$245.95 per copy (in United States).
Price subject to change without prior notice.

0069

RSMeans

Assemblies Cost Data

34th Annual Edition

2009

R.S. Means Company, Inc.
Construction Publishers & Consultants
63 Smiths Lane
Kingston, MA 02364-3008
(781) 422-5000

Copyright©2008 by R.S. Means Company, Inc.
All rights reserved.

Printed in the United States of America.
ISSN 0068-3531
ISBN 978-0-87629-094-1

Senior Editor
Barbara Balboni

Contributing Editors
Christopher Babbitt
Ted Baker
Robert A. Bastoni
John H. Chiang, PE
Gary W. Christensen
David G. Drain, PE
Cheryl Elsmore
Robert J. Kuchta
Robert C. McNichols
Melville J. Mossman, PE
Jeannene D. Murphy
Stephen C. Plotner
Eugene R. Spencer
Marshall J. Stetson
Phillip R. Waier, PE

Senior Vice President & General Manager
John Ware

Vice President of Operations
Dave Walsh

Vice President of Sales & Marketing
Sev Ritchie

Vice President of Direct Response
John M. Shea

Director of Product Development
Thomas J. Dion

Engineering Manager
Bob Mewis, CCC

Production Manager
Michael Kokernak

Production Coordinator
Jill Goodman

Technical Support
Jonathan Forgit
Mary Lou Geary
Roger Hancock
Gary L. Hoitt
Genevieve Medeiros
Debbie Panarelli
Paula Reale-Camelio
Kathryn S. Rodriguez
Sheryl A. Rose

Book & Cover Design
Norman R. Forgit

This book is printed on recycled paper (10% PCW cover, 20% PCW text), certified by the Sustainable Forestry Initiative (SFI*) and Forest Stewardship Counsil (FSC®), using soy-based printing ink. This book is recyclable.

Foreword

Our Mission

Since 1942, RSMeans has been actively engaged in construction cost publishing and consulting throughout North America.

Today, over 60 years after RSMeans began, our primary objective remains the same: to provide you, the construction and facilities professional, with the most current and comprehensive construction cost data possible.

Whether you are a contractor, an owner, an architect, an engineer, a facilities manager, or anyone else who needs a reliable construction cost estimate, you'll find this publication to be a highly useful and necessary tool.

With the constant flow of new construction methods and materials today, it's difficult to find the time to look at and evaluate all the different construction cost possibilities. In addition, because labor and material costs keep changing, last year's cost information is not a reliable basis for today's estimate or budget.

That's why so many construction professionals turn to RSMeans. We keep track of the costs for you, along with a wide range of other key information, from city cost indexes . . . to productivity rates . . . to crew composition . . . to contractor's overhead and profit rates.

RSMeans performs these functions by collecting data from all facets of the industry and organizing it in a format that is instantly accessible to you. From the preliminary budget to the detailed unit price estimate, you'll find the data in this book useful for all phases of construction cost determination.

The Staff, the Organization, and Our Services

When you purchase one of RSMeans' publications, you are, in effect, hiring the services of a full-time staff of construction and engineering professionals.

Our thoroughly experienced and highly qualified staff works daily at collecting, analyzing, and disseminating comprehensive cost information for your needs. These staff members have years of practical construction experience and engineering training prior to joining the firm. As a result, you can count on them not only for the cost figures, but also for additional background reference information that will help you create a realistic estimate.

The RSMeans organization is always prepared to help you solve construction problems through its four major divisions: Construction and Cost Data Publishing, Electronic Products and Services, Consulting and Business Solutions, and Professional Development Services.

Besides a full array of construction cost estimating books, RSMeans also publishes a number of other reference works for the construction industry. Subjects include construction estimating and project and business management; special topics such as HVAC, roofing, plumbing, and hazardous waste remediation; and a library of facility management references.

In addition, you can access all of our construction cost data electronically using *Means CostWorks®* CD or on the Web using Means CostWorks.com.

What's more, you can increase your knowledge and improve your construction estimating and management performance with an RSMeans Construction Seminar or In-House Training Program. These two-day seminar programs offer unparalleled opportunities for everyone in your organization to get updated on a wide variety of construction-related issues.

RSMeans is also a worldwide provider of construction cost management and analysis services for commercial and government owners.

In short, RSMeans can provide you with the tools and expertise for constructing accurate and dependable construction estimates and budgets in a variety of ways.

Robert Snow Means Established a Tradition of Quality That Continues Today

Robert Snow Means spent years building RSMeans, making certain he always delivered a quality product.

Today, at RSMeans, we do more than talk about the quality of our data and the usefulness of our books. We stand behind all of our data, from historical cost indexes to construction materials and techniques to current costs.

If you have any questions about our products or services, please call us toll-free at 1-800-334-3509. Our customer service representatives will be happy to assist you. You can also visit our Web site at www.rsmeans.com.

Table of Contents

Related RSMeans Products and Services

The engineers at RSMeans, suggest the following products and services as companion information resources to *RSMeans Assemblies Cost Data:*

Construction Cost Data Books
Building Construction Cost Data 2009
Facilities Construction Cost Data 2009
Square Foot Costs 2009

Reference Books
ADA Compliance Pricing Guide, 2nd Ed.
Building Security: Strategies & Costs
Designing & Building with the IBC, 2nd Ed.
Estimating Building Costs
Estimating Handbook, 2nd Ed.
Green Building: Project Planning & Estimating, 2nd Ed.
How to Estimate with Means Data and CostWorks, 3rd Ed.
Plan Reading & Material Takeoff
Square Foot & Assemblies Estimating Methods, 3rd Ed.

Seminars and In-House Training
Means CostWorks Training
Means Data for Job Order Contracting (JOC)
Plan Reading & Material Takeoff
Scheduling & Project Management
Scope of Work for Facilities Estimating
Unit Price Estimating

RSMeans on the Internet
Visit RSMeans at **www.rsmeans.com.** The site contains useful interactive cost and reference material. Request or download **FREE** estimating software demos. Visit our bookstore for convenient ordering and to learn more about new publications and companion products.

RSMeans Electronic Data
Get the information found in RSMeans cost books electronically on *Means CostWorks®* CD or on the Web.

RSMeans Business Solutions
Engineers and Analysts offer research studies, benchmark analysis, predictive cost modeling, analytics, job order contracting, and real property management consultation, as well as custom-designed, web-based dashboards and calculators that apply RCD/RSMeans extensive databases. Clients include federal government agencies, architects, construction management firms, and institutional organizations such as school systems, health care facilities, associations, and corporations.

RSMeans for Job Order Contracting (JOC)
Best practice JOC tools for cost estimating and project management to help streamline delivery processes for renovation projects. Renovation is a $147 billion market in the U.S., and includes projects in school districts, municipalities, health care facilities, colleges and universities, and corporations.

• RSMeans Engineers consult in contracting methods and conduct JOC Facility Audits

• JOCWorks™ Software (Basic, Advanced, PRO levels)

• RSMeans Job Order Contracting Cost Data for the entire U.S.

Construction Costs for Software Applications
Over 25 unit price and assemblies cost databases are available through a number of leading estimating and facilities management software providers (listed below). For more information see the "Other RSMeans Products" pages at the back of this publication.
MeansData™ is also available to federal, state, and local government agencies as multi-year, multi-seat licenses.

• 4Clicks-Solutions, LLC
• Aepco, Inc.
• Applied Flow Technology
• ArenaSoft Estimating
• Ares Corporation
• Beck
• BSD – Building Systems Design, Inc.
• CMS – Construction Management Software
• Corecon Technologies, Inc.
• CorVet Systems
• Earth Tech
• Estimating Systems, Inc.
• HCSS
• Maximus Asset Solutions
• MC^2 – Management Computer Controls
• Sage Timberline Office
• Shaw Beneco Enterprises, Inc.
• US Cost, Inc.
• VFA – Vanderweil Facility Advisers
• WinEstimator, Inc.

How the Book Is Built: An Overview

A Powerful Construction Tool

You have in your hands one of the most powerful construction tools available today. A successful project is built on the foundation of an accurate and dependable estimate. This book will enable you to construct just such an estimate.

For the casual user the book is designed to be:

- quickly and easily understood so you can get right to your estimate.
- filled with valuable information so you can understand the necessary factors that go into the cost estimate.

For the regular user, the book is designed to be:

- a handy desk reference that can be quickly referred to for key costs.
- a comprehensive, fully reliable source of current construction costs so you'll be prepared to estimate any project.
- a source book for preliminary project cost, product selections, and alternate materials and methods.

To meet all of these requirements we have organized the book into the following clearly defined sections.

How To Use the Book: *The Details*

This section contains an in-depth explanation of how the book is arranged ... and how you can use it to determine a reliable construction cost estimate. It includes information about how we develop our cost figures and how to completely prepare your estimate.

Assemblies Section

The cost data in this section has been organized in an "Assemblies" format. These assemblies are the functional elements of a building and are arranged according to the 7 divisions of the UNIFORMAT II classification system. For a complete explanation of a typical "Assemblies" page, see "How To Use the Assemblies Cost Tables."

Reference Section

This section includes information on Reference Tables, Historical Cost Indexes, City Cost Indexes, Square Foot Costs, Reference Aids, Estimating Forms, and a listing of Abbreviations. It is visually identified by a vertical gray bar on the page edges.

Reference Tables: At the beginning of selected major classifications in the Assemblies Section are "reference numbers" shown in bold squares. These numbers refer you to related information in the Reference Section.

In this section, you'll find reference tables, explanations, and estimating information that support how we develop the assemblies data. Also included are alternate pricing methods, technical data, and estimating procedures, along with information on design and economy in construction. You'll also find helpful tips on what to expect and what to avoid when estimating and constructing your project.

It is recommended that you refer to this section if a "reference number" appears within the assembly you are estimating.

Historical Cost Indexes: These indexes provide you with data to adjust construction costs over time.

City Cost Indexes: All costs in this book are U.S. national averages. Costs vary because of the regional economy. You can adjust the national average costs in this book to 316 major cities throughout the U.S. and Canada by using the data in this section.

Location Factors: You can adjust total project costs to over 900 locations throughout the U.S. and Canada by using the data in this section.

Square Foot Costs: This section contains costs for 59 different building types that allow you to make a rough estimate for the overall cost of a project or its major components.

Reference Aids: This section provides design criteria, code requirements, and many other reference aids useful in designing a building in various geographic locations of the country. Trade-offs by weight, material, thermal conductivity, and fire rating can be developed from this information. Design loads and capacities are based on the requirements of several nationally used building codes. In some cases, local building codes may have more stringent requirements, which would influence the final in-place cost.

Estimating Forms: Detailed estimating forms and checklists are included to be used when estimating on an "Assemblies" basis.

Abbreviations: A listing of the abbreviations used throughout this book, along with the terms they represent, is included.

Index

A comprehensive listing of all terms and subjects in this book will help you quickly find what you need.

The Scope of This Book

This book is designed to be as comprehensive and as easy to use as possible. To that end we have made certain assumptions and limited its scope in two key ways:

1. We have established material prices based on a national average.
2. We have computed labor costs based on a 30-city national average of union wage rates.

For a more detailed explanation of how the cost data is developed, see "How To Use the Book: The Details."

Project Size

This book is aimed primarily at commercial and industrial projects costing $1,000,000 and up, or large multi-family housing projects. Costs are primarily for new construction or major renovation of buildings rather than repairs or minor alterations.

With reasonable exercise of judgment the figures can be used for any building work. However, for civil engineering structures such as bridges, dams, highways, or the like, please refer to RSMeans Heavy Construction Cost Data.

How to Use the Book: The Details

What's Behind the Numbers? The Development of Cost Data

The staff at RSMeans continuously monitors developments in the construction industry in order to ensure reliable, thorough, and up-to-date cost information.

While *overall* construction costs may vary relative to general economic conditions, price fluctuations within the industry are dependent upon many factors. Individual price variations may, in fact, be opposite to overall economic trends. Therefore, costs are continually monitored and complete updates are published yearly. Also, new items are frequently added in response to changes in materials and methods.

Costs—$ (U.S.)

All costs represent U.S. national averages and are given in U.S. dollars. The RSMeans City Cost Indexes can be used to adjust costs to a particular location. The City Cost Indexes for Canada can be used to adjust U.S. national averages to local costs in Canadian dollars. No exchange rate conversion is necessary.

Material Costs

The RSMeans staff contacts manufacturers, dealers, distributors, and contractors all across the U.S. and Canada to determine national average material costs. If you have access to current material costs for your specific location, you may wish to make adjustments to reflect differences from the national average. Included within material costs are fasteners for a normal installation. RSMeans engineers use manufacturers' recommendations, written specifications, and/or standard construction practice for size and spacing of fasteners. Adjustments to material costs may be required for your specific application or location. Material costs do not include sales tax.

Labor Costs

Labor costs are based on the average of wage rates from 30 major U.S. cities. Rates are determined from labor union agreements or prevailing wages for construction trades for the current year. Rates, along with overhead and profit markups, are listed on the inside back cover of this book.

- If wage rates in your area vary from those used in this book, or if rate increases are expected within a given year, labor costs should be adjusted accordingly.

Labor costs reflect productivity based on actual working conditions. In addition to actual installation, these figures include time spent during a normal workday on tasks such as material receiving and handling, mobilization at site, site movement, breaks, and cleanup.

Productivity data is developed over an extended period so as not to be influenced by abnormal variations and reflects a typical average.

Equipment Costs

Equipment costs include not only rental, but also operating costs for equipment under normal use. The operating costs include parts and labor for routine servicing such as repair and replacement of pumps, filters, and worn lines. Normal operating expendables, such as fuel, lubricants, tires and electricity (where applicable), are also included. Extraordinary operating expendables with highly variable wear patterns, such as diamond bits and blades, are excluded. These costs are included under materials. Equipment rental rates are obtained from industry sources throughout North America—contractors, suppliers, dealers, manufacturers, and distributors.

Equipment costs do not include operators' wages, nor do they include the cost to move equipment to a job site (mobilization) or from a job site (demobilization).

General Conditions

Prices given in this book include the Installing Contractor's overhead and profit (O&P). General Conditions, when applicable, should also be added to the Total Cost including O&P. The costs for General Conditions are listed in the Reference Section of this book. General Conditions for the *Installing Contractor* may range from 0% to 10% of the Total Cost including O&P. For the *General* or *Prime* Contractor, costs for General Conditions may range from 5% to 15% of the Total Cost, including O&P, with a figure of 10% as the most typical allowance.

Overhead and Profit

Total Cost, which includes O&P for the *Installing Contractor*, is shown in the last column on the Assemblies Price pages of this book. This figure is the arithmetic sum of the two columns labelled Material and Installation. The value in the Material column is the bare material cost plus 10% for profit. The value in the Installation column is the bare labor cost plus total overhead and profit added to the bare equipment cost plus 10% for profit. Details for the calculation of Overhead and Profit on labor are shown on the inside back cover and in the Reference Section of this book. (See the "How To Use the Assemblies Pages" for an example of this calculation.)

Factors Affecting Costs

Costs can vary depending upon a number of variables. Here's how we have handled the main factors affecting costs.

Quality—The prices for materials and the workmanship upon which productivity is based represent sound construction work. They are also in line with U.S. government specifications.

Overtime—We have made no allowance for overtime. If you anticipate premium time or work beyond normal working hours, be sure to make an appropriate adjustment to your labor costs.

Productivity—The productivities used in calculating labor costs are based on working an eight-hour day in daylight hours in moderate temperatures. For work that extends beyond normal work hours or is performed under adverse conditions, productivity may decrease.

Size of Project—The size, scope of work, and type of construction project will have a significant impact on cost. Economies of scale can reduce costs for large projects. Costs can often run higher for small projects. Costs in this book are intended for the size and type of project as previously described in "How the Book Is Built: An Overview." Costs for projects of a significantly different size or type should be adjusted accordingly.

Location—Material prices in this book are for metropolitan areas. However, in dense urban areas, traffic and site storage limitations may increase costs. Beyond a 20-mile radius of large cities, extra trucking or transportation charges may also increase the material costs slightly. On the other hand, lower wage rates may be in effect. Be sure to consider both of these factors when preparing an estimate, particularly if the job site is located in a central city or remote rural location.

In addition, highly specialized subcontract items may require travel and per-diem expenses for mechanics.

Other Factors—

- season of year
- contractor management
- weather conditions
- local union restrictions
- building code requirements
- availability of:
  - adequate energy
  - skilled labor
  - building materials
- owner's special requirements/restrictions
- safety requirements
- environmental considerations

Unpredictable Factors—General business conditions influence "in-place" costs of all items. Substitute materials and construction methods may have to be employed. These may affect the installed cost and/or life cycle costs. Such factors may be difficult to evaluate and cannot necessarily be predicted on the basis of the job's location in a particular section of the country. Thus, where these factors apply, you may find significant, but unavoidable cost variations for which you will have to apply a measure of judgment to your estimate.

Contingencies—Estimates that include an allowance for contingencies have a margin to allow for unforeseen construction difficulties. On alterations or repair jobs, 20% is not an unreasonable allowance to make. If drawings are final and only field contingencies are being considered, 2% to 3% is probably sufficient, and often nothing need be added. As far as the contract is concerned, future changes in plans can be covered by extras. The contractor should consider inflationary price trends and possible material shortages during the course of the job. Escalation factors are dependent upon both economic conditions and the anticipated time between the estimate and actual construction. If drawings are not complete or approved, or if a budget is required before proceeding with a project, it is wise to add 5% to 10%. Contingencies are a matter of judgment.

Rounding of Costs

In general, all unit prices in excess of $5.00 have been rounded to make them easier to use and still maintain adequate precision of the results. The rounding rules we have chosen are in the following table.

Prices from . . .	Rounded to the nearest . . .
$.01 to $5.00	$.01
$5.01 to $20.00	$.05
$20.01 to $100.00	$.50
$100.01 to $300.00	$1.00
$300.01 to $1,000.00	$5.00
$1,000.01 to $10,000.00	$25.00
$10,000.01 to $50,000.00	$100.00
$50,000.01 and above	$500.00

Final Checklist

Estimating can be a straightforward process provided you remember the basics. Here's a checklist of some of the steps you should remember to complete before finalizing your estimate.

Did you remember to . . .

- factor in the City Cost Index for your locale?
- take into consideration which items have been marked up and by how much?
- mark up the entire estimate sufficiently for your purposes?
- read the background information on techniques and technical matters that could impact your project time span and cost?
- include all components of your project in the final estimate?
- double check your figures for accuracy?
- call RSMeans if you have any questions about your estimate or the data you've found in our publications?

Remember, RSMeans stands behind its publications. If you have any questions about your estimate, about the costs you've used from our books, or even about the technical aspects of the job that may affect your estimate, feel free to call the RSMeans editors at 1-800-334-3509.

Assemblies Estimating

The grouping of several different trades into building components or broad building elements is the "Systems" or "Assemblies" method of estimating. This method allows the estimator or designer to make quick comparisons of systems in various combinations within predetermined guidelines. Systems which are best suited to accommodate budget, code, load, insulation, fireproofing, acoustics, energy considerations, and the owner's special requirements can quickly be determined. This method can also be used to help match existing construction.

In order to understand how a Systems estimate is assembled, it is a good idea to compare a Unit Price estimate with a Systems estimate. In a Unit Price estimate, each item is normally included along the guidelines of the 50-division MasterFormat of the Construction Specifications Institute, Inc. In a Systems estimate, these same items are allocated to one of seven major group elements in the UNIFORMAT II organization. Certain items that were formerly grouped into a single trade breakdown must now be allocated among two or more systems.

An example of this difference would be concrete. In a Unit Price estimate, all the concrete items on a job would be priced in the "Concrete" section of the estimate, CSI Division 03. In a Systems estimate, concrete is found in a number of locations. For instance, concrete is used in all of these systems: Division A10, Foundations; Division A20, Basement Construction; Division B10, Superstructure; and Division B20, Exterior Closure.

Conversely, other items that are listed in separate trade breakdowns in a Unit Price estimate are combined into one division in the Systems estimate. For example, interior partitions might include two CSI divisions: Division 06, Wood Stud Wall; and Division 09, Lath, Plaster and Paint. In the UNIFORMAT II Systems Estimate, these items are all combined in Division C, Interior Construction.

This re-allocation of the familiar items from the CSI format may, at first, seem confusing, but once the concept is understood, the resultant increase in estimating speed is well worth the initial familiarization required.

Systems or Assemblies estimating is not a substitute for Unit Price estimating. It is normally done during the earlier conceptual stage before plans have been completed or when preparing a budget. This enables the designer to bring in the project within the owner's budget. During the actual initial design process, the designer will be forced to make important decisions and "trade-offs" for each of the various systems. Some of the trade-offs can include:

a. Price of each system
b. Appearance, quality and compatibility
c. Story height
d. Clear span
e. Complications and restrictions
f. Thermal characteristics
g. Life-cycle costs
h. Acoustical characteristics
i. Fireproofing characteristics
j. Special owner's requirements in excess of code requirements
k. Code
l. Load

Before starting a Systems Estimate, gather all the information possible pertaining to the project.

Information can be gathered from:
1. Code Requirements (see Reference Section L of this book)
2. Owner's Requirements
3. Preliminary Assumptions
4. Site Inspection and Investigation

Since the Foundation and Substructure design and price are functions of the Superstructure and the site, it is advisable to start the estimate with the Superstructure. Follow this with the Foundation and Substructure, and then the other Systems in the sequence as applicable to your project.

Assemblies Section

Table of Contents

Table of Contents

Table of Contents

Table of Contents

How to Use the Assemblies Cost Tables

The following is a detailed explanation of a sample Assemblies Cost Table. Most Assembly Tables are separated into three parts: 1) an illustration of the system to be estimated; 2) the components and related costs of a typical system; and 3) the costs for similar systems with dimensional and/or size variations. Next to each bold number below is the described item with the appropriate component of the sample entry following in parentheses. In most cases, if the work is to be subcontracted, the general contractor will need to add an additional markup (RSMeans suggests using 10%) to the "Total" figures.

1 System/Line Numbers (B1010-222-1700)

Each Assemblies Cost Line has been assigned a unique identification number based on the UNIFORMAT II classification system.

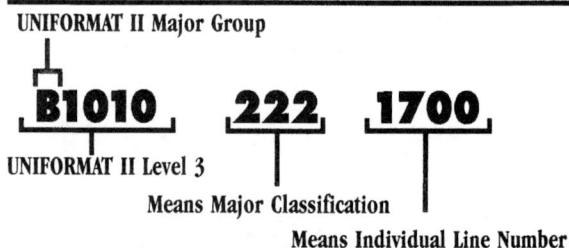

UNIFORMAT II Major Group

B1010   222   1700

UNIFORMAT II Level 3

Means Major Classification

Means Individual Line Number

B10 Superstructure

B1010 Floor Construction 1

2

General: Flat Slab: Solid uniform depth concrete two-way slabs with drop panels at columns and no column capitals.

Design and Pricing Assumptions:
Concrete f'c = 3 KSI, placed by concrete pump.
Reinforcement, fy = 60 KSI.
Forms, four use.
Finish, steel trowel.
Curing, spray on membrane.
Based on 4 bay x 4 bay structure.

System Components	QUANTITY	UNIT	COST PER S.F.		
			MAT.	INST.	TOTAL
SYSTEM B1010 222 1700					
15'X15' BAY 40 PSF S. LOAD, 12" MIN. COL. 6" SLAB, 1-1/2" DROP, 117 PSF					
Forms in place, flat slab with drop panels, to 15' high, 4 uses	.993	S.F.	1.70	5.26	6.96
Forms in place, exterior spandrel, 12" wide, 4 uses	.034	SFCA	.05	.32	.37
Reinforcing in place, elevated slabs #4 to #7	1.588	Lb.	1.41	.64	2.05
Concrete ready mix, regular weight, 3000 psi	.513	C.F.	2.11		2.11
Place and vibrate concrete, elevated slab, 6" to 10" pump	.513	C.F.		.62	.62
Finish floor, monolithic steel trowel finish for finish floor	1.000	S.F.		.78	.78
Cure with sprayed membrane curing compound	.010	C.S.F.	.06	.08	.14
TOTAL			5.33	7.70	13.03

B1010 222		Cast in Place Flat Slab with Drop Panels						
	BAY SIZE (FT.)	SUPERIMPOSED LOAD (P.S.F.)	MINIMUM COL. SIZE (IN.)	SLAB & DROP (IN.)	TOTAL LOAD (P.S.F.)	COST PER S.F.		
						MAT.	INST.	TOTAL
1700	15 x 15	40	12	6 - 1-1/2	117	5.35	7.70	13.05
1720		75	12	6 - 2-1/2	153	5.50	7.75	13.25
1760	RB1010 -010	125	14	6 - 3-1/2	205	5.85	7.90	13.75
1780		200	16	6 - 4-1/2	281	6.25	8.10	14.35
1840	15 x 20	40	12	6-1/2 - 2	124	5.70	7.80	13.50
1860		75	14	6-1/2 - 4	162	6.05	8	14.05
1880	RB1010 -100	125	16	6-1/2 - 5	213	6.50	8.20	14.70
1900		200	18	6-1/2 - 6	293	6.80	8.35	15.15
1960	20 x 20	40	12	7 - 3	132	6	7.95	13.95
1980		75	16	7 - 4	168	6.50	8.15	14.65

2 Illustration

At the top of most assembly tables is an illustration, a brief description, and the design criteria used to develop the cost.

3 System Components

The components of a typical system are listed separately to show what has been included in the development of the total system price. The table below contains prices for other similar systems with dimensional and/or size variations.

4 Quantity

This is the number of line item units required for one system unit. For example, we assume that it will take 1.588 pounds of reinforcing on a square foot basis.

5 Unit of Measure for Each Item

The abbreviated designation indicates the unit of measure, as defined by industry standards, upon which the price of the component is based. For example, reinforcing is priced by lb. (pound) while concrete is priced by C.F. (cubic foot).

6 Unit of Measure for Each System (cost per S.F.)

Costs shown in the three right-hand columns have been adjusted by the component quantity and unit of measure for the entire system. In this example, "Cost per S.F." is the unit of measure for this system or "assembly."

7 Reference Number Information

RB1010 -100

You'll see reference numbers shown in shaded boxes at the beginning of some sections. These refer to related items in the Reference Section, visually identified by a vertical gray bar on the page edges.

The relation may be: (1) an estimating procedure that should be read before estimating, (2) an alternate pricing method, or (3) technical information.

The "R" designates the Reference Section. The letters and numbers refer to the UNIFORMAT II classification system.

Example: The bold number above is directing you to refer to the reference number RB1010-100. This particular reference number shows comparative costs of floor systems.

8 Materials (5.35)

This column contains the Materials Cost of each component. These cost figures are bare costs plus 10% for profit.

9 Installation (7.70)

Installation includes labor and equipment plus the installing contractor's overhead and profit. Equipment costs are the bare rental costs plus 10% for profit. The labor overhead and profit is defined on the inside back cover of this book.

10 Total (13.05)

The figure in this column is the sum of the material and installation costs.

Material Cost	+	Installation Cost	=	Total
$5.35	+	$7.70	=	$13.05

A Substructure

A1010 Standard Foundations

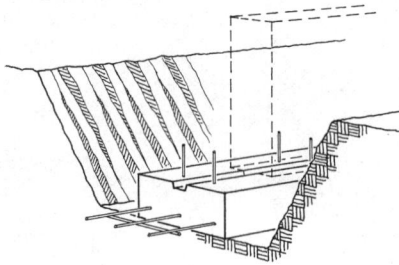

The Strip Footing System includes: excavation; hand trim; all forms needed for footing placement; forms for 2″ x 6″ keyway (four uses); dowels; and 3,000 p.s.i. concrete.

The footing size required varies for different soils. Soil bearing capacities are listed for 3 KSF and 6 KSF. Depths of the system range from 8″ to 24″. Widths range from 16″ to 96″. Smaller strip footings may not require reinforcement.

Please see the reference section for further design and cost information.

System Components			COST PER L.F.		
	QUANTITY	UNIT	MAT.	INST.	TOTAL
SYSTEM A1010 110 2500					
STRIP FOOTING, LOAD 5.1KLF, SOIL CAP. 3 KSF, 24″ WIDE X12″ DEEP, REINF.					
Trench excavation	.148	C.Y.		1.20	1.20
Hand trim	2.000	S.F.		1.66	1.66
Compacted backfill	.074	C.Y.		.24	.24
Formwork, 4 uses	2.000	S.F.	5.34	7.74	13.08
Keyway form, 4 uses	1.000	L.F.	.29	.99	1.28
Reinforcing, fy = 60000 psi	3.000	Lb.	2.67	1.68	4.35
Dowels	2.000	Ea.	2.50	4.86	7.36
Concrete, f'c = 3000 psi	.074	C.Y.	8.21		8.21
Place concrete, direct chute	.074	C.Y.		1.56	1.56
Screed finish	2.000	S.F.		.64	.64
TOTAL			19.01	20.57	39.58

A1010 110	Strip Footings		COST PER L.F.		
			MAT.	INST.	TOTAL
2100	Strip footing, load 2.6 KLF, soil capacity 3 KSF, 16″ wide x 8″ deep plain		9.65	12.30	21.95
2300	Load 3.9 KLF, soil capacity, 3 KSF, 24″ wide x 8″ deep, plain		11.55	13.50	25.05
2500	Load 5.1KLF, soil capacity 3 KSF, 24″ wide x 12″ deep, reinf.	RA1010 -140	19	20.50	39.50
2700	Load 11.1KLF, soil capacity 6 KSF, 24″ wide x 12″ deep, reinf.		19	20.50	39.50
2900	Load 6.8 KLF, soil capacity 3 KSF, 32″ wide x 12″ deep, reinf.		23	22.50	45.50
3100	Load 14.8 KLF, soil capacity 6 KSF, 32″ wide x 12″ deep, reinf.		23	22.50	45.50
3300	Load 9.3 KLF, soil capacity 3 KSF, 40″ wide x 12″ deep, reinf.		26.50	24	50.50
3500	Load 18.4 KLF, soil capacity 6 KSF, 40″ wide x 12″ deep, reinf.		27	24.50	51.50
3700	Load 10.1KLF, soil capacity 3 KSF, 48″ wide x 12″ deep, reinf.		29	26.50	55.50
3900	Load 22.1KLF, soil capacity 6 KSF, 48″ wide x 12″ deep, reinf.		31.50	28	59.50
4100	Load 11.8KLF, soil capacity 3 KSF, 56″ wide x 12″ deep, reinf.		34	29	63
4300	Load 25.8KLF, soil capacity 6 KSF, 56″ wide x 12″ deep, reinf.		38	31.50	69.50
4500	Load 10KLF, soil capacity 3 KSF, 48″ wide x 16″ deep, reinf.		36	30.50	66.50
4700	Load 22KLF, soil capacity 6 KSF, 48″ wide, 16″ deep, reinf.		37.50	31	68.50
4900	Load 11.6KLF, soil capacity 3 KSF, 56″ wide x 16″ deep, reinf.		41	43.50	84.50
5100	Load 25.6KLF, soil capacity 6 KSF, 56″ wide x 16″ deep, reinf.		44	45.50	89.50
5300	Load 13.3KLF, soil capacity 3 KSF, 64″ wide x 16″ deep, reinf.		47	36.50	83.50
5500	Load 29.3KLF, soil capacity 6 KSF, 64″ wide x 16″ deep, reinf.		51.50	39	90.50
5700	Load 15KLF, soil capacity 3 KSF, 72″ wide x 20″ deep, reinf.		61.50	43	104.50
5900	Load 33KLF, soil capacity 6 KSF, 72″ wide x 20″ deep, reinf.		66.50	46.50	113
6100	Load 18.3KLF, soil capacity 3 KSF, 88″ wide x 24″ deep, reinf.		86	55	141
6300	Load 40.3KLF, soil capacity 6 KSF, 88″ wide x 24″ deep, reinf.		96	61.50	157.50
6500	Load 20KLF, soil capacity 3 KSF, 96″ wide x 24″ deep, reinf.		93.50	58.50	152
6700	Load 44 KLF, soil capacity 6 KSF, 96″ wide x 24″ deep, reinf.		101	63.50	164.50

A10 Foundations

A1010 Standard Foundations

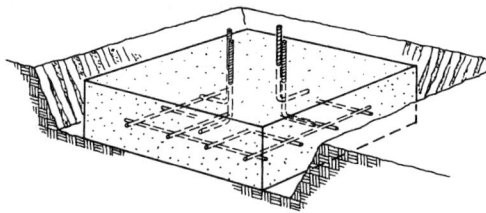

The Spread Footing System includes: excavation; backfill; forms (four uses); all reinforcement; 3,000 p.s.i. concrete (chute placed); and screed finish.

Footing systems are priced per individual unit. The Expanded System Listing at the bottom shows footings that range from 3′ square x 12″ deep, to 18′ square x 52″ deep. It is assumed that excavation is done by a truck mounted hydraulic excavator with an operator and oiler.

Backfill is with a dozer, and compaction by air tamp. The excavation and backfill equipment is assumed to operate at 30 C.Y. per hour.

Please see the reference section for further design and cost information.

System Components	QUANTITY	UNIT	COST EACH		
			MAT.	INST.	TOTAL
SYSTEM A1010 210 7100					
SPREAD FOOTINGS, LOAD 25K, SOIL CAPACITY 3 KSF, 3′ SQ X 12″ DEEP					
Bulk excavation	.590	C.Y.		4.81	4.81
Hand trim	9.000	S.F.		7.47	7.47
Compacted backfill	.260	C.Y.		.87	.87
Formwork, 4 uses	12.000	S.F.	9.24	54.48	63.72
Reinforcing, fy = 60,000 psi	.006	Ton	9.75	6.60	16.35
Dowel or anchor bolt templates	6.000	L.F.	4.32	22.56	26.88
Concrete, f'c = 3,000 psi	.330	C.Y.	36.63		36.63
Place concrete, direct chute	.330	C.Y.		6.93	6.93
Screed finish	9.000	S.F.		2.43	2.43
TOTAL			59.94	106.15	166.09

A1010 210	Spread Footings		COST EACH		
			MAT.	INST.	TOTAL
7090	Spread footings, 3000 psi concrete, chute delivered				
7100	Load 25K, soil capacity 3 KSF, 3′-0″ sq. x 12″ deep		60	107	167
7150	Load 50K, soil capacity 3 KSF, 4′-6″ sq. x 12″ deep		131	183	314
7200	Load 50K, soil capacity 6 KSF, 3′-0″ sq. x 12″ deep	RA1010-120	60	107	167
7250	Load 75K, soil capacity 3 KSF, 5′-6″ sq. x 13″ deep		211	258	469
7300	Load 75K, soil capacity 6 KSF, 4′-0″ sq. x 12″ deep		106	157	263
7350	Load 100K, soil capacity 3 KSF, 6′-0″ sq. x 14″ deep		268	310	578
7410	Load 100K, soil capacity 6 KSF, 4′-6″ sq. x 15″ deep		162	215	377
7450	Load 125K, soil capacity 3 KSF, 7′-0″ sq. x 17″ deep		425	445	870
7500	Load 125K, soil capacity 6 KSF, 5′-0″ sq. x 16″ deep		210	258	468
7550	Load 150K, soil capacity 3 KSF 7′-6″ sq. x 18″ deep		515	515	1,030
7610	Load 150K, soil capacity 6 KSF, 5′-6″ sq. x 18″ deep		280	325	605
7650	Load 200K, soil capacity 3 KSF, 8′-6″ sq. x 20″ deep		740	685	1,425
7700	Load 200K, soil capacity 6 KSF, 6′-0″ sq. x 20″ deep		370	400	770
7750	Load 300K, soil capacity 3 KSF, 10′-6″ sq. x 25″ deep		1,350	1,125	2,475
7810	Load 300K, soil capacity 6 KSF, 7′-6″ sq. x 25″ deep		705	675	1,380
7850	Load 400K, soil capacity 3 KSF, 12′-6″ sq. x 28″ deep		2,175	1,675	3,850
7900	Load 400K, soil capacity 6 KSF, 8′-6″ sq. x 27″ deep		985	880	1,865
7950	Load 500K, soil capacity 3 KSF, 14′-0″ sq. x 31″ deep		3,025	2,175	5,200
8010	Load 500K, soil capacity 6 KSF, 9′-6″ sq. x 30″ deep		1,375	1,150	2,525

A10 Foundations

A1010 Standard Foundations

A1010 210	Spread Footings	COST EACH		
		MAT.	INST.	TOTAL
8050	Load 600K, soil capacity 3 KSF, 16'-0" sq. x 35" deep	4,450	3,025	7,475
8100	Load 600K, soil capacity 6 KSF, 10'-6" sq. x 33" deep	1,850	1,475	3,325
8150	Load 700K, soil capacity 3 KSF, 17'-0" sq. x 37" deep	5,225	3,425	8,650
8200	Load 700K, soil capacity 6 KSF, 11'-6" sq. x 36" deep	2,375	1,800	4,175
8250	Load 800K, soil capacity 3 KSF, 18'-0" sq. x 39" deep	6,175	3,975	10,150
8300	Load 800K, soil capacity 6 KSF, 12'-0" sq. x 37" deep	2,650	2,000	4,650
8350	Load 900K, soil capacity 3 KSF, 19'-0" sq. x 40" deep	7,225	4,575	11,800
8400	Load 900K, soil capacity 6 KSF, 13'-0" sq. x 39" deep	3,275	2,375	5,650
8450	Load 1000K, soil capacity 3 KSF, 20'-0" sq. x 42" deep	8,325	5,150	13,475
8500	Load 1000K, soil capacity 6 KSF, 13'-6" sq. x 41" deep	3,725	2,625	6,350
8550	Load 1200K, soil capacity 6 KSF, 15'-0" sq. x 48" deep	4,975	3,350	8,325
8600	Load 1400K, soil capacity 6 KSF, 16'-0" sq. x 47" deep	6,075	4,000	10,075
8650	Load 1600K, soil capacity 6 KSF, 18'-0" sq. x 52" deep	8,450	5,300	13,750
8700				

A1010 Standard Foundations

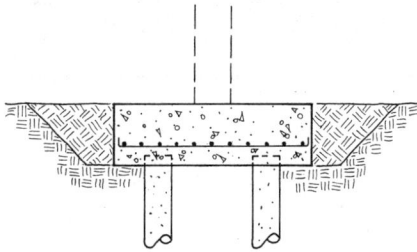

These pile cap systems include excavation with a truck mounted hydraulic excavator, hand trimming, compacted backfill, forms for concrete, templates for dowels or anchor bolts, reinforcing steel and concrete placed and screeded.

Pile embedment is assumed as 6″. Design is consistent with the *Concrete Reinforcing Steel Institute Handbook* f'c = 3000 psi, fy = 60,000.

Please see the reference section for further design and cost information.

System Components	QUANTITY	UNIT	COST EACH		
			MAT.	INST.	TOTAL
SYSTEM A1010 250 5100					
CAP FOR 2 PILES, 6′-6″X3′-6″X20″, 15 TON PILE, 8″ MIN. COL., 45K COL. LOAD					
Excavation, bulk, hyd excavator, truck mtd. 30″ bucket 1/2 CY	2.890	C.Y.		23.59	23.59
Trim sides and bottom of trench, regular soil	23.000	S.F.		19.09	19.09
Dozer backfill & roller compaction	1.500	C.Y.		5.02	5.02
Forms in place pile cap, square or rectangular, 4 uses	33.000	SFCA	29.04	162.03	191.07
Templates for dowels or anchor bolts	8.000	Ea.	5.76	30.08	35.84
Reinforcing in place footings, #8 to #14	.025	Ton	38.13	16.25	54.38
Concrete ready mix, regular weight, 3000 psi	1.400	C.Y.	155.40		155.40
Place and vibrate concrete for pile caps, under 5CY, direct chute	1.400	C.Y.		38.68	38.68
Monolithic screed finish	23.000	S.F.		6.21	6.21
TOTAL			228.33	300.95	529.28

A1010 250		Pile Caps						
	NO. PILES	SIZE FT-IN X FT-IN X IN	PILE CAPACITY (TON)	COLUMN SIZE (IN)	COLUMN LOAD (K)	COST EACH		
						MAT.	INST.	TOTAL
5100	2	6-6x3-6x20	15	8	45	228	300	528
5150	RA1010 -330	26	40	8	155	275	370	645
5200		34	80	11	314	380	475	855
5250		37	120	14	473	405	505	910
5300	3	5-6x5-1x23	15	8	75	285	350	635
5350		28	40	10	232	310	395	705
5400		32	80	14	471	365	450	815
5450		38	120	17	709	415	510	925
5500	4	5-6x5-6x18	15	10	103	325	350	675
5550		30	40	11	308	460	500	960
5600		36	80	16	626	540	590	1,130
5650		38	120	19	945	570	610	1,180
5700	6	8-6x5-6x18	15	12	156	570	505	1,075
5750		37	40	14	458	860	745	1,605
5800		40	80	19	936	1,000	835	1,835
5850		45	120	24	1413	1,125	925	2,050
5900	8	8-6x7-9x19	15	12	205	880	685	1,565
5950		36	40	16	610	1,150	900	2,050
6000		44	80	22	1243	1,450	1,075	2,525
6050		47	120	27	1881	1,575	1,175	2,750

A1010 Standard Foundations

| A1010 250 | | | | | | | Pile Caps | | |

	NO. PILES	SIZE FT-IN X FT-IN X IN	PILE CAPACITY (TON)	COLUMN SIZE (IN)	COLUMN LOAD (K)	COST EACH		
						MAT.	INST.	TOTAL
6100	10	11-6x7-9x21	15	14	250	1,225	840	2,065
6150		39	40	17	756	1,750	1,200	2,950
6200		47	80	25	1547	2,150	1,425	3,575
6250		49	120	31	2345	2,275	1,500	3,775
6300	12	11-6x8-6x22	15	15	316	1,600	1,025	2,625
6350		49	40	19	900	2,275	1,525	3,800
6400		52	80	27	1856	2,575	1,675	4,250
6450		55	120	34	2812	2,800	1,800	4,600
6500	14	11-6x10-9x24	15	16	345	2,050	1,250	3,300
6550		41	40	21	1056	2,500	1,575	4,075
6600		55	80	29	2155	3,225	1,975	5,200
6700	16	11-6x11-6x26	15	18	400	2,525	1,450	3,975
6750		48	40	22	1200	3,050	1,850	4,900
6800		60	80	31	2460	3,775	2,250	6,025
6900	18	13-0x11-6x28	15	20	450	2,825	1,600	4,425
6950		49	40	23	1349	3,575	2,075	5,650
7000		56	80	33	2776	4,275	2,425	6,700
7100	20	14-6x11-6x30	15	20	510	3,475	1,900	5,375
7150		52	40	24	1491	4,325	2,450	6,775

A1010 Standard Foundations

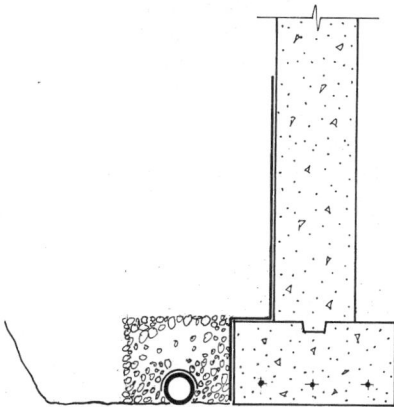

General: Footing drains can be placed either inside or outside of foundation walls depending upon the source of water to be intercepted. If the source of subsurface water is principally from grade or a subsurface stream above the bottom of the footing, outside drains should be used. For high water tables, use inside drains or both inside and outside.

The effectiveness of underdrains depends on good waterproofing. This must be carefully installed and protected during construction.

Costs below include the labor and materials for the pipe and 6" only of gravel or crushed stone around pipe. Excavation and backfill are not included.

System Components	QUANTITY	UNIT	COST PER L.F.		
			MAT.	INST.	TOTAL
SYSTEM A1010 310 1000					
FOUNDATION UNDERDRAIN, OUTSIDE ONLY, PVC 4" DIAM.					
PVC pipe 4" diam. S.D.R. 35	1.000	L.F.	1.83	3.51	5.34
Pipe bedding, graded gravel 3/4" to 1/2"	.070	C.Y.	3.36	.73	4.09
TOTAL			5.19	4.24	9.43

A1010 310	Foundation Underdrain	COST PER L.F.		
		MAT.	INST.	TOTAL
1000	Foundation underdrain, outside only, PVC, 4" diameter	5.20	4.24	9.44
1100	6" diameter	8.05	4.70	12.75
1400	Porous concrete, 6" diameter	9.05	5.10	14.15
1450	8" diameter	11.15	6.65	17.80
1500	12" diameter	20	7.70	27.70
1600	Corrugated metal, 16 ga. asphalt coated, 6" diameter	9.45	8.25	17.70
1650	8" diameter	13.15	8.65	21.80
1700	10" diameter	16.05	9.10	25.15
3000	Outside and inside, PVC, 4" diameter	10.40	8.50	18.90
3100	6" diameter	16.10	9.40	25.50
3400	Porous concrete, 6" diameter	18.10	10.25	28.35
3450	8" diameter	22.50	13.35	35.85
3500	12" diameter	40	15.35	55.35
3600	Corrugated metal, 16 ga., asphalt coated, 6" diameter	18.95	16.50	35.45
3650	8" diameter	26.50	17.35	43.85
3700	10" diameter	32	18.10	50.10

A10 Foundations

A1010 Standard Foundations

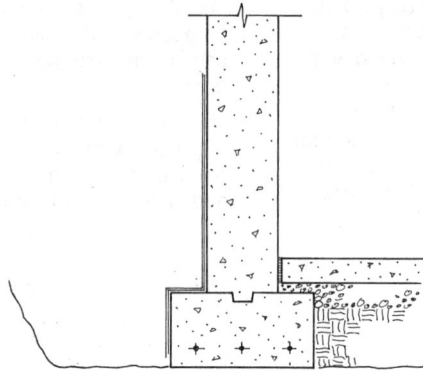

General: Apply foundation wall dampproofing over clean concrete giving particular attention to the joint between the wall and the footing. Use care in backfilling to prevent damage to the dampproofing.

Costs for four types of dampproofing are listed below.

System Components	QUANTITY	UNIT	COST PER L.F. MAT.	COST PER L.F. INST.	COST PER L.F. TOTAL
SYSTEM A1010 320 1000					
FOUNDATION DAMPPROOFING, BITUMINOUS, 1 COAT, 4' HIGH					
Bituminous asphalt dampproofing brushed on below grade, 1 coat	4.000	S.F.	.72	2.76	3.48
Labor for protection of dampproofing during backfilling	4.000	S.F.		1.17	1.17
TOTAL			.72	3.93	4.65

A1010 320	Foundation Dampproofing	MAT.	INST.	TOTAL
1000	Foundation dampproofing, bituminous, 1 coat, 4' high	.72	3.93	4.65
1400	8' high	1.44	7.85	9.29
1800	12' high	2.16	12.20	14.36
2000	2 coats, 4' high	1.44	4.85	6.29
2400	8' high	2.88	9.70	12.58
2800	12' high	4.32	14.95	19.27
3000	Asphalt with fibers, 1/16" thick, 4' high	1.64	4.85	6.49
3400	8' high	3.28	9.70	12.98
3800	12' high	4.92	14.95	19.87
4000	1/8" thick, 4' high	2.88	5.75	8.63
4400	8' high	5.75	11.55	17.30
4800	12' high	8.65	17.70	26.35
5000	Asphalt coated board and mastic, 1/4" thick, 4' high	3.48	5.30	8.78
5400	8' high	6.95	10.60	17.55
5800	12' high	10.45	16.25	26.70
6000	1/2" thick, 4' high	5.20	7.35	12.55
6400	8' high	10.45	14.70	25.15
6800	12' high	15.65	22.50	38.15
7000	Cementitious coating, on walls, 1/8" thick coating, 4' high	7.10	7.50	14.60
7400	8' high	14.25	14.95	29.20
7800	12' high	21.50	22.50	44
8000	Cementitious/metallic slurry, 2 coat, 1/4" thick, 2' high	26.50	181	207.50
8400	4' high	53	360	413
8800	6' high	79.50	545	624.50

A1020 Special Foundations

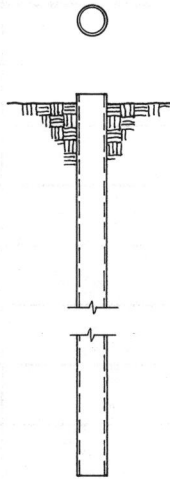

The Cast-in-Place Concrete Pile System includes: a defined number of 4,000 p.s.i. concrete piles with thin-wall, straight-sided, steel shells that have a standard steel plate driving point. An allowance for cutoffs is included.

The Expanded System Listing shows costs per cluster of piles. Clusters range from one pile to twenty piles. Loads vary from 50 Kips to 1,600 Kips. Both end-bearing and friction-type piles are shown.

Please see the reference section for cost of mobilization of the pile driving equipment and other design and cost information.

System Components	QUANTITY	UNIT	COST EACH		
			MAT.	INST.	TOTAL
SYSTEM A1020 110 2220					
CIP SHELL CONCRETE PILE, 25′ LONG, 50K LOAD, END BEARING, 1 PILE					
Cast in place piles, end bearing, no mobil	27.000	V.L.F.	864	266.76	1,130.76
Steel pipe pile standard point, 12″ or 14″ diameter pile	1.000	Ea.	72	78.50	150.50
Pile cutoff, conc. pile with thin steel shell	1.000	Ea.		13.15	13.15
TOTAL			936	358.41	1,294.41

A1020 110	C.I.P. Concrete Piles		COST EACH		
			MAT.	INST.	TOTAL
2220	CIP shell concrete pile, 25′ long, 50K load, end bearing, 1 pile		935	360	1,295
2240	100K load, end bearing, 2 pile cluster		1,875	715	2,590
2260	200K load, end bearing, 4 pile cluster	RA1020 -100	3,750	1,425	5,175
2280	400K load, end bearing, 7 pile cluster		6,550	2,500	9,050
2300	10 pile cluster		9,350	3,575	12,925
2320	800K load, end bearing, 13 pile cluster		18,600	6,550	25,150
2340	17 pile cluster		24,400	8,550	32,950
2360	1200K load, end bearing, 14 pile cluster		20,100	7,050	27,150
2380	19 pile cluster		27,300	9,575	36,875
2400	1600K load, end bearing, 19 pile cluster		27,300	9,575	36,875
2420	50′ long, 50K load, end bearing, 1 pile		1,775	615	2,390
2440	Friction type, 2 pile cluster		3,375	1,225	4,600
2460	3 pile cluster		5,075	1,850	6,925
2480	100K load, end bearing, 2 pile cluster		3,525	1,225	4,750
2500	Friction type, 4 pile cluster		6,750	2,450	9,200
2520	6 pile cluster		10,100	3,700	13,800
2540	200K load, end bearing, 4 pile cluster		7,075	2,450	9,525
2560	Friction type, 8 pile cluster		13,500	4,925	18,425
2580	10 pile cluster		16,900	6,150	23,050
2600	400K load, end bearing, 7 pile cluster		12,400	4,300	16,700
2620	Friction type, 16 pile cluster		27,000	9,850	36,850
2640	19 pile cluster		32,100	11,700	43,800
2660	800K load, end bearing, 14 pile cluster		38,600	11,600	50,200
2680	20 pile cluster		55,000	16,500	71,500
2700	1200K load, end bearing, 15 pile cluster		41,400	12,400	53,800
2720	1600K load, end bearing, 20 pile cluster		55,000	16,500	71,500

A10 Foundations

A1020 Special Foundations

A1020 110	C.I.P. Concrete Piles	COST EACH		
		MAT.	INST.	TOTAL
3740	75' long, 50K load, end bearing, 1 pile	2,725	1,050	3,775
3760	Friction type, 2 pile cluster	5,200	2,075	7,275
3780	3 pile cluster	7,800	3,125	10,925
3800	100K load, end bearing, 2 pile cluster	5,450	2,075	7,525
3820	Friction type, 3 pile cluster	7,800	3,125	10,925
3840	5 pile cluster	13,000	5,200	18,200
3860	200K load, end bearing, 4 pile cluster	10,900	4,175	15,075
3880	6 pile cluster	16,300	6,225	22,525
3900	Friction type, 6 pile cluster	15,600	6,225	21,825
3910	7 pile cluster	18,200	7,275	25,475
3920	400K load, end bearing, 7 pile cluster	19,100	7,275	26,375
3930	11 pile cluster	29,900	11,400	41,300
3940	Friction type, 12 pile cluster	31,200	12,500	43,700
3950	14 pile cluster	36,400	14,600	51,000
3960	800K load, end bearing, 15 pile cluster	63,500	20,300	83,800
3970	20 pile cluster	84,500	27,100	111,600
3980	1200K load, end bearing, 17 pile cluster	72,000	23,000	95,000
3990				

A1020 Special Foundations

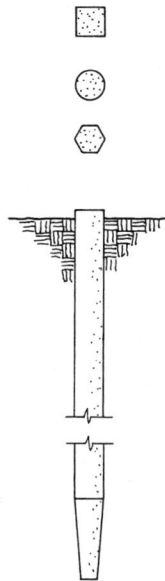

The Precast Concrete Pile System includes: pre-stressed concrete piles; standard steel driving point; and an allowance for cutoffs.

The Expanded System Listing shows costs per cluster of piles. Clusters range from one pile to twenty piles. Loads vary from 50 Kips to 1,600 Kips. Both end-bearing and friction type piles are listed.

Please see the reference section for cost of mobilization of the pile driving equipment and other design and cost information.

System Components	QUANTITY	UNIT	COST EACH		
			MAT.	INST.	TOTAL
SYSTEM A1020 120 2220					
PRECAST CONCRETE PILE, 50′ LONG, 50K LOAD, END BEARING, 1 PILE					
Precast, prestressed conc. piles, 10″ square, no mobil.	53.000	V.L.F.	840.05	448.91	1,288.96
Steel pipe pile standard point, 8″ to 10″ diameter	1.000	Ea.	52	69	121
Piling special costs cutoffs concrete piles plain	1.000	Ea.		91	91
TOTAL			892.05	608.91	1,500.96

A1020 120	Precast Concrete Piles		COST EACH		
			MAT.	INST.	TOTAL
2220	Precast conc pile, 50′ long, 50K load, end bearing, 1 pile		890	605	1,495
2240	Friction type, 2 pile cluster		2,250	1,250	3,500
2260	4 pile cluster		4,525	2,525	7,050
2280	100K load, end bearing, 2 pile cluster	RA1020 -100	1,775	1,225	3,000
2300	Friction type, 2 pile cluster		2,250	1,250	3,500
2320	4 pile cluster		4,525	2,525	7,050
2340	7 pile cluster		7,900	4,400	12,300
2360	200K load, end bearing, 3 pile cluster		2,675	1,825	4,500
2380	4 pile cluster		3,575	2,425	6,000
2400	Friction type, 8 pile cluster		9,025	5,025	14,050
2420	9 pile cluster		10,200	5,675	15,875
2440	14 pile cluster		15,800	8,850	24,650
2460	400K load, end bearing, 6 pile cluster		5,350	3,650	9,000
2480	8 pile cluster		7,125	4,875	12,000
2500	Friction type, 14 pile cluster		15,800	8,850	24,650
2520	16 pile cluster		18,100	10,100	28,200
2540	18 pile cluster		20,300	11,400	31,700
2560	800K load, end bearing, 12 pile cluster		11,600	7,900	19,500
2580	16 pile cluster		14,300	9,750	24,050
2600	1200K load, end bearing, 19 pile cluster		48,500	14,200	62,700
2620	20 pile cluster		51,000	15,000	66,000
2640	1600K load, end bearing, 19 pile cluster		48,500	14,200	62,700
4660	100′ long, 50K load, end bearing, 1 pile		1,725	1,050	2,775
4680	Friction type, 1 pile		2,175	1,075	3,250

A1020 Special Foundations

A1020 120	Precast Concrete Piles	COST EACH		
		MAT.	INST.	TOTAL
4700	2 pile cluster	4,325	2,175	6,500
4720	100K load, end bearing, 2 pile cluster	3,425	2,100	5,525
4740	Friction type, 2 pile cluster	4,325	2,175	6,500
4760	3 pile cluster	6,500	3,250	9,750
4780	4 pile cluster	8,675	4,325	13,000
4800	200K load, end bearing, 3 pile cluster	5,150	3,150	8,300
4820	4 pile cluster	6,875	4,200	11,075
4840	Friction type, 3 pile cluster	6,500	3,250	9,750
4860	5 pile cluster	10,800	5,425	16,225
4880	400K load, end bearing, 6 pile cluster	10,300	6,300	16,600
4900	8 pile cluster	13,700	8,400	22,100
4910	Friction type, 8 pile cluster	17,300	8,675	25,975
4920	10 pile cluster	21,700	10,900	32,600
4930	800K load, end bearing, 13 pile cluster	22,300	13,600	35,900
4940	16 pile cluster	27,500	16,800	44,300
4950	1200K load, end bearing, 19 pile cluster	94,500	24,700	119,200
4960	20 pile cluster	99,500	26,000	125,500
4970	1600K load, end bearing, 19 pile cluster	94,500	24,700	119,200

A10 Foundations

A1020 Special Foundations

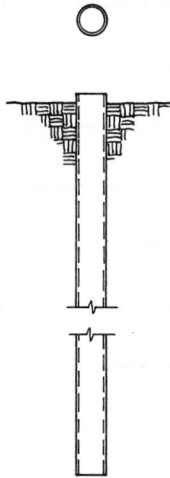

The Steel Pipe Pile System includes: steel pipe sections filled with 4,000 p.s.i. concrete; a standard steel driving point; splices when required and an allowance for cutoffs.

The Expanded System Listing shows costs per cluster of piles. Clusters range from one pile to twenty piles. Loads vary from 50 Kips to 1,600 Kips. Both end-bearing and friction-type piles are shown.

Please see the reference section for cost of mobilization of the pile driving equipment and other design and cost information.

System Components	QUANTITY	UNIT	COST EACH		
			MAT.	INST.	TOTAL
SYSTEM A1020 130 2220					
CONC. FILL STEEL PIPE PILE, 50' LONG, 50K LOAD, END BEARING, 1 PILE					
Piles, steel, pipe, conc. filled, 12" diameter	53.000	V.L.F.	1,722.50	756.84	2,479.34
Steel pipe pile, standard point, for 12" or 14" diameter pipe	1.000	Ea.	144	157	301
Pile cut off, concrete pile, thin steel shell	1.000	Ea.		13.15	13.15
TOTAL			1,866.50	926.99	2,793.49

A1020 130	Steel Pipe Piles	COST EACH		
		MAT.	INST.	TOTAL
2220	Conc. fill steel pipe pile, 50' long, 50K load, end bearing, 1 pile	1,875	930	2,805
2240	Friction type, 2 pile cluster	3,725	1,850	5,575
2250	100K load, end bearing, 2 pile cluster	3,725	1,850	5,575
2260	3 pile cluster	5,600	2,775	8,375
2300	Friction type, 4 pile cluster	7,475	3,725	11,200
2320	5 pile cluster	9,325	4,650	13,975
2340	10 pile cluster	18,700	9,275	27,975
2360	200K load, end bearing, 3 pile cluster	5,600	2,775	8,375
2380	4 pile cluster	7,475	3,725	11,200
2400	Friction type, 4 pile cluster	7,475	3,725	11,200
2420	8 pile cluster	14,900	7,400	22,300
2440	9 pile cluster	16,800	8,325	25,125
2460	400K load, end bearing, 6 pile cluster	11,200	5,575	16,775
2480	7 pile cluster	13,100	6,475	19,575
2500	Friction type, 9 pile cluster	16,800	8,325	25,125
2520	16 pile cluster	29,900	14,900	44,800
2540	19 pile cluster	35,500	17,600	53,100
2560	800K load, end bearing, 11 pile cluster	20,500	10,200	30,700
2580	14 pile cluster	26,100	13,000	39,100
2600	15 pile cluster	28,000	13,900	41,900
2620	Friction type, 17 pile cluster	31,700	15,700	47,400
2640	1200K load, end bearing, 16 pile cluster	29,900	14,900	44,800
2660	20 pile cluster	37,300	18,600	55,900
2680	1600K load, end bearing, 17 pile cluster	31,700	15,700	47,400
3700	100' long, 50K load, end bearing, 1 pile	3,675	1,850	5,525
3720	Friction type, 1 pile	3,675	1,850	5,525

Note: RA1020 -100 (reference box beside line 2260)

13

A10 Foundations

A1020 Special Foundations

A1020 130	Steel Pipe Piles	COST EACH		
		MAT.	INST.	TOTAL
3740	2 pile cluster	7,350	3,675	11,025
3760	100K load, end bearing, 2 pile cluster	7,350	3,675	11,025
3780	Friction type, 2 pile cluster	7,350	3,675	11,025
3800	3 pile cluster	11,000	5,525	16,525
3820	200K load, end bearing, 3 pile cluster	11,000	5,525	16,525
3840	4 pile cluster	14,700	7,350	22,050
3860	Friction type, 3 pile cluster	11,000	5,525	16,525
3880	4 pile cluster	14,700	7,350	22,050
3900	400K load, end bearing, 6 pile cluster	22,100	11,000	33,100
3910	7 pile cluster	25,700	12,900	38,600
3920	Friction type, 5 pile cluster	18,400	9,175	27,575
3930	8 pile cluster	29,400	14,700	44,100
3940	800K load, end bearing, 11 pile cluster	40,500	20,200	60,700
3950	14 pile cluster	51,500	25,700	77,200
3960	15 pile cluster	55,000	27,600	82,600
3970	1200K load, end bearing, 16 pile cluster	59,000	29,400	88,400
3980	20 pile cluster	73,500	36,700	110,200
3990	1600K load, end bearing, 17 pile cluster	62,500	31,300	93,800

A1020 Special Foundations

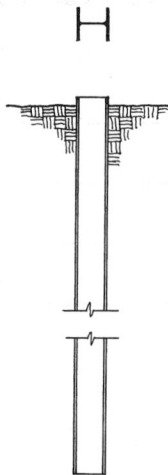

H

A Steel "H" Pile System includes: steel H sections; heavy duty driving point; splices where applicable and allowance for cutoffs.

The Expanded System Listing shows costs per cluster of piles. Clusters range from one pile to seventeen piles. Loads vary from 50 Kips to 2,000 Kips. All loads for Steel H Pile systems are given in terms of end bearing capacity.

Steel sections range from 10″ x 10″ to 14″ x 14″ in the Expanded System Listing. The 14″ x 14″ steel section is used for all H piles used in applications requiring a working load over 800 Kips.

Please see the reference section for cost of mobilization of the pile driving equipment and other design and cost information.

System Components	QUANTITY	UNIT	COST EACH		
			MAT.	INST.	TOTAL
SYSTEM A1020 140 2220					
STEEL H PILES, 50′ LONG, 100K LOAD, END BEARING, 1 PILE					
Steel H piles 10″ x 10″, 42 #/L.F.	53.000	V.L.F.	1,192.50	515.69	1,708.19
Heavy duty points, not in leads, 10″ wide	1.000	Ea.	194	159	353
Pile cut off, steel pipe or H piles	1.000	Ea.		26.50	26.50
TOTAL			1,386.50	701.19	2,087.69

A1020 140	Steel H Piles	COST EACH		
		MAT.	INST.	TOTAL
2220	Steel H piles, 50′ long, 100K load, end bearing, 1 pile	1,375	705	2,080
2260	2 pile cluster	2,775	1,400	4,175
2280	200K load, end bearing, 2 pile cluster	2,775	1,400	4,175
2300	3 pile cluster	4,150	2,100	6,250
2320	400K load, end bearing, 3 pile cluster	4,150	2,100	6,250
2340	4 pile cluster	5,550	2,800	8,350
2360	6 pile cluster	8,325	4,200	12,525
2380	800K load, end bearing, 5 pile cluster	6,925	3,525	10,450
2400	7 pile cluster	9,700	4,900	14,600
2420	12 pile cluster	16,600	8,425	25,025
2440	1200K load, end bearing, 8 pile cluster	11,100	5,600	16,700
2460	11 pile cluster	15,300	7,700	23,000
2480	17 pile cluster	23,600	11,900	35,500
2500	1600K load, end bearing, 10 pile cluster	17,200	7,350	24,550
2520	14 pile cluster	24,100	10,300	34,400
2540	2000K load, end bearing, 12 pile cluster	20,700	8,800	29,500
2560	18 pile cluster	31,000	13,300	44,300
2580				
3580	100′ long, 50K load, end bearing, 1 pile	4,575	1,650	6,225
3600	100K load, end bearing, 1 pile	4,575	1,650	6,225
3620	2 pile cluster	9,150	3,275	12,425
3640	200K load, end bearing, 2 pile cluster	9,150	3,275	12,425
3660	3 pile cluster	13,700	4,900	18,600
3680	400K load, end bearing, 3 pile cluster	13,700	4,900	18,600
3700	4 pile cluster	18,300	6,525	24,825
3720	6 pile cluster	27,500	9,825	37,325

RA1020 -100

A10 Foundations

A1020 Special Foundations

A1020 140	Steel H Piles	COST EACH		
		MAT.	INST.	TOTAL
3740	800K load, end bearing, 5 pile cluster	22,900	8,175	31,075
3760	7 pile cluster	32,100	11,400	43,500
3780	12 pile cluster	55,000	19,600	74,600
3800	1200K load, end bearing, 8 pile cluster	36,600	13,100	49,700
3820	11 pile cluster	50,500	18,000	68,500
3840	17 pile cluster	78,000	27,800	105,800
3860	1600K load, end bearing, 10 pile cluster	45,800	16,300	62,100
3880	14 pile cluster	64,000	22,900	86,900
3900	2000K load, end bearing, 12 pile cluster	55,000	19,600	74,600
3920	18 pile cluster	82,500	29,400	111,900

A1020 Special Foundations

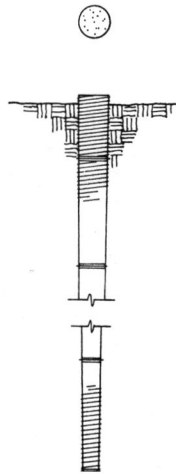

The Step Tapered Steel Pile System includes: step tapered piles filled with 4,000 p.s.i. concrete. The cost for splices and pile cutoffs is included.

The Expanded System Listing shows costs per cluster of piles. Clusters range from one pile to twenty-four piles. Both end bearing piles and friction piles are listed. Loads vary from 50 Kips to 1,600 Kips.

Please see the reference section for cost of mobilization of the pile driving equipment and other design and cost information.

System Components	QUANTITY	UNIT	COST EACH		
			MAT.	INST.	TOTAL
SYSTEM A1020 150 1000					
STEEL PILE, STEP TAPERED, 50' LONG, 50K LOAD, END BEARING, 1 PILE					
Steel shell step tapered conc. filled piles, 8" tip 60 ton capacity to 60'	53.000	V.L.F.	667.80	425.06	1,092.86
Pile cutoff, steel pipe or H piles	1.000	Ea.		26.50	26.50
TOTAL			667.80	451.56	1,119.36

A1020 150	Step-Tapered Steel Piles		COST EACH		
			MAT.	INST.	TOTAL
1000	Steel pile, step tapered, 50' long, 50K load, end bearing, 1 pile		670	455	1,125
1200	Friction type, 3 pile cluster		2,000	1,350	3,350
1400	100K load, end bearing, 2 pile cluster	RA1020 -100	1,325	905	2,230
1600	Friction type, 4 pile cluster		2,675	1,800	4,475
1800	200K load, end bearing, 4 pile cluster		2,675	1,800	4,475
2000	Friction type, 6 pile cluster		4,000	2,700	6,700
2200	400K load, end bearing, 7 pile cluster		4,675	3,175	7,850
2400	Friction type, 10 pile cluster		6,675	4,525	11,200
2600	800K load, end bearing, 14 pile cluster		9,350	6,325	15,675
2800	Friction type, 18 pile cluster		12,000	8,125	20,125
3000	1200K load, end bearing, 16 pile cluster		12,000	7,725	19,725
3200	Friction type, 21 pile cluster		15,700	10,100	25,800
3400	1600K load, end bearing, 18 pile cluster		13,500	8,675	22,175
3600	Friction type, 24 pile cluster		17,900	11,600	29,500
5000	100' long, 50K load, end bearing, 1 pile		1,350	900	2,250
5200	Friction type, 2 pile cluster		2,700	1,800	4,500
5400	100K load, end bearing, 2 pile cluster		2,700	1,800	4,500
5600	Friction type, 3 pile cluster		4,025	2,700	6,725
5800	200K load, end bearing, 4 pile cluster		5,375	3,600	8,975
6000	Friction type, 5 pile cluster		6,725	4,500	11,225
6200	400K load, end bearing, 7 pile cluster		9,400	6,300	15,700
6400	Friction type, 8 pile cluster		10,800	7,175	17,975
6600	800K load, end bearing, 15 pile cluster		20,200	13,500	33,700
6800	Friction type, 16 pile cluster		21,500	14,400	35,900
7000	1200K load, end bearing, 17 pile cluster		39,600	23,500	63,100
7200	Friction type, 19 pile cluster		29,700	17,800	47,500
7400	1600K load, end bearing, 20 pile cluster		31,300	18,800	50,100
7600	Friction type, 22 pile cluster		34,400	20,700	55,100

A10 Foundations

A1020 Special Foundations

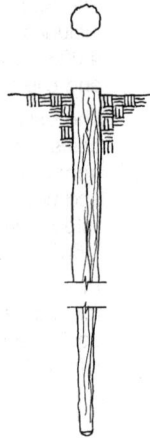

The Treated Wood Pile System includes: creosoted wood piles; a standard steel driving point; and an allowance for cutoffs.

The Expanded System Listing shows costs per cluster of piles. Clusters range from three piles to twenty piles. Loads vary from 50 Kips to 400 Kips. Both end-bearing and friction type piles are listed.

Please see the reference section for cost of mobilization of the pile driving equipment and other design and cost information.

System Components	QUANTITY	UNIT	COST EACH		
			MAT.	INST.	TOTAL
SYSTEM A1020 160 2220					
WOOD PILES, 25' LONG, 50K LOAD, END BEARING, 3 PILE CLUSTER					
Wood piles, treated, 12" butt, 8" tip, up to 30' long	81.000	V.L.F.	951.75	769.50	1,721.25
Point for driving wood piles	3.000	Ea.	72	75	147
Pile cutoff, wood piles	3.000	Ea.		39.45	39.45
TOTAL			1,023.75	883.95	1,907.70

A1020 160	Treated Wood Piles	COST EACH		
		MAT.	INST.	TOTAL
2220	Wood piles, 25' long, 50K load, end bearing, 3 pile cluster	1,025	885	1,910
2240	Friction type, 3 pile cluster	950	810	1,760
2260	5 pile cluster	1,575	1,350	2,925
2280	100K load, end bearing, 4 pile cluster	1,375	1,175	2,550
2300	5 pile cluster	1,700	1,475	3,175
2320	6 pile cluster	2,050	1,775	3,825
2340	Friction type, 5 pile cluster	1,575	1,350	2,925
2360	6 pile cluster	1,900	1,625	3,525
2380	10 pile cluster	3,175	2,675	5,850
2400	200K load, end bearing, 8 pile cluster	2,725	2,375	5,100
2420	10 pile cluster	3,425	2,925	6,350
2440	12 pile cluster	4,100	3,525	7,625
2460	Friction type, 10 pile cluster	3,175	2,675	5,850
2480	400K load, end bearing, 16 pile cluster	5,450	4,725	10,175
2500	20 pile cluster	6,825	5,900	12,725
4520	50' long, 50K load, end bearing, 3 pile cluster	2,225	1,300	3,525
4540	4 pile cluster	2,950	1,750	4,700
4560	Friction type, 2 pile cluster	1,400	800	2,200
4580	3 pile cluster	2,125	1,200	3,325
4600	100K load, end bearing, 5 pile cluster	3,700	2,175	5,875
4620	8 pile cluster	5,925	3,475	9,400
4640	Friction type, 3 pile cluster	2,125	1,200	3,325
4660	5 pile cluster	3,525	2,000	5,525
4680	200K load, end bearing, 9 pile cluster	6,675	3,925	10,600
4700	10 pile cluster	7,400	4,375	11,775
4720	15 pile cluster	11,100	6,550	17,650
4740	Friction type, 5 pile cluster	3,525	2,000	5,525
4760	6 pile cluster	4,225	2,375	6,600

RA1020 -100

A10 Foundations

A1020 Special Foundations

A1020 160	Treated Wood Piles	COST EACH		
		MAT.	INST.	TOTAL
4780	10 pile cluster	7,050	4,000	11,050
4800	400K load, end bearing, 18 pile cluster	13,300	7,850	21,150
4820	20 pile cluster	14,800	8,725	23,525
4840	Friction type, 9 pile cluster	6,350	3,575	9,925
4860	10 pile cluster	7,050	4,000	11,050
9000	Add for boot for driving tip, each pile	24	18.50	42.50

A1020 Special Foundations

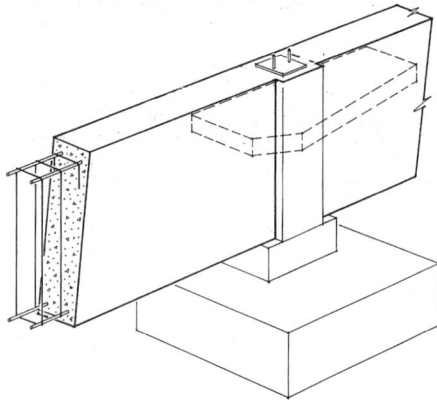

The Grade Beam System includes: excavation with a truck mounted backhoe; hand trim; backfill; forms (four uses); reinforcing steel; and 3,000 p.s.i. concrete placed from chute.

Superimposed loads vary in the listing from 8 Kips per linear foot (KLF) to 50 KLF. In the Expanded System Listing, the span of the beams varies from 15' to 40'. Depth varies from 28" to 52". Width varies from 12" to 28".

Please see the reference section for further design and cost information.

System Components	QUANTITY	UNIT	COST PER L.F.		
			MAT.	INST.	TOTAL
SYSTEM A1020 210 2220					
GRADE BEAM, 15' SPAN, 28" DEEP, 12" WIDE, 8 KLF LOAD					
Excavation, trench, hydraulic backhoe, 3/8 CY bucket	.260	C.Y.		2.11	2.11
Trim sides and bottom of trench, regular soil	2.000	S.F.		1.66	1.66
Backfill, by hand, compaction in 6" layers, using vibrating plate	.170	C.Y.		1.21	1.21
Forms in place, grade beam, 4 uses	4.700	SFCA	5.45	22.51	27.96
Reinforcing in place, beams & girders, #8 to #14	.019	Ton	32.30	16.44	48.74
Concrete ready mix, regular weight, 3000 psi	.090	C.Y.	9.99		9.99
Place and vibrate conc. for grade beam, direct chute	.090	C.Y.		1.49	1.49
TOTAL			47.74	45.42	93.16

A1020 210	Grade Beams		COST PER L.F.		
			MAT.	INST.	TOTAL
2220	Grade beam, 15' span, 28" deep, 12" wide, 8 KLF load		47.50	45.50	93
2240	14" wide, 12 KLF load		49	46	95
2260	40" deep, 12" wide, 16 KLF load	RA1020	46.50	56	102.50
2280	20 KLF load	-230	55	60.50	115.50
2300	52" deep, 12" wide, 30 KLF load		62	76	138
2320	40 KLF load		79	84.50	163.50
2340	50 KLF load		96	92.50	188.50
3360	20' span, 28" deep, 12" wide, 2 KLF load		27.50	35.50	63
3380	16" wide, 4 KLF load		37	39.50	76.50
3400	40" deep, 12" wide, 8 KLF load		46.50	56	102.50
3420	12 KLF load		62	63	125
3440	14" wide, 16 KLF load		76	70.50	146.50
3460	52" deep, 12" wide, 20 KLF load		80.50	85.50	166
3480	14" wide, 30 KLF load		114	102	216
3500	20" wide, 40 KLF load		132	108	240
3520	24" wide, 50 KLF load		165	123	288
4540	30' span, 28" deep, 12" wide, 1 KLF load		29	36	65
4560	14" wide, 2 KLF load		50	50	100
4580	40" deep, 12" wide, 4 KLF load		56	59.50	115.50
4600	18" wide, 8 KLF load		83.50	72.50	156
4620	52" deep, 14" wide, 12 KLF load		109	99	208
4640	20" wide, 16 KLF load		132	108	240
4660	24" wide, 20 KLF load		165	124	289
4680	36" wide, 30 KLF load		237	154	391
4700	48" wide, 40 KLF load		310	188	498
5720	40' span, 40" deep, 12" wide, 1 KLF load		40	52.50	92.50

A10 Foundations

A1020   Special Foundations

A1020 210	Grade Beams	COST PER L.F.		
		MAT.	INST.	TOTAL
5740	2 KLF load	51.50	59	110.50
5760	52″ deep, 12″ wide, 4 KLF load	79	84.50	163.50
5780	20″ wide, 8 KLF load	127	106	233
5800	28″ wide, 12 KLF load	175	127	302
5820	38″ wide, 16 KLF load	242	156	398
5840	46″ wide, 20 KLF load	320	192	512

A1020 Special Foundations

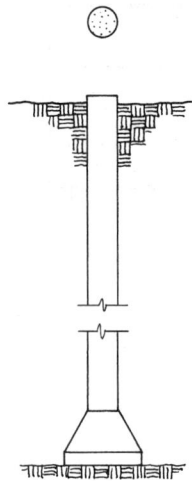

Caisson Systems are listed for three applications: stable ground, wet ground and soft rock. Concrete used is 3,000 p.s.i. placed from chute. Included are a bell at the bottom of the caisson shaft (if applicable) along with required excavation and disposal of excess excavated material up to two miles from job site.

The Expanded System lists cost per caisson. End-bearing loads vary from 200 Kips to 3,200 Kips. The dimensions of the caissons range from 2' x 50' to 7' x 200'.

Please see the reference section for further design and cost information.

System Components	QUANTITY	UNIT	COST EACH		
			MAT.	INST.	TOTAL
SYSTEM A1020 310 2200					
CAISSON, STABLE GROUND, 3000 PSI CONC., 10KSF BRNG, 200K LOAD, 2'X50'					
Caissons, drilled, to 50', 24" shaft diameter, .116 C.Y./L.F.	50.000	V.L.F.	892.50	1,402.50	2,295
Reinforcing in place, columns, #3 to #7	.060	Ton	102	93	195
Caisson bell excavation and concrete, 4' diameter .444 CY	1.000	Ea.	49.50	266	315.50
Load & haul excess excavation, 2 miles	6.240	C.Y.		34.32	34.32
TOTAL			1,044	1,795.82	2,839.82

A1020 310	Caissons		COST EACH		
			MAT.	INST.	TOTAL
2200	Caisson, stable ground, 3000 PSI conc, 10KSF brng, 200K load, 2'x50'		1,050	1,800	2,850
2400	400K load, 2'-6"x50'-0"		1,700	2,875	4,575
2600	800K load, 3'-0"x100'-0"	RA1020 -200	4,600	6,825	11,425
2800	1200K load, 4'-0"x100'-0"		8,025	8,650	16,675
3000	1600K load, 5'-0"x150'-0"		18,100	13,400	31,500
3200	2400K load, 6'-0"x150'-0"		26,500	17,400	43,900
3400	3200K load, 7'-0"x200'-0"		47,400	25,200	72,600
5000	Wet ground, 3000 PSI conc., 10 KSF brng, 200K load, 2'-0"x50'-0"		910	2,525	3,435
5200	400K load, 2'-6"x50'-0"		1,500	4,325	5,825
5400	800K load, 3'-0"x100'-0"		4,000	11,600	15,600
5600	1200K load, 4'-0"x100'-0"		6,950	17,300	24,250
5800	1600K load, 5'-0"x150'-0"		15,600	37,300	52,900
6000	2400K load, 6'-0"x150'-0"		22,800	45,800	68,600
6200	3200K load, 7'-0"x200'-0"		40,800	73,500	114,300
7800	Soft rock, 3000 PSI conc., 10 KSF brng, 200K load, 2'-0"x50'-0"		910	13,700	14,610
8000	400K load, 2'-6"x50'-0"		1,500	22,200	23,700
8200	800K load, 3'-0"x100'-0"		4,000	58,000	62,000
8400	1200K load, 4'-0"x100'-0"		6,950	85,000	91,950
8600	1600K load, 5'-0"x150'-0"		15,600	175,000	190,600
8800	2400K load, 6'-0"x150'-0"		22,800	208,000	230,800
9000	3200K load, 7'-0"x200'-0"		40,800	332,000	372,800

A10 Foundations

A1020 Special Foundations

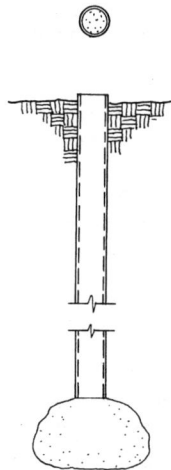

Pressure Injected Piles are usually uncased up to 25' and cased over 25' depending on soil conditions.

These costs include excavation and hauling of excess materials; steel casing over 25'; reinforcement; 3,000 p.s.i. concrete; plus mobilization and demobilization of equipment for a distance of up to fifty miles to and from the job site.

The Expanded System lists cost per cluster of piles. Clusters range from one pile to eight piles. End-bearing loads range from 50 Kips to 1,600 Kips.

Please see the reference section for further design and cost information.

System Components	QUANTITY	UNIT	COST EACH		
			MAT.	INST.	TOTAL
SYSTEM A1020 710 4200 **PRESSURE INJECTED FOOTING, END BEARING, 50' LONG, 50K LOAD, 1 PILE**					
Pressure injected footings, cased, 30-60 ton cap., 12" diameter	50.000	V.L.F.	855	1,265	2,120
Pile cutoff, concrete pile with thin steel shell	1.000	Ea.		13.15	13.15
TOTAL			855	1,278.15	2,133.15

A1020 710	Pressure Injected Footings		COST EACH		
			MAT.	INST.	TOTAL
2200	Pressure injected footing, end bearing, 25' long, 50K load, 1 pile		395	900	1,295
2400	100K load, 1 pile		395	900	1,295
2600	2 pile cluster	RA1020	795	1,800	2,595
2800	200K load, 2 pile cluster	-100	795	1,800	2,595
3200	400K load, 4 pile cluster		1,575	3,575	5,150
3400	7 pile cluster		2,775	6,275	9,050
3800	1200K load, 6 pile cluster		3,300	6,800	10,100
4000	1600K load, 7 pile cluster		3,850	7,950	11,800
4200	50' long, 50K load, 1 pile		855	1,275	2,130
4400	100K load, 1 pile		1,475	1,275	2,750
4600	2 pile cluster		2,950	2,575	5,525
4800	200K load, 2 pile cluster		2,950	2,575	5,525
5000	4 pile cluster		5,900	5,100	11,000
5200	400K load, 4 pile cluster		5,900	5,100	11,000
5400	8 pile cluster		11,800	10,200	22,000
5600	800K load, 7 pile cluster		10,300	8,950	19,250
5800	1200K load, 6 pile cluster		9,450	7,675	17,125
6000	1600K load, 7 pile cluster		11,000	8,950	19,950

23

A1030   Slab on Grade

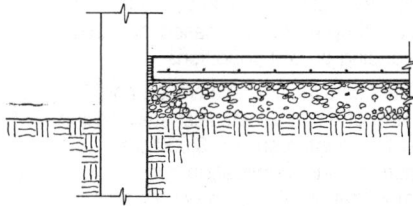

There are four types of Slab on Grade Systems listed: Non-industrial, Light industrial, Industrial and Heavy industrial. Each type is listed two ways: reinforced and non-reinforced. A Slab on Grade system includes three passes with a grader; 6" of compacted gravel fill; polyethylene vapor barrier; 3500 p.s.i. concrete placed by chute; bituminous fiber expansion joint; all necessary edge forms (4 uses); steel trowel finish; and sprayed-on membrane curing compound.

The Expanded System Listing shows costs on a per square foot basis. Thicknesses of the slabs range from 4" to 8". Non-industrial applications are for foot traffic only with negligible abrasion. Light industrial applications are for pneumatic wheels and light abrasion. Industrial applications are for solid rubber wheels and moderate abrasion. Heavy industrial applications are for steel wheels and severe abrasion. All slabs are either shown unreinforced or reinforced with welded wire fabric.

System Components	QUANTITY	UNIT	COST PER S.F.		
			MAT.	INST.	TOTAL
SYSTEM A1030 120 2220					
SLAB ON GRADE, 4" THICK, NON INDUSTRIAL, NON REINFORCED					
Fine grade, 3 passes with grader and roller	.110	S.Y.		.41	.41
Gravel under floor slab, 6" deep, compacted	1.000	S.F.	.28	.27	.55
Polyethylene vapor barrier, standard, .006" thick	1.000	S.F.	.06	.13	.19
Concrete ready mix, regular weight, 3500 psi	.012	C.Y.	1.37		1.37
Place and vibrate concrete for slab on grade, 4" thick, direct chute	.012	C.Y.		.27	.27
Expansion joint, premolded bituminous fiber, 1/2" x 6"	.100	L.F.	.10	.29	.39
Edge forms in place for slab on grade to 6" high, 4 uses	.030	L.F.	.01	.09	.10
Cure with sprayed membrane curing compound	1.000	S.F.	.06	.08	.14
Finishing floor, monolithic steel trowel	1.000	S.F.		.78	.78
TOTAL			1.88	2.32	4.20

A1030 120	Plain & Reinforced		COST PER S.F.		
			MAT.	INST.	TOTAL
2220	Slab on grade, 4" thick, non industrial, non reinforced		1.88	2.32	4.20
2240	Reinforced		2.08	2.66	4.74
2260	Light industrial, non reinforced	RA1030 -200	2.46	2.85	5.31
2280	Reinforced		2.66	3.19	5.85
2300	Industrial, non reinforced		3.08	6	9.08
2320	Reinforced		3.28	6.30	9.58
3340	5" thick, non industrial, non reinforced		2.22	2.39	4.61
3360	Reinforced		2.42	2.73	5.15
3380	Light industrial, non reinforced		2.81	2.92	5.73
3400	Reinforced		3.01	3.26	6.27
3420	Heavy industrial, non reinforced		4.05	7.15	11.20
3440	Reinforced		4.25	7.55	11.80
4460	6" thick, non industrial, non reinforced		2.68	2.34	5.02
4480	Reinforced		3.03	2.79	5.82
4500	Light industrial, non reinforced		3.27	2.87	6.14
4520	Reinforced		3.82	3.47	7.29
4540	Heavy industrial, non reinforced		4.53	7.25	11.78
4560	Reinforced		4.88	7.70	12.58
5580	7" thick, non industrial, non reinforced		3.03	2.42	5.45
5600	Reinforced		3.46	2.90	6.36
5620	Light industrial, non reinforced		3.63	2.95	6.58
5640	Reinforced		4.06	3.43	7.49
5660	Heavy industrial, non reinforced		4.89	7.15	12.04
5680	Reinforced		5.25	7.60	12.85

A10 Foundations

A1030 Slab on Grade

A1030 120	Plain & Reinforced	COST PER S.F.		
		MAT.	INST.	TOTAL
6700	8" thick, non industrial, non reinforced	3.37	2.47	5.84
6720	Reinforced	3.73	2.87	6.60
6740	Light industrial, non reinforced	3.98	3	6.98
6760	Reinforced	4.34	3.40	7.74
6780	Heavy industrial, non reinforced	5.25	7.25	12.50
6800	Reinforced	5.75	7.65	13.40

A2010 Basement Excavation

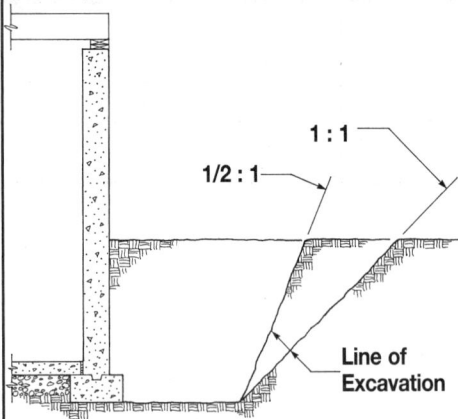

Pricing Assumptions: Two-thirds of excavation is by 2-1/2 C.Y. wheel mounted front end loader and one-third by 1-1/2 C.Y. hydraulic excavator.

Two-mile round trip haul by 12 C.Y. tandem trucks is included for excavation wasted and storage of suitable fill from excavated soil. For excavation in clay, all is wasted and the cost of suitable backfill with two-mile haul is included.

Sand and gravel assumes 15% swell and compaction; common earth assumes 25% swell and 15% compaction; clay assumes 40% swell and 15% compaction (non-clay).

In general, the following items are accounted for in the costs in the table below.

1. Excavation for building or other structure to depth and extent indicated.
2. Backfill compacted in place.
3. Haul of excavated waste.
4. Replacement of unsuitable material with bank run gravel.

Note: Additional excavation and fill beyond this line of general excavation for the building (as required for isolated spread footings, strip footings, etc.) are included in the cost of the appropriate component systems.

System Components	QUANTITY	UNIT	COST PER S.F.		
			MAT.	INST.	TOTAL
SYSTEM A2010 110 2280					
EXCAVATE & FILL, 1000 S.F., 8′ DEEP, SAND, ON SITE STORAGE					
Excavating bulk shovel, 1.5 C.Y. bucket, 150 cy/hr	.262	C.Y.		.44	.44
Excavation, front end loader, 2-1/2 C.Y.	.523	C.Y.		1.06	1.06
Haul earth, 12 C.Y. dump truck	.341	L.C.Y.		1.85	1.85
Backfill, dozer bulk push 300′, including compaction	.562	C.Y.		1.88	1.88
TOTAL				5.23	5.23

A2010 110	Building Excavation & Backfill		COST PER S.F.		
			MAT.	INST.	TOTAL
2220	Excav & fill, 1000 S.F. 4′ sand, gravel, or common earth, on site storage			.88	.88
2240	Off site storage			1.31	1.31
2260	Clay excavation, bank run gravel borrow for backfill		3	1.97	4.97
2280	8′ deep, sand, gravel, or common earth, on site storage	RG1010 -010		5.25	5.25
2300	Off site storage			11.30	11.30
2320	Clay excavation, bank run gravel borrow for backfill		12.20	9.25	21.45
2340	16′ deep, sand, gravel, or common earth, on site storage	RG1030 -400		14.35	14.35
2350	Off site storage			28	28
2360	Clay excavation, bank run gravel borrow for backfill		33.50	24	57.50
3380	4000 S.F.,4′ deep, sand, gravel, or common earth, on site storage			.46	.46
3400	Off site storage			.90	.90
3420	Clay excavation, bank run gravel borrow for backfill		1.53	1	2.53
3440	8′ deep, sand, gravel, or common earth, on site storage			3.66	3.66
3460	Off site storage			6.35	6.35
3480	Clay excavation, bank run gravel borrow for backfill		5.55	5.55	11.10
3500	16′ deep, sand, gravel, or common earth, on site storage			8.80	8.80
3520	Off site storage			17.25	17.25
3540	Clay, excavation, bank run gravel borrow for backfill		14.85	13.20	28.05
4560	10,000 S.F., 4′ deep, sand, gravel, or common earth, on site storage			.26	.26
4580	Off site storage			.52	.52
4600	Clay excavation, bank run gravel borrow for backfill		.93	.61	1.54
4620	8′ deep, sand, gravel, or common earth, on site storage			3.17	3.17
4640	Off site storage			4.79	4.79
4660	Clay excavation, bank run gravel borrow for backfill		3.36	4.31	7.67
4680	16′ deep, sand, gravel, or common earth, on site storage			7.15	7.15
4700	Off site storage			12.10	12.10
4720	Clay excavation, bank run gravel borrow for backfill		8.90	9.80	18.70
5740	30,000 S.F., 4′ deep, sand, gravel, or common earth, on site storage			.17	.17

A20 Basement Construction

A2010 Basement Excavation

A2010 110	Building Excavation & Backfill	COST PER S.F.		
		MAT.	INST.	TOTAL
5760	Off site storage		.33	.33
5780	Clay excavation, bank run gravel borrow for backfill	.54	.35	.89
5800	8' deep, sand & gravel, or common earth, on site storage		2.82	2.82
5820	Off site storage		3.72	3.72
5840	Clay excavation, bank run gravel borrow for backfill	1.89	3.48	5.37
5860	16' deep, sand, gravel, or common earth, on site storage		6.10	6.10
5880	Off site storage		8.80	8.80
5900	Clay excavation, bank run gravel borrow for backfill	4.95	7.60	12.55
6910	100,000 S.F., 4' deep, sand, gravel, or common earth, on site storage		.07	.07
6920	Off site storage		.16	.16
6930	Clay excavation, bank run gravel borrow for backfill	.27	.19	.46
6940	8' deep, sand, gravel, or common earth, on site storage		2.63	2.63
6950	Off site storage		3.12	3.12
6960	Clay excavation, bank run gravel borrow for backfill	1.02	2.99	4.01
6970	16' deep, sand, gravel, or common earth, on site storage		5.50	5.50
6980	Off site storage		6.95	6.95
6990	Clay excavation, bank run gravel borrow for backfill	2.67	6.35	9.02

A2020 Basement Walls

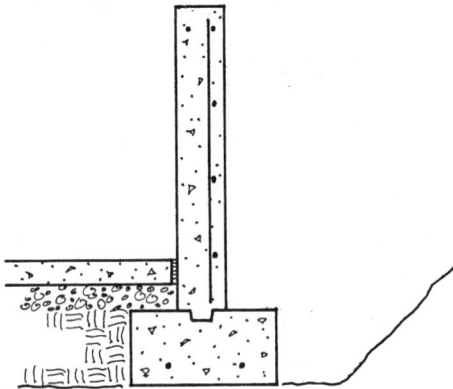

The Foundation Bearing Wall System includes: forms up to 16' high (four uses); 3,000 p.s.i. concrete placed and vibrated; and form removal with breaking form ties and patching walls. The wall systems list walls from 6″ to 16″ thick and are designed with minimum reinforcement.

Excavation and backfill are not included.

Please see the reference section for further design and cost information.

System Components				QUANTITY	UNIT	COST PER L.F.		
						MAT.	INST.	TOTAL
SYSTEM A2020 110 1500								
FOUNDATION WALL, CAST IN PLACE, DIRECT CHUTE, 4' HIGH, 6″ THICK								
Formwork				8.000	SFCA	7.28	38.56	45.84
Reinforcing				3.300	Lb.	2.68	1.28	3.96
Unloading & sorting reinforcing				3.300	Lb.		.07	.07
Concrete, 3,000 psi				.074	C.Y.	8.21		8.21
Place concrete, direct chute				.074	C.Y.		2.05	2.05
Finish walls, break ties and patch voids, one side				4.000	S.F.	.12	3.32	3.44
			TOTAL			18.29	45.28	63.57

A2020 110		Walls, Cast in Place						
	WALL HEIGHT (FT.)	PLACING METHOD	CONCRETE (C.Y. per L.F.)	REINFORCING (LBS. per L.F.)	WALL THICKNESS (IN.)	COST PER L.F.		
						MAT.	INST.	TOTAL
1500	4'	direct chute	.074	3.3	6	18.30	45	63.30
1520			.099	4.8	8	22.50	46.50	69
1540			.123	6.0	10	26	47.50	73.50
1560			.148	7.2	12	29.50	48.50	78
1580	RA2020 -210		.173	8.1	14	33	49	82
1600			.197	9.44	16	37	50.50	87.50
1700	4'	pumped	.074	3.3	6	18.30	46	64.30
1720			.099	4.8	8	22.50	48	70.50
1740			.123	6.0	10	26	49	75
1760			.148	7.2	12	29.50	50.50	80
1780			.173	8.1	14	33	51.50	84.50
1800			.197	9.44	16	37	52.50	89.50
3000	6'	direct chute	.111	4.95	6	27.50	68	95.50
3020			.149	7.20	8	33.50	70	103.50
3040			.184	9.00	10	39	71	110
3060			.222	10.8	12	44.50	72.50	117
3080			.260	12.15	14	50	74	124
3100			.300	14.39	16	56	75.50	131.50

A2020 Basement Walls

A2020 110	Walls, Cast in Place

	WALL HEIGHT (FT.)	PLACING METHOD	CONCRETE (C.Y. per L.F.)	REINFORCING (LBS. per L.F.)	WALL THICKNESS (IN.)	COST PER L.F.		
						MAT.	INST.	TOTAL
3200	6'	pumped	.111	4.95	6	27.50	69.50	97
3220			.149	7.20	8	33.50	72	105.50
3240			.184	9.00	10	39	73.50	112.50
3260			.222	10.8	12	44.50	76	120.50
3280			.260	12.15	14	50	77	127
3300			.300	14.39	16	56	79	135
5000	8'	direct chute	.148	6.6	6	36.50	90.50	127
5020			.199	9.6	8	44.50	93	137.50
5040			.250	12	10	52.50	95	147.50
5060			.296	14.39	12	59.50	97	156.50
5080			.347	16.19	14	62	97.50	159.50
5100			.394	19.19	16	74	101	175
5200	8'	pumped	.148	6.6	6	36.50	93	129.50
5220			.199	9.6	8	44.50	96	140.50
5240			.250	12	10	52.50	98	150.50
5260			.296	14.39	12	59.50	101	160.50
5280			.347	16.19	14	62	101	163
5300			.394	19.19	16	74	105	179
6020	10'	direct chute	.248	12	8	56	116	172
6040			.307	14.99	10	65	118	183
6060			.370	17.99	12	74	121	195
6080			.433	20.24	14	83	123	206
6100			.493	23.99	16	92.50	126	218.50
6220	10'	pumped	.248	12	8	56	120	176
6240			.307	14.99	10	65	123	188
6260			.370	17.99	12	74	126	200
6280			.433	20.24	14	83	128	211
6300			.493	23.99	16	92.50	132	224.50
7220	12'	pumped	.298	14.39	8	67	144	211
7240			.369	17.99	10	78	147	225
7260			.444	21.59	12	89	152	241
7280			.52	24.29	14	99.50	154	253.50
7300			.591	28.79	16	111	158	269
7420	12'	crane & bucket	.298	14.39	8	67	151	218
7440			.369	17.99	10	78	154	232
7460			.444	21.59	12	89	160	249
7480			.52	24.29	14	99.50	163	262.50
7500			.591	28.79	16	111	171	282
8220	14'	pumped	.347	16.79	8	78	168	246
8240			.43	20.99	10	90.50	171	261.50
8260			.519	25.19	12	104	176	280
8280			.607	28.33	14	116	180	296
8300			.69	33.59	16	130	184	314
8420	14'	crane & bucket	.347	16.79	8	78	176	254
8440			.43	20.99	10	90.50	179	269.50
8460			.519	25.19	12	104	187	291
8480			.607	28.33	14	116	191	307
8500			.69	33.59	16	130	198	328
9220	16'	pumped	.397	19.19	8	89.50	192	281.50
9240			.492	23.99	10	104	196	300
9260			.593	28.79	12	119	202	321
9280			.693	32.39	14	133	205	338
9300			.788	38.38	16	148	211	359

A20 Basement Construction

A2020 Basement Walls

A2020 110	Walls, Cast in Place

	WALL HEIGHT (FT.)	PLACING METHOD	CONCRETE (C.Y. per L.F.)	REINFORCING (LBS. per L.F.)	WALL THICKNESS (IN.)	COST PER L.F.		
						MAT.	INST.	TOTAL
9420	16'	crane & bucket	.397	19.19	8	89.50	201	290.50
9440			.492	23.99	10	104	205	309
9460			.593	28.79	12	119	213	332
9480			.693	32.39	14	133	218	351
9500			.788	38.38	16	148	225	373

A20 Basement Construction

A2020 Basement Walls

A2020 220	Subdrainage Piping	COST PER L.F.		
		MAT.	INST.	TOTAL
2000	Piping, excavation & backfill excluded, PVC, perforated			
2110	3" diameter	1.83	3.51	5.34
2130	4" diameter	1.83	3.51	5.34
2140	5" diameter	3.74	3.76	7.50
2150	6" diameter	3.74	3.76	7.50
3000	Metal alum. or steel, perforated asphalt coated			
3150	6" diameter	5.15	7.30	12.45
3160	8" diameter	7.85	7.50	15.35
3170	10" diameter	9.80	7.70	17.50
3180	12" diameter	11	9.75	20.75
3220	18" diameter	16.45	13.55	30
4000	Porous wall concrete			
4130	4" diameter	3.63	3.92	7.55
4150	6" diameter	4.72	4.17	8.89
4160	8" diameter	5.85	5.50	11.35
4180	12" diameter	12.35	6	18.35
4200	15" diameter	14.15	7.45	21.60
4220	18" diameter	18.70	10.35	29.05

CONCRETE COLUMNS

General: It is desirable for purposes of consistency and simplicity to maintain constant column sizes throughout the building height. To do this, concrete strength may be varied (higher strength concrete at lower stories and lower strength concrete at upper stories), as well as varying the amount of reinforcing.

The first portion of the table provides probable minimum column sizes with related costs and weights per lineal foot of story height for bottom level columns.

The second portion of the table provides costs by column size for top level columns with minimum code reinforcement. Probable maximum loads for these columns are also given.

How to Use Table:

1. Enter the second portion (minimum reinforcing) of the table with the minimum allowable column size from the selected cast in place floor system.

   If the total load on the column does not exceed the allowable working load shown, use the cost per L.F. multiplied by the length of columns required to obtain the column cost.

2. If the total load on the column exceeds the allowable working load shown in the second portion of the table, enter the first portion of the

table with the total load on the column and the minimum allowable column size from the selected cast in place floor system.

Select a cost per L.F. for bottom level columns by total load or minimum allowable column size.

Select a cost per L.F. for top level columns using the column size required for bottom level columns from the second portion of the table.

$$\frac{\text{Btm.} + \text{Top Col. Costs/L.F.}}{2} = \text{Avg. Col. Cost/L.F.}$$

Column Cost = Average Col. Cost/L.F. x Length of Cols. Required.

See reference section to determine total loads.

Design and Pricing Assumptions:
Normal wt. concrete, f'c = 4 or 6 KSI, placed by pump.
Steel, fy = 60 KSI, spliced every other level.
Minimum design eccentricity of 0.1t.
Assumed load level depth is 8" (weights prorated to full story basis).
Gravity loads only (no frame or lateral loads included).

Please see the reference section for further design and cost information.

System Components			COST PER V.L.F.		
	QUANTITY	UNIT	MAT.	INST.	TOTAL
SYSTEM B1010 201 1050					
ROUND TIED COLUMNS, 4 KSI CONCRETE, 100K MAX. LOAD, 10' STORY, 12" SIZE					
Forms in place, columns, round fiber tube, 12" diam 1 use	1.000	L.F.	2.56	12.50	15.06
Reinforcing in place, column ties	1.393	Lb.	6.84	6	12.84
Concrete ready mix, regular weight, 4000 psi	.029	C.Y.	3.36		3.36
Placing concrete, incl. vibrating, 12" sq./round columns, pumped	.029	C.Y.		2.03	2.03
Finish, burlap rub w/grout	3.140	S.F.	.09	3.14	3.23
TOTAL			12.85	23.67	36.52

B1010 201		C.I.P. Column - Round Tied						
	LOAD (KIPS)	STORY HEIGHT (FT.)	COLUMN SIZE (IN.)	COLUMN WEIGHT (P.L.F.)	CONCRETE STRENGTH (PSI)	COST PER V.L.F.		
						MAT.	INST.	TOTAL
1050	100	10	12	110	4000	12.85	24	36.85
1060	RB1010 -112	12	12	111	4000	13.20	24	37.20
1070		14	12	112	4000	13.55	24.50	38.05
1080	150	10	12	110	4000	14.55	25.50	40.05
1090		12	12	111	4000	14.90	25.50	40.40
1100		14	14	153	4000	17.60	27.50	45.10
1120	200	10	14	150	4000	18.05	28	46.05
1140		12	14	152	4000	18.50	28.50	47

B1010 Floor Construction

B1010 201	C.I.P. Column - Round Tied

	LOAD (KIPS)	STORY HEIGHT (FT.)	COLUMN SIZE (IN.)	COLUMN WEIGHT (P.L.F.)	CONCRETE STRENGTH (PSI)	COST PER V.L.F.		
						MAT.	INST.	TOTAL
1160		14	14	153	4000	19	28.50	47.50
1180	300	10	16	194	4000	22.50	30.50	53
1190		12	18	250	4000	29.50	35	64.50
1200		14	18	252	4000	30.50	35.50	66
1220	400	10	20	306	4000	35	40	75
1230		12	20	310	4000	36	40.50	76.50
1260		14	20	313	4000	37	41.50	78.50
1280	500	10	22	368	4000	43	45	88
1300		12	22	372	4000	44.50	46	90.50
1325		14	22	375	4000	45.50	47	92.50
1350	600	10	24	439	4000	50	50	100
1375		12	24	445	4000	51.50	51	102.50
1400		14	24	448	4000	52.50	52.50	105
1420	700	10	26	517	4000	61	56.50	117.50
1430		12	26	524	4000	62.50	58	120.50
1450		14	26	528	4000	64	59.50	123.50
1460	800	10	28	596	4000	68.50	62	130.50
1480		12	28	604	4000	70.50	64	134.50
1490		14	28	609	4000	72.50	65.50	138
1500	900	10	28	596	4000	74	67	141
1510		12	28	604	4000	76	69	145
1520		14	28	609	4000	78	70.50	148.50
1530	1000	10	30	687	4000	81	69.50	150.50
1540		12	30	695	4000	83	71.50	154.50
1620		14	30	701	4000	85.50	73	158.50
1640	100	10	12	110	6000	13.50	23.50	37
1660		12	12	111	6000	13.90	24	37.90
1680		14	12	112	6000	14.20	24.50	38.70
1700	150	10	12	110	6000	14.90	25	39.90
1710		12	12	111	6000	15.25	25.50	40.75
1720		14	12	112	6000	15.60	25.50	41.10
1730	200	10	12	110	6000	15.90	26	41.90
1740		12	12	111	6000	16.25	26	42.25
1760		14	12	112	6000	16.60	26.50	43.10
1780	300	10	14	150	6000	19.45	28.50	47.95
1790		12	14	152	6000	19.90	28.50	48.40
1800		14	16	153	6000	23	30	53
1810	400	10	16	194	6000	25	31.50	56.50
1820		12	16	196	6000	26	32.50	58.50
1830		14	18	252	6000	31	35	66
1850	500	10	18	247	6000	33	36.50	69.50
1870		12	18	250	6000	33.50	37	70.50
1880		14	20	252	6000	37	40	77
1890	600	10	20	306	6000	39	41.50	80.50
1900		12	20	310	6000	40	42.50	82.50
1905		14	20	313	6000	41	43	84
1910	700	10	22	368	6000	45.50	45	90.50
1915		12	22	372	6000	46.50	46	92.50
1920		14	22	375	6000	48	47	95
1925	800	10	24	439	6000	52.50	50	102.50
1930		12	24	445	6000	54	51	105
1935		14	24	448	6000	55.50	52.50	108

B10 Superstructure

B1010 Floor Construction

B1010 201		C.I.P. Column - Round Tied						

	LOAD (KIPS)	STORY HEIGHT (FT.)	COLUMN SIZE (IN.)	COLUMN WEIGHT (P.L.F.)	CONCRETE STRENGTH (PSI)	COST PER V.L.F.		
						MAT.	INST.	TOTAL
1940	900	10	24	439	6000	56.50	53.50	110
1945		12	24	445	6000	58	55	113
1970	1000	10	26	517	6000	66	58	124
1980		12	26	524	6000	67.50	59.50	127
1995		14	26	528	6000	69	60.50	129.50

B1010 202		C.I.P. Columns, Round Tied - Minimum Reinforcing						

	LOAD (KIPS)	STORY HEIGHT (FT.)	COLUMN SIZE (IN.)	COLUMN WEIGHT (P.L.F.)	CONCRETE STRENGTH (PSI)	COST PER V.L.F.		
						MAT.	INST.	TOTAL
2500	100	10-14	12	107	4000	11.15	22.50	33.65
2510	200	10-14	16	190	4000	19.55	28	47.55
2520	400	10-14	20	295	4000	30.50	35.50	66
2530	600	10-14	24	425	4000	43	44	87
2540	800	10-14	28	580	4000	59.50	54	113.50
2550	1100	10-14	32	755	4000	77	63	140
2560	1400	10-14	36	960	4000	93.50	74	167.50
2570								

B1010 Floor Construction

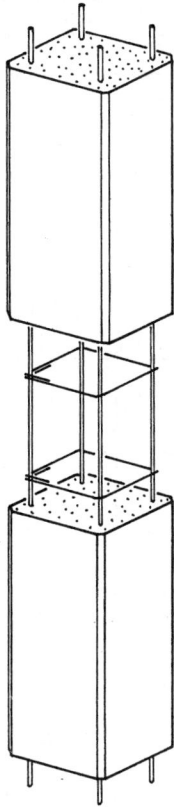

CONCRETE COLUMNS

General: It is desirable for purposes of consistency and simplicity to maintain constant column sizes throughout the building height. To do this, concrete strength may be varied (higher strength concrete at lower stories and lower strength concrete at upper stories), as well as varying the amount of reinforcing.

The first portion of the table provides probable minimum column sizes with related costs and weights per lineal foot of story height for bottom level columns.

The second portion of the table provides costs by column size for top level columns with minimum code reinforcement. Probable maximum loads for these columns are also given.

How to Use Table:

1. Enter the second portion (minimum reinforcing) of the table with the minimum allowable column size from the selected cast in place floor system.

   If the total load on the column does not exceed the allowable working load shown, use the cost per L.F. multiplied by the length of columns required to obtain the column cost.

2. If the total load on the column exceeds the allowable working load shown in the second portion of the table, enter the first portion of the

table with the total load on the column and the minimum allowable column size from the selected cast in place floor system.

Select a cost per L.F. for bottom level columns by total load or minimum allowable column size.

Select a cost per L.F. for top level columns using the column size required for bottom level columns from the second portion of the table.

$$\frac{\text{Btm.} + \text{Top Col. Costs/L.F.}}{2} = \text{Avg. Col. Cost/L.F}$$

Column Cost = Average Col. Cost/L.F. x Length of Cols. Required.

See reference section in back of book to determine total loads.

Design and Pricing Assumptions:

Normal wt. concrete, f'c = 4 or 6 KSI, placed by pump.
Steel, fy = 60 KSI, spliced every other level.
Minimum design eccentricity of 0.1t.
Assumed load level depth is 8″ (weights prorated to full story basis).
Gravity loads only (no frame or lateral loads included).

Please see the reference section for further design and cost information.

System Components			COST PER V.L.F.		
	QUANTITY	UNIT	MAT.	INST.	TOTAL
SYSTEM B1010 203 0640					
SQUARE COLUMNS, 100K LOAD,10′ STORY, 10″ SQUARE					
Forms in place, columns, plywood, 10″ x 10″, 4 uses	3.323	SFCA	2.70	27.83	30.53
Chamfer strip,wood, 3/4″ wide	4.000	L.F.	.92	3.76	4.68
Reinforcing in place, column ties	1.405	Lb.	6.05	5.30	11.35
Concrete ready mix, regular weight, 4000 psi	.026	C.Y.	3.02		3.02
Placing concrete, incl. vibrating, 12″ sq./round columns, pumped	.026	C.Y.		1.82	1.82
Finish, break ties, patch voids, burlap rub w/grout	3.323	S.F.	.10	3.33	3.43
TOTAL			12.79	42.04	54.83

B1010 203		C.I.P. Column, Square Tied						
	LOAD (KIPS)	STORY HEIGHT (FT.)	COLUMN SIZE (IN.)	COLUMN WEIGHT (P.L.F.)	CONCRETE STRENGTH (PSI)	COST PER V.L.F.		
						MAT.	INST.	TOTAL
0640	100	10	10	96	4000	12.80	42	54.80
0680	RB1010 -112	12	10	97	4000	13.10	42.50	55.60
0700		14	12	142	4000	17.30	51.50	68.80
0710								

B10 Superstructure

B1010 Floor Construction

| B1010 203 | C.I.P. Column, Square Tied |

	LOAD (KIPS)	STORY HEIGHT (FT.)	COLUMN SIZE (IN.)	COLUMN WEIGHT (P.L.F.)	CONCRETE STRENGTH (PSI)	COST PER V.L.F.		
						MAT.	INST.	TOTAL
0740	150	10	10	96	4000	15.80	45	60.80
0780		12	12	142	4000	18.15	52	70.15
0800		14	12	143	4000	18.60	52.50	71.10
0840	200	10	12	140	4000	19.45	53.50	72.95
0860		12	12	142	4000	19.90	53.50	73.40
0900		14	14	196	4000	22.50	59.50	82
0920	300	10	14	192	4000	24.50	61	85.50
0960		12	14	194	4000	25	61.50	86.50
0980		14	16	253	4000	28.50	68	96.50
1020	400	10	16	248	4000	31	70	101
1060		12	16	251	4000	31.50	71	102.50
1080		14	16	253	4000	32.50	71.50	104
1200	500	10	18	315	4000	35.50	78.50	114
1250		12	20	394	4000	43	89	132
1300		14	20	397	4000	44.50	90.50	135
1350	600	10	20	388	4000	45.50	91.50	137
1400		12	20	394	4000	47	92.50	139.50
1600		14	20	397	4000	48	93.50	141.50
1900	700	10	20	388	4000	52.50	101	153.50
2100		12	22	474	4000	53	101	154
2300		14	22	478	4000	54.50	103	157.50
2600	800	10	22	388	4000	55.50	104	159.50
2900		12	22	474	4000	57	105	162
3200		14	22	478	4000	58.50	106	164.50
3400	900	10	24	560	4000	62	113	175
3800		12	24	567	4000	64	115	179
4000		14	24	571	4000	65.50	116	181.50
4250	1000	10	24	560	4000	69	119	188
4500		12	26	667	4000	72	125	197
4750		14	26	673	4000	74	127	201
5600	100	10	10	96	6000	13.35	42	55.35
5800		12	10	97	6000	13.65	42.50	56.15
6000		14	12	142	6000	18.15	51.50	69.65
6200	150	10	10	96	6000	16.40	44.50	60.90
6400		12	12	98	6000	19	52	71
6600		14	12	143	6000	19.45	52.50	71.95
6800	200	10	12	140	6000	20.50	53.50	74
7000		12	12	142	6000	21	53.50	74.50
7100		14	14	196	6000	24	59.50	83.50
7300	300	10	14	192	6000	25	60.50	85.50
7500		12	14	194	6000	25.50	61	86.50
7600		14	14	196	6000	26	61.50	87.50
7700	400	10	14	192	6000	27.50	62.50	90
7800		12	14	194	6000	28	63	91
7900		14	16	253	6000	31	68	99
8000	500	10	16	248	6000	32.50	70	102.50
8050		12	16	251	6000	33	71	104
8100		14	16	253	6000	34	71.50	105.50
8200	600	10	18	315	6000	37	79	116
8300		12	18	319	6000	38	80	118
8400		14	18	321	6000	39	80.50	119.50

38

B10 Superstructure

B1010 Floor Construction

B1010 203				C.I.P. Column, Square Tied				

	LOAD (KIPS)	STORY HEIGHT (FT.)	COLUMN SIZE (IN.)	COLUMN WEIGHT (P.L.F.)	CONCRETE STRENGTH (PSI)	COST PER V.L.F.		
						MAT.	INST.	TOTAL
8500	700	10	18	315	6000	40	81.50	121.50
8600		12	18	319	6000	41	82.50	123.50
8700		14	18	321	6000	42	83.50	125.50
8800	800	10	20	388	6000	45.50	89	134.50
8900		12	20	394	6000	47	90.50	137.50
9000		14	20	397	6000	48	91.50	139.50
9100	900	10	20	388	6000	50.50	93.50	144
9300		12	20	394	6000	51.50	94.50	146
9600		14	20	397	6000	53	95.50	148.50
9800	1000	10	22	469	6000	57	103	160
9840		12	22	474	6000	58.50	104	162.50
9900		14	22	478	6000	60	105	165

B1010 204				C.I.P. Column, Square Tied-Minimum Reinforcing				

	LOAD (KIPS)	STORY HEIGHT (FT.)	COLUMN SIZE (IN.)	COLUMN WEIGHT (P.L.F.)	CONCRETE STRENGTH (PSI)	COST PER V.L.F.		
						MAT.	INST.	TOTAL
9913	150	10-14	12	135	4000	15.10	49.50	64.60
9918	300	10-14	16	240	4000	24.50	65	89.50
9924	500	10-14	20	375	4000	37	84	121
9930	700	10-14	24	540	4000	51.50	104	155.50
9936	1000	10-14	28	740	4000	68.50	127	195.50
9942	1400	10-14	32	965	4000	85	143	228
9948	1800	10-14	36	1220	4000	107	167	274
9954	2300	10-14	40	1505	4000	130	193	323

B10 Superstructure

B1010 Floor Construction

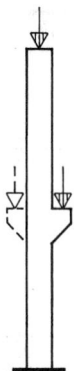

Concentric Load

Eccentric Load

General: Data presented here is for plant produced members transported 50 miles to 100 miles to the site and erected.

Design and pricing assumptions:
Normal wt. concrete, f'c = 5 KSI

Main reinforcement, fy = 60 KSI
Ties, fy = 40 KSI

Minimum design eccentricity, 0.1t.

Concrete encased structural steel haunches are assumed where practical; otherwise galvanized rebar haunches are assumed.

Base plates are integral with columns.

Foundation anchor bolts, nuts and washers are included in price.

System Components	QUANTITY	UNIT	COST PER V.L.F.		
			MAT.	INST.	TOTAL
SYSTEM B1010 206 0560					
PRECAST TIED COLUMN, CONCENTRIC LOADING, 100 K MAX.					
TWO STORY - 10' PER STORY, 5 KSI CONCRETE, 12"X12"					
Precast column, two story-10'/story, 5 KSI conc., 12"x12"	1.000	Ea.	116		116
Anchor bolts, 3/4" diameter x 24" long	.200	Ea.	.87		.87
Steel bearing plates; top, bottom, haunches	3.250	Lb.	5.92		5.92
Erection crew	.013	Hr.		11.28	11.28
TOTAL			122.79	11.28	134.07

B1010 206		Tied, Concentric Loaded Precast Concrete Columns						
	LOAD (KIPS)	STORY HEIGHT (FT.)	COLUMN SIZE (IN.)	COLUMN WEIGHT (P.L.F.)	LOAD LEVELS	COST PER V.L.F.		
						MAT.	INST.	TOTAL
0560	100	10	12x12	164	2	123	11.25	134.25
0570		12	12x12	162	2	117	9.40	126.40
0580		14	12x12	161	2	117	9.40	126.40
0590	150	10	12x12	166	3	117	9.40	126.40
0600		12	12x12	169	3	116	8.45	124.45
0610		14	12x12	162	3	116	8.45	124.45
0620	200	10	12x12	168	4	117	10.35	127.35
0630		12	12x12	170	4	117	9.40	126.40
0640		14	12x12	220	4	116	9.40	125.40
0680	300	10	14x14	225	3	159	9.40	168.40
0690		12	14x14	225	3	159	8.45	167.45
0700		14	14x14	250	3	159	8.45	167.45
0710	400	10	16x16	255	4	160	10.35	170.35
0720		12	16x16	295	4	160	9.40	169.40
0750		14	16x16	305	4	128	9.40	137.40
0790	450	10	16x16	320	3	160	9.40	169.40
0800		12	16x16	315	3	160	8.45	168.45
0810		14	16x16	330	3	159	8.45	167.45
0820	600	10	18x18	405	4	202	10.35	212.35
0830		12	18x18	395	4	202	9.40	211.40
0840		14	18x18	410	4	202	9.40	211.40
0910	800	10	20x20	495	4	202	10.35	212.35
0920		12	20x20	505	4	202	9.40	211.40
0930		14	20x20	510	4	202	9.40	211.40

B1010 Floor Construction

B1010 206 — Tied, Concentric Loaded Precast Concrete Columns

	LOAD (KIPS)	STORY HEIGHT (FT.)	COLUMN SIZE (IN.)	COLUMN WEIGHT (P.L.F.)	LOAD LEVELS	COST PER V.L.F.		
						MAT.	INST.	TOTAL
0970	900	10	22x22	625	3	245	9.40	254.40
0980		12	22x22	610	3	245	8.45	253.45
0990		14	22x22	605	3	246	8.45	254.45

B1010 207 — Tied, Eccentric Loaded Precast Concrete Columns

	LOAD (KIPS)	STORY HEIGHT (FT.)	COLUMN SIZE (IN.)	COLUMN WEIGHT (P.L.F.)	LOAD LEVELS	COST PER V.L.F.		
						MAT.	INST.	TOTAL
1130	100	10	12x12	161	2	115	11.25	126.25
1140		12	12x12	159	2	114	9.40	123.40
1150		14	12x12	159	2	114	9.40	123.40
1160	150	10	12x12	161	3	113	9.40	122.40
1170		12	12x12	160	3	113	8.45	121.45
1180		14	12x12	159	3	113	8.45	121.45
1190	200	10	12x12	161	4	112	10.35	122.35
1200		12	12x12	160	4	112	9.40	121.40
1210		14	12x12	177	4	112	9.40	121.40
1250	300	10	12x12	185	3	154	9.40	163.40
1260		12	14x14	215	3	154	8.45	162.45
1270		14	14x14	215	3	154	8.45	162.45
1280	400	10	14x14	235	4	152	10.35	162.35
1290		12	16x16	285	4	152	9.40	161.40
1300		14	16x16	295	4	152	9.40	161.40
1360	450	10	14x14	245	3	155	9.40	164.40
1370		12	16x16	285	3	155	8.45	163.45
1380		14	16x16	290	3	155	8.45	163.45
1390	600	10	18x18	385	4	194	10.35	204.35
1400		12	18x18	380	4	194	9.40	203.40
1410		14	18x18	375	4	193	9.40	202.40
1480	800	10	20x20	490	4	194	10.35	204.35
1490		12	20x20	480	4	195	9.40	204.40
1500		14	20x20	475	4	195	9.40	204.40

B10 Superstructure

B1010 Floor Construction

I (A) Wide Flange

○ (B) Pipe

◉ (C) Pipe, Concrete Filled

▢ (D) Square Tube

▣ (E) Square Tube Concrete Filled

▭ (F) Rectangular Tube

▨ (G) Rectangular Tube, Concrete Filled

General: The following pages provide data for seven types of steel columns: wide flange, round pipe, round pipe concrete filled, square tube, square tube concrete filled, rectangular tube and rectangular tube concrete filled.

Design Assumptions: Loads are concentric; wide flange and round pipe bearing capacity is for 36 KSI steel. Square and rectangular tubing bearing capacity is for 46 KSI steel.

The effective length factor K=1.1 is used for determining column values in the tables. K=1.1 is within a frequently used range for pinned connections with cross bracing.

How To Use Tables:
a. Steel columns usually extend through two or more stories to minimize splices. Determine floors with splices.
b. Enter Table No. below with load to column at the splice. Use the unsupported height.
c. Determine the column type desired by price or design.

Cost:
a. Multiply number of columns at the desired level by the total height of the column by the cost/VLF.
b. Repeat the above for all tiers.

Please see the reference section for further design and cost information.

B1010 208 — Steel Columns

	LOAD (KIPS)	UNSUPPORTED HEIGHT (FT.)	WEIGHT (P.L.F.)	SIZE (IN.)	TYPE	COST PER V.L.F. MAT.	COST PER V.L.F. INST.	COST PER V.L.F. TOTAL
1000	25	10	13	4	A	26	10.15	36.15
1020	RB1010 -130		7.58	3	B	15.15	10.15	25.30
1040			15	3-1/2	C	18.40	10.15	28.55
1060			6.87	3	D	13.70	10.15	23.85
1080			15	3	E	17.80	10.15	27.95
1100			8.15	4x3	F	16.25	10.15	26.40
1120			20	4x3	G	21.50	10.15	31.65
1200		16	16	5	A	29.50	7.60	37.10
1220			10.79	4	B	19.95	7.60	27.55
1240			36	5-1/2	C	27.50	7.60	35.10
1260			11.97	5	D	22	7.60	29.60
1280			36	5	E	29.50	7.60	37.10
1300			11.97	6x4	F	22	7.60	29.60
1320			64	8x6	G	42.50	7.60	50.10
1400		20	20	6	A	35	7.60	42.60
1420			14.62	5	B	25.50	7.60	33.10
1440			49	6-5/8	C	34	7.60	41.60
1460			11.97	5	D	21	7.60	28.60
1480			49	6	E	34	7.60	41.60
1500			14.53	7x5	F	25.50	7.60	33.10
1520			64	8x6	G	40.50	7.60	48.10
1600	50	10	16	5	A	32	10.15	42.15
1620			14.62	5	B	29	10.15	39.15
1640			24	4-1/2	C	22	10.15	32.15
1660			12.21	4	D	24.50	10.15	34.65
1680			25	4	E	25	10.15	35.15
1700			11.97	6x4	F	24	10.15	34.15
1720			28	6x3	G	28	10.15	38.15

B1010 Floor Construction

B1010 208	Steel Columns

	LOAD (KIPS)	UNSUPPORTED HEIGHT (FT.)	WEIGHT (P.L.F.)	SIZE (IN.)	TYPE	COST PER V.L.F.		
						MAT.	INST.	TOTAL
1800		16	24	8	A	44.50	7.60	52.10
1820			18.97	6	B	35	7.60	42.60
1840			36	5-1/2	C	27.50	7.60	35.10
1860			14.63	6	D	27	7.60	34.60
1880			36	5	E	29.50	7.60	37.10
1900			14.53	7x5	F	27	7.60	34.60
1920			64	8x6	G	42.50	7.60	50.10
1940								
2000	50	20	28	8	A	49	7.60	56.60
2020			18.97	6	B	33	7.60	40.60
2040			49	6-5/8	C	34	7.60	41.60
2060			19.02	6	D	33.50	7.60	41.10
2080			49	6	E	34	7.60	41.60
2100			22.42	8x6	F	39	7.60	46.60
2120			64	8x6	G	40.50	7.60	48.10
2200	75	10	20	6	A	40	10.15	50.15
2220			18.97	6	B	38	10.15	48.15
2240			36	4-1/2	C	55	10.15	65.15
2260			14.53	6	D	29	10.15	39.15
2280			28	4	E	24.50	10.15	34.65
2300			14.33	7x5	F	28.50	10.15	38.65
2320			35	6x4	G	32	10.15	42.15
2400		16	31	8	A	57.50	7.60	65.10
2420			28.55	8	B	53	7.60	60.60
2440			49	6-5/8	C	36	7.60	43.60
2460			17.08	7	D	31.50	7.60	39.10
2480			36	5	E	29.50	7.60	37.10
2500			23.34	7x5	F	43	7.60	50.60
2520			64	8x6	G	42.50	7.60	50.10
2600		20	31	8	A	54	7.60	61.60
2620			28.55	8	B	50	7.60	57.60
2640			81	8-5/8	C	51.50	7.60	59.10
2660			22.42	7	D	39	7.60	46.60
2680			49	6	E	34	7.60	41.60
2700			22.42	8x6	F	39	7.60	46.60
2720			64	8x6	G	40.50	7.60	48.10
2800	100	10	24	8	A	48	10.15	58.15
2820			28.57	6	B	57	10.15	67.15
2840			35	4-1/2	C	55	10.15	65.15
2860			17.08	7	D	34	10.15	44.15
2880			36	5	E	32	10.15	42.15
2900			19.02	7x5	F	38	10.15	48.15
2920			46	8x4	G	39	10.15	49.15
3000		16	31	8	A	57.50	7.60	65.10
3020			28.55	8	B	53	7.60	60.60
3040			56	6-5/8	C	53.50	7.60	61.10
3060			22.42	7	D	41.50	7.60	49.10
3080			49	6	E	36	7.60	43.60
3100			22.42	8x6	F	41.50	7.60	49.10
3120			64	8x6	G	42.50	7.60	50.10

B10 Superstructure

B1010 Floor Construction

B1010 208				Steel Columns				

	LOAD (KIPS)	UNSUPPORTED HEIGHT (FT.)	WEIGHT (P.L.F.)	SIZE (IN.)	TYPE	COST PER V.L.F.		
						MAT.	INST.	TOTAL
3200	100	20	40	8	A	70	7.60	77.60
3220			28.55	8	B	50	7.60	57.60
3240			81	8-5/8	C	51.50	7.60	59.10
3260			25.82	8	D	45	7.60	52.60
3280			66	7	E	40.50	7.60	48.10
3300			27.59	8x6	F	48.50	7.60	56.10
3320			70	8x6	G	58	7.60	65.60
3400	125	10	31	8	A	62	10.15	72.15
3420			28.57	6	B	57	10.15	67.15
3440			81	8	C	58.50	10.15	68.65
3460			22.42	7	D	45	10.15	55.15
3480			49	6	E	39	10.15	49.15
3500			22.42	8x6	F	45	10.15	55.15
3520			64	8x6	G	46	10.15	56.15
3600		16	40	8	A	74	7.60	81.60
3620			28.55	8	B	53	7.60	60.60
3640			81	8	C	54.50	7.60	62.10
3660			25.82	8	D	47.50	7.60	55.10
3680			66	7	E	43	7.60	50.60
3700			27.59	8x6	F	51	7.60	58.60
3720			64	8x6	G	42.50	7.60	50.10
3800		20	48	8	A	84	7.60	91.60
3820			40.48	10	B	71	7.60	78.60
3840			81	8	C	51.50	7.60	59.10
3860			25.82	8	D	45	7.60	52.60
3880			66	7	E	40.50	7.60	48.10
3900			37.59	10x6	F	65.50	7.60	73.10
3920			60	8x6	G	58	7.60	65.60
4000	150	10	35	8	A	70	10.15	80.15
4020			40.48	10	B	81	10.15	91.15
4040			81	8-5/8	C	58.50	10.15	68.65
4060			25.82	8	D	51.50	10.15	61.65
4080			66	7	E	46	10.15	56.15
4100			27.48	7x5	F	55	10.15	65.15
4120			64	8x6	G	46	10.15	56.15
4200		16	45	10	A	83	7.60	90.60
4220			40.48	10	B	75	7.60	82.60
4240			81	8-5/8	C	54.50	7.60	62.10
4260			31.84	8	D	59	7.60	66.60
4280			66	7	E	43	7.60	50.60
4300			37.69	10x6	F	69.50	7.60	77.10
4320			70	8x6	G	61.50	7.60	69.10
4400		20	49	10	A	85.50	7.60	93.10
4420			40.48	10	B	71	7.60	78.60
4440			123	10-3/4	C	73.50	7.60	81.10
4460			31.84	8	D	55.50	7.60	63.10
4480			82	8	E	47	7.60	54.60
4500			37.69	10x6	F	66	7.60	73.60
4520			86	10x6	G	57	7.60	64.60

B10 Superstructure

B1010 Floor Construction

B1010 208 — Steel Columns

	LOAD (KIPS)	UNSUPPORTED HEIGHT (FT.)	WEIGHT (P.L.F.)	SIZE (IN.)	TYPE	COST PER V.L.F.		
						MAT.	INST.	TOTAL
4600	200	10	45	10	A	90	10.15	100.15
4620			40.48	10	B	81	10.15	91.15
4640			81	8-5/8	C	58.50	10.15	68.65
4660			31.84	8	D	63.50	10.15	73.65
4680			82	8	E	53.50	10.15	63.65
4700			37.69	10x6	F	75.50	10.15	85.65
4720			70	8x6	G	66	10.15	76.15
4800		16	49	10	A	90.50	7.60	98.10
4820			49.56	12	B	91.50	7.60	99.10
4840			123	10-3/4	C	77.50	7.60	85.10
4860			37.60	8	D	69.50	7.60	77.10
4880			90	8	E	71.50	7.60	79.10
4900			42.79	12x6	F	79	7.60	86.60
4920			85	10x6	G	71	7.60	78.60
5000		20	58	12	A	101	7.60	108.60
5020			49.56	12	B	86.50	7.60	94.10
5040			123	10-3/4	C	73.50	7.60	81.10
5060			40.35	10	D	70.50	7.60	78.10
5080			90	8	E	67.50	7.60	75.10
5100			47.90	12x8	F	84	7.60	91.60
5120			93	10x6	G	87	7.60	94.60
5200	300	10	61	14	A	122	10.15	132.15
5220			65.42	12	B	131	10.15	141.15
5240			169	12-3/4	C	103	10.15	113.15
5260			47.90	10	D	95.50	10.15	105.65
5280			90	8	E	76.50	10.15	86.65
5300			47.90	12x8	F	95.50	10.15	105.65
5320			86	10x6	G	99	10.15	109.15
5400		16	72	12	A	133	7.60	140.60
5420			65.42	12	B	121	7.60	128.60
5440			169	12-3/4	C	95	7.60	102.60
5460			58.10	12	D	107	7.60	114.60
5480			135	10	E	91	7.60	98.60
5500			58.10	14x10	F	107	7.60	114.60
5600		20	79	12	A	138	7.60	145.60
5620			65.42	12	B	114	7.60	121.60
5640			169	12-3/4	C	90	7.60	97.60
5660			58.10	12	D	102	7.60	109.60
5680			135	10	E	86.50	7.60	94.10
5700			58.10	14x10	F	102	7.60	109.60
5800	400	10	79	12	A	158	10.15	168.15
5840			178	12-3/4	C	134	10.15	144.15
5860			68.31	14	D	136	10.15	146.15
5880			135	10	E	98.50	10.15	108.65
5900			62.46	14x10	F	125	10.15	135.15
6000		16	87	12	A	161	7.60	168.60
6040			178	12-3/4	C	124	7.60	131.60
6060			68.31	14	D	126	7.60	133.60
6080			145	10	E	118	7.60	125.60
6100			76.07	14x10	F	141	7.60	148.60

B10 Superstructure

B1010 Floor Construction

B1010 208					Steel Columns			

	LOAD (KIPS)	UNSUPPORTED HEIGHT (FT.)	WEIGHT (P.L.F.)	SIZE (IN.)	TYPE	COST PER V.L.F.		
						MAT.	INST.	TOTAL
6200	400	20	90	14	A	157	7.60	164.60
6240			178	12-3/4	C	118	7.60	125.60
6260			68.31	14	D	119	7.60	126.60
6280			145	10	E	112	7.60	119.60
6300			76.07	14x10	F	133	7.60	140.60
6400	500	10	99	14	A	198	10.15	208.15
6460			76.07	12	D	152	10.15	162.15
6480			145	10	E	127	10.15	137.15
6500			76.07	14x10	F	152	10.15	162.15
6600		16	109	14	A	201	7.60	208.60
6660			89.68	14	D	166	7.60	173.60
6700			89.68	16x12	F	166	7.60	173.60
6800		20	120	12	A	210	7.60	217.60
6860			89.68	14	D	157	7.60	164.60
6900			89.68	16x12	F	157	7.60	164.60
7000	600	10	120	12	A	240	10.15	250.15
7060			89.68	14	D	179	10.15	189.15
7100			89.68	16x12	F	179	10.15	189.15
7200		16	132	14	A	244	7.60	251.60
7260			103.30	16	D	191	7.60	198.60
7400		20	132	14	A	231	7.60	238.60
7460			103.30	16	D	181	7.60	188.60
7600	700	10	136	12	A	272	10.15	282.15
7660			103.30	16	D	206	10.15	216.15
7800		16	145	14	A	268	7.60	275.60
7860			103.30	16	D	191	7.60	198.60
8000		20	145	14	A	254	7.60	261.60
8060			103.30	16	D	181	7.60	188.60
8200	800	10	145	14	A	289	10.15	299.15
8300		16	159	14	A	294	7.60	301.60
8400		20	176	14	A	310	7.60	317.60
8800	900	10	159	14	A	315	10.15	325.15
8900		16	176	14	A	325	7.60	332.60
9000		20	193	14	A	340	7.60	347.60
9100	1000	10	176	14	A	350	10.15	360.15
9200		16	193	14	A	355	7.60	362.60
9300		20	211	14	A	370	7.60	377.60

B1010 Floor Construction

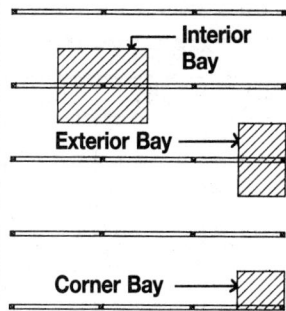

- Interior Bay
- Exterior Bay →
- Corner Bay →

Description: Table below lists costs of columns per S.F. of bay for wood columns of various sizes and unsupported heights and the maximum allowable total load per S.F. by bay size.

Design Assumptions: Columns are concentrically loaded and are not subject to bending.

Fiber stress (f) is 1200 psi maximum.

Modulus of elasticity is 1,760,000. Use table to factor load capacity figures for modulus of elasticity other than 1,760,000.

The cost of columns per S.F. of exterior bay is proportional to the area supported. For exterior bays, multiply the costs below by two. For corner bays, multiply the cost by four.

Modulus of Elasticity	Factor
1,210,000 psi	0.69
1,320,000 psi	0.75
1,430,000 psi	0.81
1,540,000 psi	0.87
1,650,000 psi	0.94
1,760,000 psi	1.00

B1010 210	Wood Columns						
	NOMINAL COLUMN SIZE (IN.)	BAY SIZE (FT.)	UNSUPPORTED HEIGHT (FT.)	MATERIAL (BF per M.S.F.)	TOTAL LOAD (P.S.F.)	COST PER S.F.		
						MAT.	INST.	TOTAL
1000	4 x 4	10 x 8	8	133	100	.19	.22	.41
1050			10	167	60	.24	.28	.52
1100			12	200	40	.29	.33	.62
1150			14	233	30	.34	.38	.72
1200		10 x 10	8	106	80	.15	.17	.32
1250			10	133	50	.19	.22	.41
1300			12	160	30	.23	.26	.49
1350			14	187	20	.27	.31	.58
1400		10 x 15	8	71	50	.10	.12	.22
1450			10	88	30	.13	.15	.28
1500			12	107	20	.16	.18	.34
1550			14	124	15	.18	.20	.38
1600		15 x 15	8	47	30	.07	.08	.15
1650			10	59	15	.09	.10	.19
1800		15 x 20	8	35	15	.05	.06	.11
1900								
2000	6 x 6	10 x 15	8	160	230	.28	.24	.52
2050			10	200	210	.36	.31	.67
2100			12	240	140	.43	.37	.80
2150			14	280	100	.50	.43	.93
2200		15 x 15	8	107	150	.19	.16	.35
2250			10	133	140	.24	.20	.44
2300			12	160	90	.28	.24	.52
2350			14	187	60	.33	.29	.62
2400		15 x 20	8	80	110	.14	.12	.26
2450			10	100	100	.18	.15	.33
2500			12	120	70	.21	.18	.39
2550			14	140	50	.25	.21	.46
2600		20 x 20	8	60	80	.11	.09	.20
2650			10	75	70	.13	.11	.24
2700			12	90	50	.16	.14	.30
2750			14	105	30	.19	.16	.35
2800		20 x 25	8	48	60	.09	.07	.16
2850			10	60	50	.11	.09	.20
2900			12	72	40	.13	.11	.24
2950			14	84	20	.15	.13	.28

B1010 Floor Construction

B1010 210	Wood Columns

	NOMINAL COLUMN SIZE (IN.)	BAY SIZE (FT.)	UNSUPPORTED HEIGHT (FT.)	MATERIAL (BF per M.S.F.)	TOTAL LOAD (P.S.F.)	COST PER S.F.		
						MAT.	INST.	TOTAL
3000		25 x 25	8	38	50	.07	.06	.13
3050			10	48	40	.09	.07	.16
3100			12	58	30	.10	.09	.19
3150			14	67	20	.12	.10	.22
3400	8 x 8	20 x 20	8	107	160	.24	.15	.39
3450			10	133	160	.30	.19	.49
3500			12	160	160	.36	.23	.59
3550			14	187	140	.27	.31	.58
3600		20 x 25	8	85	130	.12	.14	.26
3650			10	107	130	.16	.18	.34
3700			12	128	130	.32	.17	.49
3750			14	149	110	.37	.20	.57
3800		25 x 25	8	68	100	.17	.09	.26
3850			10	85	100	.21	.11	.32
3900			12	102	100	.26	.14	.40
3950			14	119	90	.30	.16	.46
4200	10 x 10	20 x 25	8	133	210	.33	.18	.51
4250			10	167	210	.42	.22	.64
4300			12	200	210	.50	.27	.77
4350			14	233	210	.58	.31	.89
4400		25 x 25	8	107	160	.27	.14	.41
4450			10	133	160	.33	.18	.51
4500			12	160	160	.40	.21	.61
4550			14	187	160	.47	.25	.72
4700	12 x 12	20 x 25	8	192	310	.49	.24	.73
4750			10	240	310	.61	.30	.91
4800			12	288	310	.73	.36	1.09
4850			14	336	310	.86	.42	1.28
4900		25 x 25	8	154	240	.39	.19	.58
4950			10	192	240	.49	.24	.73
5000			12	230	240	.59	.29	.88
5050			14	269	240	.69	.34	1.03

B10 Superstructure

B1010 Floor Construction

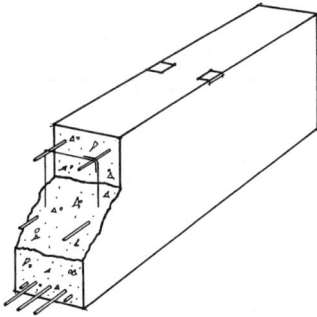

General: Beams priced in the following table are plant produced prestressed members transported to the site and erected.

Pricing Assumptions: Prices are based upon 10,000 S.F. to 20,000 S.F. projects and 50 mile to 100 mile transport.

Normal steel for connections is included in price. Deduct 20% from prices for Southern states. Add 10% to prices for Western states.

Design Assumptions: Normal weight concrete to 150 lbs./C.F. f'c = 5 KSI. Prestressing Steel: 250 KSI straight strand. Non-Prestressing Steel: fy = 60 KSI.

System Components	QUANTITY	UNIT	COST PER L.F.		
			MAT.	INST.	TOTAL
SYSTEM B1010 213 2200					
RECTANGULAR PRECAST BEAM, 15' SPAN, 12"X16"					
Precast beam, rectangular	1.000	L.F.	53.50		53.50
Delivery, 30 mile radius	.006	Hr.		2.13	2.13
Erection crew	.019	Hr.		17.86	17.86
TOTAL			53.50	19.99	73.49

B1010 213		Rectangular Precast Beams						
	SPAN (FT.)	SUPERIMPOSED LOAD (K.L.F.)	SIZE W X D (IN.)	BEAM WEIGHT (P.L.F.)	TOTAL LOAD (K.L.F.)	COST PER L.F.		
						MAT.	INST.	TOTAL
2200	15	2.32	12x16	200	2.52	53.50	20	73.50
2250		3.80	12x20	250	4.05	82	17.85	99.85
2300		5.60	12x24	300	5.90	92	18.80	110.80
2350		7.65	12x28	350	8.00	106	18.80	124.80
2400		5.85	18x20	375	6.73	98	18.80	116.80
2450		8.26	18x24	450	8.71	109	19.75	128.75
2500		11.39	18x28	525	11.91	124	19.75	143.75
2550		18.90	18x36	675	19.58	148	19.75	167.75
2600		10.78	24x24	600	11.38	129	19.75	148.75
2700		15.12	24x28	700	15.82	146	19.75	165.75
2750		25.23	24x36	900	26.13	174	20.50	194.50
2800	20	1.22	12x16	200	1.44	71.50	13.15	84.65
2850		2.03	12x20	250	2.28	81.50	14.10	95.60
2900		3.02	12x24	300	3.32	92	14.10	106.10
2950		4.15	12x28	350	4.50	106	14.10	120.10
3000		6.85	12x36	450	7.30	126	15.05	141.05
3050		3.13	18x20	375	3.50	97.50	14.10	111.60
3100		4.45	18x24	450	4.90	109	15.05	124.05
3150		6.18	18x28	525	6.70	90	15.05	105.05
3200		10.33	18x36	675	11.00	148	15.95	163.95
3400		15.48	18x44	825	16.30	174	16.95	190.95
3450		21.63	18x52	975	22.60	202	16.95	218.95
3500		5.80	24x24	600	6.40	129	15.05	144.05
3600		8.20	24x28	700	8.90	130	15.95	145.95
3700		13.80	24x36	900	14.70	148	16.95	164.95
3800		20.70	24x44	1100	21.80	204	23.50	227.50
3850		29.20	24x52	1300	30.50	236	23.50	259.50

B1010 Floor Construction

B1010 213	Rectangular Precast Beams

	SPAN (FT.)	SUPERIMPOSED LOAD (K.L.F.)	SIZE W X D (IN.)	BEAM WEIGHT (P.L.F.)	TOTAL LOAD (K.L.F.)	COST PER L.F.		
						MAT.	INST.	TOTAL
3900	25	1.82	12x24	300	2.12	94	11.25	105.25
4000		2.53	12x28	350	2.88	100	12.25	112.25
4050		5.18	12x36	450	5.63	127	12.25	139.25
4100		6.55	12x44	550	7.10	146	12.25	158.25
4150		9.08	12x52	650	9.73	171	14.10	185.10
4200		1.86	18x20	375	2.24	99	12.25	111.25
4250		2.69	18x24	450	3.14	111	12.25	123.25
4300		3.76	18x28	525	4.29	91.50	12.25	103.75
4350		6.37	18x36	675	7.05	150	14.10	164.10
4400		9.60	18x44	872	10.43	177	18.80	195.80
4450		13.49	18x52	975	14.47	178	18.80	196.80
4500		18.40	18x60	1125	19.53	228	18.80	246.80
4600		3.50	24x24	600	4.10	131	14.10	145.10
4700		5.00	24x28	700	5.70	148	14.10	162.10
4800		8.50	24x36	900	9.40	176	18.80	194.80
4900		12.85	24x44	1100	13.95	206	18.80	224.80
5000		18.22	24x52	1300	19.52	252	25.50	277.50
5050		24.48	24x60	1500	26.00	275	25.50	300.50
5100	30	1.65	12x28	350	2.00	108	10.35	118.35
5150		2.79	12x36	450	3.24	128	10.35	138.35
5200		4.38	12x44	550	4.93	171	11.25	182.25
5250		6.10	12x52	650	6.75	174	11.25	185.25
5300		8.36	12x60	750	9.11	197	15.95	212.95
5400		1.72	18x24	450	2.17	111	10.35	121.35
5450		2.45	18x28	525	2.98	127	11.25	138.25
5750		4.21	18x36	675	4.89	152	15.95	167.95
6000		6.42	18x44	825	7.25	177	15.95	192.95
6250		9.07	18x52	975	10.05	206	15.95	221.95
6500		12.43	18x60	1125	13.56	242	20.50	262.50
6750		2.24	24x24	600	2.84	131	11.25	142.25
7000		3.26	24x28	700	3.96	149	15.95	164.95
7100		5.63	24x36	900	6.53	177	15.95	192.95
7200	35	8.59	24x44	1100	9.69	207	17.85	224.85
7300		12.25	24x52	1300	13.55	241	17.85	258.85
7400		16.54	24x60	1500	18.04	270	17.85	287.85
7500	40	2.23	12x44	550	2.78	151	17.85	168.85
7600		3.15	12x52	650	3.80	177	17.85	194.85
7700		4.38	12x60	750	5.13	203	17.85	220.85
7800		2.08	18x36	675	2.76	153	17.85	170.85
7900		3.25	18x44	825	4.08	186	21.50	207.50
8000		4.68	18x52	975	5.66	215	21.50	236.50
8100		6.50	18x60	1175	7.63	237	21.50	258.50
8200		2.78	24x36	900	3.60	185	21.50	206.50
8300		4.35	24x44	1100	5.45	211	21.50	232.50
8400		6.35	24x52	1300	7.65	244	23.50	267.50
8500		8.65	24x60	1500	10.15	273	23.50	296.50

B10 Superstructure

B1010 Floor Construction

| B1010 213 | | Rectangular Precast Beams | | | | | | |

	SPAN (FT.)	SUPERIMPOSED LOAD (K.L.F.)	SIZE W X D (IN.)	BEAM WEIGHT (P.L.F.)	TOTAL LOAD (K.L.F.)	COST PER L.F.		
						MAT.	INST.	TOTAL
8600	45	2.35	12x52	650	3.00	179	16.95	195.95
8800		3.29	12x60	750	4.04	207	18.80	225.80
9000		2.39	18x44	825	3.22	183	18.80	201.80
9020		3.49	18x52	975	4.47	213	18.80	231.80
9200		4.90	18x60	1125	6.03	238	21.50	259.50
9250		3.21	24x44	1100	4.31	213	21.50	234.50
9300		4.72	24x52	1300	6.02	242	21.50	263.50
9500		6.52	24x60	1500	8.02	277	21.50	298.50
9600	50	2.64	18x52	975	3.62	215	18.80	233.80
9700		3.75	18x60	1125	4.88	242	18.80	260.80
9750		3.58	24x52	1300	4.88	375	18.80	393.80
9800		5.00	24x60	1500	6.50	280	18.80	298.80
9950	55	3.86	24x60	1500	5.36	281	16.95	297.95

B1010 Floor Construction

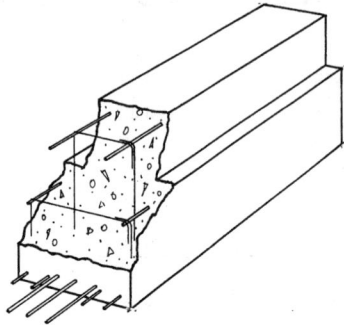

General: Beams priced in the following table are plant produced prestressed members transported to the site and erected.

Pricing Assumptions: Prices are based upon 10,000 S.F. to 20,000 S.F. projects and 50 mile to 100 mile transport.

Normal steel for connections is included in price. Deduct 20% from prices for Southern states. Add 10% to prices for Western states.

Design Assumptions: Normal weight concrete to 150 lbs./C.F. f'c = 5 KSI. Prestressing Steel: 250 KSI straight strand. Non-Prestressing Steel: fy = 60 KSI.

System Components	QUANTITY	UNIT	COST PER L.F.		
			MAT.	INST.	TOTAL
SYSTEM B1010 214 2300					
"T" SHAPED PRECAST BEAM, 15' SPAN, 12"X16"					
Precast beam, "T" shaped, 12"x16"	1.000	L.F.	166		166
Delivery, 30 mile radius	.009	Hr.		2.13	2.13
Erection crew	.019	Hr.		17.86	17.86
TOTAL			166	19.99	185.99

B1010 214		"T" Shaped Precast Beams						
	SPAN (FT.)	SUPERIMPOSED LOAD (K.L.F.)	SIZE W X D (IN.)	BEAM WEIGHT (P.L.F.)	TOTAL LOAD (K.L.F.)	COST PER L.F.		
						MAT.	INST.	TOTAL
2300	15	2.8	12x16	260	3.06	166	20	186
2350		4.33	12x20	355	4.69	161	18.80	179.80
2400		6.17	12x24	445	6.62	190	19.75	209.75
2500		8.37	12x28	515	8.89	204	19.75	223.75
2550		13.54	12x36	680	14.22	230	19.75	249.75
2600		8.83	18x24	595	9.43	208	19.75	227.75
2650		12.11	18x28	690	12.8	223	19.75	242.75
2700		19.72	18x36	905	24.63	253	20.50	273.50
2800		11.52	24x24	745	12.27	228	19.75	247.75
2900		15.85	24x28	865	16.72	246	20.50	266.50
3000		26.07	24x36	1130	27.20	280	22.50	302.50

B1010 Floor Construction

B1010 214	"T" Shaped Precast Beams

	SPAN (FT.)	SUPERIMPOSED LOAD (K.L.F.)	SIZE W X D (IN.)	BEAM WEIGHT (P.L.F.)	TOTAL LOAD (K.L.F.)	COST PER L.F.		
						MAT.	INST.	TOTAL
3100	20	1.46	12x16	260	1.72	166	14.10	180.10
3200		2.28	12x20	355	2.64	161	14.10	175.10
3300		3.28	12x24	445	3.73	190	15.05	205.05
3400		4.49	12x28	515	5.00	204	15.05	219.05
3500		7.32	12x36	680	8.00	230	15.95	245.95
3600		11.26	12x44	840	12.10	257	16.95	273.95
3700		4.70	18x24	595	5.30	208	15.05	223.05
3800		6.51	18x28	690	7.20	223	15.95	238.95
3900		10.7	18x36	905	11.61	253	16.95	269.95
4300		16.19	18x44	1115	17.31	289	23.50	312.50
4400		22.77	18x52	1330	24.10	320	23.50	343.50
4500		6.15	24x24	745	6.90	228	15.95	243.95
4600		8.54	24x28	865	9.41	246	16.95	262.95
4700		14.17	24x36	1130	15.30	280	23.50	303.50
4800		21.41	24x44	1390	22.80	320	23.50	343.50
4900		30.25	24x52	1655	31.91	365	31	396
5000	25	2.68	12x28	515	3.2	207	12.25	219.25
5050		4.44	12x36	680	5.12	232	14.10	246.10
5100		6.90	12x44	840	7.74	262	18.80	280.80
5200		9.75	12x52	1005	10.76	292	18.80	310.80
5300		13.43	12x60	1165	14.60	320	18.80	338.80
5350		3.92	18x28	690	4.61	226	14.10	240.10
5400		6.52	18x36	905	7.43	257	18.80	275.80
5500		9.96	18x44	1115	11.08	233	18.80	251.80
5600		14.09	18x52	1330	15.42	330	25.50	355.50
5650		19.39	18x60	1540	20.93	360	25.50	385.50
5700		3.67	24x24	745	4.42	232	14.10	246.10
5750		5.15	24x28	865	6.02	250	18.80	268.80
5800		8.66	24x36	1130	9.79	222	18.80	240.80
5850		13.20	24x44	1390	14.59	330	25.50	355.50
5900		18.76	24x52	1655	20.42	365	25.50	390.50
5950		25.35	24x60	1916	27.27	395	25.50	420.50
6000	30	2.88	12x36	680	3.56	235	15.95	250.95
6100		4.54	12x44	840	5.38	270	15.95	285.95
6200		6.46	12x52	1005	7.47	305	20.50	325.50
6250		8.97	12x60	1165	10.14	330	20.50	350.50
6300		4.25	18x36	905	5.16	258	15.95	273.95
6350		6.57	18x44	1115	7.69	298	20.50	318.50
6400		9.38	18x52	1330	10.71	330	20.50	350.50
6500		13.00	18x60	1540	14.54	365	20.50	385.50
6700		3.31	24x28	865	4.18	251	15.95	266.95
6750		5.67	24x36	1130	6.80	289	20.50	309.50
6800		8.74	24x44	1390	10.13	325	20.50	345.50
6850		12.52	24x52	1655	14.18	310	20.50	330.50
6900		17.00	24x60	1215	18.92	400	20.50	420.50

B1010 Floor Construction

B1010 214	"T" Shaped Precast Beams

	SPAN (FT.)	SUPERIMPOSED LOAD (K.L.F.)	SIZE W X D (IN.)	BEAM WEIGHT (P.L.F.)	TOTAL LOAD (K.L.F.)	COST PER L.F.		
						MAT.	INST.	TOTAL
7000	35	3.11	12x44	840	3.95	264	13.15	277.15
7100		4.48	12x52	1005	5.49	305	17.85	322.85
7200		6.28	12x60	1165	7.45	330	17.85	347.85
7300		4.53	18x44	1115	5.65	298	17.85	315.85
7500		6.54	18x52	1330	7.87	330	17.85	347.85
7600		9.14	18x60	1540	10.68	350	22.50	372.50
7700		3.87	24x36	1130	5.00	231	17.85	248.85
7800		6.05	24x44	1390	7.44	330	17.85	347.85
7900		8.76	24x52	1655	10.42	365	22.50	387.50
8000		12.00	24x60	1915	13.92	405	22.50	427.50
8100	40	3.19	12x52	1005	4.2	310	21.50	331.50
8200		4.53	12x60	1165	5.7	340	21.50	361.50
8300		4.70	18x52	1330	6.03	340	21.50	361.50
8400		6.64	18x60	1540	8.18	375	21.50	396.50
8600		4.31	24x44	1390	5.7	325	21.50	346.50
8700		6.32	24x52	1655	7.98	375	21.50	396.50
8800		8.74	24x60	1915	10.66	415	21.50	436.50
8900	45	3.34	12x60	1165	4.51	345	18.80	363.80
9000		4.92	18x60	1540	6.46	380	18.80	398.80
9250		4.64	24x52	1655	6.30	385	18.80	403.80
9900		6.5	24x60	1915	8.42	420	18.80	438.80
9950	50	4.9	24x60	1915	6.82	425	22.50	447.50

B10 Superstructure

B1010 Floor Construction

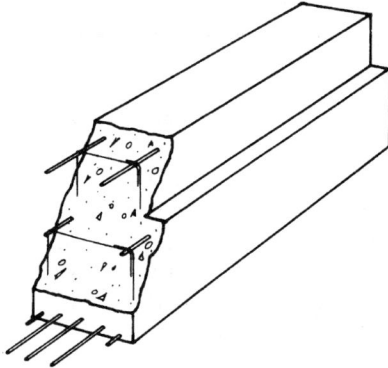

General: Beams priced in the following table are plant produced prestressed members transported to the site and erected.

Pricing Assumptions: Prices are based upon 10,000 S.F. to 20,000 S.F. projects and 50 mile to 100 mile transport.

Normal steel for connections is included in price. Deduct 20% from prices for Southern states. Add 10% to prices for Western states.

Design Assumptions: Normal weight concrete to 150 lbs./C.F. f'c = 5 KSI. Prestressing Steel: 250 KSI straight strand. Non-Prestressing Steel: fy = 60 KSI.

System Components	QUANTITY	UNIT	COST PER L.F.		
			MAT.	INST.	TOTAL
SYSTEM B1010 215 2250					
"L" SHAPED PRECAST BEAM, 15' SPAN, 12"X16"					
Precast beam, "L" shaped, 12"x16"	1.000	L.F.	122		122
Delivery, 30 mile radius	.008	Hr.		2.13	2.13
Erection crew	.019	Hr.		17.86	17.86
TOTAL			122	19.99	141.99

B1010 215	"L" Shaped Precast Beams							
	SPAN (FT.)	SUPERIMPOSED LOAD (K.L.F.)	SIZE W X D (IN.)	BEAM WEIGHT (P.L.F.)	TOTAL LOAD (K.L.F.)	COST PER L.F.		
						MAT.	INST.	TOTAL
2250	15	2.58	12x16	230	2.81	122	20	142
2300		4.10	12x20	300	4.40	132	18.80	150.80
2400		5.92	12x24	370	6.29	144	18.80	162.80
2450		8.09	12x28	435	8.53	156	18.80	174.80
2500		12.95	12x36	565	13.52	179	19.75	198.75
2600		8.55	18x24	520	9.07	161	19.75	180.75
2650		11.83	18x28	610	12.44	175	19.75	194.75
2700		19.30	18x36	790	20.09	202	19.75	221.75
2750		11.24	24x24	670	11.91	182	19.75	201.75
2800		15.40	24x28	780	16.18	198	19.75	217.75
2850		25.65	24x36	1015	26.67	229	22.50	251.50

B1010   Floor Construction

B1010 215	"L" Shaped Precast Beams

	SPAN (FT.)	SUPERIMPOSED LOAD (K.L.F.)	SIZE W X D (IN.)	BEAM WEIGHT (P.L.F.)	TOTAL LOAD (K.L.F.)	COST PER L.F.		
						MAT.	INST.	TOTAL
2900	20	2.18	12x20	300	2.48	132	14.10	146.10
2950		3.17	12x24	370	3.54	144	14.10	158.10
3000		4.37	12x28	435	4.81	156	15.05	171.05
3100		7.04	12x36	565	7.60	179	15.05	194.05
3150		10.80	12x44	695	11.50	203	15.95	218.95
3200		15.08	12x52	825	15.91	231	16.95	247.95
3250		4.58	18x24	520	5.10	161	15.05	176.05
3300		6.39	18x28	610	7.00	176	15.95	191.95
3350		10.51	18x36	790	11.30	203	16.95	219.95
3400		15.73	18x44	970	16.70	233	16.95	249.95
3550		22.15	18x52	1150	23.30	267	23.50	290.50
3600		6.03	24x24	370	6.40	178	14.10	192.10
3650		8.32	24x28	780	9.10	199	16.95	215.95
3700		13.98	24x36	1015	15.00	230	23.50	253.50
3800		21.05	24x44	1245	22.30	264	23.50	287.50
3900		29.73	24x52	1475	31.21	300	23.50	323.50
4000	25	2.64	12x28	435	3.08	157	11.25	168.25
4100		4.30	12x36	565	4.87	179	12.25	191.25
4200		6.65	12x44	695	7.35	202	12.25	214.25
4250		9.35	12x52	875	10.18	231	14.10	245.10
4300		12.81	12x60	950	13.76	259	14.10	273.10
4350		2.74	18x24	570	3.76	169	18.80	187.80
4400		3.67	18x28	610	4.28	177	12.25	189.25
4450		6.44	18x36	790	7.23	203	14.10	217.10
4500		9.72	18x44	970	10.69	237	18.80	255.80
4600		13.76	18x52	1150	14.91	260	12.25	272.25
4700		18.33	18x60	1330	20.16	300	25.50	325.50
4800		3.62	24x24	370	3.99	180	15.05	195.05
4900		5.04	24x28	780	5.82	201	14.10	215.10
5000		8.58	24x36	1015	9.60	232	18.80	250.80
5100		13.00	24x44	1245	14.25	275	25.50	300.50
5200		18.50	24x52	1475	19.98	310	25.50	335.50
5300	30	2.80	12x36	565	3.37	184	11.25	195.25
5400		4.42	12x44	695	5.12	207	15.95	222.95
5500		6.24	12x52	875	7.07	236	15.95	251.95
5600		8.60	12x60	950	9.55	267	15.95	282.95
5700		2.50	18x28	610	3.11	179	11.25	190.25
5800		4.23	18x36	790	5.02	207	15.95	222.95
5900		6.45	18x44	970	7.42	237	15.95	252.95
6000		9.20	18x52	1150	10.35	275	20.50	295.50
6100		12.67	18x60	1330	14.00	305	20.50	325.50
6200		3.26	24x28	780	4.04	203	15.95	218.95
6300		5.65	24x36	1015	6.67	239	20.50	259.50
6400		8.66	24x44	1245	9.90	272	20.50	292.50

B1010 Floor Construction

B1010 215	"L" Shaped Precast Beams

	SPAN (FT.)	SUPERIMPOSED LOAD (K.L.F.)	SIZE W X D (IN.)	BEAM WEIGHT (P.L.F.)	TOTAL LOAD (K.L.F.)	COST PER L.F.		
						MAT.	INST.	TOTAL
6500	35	3.00	12x44	695	3.70	210	13.15	223.15
6600		4.27	12x52	825	5.20	238	13.15	251.15
6700		6.00	12x60	950	6.95	243	17.85	260.85
6800		2.90	18x36	790	3.69	209	13.15	222.15
6900		4.48	18x44	970	5.45	245	17.85	262.85
7000		6.45	18x52	1150	7.60	276	17.85	293.85
7100		8.95	18x60	1330	10.28	305	17.85	322.85
7200		3.88	24x36	1015	4.90	242	17.85	259.85
7300		6.03	24x44	1245	7.28	274	17.85	291.85
7400		8.71	24x52	1475	10.19	310	22.50	332.50
7500	40	3.15	12x52	825	3.98	243	21.50	264.50
7600		4.43	12x60	950	5.38	277	21.50	298.50
7700		3.20	18x44	970	4.17	246	21.50	267.50
7800		4.68	18x52	1150	5.83	279	21.50	300.50
7900		6.55	18x60	1330	7.88	310	21.50	331.50
8000		4.43	24x44	1245	5.68	277	21.50	298.50
8100		6.33	24x52	1475	7.81	315	21.50	336.50
8200	45	3.30	12x60	950	4.25	279	18.80	297.80
8300		3.45	18x52	1150	4.60	282	18.80	300.80
8500		4.44	18x60	1330	5.77	315	18.80	333.80
8750		3.16	24x44	1245	4.41	280	18.80	298.80
9000		4.68	24x52	1475	6.16	320	18.80	338.80
9900	50	3.71	18x60	1330	5.04	315	17.85	332.85
9950		3.82	24x52	1475	5.30	315	17.85	332.85

B10 Superstructure

B1010 Floor Construction

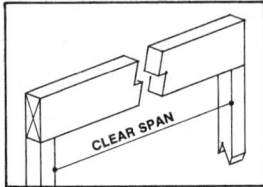

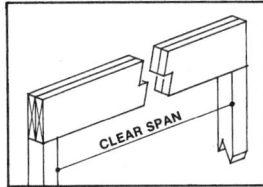

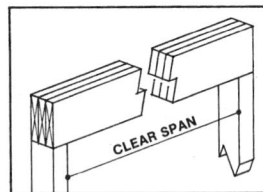

Description: Table below lists the cost per L.F. and total uniform load allowable for various size beams at various spans.

Design Assumptions: Fiber strength (f) is 1,000 PSI. Maximum deflection does not exceed 1/360 the span of the beam. Modulus of elasticity (E) is 1,100,000 psi. Anything less will result in excessive deflection in the longer members so that spans must be reduced or member size increased.

The total loads are in pounds per foot.

The span is in feet and is the unsupported clear span.

The costs are based on net quantities and do not include any allowance for waste or overlap.

The member sizes are from 2″ x 6″ to 4″ x 12″ and include one, two and three pieces of each member size to arrive at the costs, spans and loading. If more than a three-piece beam is required, its cost may be calculated directly. A four-piece 2″ x 6″ would cost two times as much as a two-piece 2″ x 6″ beam. The maximum span and total load cannot be doubled, however.

B1010 216		Wood Beams						
	BEAM SIZE (IN.)	TOTAL LOAD (P.L.F.) 6′ SPAN	TOTAL LOAD (P.L.F.) 10′ SPAN	TOTAL LOAD (P.L.F.) 14′ SPAN	TOTAL LOAD (P.L.F.) 18′ SPAN	COST PER L.F.		
						MAT.	INST.	TOTAL
1000	2 - 2 x 6	238				1.15	3.80	4.95
1050	3 - 2 x 6	357	RB1010 -240			1.73	5.70	7.43
1100	2 - 2 x 8	314				1.56	4.40	5.96
1150	3 - 2 x 8	471	232			2.34	6.60	8.94
1200	2 - 2 x 10	400	240			2.79	4.58	7.37
1250	3 - 2 x 10	600	360			4.18	6.90	11.08
1300	2 - 2 x 12	487	292	208		3.04	5	8.04
1350	3 - 2 x 12	731	438	313		4.56	7.50	12.06
1400	2 - 2 x 14	574	344	246		4.67	5.15	9.82
1450	3 - 2 x 14	861	516	369	243	7	7.70	14.70
1550	2 - 3 x 6	397				3.38	4.28	7.66
1600	3 - 3 x 6	595				5.05	6.40	11.45
1650	1 - 3 x 8	261				2.30	2.50	4.80
1700	2 - 3 x 8	523	258			4.60	5	9.60
1750	3 - 3 x 8	785	388			6.90	7.50	14.40
1800	1 - 3 x 10	334	200			3.25	2.75	6
1850	2 - 3 x 10	668	400			6.50	5.50	12
1900	3 - 3 x 10	1,002	600	293		9.75	8.25	18
1950	1 - 3 x 12	406	243			3.90	2.97	6.87
2000	2 - 3 x 12	812	487	348		7.80	5.95	13.75
2050	3 - 3 x 12	1,218	731	522	248	11.70	8.90	20.60
2100	1 - 4 x 6	278				2.55	2.50	5.05
2150	2 - 4 x 6	556				5.10	5	10.10
2200	1 - 4 x 8	366				3.07	2.93	6
2250	2 - 4 x 8	733	362			6.15	5.85	12
2300	1 - 4 x 10	467	280			4.08	3.30	7.38
2350	2 - 4 x 10	935	560			8.15	6.60	14.75
2400	1 - 4 x 12	568	341			5.20	3.60	8.80
2450	2 - 4 x 12	1,137	682		232	10.40	7.20	17.60

B1010 Floor Construction

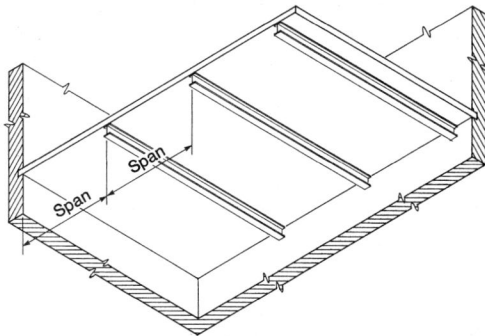

General: Solid concrete slabs of uniform depth reinforced for flexure in one direction and for temperature and shrinkage in the other direction.

Design and Pricing Assumptions:
Concrete f'c = 3 KSI normal weight placed by pump.
Reinforcement fy = 60 KSI.
Deflection ≤ span/360.
Forms, four uses hung from steel beams plus edge forms.
Steel trowel finish for finish floor and cure.

System Components	QUANTITY	UNIT	COST PER S.F.		
			MAT.	INST.	TOTAL
SYSTEM B1010 217 2000					
6'-0" SINGLE SPAN, 4" SLAB DEPTH, 40 PSF SUPERIMPOSED LOAD					
Forms in place, floor slab, with 1-way joist pans, 4 uses	1.000	S.F.	3.12	5.10	8.22
Edge forms to 6" high on elevated slab, 4 uses	.080	L.F.	.06	1.25	1.31
Reinforcing in place, elevated slabs #4 to #7	1.030	Lb.	.92	.41	1.33
Concrete ready mix, regular weight, 3000 psi	.330	C.F.	1.37		1.37
Place and vibrate concrete, elevated slab less than 6", pumped	.330	C.F.		.47	.47
Finishing floor, monolithic steel trowel finish for finish floor	1.000	S.F.		.78	.78
Curing with sprayed membrane curing compound	1.000	S.F.	.06	.08	.14
TOTAL			5.53	8.09	13.62

B1010 217		Cast in Place Slabs, One Way						
	SLAB DESIGN & SPAN (FT.)	SUPERIMPOSED LOAD (P.S.F.)	THICKNESS (IN.)	TOTAL LOAD (P.S.F.)		COST PER S.F.		
						MAT.	INST.	TOTAL
2000	Single 6	40	4	90	RB1010-010	5.55	8.10	13.65
2100		75	4	125		5.55	8.10	13.65
2200		125	4	175	RB1010-100	5.55	8.10	13.65
2300		200	4	250		5.85	8.25	14.10
2500	Single 8	40	4	90		5.50	7.15	12.65
2600		75	4	125		5.50	7.80	13.30
2700		125	4-1/2	181		5.70	7.85	13.55
2800		200	5	262		6.35	8.10	14.45
3000	Single 10	40	4	90		5.75	7.70	13.45
3100		75	4	125		5.75	7.70	13.45
3200		125	5	188		6.30	7.90	14.20
3300		200	7-1/2	293		8.15	8.90	17.05
3500	Single 15	40	5-1/2	90		6.75	7.85	14.60
3600		75	6-1/2	156		7.45	8.25	15.70
3700		125	7-1/2	219		8	8.45	16.45
3800		200	8-1/2	306		8.60	8.65	17.25
4000	Single 20	40	7-1/2	115		7.95	8.30	16.25
4100		75	9	200		8.55	8.25	16.80
4200		125	10	250		9.45	8.60	18.05
4300		200	10	324		10.30	9	19.30

B10 Superstructure

B1010 Floor Construction

B1010 217	Cast in Place Slabs, One Way

	SLAB DESIGN & SPAN (FT.)	SUPERIMPOSED LOAD (P.S.F.)	THICKNESS (IN.)	TOTAL LOAD (P.S.F.)		COST PER S.F.		
						MAT.	INST.	TOTAL
4500	Multi 6	40	4	90		5.65	7.70	13.35
4600		75	4	125		5.65	7.70	13.35
4700		125	4	175		5.65	7.70	13.35
4800		200	4	250		5.65	7.70	13.35
5000	Multi 8	40	4	90		5.65	7.50	13.15
5100		75	4	125		5.65	7.50	13.15
5200		125	4	175		5.65	7.50	13.15
5300		200	4	250		5.65	7.50	13.15
5500	Multi 10	40	4	90		5.65	7.40	13.05
5600		75	4	125		5.65	7.45	13.10
5700		125	4	175		5.90	7.55	13.45
5800		200	5	263		6.45	7.85	14.30
6600	Multi 15	40	5	103		6.40	7.65	14.05
6800		75	5	138		6.75	7.80	14.55
7000		125	6-1/2	206		7.40	8.05	15.45
7100		200	6-1/2	281		7.65	8.15	15.80
7500	Multi 20	40	5-1/2	109		7.40	7.95	15.35
7600		75	6-1/2	156		8	8.20	16.20
7800		125	9	238		9.85	8.95	18.80
8000		200	9	313		9.85	8.95	18.80

B10 Superstructure

B1010 Floor Construction

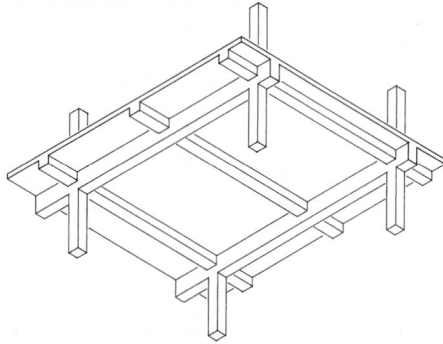

General: Solid concrete one-way slab cast monolithically with reinforced concrete support beams and girders.

Design and Pricing Assumptions:
Concrete f'c = 3 KSI, normal weight, placed by concrete pump.
Reinforcement, fy = 60 KSI.
Forms, four use.
Finish, steel trowel.
Curing, spray on membrane.
Based on 4 bay x 4 bay structure.

System Components	QUANTITY	UNIT	COST PER S.F.		
			MAT.	INST.	TOTAL
SYSTEM B1010 219 3000					
BM. & SLAB ONE WAY 15′X15′BAY, 40 PSF S.LOAD,12″ MIN. COL.					
Forms in place, flat plate to 15′ high, 4 uses	.858	S.F.	1.39	4.42	5.81
Forms in place, exterior spandrel, 12″ wide, 4 uses	.142	SFCA	.19	1.33	1.52
Forms in place, interior beam. 12″ wide, 4 uses	.306	SFCA	.47	2.36	2.83
Reinforcing in place, elevated slabs #4 to #7	1.600	Lb.	1.42	.64	2.06
Concrete ready mix, regular weight, 3000 psi	.410	C.F.	1.69		1.69
Place and vibrate concrete, elevated slab less than 6″, pump	.410	C.F.		.58	.58
Finish floor, monolithic steel trowel finish for finish floor	1.000	S.F.		.78	.78
Cure with sprayed membrane curing compound	.010	C.S.F.	.06	.08	.14
TOTAL			5.22	10.19	15.41

B1010 219			Cast in Place Beam & Slab, One Way					
	BAY SIZE (FT.)	SUPERIMPOSED LOAD (P.S.F.)	MINIMUM COL. SIZE (IN.)	SLAB THICKNESS (IN.)	TOTAL LOAD (P.S.F.)	COST PER S.F.		
						MAT.	INST.	TOTAL
3000	15x15	40	12	4	120	5.20	10.20	15.40
3100		75	12	4	138	5.35	10.25	15.60
3200	RB1010 -010	125	12	4	188	5.55	10.35	15.90
3300		200	14	4	266	5.95	10.70	16.65
3600	15x20	40	12	4	102	5.40	10.10	15.50
3700		75	12	4	140	5.70	10.40	16.10
3800	RB1010 -100	125	14	4	192	6.05	10.75	16.80
3900		200	16	4	272	6.85	11.45	18.30
4200	20x20	40	12	5	115	5.90	9.85	15.75
4300		75	14	5	154	6.50	10.65	17.15
4400		125	16	5	206	6.75	11.20	17.95
4500		200	18	5	287	7.75	12	19.75
5000	20x25	40	12	5-1/2	121	6.15	9.90	16.05
5100		75	14	5-1/2	160	6.85	10.75	17.60
5200		125	16	5-1/2	215	7.40	11.35	18.75
5300		200	18	5-1/2	294	8.20	12.15	20.35
5500	25x25	40	12	6	129	6.45	9.70	16.15
5600		75	16	6	171	7.15	10.40	17.55
5700		125	18	6	227	8.55	11.95	20.50
5800		200	2	6	300	9.60	12.80	22.40
6500	25x30	40	14	6-1/2	132	6.60	9.90	16.50
6600		75	16	6-1/2	172	7.25	10.50	17.75
6700		125	18	6-1/2	231	8.70	11.90	20.60
6800		200	20	6-1/2	312	9.65	12.90	22.55

B10 Superstructure

B1010 Floor Construction

B1010 219		Cast in Place Beam & Slab, One Way						
	BAY SIZE (FT.)	SUPERIMPOSED LOAD (P.S.F.)	MINIMUM COL. SIZE (IN.)	SLAB THICKNESS (IN.)	TOTAL LOAD (P.S.F.)	COST PER S.F.		
						MAT.	INST.	TOTAL
7000	30x30	40	14	7-1/2	150	7.70	10.65	18.35
7100		75	18	7-1/2	191	8.65	11.20	19.85
7300		125	20	7-1/2	245	9.30	11.95	21.25
7400		200	24	7-1/2	328	10.55	13.25	23.80
7500	30x35	40	16	8	158	8.10	11	19.10
7600		75	18	8	196	8.80	11.35	20.15
7700		125	22	8	254	10.05	12.60	22.65
7800		200	26	8	332	11.15	13.05	24.20
8000	35x35	40	16	9	169	9.20	11.35	20.55
8200		75	20	9	213	10.05	12.35	22.40
8400		125	24	9	272	11.25	12.85	24.10
8600		200	26	9	355	12.55	13.75	26.30
9000	35x40	40	18	9	174	9.45	11.55	21
9300		75	22	9	214	10.35	12.45	22.80
9400		125	26	9	273	11.40	12.95	24.35
9600		200	30	9	355	12.70	13.80	26.50

B1010 Floor Construction

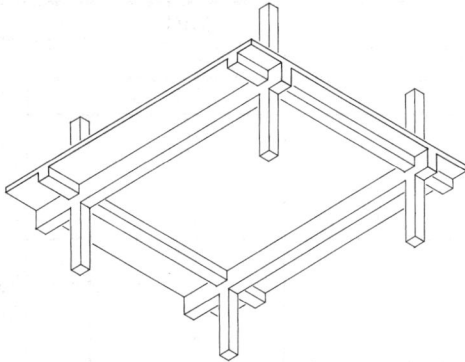

General: Solid concrete two-way slab cast monolithically with reinforced concrete support beams and girders.

Design and Pricing Assumptions:
Concrete f'c = 3 KSI, normal weight, placed by concrete pump.
Reinforcement, fy = 60 KSI.
Forms, four use.
Finish, steel trowel.
Curing, spray on membrane.
Based on 4 bay x 4 bay structure.

System Components			COST PER S.F.		
	QUANTITY	UNIT	MAT.	INST.	TOTAL
SYSTEM B1010 220 2000					
15'X 15' BAY, 40 PSF, S. LOAD 12" MIN. COL					
Forms in place, flat plate to 15' high, 4 uses	.894	S.F.	1.45	4.60	6.05
Forms in place, exterior spandrel, 12" wide, 4 uses	.142	SFCA	.19	1.33	1.52
Forms in place, interior beam. 12" wide, 4 uses	.178	SFCA	.27	1.37	1.64
Reinforcing in place, elevated slabs #4 to #7	1.598	Lb.	1.42	.64	2.06
Concrete ready mix, regular weight, 3000 psi	.437	C.F.	1.80		1.80
Place and vibrate concrete, elevated slab less than 6", pump	.437	C.F.		.62	.62
Finish floor, monolithic steel trowel finish for finish floor	1.000	S.F.		.78	.78
Cure with sprayed membrane curing compound	.010	C.S.F.	.06	.08	.14
TOTAL			5.19	9.42	14.61

B1010 220		Cast in Place Beam & Slab, Two Way						
	BAY SIZE (FT.)	SUPERIMPOSED LOAD (P.S.F.)	MINIMUM COL. SIZE (IN.)	SLAB THICKNESS (IN.)	TOTAL LOAD (P.S.F.)	COST PER S.F.		
						MAT.	INST.	TOTAL
2000	15x15	40	12	4-1/2	106	5.20	9.40	14.60
2200	RB1010 -010	75	12	4-1/2	143	5.35	9.50	14.85
2250		125	12	4-1/2	195	5.90	9.80	15.70
2300		200	14	4-1/2	274	6.45	10.40	16.85
2400	15 x 20	40	12	5-1/2	120	5.95	9.40	15.35
2600	RB1010 -100	75	12	5-1/2	159	6.60	10.05	16.65
2800		125	14	5-1/2	213	7.20	10.85	18.05
2900		200	16	5-1/2	294	7.95	11.60	19.55
3100	20 x 20	40	12	6	132	6.65	9.55	16.20
3300		75	14	6	167	7.15	10.45	17.60
3400		125	16	6	220	7.40	10.80	18.20
3600		200	18	6	301	8.40	11.55	19.95
4000	20 x 25	40	12	7	141	7.10	9.95	17.05
4300		75	14	7	181	8.40	10.90	19.30
4500		125	16	7	236	8.50	11.20	19.70
4700		200	18	7	317	9.55	12.10	21.65
5100	25 x 25	40	12	7-1/2	149	7.40	10.10	17.50
5200		75	16	7-1/2	185	8.25	10.80	19.05
5300		125	18	7-1/2	250	8.90	11.75	20.65
5400		200	20	7-1/2	332	11.15	12.75	23.90
6000	25 x 30	40	14	8-1/2	165	8.30	10.10	18.40
6200		75	16	8-1/2	201	9.05	11.05	20.10
6400		125	18	8-1/2	259	10	11.90	21.90

B1010 Floor Construction

B1010 220	Cast in Place Beam & Slab, Two Way

	BAY SIZE (FT.)	SUPERIMPOSED LOAD (P.S.F.)	MINIMUM COL. SIZE (IN.)	SLAB THICKNESS (IN.)	TOTAL LOAD (P.S.F.)	COST PER S.F.		
						MAT.	INST.	TOTAL
6600		200	20	8-1/2	341	11.55	12.60	24.15
7000	30 x 30	40	14	9	170	9.15	10.80	19.95
7100		75	18	9	212	10.15	11.65	21.80
7300		125	20	9	267	11.05	12.10	23.15
7400		200	24	9	351	12.25	12.80	25.05
7600	30 x 35	40	16	10	188	10.05	11.45	21.50
7700		75	18	10	225	10.80	11.90	22.70
8000		125	22	10	282	12.15	12.80	24.95
8200		200	26	10	371	13.45	13.20	26.65
8500	35 x 35	40	16	10-1/2	193	10.90	11.70	22.60
8600		75	20	10-1/2	233	11.55	12.15	23.70
9000		125	24	10-1/2	287	13.05	13.05	26.10
9100		200	26	10-1/2	370	13.65	13.85	27.50
9300	35 x 40	40	18	11-1/2	208	11.85	12.10	23.95
9500		75	22	11-1/2	247	12.55	12.40	24.95
9800		125	24	11-1/2	302	13.55	12.85	26.40
9900		200	26	11-1/2	383	14.70	13.75	28.45

B1010   Floor Construction

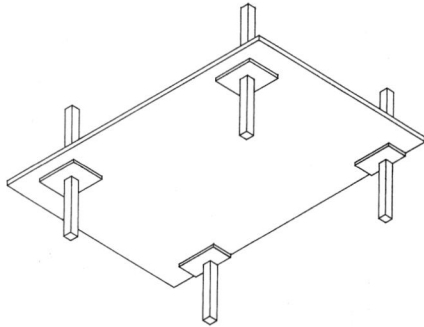

General: Flat Slab: Solid uniform depth concrete two-way slabs with drop panels at columns and no column capitals.

Design and Pricing Assumptions:
Concrete f'c = 3 KSI, placed by concrete pump.
Reinforcement, fy = 60 KSI.
Forms, four use.
Finish, steel trowel.
Curing, spray on membrane.
Based on 4 bay x 4 bay structure.

System Components	QUANTITY	UNIT	COST PER S.F. MAT.	COST PER S.F. INST.	COST PER S.F. TOTAL
SYSTEM B1010 222 1700					
15'X15' BAY 40 PSF S. LOAD, 12" MIN. COL. 6" SLAB, 1-1/2" DROP, 117 PSF					
Forms in place, flat slab with drop panels, to 15' high, 4 uses	.993	S.F.	1.70	5.26	6.96
Forms in place, exterior spandrel, 12" wide, 4 uses	.034	SFCA	.05	.32	.37
Reinforcing in place, elevated slabs #4 to #7	1.588	Lb.	1.41	.64	2.05
Concrete ready mix, regular weight, 3000 psi	.513	C.F.	2.11		2.11
Place and vibrate concrete, elevated slab, 6" to 10" pump	.513	C.F.		.62	.62
Finish floor, monolithic steel trowel finish for finish floor	1.000	S.F.		.78	.78
Cure with sprayed membrane curing compound	.010	C.S.F.	.06	.08	.14
TOTAL			5.33	7.70	13.03

B1010 222		Cast in Place Flat Slab with Drop Panels							
	BAY SIZE (FT.)	SUPERIMPOSED LOAD (P.S.F.)	MINIMUM COL. SIZE (IN.)	SLAB & DROP (IN.)	TOTAL LOAD (P.S.F.)	COST PER S.F. MAT.	COST PER S.F. INST.	COST PER S.F. TOTAL
1700	15 x 15	40	12	6 - 1-1/2	117	5.35	7.70	13.05
1720	RB1010 -010	75	12	6 - 2-1/2	153	5.50	7.75	13.25
1760		125	14	6 - 3-1/2	205	5.85	7.90	13.75
1780		200	16	6 - 4-1/2	281	6.25	8.10	14.35
1840	15 x 20	40	12	6-1/2 - 2	124	5.70	7.80	13.50
1860	RB1010 -100	75	14	6-1/2 - 4	162	6.05	8	14.05
1880		125	16	6-1/2 - 5	213	6.50	8.20	14.70
1900		200	18	6-1/2 - 6	293	6.80	8.35	15.15
1960	20 x 20	40	12	7 - 3	132	6	7.95	13.95
1980		75	16	7 - 4	168	6.50	8.15	14.65
2000		125	18	7 - 6	221	7.30	8.40	15.70
2100		200	20	8 - 6-1/2	309	7.50	8.55	16.05
2300	20 x 25	40	12	8 - 5	147	6.80	8.25	15.05
2400		75	18	8 - 6-1/2	184	7.50	8.55	16.05
2600		125	20	8 - 8	236	8.35	8.90	17.25
2800		200	22	8-1/2 - 8-1/2	323	8.75	9.15	17.90
3200	25 x 25	40	12	8-1/2 - 5-1/2	154	7.15	8.30	15.45
3400		75	18	8-1/2 - 7	191	7.70	8.60	16.30
4000		125	20	8-1/2 - 8-1/2	243	8.50	8.95	17.45
4400		200	24	9 - 8-1/2	329	8.95	9.15	18.10
5000	25 x 30	40	14	9-1/2 - 7	168	7.85	8.65	16.50
5200		75	18	9-1/2 - 7	203	8.60	9	17.60
5600		125	22	9-1/2 - 8	256	9.15	9.20	18.35
5800		200	24	10 - 10	342	9.80	9.50	19.30

B1010 Floor Construction

| B1010 222 | Cast in Place Flat Slab with Drop Panels |

	BAY SIZE (FT.)	SUPERIMPOSED LOAD (P.S.F.)	MINIMUM COL. SIZE (IN.)	SLAB & DROP (IN.)	TOTAL LOAD (P.S.F.)	COST PER S.F.		
						MAT.	INST.	TOTAL
6400	30 x 30	40	14	10-1/2 - 7-1/2	182	8.55	8.90	17.45
6600		75	18	10-1/2 - 7-1/2	217	9.35	9.25	18.60
6800		125	22	10-1/2 - 9	269	9.90	9.45	19.35
7000		200	26	11 - 11	359	10.70	9.85	20.55
7400	30 x 35	40	16	11-1/2 - 9	196	9.40	9.25	18.65
7900		75	20	11-1/2 - 9	231	10.25	9.65	19.90
8000		125	24	11-1/2 - 11	284	10.80	9.85	20.65
9000	35 x 35	40	16	12 - 9	202	9.65	9.30	18.95
9400		75	20	12 - 11	240	10.65	9.75	20.40
9600		125	24	12 - 11	290	11.10	9.95	21.05

B1010 Floor Construction

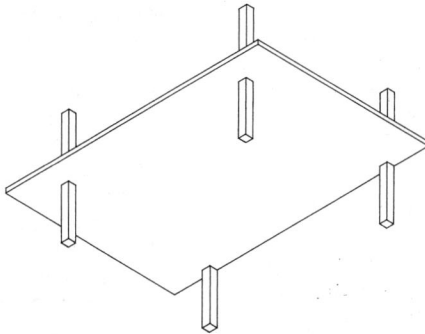

General: Flat Plates: Solid uniform depth concrete two-way slab without drops or interior beams. Primary design limit is shear at columns.

Design and Pricing Assumptions:
Concrete f'c to 4 KSI, placed by concrete pump.
Reinforcement, fy = 60 KSI.
Forms, four use.
Finish, steel trowel.
Curing, spray on membrane.
Based on 4 bay x 4 bay structure.

System Components	QUANTITY	UNIT	COST PER S.F.		
			MAT.	INST.	TOTAL
SYSTEM B1010 223 2000					
15'X15' BAY 40 PSF S. LOAD, 12" MIN. COL.					
Forms in place, flat plate to 15' high, 4 uses	.992	S.F.	1.61	5.11	6.72
Edge forms to 6" high on elevated slab, 4 uses	.065	L.F.	.01	.24	.25
Reinforcing in place, elevated slabs #4 to #7	1.706	Lb.	1.52	.68	2.20
Concrete ready mix, regular weight, 3000 psi	.459	C.F.	1.89		1.89
Place and vibrate concrete, elevated slab less than 6", pump	.459	C.F.		.64	.64
Finish floor, monolithic steel trowel finish for finish floor	1.000	S.F.		.78	.78
Cure with sprayed membrane curing compound	.010	C.S.F.	.06	.08	.14
TOTAL			5.09	7.53	12.62

B1010 223			Cast in Place Flat Plate						
	BAY SIZE (FT.)	SUPERIMPOSED LOAD (P.S.F.)	MINIMUM COL. SIZE (IN.)	SLAB THICKNESS (IN.)	TOTAL LOAD (P.S.F.)	COST PER S.F.		
						MAT.	INST.	TOTAL
2000	15 x 15	40	12	5-1/2	109	5.10	7.50	12.60
2200		75	14	5-1/2	144	5.15	7.50	12.65
2400	RB1010 -010	125	20	5-1/2	194	5.45	7.60	13.05
2600		175	22	5-1/2	244	5.60	7.65	13.25
3000	15 x 20	40	14	7	127	5.85	7.65	13.50
3400	RB1010 -100	75	16	7-1/2	169	6.30	7.80	14.10
3600		125	22	8-1/2	231	6.95	8	14.95
3800		175	24	8-1/2	281	7.05	8	15.05
4200	20 x 20	40	16	7	127	5.85	7.65	13.50
4400		75	20	7-1/2	175	6.40	7.80	14.20
4600		125	24	8-1/2	231	7	8	15
5000		175	24	8-1/2	281	7.05	8.05	15.10
5600	20 x 25	40	18	8-1/2	146	6.90	8	14.90
6000		75	20	9	188	7.15	8.05	15.20
6400		125	26	9-1/2	244	7.85	8.35	16.20
6600		175	30	10	300	8.15	8.45	16.60
7000	25 x 25	40	20	9	152	7.15	8.05	15.20
7400		75	24	9-1/2	194	7.60	8.30	15.90
7600		125	30	10	250	8.20	8.50	16.70
8000								

B1010 Floor Construction

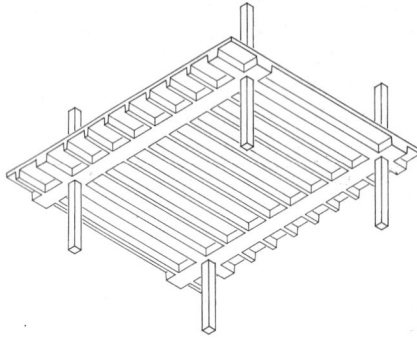

General: Combination of thin concrete slab and monolithic ribs at uniform spacing to reduce dead weight and increase rigidity. The ribs (or joists) are arranged parallel in one direction between supports.

Square end joists simplify forming. Tapered ends can increase span or provide for heavy load.

Costs for multiple span joists are provided in this section. Single span joist costs are not provided here.

Design and Pricing Assumptions:
Concrete f'c = 4 KSI, normal weight placed by concrete pump.
Reinforcement, fy = 60 KSI.
Forms, four use.
4-1/2" slab.
30" pans, sq. ends (except for shear req.).
6" rib thickness.
Distribution ribs as required.
Finish, steel trowel.
Curing, spray on membrane.
Based on 4 bay x 4 bay structure.

System Components			COST PER S.F.		
	QUANTITY	UNIT	MAT.	INST.	TOTAL
SYSTEM B1010 226 2000					
15'X15' BAY, 40 PSF, S. LOAD 12" MIN. COLUMN					
Forms in place, floor slab with 30" metal pans, 4 use	.905	S.F.	3.24	5.25	8.49
Forms in place, exterior spandrel, 12" wide, 4 uses	.170	SFCA	.23	1.59	1.82
Forms in place, interior beam. 12" wide, 4 uses	.095	SFCA	.15	.73	.88
Edge forms, to 6" high on elevated slab, 1 use	.010	L.F.	.01	.06	.07
Reinforcing in place, elevated slabs #4 to #7	.628	Lb.	.56	.25	.81
Concrete ready mix, regular weight, 4000 psi	.555	C.F.	2.39		2.39
Place and vibrate concrete, elevated slab, 6" to 10" pump	.555	C.F.		.67	.67
Finish floor, monolithic steel trowel finish for finish floor	1.000	S.F.		.78	.78
Cure with sprayed membrane curing compound	.010	S.F.	.06	.08	.14
TOTAL			6.64	9.41	16.05

B1010 226			Cast in Place Multispan Joist Slab					
	BAY SIZE (FT.)	SUPERIMPOSED LOAD (P.S.F.)	MINIMUM COL. SIZE (IN.)	RIB DEPTH (IN.)	TOTAL LOAD (P.S.F.)	COST PER S.F.		
						MAT.	INST.	TOTAL
2000	15 x 15	40	12	8	115	6.65	9.40	16.05
2100	RB1010 -010	75	12	8	150	6.70	9.40	16.10
2200		125	12	8	200	6.90	9.50	16.40
2300		200	14	8	275	7.15	9.90	17.05
2600	15 x 20	40	12	8	115	6.80	9.40	16.20
2800	RB1010 -100	75	12	8	150	7	9.90	16.90
3000		125	14	8	200	7.30	10.05	17.35
3300		200	16	8	275	7.75	10.30	18.05
3600	20 x 20	40	12	10	120	7	9.25	16.25
3900		75	14	10	155	7.35	9.80	17.15
4000		125	16	10	205	7.45	10	17.45
4100		200	18	10	280	7.90	10.45	18.35
4300	20 x 25	40	12	10	120	7	9.40	16.40
4400		75	14	10	155	7.40	9.90	17.30
4500		125	16	10	205	7.85	10.40	18.25
4600		200	18	12	280	8.35	10.95	19.30
4700	25 x 25	40	12	12	125	7	9.20	16.20
4800		75	16	12	160	7.55	9.70	17.25
4900		125	18	12	210	8.60	10.65	19.25
5000		200	20	14	291	9.15	10.90	20.05

B1010 Floor Construction

B1010 226	Cast in Place Multispan Joist Slab

	BAY SIZE (FT.)	SUPERIMPOSED LOAD (P.S.F.)	MINIMUM COL. SIZE (IN.)	RIB DEPTH (IN.)	TOTAL LOAD (P.S.F.)	COST PER S.F.		
						MAT.	INST.	TOTAL
5400	25 x 30	40	14	12	125	7.40	9.60	17
5600		75	16	12	160	7.75	9.90	17.65
5800		125	18	12	210	8.45	10.60	19.05
6000		200	20	14	291	9.20	11.05	20.25
6200	30 x 30	40	14	14	131	7.85	9.70	17.55
6400		75	18	14	166	8.15	10	18.15
6600		125	20	14	216	8.75	10.55	19.30
6700		200	24	16	297	9.45	10.95	20.40
6900	30 x 35	40	16	14	131	8.05	10.10	18.15
7000		75	18	14	166	8.30	10.15	18.45
7100		125	22	14	216	8.65	10.75	19.40
7200		200	26	16	297	9.70	11.35	21.05
7400	35 x 35	40	16	16	137	8.35	10	18.35
7500		75	20	16	172	8.95	10.40	19.35
7600		125	24	16	222	9	10.45	19.45
7700		200	26	20	309	9.85	11.10	20.95
8000	35 x 40	40	18	16	137	8.65	10.40	19.05
8100		75	22	16	172	9.10	10.85	19.95
8300		125	26	16	222	9.25	10.70	19.95
8400		200	30	20	309	10.20	11.25	21.45
8750	40 x 40	40	18	20	149	9.30	10.40	19.70
8800		75	24	20	184	9.65	10.60	20.25
8900		125	26	20	234	10.20	11	21.20
9100	40 x 45	40	20	20	149	9.80	10.70	20.50
9500		75	24	20	184	9.90	10.80	20.70
9800		125	28	20	234	10.30	11.20	21.50

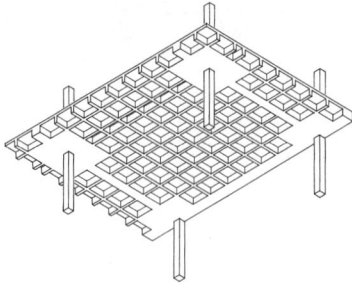

General: Waffle slabs are basically flat slabs with hollowed out domes on bottom side to reduce weight. Solid concrete heads at columns function as drops without increasing depth. The concrete ribs function as two-way right angle joist.

Joists are formed with standard sized domes. Thin slabs cover domes and are usually reinforced with welded wire fabric. Ribs have bottom steel and may have stirrups for shear.

Design and Pricing Assumptions:
Concrete f'c = 4 KSI, normal weight placed by concrete pump.
Reinforcement, fy = 60 KSI.
Forms, four use.
  4-1/2" slab.
  30" x 30" voids.
  6" wide ribs.
  (ribs @ 36" O.C.).
  Rib depth filler beams as required.
  Solid concrete heads at columns.
Finish, steel trowel.
Curing, spray on membrane.
Based on 4 bay x 4 bay structure.

System Components	QUANTITY	UNIT	COST PER S.F. MAT.	COST PER S.F. INST.	COST PER S.F. TOTAL
SYSTEM B1010 227 3900					
20X20' BAY, 40 PSF S. LOAD, 12" MIN. COLUMN					
Forms in place, floor slab with 2-way joist domes, 4 uses	1.000	S.F.	1.62	6.90	8.52
Edge forms, to 6" high on elevated slab, 1 use	.052	SFCA	.03	.30	.33
Forms in place, bulkhead for slab with keyway, 1 use, 3 piece	.010	L.F.	.03	.06	.09
Reinforcing in place, elevated slabs #4 to #7	1.580	Lb.	1.41	.63	2.04
Welded wire fabric rolls, 6 x 6 - W4 x W4 (4 x 4) 58 lb./c.s.f	1.000	S.F.	.51	.43	.94
Concrete ready mix, regular weight, 4000 psi	.690	C.F.	2.97		2.97
Place and vibrate concrete, elevated slab, over 10", pump	.690	C.F.		.87	.87
Finish floor, monolithic steel trowel finish for finish floor	1.000	S.F.		.78	.78
Cure with sprayed membrane curing compound	.010	C.S.F.	.06	.08	.14
TOTAL			6.63	10.05	16.68

B1010 227		Cast in Place Waffle Slab						
	BAY SIZE (FT.)	SUPERIMPOSED LOAD (P.S.F.)	MINIMUM COL. SIZE (IN.)	RIB DEPTH (IN.)	TOTAL LOAD (P.S.F.)	COST PER S.F. MAT.	COST PER S.F. INST.	COST PER S.F. TOTAL
3900	20 x 20	40	12	8	144	6.65	10.05	16.70
4000	RB1010 -010	75	12	8	179	6.85	10.15	17
4100		125	16	8	229	7.15	10.25	17.40
4200		200	18	8	304	7.90	10.60	18.50
4400	20 x 25	40	12	8	146	6.85	10.10	16.95
4500	RB1010 -100	75	14	8	181	7.20	10.25	17.45
4600		125	16	8	231	7.45	10.35	17.80
4700		200	18	8	306	8.10	10.65	18.75
4900	25 x 25	40	12	10	150	7.15	10.20	17.35
5000		75	16	10	185	7.55	10.40	17.95
5300		125	18	10	235	7.95	10.60	18.55
5500		200	20	10	310	8.40	10.75	19.15
5700	25 x 30	40	14	10	154	7.35	10.30	17.65
5800		75	16	10	189	7.75	10.50	18.25
5900		125	18	10	239	8.15	10.65	18.80
6000		200	20	12	329	9.35	11.10	20.45
6400	30 x 30	40	14	12	169	8	10.50	18.50
6500		75	18	12	204	8.35	10.65	19
6600		125	20	12	254	8.60	10.75	19.35
6700		200	24	12	329	9.95	11.35	21.30

B10 Superstructure

B1010 Floor Construction

| B1010 227 | | Cast in Place Waffle Slab | | | | | | |

	BAY SIZE (FT.)	SUPERIMPOSED LOAD (P.S.F.)	MINIMUM COL. SIZE (IN.)	RIB DEPTH (IN.)	TOTAL LOAD (P.S.F.)	COST PER S.F.		
						MAT.	INST.	TOTAL
6900	30 x 35	40	16	12	169	8.25	10.60	18.85
7000		75	18	12	204	8.25	10.60	18.85
7100		125	22	12	254	9.05	10.95	20
7200		200	26	14	334	10.55	11.60	22.15
7400	35 x 35	40	16	14	174	8.90	10.85	19.75
7500		75	20	14	209	9.30	11.05	20.35
7600		125	24	14	259	9.65	11.20	20.85
7700		200	26	16	346	10.85	11.75	22.60
8000	35 x 40	40	18	14	176	9.35	11.05	20.40
8300		75	22	14	211	9.90	11.30	21.20
8500		125	26	16	271	10.50	11.55	22.05
8750		200	30	20	372	11.95	12.15	24.10
9200	40 x 40	40	18	14	176	9.90	11.30	21.20
9400		75	24	14	211	10.55	11.60	22.15
9500		125	26	16	271	10.75	11.65	22.40
9700	40 x 45	40	20	16	186	10.30	11.45	21.75
9800		75	24	16	221	10.95	11.75	22.70
9900		125	28	16	271	11.30	11.90	23.20

B1010 Floor Construction

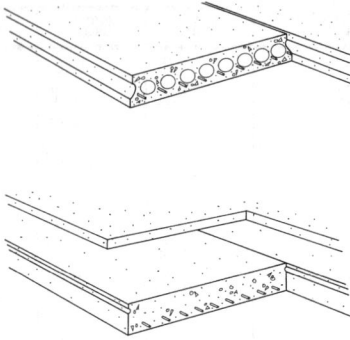

General: Units priced here are for plant produced prestressed members, transported to site and erected.

Normal weight concrete is most frequently used. Lightweight concrete may be used to reduce dead weight.

Structural topping is sometimes used on floors: insulating concrete or rigid insulation on roofs.

Camber and deflection may limit use by depth considerations.

Prices are based upon 10,000 S.F. to 20,000 S.F. projects, and 50 mile to 100 mile transport.

Concrete is f'c = 5 KSI and Steel is fy = 250 or 300 KSI

Note: Deduct from prices 20% for Southern states. Add to prices 10% for Western states.

Description of Table: Enter table at span and load. Most economical sections will generally consist of normal weight concrete without topping. If acceptable, note this price, depth and weight. For topping and/or lightweight concrete, note appropriate data.

Generally used on masonry and concrete bearing or reinforced concrete and steel framed structures.

The solid 4" slabs are used for light loads and short spans. The 6" to 12" thick hollow core units are used for longer spans and heavier loads. Cores may carry utilities.

Topping is used structurally for loads or rigidity and architecturally to level or slope surface.

Camber and deflection and change in direction of spans must be considered (door openings, etc.), especially untopped.

System Components			COST PER S.F.		
	QUANTITY	UNIT	MAT.	INST.	TOTAL
SYSTEM B1010 230 2000					
10' SPAN, 40 LBS S.F. WORKING LOAD, 2" TOPPING					
Precast prestressed concrete roof/floor slabs 4" thick, grouted	1.000	S.F.	6.25	3.14	9.39
Edge forms to 6" high on elevated slab, 4 uses	.100	L.F.	.02	.38	.40
Welded wire fabric 6 x 6 - W1.4 x W1.4 (10 x 10), 21 lb/csf, 10% lap	.010	C.S.F.	.20	.34	.54
Concrete ready mix, regular weight, 3000 psi	.170	C.F.	.70		.70
Place and vibrate concrete, elevated slab less than 6", pumped	.170	C.F.		.24	.24
Finishing floor, monolithic steel trowel finish for resilient tile	1.000	S.F.		1.02	1.02
Curing with sprayed membrane curing compound	.010	C.S.F.	.06	.08	.14
TOTAL			7.23	5.20	12.43

B1010 229		Precast Plank with No Topping						
	SPAN (FT.)	SUPERIMPOSED LOAD (P.S.F.)	TOTAL DEPTH (IN.)	DEAD LOAD (P.S.F.)	TOTAL LOAD (P.S.F.)	COST PER S.F.		
						MAT.	INST.	TOTAL
0720	10	40	4	50	90	6.25	3.14	9.39
0750	RB1010 -010	75	6	50	125	7.35	2.68	10.03
0770		100	6	50	150	7.35	2.68	10.03
0800	15	40	6	50	90	7.35	2.68	10.03
0820	RB1010 -100	75	6	50	125	7.35	2.68	10.03
0850		100	6	50	150	7.35	2.68	10.03
0875	20	40	6	50	90	7.35	2.68	10.03
0900		75	6	50	125	7.35	2.68	10.03
0920		100	6	50	150	7.35	2.68	10.03
0950	25	40	6	50	90	7.35	2.68	10.03
0970		75	8	55	130	8.05	2.35	10.40
1000		100	8	55	155	8.05	2.35	10.40
1200	30	40	8	55	95	8.05	2.35	10.40
1300		75	8	55	130	8.05	2.35	10.40
1400		100	10	70	170	8.50	2.09	10.59
1500	40	40	10	70	110	8.50	2.09	10.59
1600		75	12	70	145	9.15	1.88	11.03

B10 Superstructure

B1010 Floor Construction

B1010 229 — Precast Plank with No Topping

	SPAN (FT.)	SUPERIMPOSED LOAD (P.S.F.)	TOTAL DEPTH (IN.)	DEAD LOAD (P.S.F.)	TOTAL LOAD (P.S.F.)	COST PER S.F.		
						MAT.	INST.	TOTAL
1700	45	40	12	70	110	9.15	1.88	11.03

B1010 230 — Precast Plank with 2″ Concrete Topping

	SPAN (FT.)	SUPERIMPOSED LOAD (P.S.F.)	TOTAL DEPTH (IN.)	DEAD LOAD (P.S.F.)	TOTAL LOAD (P.S.F.)	COST PER S.F.		
						MAT.	INST.	TOTAL
2000	10	40	6	75	115	7.25	5.20	12.45
2100		75	8	75	150	8.35	4.74	13.09
2200		100	8	75	175	8.35	4.74	13.09
2500	15	40	8	75	115	8.35	4.74	13.09
2600		75	8	75	150	8.35	4.74	13.09
2700		100	8	75	175	8.35	4.74	13.09
2800	20	40	8	75	115	8.35	4.74	13.09
2900		75	8	75	150	8.35	4.74	13.09
3000		100	8	75	175	8.35	4.74	13.09
3100	25	40	8	75	115	8.35	4.74	13.09
3200		75	8	75	150	8.35	4.74	13.09
3300		100	10	80	180	9.05	4.41	13.46
3400	30	40	10	80	120	9.05	4.41	13.46
3500		75	10	80	155	9.05	4.41	13.46
3600		100	10	80	180	9.05	4.41	13.46
3700	35	40	12	95	135	9.50	4.15	13.65
3800		75	12	95	170	9.50	4.15	13.65
3900		100	14	95	195	10.15	3.94	14.09
4000	40	40	12	95	135	9.50	4.15	13.65
4500		75	14	95	170	10.15	3.94	14.09
5000	45	40	14	95	135	10.15	3.94	14.09

B1010 Floor Construction

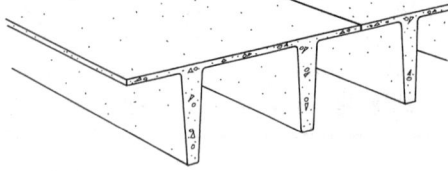

Most widely used for moderate span floors and roofs. At shorter spans, they tend to be competitive with hollow core slabs. They are also used as wall panels.

System Components			COST PER S.F.		
	QUANTITY	UNIT	MAT.	INST.	TOTAL
SYSTEM B1010 235 6700					
PRECAST, DOUBLE "T",2" TOPPING,30' SPAN,30 PSF SUP. LOAD, 18" X 8'					
Double "T" beams, reg. wt, 18" x 8' w, 30' span	1.000	S.F.	8.85	1.56	10.41
Edge forms to 6" high on elevated slab, 4 uses	.050	L.F.	.01	.19	.20
Concrete ready mix, regular weight, 3000 psi	.250	C.F.	1.03		1.03
Place and vibrate concrete, elevated slab less than 6", pumped	.250	C.F.		.35	.35
Finishing floor, monolithic steel trowel finish for finish floor	1.000	S.F.		.78	.78
Curing with sprayed membrane curing compound	.010	C.S.F.	.06	.08	.14
TOTAL			9.95	2.96	12.91

B1010 234 — Precast Double "T" Beams with No Topping

	SPAN (FT.)	SUPERIMPOSED LOAD (P.S.F.)	DBL. "T" SIZE D (IN.) W (FT.)	CONCRETE "T" TYPE	TOTAL LOAD (P.S.F.)	MAT.	INST.	TOTAL
1500	30	30	18x8	Reg. Wt.	92	8.85	1.56	10.41
1600		40	18x8	Reg. Wt.	102	9	2.01	11.01
1700	RB1010 -010	50	18x8	Reg. Wt	112	9	2.01	11.01
1800		75	18x8	Reg. Wt.	137	9.05	2.09	11.14
1900		100	18x8	Reg. Wt.	162	9.05	2.09	11.14
2000	40	30	20x8	Reg. Wt.	87	6.80	1.30	8.10
2100		40	20x8	Reg. Wt.	97	6.90	1.56	8.46
2200	RB1010 -100	50	20x8	Reg. Wt.	107	6.90	1.56	8.46
2300		75	20x8	Reg. Wt.	132	6.95	1.67	8.62
2400		100	20x8	Reg. Wt.	157	7.05	2.05	9.10
2500	50	30	24x8	Reg. Wt.	103	7.25	1.18	8.43
2600		40	24x8	Reg. Wt.	113	7.35	1.44	8.79
2700		50	24x8	Reg. Wt.	123	7.40	1.53	8.93
2800		75	24x8	Reg. Wt.	148	7.40	1.55	8.95
2900		100	24x8	Reg. Wt.	173	7.50	1.92	9.42
3000	60	30	24x8	Reg. Wt.	82	7.40	1.54	8.94
3100		40	32x10	Reg. Wt.	104	8.65	1.26	9.91
3150		50	32x10	Reg. Wt.	114	8.55	1.10	9.65
3200		75	32x10	Reg. Wt.	139	8.60	1.19	9.79
3250		100	32x10	Reg. Wt.	164	8.70	1.49	10.19
3300	70	30	32x10	Reg. Wt.	94	8.60	1.17	9.77
3350		40	32x10	Reg. Wt.	104	8.60	1.19	9.79
3400		50	32x10	Reg. Wt.	114	8.70	1.49	10.19
3450		75	32x10	Reg. Wt.	139	8.80	1.78	10.58
3500		100	32x10	Reg. Wt.	164	9.05	2.37	11.42
3550	80	30	32x10	Reg. Wt.	94	8.70	1.49	10.19
3600		40	32x10	Reg. Wt.	104	8.95	2.07	11.02
3900		50	32x10	Reg. Wt.	114	9.05	2.37	11.42

B10 Superstructure

B1010 Floor Construction

B1010 234 — Precast Double "T" Beams with No Topping

	SPAN (FT.)	SUPERIMPOSED LOAD (P.S.F.)	DBL. "T" SIZE D (IN.) W (FT.)	CONCRETE "T" TYPE	TOTAL LOAD (P.S.F.)	COST PER S.F.		
						MAT.	INST.	TOTAL
4300	50	30	20x8	Lt. Wt.	66	8.60	1.44	10.04
4400		40	20x8	Lt. Wt.	76	9.20	1.30	10.50
4500		50	20x8	Lt. Wt.	86	9.30	1.56	10.86
4600		75	20x8	Lt. Wt.	111	9.40	1.78	11.18
4700		100	20x8	Lt. Wt.	136	9.55	2.15	11.70
4800	60	30	24x8	Lt. Wt.	70	9.25	1.40	10.65
4900		40	32x10	Lt. Wt.	88	10.05	.99	11.04
5000		50	32x10	Lt. Wt.	98	10.15	1.20	11.35
5200		75	32x10	Lt. Wt.	123	10.20	1.38	11.58
5400		100	32x10	Lt. Wt.	148	10.30	1.68	11.98
5600	70	30	32x10	Lt. Wt.	78	10.05	.99	11.04
5750		40	32x10	Lt. Wt.	88	10.15	1.19	11.34
5900		50	32x10	Lt. Wt.	98	10.20	1.38	11.58
6000		75	32x10	Lt. Wt.	123	10.30	1.67	11.97
6100		100	32x10	Lt. Wt.	148	10.55	2.26	12.81
6200	80	30	32x10	Lt. Wt.	78	10.20	1.38	11.58
6300		40	32x10	Lt. Wt.	88	10.30	1.67	11.97
6400		50	32x10	Lt. Wt.	98	10.45	1.96	12.41

B1010 235 — Precast Double "T" Beams With 2" Topping

	SPAN (FT.)	SUPERIMPOSED LOAD (P.S.F.)	DBL. "T" SIZE D (IN.) W (FT.)	CONCRETE "T" TYPE	TOTAL LOAD (P.S.F.)	COST PER S.F.		
						MAT.	INST.	TOTAL
6700	30	30	18x8	Reg. Wt.	117	9.95	2.96	12.91
6750		40	18x8	Reg. Wt.	127	9.95	2.96	12.91
6800		50	18x8	Reg. Wt.	137	10.05	3.22	13.27
6900		75	18x8	Reg. Wt.	162	10.05	3.22	13.27
7000		100	18x8	Reg. Wt.	187	10.10	3.34	13.44
7100	40	30	18x8	Reg. Wt.	120	7.85	2.84	10.69
7200		40	20x8	Reg. Wt.	130	7.90	2.70	10.60
7300		50	20x8	Reg. Wt.	140	8	2.96	10.96
7400		75	20x8	Reg. Wt.	165	8.05	3.07	11.12
7500		100	20x8	Reg. Wt.	190	8.15	3.45	11.60
7550	50	30	24x8	Reg. Wt.	120	8.45	2.84	11.29
7600		40	24x8	Reg. Wt.	130	8.50	2.93	11.43
7750		50	24x8	Reg. Wt.	140	8.50	2.95	11.45
7800		75	24x8	Reg. Wt.	165	8.60	3.32	11.92
7900		100	32x10	Reg. Wt.	189	9.75	2.71	12.46
8000	60	30	32x10	Reg. Wt.	118	9.60	2.30	11.90
8100		40	32x10	Reg. Wt.	129	9.65	2.50	12.15
8200		50	32x10	Reg. Wt.	139	9.75	2.71	12.46
8300		75	32x10	Reg. Wt.	164	9.80	2.89	12.69
8350		100	32x10	Reg. Wt.	189	9.90	3.19	13.09
8400	70	30	32x10	Reg. Wt.	119	9.70	2.59	12.29
8450		40	32x10	Reg. Wt.	129	9.80	2.89	12.69
8500		50	32x10	Reg. Wt.	139	9.90	3.18	13.08
8550		75	32x10	Reg. Wt.	164	10.15	3.77	13.92
8600	80	30	32x10	Reg. Wt.	119	10.15	3.77	13.92
8800	50	30	20x8	Lt. Wt.	105	9.75	2.95	12.70
8850		40	24x8	Lt. Wt.	121	9.80	3.10	12.90
8900		50	24x8	Lt. Wt.	131	9.90	3.32	13.22
8950		75	24x8	Lt. Wt.	156	9.90	3.32	13.22
9000		100	24x8	Lt. Wt.	181	10	3.69	13.69

B10 Superstructure

B1010 Floor Construction

| B1010 235 | | | | Precast Double "T" Beams With 2" Topping | | | | |

	SPAN (FT.)	SUPERIMPOSED LOAD (P.S.F.)	DBL. "T" SIZE D (IN.) W (FT.)	CONCRETE "T" TYPE	TOTAL LOAD (P.S.F.)	COST PER S.F.		
						MAT.	INST.	TOTAL
9200	60	30	32x10	Lt. Wt.	103	11.15	2.39	13.54
9300		40	32x10	Lt. Wt.	113	11.25	2.60	13.85
9350		50	32x10	Lt. Wt.	123	11.25	2.60	13.85
9400		75	32x10	Lt. Wt.	148	11.30	2.78	14.08
9450		100	32x10	Lt. Wt.	173	11.40	3.08	14.48
9500	70	30	32x10	Lt. Wt.	103	11.30	2.78	14.08
9550		40	32x10	Lt. Wt.	113	11.30	2.78	14.08
9600		50	32x10	Lt. Wt.	123	11.40	3.07	14.47
9650		75	32x10	Lt. Wt.	148	11.55	3.36	14.91
9700	80	30	32x10	Lt. Wt.	103	11.40	3.07	14.47
9800		40	32x10	Lt. Wt.	113	11.55	3.36	14.91
9900		50	32x10	Lt. Wt.	123	11.65	3.66	15.31

B10 Superstructure

B1010 Floor Construction

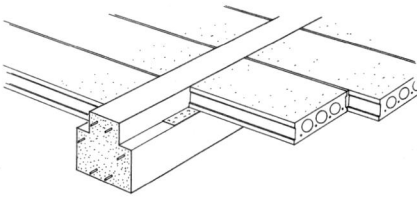

General: Units priced here are for plant produced prestressed members transported to the site and erected.

System has precast prestressed concrete beams and precast, prestressed hollow core slabs spanning the longer direction when applicable.

Camber and deflection must be considered when using untopped hollow core slabs.

Design and Pricing Assumptions:
Prices are based on 10,000 S.F. to 20,000 S.F. projects and 50 mile to 100 mile transport.

Concrete is f'c 5 KSI and prestressing steel is fy = 250 or 300 KSI.

System Components	QUANTITY	UNIT	COST PER S.F.		
			MAT.	INST.	TOTAL
SYSTEM B1010 236 5200					
20'X20' BAY, 6" DEPTH, 40 PSF S. LOAD, 100PSF TOTAL LOAD					
12" x 20" precast "T" beam, 20' span	.038	L.F.		.94	.94
Installation labor and equipment	.038	L.F.	6.12		6.12
12" x 20" precast "L" beam, 20' span	.025	L.F.		.94	.94
Installation labor and equipment	.025	L.F.	3.30		3.30
Precast prestressed concrete roof/floor slabs 6" deep, grouted	1.000	S.F.	7.35	2.68	10.03
TOTAL			16.77	4.56	21.33

B1010 236		**Precast Beam & Plank with No Topping**						
	BAY SIZE (FT.)	SUPERIMPOSED LOAD (P.S.F.)	PLANK THICKNESS (IN.)	TOTAL DEPTH (IN.)	TOTAL LOAD (P.S.F.)	COST PER S.F.		
						MAT.	INST.	TOTAL
5200	20x20	40	6	20	110	16.75	4.56	21.31
5400		75	6	24	148	17.85	4.56	22.41
5600		100	6	24	173	17.85	4.56	22.41
5800	20x25	40	8	24	113	16.40	4.23	20.63
6000		75	8	24	149	16.40	4.23	20.63
6100		100	8	36	183	17.60	4.23	21.83
6200	25x25	40	8	28	118	17.40	4.23	21.63
6400		75	8	36	158	18.15	4.23	22.38
6500		100	8	36	183	18.15	4.23	22.38
7000	25x30	40	8	36	110	16.50	4.23	20.73
7200		75	10	36	159	17.35	3.97	21.32
7400		100	12	36	188	18.75	3.76	22.51
7600	30x30	40	8	36	121	17.05	4.23	21.28
8000		75	10	44	140	17.95	4.23	22.18
8250		100	12	52	206	20.50	3.76	24.26
8500	30x35	40	12	44	135	17.40	3.76	21.16
8750		75	12	52	176	18.45	3.76	22.21
9000	35x35	40	12	52	141	18.50	3.76	22.26
9250		75	12	60	181	19.40	3.76	23.16
9500	35x40	40	12	52	137	17.70	4.70	22.40
9750	40x40	40	12	60	141	18.70	4.70	23.40

B10 Superstructure

B1010 Floor Construction

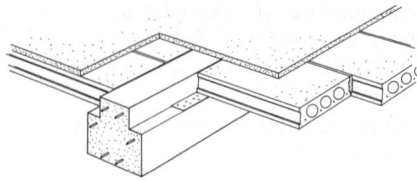

General: Beams and hollow core slabs priced here are for plant produced prestressed members transported to the site and erected.

The 2″ structural topping is applied after the beams and hollow core slabs are in place and is reinforced with W.W.F.

Design and Pricing Assumptions:
Prices are based on 10,000 S.F. to 20,000 S.F. projects and 50 mile to 100 mile transport.

Concrete for prestressed members is f'c 5 KSI.

Concrete for topping is f'c 3000 PSI and placed by pump.

Prestressing steel is fy = 250 or 300 KSI.

W.W.F. is 6 x 6 – W1.4 x W1.4 (10 x 10).

System Components	QUANTITY	UNIT	COST PER S.F.		
			MAT.	INST.	TOTAL
SYSTEM B1010 238 4300					
20′X20′ BAY, 6″ PLANK, 40 PSF S. LOAD, 135 PSF TOTAL LOAD					
12″ x 20″ precast "T" beam, 20′ span	.038	L.F.		.94	.94
Installation labor and equipment	.038	L.F.	6.12		6.12
12″ x 20″ precast "L" beam, 20′ span	.025	L.F.		.94	.94
Installation labor and equipment	.025	L.F.	3.30		3.30
Precast prestressed concrete roof/floor slabs 6″ deep, grouted	1.000	S.F.	7.35	2.68	10.03
Edge forms to 6″ high on elevated slab, 4 uses	.050	L.F.	.01	.19	.20
Forms in place, bulkhead for slab with keyway, 1 use, 2 piece	.013	L.F.	.03	.08	.11
Welded wire fabric rolls, 6 x 6 - W1.4 x W1.4 (10 x 10), 21 lb/csf	.010	C.S.F.	.20	.34	.54
Concrete ready mix, regular weight, 3000 psi	.170	C.F.	.70		.70
Place and vibrate concrete, elevated slab less than 6″, pump	.170	C.F.		.24	.24
Finish floor, monolithic steel trowel finish for finish floor	1.000	S.F.		.78	.78
Cure with sprayed membrane curing compound	.010	C.S.F.	.06	.08	.14
TOTAL			17.77	6.27	24.04

B1010 238		Precast Beam & Plank with 2″ Topping						
	BAY SIZE (FT.)	SUPERIMPOSED LOAD (P.S.F.)	PLANK THICKNESS (IN.)	TOTAL DEPTH (IN.)	TOTAL LOAD (P.S.F.)	COST PER S.F.		
						MAT.	INST.	TOTAL
4300	20x20	40	6	22	135	17.75	6.25	24
4400	RB1010 -010	75	6	24	173	18.85	6.25	25.10
4500		100	6	28	200	19.40	6.25	25.65
4600	20x25	40	6	26	134	16.70	6.25	22.95
5000	RB1010 -100	75	8	30	177	18.05	5.90	23.95
5200		100	8	30	202	18.05	5.90	23.95
5400	25x25	40	6	38	143	18.45	6.20	24.65
5600		75	8	38	183	18.45	6.20	24.65
6000		100	8	46	216	20.50	5.90	26.40
6200	25x30	40	8	38	144	17.50	5.85	23.35
6400		75	10	46	200	19.10	5.60	24.70
6600		100	10	46	225	19.10	5.60	24.70
7000	30x30	40	8	46	150	18.90	5.85	24.75
7200		75	10	54	181	20	5.85	25.85
7600		100	10	54	231	20	5.85	25.85
7800	30x35	40	10	54	166	18.80	5.55	24.35
8000		75	12	54	200	19.15	5.35	24.50
8200	35x35	40	10	62	170	19.35	5.55	24.90
9300		75	12	62	206	20.50	6.30	26.80
9500	35x40	40	12	62	167	19.50	6.25	25.75
9600	40x40	40	12	62	173	20.50	6.25	26.75

B1010 Floor Construction

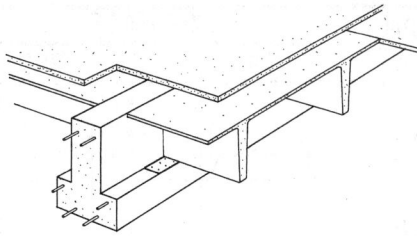

General: Beams and double tees priced here are for plant produced prestressed members transported to the site and erected.

The 2″ structural topping is applied after the beams and double tees are in place and is reinforced with W.W.F.

Design and Pricing Assumptions:
Prices are based on 10,000 S.F. to 20,000 S.F. projects and 50 mile to 100 mile transport.

Concrete for prestressed members is f'c 5 KSI.

Concrete for topping is f'c 3000 PSI and placed by pump.

Prestressing steel is fy = 250 or 300 KSI.

W.W.F. is 6 x 6 – W1.4 x W1.4 (10x10).

System Components	QUANTITY	UNIT	COST PER S.F. MAT.	INST.	TOTAL
SYSTEM B1010 239 3000					
25'X30' BAY 38″ DEPTH, 130 PSF T.L., 2″ TOPPING					
12″ x 36″ precast "T" beam, 25' span	.025	L.F.		.94	.94
Installation labor and equipment	.025	L.F.	5.80		5.80
Precast beam, "L" shaped, 12″ x 20″	.017	L.F.		.94	.94
Installation labor and equipment	.017	L.F.	2.67		2.67
Double T, standard weight, 16″ x 8' w, 25' span	.989	S.F.	7.83	1.55	9.38
Edge forms to 6″ high on elevated slab, 4 uses	.037	L.F.	.01	.14	.15
Forms in place, bulkhead for slab with keyway, 1 use, 2 piece	.010	L.F.	.03	.06	.09
Welded wire fabric rolls, 6 x 6 - W1.4 x W1.4 (10 x 10), 21 lb/csf	1.000	S.F.	.20	.34	.54
Concrete ready mix, regular weight, 3000 psi	.170	C.F.	.70		.70
Place and vibrate concrete, elevated slab less than 6″, pumped	.170	C.F.		.24	.24
Finishing floor, monolithic steel trowel finish for finish floor	1.000	S.F.		.78	.78
Curing with sprayed membrane curing compound	.010	S.F.	.06	.08	.14
TOTAL			17.30	5.07	22.37

B1010 239		Precast Double "T" & 2″ Topping on Precast Beams							
	BAY SIZE (FT.)	SUPERIMPOSED LOAD (P.S.F.)	DEPTH (IN.)			TOTAL LOAD (P.S.F.)	COST PER S.F. MAT.	INST.	TOTAL
3000	25x30	40	38			130	17.30	5.05	22.35
3100	RB1010 -100	75	38			168	17.30	5.05	22.35
3300		100	46			196	18.05	5.05	23.10
3600	30x30	40	46			150	18.70	5.05	23.75
3750		75	46			174	18.70	5.05	23.75
4000		100	54			203	19.60	5.05	24.65
4100	30x40	40	46			136	15.25	4.77	20.02
4300		75	54			173	15.90	4.77	20.67
4400		100	62			204	16.70	4.77	21.47
4600	30x50	40	54			138	14.60	4.61	19.21
4800		75	54			181	15.20	4.61	19.81
5000		100	54			219	16.75	4.35	21.10
5200	30x60	40	62			151	15.40	4.36	19.76
5400		75	62			192	16.10	4.36	20.46
5600		100	62			215	16.10	4.36	20.46
5800	35x40	40	54			139	16.25	4.75	21
6000		75	62			179	17.60	4.62	22.22
6250		100	62			212	18	4.62	22.62
6500	35x50	40	62			142	15.50	4.61	20.11
6750		75	62			186	16.20	4.61	20.81
7300		100	62			231	18.60	4.35	22.95

B10 Superstructure

B1010 Floor Construction

B1010 239	Precast Double "T" & 2" Topping on Precast Beams

	BAY SIZE (FT.)	SUPERIMPOSED LOAD (P.S.F.)	DEPTH (IN.)		TOTAL LOAD (P.S.F.)	COST PER S.F.		
						MAT.	INST.	TOTAL
7600	35x60	40	54		154	17.45	4.23	21.68
7750		75	54		179	17.90	4.23	22.13
8000		100	62		224	18.70	4.23	22.93
8250	40x40	40	62		145	17.85	5.70	23.55
8400		75	62		187	18.50	5.70	24.20
8750		100	62		223	19.70	5.70	25.40
9000	40x50	40	62		151	16.60	5.55	22.15
9300		75	62		193	17.50	4.75	22.25
9800	40x60	40	62		164	21.50	4.93	26.43

B1010 Floor Construction

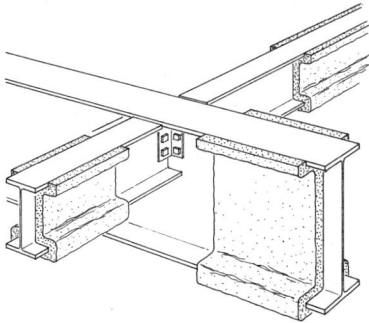

General: The following table is based upon structural W shape beam and girder framing. Non-composite action is assumed between beams and decking. Deck costs not included.

The deck spans the short direction. The steel beams and girders are fireproofed with sprayed fiber fireproofing.

Design and Pricing Assumptions: Structural steel is A36, with high strength A325 bolts.

Fireproofing is sprayed fiber (non-asbestos).

Total load includes steel, deck & live load.

Spandrels are assumed the same as interior beams and girders to allow for exterior wall loads and bracing or moment connections. No columns included in price.

See Tables B1010 528 and B1020 128 for metal deck costs.

System Components	QUANTITY	UNIT	COST PER S.F.		
			MAT.	INST.	TOTAL
SYSTEM B1010 241 1350 **15'X20' BAY, 40 P.S.F. L.L., 12" DEPTH, .535 P.S.F. FIREPROOF, 50 PSF T.LOAD**					
Structural steel	3.200	Lb.	5.38	1.28	6.66
Spray mineral fiber/cement for fire proof., 1" thick on beams	.535	S.F.	.31	.48	.79
TOTAL			5.69	1.76	7.45

B1010 241 — W Shape Beams & Girders

	BAY SIZE (FT.) BEAM X GIRD	SUPERIMPOSED LOAD (P.S.F.)	STEEL FRAMING DEPTH (IN.)	FIREPROOFING (S.F. PER S.F.)	TOTAL LOAD (P.S.F.)	COST PER S.F.		
						MAT.	INST.	TOTAL
1350	15x20	40	12	.535	50	5.70	1.76	7.46
1400		40	16	.65	90	7.45	2.27	9.72
1450		75	18	.694	125	9.80	2.86	12.66
1500		125	24	.796	175	13.55	3.98	17.53
1550		200	24	.89	263	15.35	3.68	19.03
1600	20x15	40	14	.659	50	5.75	1.87	7.62
1650		40	14	.69	90	7.80	2.38	10.18
1700		75	14	.806	125	9.55	2.88	12.43
1800		125	16	.86	175	11.25	3.47	14.72
1900		200	18	1.00	250	13.35	3.25	16.60
2000	20x20	40	12	.55	50	6.35	1.94	8.29
2050		40	14	.579	90	8.75	2.52	11.27
2100		75	16	.672	125	10.45	3	13.45
2150		125	16	.714	175	12.50	3.67	16.17
2200		200	24	.841	263	15.65	3.71	19.36
2300	20x20	40	14	.67	50	6.45	2.04	8.49
2400		40	14	.718	90	8.80	2.64	11.44
2500		75	18	.751	125	10.20	3	13.20
2550		125	21	.879	175	13.95	4.15	18.10
2600		200	21	.976	250	16.75	4.01	20.76
2650	20x20	40	14	.746	50	6.50	2.11	8.61
2700		40	14	.839	90	8.90	2.76	11.66
2750		75	18	.894	125	11.25	3.36	14.61
2800		125	21	.959	175	15	4.48	19.48
2850		200	21	1.10	250	18.45	4.44	22.89
2900	20x25	40	16	.53	50	7.05	2.08	9.13
2950		40	18	.621	96	11.10	3.12	14.22
3000		75	18	.651	131	12.80	3.55	16.35
3050		125	24	.77	200	16.90	4.80	21.70

B1010 Floor Construction

B1010 241		W Shape Beams & Girders						

	BAY SIZE (FT.) BEAM X GIRD	SUPERIMPOSED LOAD (P.S.F.)	STEEL FRAMING DEPTH (IN.)	FIREPROOFING (S.F. PER S.F.)	TOTAL LOAD (P.S.F.)	COST PER S.F.		
						MAT.	INST.	TOTAL
3100	20x25	200	27	.855	275	19.30	4.42	23.72
3300	20x25	40	14	.608	50	7.05	2.14	9.19
3350		40	21	.751	90	9.85	2.92	12.77
3400		75	24	.793	125	12.20	3.51	15.71
3450		125	24	.846	175	14.60	4.30	18.90
3500		200	24	.947	256	18.35	4.31	22.66
3550	20x25	40	14	.72	50	8.15	2.48	10.63
3600		40	21	.802	90	9.90	2.96	12.86
3650		75	24	.924	125	12.65	3.71	16.36
3700		125	24	.964	175	15.35	4.56	19.91
3750		200	27	1.09	250	19.45	4.63	24.08
3800	25x20	40	12	.512	50	7	2.06	9.06
3850		40	16	.653	90	9.45	2.75	12.20
3900		75	18	.726	125	12.20	3.46	15.66
4000		125	21	.827	175	15.25	4.43	19.68
4100		200	24	.928	250	18.70	4.35	23.05
4200	25x20	40	12	.65	50	7.10	2.19	9.29
4300		40	18	.702	90	10.50	3.03	13.53
4400		75	21	.829	125	12.25	3.54	15.79
4500		125	24	.914	175	15	4.43	19.43
4600		200	24	1.015	250	18.40	4.36	22.76
4700	25x20	40	14	.769	50	7.50	2.37	9.87
4800		40	16	.938	90	11.30	3.40	14.70
4900		75	18	.969	125	13.65	3.99	17.64
5000		125	24	1.136	175	19.15	5.65	24.80
5100		200	24	1.239	250	25	5.80	30.80
5200	25x25	40	18	.486	50	7.65	2.19	9.84
5300		40	18	.592	96	11.45	3.17	14.62
5400		75	21	.668	131	13.85	3.80	17.65
5450		125	24	.738	191	18.25	5.10	23.35
5500		200	30	.861	272	21.50	4.81	26.31
5550	25x25	40	18	.597	50	7.40	2.21	9.61
5600		40	18	.704	90	11.85	3.35	15.20
5650		75	21	.777	125	13.55	3.82	17.37
5700		125	24	.865	175	17.30	4.98	22.28
5750		200	27	.96	250	21	4.84	25.84
5800	25x25	40	18	.71	50	8.15	2.48	10.63
5850		40	21	.767	90	11.85	3.41	15.26
5900		75	24	.887	125	14.30	4.08	18.38
5950		125	24	.972	175	18.05	5.25	23.30
6000		200	30	1.10	250	22	5.15	27.15
6050	25x30	40	24	.547	50	9.75	2.73	12.48
6100		40	24	.629	103	13.80	3.77	17.57
6150		75	30	.726	138	16.55	4.50	21.05
6200		125	30	.751	206	19.60	5.45	25.05
6250		200	33	.868	281	23.50	5.25	28.75
6300	25x30	40	21	.568	50	8.75	2.51	11.26
6350		40	21	.694	90	11.80	3.34	15.14
6400		75	24	.776	125	15.25	4.22	19.47
6450		125	30	.904	175	18.35	5.25	23.60
6500		200	33	1.008	263	22	5.05	27.05

B1010 Floor Construction

| B1010 241 | W Shape Beams & Girders |

	BAY SIZE (FT.) BEAM X GIRD	SUPERIMPOSED LOAD (P.S.F.)	STEEL FRAMING DEPTH (IN.)	FIREPROOFING (S.F. PER S.F.)	TOTAL LOAD (P.S.F.)	COST PER S.F.		
						MAT.	INST.	TOTAL
6550	25x30	40	16	.632	50	9.10	2.65	11.75
6600		40	21	.76	90	12.55	3.57	16.12
6650		75	24	.857	125	14.95	4.21	19.16
6700		125	30	.983	175	18.70	5.40	24.10
6750		200	33	1.11	250	23.50	5.50	29
6800	30x25	40	16	.532	50	8.35	2.40	10.75
6850		40	21	.672	96	12.80	3.56	16.36
6900		75	24	.702	131	15.20	4.15	19.35
6950		125	27	1.020	175	19.75	5.50	25.25
7000		200	30	1.160	250	25	6.85	31.85
7100	30x25	40	18	.569	50	8.75	2.51	11.26
7150		40	24	.740	90	12.20	3.47	15.67
7200		75	24	.787	125	15.25	4.23	19.48
7300		125	24	.874	175	19	5.40	24.40
7400		200	30	1.013	250	23.50	5.30	28.80
7450	30x25	40	16	.637	50	9.10	2.66	11.76
7500		40	24	.839	90	12.90	3.72	16.62
7550		75	24	.919	125	15.65	4.42	20.07
7600		125	27	1.02	175	19.75	5.70	25.45
7650		200	30	1.160	250	25	5.70	30.70
7700	30x30	40	21	.52	50	9.35	2.63	11.98
7750		40	24	.629	103	14.45	3.93	18.38
7800		75	30	.715	138	17.20	4.64	21.84
7850		125	36	.822	206	22.50	6.30	28.80
7900		200	36	.878	281	25.50	5.60	31.10
7950	30x30	40	24	.619	50	9.75	2.80	12.55
8000		40	24	.706	90	13.20	3.67	16.87
8020		75	27	.818	125	15.60	4.33	19.93
8040		125	30	.910	175	20	5.70	25.70
8060		200	33	.999	263	24.50	5.55	30.05
8080	30x30	40	18	.631	50	10.45	2.97	13.42
8100		40	24	.805	90	14.25	4	18.25
8120		75	27	.899	125	17	4.73	21.73
8150		125	30	1.010	175	21	6.05	27.05
8200		200	36	1.148	250	25	5.75	30.75
8250	30x35	40	21	.508	50	10.70	2.94	13.64
8300		40	24	.651	109	15.85	4.27	20.12
8350		75	33	.732	150	19.25	5.15	24.40
8400		125	36	.802	225	24	6.60	30.60
8450		200	36	.888	300	31.50	6.85	38.35
8500	30x35	40	24	.554	50	9.40	2.66	12.06
8520		40	24	.655	90	13.80	3.79	17.59
8540		75	30	.751	125	17.25	4.68	21.93
8600		125	33	.845	175	21	5.90	26.90
8650		200	36	.936	263	27.50	6.10	33.60
8700	30x35	40	21	.644	50	10.10	2.90	13
8720		40	24	.733	90	14.55	4.02	18.57
8740		75	30	.833	125	18.30	4.98	23.28
8760		125	36	.941	175	21	5.95	26.95
8780		200	36	1.03	250	28	6.25	34.25

B1010 Floor Construction

B1010 241		W Shape Beams & Girders						
	BAY SIZE (FT.) BEAM X GIRD	SUPERIMPOSED LOAD (P.S.F.)	STEEL FRAMING DEPTH (IN.)	FIREPROOFING (S.F. PER S.F.)	TOTAL LOAD (P.S.F.)	COST PER S.F.		
						MAT.	INST.	TOTAL
8800	35x30	40	24	.540	50	10.40	2.89	13.29
8850		40	30	.670	103	15.50	4.20	19.70
8900		75	33	.748	138	18.90	5.10	24
8950		125	36	.824	206	23.50	6.45	29.95
8980		200	36	.874	281	28	6.15	34.15
9000	35x30	40	24	.619	50	10.10	2.88	12.98
9050		40	24	.754	90	14.90	4.12	19.02
9100		75	27	.844	125	17.65	4.84	22.49
9200		125	30	.856	175	21.50	6.05	27.55
9250		200	33	.953	263	24	5.40	29.40
9300	35x30	40	24	.705	50	10.85	3.11	13.96
9350		40	24	.833	90	15.25	4.26	19.51
9400		75	30	.963	125	18.70	5.20	23.90
9450		125	33	1.078	175	23.50	6.70	30.20
9500		200	36	1.172	250	28.50	6.45	34.95
9550	35x35	40	27	.560	50	11.05	3.06	14.11
9600		40	36	.706	109	20	5.35	25.35
9650		75	36	.750	150	21	5.55	26.55
9820		125	36	.797	225	28	7.65	35.65
9840		200	36	.914	300	33	7.15	40.15
9860	35x35	40	24	.580	50	10.10	2.84	12.94
9880		40	30	.705	90	15.55	4.23	19.78
9890		75	33	.794	125	18.95	5.10	24.05
9900		125	36	.878	175	23.50	6.50	30
9920		200	36	.950	263	29	6.35	35.35
9930	35x35	40	24	.689	50	10.80	3.10	13.90
9940		40	30	.787	90	15.90	4.39	20.29
9960		75	33	.871	125	20.50	5.50	26
9970		125	36	.949	175	22.50	6.35	28.85
9980		200	36	1.060	250	31	6.85	37.85

B10 Superstructure

B1010 Floor Construction

Description: Table below lists costs for light gauge CEE or PUNCHED DOUBLE joists to suit the span and loading with the minimum thickness subfloor required by the joist spacing.

Design Assumptions:

Maximum live load deflection is 1/360 of the clear span.

Maximum total load deflection is 1/240 of the clear span.

Bending strength is 20,000 psi.

8% allowance has been added to framing quantities for overlaps, double joists at openings under partitions, etc.; 5% added to glued & nailed subfloor for waste.

Maximum span is in feet and is the unsupported clear span.

System Components	QUANTITY	UNIT	COST PER S.F.		
			MAT.	INST.	TOTAL
SYSTEM B1010 244 1500					
LIGHT GAUGE STL & PLYWOOD FLR SYS,40 PSF S.LOAD,15'SPAN,8"DEPTH,16" O.C.					
Light gauge steel, 8" deep, 16" O.C., 12 ga.	1.870	Lb.	2.45	1.17	3.62
Subfloor plywood CDX 5/8"	1.050	S.F.	.78	.77	1.55
TOTAL			3.23	1.94	5.17

B1010 244		Light Gauge Steel Floor Systems						
	SPAN (FT.)	SUPERIMPOSED LOAD (P.S.F.)	FRAMING DEPTH (IN.)	FRAMING SPAC. (IN.)	TOTAL LOAD (P.S.F.)	COST PER S.F.		
						MAT.	INST.	TOTAL
1500	15	40	8	16	54	3.23	1.94	5.17
1550			8	24	54	3.37	1.99	5.36
1600		65	10	16	80	4.33	2.38	6.71
1650			10	24	80	4.25	2.32	6.57
1700		75	10	16	90	4.33	2.38	6.71
1750			10	24	90	4.25	2.32	6.57
1800		100	10	16	116	5.75	3.02	8.77
1850			10	24	116	4.25	2.32	6.57
1900		125	10	16	141	5.75	3.02	8.77
1950			10	24	141	4.25	2.32	6.57
2500	20	40	8	16	55	3.75	2.20	5.95
2550			8	24	55	3.37	1.99	5.36
2600		65	8	16	80	4.40	2.51	6.91
2650			10	24	80	4.26	2.34	6.60
2700		75	10	16	90	4.07	2.26	6.33
2750			12	24	90	4.79	2.02	6.81
2800		100	10	16	115	4.07	2.26	6.33
2850			10	24	116	5.55	2.60	8.15
2900		125	12	16	142	6.55	2.56	9.11
2950			12	24	141	6.05	2.78	8.83
3500	25	40	10	16	55	4.33	2.38	6.71
3550			10	24	55	4.25	2.32	6.57
3600		65	10	16	81	5.75	3.02	8.77
3650			12	24	81	5.35	2.19	7.54
3700		75	12	16	92	6.50	2.55	9.05
3750			12	24	91	6.05	2.78	8.83
3800		100	12	16	117	7.40	2.82	10.22
3850		125	12	16	143	8.50	3.73	12.23
4500	30	40	12	16	57	6.50	2.55	9.05
4550			12	24	56	5.30	2.18	7.48
4600		65	12	16	82	7.35	2.81	10.16
4650		75	12	16	92	7.35	2.81	10.16

85

B1010 Floor Construction

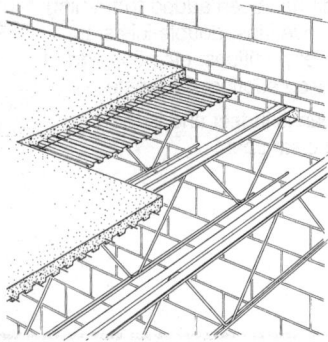

Description: Table below lists cost per S.F. for a floor system on bearing walls using open web steel joists, galvanized steel slab form and 2-1/2″ concrete slab reinforced with welded wire fabric.

Design and Pricing Assumptions:
Concrete f'c = 3 KSI placed by pump.
WWF 6 x 6 – W1.4 x W1.4 (10 x 10)
Joists are spaced as shown.
Slab form is 28 gauge galvanized.
Joists costs include appropriate bridging. Deflection is limited to 1/360 of the span. Screeds and steel trowel finish.

Design Loads	Min.	Max.
Joists	3.0 PSF	7.6 PSF
Slab Form	1.0	1.0
2-1/2″ Concrete	27.0	27.0
Ceiling	3.0	3.0
Misc.	9.0	9.4
	43.0 PSF	48.0 PSF

System Components	QUANTITY	UNIT	COST PER S.F.		
			MAT.	INST.	TOTAL
SYSTEM B1010 246 1050					
SPAN 20′ S.LOAD 40 PSF, JOIST SPACING 2′-0 O.C., 14-1/2″ DEPTH					
Open web joists	2.800	Lb.	3.14	.95	4.09
Slab form, galvanized steel 9/16″ deep, 28 gauge	1.050	S.F.	2.51	1.02	3.53
Welded wire fabric rolls, 6 x 6 - W1.4 x W1.4 (10 x 10), 21 lb/csf	1.000	S.F.	.20	.34	.54
Concrete ready mix, regular weight, 3000 psi	.210	C.F.	.86		.86
Place and vibrate concrete, elevated slab less than 6″, pumped	.210	C.F.		.30	.30
Finishing floor, monolithic steel trowel finish for finish floor	1.000	S.F.		.78	.78
Curing with sprayed membrane curing compound	.010	C.S.F.	.06	.08	.14
TOTAL			6.77	3.47	10.24

B1010 246		**Deck & Joists on Bearing Walls**						
	SPAN (FT.)	SUPERIMPOSED LOAD (P.S.F.)	JOIST SPACING FT. - IN.	DEPTH (IN.)		COST PER S.F.		
						MAT.	INST.	TOTAL
1050	20	40	2-0	14-1/2	83	6.75	3.47	10.22
1070		65	2-0	16-1/2	109	7.35	3.68	11.03
1100		75	2-0	16-1/2	119	7.35	3.68	11.03
1120		100	2-0	18-1/2	145	7.40	3.69	11.09
1150		125	1-9	18-1/2	170	9.10	4.40	13.50
1170	25	40	2-0	18-1/2	84	8.20	4.09	12.29
1200		65	2-0	20-1/2	109	8.50	4.20	12.70
1220		75	2-0	20-1/2	119	8.20	3.94	12.14
1250		100	2-0	22-1/2	145	8.35	3.98	12.33
1270		125	1-9	22-1/2	170	9.60	4.36	13.96
1300	30	40	2-0	22-1/2	84	8.35	3.58	11.93
1320		65	2-0	24-1/2	110	9.10	3.74	12.84
1350		75	2-0	26-1/2	121	9.45	3.82	13.27
1370		100	2-0	26-1/2	146	10.10	3.95	14.05
1400		125	2-0	24-1/2	172	11	5.20	16.20
1420	35	40	2-0	26-1/2	85	10.40	4.97	15.37
1450		65	2-0	28-1/2	111	10.75	5.10	15.85
1470		75	2-0	28-1/2	121	10.75	5.10	15.85
1500		100	1-11	28-1/2	147	11.70	5.45	17.15
1520		125	1-8	28-1/2	172	12.85	5.85	18.70
1550 1560	5/8″ gyp. fireproof. On metal furring, add					.83	2.58	3.41

B1010   Floor Construction

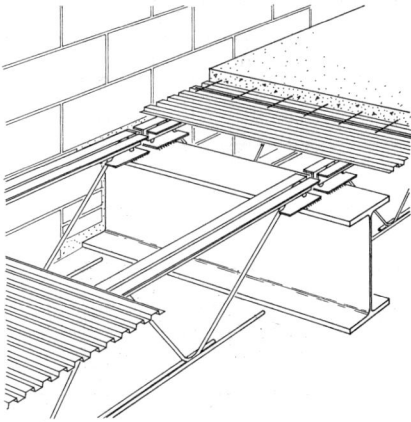

Table below lists costs for a floor system on exterior bearing walls and interior columns and beams using open web steel joists, galvanized steel slab form, 2-1/2" concrete slab reinforced with welded wire fabric.

Design and Pricing Assumptions:
Structural Steel is A36.
Concrete f'c = 3 KSI placed by pump.
WWF 6 x 6 – W1.4 x W1.4 (10 x 10)
Columns are 12' high.
Building is 4 bays long by 4 bays wide.
Joists are 2' O.C. ± and span the long direction of the bay.

Joists at columns have bottom chords extended and are connected to columns.

Slab form is 28 gauge galvanized.
Column costs in table are for columns to support 1 floor plus roof loading in a 2-story building; however, column costs are from ground floor to 2nd floor only. Joist costs include appropriate bridging. Deflection is limited to 1/360 of the span. Screeds and steel trowel finish.

Design Loads	Min.	Max.
S.S & Joists	4.4 PSF	11.5 PSF
Slab Form	1.0	1.0
2-1/2" Concrete	27.0	27.0
Ceiling	3.0	3.0
Misc.	7.6	5.5
	43.0 PSF	48.0 PSF

System Components			COST PER S.F.		
	QUANTITY	UNIT	MAT.	INST.	TOTAL
SYSTEM B1010 248 1200					
15'X20' BAY, W.F. STEEL, STEEL JOISTS, SLAB FORM, CONCRETE SLAB					
Structural steel	1.248	Lb.	1.55	.37	1.92
Open web joists	3.140	Lb.	3.52	1.07	4.59
Slab form, galvanized steel 9/16" deep, 28 gauge	1.020	S.F.	1.70	.69	2.39
Welded wire fabric 6x6 - W1.4 x W1.4 (10 x 10), 21 lb/CSF roll, 10% lap	1.000	S.F.	.20	.34	.54
Concrete ready mix, regular weight, 3000 psi	.210	C.F.	.86		.86
Place and vibrate concrete, elevated slab less than 6", pumped	.210	C.F.		.30	.30
Finishing floor, monolithic steel trowel finish for finish floor	1.000	S.F.		.78	.78
Curing with sprayed membrane curing compound	.010	C.S.F.	.06	.08	.14
TOTAL			7.89	3.63	11.52

B1010 248		Steel Joists on Beam & Wall						
	BAY SIZE (FT.)	SUPERIMPOSED LOAD (P.S.F.)	DEPTH (IN.)	TOTAL LOAD (P.S.F.)	COLUMN ADD	COST PER S.F.		
						MAT.	INST.	TOTAL
1200	15x20 RB1010 -100	40	17	83		7.90	3.63	11.53
1210					columns	.51	.12	.63
1220	15x20	65	19	108		8.10	3.71	11.81
1230					columns	.51	.12	.63
1250	15x20	75	19	119		8.65	3.87	12.52
1260					columns	.61	.15	.76
1270	15x20	100	19	144		10	4.46	14.46
1280					columns	.61	.15	.76
1300	15x20	125	19	170		10.30	4.53	14.83
1310					columns	.82	.20	1.02
1350	20x20	40	19	83		8.10	3.69	11.79
1360					columns	.46	.11	.57
1370	20x20	65	23	109		9.05	3.95	13
1380					columns	.61	.15	.76
1400	20x20	75	23	119		9.35	4.03	13.38
1410					columns	.61	.15	.76
1420	20x20	100	23	144		9.65	4.11	13.76
1430					columns	.61	.15	.76
1450	20x20	125	23	170		11.80	4.94	16.74
1460					columns	.74	.17	.91

B1010 Floor Construction

B1010 248	Steel Joists on Beam & Wall

	BAY SIZE (FT.)	SUPERIMPOSED LOAD (P.S.F.)	DEPTH (IN.)	TOTAL LOAD (P.S.F.)	COLUMN ADD	COST PER S.F.		
						MAT.	INST.	TOTAL
1500	20x25	40	23	84		9.35	4.26	13.61
1510					columns	.49	.12	.61
1520	20x25	65	26	110		10.15	4.47	14.62
1530					columns	.49	.12	.61
1550	20x25	75	26	120		10.85	4.72	15.57
1560					columns	.59	.15	.74
1570	20x25	100	26	145		10.40	4.35	14.75
1580					columns	.59	.15	.74
1670	20x25	125	29	170		11.90	4.78	16.68
1680					columns	.69	.16	.85
1720	25x25	40	23	84		10	4.41	14.41
1730					columns	.47	.12	.59
1750	25x25	65	29	110		10.55	4.55	15.10
1760					columns	.47	.12	.59
1770	25x25	75	26	120		10.90	4.46	15.36
1780					columns	.55	.13	.68
1800	25x25	100	29	145		12.35	4.88	17.23
1810					columns	.55	.13	.68
1820	25x25	125	29	170		12.95	5.05	18
1830					columns	.61	.15	.76
1870	25x30	40	29	84		10.25	4.59	14.84
1880					columns	.46	.11	.57
1900	25x30	65	29	110		10.95	4.05	15
1910					columns	.46	.11	.57
1920	25x30	75	29	120		11.55	4.20	15.75
1930					columns	.50	.12	.62
1950	25x30	100	29	145		12.55	4.43	16.98
1960					columns	.50	.12	.62
1970	25x30	125	32	170		13.55	5.60	19.15
1980					columns	.57	.13	.70
2020	30x30	40	29	84		10.70	4.01	14.71
2030					columns	.42	.11	.53
2050	30x30	65	29	110		12.15	4.34	16.49
2060					columns	.42	.11	.53
2070	30x30	75	32	120		12.50	4.42	16.92
2080					columns	.49	.12	.61
2100	30x30	100	35	145		13.90	4.73	18.63
2110					columns	.57	.13	.70
2120	30x30	125	35	172		15.20	6.10	21.30
2130					columns	.68	.16	.84
2170	30x35	40	29	85		12.15	4.33	16.48
2180					columns	.41	.09	.50
2200	30x35	65	29	111		13.70	5.70	19.40
2210					columns	.47	.11	.58
2220	30x35	75	32	121		13.65	5.70	19.35
2230					columns	.48	.12	.60
2250	30x35	100	35	148		14.80	4.92	19.72
2260					columns	.58	.15	.73
2270	30x35	125	38	173		16.65	5.35	22
2280					columns	.59	.15	.74
2320	35x35	40	32	85		12.50	4.42	16.92
2330					columns	.42	.11	.53

B10 Superstructure

B1010 Floor Construction

B1010 248	Steel Joists on Beam & Wall

	BAY SIZE (FT.)	SUPERIMPOSED LOAD (P.S.F.)	DEPTH (IN.)	TOTAL LOAD (P.S.F.)	COLUMN ADD	COST PER S.F.		
						MAT.	INST.	TOTAL
2350	35x35	65	35	111		14.45	5.90	20.35
2360					columns	.50	.12	.62
2370	35x35	75	35	121		14.75	6	20.75
2380					columns	.50	.12	.62
2400	35x35	100	38	148		15.40	6.20	21.60
2410					columns	.62	.15	.77
2420	35x35	125	41	173		17.55	6.80	24.35
2430					columns	.63	.15	.78
2460	5/8 gyp. fireproof.							
2475	On metal furring, add					.83	2.58	3.41

B1010 Floor Construction

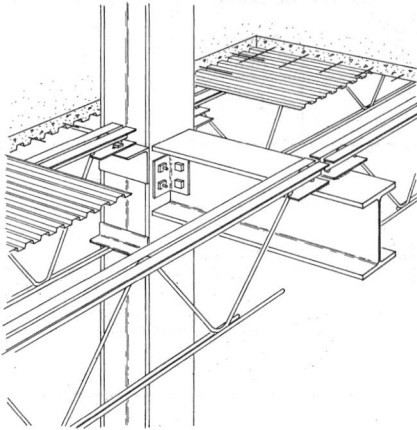

Table below lists costs for a floor system on steel columns and beams using open web steel joists, galvanized steel slab form, and 2-1/2″ concrete slab reinforced with welded wire fabric.

Design and Pricing Assumptions:
Structural Steel is A36.
Concrete f'c = 3 KSI placed by pump.
WWF 6 x 6 – W1.4 x W1.4 (10 x 10)
Columns are 12′ high.
Building is 4 bays long by 4 bays wide.
Joists are 2′ O.C. ± and span the long direction of the bay.

Joists at columns have bottom chords extended and are connected to columns.

Slab form is 28 gauge galvanized. Column costs in table are for columns to support 1 floor plus roof loading in a 2-story building; however, column costs are from ground floor to 2nd floor only. Joist costs include appropriate bridging. Deflection is limited to 1/360 of the span. Screeds and steel trowel finish.

Design Loads	Min.	Max.
S.S. & Joists	6.3 PSF	15.3 PSF
Slab Form	1.0	1.0
2-1/2″ Concrete	27.0	27.0
Ceiling	3.0	3.0
Misc.	5.7	1.7
	43.0 PSF	48.0 PSF

System Components			QUANTITY	UNIT	COST PER S.F. MAT.	COST PER S.F. INST.	COST PER S.F. TOTAL
SYSTEM B1010 250 2350							
15′X20′BAY 40 PSF S. LOAD, 17″ DEPTH, 83 PSF TOTAL LOAD							
Structural steel			1.974	Lb.	3.26	.79	4.05
Open web joists			3.140	Lb.	3.52	1.07	4.59
Slab form, galvanized steel 9/16″ deep, 28 gauge			1.020	S.F.	1.70	.69	2.39
Welded wire fabric rolls, 6 x 6 - W1.4 x W1.4 (10 x 10), 21 lb/csf			1.000	S.F.	.20	.34	.54
Concrete ready mix, regular weight, 3000 psi			.210	C.F.	.86		.86
Place and vibrate concrete, elevated slab less than 6″, pumped			.210	C.F.		.30	.30
Finishing floor, monolithic steel trowel finish for finish floor			1.000	S.F.		.78	.78
Curing with sprayed membrane curing compound			.010	S.F.	.06	.08	.14
		TOTAL			9.60	4.05	13.65

B1010 250	Steel Joists, Beams & Slab on Columns							
	BAY SIZE (FT.)	SUPERIMPOSED LOAD (P.S.F.)	DEPTH (IN.)	TOTAL LOAD (P.S.F.)	COLUMN ADD	COST PER S.F. MAT.	COST PER S.F. INST.	COST PER S.F. TOTAL
2350	15x20 RB1010 -100	40	17	83		9.60	4.05	13.65
2400					column	1.56	.37	1.93
2450	15x20	65	19	108		10.65	4.31	14.96
2500					column	1.56	.37	1.93
2550	15x20	75	19	119		11.10	4.44	15.54
2600					column	1.70	.41	2.11
2650	15x20	100	19	144		11.80	4.63	16.43
2700					column	1.70	.41	2.11
2750	15x20	125	19	170		13.25	5.25	18.50
2800					column	2.27	.55	2.82
2850	20x20	40	19	83		10.45	4.24	14.69
2900					column	1.28	.31	1.59
2950	20x20	65	23	109		11.55	4.54	16.09
3000					column	1.70	.41	2.11
3100	20x20	75	26	119		12.20	4.70	16.90
3200					column	1.70	.41	2.11
3400	20x20	100	23	144		12.65	4.81	17.46
3450					column	1.70	.41	2.11
3500	20x20	125	23	170		14.10	5.20	19.30
3600					column	2.04	.49	2.53

B1010 Floor Construction

B1010 250		Steel Joists, Beams & Slab on Columns						

	BAY SIZE (FT.)	SUPERIMPOSED LOAD (P.S.F.)	DEPTH (IN.)	TOTAL LOAD (P.S.F.)	COLUMN ADD	COST PER S.F.		
						MAT.	INST.	TOTAL
3700	20x25	40	44	83		12	4.87	16.87
3800					column	1.36	.33	1.69
3900	20x25	65	26	110		13.10	5.15	18.25
4000					column	1.36	.33	1.69
4100	20x25	75	26	120		12.85	4.91	17.76
4200					column	1.63	.40	2.03
4300	20x25	100	26	145		13.65	5.10	18.75
4400					column	1.63	.40	2.03
4500	20x25	125	29	170		15.25	5.55	20.80
4600					column	1.91	.47	2.38
4700	25x25	40	23	84		12.90	5.05	17.95
4800					column	1.31	.32	1.63
4900	25x25	65	29	110		13.65	5.30	18.95
5000					column	1.31	.32	1.63
5100	25x25	75	26	120		14.25	5.25	19.50
5200					column	1.52	.37	1.89
5300	25x25	100	29	145		15.90	5.70	21.60
5400					column	1.52	.37	1.89
5500	25x25	125	32	170		16.80	5.95	22.75
5600					column	1.69	.41	2.10
5700	25x30	40	29	84		13.40	5.30	18.70
5800					column	1.27	.31	1.58
5900	25x30	65	29	110		13.95	5.55	19.50
6000					column	1.27	.31	1.58
6050	25x30	75	29	120		15	5	20
6100					column	1.41	.35	1.76
6150	25x30	100	29	145		16.30	5.30	21.60
6200					column	1.41	.35	1.76
6250	25x30	125	32	170		17.55	6.55	24.10
6300					column	1.62	.39	2.01
6350	30x30	40	29	84		13.95	4.78	18.73
6400					column	1.17	.28	1.45
6500	30x30	65	29	110		15.90	5.25	21.15
6600					column	1.17	.28	1.45
6700	30x30	75	32	120		16.25	5.30	21.55
6800					column	1.35	.33	1.68
6900	30x30	100	35	145		18.05	5.75	23.80
7000					column	1.57	.39	1.96
7100	30x30	125	35	172		19.80	7.15	26.95
7200					column	1.75	.43	2.18
7300	30x35	40	29	85		15.80	5.20	21
7400					column	1	.24	1.24
7500	30x35	65	29	111		17.60	6.60	24.20
7600					column	1.30	.32	1.62
7700	30x35	75	32	121		17.60	6.60	24.20
7800					column	1.32	.32	1.64
7900	30x35	100	35	148		19.10	5.95	25.05
8000					column	1.62	.39	2.01
8100	30x35	125	38	173		21	6.45	27.45
8200					column	1.65	.40	2.05
8300	35x35	40	32	85		16.25	5.30	21.55
8400					column	1.16	.28	1.44

B10 Superstructure

B1010 Floor Construction

B1010 250 — Steel Joists, Beams & Slab on Columns

	BAY SIZE (FT.)	SUPERIMPOSED LOAD (P.S.F.)	DEPTH (IN.)	TOTAL LOAD (P.S.F.)	COLUMN ADD	COST PER S.F.		
						MAT.	INST.	TOTAL
8500	35x35	65	35	111		18.55	6.90	25.45
8600					column	1.39	.33	1.72
9300	35x35	75	38	121		19.05	7	26.05
9400					column	1.39	.33	1.72
9500	35x35	100	38	148		20.50	7.40	27.90
9600					column	1.71	.41	2.12
9750	35x35	125	41	173		22	6.65	28.65
9800					column	1.75	.43	2.18
9810	5/8 gyp. fireproof.							
9815	On metal furring, add					.83	2.58	3.41

B1010 Floor Construction

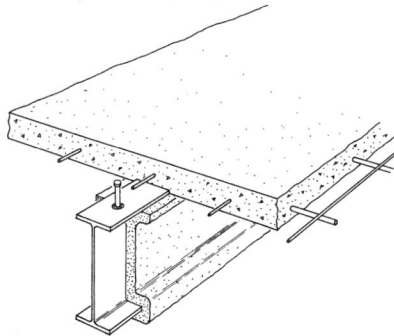

General: Composite construction of W shape flange beams and concrete slabs is most efficiently used when loads are heavy and spans are moderately long. It is stiffer with less deflection than non-composite construction of similar depth and spans.

In practice, composite construction is typically shallower in depth than non-composite would be.

Design Assumptions:
Steel, fy = 36 KSI
Beams unshored during construction
Deflection limited to span/360
Shear connectors, welded studs
Concrete, f'c = 3 KSI

System Components	QUANTITY	UNIT	COST PER S.F.		
			MAT.	INST.	TOTAL
SYSTEM B1010 252 3800					
20X25 BAY, 40 PSF S. LOAD, 4" THICK SLAB, 20" TOTAL DEPTH					
Structural steel	3.820	Lb.	6.42	1.53	7.95
Welded shear connectors 3/4" diameter 3-3/8" long	.140	Ea.	.08	.25	.33
Forms in place, floor slab, with 1-way joist pans, 4 uses	1.000	S.F.	3.12	5.10	8.22
Edge forms to 6" high on elevated slab, 4 uses	.050	L.F.	.01	.19	.20
Reinforcing in place, elevated slabs #4 to #7	1.190	Lb.	1.06	.48	1.54
Concrete ready mix, regular weight, 3000 psi	.330	C.F.	1.36		1.36
Place and vibrate concrete, elevated slab less than 6", pumped	.330	C.F.		.47	.47
Finishing floor, monolithic steel trowel finish for resilient tile	1.000	S.F.		1.02	1.02
Curing with sprayed membrane curing compound	.010	S.F.	.06	.08	.14
Spray mineral fiber/cement for fire proof, 1" thick on beams	.520	S.F.	.30	.47	.77
TOTAL			12.41	9.59	22

B1010 252		Composite Beam & Cast in Place Slab						
	BAY SIZE (FT.)	SUPERIMPOSED LOAD (P.S.F.)	SLAB THICKNESS (IN.)	TOTAL DEPTH (FT.-IN.)	TOTAL LOAD (P.S.F.)	COST PER S.F.		
						MAT.	INST.	TOTAL
3800	20x25	40	4	1 - 8	94	12.40	9.60	22
3900	RB1010 -100	75	4	1 - 8	130	14.35	10.20	24.55
4000		125	4	1 - 10	181	16.65	10.90	27.55
4100		200	5	2 - 2	272	21.50	12.40	33.90
4200	25x25	40	4-1/2	1 - 8-1/2	99	12.95	9.70	22.65
4300		75	4-1/2	1 - 10-1/2	136	15.55	10.45	26
4400		125	5-1/2	2 - 0-1/2	200	18.25	11.40	29.65
4500		200	5-1/2	2 - 2-1/2	278	22.50	12.60	35.10
4600	25x30	40	4-1/2	1 - 8-1/2	100	14.40	10.15	24.55
4700		75	4-1/2	1 - 10-1/2	136	16.80	10.85	27.65
4800		125	5-1/2	2 - 2-1/2	202	20.50	11.95	32.45
4900		200	5-1/2	2 - 5-1/2	279	24.50	13.10	37.60
5000	30x30	40	4	1 - 8	95	14.40	10.10	24.50
5200		75	4	2 - 1	131	16.60	10.70	27.30
5400		125	4	2 - 4	183	20.50	11.85	32.35
5600		200	5	2 - 10	274	26	13.45	39.45

B1010 Floor Construction

B1010 252		Composite Beam & Cast in Place Slab						
	BAY SIZE (FT.)	SUPERIMPOSED LOAD (P.S.F.)	SLAB THICKNESS (IN.)	TOTAL DEPTH (FT.-IN.)	TOTAL LOAD (P.S.F.)	COST PER S.F.		
						MAT.	INST.	TOTAL
5800	30x35	40	4	2 - 1	95	15.25	10.30	25.55
6000		75	4	2 - 4	139	18.25	11.15	29.40
6250		125	4	2 - 4	185	23.50	12.65	36.15
6500		200	5	2 - 11	276	28.50	14.15	42.65
7000	35x30	40	4	2 - 1	95	15.65	10.45	26.10
7200		75	4	2 - 4	139	18.55	11.30	29.85
7400		125	4	2 - 4	185	23.50	12.65	36.15
7600		200	5	2 - 11	276	28.50	14.15	42.65
8000	35x35	40	4	2 - 1	96	16.05	10.55	26.60
8250		75	4	2 - 4	133	19	11.40	30.40
8500		125	4	2 - 10	185	23	12.45	35.45
8750		200	5	3 - 5	276	29.50	14.35	43.85
9000	35x40	40	4	2 - 4	97	17.80	11	28.80
9250		75	4	2 - 4	134	21	11.75	32.75
9500		125	4	2 - 7	186	25	13	38
9750		200	5	3 - 5	278	32.50	15.05	47.55

B1010 Floor Construction

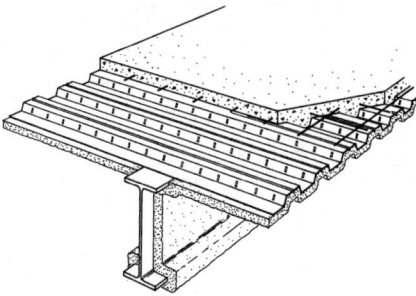

Description: Table below lists costs per S.F. for floors using steel beams and girders, composite steel deck, concrete slab reinforced with W.W.F. and sprayed fiber fireproofing (non-asbestos) on the steel beams and girders and on the steel deck.

Design and Pricing Assumptions:
Structural Steel is A36, high strength bolted.

Composite steel deck varies from 2"–20 gauge to 3"–16 gauge galvanized. WWF 6 x 6 – W1.4 x W1.4 (10 x 10) Concrete f'c = 3 KSI. Steel trowel finish and cure.

Spandrels are assumed the same weight as interior beams and girders to allow for exterior wall loads and bracing or moment connections.

System Components	QUANTITY	UNIT	COST PER S.F.		
			MAT.	INST.	TOTAL
SYSTEM B1010 254 0540					
W SHAPE BEAMS & DECK, FIREPROOFED, 15'X20', 5" SLAB, 40 PSF LOAD					
Structural steel	4.470	Lb.	6.92	1.65	8.57
Metal decking, non-cellular composite, galv 3" deep, 20 gage	1.050	S.F.	3.44	.95	4.39
Sheet metal edge closure form, 12", w/2 bends, 18 ga, galv	.058	L.F.	.34	.13	.47
Welded wire fabric 6 x 6 - W1.4 x W1.4 (10 x 10), 21 lb/CSF roll, 10% lap	1.000	S.F.	.20	.34	.54
Concrete ready mix, regular weight, 3000 psi	.011	C.Y.	1.22		1.22
Place and vibrate concrete, elevated slab less than 6", pumped	.011	C.Y.		.33	.33
Finishing floor, monolithic steel trowel finish for finish floor	1.000	S.F.		.78	.78
Curing with sprayed membrane curing compound	.010	C.S.F.	.06	.08	.14
Sprayed mineral fiber/cement for fireproof, 1" thick on decks	1.000	S.F.	.87	1.08	1.95
Sprayed mineral fiber/cement for fireproof, 1" thick on beams	.615	S.F.	.36	.56	.92
TOTAL			13.41	5.90	19.31

B1010 254		W Shape, Composite Deck, & Slab						
	BAY SIZE (FT.) BEAM X GIRD	SUPERIMPOSED LOAD (P.S.F.)	SLAB THICKNESS (IN.)	TOTAL DEPTH (FT.-IN.)	TOTAL LOAD (P.S.F.)	COST PER S.F.		
						MAT.	INST.	TOTAL
0540	15x20	40	5	1-7	89	13.40	5.90	19.30
0560	RB1010 -100	75	5	1-9	125	14.75	6.30	21.05
0580		125	5	1-11	176	17.25	6.90	24.15
0600		200	5	2-2	254	21.50	7.85	29.35
0620	20x15	40	4	1-4	89	13.50	6.05	19.55
0640		75	4	1-4	119	15.10	6.30	21.40
0660		125	4	1-6	170	16.85	6.70	23.55
0680		200	4	2-1	247	19.80	7.55	27.35
0700	20x20	40	5	1-9	90	14.60	6.15	20.75
0720		75	5	1-9	126	16.50	6.70	23.20
0740		125	5	1-11	177	18.75	7.30	26.05
0760		200	5	2-5	255	22.50	8.15	30.65
0780	20x25	40	5	1-9	90	14.90	6.15	21.05
0800		75	5	1-9	127	18.60	7.05	25.65
0820		125	5	1-11	180	22.50	7.75	30.25
0840		200	5	2-5	256	24.50	8.80	33.30
0860	25x20	40	5	1-11	90	15.10	6.30	21.40
0880		75	5	2-5	127	17.65	6.95	24.60
0900		125	5	2-5	178	19.95	7.60	27.55
0920		200	5	2-5	257	26	9	35
0940	25x25	40	5	1-11	91	16.90	6.55	23.45
0960		75	5	2-5	178	19.50	7.25	26.75
0980		125	5	2-5	181	24	8.10	32.10
1000		200	5-1/2	2-8-1/2	263	27	9.45	36.45

B10 Superstructure

B1010 Floor Construction

B1010 254			W Shape, Composite Deck, & Slab					

	BAY SIZE (FT.) BEAM X GIRD	SUPERIMPOSED LOAD (P.S.F.)	SLAB THICKNESS (IN.)	TOTAL DEPTH (FT.-IN.)	TOTAL LOAD (P.S.F.)	COST PER S.F.		
						MAT.	INST.	TOTAL
1400	25x30	40	5	2-5	91	16.70	6.75	23.45
1500		75	5	2-5	128	20.50	7.55	28.05
1600		125	5	2-8	180	23	8.40	31.40
1700		200	5	2-11	259	29	9.70	38.70
1800	30x25	40	5	2-5	92	17.75	7.05	24.80
1900		75	5	2-5	129	21	7.90	28.90
2000		125	5	2-8	181	24	8.65	32.65
2100		200	5-1/2	2-11	200	29	10.05	39.05
2200	30x30	40	5	2-2	92	18.55	7.15	25.70
2300		75	5	2-5	129	22	8	30
2400		125	5	2-11	182	26	9.05	35.05
2500		200	5	3-2	263	35	11.40	46.40
2600	30x35	40	5	2-5	94	20.50	7.60	28.10
2700		75	5	2-11	131	23.50	8.45	31.95
2800		125	5	3-2	183	28	9.50	37.50
2900		200	5-1/2	3-5-1/2	268	34	10.95	44.95
3400	35x30	40	5	2-5	93	19.50	7.50	27
3500		75	5	2-8	130	23.50	8.45	31.95
3600		125	5	2-11	183	28	9.60	37.60
3700		200	5	3-5	262	34.50	11.20	45.70
3800	35x35	40	5	2-8	94	21	7.60	28.60
3900		75	5	2-11	131	24.50	8.45	32.95
4000		125	5	3-5	184	29	9.70	38.70
4100		200	5-1/2	3-5-1/2	270	38	11.55	49.55
4200	35x40	40	5	2-11	94	21.50	8	29.50
4300		75	5	3-2	131	25	8.95	33.95
4400		125	5	3-5	184	30	10.10	40.10
4500		200	5	3-5-1/2	264	38.50	12.15	50.65

B1010 Floor Construction

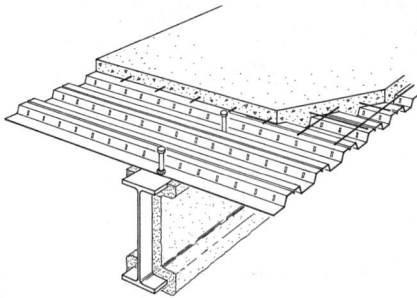

Description: Table below lists costs ($/S.F.) for a floor system using composite steel beams with welded shear studs, composite steel deck, and light weight concrete slab reinforced with W.W.F. Price includes sprayed fiber fireproofing on steel beams.

Design and Pricing Assumptions:
Structural steel is A36, high strength bolted.
Composite steel deck varies from 22 gauge to 16 gauge, galvanized.

Shear Studs are 3/4″.
W.W.F., 6 x 6 – W1.4 x W1.4 (10 x 10)
Concrete f'c = 3 KSI, lightweight.
Steel trowel finish and cure.
Fireproofing is sprayed fiber (non-asbestos).

Spandrels are assumed the same as interior beams and girders to allow for exterior wall loads and bracing or moment connections.

System Components	QUANTITY	UNIT	COST PER S.F.		
			MAT.	INST.	TOTAL
SYSTEM B1010 256 2400					
20X25 BAY, 40 PSF S. LOAD, 5-1/2″ SLAB, 17-1/2″ TOTAL THICKNESS					
Structural steel	4.320	Lb.	7.26	1.73	8.99
Welded shear connectors 3/4″ diameter 4-7/8″ long	.163	Ea.	.12	.30	.42
Metal decking, non-cellular composite, galv. 3″ deep, 22 gauge	1.050	S.F.	3.08	.90	3.98
Sheet metal edge closure form, 12″, w/2 bends, 18 ga, galv	.045	L.F.	.26	.10	.36
Welded wire fabric rolls, 6 x 6 - W1.4 x W1.4 (10 x 10), 21 lb/csf	1.000	S.F.	.20	.34	.54
Concrete ready mix, light weight, 3,000 PSI	.333	C.F.	2.58		2.58
Place and vibrate concrete, elevated slab less than 6″, pumped	.333	C.F.		.47	.47
Finishing floor, monolithic steel trowel finish for finish floor	1.000	S.F.		.78	.78
Curing with sprayed membrane curing compound	.010	C.S.F.	.06	.08	.14
Shores, erect and strip vertical to 10′ high	.020	Ea.		.38	.38
Sprayed mineral fiber/cement for fireproof, 1″ thick on beams	.483	S.F.	.28	.43	.71
TOTAL			13.84	5.51	19.35

B1010 256		Composite Beams, Deck & Slab						
	BAY SIZE (FT.)	SUPERIMPOSED LOAD (P.S.F.)	SLAB THICKNESS (IN.)	TOTAL DEPTH (FT.-IN.)	TOTAL LOAD (P.S.F.)	COST PER S.F.		
						MAT.	INST.	TOTAL
2400	20x25	40	5-1/2	1 - 5-1/2	80	13.85	5.50	19.35
2500		75	5-1/2	1 - 9-1/2	115	14.40	5.55	19.95
2750	RB1010 -100	125	5-1/2	1 - 9-1/2	167	17.70	6.50	24.20
2900		200	6-1/4	1 - 11-1/2	251	19.85	7	26.85
3000	25x25	40	5-1/2	1 - 9-1/2	82	13.70	5.25	18.95
3100		75	5-1/2	1 - 11-1/2	118	15.30	5.35	20.65
3200		125	5-1/2	2 - 2-1/2	169	15.95	5.75	21.70
3300		200	6-1/4	2 - 6-1/4	252	22	6.70	28.70
3400	25x30	40	5-1/2	1 - 11-1/2	83	14	5.20	19.20
3600		75	5-1/2	1 - 11-1/2	119	15.10	5.25	20.35
3900		125	5-1/2	1 - 11-1/2	170	17.60	5.95	23.55
4000		200	6-1/4	2 - 6-1/4	252	22	6.80	28.80
4200	30x30	40	5-1/2	1 - 11-1/2	81	13.95	5.40	19.35
4400		75	5-1/2	2 - 2-1/2	116	15.15	5.60	20.75
4500		125	5-1/2	2 - 5-1/2	168	18.40	6.30	24.70
4700		200	6-1/4	2 - 9-1/4	252	22	7.30	29.30
4900	30x35	40	5-1/2	2 - 2-1/2	82	14.65	5.55	20.20
5100		75	5-1/2	2 - 5-1/2	117	16.05	5.70	21.75
5300		125	5-1/2	2 - 5-1/2	169	19	6.45	25.45
5500		200	6-1/4	2 - 9-1/4	254	22	7.35	29.35
5750	35x35	40	5-1/2	2 - 5-1/2	84	15.75	5.55	21.30
6000		75	5-1/2	2 - 5-1/2	121	18	5.95	23.95

B10 Superstructure

B1010 Floor Construction

B1010 256						Composite Beams, Deck & Slab		
	BAY SIZE (FT.)	SUPERIMPOSED LOAD (P.S.F.)	SLAB THICKNESS (IN.)	TOTAL DEPTH (FT.-IN.)	TOTAL LOAD (P.S.F.)	COST PER S.F.		
						MAT.	INST.	TOTAL
7000		125	5-1/2	2 - 8-1/2	170	21	6.85	27.85
7200		200	5-1/2	2 - 11-1/2	254	24.50	7.55	32.05
7400	35x40	40	5-1/2	2 - 5-1/2	85	17.40	6	23.40
7600		75	5-1/2	2 - 5-1/2	121	18.95	6.20	25.15
8000		125	5-1/2	2 - 5-1/2	171	21.50	6.95	28.45
9000		200	5-1/2	2 - 11-1/2	255	26.50	7.90	34.40

B10 Superstructure

B1010 Floor Construction

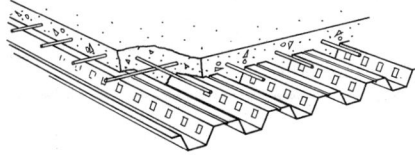

How to Use Tables: Enter any table at superimposed load and support spacing for deck.

Cellular decking tends to be stiffer and has the potential for utility integration with deck structure.

System Components			QUANTITY	UNIT	COST PER S.F.		
					MAT.	INST.	TOTAL
SYSTEM B1010 258 0900							
SPAN 6′, LOAD 125 PSF, DECK 1-1/2″-22 GA., 4″ SLAB							
Metal decking, open, galv., 1-1/2″ deep, 22 ga., over 50 sq.			1.050	S.F.	2.14	.57	2.71
Welded wire fabric rolls, 6 x 6 - W1.4 x W1.4 (10 x 10), 21 lb/csf			1.000	S.F.	.20	.34	.54
Edge forms to 6″ high on elevated slab, 4 uses			.050	L.F.	.01	.19	.20
Concrete ready mix, regular weight, 3000 psi			.253	C.F.	1		1
Place and vibrate concrete, elevated slab less than 6″, pumped			.253	C.F.		.27	.27
Finishing floor, monolithic steel trowel finish for finish floor			1.000	S.F.		.78	.78
Curing with sprayed membrane curing compound			.010	C.S.F.	.06	.08	.14
		TOTAL			3.41	2.23	5.64

B1010 258		Metal Deck/Concrete Fill						
	SUPERIMPOSED LOAD (P.S.F.)	DECK SPAN (FT.)	DECK GAGE DEPTH	SLAB THICKNESS (IN.)	TOTAL LOAD (P.S.F.)	COST PER S.F.		
						MAT.	INST.	TOTAL
0900	125	6	22 1-1/2	4	164	3.41	2.23	5.64
0920		7	20 1-1/2	4	164	3.81	2.35	6.16
0950		8	20 1-1/2	4	165	3.81	2.35	6.16
0970		9	18 1-1/2	4	165	4.55	2.35	6.90
1000		10	18 2	4	165	4.42	2.49	6.91
1020		11	18 3	5	169	5.50	2.69	8.19
1050	150	6	22 1-1/2	4	189	3.43	2.26	5.69
1070		7	22 1-1/2	4	189	3.43	2.26	5.69
1100		8	20 1-1/2	4	190	3.81	2.35	6.16
1120		9	18 1-1/2	4	190	4.55	2.35	6.90
1150		10	18 3	5	190	5.50	2.69	8.19
1170		11	18 3	5-1/2	194	5.50	2.69	8.19
1200	200	6	22 1-1/2	4	239	3.43	2.26	5.69
1220		7	22 1-1/2	4	239	3.43	2.26	5.69
1250		8	20 1-1/2	4	239	3.81	2.35	6.16
1270		9	18 2	4	241	5.25	2.53	7.78
1300		10	16 2	4	240	6.25	2.59	8.84
1320	250	6	22 1-1/2	4	289	3.43	2.26	5.69
1350		7	22 1-1/2	4	289	3.43	2.26	5.69
1370		8	18 2	4	290	5.25	2.53	7.78
1400		9	16 2	4	290	6.25	2.59	8.84

B10 Superstructure

B1010 Floor Construction

B1010 260	Cellular Composite Deck							
	SUPERIMPOSED LOAD (P.S.F.)	DECK SPAN (FT.)	DECK & PLATE GAUGE	SLAB THICKNESS (IN.)	TOTAL LOAD (P.S.F.)	COST PER S.F.		
						MAT.	INST.	TOTAL
1450	150	10	20 20	5	195	14.25	3.77	18.02
1470		11	20 20	5	195	14.25	3.77	18.02
1500		12	20 20	5	195	14.25	3.77	18.02
1520		13	20 20	5	195	14.25	3.77	18.02
1570	200	10	20 20	5	250	14.25	3.77	18.02
1650		11	20 20	5	250	14.25	3.77	18.02
1670		12	20 20	5	250	14.25	3.77	18.02
1700		13	18 20	5	250	17	3.82	20.82

B1010 Floor Construction

Description: Table below lists the S.F. costs for wood joists and a minimum thickness plywood subfloor.

Design Assumptions: 10% allowance has been added to framing quantities for overlaps, waste, double joists at openings or under partitions, etc. 5% added to subfloor for waste.

System Components	QUANTITY	UNIT	COST PER S.F.		
			MAT.	INST.	TOTAL
SYSTEM B1010 261 2500					
WOOD JOISTS 2"X6", 12" O.C.					
Framing joists, fir, 2"x6"	1.100	B.F.	.63	.87	1.50
Subfloor plywood CDX 1/2"	1.050	S.F.	.54	.69	1.23
TOTAL			1.17	1.56	2.73

B1010 261	Wood Joist	COST PER S.F.		
		MAT.	INST.	TOTAL
2500	Wood joists, 2"x6", 12" O.C.	1.17	1.56	2.73
2550	16" O.C.	1.02	1.35	2.37
2600	24" O.C.	1.28	1.27	2.55
2900	2"x8", 12" O.C.	1.39	1.68	3.07
2950	16" O.C.	1.18	1.44	2.62
3000	24" O.C.	1.39	1.33	2.72
3300	2"x10", 12" O.C.	1.75	1.91	3.66
3350	16" O.C.	1.44	1.61	3.05
3400	24" O.C.	1.56	1.44	3
3700	2"x12", 12" O.C.	2.21	1.93	4.14
3750	16" O.C.	1.79	1.62	3.41
3800	24" O.C.	1.80	1.45	3.25
4100	2"x14", 12" O.C.	3.10	2.11	5.21
4150	16" O.C.	2.47	1.76	4.23
4200	24" O.C.	2.25	1.55	3.80
4500	3"x6", 12" O.C.	2.40	1.87	4.27
4550	16" O.C.	1.94	1.58	3.52
4600	24" O.C.	1.89	1.42	3.31
4900	3"x8", 12" O.C.	3.07	1.83	4.90
4950	16" O.C.	2.44	1.55	3.99
5000	24" O.C.	2.23	1.40	3.63
5300	3"x10", 12" O.C.	4.12	2.09	6.21
5350	16" O.C.	3.23	1.75	4.98
5400	24" O.C.	2.75	1.53	4.28
5700	3"x12", 12" O.C.	4.29	1.82	6.11
5750	16" O.C.	3.76	2.05	5.81
5800	24" O.C.	3.11	1.74	4.85
6100	4"x6", 12" O.C.	3.40	1.90	5.30
6150	16" O.C.	2.69	1.60	4.29
6200	24" O.C.	2.39	1.44	3.83

B1010 Floor Construction

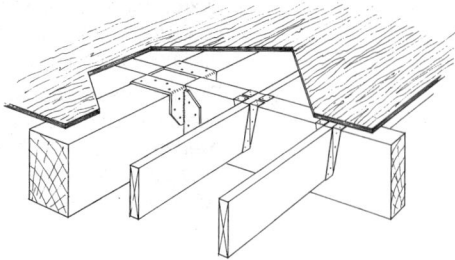

Description: Table lists the S.F. costs, total load, and member sizes, for various bay sizes and loading conditions.

Design Assumptions: Dead load = girder, beams, and joist weight plus 3/4" plywood floor.

Maximum deflection is 1/360 of the clear span.

Lumber is stress grade f(w) = 1,800 PSI

System Components		QUANTITY	UNIT	COST PER S.F.		
				MAT.	INST.	TOTAL
SYSTEM B1010 264 2000						
15' X 15' BAY, S. LOAD 40 P.S.F.						
Beams and girders, structural grade, 8" x 12"		.730	B.F.	1.82	.21	2.03
Framing joists, fir 4" x 12"		.660	B.F.	.86	.36	1.22
Framing joists, 2" x 6"		.840	B.F.	.48	.67	1.15
Beam to girder saddles		.510	Lb.	3.22	.65	3.87
Column caps		.510	Lb.	.87	.09	.96
Drilling, bolt holes		.510	Lb.		.32	.32
Machine bolts		.510	Lb.	.13	.14	.27
Joist hangers 18 ga.		.213	Ea.	.28	.64	.92
Subfloor plywood CDX 3/4"		1.050	S.F.	.96	.83	1.79
	TOTAL			8.62	3.91	12.53

B1010 264			Wood Beam & Joist					
	BAY SIZE (FT.)	SUPERIMPOSED LOAD (P.S.F.)	GIRDER BEAM (IN.)	JOISTS (IN.)	TOTAL LOAD (P.S.F.)	COST PER S.F.		
						MAT.	INST.	TOTAL
2000	15x15	40	8 x 12 4 x 12	2 x 6 @ 16	53	8.60	3.91	12.51
2050	RB1010 -100	75	8 x 16 4 x 16	2 x 8 @ 16	90	11.25	4.28	15.53
2100		125	12 x 16 6 x 16	2 x 8 @ 12	144	17.20	5.45	22.65
2150		200	14 x 22 12 x 16	2 x 10 @ 12	227	35	8.35	43.35
2500	15x20	40	10 x 16 8 x 12	2 x 6 @ 16	58	11.30	3.98	15.28
2550		75	12 x 14 8 x 14	2 x 8 @ 16	96	14.80	4.61	19.41
2600		125	10 x 18 12 x 14	2 x 8 @ 12	152	21	6.20	27.20
2650		200	14 x 20 14 x 16	2 x 10 @ 12	234	31.50	7.60	39.10
3000	20x20	40	10 x 14 10 x 12	2 x 8 @ 16	63	11	3.93	14.93
3050		75	12 x 16 8 x 16	2 x 10 @ 16	102	14.80	4.39	19.19
3100		125	14 x 22 12 x 16	2 x 10 @ 12	163	29.50	7.05	36.55

B1010 Floor Construction

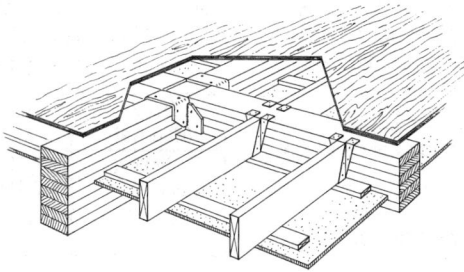

Description: Table below lists the S.F. costs, total load, and member sizes, for various bay sizes and loading conditions.

Design Assumptions: Dead load = girder, beams, and joist weight plus 3/4″ plywood floor.

Maximum deflection is 1/360 of the clear span.

Lumber is stress grade f(w) = 2,500 PSI

System Components			COST PER S.F.		
	QUANTITY	UNIT	MAT.	INST.	TOTAL
SYSTEM B1010 265 2000					
15′X15′ BAY, 40 PSF L.L., JOISTS 2X6, @ 16″ O.C.					
Laminated framing, straight beams	1.000	B.F.	3.20	.95	4.15
Framing joists,2″x6″	.840	B.F.	.48	.67	1.15
Saddles	3.375	Lb.	3.22	.65	3.87
Column caps	3.375	Lb.	.87	.09	.96
Drilling, bolt holes	3.375	Lb.		.30	.30
Machine bolts	3.375	Lb.	.12	.12	.24
Joist & beam hangers 18 ga.	.213	Ea.	.28	.64	.92
Subfloor plywood CDX 3/4″	1.050	S.F.	.96	.83	1.79
TOTAL			9.13	4.25	13.38

B1010 265		Laminated Wood Floor Beams						
	BAY SIZE (FT.)	SUPERIMPOSED LOAD (P.S.F.)	GIRDER (IN.)	BEAM (IN.)	TOTAL LOAD (P.S.F.)	COST PER S.F.		
						MAT.	INST.	TOTAL
2000	15x15	40	5 x 12-3/8	5 x 8-1/4	54	9.15	4.25	13.40
2050	RB1010 -100	75	5 x 16-1/2	5 x 11	92	11.70	4.93	16.63
2100		125	8-5/8 x 15-1/8	8-5/8 x 11	147	17.35	6.90	24.25
2150		200	8-5/8 x 22	8-5/8 x 16-1/2	128	32	10.05	42.05
2500	15x20	40	5 x 13-3/4	5 x 11	55	9.20	4.20	13.40
2550		75	5 x 17-3/8	5 x 15-1/8	92	13.20	5.10	18.30
2600		125	8-5/8 x 17-7/8	8-5/8 x 13-3/4	148	20.50	7.40	27.90
2650		200	8-5/8 x 22	8-5/8 x 17-7/8	226	28	9.20	37.20
3000	20x20	40	5 x 16-1/2	5 x 13-3/4	55	9.50	4.11	13.61
3050		75	5 x 20-5/8	5 x 16-1/2	92	13.10	4.95	18.05
3100		125	8-5/8 x 22	8-5/8 x 15-1/8	148	20	6.90	26.90
3150		200	8-5/8 x 28-7/8	8-5/8 x 20-5/8	229	30	9.30	39.30

B1010 Floor Construction

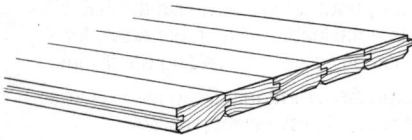

Description: Table below lists the S.F. costs and maximum spans for commonly used wood decking materials for various loading conditions.

Design Assumptions: Total applied load is the superimposed load plus dead load.

Maximum deflection is 1/180 of the clear span which is not suitable if plaster ceilings will be supported by the roof or floor.

Modulus of elasticity (E) and fiber strength (f) are as shown in the table to the right.

No allowance for waste has been included.

Supporting beams or purlins are not included in costs below.

Decking Material Characteristics

Deck Material	Modulus of Elasticity	Fiber Strength
Cedar	1,100,000	1000 psi
Douglas fir	1,760,000	1200 psi
Hemlock	1,600,000	1200 psi
White spruce	1,320,000	1200 psi

B1010 266	Wood Deck						
	NOMINAL THK (IN.)	**TYPE WOOD**	**MAXIMUM SPAN**			**COST PER S.F.**		
			MAXIMUM SPAN 40 P.S.F. LOAD	**MAXIMUM SPAN 100 P.S.F.**	**MAXIMUM SPAN 250 P.S.F. LOAD**	**MAT.**	**INST.**	**TOTAL**
1000	2	cedar	6.5	5	3.5	4.79	2.83	7.62
1050		douglas fir	8	6	4	1.32	2.83	4.15
1100		hemlock	7.5	5.5	4	1.71	2.83	4.54
1150		white spruce	7	5	3.5	1.64	2.83	4.47
1200	3	cedar	11	8	6	6.85	3.10	9.95
1250		douglas fir	13	10	7	2.56	3.10	5.66
1300		hemlock	12	9	7	2.56	3.10	5.66
1350		white spruce	11	8	6	2.46	3.10	5.56
1400	4	cedar	15	11	8	9.20	3.96	13.16
1450		douglas fir	18	13	10	3.43	3.96	7.39
1500		hemlock	17	13	9	3.42	3.96	7.38
1550		white spruce	16	12	8	3.29	3.96	7.25
1600	6	cedar	23	17	13	15	4.96	19.96
1650		douglas fir	24+	21	15	5.15	4.96	10.11
1700		hemlock	24+	20	15	5.15	4.96	10.11
1750		white spruce	24	18	13	4.09	4.96	9.05

B10 Superstructure

B1010 Floor Construction

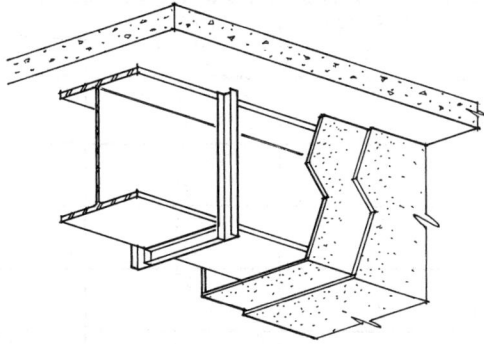

The table below lists fireproofing costs for steel beams by type, beam size, thickness and fire rating. Weights listed are for the fireproofing material only.

System Components	QUANTITY	UNIT	COST PER L.F.		
			MAT.	INST.	TOTAL
SYSTEM B1010 710 1300					
FIREPROOFING, 5/8″ F.R. GYP. BOARD, 12″X 4″ BEAM, 2″ THICK, 2 HR. F.R.					
Corner bead for drywall, 1-1/4″ x 1-1/4″, galvanized	.020	C.L.F.	.40	2.84	3.24
L bead for drywall, galvanized	.020	C.L.F.	.43	3.30	3.73
Furring, beams & columns, 3/4″ galv. channels, 24″ O.C.	2.330	S.F.	.54	5.29	5.83
Drywall on beam, no finish, 2 layers at 5/8″ thick	3.000	S.F.	2.67	9.90	12.57
Drywall, taping and finishing joints, add	3.000	S.F.	.15	1.50	1.65
TOTAL			4.19	22.83	27.02

B1010 710 — Steel Beam Fireproofing

	ENCASEMENT SYSTEM	BEAM SIZE (IN.)	THICKNESS (IN.)	FIRE RATING (HRS.)	WEIGHT (P.L.F.)	COST PER L.F.		
						MAT.	INST.	TOTAL
0400	Concrete	12x4	1	1	77	7.65	24.50	32.15
0450	3000 PSI		1-1/2	2	93	8.80	26.50	35.30
0500			2	3	121	9.85	29	38.85
0550		14x5	1	1	100	10	30	40
0600			1-1/2	2	122	11.60	33.50	45.10
0650			2	3	142	13.20	36.50	49.70
0700		16x7	1	1	147	11.15	32	43.15
0750			1-1/2	2	169	11.80	33	44.80
0800			2	3	195	13.50	36	49.50
0850		18x7-1/2	1	1	172	13.10	36.50	49.60
0900			1-1/2	2	196	14.85	40.50	55.35
0950			2	3	225	16.65	44.50	61.15
1000		24x9	1	1	264	16.85	44.50	61.35
1050			1-1/2	2	295	18.25	47	65.25
1100			2	3	328	19.70	50	69.70
1150		30x10-1/2	1	1	366	22	56	78
1200			1-1/2	2	404	24	59.50	83.50
1250			2	3	449	25.50	63	88.50
1300	5/8″ fire rated	12x4	2	2	15	4.19	23	27.19
1350	Gypsum board		2-5/8	3	24	5.85	28.50	34.35
1400		14x5	2	2	17	4.50	23.50	28
1450			2-5/8	3	27	6.45	30	36.45
1500		16x7	2	2	20	5.05	26.50	31.55
1550			2-5/8	3	31	6.50	26	32.50
1600		18x7-1/2	2	2	22	5.45	28.50	33.95
1650			2-5/8	3	34	7.75	36	43.75

B1010 Floor Construction

B1010 710	Steel Beam Fireproofing

	ENCASEMENT SYSTEM	BEAM SIZE (IN.)	THICKNESS (IN.)	FIRE RATING (HRS.)	WEIGHT (P.L.F.)	COST PER L.F.		
						MAT.	INST.	TOTAL
1700	5/8" fire rated	24x9	2	2	27	6.75	35	41.75
1750	Gypsum board		2-5/8	3	42	9.55	44	53.55
1800		30x10-1/2	2	2	33	7.95	41	48.95
1850			2-5/8	3	51	11.25	51.50	62.75
1900	Gypsum	12x4	1-1/8	3	18	6	17.30	23.30
1950	Plaster on	14x5	1-1/8	3	21	6.80	19.50	26.30
2000	Metal lath	16x7	1-1/8	3	25	7.95	22.50	30.45
2050		18x7-1/2	1-1/8	3	32	9	25	34
2100		24x9	1-1/8	3	35	11.10	30.50	41.60
2150		30x10-1/2	1-1/8	3	44	13.35	36.50	49.85
2200	Perlite plaster	12x4	1-1/8	2	16	5.20	19.25	24.45
2250	On metal lath		1-1/4	3	20	5.50	20	25.50
2300			1-1/2	4	22	5.90	21.50	27.40
2350		14x5	1-1/8	2	18	6.75	23.50	30.25
2400			1-1/4	3	23	7.15	24.50	31.65
2450			1-1/2	4	25	7.90	26.50	34.40
2500		16x7	1-1/8	2	21	6.25	23.50	29.75
2550			1-1/4	3	26	6.55	24.50	31.05
2600			1-1/2	4	29	6.95	25.50	32.45
2650		18x7-1/2	1-1/8	2	26	7.75	27.50	35.25
2700			1-1/4	3	33	8.10	29	37.10
2750			1-1/2	4	36	8.90	31	39.90
2800		24x9	1-1/8	2	30	7.95	31	38.95
2850			1-1/4	3	38	8.30	32	40.30
2900			1-1/2	4	41	8.70	33	41.70
2950		30x10-1/2	1-1/8	2	37	10.30	39	49.30
3000			1-1/4	3	46	10.70	40	50.70
3050			1-1/2	4	51	11.50	41.50	53
3100	Sprayed Fiber	12x4	5/8	1	12	1.26	1.96	3.22
3150	(non asbestos)		1-1/8	2	21	2.26	3.51	5.77
3200			1-1/4	3	22	2.52	3.90	6.42
3250		14x5	5/8	1	14	1.26	1.96	3.22
3300			1-1/8	2	26	2.26	3.51	5.77
3350			1-1/4	3	29	2.52	3.90	6.42
3400		16x7	5/8	1	18	1.59	2.47	4.06
3450			1-1/8	2	32	2.86	4.44	7.30
3500			1-1/4	3	36	3.18	4.94	8.12
3550		18x7-1/2	5/8	1	20	1.76	2.73	4.49
3600			1-1/8	2	35	3.16	4.91	8.07
3650			1-1/4	3	39	3.49	5.40	8.89
3700		24x9	5/8	1	25	2.22	3.44	5.66
3750			1-1/8	2	45	4.02	6.20	10.22
3800			1-1/4	3	50	4.47	6.95	11.42
3850		30x10-1/2	5/8	1	31	2.69	4.18	6.87
3900			1-1/8	2	55	4.84	7.50	12.34
3950			1-1/4	3	61	5.40	8.35	13.75
4000	On Decking	Flat	1		6	.58	.57	1.15
4050	Per S.F.	Corrugated	1		7	.87	1.08	1.95
4100		Fluted	1		7	.87	1.08	1.95

B10 Superstructure

B1010 Floor Construction

Listed below are costs per V.L.F. for fireproofing by material, column size, thickness and fire rating. Weights listed are for the fireproofing material only.

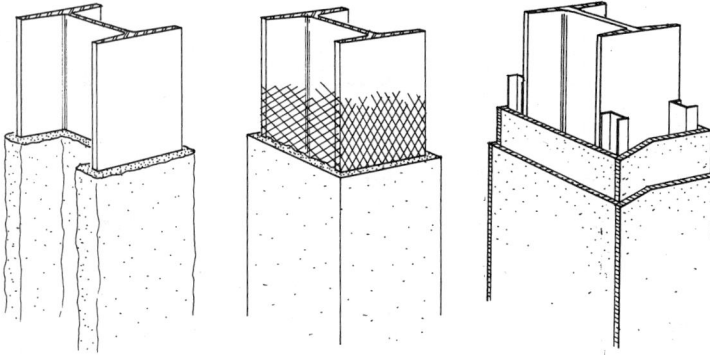

System Components	QUANTITY	UNIT	COST PER V.L.F.		
			MAT.	INST.	TOTAL
SYSTEM B1010 720 3000					
CONCRETE FIREPROOFING, 8″ STEEL COLUMN, 1″ THICK, 1 HR. FIRE RATING					
Forms in place, columns, plywood, 4 uses	3.330	SFCA	2.16	24.98	27.14
Welded wire fabric, 2 x 2 #14 galv. 21 lb./C.S.F., column wrap	2.700	S.F.	1.97	4.83	6.80
Concrete ready mix, regular weight, 3000 psi	.621	C.F.	2.55		2.55
Place and vibrate concrete, 12″ sq./round columns, pumped	.621	C.F.		1.61	1.61
TOTAL			6.68	31.42	38.10

B1010 720 — Steel Column Fireproofing

	ENCASEMENT SYSTEM	COLUMN SIZE (IN.)	THICKNESS (IN.)	FIRE RATING (HRS.)	WEIGHT (P.L.F.)	COST PER V.L.F.		
						MAT.	INST.	TOTAL
3000	Concrete	8	1	1	110	6.70	31.50	38.20
3050			1-1/2	2	133	7.70	35	42.70
3100			2	3	145	8.55	38	46.55
3150		10	1	1	145	8.85	39	47.85
3200			1-1/2	2	168	9.75	41.50	51.25
3250			2	3	196	10.75	44.50	55.25
3300		14	1	1	258	11.20	45.50	56.70
3350			1-1/2	2	294	12.05	48.50	60.55
3400			2	3	325	13.15	52	65.15
3450	Gypsum board	8	1/2	2	8	3.18	19.25	22.43
3500	1/2″ fire rated	10	1/2	2	11	3.35	20	23.35
3550	1 layer	14	1/2	2	18	3.44	20.50	23.94
3600	Gypsum board	8	1	3	14	4.43	24.50	28.93
3650	1/2″ fire rated	10	1	3	17	4.76	26	30.76
3700	2 layers	14	1	3	22	4.93	27	31.93
3750	Gypsum board	8	1-1/2	3	23	5.90	31	36.90
3800	1/2″ fire rated	10	1-1/2	3	27	6.65	34.50	41.15
3850	3 layers	14	1-1/2	3	35	7.40	37.50	44.90
3900	Sprayed fiber	8	1-1/2	2	6.3	3.93	6.10	10.03
3950	Direct application		2	3	8.3	5.40	8.40	13.80
4000			2-1/2	4	10.4	7	10.90	17.90
4050		10	1-1/2	2	7.9	4.75	7.40	12.15
4100			2	3	10.5	6.50	10.10	16.60
4150			2-1/2	4	13.1	8.40	13	21.40

107

B1010 720			Steel Column Fireproofing				

	ENCASEMENT SYSTEM	COLUMN SIZE (IN.)	THICKNESS (IN.)	FIRE RATING (HRS.)	WEIGHT (P.L.F.)	COST PER V.L.F.		
						MAT.	INST.	TOTAL
4200		14	1-1/2	2	10.8	5.90	9.15	15.05
4250			2	3	14.5	8.05	12.50	20.55
4300			2-1/2	4	18	10.30	16	26.30
4350	3/4" gypsum plaster	8	3/4	1	23	7.60	25	32.60
4400	On metal lath	10	3/4	1	28	8.60	28	36.60
4450		14	3/4	1	38	10.45	35	45.45
4500	Perlite plaster	8	1	2	18	8.15	27	35.15
4550	On metal lath		1-3/8	3	23	8.85	29.50	38.35
4600			1-3/4	4	35	10.40	36	46.40
4650	Perlite plaster	10	1	2	21	9.30	31.50	40.80
4700			1-3/8	3	27	10.45	36	46.45
4750			1-3/4	4	41	11.70	41	52.70
4800		14	1	2	29	11.60	39	50.60
4850			1-3/8	3	35	12.95	45	57.95
4900			1-3/4	4	53	13.45	46.50	59.95
4950	1/2 gypsum plaster	8	7/8	1	13	7.50	20.50	28
5000	On 3/8" gypsum lath	10	7/8	1	16	8.95	24	32.95
5050		14	7/8	1	21	10.95	29.50	40.45
5100	5/8" gypsum plaster	8	1	1-1/2	20	7.40	22.50	29.90
5150	On 3/8" gypsum lath	10	1	1-1/2	24	8.80	26.50	35.30
5200		14	1	1-1/2	33	10.75	32.50	43.25
5250	1"perlite plaster	8	1-3/8	2	23	8.10	23.50	31.60
5300	On 3/8" gypsum lath	10	1-3/8	2	28	9.30	27	36.30
5350		14	1-3/8	2	37	11.60	34	45.60
5400	1-3/8" perlite plaster	8	1-3/4	3	27	8.50	26	34.50
5450	On 3/8" gypsum lath	10	1-3/4	3	33	9.75	30.50	40.25
5500		14	1-3/4	3	43	12.15	38	50.15
5550	Concrete masonry	8	4-3/4	4	126	10.65	29	39.65
5600	Units 4" thick	10	4-3/4	4	166	13.35	36.50	49.85
5650	75% solid	14	4-3/4	4	262	16	43.50	59.50

B1020 Roof Construction

The table below lists prices per S.F. for roof rafters and sheathing by nominal size and spacing. Sheathing is 5/16″ CDX for 12″ and 16″ spacing and 3/8″ CDX for 24″ spacing.

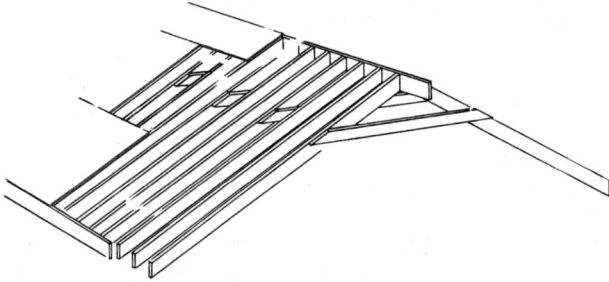

Factors for Converting Inclined to Horizontal

Roof Slope	Approx. Angle	Factor	Roof Slope	Approx. Angle	Factor
Flat	0°	1.000	12 in 12	45.0°	1.414
1 in 12	4.8°	1.003	13 in 12	47.3°	1.474
2 in 12	9.5°	1.014	14 in 12	49.4°	1.537
3 in 12	14.0°	1.031	15 in 12	51.3°	1.601
4 in 12	18.4°	1.054	16 in 12	53.1°	1.667
5 in 12	22.6°	1.083	17 in 12	54.8°	1.734
6 in 12	26.6°	1.118	18 in 12	56.3°	1.803
7 in 12	30.3°	1.158	19 in 12	57.7°	1.873
8 in 12	33.7°	1.202	20 in 12	59.0°	1.943
9 in 12	36.9°	1.250	21 in 12	60.3°	2.015
10 in 12	39.8°	1.302	22 in 12	61.4°	2.088
11 in 12	42.5°	1.357	23 in 12	62.4°	2.162

System Components		QUANTITY	UNIT	COST PER S.F.		
				MAT.	INST.	TOTAL
SYSTEM B1020 102 2500						
RAFTER 2″X4″, 12″ O.C.						
	Framing joists, fir 2″x 4″	.730	B.F.	.35	.88	1.23
	Sheathing plywood on roof CDX 5/16″	1.050	S.F.	.61	.65	1.26
	TOTAL			.96	1.53	2.49

B1020 102	Wood/Flat or Pitched	COST PER S.F.		
		MAT.	INST.	TOTAL
2500	Flat rafter, 2″x4″, 12″ O.C.	.96	1.53	2.49
2550	16″ O.C.	.87	1.31	2.18
2600	24″ O.C.	.69	1.12	1.81
2900	2″x6″, 12″ O.C.	1.24	1.52	2.76
2950	16″ O.C.	1.09	1.31	2.40
3000	24″ O.C.	.83	1.12	1.95
3300	2″x8″, 12″ O.C.	1.46	1.64	3.10
3350	16″ O.C.	1.25	1.40	2.65
3400	24″ O.C.	.94	1.18	2.12
3700	2″x10″, 12″ O.C.	1.82	1.87	3.69
3750	16″ O.C.	1.51	1.57	3.08
3800	24″ O.C.	1.11	1.29	2.40
4100	2″x12″, 12″ O.C.	2.28	1.89	4.17
4150	16″ O.C.	1.86	1.58	3.44
4200	24″ O.C.	1.35	1.30	2.65
4500	2″x14″, 12″ O.C.	3.17	2.76	5.93
4550	16″ O.C.	2.54	2.24	4.78
4600	24″ O.C.	1.80	1.74	3.54
4900	3″x6″, 12″ O.C.	2.47	1.62	4.09
4950	16″ O.C.	2.01	1.38	3.39
5000	24″ O.C.	1.44	1.17	2.61
5300	3″x8″, 12″ O.C.	3.14	1.79	4.93
5350	16″ O.C.	2.51	1.51	4.02
5400	24″ O.C.	1.78	1.25	3.03
5700	3″x10″, 12″ O.C.	4.19	2.05	6.24
5750	16″ O.C.	3.30	1.71	5.01
5800	24″ O.C.	2.30	1.38	3.68
6100	3″x12″, 12″ O.C.	4.90	2.47	7.37
6150	16″ O.C.	3.83	2.01	5.84
6200	24″ O.C.	2.66	1.59	4.25

B1020 Roof Construction

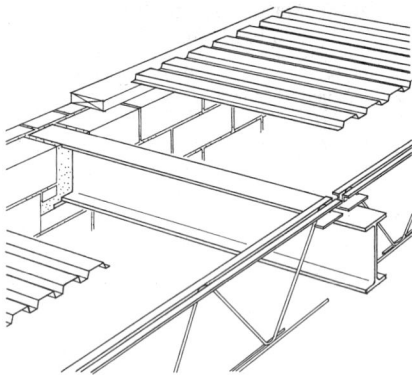

Table below lists the cost per S.F. for a roof system with steel columns, beams, and deck using open web steel joists and 1-1/2" galvanized metal deck. Perimeter of system is supported on bearing walls.

Design and Pricing Assumptions:
Columns are 18' high.
Joists are 5'-0" O.C. and span the long direction of the bay.

Joists at columns have bottom chords extended and are connected to columns. Column costs are not included but are listed separately per S.F. of floor.

Roof deck is 1-1/2", 22 gauge galvanized steel. Joist cost includes appropriate bridging. Deflection is limited to 1/240 of the span. Fireproofing is not included.

Costs/S.F. are based on a building 4 bays long and 4 bays wide.

Design Loads	Min.	Max.
Joists & Beams	3 PSF	5 PSF
Deck	2	2
Insulation	3	3
Roofing	6	6
Misc.	6	6
Total Dead Load	20 PSF	22 PSF

System Components		QUANTITY	UNIT	COST PER S.F.		
				MAT.	INST.	TOTAL
SYSTEM B1020 108 1200						
METAL DECK & STEEL JOISTS,15'X20' BAY,20 PSF S. LOAD,16" DP,40 PSF T.LOAD						
Structural steel		.488	Lb.	.81	.16	.97
Open web joists		1.022	Lb.	1.14	.35	1.49
Metal decking, open, galvanized, 1-1/2" deep, 22 gauge		1.050	S.F.	2.14	.57	2.71
	TOTAL			4.09	1.08	5.17

B1020 108 — Steel Joists, Beams & Deck on Columns & Walls

	BAY SIZE (FT.)	SUPERIMPOSED LOAD (P.S.F.)	DEPTH (IN.)	TOTAL LOAD (P.S.F.)	COLUMN ADD	COST PER S.F.		
						MAT.	INST.	TOTAL
1200	15x20	20	16	40		4.09	1.08	5.17
1300					columns	.92	.17	1.09
1400		30	16	50		4.44	1.15	5.59
1500					columns	.92	.17	1.09
1600		40	18	60		4.80	1.24	6.04
1700					columns	.92	.17	1.09
1800	20x20	20	16	40		4.81	1.21	6.02
1900					columns	.69	.13	.82
2000		30	18	50		4.81	1.21	6.02
2100					columns	.69	.13	.82
2200		40	18	60		5.35	1.37	6.72
2300					columns	.69	.13	.82
2400	20x25	20	18	40		4.81	1.25	6.06
2500					columns	.55	.11	.66
2600		30	18	50		5.35	1.46	6.81
2700					columns	.55	.11	.66
2800		40	20	60		5.35	1.39	6.74
2900					columns	.74	.15	.89
3000	25x25	20	18	40		5.15	1.31	6.46
3100					columns	.44	.08	.52
3200		30	22	50		5.85	1.55	7.40
3300					columns	.59	.12	.71
3400		40	20	60		6.15	1.55	7.70
3500					columns	.59	.12	.71

B1020 Roof Construction

B1020 108	Steel Joists, Beams & Deck on Columns & Walls

	BAY SIZE (FT.)	SUPERIMPOSED LOAD (P.S.F.)	DEPTH (IN.)	TOTAL LOAD (P.S.F.)	COLUMN ADD	COST PER S.F.		
						MAT.	INST.	TOTAL
3600	25x30	20	22	40		5.50	1.27	6.77
3700					columns	.49	.09	.58
3800		30	20	50		5.95	1.64	7.59
3900					columns	.49	.09	.58
4000		40	25	60		6.45	1.45	7.90
4100					columns	.59	.12	.71
4200	30x30	20	25	42		6	1.37	7.37
4300					columns	.41	.08	.49
4400		30	22	52		6.55	1.48	8.03
4500					columns	.49	.09	.58
4600		40	28	62		6.80	1.52	8.32
4700					columns	.49	.09	.58
4800	30x35	20	22	42		6.25	1.41	7.66
4900					columns	.42	.08	.50
5000		30	28	52		6.55	1.47	8.02
5100					columns	.42	.08	.50
5200		40	25	62		7.10	1.60	8.70
5300					columns	.49	.09	.58
5400	35x35	20	28	42		6.25	1.42	7.67
5500					columns	.36	.07	.43
5600		30	25	52		7.55	1.68	9.23
5700					columns	.42	.08	.50
5800		40	28	62		7.65	1.70	9.35
5900					columns	.47	.09	.56

B1020 Roof Construction

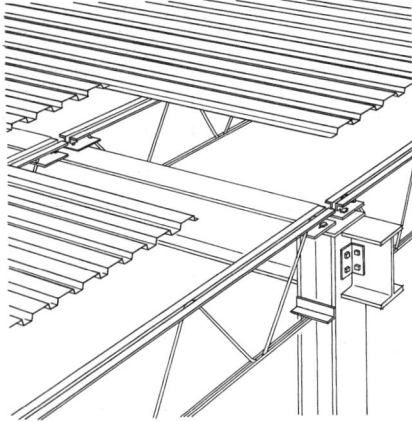

Description: Table below lists the cost per S.F. for a roof system with steel columns, beams, and deck, using open web steel joists and 1-1/2″ galvanized metal deck.

Roof deck is 1-1/2″, 22 gauge galvanized steel. Joist cost includes appropriate bridging. Deflection is limited to 1/240 of the span. Fireproofing is not included.

Design and Pricing Assumptions:
 Columns are 18′ high.
 Building is 4 bays long by 4 bays wide.
 Joists are 5′-0″ O.C. and span the long direction of the bay.
Joists at columns have bottom chords extended and are connected to columns. Column costs are not included but are listed separately per S.F. of floor.

Design Loads	Min.		Max.	
Joists & Beams	3	PSF	5	PSF
Deck	2		2	
Insulation	3		3	
Roofing	6		6	
Misc.	6		6	
Total Dead Load	20	PSF	22	PSF

System Components	QUANTITY	UNIT	COST PER S.F.		
			MAT.	INST.	TOTAL
SYSTEM B1020 112 1100					
METAL DECK AND JOISTS,15′X20′ BAY,20 PSF S. LOAD					
Structural steel	.954	Lb.	1.57	.31	1.88
Open web joists	1.260	Lb.	1.41	.43	1.84
Metal decking, open, galvanized, 1-1/2″ deep, 22 gauge	1.050	S.F.	2.14	.57	2.71
TOTAL			5.12	1.31	6.43

B1020 112	Steel Joists, Beams, & Deck on Columns						
	BAY SIZE (FT.)	SUPERIMPOSED LOAD (P.S.F.)	DEPTH (IN.)	TOTAL LOAD (P.S.F.)	COLUMN ADD	COST PER S.F.		
						MAT.	INST.	TOTAL
1100	15x20	20	16	40		5.10	1.31	6.41
1200					columns	2.55	.49	3.04
1300		30	16	50		5.60	1.40	7
1400					columns	2.55	.49	3.04
1500		40	18	60		5.70	1.45	7.15
1600					columns	2.55	.49	3.04
1700	20x20	20	16	40		5.50	1.38	6.88
1800					columns	1.91	.37	2.28
1900		30	18	50		6.05	1.47	7.52
2000					columns	1.91	.37	2.28
2100		40	18	60		6.75	1.64	8.39
2200					columns	1.91	.37	2.28
2300	20x25	20	18	40		5.80	1.46	7.26
2400					columns	1.53	.29	1.82
2500		30	18	50		6.50	1.71	8.21
2600					columns	1.53	.29	1.82
2700		40	20	60		6.50	1.63	8.13
2800					columns	2.04	.40	2.44
2900	25x25	20	18	40		6.75	1.64	8.39
3000					columns	1.22	.24	1.46
3100		30	22	50		7.30	1.86	9.16
3200					columns	1.63	.32	1.95
3300		40	20	60		7.80	1.89	9.69
3400					columns	1.63	.32	1.95

B1020 Roof Construction

B1020 112	Steel Joists, Beams, & Deck on Columns

	BAY SIZE (FT.)	SUPERIMPOSED LOAD (P.S.F.)	DEPTH (IN.)	TOTAL LOAD (P.S.F.)	COLUMN ADD	COST PER S.F.		
						MAT.	INST.	TOTAL
3500	25x30	20	22	40		6.65	1.49	8.14
3600					columns	1.36	.27	1.63
3700		30	20	50		7.35	1.95	9.30
3800					columns	1.36	.27	1.63
3900		40	25	60		8	1.76	9.76
4000					columns	1.63	.32	1.95
4100	30x30	20	25	42		7.50	1.65	9.15
4200					columns	1.14	.23	1.37
4300		30	22	52		8.25	1.81	10.06
4400					columns	1.36	.27	1.63
4500		40	28	62		8.60	1.88	10.48
4600					columns	1.36	.27	1.63
4700	30x35	20	22	42		7.70	1.71	9.41
4800					columns	1.17	.23	1.40
4900		30	28	52		8.20	1.81	10.01
5000					columns	1.17	.23	1.40
5100		40	25	62		8.95	1.94	10.89
5200					columns	1.36	.27	1.63
5300	35x35	20	28	42		8.15	1.79	9.94
5400					columns	1	.20	1.20
5500		30	25	52		9.10	1.98	11.08
5600					columns	1.17	.23	1.40
5700		40	28	62		9.80	2.12	11.92
5800					columns	1.29	.25	1.54

B1020 Roof Construction

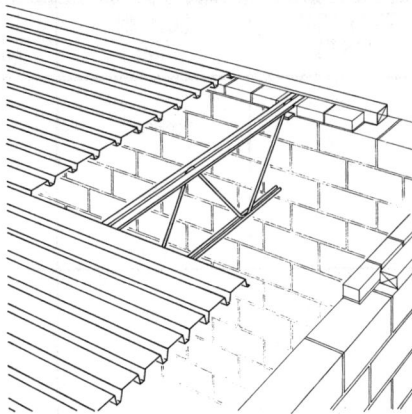

Description: Table below lists cost per S.F. for a roof system using open web steel joists and 1-1/2″ galvanized metal deck. The system is assumed supported on bearing walls or other suitable support. Costs for the supports are not included.

Design and Pricing Assumptions:
Joists are 5′-0″ O.C.
Roof deck is 1-1/2″, 22 gauge galvanized.

System Components	QUANTITY	UNIT	COST PER S.F.		
			MAT.	INST.	TOTAL
SYSTEM B1020 116 1100					
METAL DECK AND JOISTS, 20′ SPAN, 20 PSF S. LOAD					
Open web joists, horiz. bridging T.L. lot, to 30′ span	1.114	Lb.	1.25	.38	1.63
Metal decking, open type, galv 1-1/2″ deep	1.050	S.F.	2.14	.57	2.71
TOTAL			3.39	.95	4.34

B1020 116	Steel Joists & Deck on Bearing Walls							
	BAY SIZE (FT.)	SUPERIMPOSED LOAD (P.S.F.)	DEPTH (IN.)	TOTAL LOAD (P.S.F.)		COST PER S.F.		
						MAT.	INST.	TOTAL
1100	20	20	13-1/2	40		3.39	.95	4.34
1200		30	15-1/2	50		3.46	.97	4.43
1300		40	15-1/2	60		3.70	1.05	4.75
1400	25	20	17-1/2	40		3.71	1.05	4.76
1500		30	17-1/2	50		3.99	1.13	5.12
1600		40	19-1/2	60		4.04	1.15	5.19
1700	30	20	19-1/2	40		4.03	1.14	5.17
1800		30	21-1/2	50		4.12	1.17	5.29
1900		40	23-1/2	60		4.42	1.26	5.68
2000	35	20	23-1/2	40		4.38	1.06	5.44
2100		30	25-1/2	50		4.52	1.09	5.61
2200		40	25-1/2	60		4.79	1.14	5.93
2300	40	20	25-1/2	41		4.86	1.16	6.02
2400		30	25-1/2	51		5.15	1.22	6.37
2500		40	25-1/2	61		5.35	1.26	6.61
2600	45	20	27-1/2	41		5.40	1.75	7.15
2700		30	31-1/2	51		5.70	1.86	7.56
2800		40	31-1/2	61		6	1.96	7.96
2900	50	20	29-1/2	42		6	1.97	7.97
3000		30	31-1/2	52		6.55	2.16	8.71
3100		40	31-1/2	62		6.95	2.29	9.24
3200	60	20	37-1/2	42		7.70	2.04	9.74
3300		30	37-1/2	52		8.50	2.26	10.76
3400		40	37-1/2	62		8.50	2.26	10.76
3500	70	20	41-1/2	42		8.50	2.26	10.76
3600		30	41-1/2	52		9.05	2.41	11.46
3700		40	41-1/2	64		11	2.92	13.92

B1020 Roof Construction

B1020 116	Steel Joists & Deck on Bearing Walls

	BAY SIZE (FT.)	SUPERIMPOSED LOAD (P.S.F.)	DEPTH (IN.)	TOTAL LOAD (P.S.F.)		COST PER S.F.		
						MAT.	INST.	TOTAL
3800	80	20	45-1/2	44		12.30	3.19	15.49
3900		30	45-1/2	54		12.30	3.19	15.49
4000		40	45-1/2	64		13.60	3.53	17.13
4100	90	20	53-1/2	44		10.05	2.55	12.60
4200		30	53-1/2	54		10.65	2.70	13.35
4300		40	53-1/2	65		12.70	3.21	15.91
4400	100	20	57-1/2	44		10.65	2.70	13.35
4500		30	57-1/2	54		12.70	3.21	15.91
4600		40	57-1/2	65		14	3.54	17.54
4700	125	20	69-1/2	44		14.90	3.60	18.50
4800		30	69-1/2	56		17.40	4.19	21.59
4900		40	69-1/2	67		19.85	4.78	24.63

B1020 Roof Construction

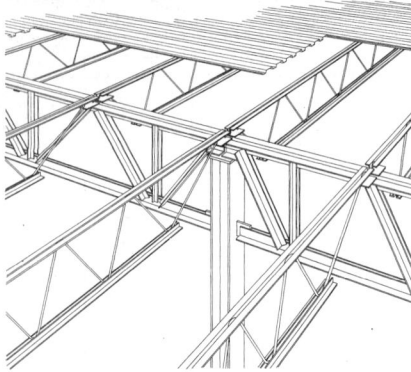

Description: Table below lists costs for a roof system supported on exterior bearing walls and interior columns. Costs include bracing, joist girders, open web steel joists and 1-1/2″ galvanized metal deck.

Design and Pricing Assumptions:
Columns are 18′ high.
Joists are 5′-0″ O.C.
Joist girders and joists have bottom chords connected to columns. Roof deck is 1-1/2″, 22 gauge galvanized steel. Costs include bridging and bracing. Deflection is limited to 1/240 of the span.

Fireproofing is not included.
Costs/S.F. are based on a building 4 bays long and 4 bays wide.
Costs for bearing walls are not included.

Column costs are not included but are listed separately per S.F. of floor.

System Components	QUANTITY	UNIT	COST PER S.F.		
			MAT.	INST.	TOTAL
SYSTEM B1020 120 2050					
30′ X 30′ BAY SIZE, 20 PSF SUPERIMPOSED LOAD					
Joist girders, 40-ton job lots, shop primer, minimum	.475	Lb.	.49	.13	.62
Open web joists, horiz. bridging, T.L. lots, 30′ to 50′ span	.067	Lb.	1.51	.38	1.89
Cross bracing, rods, shop fabricated, 1″ diameter	.062	Lb.	.12	.13	.25
Metal decking, open, galv., 1-1/2″ deep, 22 ga., over 50 sq.	1.050	S.F.	2.14	.57	2.71
TOTAL			4.26	1.21	5.47

B1020 120	Steel Joists, Joist Girders & Deck on Columns & Walls							
	BAY SIZE (FT.) GIRD X JOISTS	SUPERIMPOSED LOAD (P.S.F.)	DEPTH (IN.)	TOTAL LOAD (P.S.F.)	COLUMN ADD	COST PER S.F.		
						MAT.	INST.	TOTAL
2050	30x30	20	17-1/2	40		4.26	1.21	5.47
2100					columns	.49	.09	.58
2150		30	17-1/2	50		4.59	1.30	5.89
2200					columns	.49	.09	.58
2250		40	21-1/2	60		4.82	1.36	6.18
2300					columns	.49	.09	.58
2350	30x35	20	32-1/2	40		4.82	1.35	6.17
2400					columns	.42	.08	.50
2450		30	36-1/2	50		5.05	1.41	6.46
2500					columns	.42	.08	.50
2550		40	36-1/2	60		5.40	1.50	6.90
2600					columns	.49	.09	.58
2650	35x30	20	36-1/2	40		4.48	1.27	5.75
2700					columns	.42	.08	.50
2750		30	36-1/2	50		4.93	1.38	6.31
2800					columns	.42	.08	.50
2850		40	36-1/2	60		6.55	1.79	8.34
2900					columns	.49	.09	.58
3000	35x35	20	36-1/2	40		5.20	1.66	6.86
3050					columns	.36	.07	.43
3100		30	36-1/2	50		5.40	1.70	7.10
3150					columns	.36	.07	.43
3200		40	36-1/2	60		5.85	1.83	7.68
3250					columns	.46	.09	.55

B1020 Roof Construction

| B1020 120 | Steel Joists, Joist Girders & Deck on Columns & Walls |

	BAY SIZE (FT.) GIRD X JOISTS	SUPERIMPOSED LOAD (P.S.F.)	DEPTH (IN.)	TOTAL LOAD (P.S.F.)	COLUMN ADD	COST PER S.F.		
						MAT.	INST.	TOTAL
3300	35x40	20	36-1/2	40		5.30	1.53	6.83
3350					columns	.37	.07	.44
3400		30	36-1/2	50		5.50	1.57	7.07
3450					columns	.41	.08	.49
3500		40	36-1/2	60		6	1.69	7.69
3550					columns	.41	.08	.49
3600	40x35	20	40-1/2	40		5.15	1.48	6.63
3650					columns	.37	.07	.44
3700		30	40-1/2	50		5.35	1.54	6.89
3750					columns	.41	.08	.49
3800		40	40-1/2	60		6.25	2.08	8.33
3850					columns	.41	.08	.49
3900	40x40	20	40-1/2	41		5.90	1.97	7.87
3950					columns	.36	.07	.43
4000		30	40-1/2	51		6.10	2.02	8.12
4050					columns	.36	.07	.43
4100		40	40-1/2	61		6.60	2.14	8.74
4150					columns	.36	.07	.43
4200	40x45	20	40-1/2	41		6	2.05	8.05
4250					columns	.32	.07	.39
4300		30	40-1/2	51		6.70	2.31	9.01
4350					columns	.32	.07	.39
4400		40	40-1/2	61		7.45	2.55	10
4450					columns	.36	.07	.43
4500	45x40	20	52-1/2	41		5.95	2.04	7.99
4550					columns	.32	.07	.39
4600		30	52-1/2	51		5.90	1.78	7.68
4650					columns	.36	.07	.43
4700		40	52-1/2	61		6.75	2.31	9.06
4750					columns	.36	.07	.43
4800	45x45	20	52-1/2	41		5.65	1.94	7.59
4850					columns	.28	.05	.33
4900		30	52-1/2	51		7.05	2.40	9.45
4950					columns	.32	.07	.39
5000		40	52-1/2	61		7.20	2.46	9.66
5050					columns	.36	.07	.43
5100	45x50	20	52-1/2	41		6.60	2.26	8.86
5150					columns	.25	.05	.30
5200		30	52-1/2	51		7.50	2.55	10.05
5250					columns	.29	.05	.34
5300		40	52-1/2	61		7.95	2.72	10.67
5350					columns	.37	.07	.44
5400	50x45	20	56-1/2	41		6.40	2.17	8.57
5450					columns	.25	.05	.30
5500		30	56-1/2	51		7.20	2.43	9.63
5550					columns	.29	.05	.34
5600		40	56-1/2	61		8.15	2.78	10.93
5650					columns	.35	.07	.42

B1020 Roof Construction

B1020 120	Steel Joists, Joist Girders & Deck on Columns & Walls

	BAY SIZE (FT.) GIRD X JOISTS	SUPERIMPOSED LOAD (P.S.F.)	DEPTH (IN.)	TOTAL LOAD (P.S.F.)	COLUMN ADD	COST PER S.F.		
						MAT.	INST.	TOTAL
5700	50x50	20	56-1/2	42		7.15	2.44	9.59
5750					columns	.26	.05	.31
5800		30	56-1/2	53		7.75	2.63	10.38
5850					columns	.33	.07	.40
5900		40	59	64		9.75	2.74	12.49
5950					columns	.36	.07	.43
6000	55x50	20	60-1/2	43		7.25	2.50	9.75
6050					columns	.45	.08	.53
6100		30	64-1/2	54		8.30	2.86	11.16
6150					columns	.55	.11	.66
6200		40	67	65		10	2.84	12.84
6250					columns	.55	.11	.66
6300	60x50	20	62-1/2	43		7.20	2.48	9.68
6350					columns	.41	.08	.49
6400		30	68-1/2	54		8.30	2.84	11.14
6450					columns	.50	.09	.59
6500		40	71	65		9.95	2.82	12.77
6550					columns	.54	.11	.65

B10 Superstructure

B1020 Roof Construction

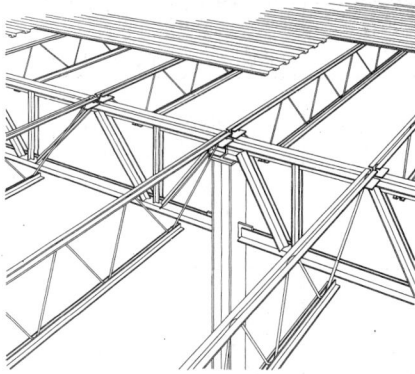

Description: Table below lists the cost per S.F. for a roof system supported on columns. Costs include joist girders, open web steel joists and 1-1/2″ galvanized metal deck.

Design and Pricing Assumptions:
Columns are 18′ high.
Joists are 5′-0″ O.C.
Joist girders and joists have bottom chords connected to columns. Roof deck is 1-1/2″, 22 gauge galvanized steel. Costs include bridging and bracing. Deflection is limited to 1/240 of the span. Fireproofing is not included.

Costs/S.F. are based on a building 4 bays long and 4 bays wide.

Costs for columns are not included, but are listed separately per S.F. of area.

System Components	QUANTITY	UNIT	COST PER S.F.		
			MAT.	INST.	TOTAL
SYSTEM B1020 124 2000					
30′X30′ BAY SIZE, 20 PSF S. LOAD, 17-1/2″ DEPTH, 40 PSF T. LOAD					
Joist girders, 40-ton job lots, shop primer, minimum	.758	Lb.	.78	.20	.98
Open web joists, spans to 30′ average	1.580	Lb.	1.78	.55	2.33
Cross bracing, rods, shop fabricated, 1″ diameter	.120	Lb.	.24	.25	.49
Metal decking, open, galv., 1-1/2″ deep, 22 ga., over 50 sq.	1.050	S.F.	2.14	.57	2.71
TOTAL			4.94	1.57	6.51

B1020 124	Steel Joists & Joist Girders on Columns							
	BAY SIZE (FT.) GIRD X JOISTS	SUPERIMPOSED LOAD (P.S.F.)	DEPTH (IN.)	TOTAL LOAD (P.S.F.)	COLUMN ADD	COST PER S.F.		
						MAT.	INST.	TOTAL
2000	30x30	20	17-1/2	40		4.94	1.57	6.51
2050					columns	1.36	.27	1.63
2100		30	17-1/2	50		5.35	1.69	7.04
2150					columns	1.36	.27	1.63
2200		40	21-1/2	60		5.50	1.74	7.24
2250					columns	1.36	.27	1.63
2300	30x35	20	32-1/2	40		5.15	1.52	6.67
2350					columns	1.17	.23	1.40
2400		30	36-1/2	50		5.45	1.58	7.03
2450					columns	1.17	.23	1.40
2500		40	36-1/2	60		5.80	1.69	7.49
2550					columns	1.36	.27	1.63
2600	35x30	20	36-1/2	40		5.05	1.59	6.64
2650					columns	1.17	.23	1.40
2700		30	36-1/2	50		5.60	1.75	7.35
2750					columns	1.17	.23	1.40
2800		40	36-1/2	60		5.70	1.79	7.49
2850					columns	1.36	.27	1.63
2900	35x35	20	36-1/2	40		5.60	1.81	7.41
2950					columns	1	.19	1.19
3000		30	36-1/2	50		5.80	1.85	7.65
3050					columns	1.17	.23	1.40
3100		40	36-1/2	60		6.30	1.99	8.29
3150					columns	1.29	.25	1.54

B1020 Roof Construction

| B1020 124 | Steel Joists & Joist Girders on Columns |

	BAY SIZE (FT.) GIRD X JOISTS	SUPERIMPOSED LOAD (P.S.F.)	DEPTH (IN.)	TOTAL LOAD (P.S.F.)	COLUMN ADD	COST PER S.F.		
						MAT.	INST.	TOTAL
3200	35x40	20	36-1/2	40		6.40	2.30	8.70
3250					columns	1.02	.20	1.22
3300		30	36-1/2	50		6.60	2.36	8.96
3350					columns	1.13	.21	1.34
3400		40	36-1/2	60		7.20	2.55	9.75
3450					columns	1.13	.21	1.34
3500	40x35	20	40-1/2	40		5.90	1.87	7.77
3550					columns	1.02	.20	1.22
3600		30	40-1/2	50		6.05	1.91	7.96
3650					columns	1.13	.21	1.34
3700		40	40-1/2	60		6.45	2.01	8.46
3750					columns	1.13	.21	1.34
3800	40x40	20	40-1/2	41		6.50	2.30	8.80
3850					columns	.99	.19	1.18
3900		30	40-1/2	51		6.70	2.37	9.07
3950					columns	.99	.19	1.18
4000		40	40-1/2	61		7.25	2.54	9.79
4050					columns	.99	.19	1.18
4100	40x45	20	40-1/2	41		6.95	2.58	9.53
4150					columns	.88	.17	1.05
4200		30	40-1/2	51		7.50	2.79	10.29
4250					columns	.88	.17	1.05
4300		40	40-1/2	61		8.35	3.06	11.41
4350					columns	.99	.19	1.18
4400	45x40	20	52-1/2	41		6.75	2.51	9.26
4450					columns	.88	.17	1.05
4500		30	52-1/2	51		7	2.58	9.58
4550					columns	.88	.17	1.05
4600		40	52-1/2	61		7.75	2.84	10.59
4650					columns	.99	.19	1.18
4700	45x45	20	52-1/2	41		7.05	2.62	9.67
4750					columns	.78	.15	.93
4800		30	52-1/2	51		7.90	2.89	10.79
4850					columns	.88	.17	1.05
4900		40	52-1/2	61		8.10	2.96	11.06
4950					columns	1.01	.20	1.21
5000	45x50	20	52-1/2	41		7.30	2.67	9.97
5050					columns	.70	.13	.83
5100		30	52-1/2	51		8.35	3.01	11.36
5150					columns	.79	.15	.94
5200		40	52-1/2	61		8.90	3.19	12.09
5250					columns	.91	.17	1.08
5300	50x45	20	56-1/2	41		7.15	2.59	9.74
5350					columns	.70	.13	.83
5400		30	56-1/2	51		8	2.89	10.89
5450					columns	.79	.15	.94
5500		40	56-1/2	61		9.15	3.26	12.41
5550					columns	.91	.17	1.08

B10 Superstructure

B1020 Roof Construction

B1020 124	Steel Joists & Joist Girders on Columns

	BAY SIZE (FT.) GIRD X JOISTS	SUPERIMPOSED LOAD (P.S.F.)	DEPTH (IN.)	TOTAL LOAD (P.S.F.)	COLUMN ADD	COST PER S.F.		
						MAT.	INST.	TOTAL
5600	50x50	20	56-1/2	42		7.95	2.88	10.83
5650					columns	.71	.13	.84
5700		30	56-1/2	53		8.65	3.11	11.76
5750					columns	.92	.17	1.09
5800		40	59	64		8.50	2.61	11.11
5850					columns	1	.19	1.19
5900	55x50	20	60-1/2	43		8	2.90	10.90
5950					columns	.74	.15	.89
6000		30	64-1/2	54		9.15	3.29	12.44
6050					columns	.91	.17	1.08
6100		40	67	65		10.80	3.23	14.03
6150					columns	.91	.17	1.08
6200	60x50	20	62-1/2	43		8	2.89	10.89
6250					columns	.68	.13	.81
6300		30	68-1/2	54		9.25	3.29	12.54
6350					columns	.83	.16	.99
6400		40	71	65		10.90	3.23	14.13
6450					columns	.90	.17	1.07

B1020 Roof Construction

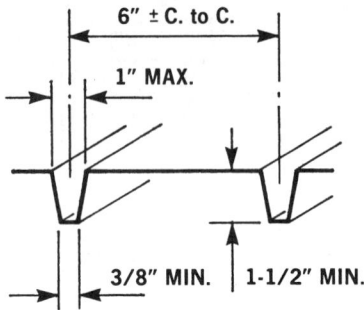

NARROW RIB DECK

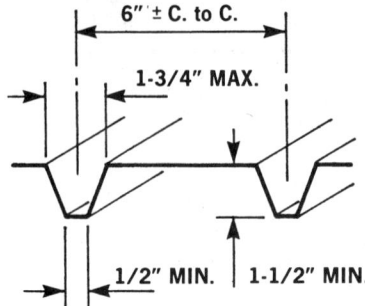

INTERMEDIATE RIB DECK

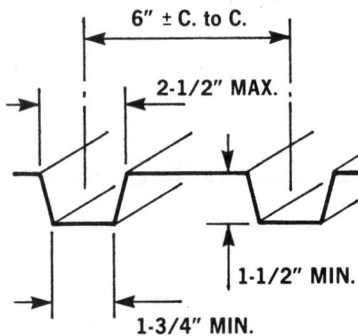

WIDE RIB DECK

General: Steel roof deck has the principal advantage of lightweight and speed of erection. It is covered by varying thicknesses of rigid insulation and roofing material.

Design Assumptions: Maximum spans shown in tables are based on construction and maintenance loads as specified by the Steel Deck Institute. Deflection is limited to 1/240 of the span.

Note: Costs shown in tables are based on projects over 5000 S.F. Under 5000 S.F. multiply costs by 1.16. For over four stories or congested site add 50% to the installation costs.

B1020 125		Narrow Rib, Single Span Steel Deck						
	TOTAL LOAD (P.S.F.)	MAX. SPAN (FT. - IN.)	DECK GAUGE	WEIGHT (P.S.F.)	DEPTH (IN.)	COST PER S.F.		
						MAT.	INST.	TOTAL
1250	85	3-10	22	1.80	1-1/2	2.04	.54	2.58
1300	66	4-10	20	2.15	1-1/2	2.40	.63	3.03
1350	60	5-11	18	2.85	1-1/2	3.10	.63	3.73
1360								

B1020 126		Narrow Rib, Multi Span Steel Deck						
	TOTAL LOAD (P.S.F.)	MAX. SPAN (FT.-IN.)	DECK GAUGE	WEIGHT (P.S.F.)	DEPTH (IN.)	COST PER S.F.		
						MAT.	INST.	TOTAL
1500	69	4-9	22	1.80	1-1//2	2.04	.54	2.58
1550	55	5-11	20	2.15	1-1/2	2.40	.63	3.03
1600	55	6-11	18	2.85	1-1/2	3.10	.63	3.73
1610								

B1020 127		Intermediate Rib, Single Span Steel Deck						
	TOTAL LOAD (P.S.F.)	MAX. SPAN (FT.-IN.)	DECK GAUGE	WEIGHT (P.S.F.)	DEPTH (IN.)	COST PER S.F.		
						MAT.	INST.	TOTAL
1750	72	4-6	22	1.75	1-1/2	2.04	.54	2.58
1800	65	5-3	20	2.10	1-1/2	2.40	.63	3.03
1850	64	6-2	18	2.80	1-1/2	3.10	.63	3.73
1860								

B1020 Roof Construction

B1020 128	Intermediate Rib, Multi Span Steel Deck							
	TOTAL LOAD (P.S.F.)	MAX. SPAN (FT.-IN.)	DECK GAUGE	WEIGHT (P.S.F.)	DEPTH (IN.)	COST PER S.F.		
						MAT.	INST.	TOTAL
2000	60	5-6	22	1.75	1-1/2	2.04	.54	2.58
2050	57	6-3	20	2.10	1-1/2	2.40	.63	3.03
2100	57	7-4	18	2.80	1-1/2	3.10	.63	3.73
2110								

B1020 129	Wide Rib, Single Span Steel Deck							
	TOTAL LOAD (P.S.F.)	MAX. SPAN (FT.-IN.)	DECK GAUGE	WEIGHT (P.S.F.)	DEPTH (IN.)	COST PER S.F.		
						MAT.	INST.	TOTAL
2250	66	5-6	22	1.80	1-1/2	2.04	.54	2.58
2300	56	6-3	20	2.15	1-1/2	2.40	.63	3.03
2350	46	7-6	18	2.85	1-1/2	3.10	.63	3.73
2375								

B1020 130	Wide Rib, Multi Span Steel Deck							
	TOTAL LOAD (P.S.F.)	MAX. SPAN (FT.-IN.)	DECK GAUGE	WEIGHT (P.S.F.)	DEPTH (IN.)	COST PER S.F.		
						MAT.	INST.	TOTAL
2500	76	6-6	22	1.80	1-1/2	2.04	.54	2.58
2550	64	7-5	20	2.15	1-1/2	2.40	.63	3.03
2600	55	8-10	18	2.80	1-1/2	3.10	.63	3.73
2610								

B1020 Roof Construction

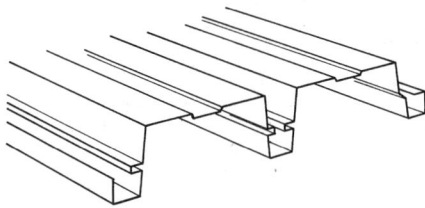

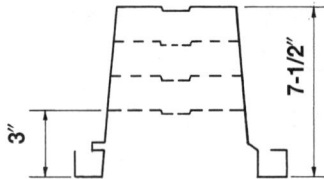

Design & Pricing Assumptions:
Costs in table below are based upon 5000 S.F. or more.
Deflection is limited to 1/240 of the span.

B1020 131		Single Span, Steel Deck						
	TOTAL LOAD (P.S.F.)	DECK SUPPORT SPACING (FT.)	DEPTH (IN.)	DECK GAUGE	WEIGHT (P.S.F.)	COST PER S.F.		
						MAT.	INST.	TOTAL
0850	30	10	3	20	2.5	4.53	.80	5.33
0900		12	3	18	3.0	5.85	.86	6.71
0950		14	3	16	4.0	7.75	.91	8.66
1000		16	4.5	20	3.0	7.25	1	8.25
1050		18	4.5	18	4.0	9.35	1.11	10.46
1100		20	4.5	16	5.0	9.70	1.60	11.30
1200		24	6.0	16	6.0	10	1.42	11.42
1300		28	7.5	16	7.0	11	1.71	12.71
1350	40	10	3.0	20	2.5	4.53	.80	5.33
1400		12	3.0	18	3.0	5.85	.86	6.71
1450		14	3.0	16	4.0	7.75	.91	8.66
1500		16	4.5	20	3.0	7.25	1	8.25
1550		18	4.5	18	4.0	9.35	1.11	10.46
1600		20	6.0	16	6.0	10	1.42	11.42
1700		24	7.5	16	7.0	11	1.71	12.71
1750	50	10	3.0	20	2.5	4.53	.80	5.33
1800		12	3.0	18	3.0	5.85	.86	6.71
1850		14	3.0	16	4.0	7.75	.91	8.66
1900		16	4.5	18	4.0	9.35	1.11	10.46
1950		18	4.5	16	5.0	9.70	1.60	11.30
2000		20	6.0	16	6.0	9.95	1.01	10.96
2100		24	7.5	16	7.0	11	1.71	12.71

B1020 132		Double Span, Steel Deck						
	TOTAL LOAD (P.S.F.)	DECK SUPPORT SPACING (FT.)	DEPTH (IN.)	DECK GAUGE	WEIGHT (P.S.F.)	COST PER S.F.		
						MAT.	INST.	TOTAL
2250	30	10	3	20	2.5	4.53	.80	5.33
2300		12	3	20	2.5	4.53	.80	5.33
2350		14	3	18	3.0	5.85	.86	6.71
2400		16	3	16	4.0	7.75	.91	8.66
2450	40	10	3	20	2.5	4.53	.80	5.33
2500		12	3	20	2.5	4.53	.80	5.33
2550		14	3	18	3.0	5.85	.86	6.71
2600		16	3	16	4.0	7.75	.91	8.66
2650	50	10	3	20	2.5	4.53	.80	5.33
2700		12	3	18	3.0	5.85	.86	6.71
2750		14	3	16	4.0	7.75	.91	8.66
2800		16	4.5	20	4.0	7.25	1	8.25

B10 Superstructure

B1020 Roof Construction

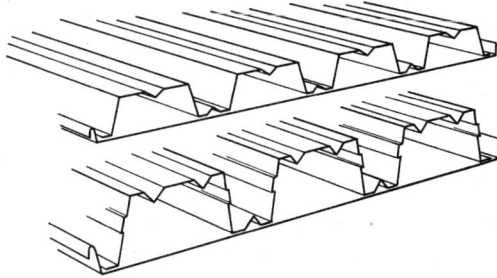

Design & Pricing Assumptions:
Costs in table below are based upon
15,000 S.F. or more.
No utility outlets or modifications included
in cost.
Deflection is limited to 1/240 of the span.

B1020 134	Cellular Steel Deck, Single Span							
	TOTAL LOAD (P.S.F.)	DECK SUPPORT SPACING (FT.)	DEPTH (IN.)	WEIGHT (P.S.F.)		COST PER S.F.		
						MAT.	INST.	TOTAL
1150	30	10	1-5/8	4		12.60	1.88	14.48
1200		12	1-5/8	5		13.15	1.96	15.11
1250		14	3	5		14.95	2.02	16.97
1300		16	3	5		14.95	2.11	17.06
1350		18	3	5		14.95	2.11	17.06
1400		20	3	6		16.80	2.22	19.02
1450	40	8	1-5/8	4		12.60	1.88	14.48
1500		10	1-5/8	5		13.15	1.96	15.11
1550		12	3	5		14.95	2.11	17.06
1600		14	3	5		14.95	2.11	17.06
1650		16	3	5		14.95	2.11	17.06
1700		18	3	6		16.80	2.22	19.02
1750	50	8	1-5/8	4		12.60	1.88	14.48
1800		10	1-5/8	5		13.15	1.96	15.11
1850		12	3	4		14.95	2.02	16.97
1900		14	3	5		14.95	2.11	17.06
1950		16	3	6		16.80	2.22	19.02
1960								

B1020 135	Cellular Steel Deck, Double Span							
	TOTAL LOAD (P.S.F.)	DECK SUPPORT SPACING (FT.)	DEPTH (IN.)	WEIGHT (P.S.F.)		COST PER S.F.		
						MAT.	INST.	TOTAL
2150	30	10	1-5/8	4		12.60	1.88	14.48
2200		12	1-5/8	5		13.15	1.96	15.11
2250		14	1-5/8	5		13.15	1.96	15.11
2300		16	3	5		14.95	2.11	17.06
2350		18	3	5		14.95	2.11	17.06
2400		20	3	6		16.80	2.22	19.02
2450	40	8	1-5/8	4		12.60	1.88	14.48
2500		10	1-5/8	5		13.15	1.96	15.11
2550		12	1-5/8	5		13.15	1.96	15.11
2600		14	3	5		14.95	2.11	17.06
2650		16	3	5		14.95	2.11	17.06
2700		18	3	6		16.80	2.22	19.02

B1020 Roof Construction

B1020 135 — Cellular Steel Deck, Double Span

	TOTAL LOAD (P.S.F.)	DECK SUPPORT SPACING (FT.)	DEPTH (IN.)	WEIGHT (P.S.F.)		COST PER S.F.		
						MAT.	INST.	TOTAL
2750	50	8	1-5/8	4		12.60	1.88	14.48
2800		10	1-5/8	5		13.15	1.96	15.11
2850		12	1-5/8	5		13.15	1.96	15.11
2900		14	3	5		14.95	2.11	17.06
2950		16	3	6		16.80	2.22	19.02
2960								

B1020 136 — Cellular Steel Deck, Triple Span

	TOTAL LOAD (P.S.F.)	DECK SUPPORT SPACING (FT.)	DEPTH (IN.)	WEIGHT (P.S.F.)		COST PER S.F.		
						MAT.	INST.	TOTAL
3200	30	10	1-5/8	4		12.60	1.88	14.48
3210								
3250	40	8	1-5/8	4		12.60	1.88	14.48
3300		10	1-5/8	4		12.60	1.88	14.48
3350		12	1-5/8	5		13.15	1.96	15.11
3400	50	8	1-5/8	4		12.60	1.88	14.48
3450		10	1-5/8	5		13.15	1.96	15.11
3500		12	1-5/8	5		13.15	1.96	15.11

B1020 310 — Canopies

	Canopies	COST PER S.F.		
		MAT.	INST.	TOTAL
0100	Canopies, wall hung, prefinished aluminum, 8' x 10'	2,350	1,500	3,850
0110	12' x 40'	10,800	3,250	14,050
0120	Canopies, wall hung, canvas awnings, with frame and lettering	82	11	93
0130	Freestanding (walkways), vinyl coated steel, 12' wide	25	7.80	32.80

B20 Exterior Enclosure

B2010 Exterior Walls

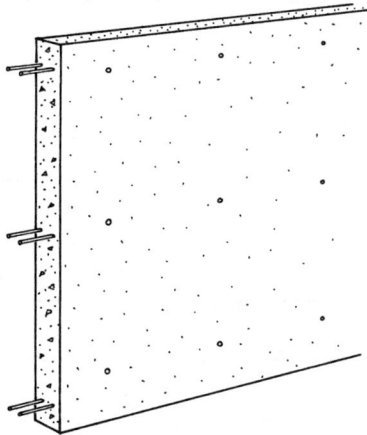

The table below describes a concrete wall system for exterior closure. There are several types of wall finishes priced from plain finish to a finish with 3/4" rustication strip.

Design Assumptions:
Conc. f'c = 3000 to 5000 psi
Reinf. fy = 60,000 psi

System Components	QUANTITY	UNIT	COST PER S.F.		
			MAT.	INST.	TOTAL
SYSTEM B2010 101 2100					
CONC. WALL, REINFORCED, 8' HIGH, 6" THICK, PLAIN FINISH, 3,000 PSI					
Forms in place, wall, job built plyform to 8' high, 4 uses	2.000	SFCA	2.10	11.50	13.60
Reinforcing in place, walls, #3 to #7	.752	Lb.	.67	.29	.96
Concrete ready mix, regular weight, 3000 psi	.018	C.Y.	2		2
Place and vibrate concrete, walls 6" thick, pump	.018	C.Y.		.75	.75
Finish wall, break ties, patch voids	2.000	S.F.	.06	1.66	1.72
TOTAL			4.83	14.20	19.03

B2010 101	Cast In Place Concrete	COST PER S.F.		
		MAT.	INST.	TOTAL
2100	Conc wall reinforced, 8' high, 6" thick, plain finish, 3000 PSI	4.83	14.20	19.03
2200	4000 PSI	4.92	14.20	19.12
2300	5000 PSI	5.05	14.20	19.25
2400	Rub concrete 1 side, 3000 PSI	4.83	16.75	21.58
2500	4000 PSI	4.92	16.75	21.67
2600	5000 PSI	5.05	16.75	21.80
2700	Aged wood liner, 3000 PSI	6	16.15	22.15
2800	4000 PSI	6.10	16.15	22.25
2900	5000 PSI	6.20	16.15	22.35
3000	Sand blast light 1 side, 3000 PSI	5.60	16.10	21.70
3100	4000 PSI	5.70	16.10	21.80
3300	5000 PSI	5.80	16.10	21.90
3400	Sand blast heavy 1 side, 3000 PSI	6.40	19.90	26.30
3500	4000 PSI	6.45	19.90	26.35
3600	5000 PSI	6.60	19.90	26.50
3700	3/4" bevel rustication strip, 3000 PSI	4.98	15.05	20.03
3800	4000 PSI	5.05	15.05	20.10
3900	5000 PSI	5.20	15.05	20.25
4000	8" thick, plain finish, 3000 PSI	5.80	14.55	20.35
4100	4000 PSI	5.90	14.55	20.45
4200	5000 PSI	6.05	14.55	20.60
4300	Rub concrete 1 side, 3000 PSI	5.80	17.15	22.95
4400	4000 PSI	5.90	17.15	23.05
4500	5000 PSI	6.05	17.15	23.20
4550	8" thick, aged wood liner, 3000 PSI	6.95	16.50	23.45
4600	4000 PSI	7.05	16.50	23.55

B2010 Exterior Walls

B2010 101	Cast In Place Concrete	COST PER S.F.		
		MAT.	INST.	TOTAL
4700	5000 PSI	7.20	16.50	23.70
4750	Sand blast light 1 side, 3000 PSI	6.55	16.50	23.05
4800	4000 PSI	6.65	16.50	23.15
4900	5000 PSI	6.80	16.50	23.30
5000	Sand blast heavy 1 side, 3000 PSI	7.35	20	27.35
5100	4000 PSI	7.45	20	27.45
5200	5000 PSI	7.60	20	27.60
5300	3/4" bevel rustication strip, 3000 PSI	5.95	15.40	21.35
5400	4000 PSI	6.05	15.40	21.45
5500	5000 PSI	6.20	15.40	21.60
5600	10" thick, plain finish, 3000 PSI	6.70	14.90	21.60
5700	4000 PSI	6.85	14.90	21.75
5800	5000 PSI	7	14.90	21.90
5900	Rub concrete 1 side, 3000 PSI	6.70	17.50	24.20
6000	4000 PSI	6.85	17.50	24.35
6100	5000 PSI	7	17.50	24.50
6200	Aged wood liner, 3000 PSI	7.85	16.85	24.70
6300	4000 PSI	8	16.85	24.85
6400	5000 PSI	8.15	16.85	25
6500	Sand blast light 1 side, 3000 PSI	7.45	16.85	24.30
6600	4000 PSI	7.60	16.85	24.45
6700	5000 PSI	7.80	16.85	24.65
6800	Sand blast heavy 1 side, 3000 PSI	8.25	20.50	28.75
6900	4000 PSI	8.40	20.50	28.90
7000	5000 PSI	8.55	20.50	29.05
7100	3/4" bevel rustication strip, 3000 PSI	6.85	15.75	22.60
7200	4000 PSI	7	15.75	22.75
7300	5000 PSI	7.15	15.75	22.90
7400	12" thick, plain finish, 3000 PSI	7.75	15.35	23.10
7500	4000 PSI	7.95	15.35	23.30
7600	5000 PSI	8.15	15.35	23.50
7700	Rub concrete 1 side, 3000 PSI	7.75	17.90	25.65
7800	4000 PSI	7.95	17.90	25.85
7900	5000 PSI	8.15	17.90	26.05
8000	Aged wood liner, 3000 PSI	8.90	17.25	26.15
8100	4000 PSI	9.10	17.25	26.35
8200	5000 PSI	9.30	17.25	26.55
8300	Sand blast light 1 side, 3000 PSI	8.50	17.25	25.75
8400	4000 PSI	8.70	17.25	25.95
8500	5000 PSI	8.90	17.25	26.15
8600	Sand blast heavy 1 side, 3000 PSI	9.30	21	30.30
8700	4000 PSI	9.50	21	30.50
8800	5000 PSI	9.70	21	30.70
8900	3/4" bevel rustication strip, 3000 PSI	7.90	16.20	24.10
9000	4000 PSI	8.10	16.20	24.30
9500	5000 PSI	8.30	16.20	24.50

B2010 Exterior Walls

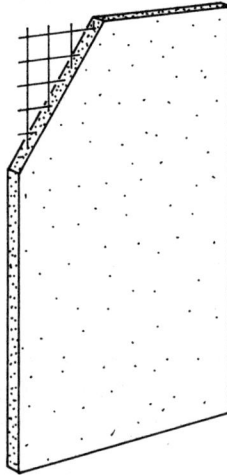

Precast concrete wall panels are either solid or insulated with plain, colored or textured finishes. Transportation is an important cost factor. Prices below are based on delivery within fifty miles of a plant. Engineering data is available from fabricators to assist with construction details. Usual minimum job size for economical use of panels is about 5000 S.F. Small jobs can double the prices below. For large, highly repetitive jobs, deduct up to 15% from the prices below.

B2010 102		Flat Precast Concrete						
	THICKNESS (IN.)	PANEL SIZE (FT.)	FINISHES	RIGID INSULATION (IN)	TYPE	COST PER S.F.		
						MAT.	INST.	TOTAL
3000	4	5x18	smooth gray	none	low rise	9.05	7.20	16.25
3050		6x18				7.55	6	13.55
3100		8x20				15.10	2.90	18
3150		12x20				14.30	2.74	17.04
3200	6	5x18	smooth gray	2	low rise	10.25	7.65	17.90
3250		6x18				8.75	6.50	15.25
3300		8x20				16.55	3.49	20.04
3350		12x20				15	3.20	18.20
3400	8	5x18	smooth gray	2	low rise	21.50	4.41	25.91
3450		6x18				20	4.20	24.20
3500		8x20				18.45	3.86	22.31
3550		12x20				16.80	3.55	20.35
3600	4	4x8	white face	none	low rise	36	4.15	40.15
3650		8x8				27	3.11	30.11
3700		10x10				23.50	2.74	26.24
3750		20x10				21.50	2.47	23.97
3800	5	4x8	white face	none	low rise	36.50	4.24	40.74
3850		8x8				27.50	3.21	30.71
3900		10x10				24.50	2.83	27.33
3950		20x20				22.50	2.60	25.10
4000	6	4x8	white face	none	low rise	38	4.39	42.39
4050		8x8				29	3.33	32.33
4100		10x10				25.50	2.94	28.44
4150		20x10				23.50	2.70	26.20
4200	6	4x8	white face	2	low rise	39.50	4.95	44.45
4250		8x8				30	3.89	33.89
4300		10x10				26.50	3.50	30
4350		20x10				23.50	2.70	26.20
4400	7	4x8	white face	none	low rise	39	4.49	43.49
4450		8x8				30	3.46	33.46
4500		10x10				27	3.10	30.10
4550		20x10				24.50	2.83	27.33
4600	7	4x8	white face	2	low rise	40	5.05	45.05
4650		8x8				31	4.02	35.02
4700		10x10				28	3.66	31.66

B20 Exterior Enclosure

B2010 Exterior Walls

B2010 102			Flat Precast Concrete					

	THICKNESS (IN.)	PANEL SIZE (FT.)	FINISHES	RIGID INSULATION (IN)	TYPE	COST PER S.F.		
						MAT.	INST.	TOTAL
4750	7	20x10	white face	2	low rise	26	3.39	29.39
4800	8	4x8	white face	none	low rise	40	4.60	44.60
4850		8x8				30.50	3.54	34.04
4900		10x10				27.50	3.19	30.69
4950		20x10				25.50	2.93	28.43
5000	8	4x8	white face	2	low rise	41	5.15	46.15
5050		8x8				32	4.10	36.10
5100		10x10				29	3.75	32.75
5150		20x10				26.50	3.49	29.99

B2010 103			Fluted Window or Mullion Precast Concrete					

	THICKNESS (IN.)	PANEL SIZE (FT.)	FINISHES	RIGID INSULATION (IN)	TYPE	COST PER S.F.		
						MAT.	INST.	TOTAL
5200	4	4x8	smooth gray	none	high rise	20.50	16.45	36.95
5250		8x8				14.80	11.75	26.55
5300		10x10				29	5.55	34.55
5350		20x10				25.50	4.87	30.37
5400	5	4x8	smooth gray	none	high rise	21	16.70	37.70
5450		8x8				15.35	12.15	27.50
5500		10x10				30	5.80	35.80
5550		20x10				26.50	5.10	31.60
5600	6	4x8	smooth gray	none	high rise	21.50	17.20	38.70
5650		8x8				15.85	12.60	28.45
5700		10x10				31	5.95	36.95
5750		20x10				27.50	5.30	32.80
5800	6	4x8	smooth gray	2	high rise	23	17.75	40.75
5850		8x8				17.10	13.15	30.25
5900		10x10				32	6.50	38.50
5950		20x10				29	5.85	34.85
6000	7	4x8	smooth gray	none	high rise	22	17.55	39.55
6050		8x8				16.30	12.90	29.20
6100		10x10				32.50	6.20	38.70
6150		20x10				28.50	5.45	33.95
6200	7	4x8	smooth gray	2	high rise	23.50	18.10	41.60
6250		8x8				17.55	13.45	31
6300		10x10				33.50	6.75	40.25
6350		20x10				29.50	6	35.50
6400	8	4x8	smooth gray	none	high rise	22.50	17.75	40.25
6450		8x8				16.70	13.25	29.95
6500		10x10				33.50	6.40	39.90
6550		20x10				30	5.70	35.70
6600	8	4x8	smooth gray	2	high rise	23.50	18.35	41.85
6650		8x8				17.95	13.80	31.75
6700		10x10				34.50	6.95	41.45
6750		20x10				31	6.30	37.30

B2010 104			Ribbed Precast Concrete					

	THICKNESS (IN.)	PANEL SIZE (FT.)	FINISHES	RIGID INSULATION (IN.)	TYPE	COST PER S.F.		
						MAT.	INST.	TOTAL
6800	4	4x8	aggregate	none	high rise	27	16.45	43.45
6850		8x8				19.75	12	31.75
6900		10x10				34	4.35	38.35

B20 Exterior Enclosure

B2010 Exterior Walls

B2010 104				Ribbed Precast Concrete				

	THICKNESS (IN.)	PANEL SIZE (FT.)	FINISHES	RIGID INSULATION(IN.)	TYPE	COST PER S.F.		
						MAT.	INST.	TOTAL
6950	4	20x10	aggregate	none	high rise	30	3.85	33.85
7000	5	4x8	aggregate	none	high rise	27.50	16.70	44.20
7050		8x8				20.50	12.35	32.85
7100		10x10				35	4.51	39.51
7150		20x10				31.50	4.03	35.53
7200	6	4x8	aggregate	none	high rise	28	17.05	45.05
7250		8x8				21	12.70	33.70
7300		10x10				36	4.66	40.66
7350		20x10				32.50	4.18	36.68
7400	6	4x8	aggregate	2	high rise	29.50	17.60	47.10
7450		8x8				22	13.25	35.25
7500		10x10				37.50	5.20	42.70
7550		20x10				33.50	4.74	38.24
7600	7	4x8	aggregate	none	high rise	28.50	17.45	45.95
7650		8x8				21.50	13	34.50
7700		10x10				37.50	4.80	42.30
7750		20x10				33.50	4.29	37.79
7800	7	4x8	aggregate	2	high rise	30	18	48
7850		8x8				22.50	13.55	36.05
7900		10x10				38.50	5.35	43.85
7950		20x10				34.50	4.85	39.35
8000	8	4x8	aggregate	none	high rise	29	17.75	46.75
8050		8x8				22	13.40	35.40
8100		10x10				38.50	4.97	43.47
8150		20x10				35	4.50	39.50
8200	8	4x8	aggregate	2	high rise	30.50	18.30	48.80
8250		8x8				23.50	13.95	37.45
8300		10x10				40	5.55	45.55
8350		20x10				36	5.05	41.05

B2010 105				Precast Concrete Specialties				

	TYPE	SIZE				COST PER L.F.		
						MAT.	INST.	TOTAL
8400	Coping, precast	6"wide				19.15	11.75	30.90
8450	Stock units	10" wide				19.15	12.60	31.75
8460		12" wide				22.50	13.55	36.05
8480		14" wide				18.80	14.70	33.50
8500	Window sills	6" wide				12.95	12.60	25.55
8550	Precast	10" wide				17.15	14.70	31.85
8600		14" wide				16.50	17.65	34.15
8610								

B2010 Exterior Walls

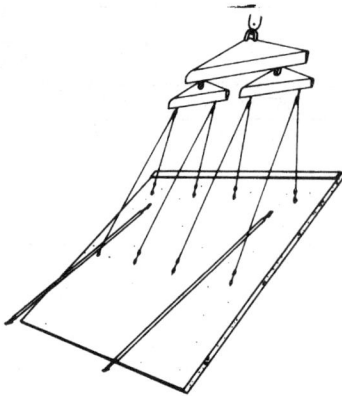

The advantage of tilt up construction is in the low cost of forms and placing of concrete and reinforcing. Tilt up has been used for several types of buildings, including warehouses, stores, offices, and schools. The panels are cast in forms on the ground, or floor slab. Most jobs use 5-1/2″ thick solid reinforced concrete panels.

Design Assumptions:
Conc. f'c = 3000 psi
Reinf. fy = 60,000

System Components	QUANTITY	UNIT	COST PER S.F.		
			MAT.	INST.	TOTAL
SYSTEM B2010 106 3200					
TILT-UP PANELS, 20′X25′, BROOM FINISH, 5-1/2″ THICK, 3000 PSI					
Apply liquid bond release agent	500.000	S.F.	10	60	70
Edge forms in place for slab on grade	120.000	L.F.	50.40	375.60	426
Reinforcing in place	.350	Ton	568.75	358.75	927.50
Footings, form braces, steel	1.000	Set	1,442		1,442
Slab lifting inserts	1.000	Set	104		104
Framing, less than 4″ angles	1.000	Set	94.05	746.70	840.75
Concrete ready mix, regular weight, 3000 psi	8.550	C.Y.	949.05		949.05
Place and vibrate concrete for slab on grade, 4″ thick, direct chute	8.550	C.Y.		192.55	192.55
Finish floor, monolithic broom finish	500.000	S.F.		350	350
Cure with curing compound, sprayed	500.000	S.F.	28.75	41.25	70
Erection crew	.058	Day		590.15	590.15
TOTAL	500.000		3,247	2,715	5,962
Cost per S.F.			6.49	5.43	11.92

B2010 106	Tilt-Up Concrete Panel	COST PER S.F.		
		MAT.	INST.	TOTAL
3200	Tilt up conc panels, broom finish, 5-1/2″ thick, 3000 PSI	6.50	5.45	11.95
3250	5000 PSI	6.55	5.35	11.90
3300	6″ thick, 3000 PSI	7.25	5.60	12.85
3350	5000 PSI	7.30	5.50	12.80
3400	7-1/2″ thick, 3000 PSI	9.55	5.80	15.35
3450	5000 PSI	9.65	5.70	15.35
3500	8″ thick, 3000 PSI	10.30	5.95	16.25
3550	5000 PSI	10.45	5.85	16.30
3700	Steel trowel finish, 5-1/2″ thick, 3000 PSI	6.50	5.50	12
3750	5000 PSI	6.55	5.40	11.95
3800	6″ thick, 3000 PSI	7.25	5.65	12.90
3850	5000 PSI	7.30	5.55	12.85
3900	7-1/2″ thick, 3000 PSI	9.55	5.85	15.40
3950	5000 PSI	9.65	5.75	15.40
4000	8″ thick, 3000 PSI	10.30	6	16.30
4050	5000 PSI	10.45	5.90	16.35
4200	Exp. aggregate finish, 5-1/2″ thick, 3000 PSI	6.90	5.60	12.50
4250	5000 PSI	6.95	5.50	12.45
4300	6″ thick, 3000 PSI	7.65	5.70	13.35
4350	5000 PSI	7.70	5.60	13.30
4400	7-1/2″ thick, 3000 PSI	10	5.90	15.90
4450	5000 PSI	10.05	5.80	15.85

B20 Exterior Enclosure

B2010 Exterior Walls

B2010 106	Tilt-Up Concrete Panel	COST PER S.F.		
		MAT.	INST.	TOTAL
4500	8" thick, 3000 PSI	10.75	6.05	16.80
4550	5000 PSI	10.85	5.95	16.80
4600	Exposed aggregate & vert. rustication 5-1/2" thick, 3000 PSI	9.10	6.85	15.95
4650	5000 PSI	9.15	6.75	15.90
4700	6" thick, 3000 PSI	9.85	7	16.85
4750	5000 PSI	9.90	6.90	16.80
4800	7-1/2" thick, 3000 PSI	12.20	7.20	19.40
4850	5000 PSI	12.25	7.10	19.35
4900	8" thick, 3000 PSI	12.95	7.35	20.30
4950	5000 PSI	13.05	7.25	20.30
5000	Vertical rib & light sandblast, 5-1/2" thick, 3000 PSI	8.40	9	17.40
5050	5000 PSI	8.45	8.90	17.35
5100	6" thick, 3000 PSI	9.15	9.15	18.30
5150	5000 PSI	9.20	9.05	18.25
5200	7-1/2" thick, 3000 PSI	11.50	9.35	20.85
5250	5000 PSI	11.55	9.25	20.80
5300	8" thick, 3000 PSI	12.20	9.50	21.70
5350	5000 PSI	12.35	9.40	21.75
6000	Broom finish w/2" polystyrene insulation, 6" thick, 3000 PSI	6.15	6.70	12.85
6050	5000 PSI	6.25	6.70	12.95
6100	Broom finish 2" fiberplank insulation, 6" thick, 3000 PSI	6.85	6.60	13.45
6150	5000 PSI	7	6.60	13.60
6200	Exposed aggregate w/2" polystyrene insulation, 6" thick, 3000 PSI	6.40	6.60	13
6250	5000 PSI	6.50	6.60	13.10
6300	Exposed aggregate 2" fiberplank insulation, 6" thick, 3000 PSI	7.10	6.55	13.65
6350	5000 PSI	7.25	6.55	13.80

B2010 Exterior Walls

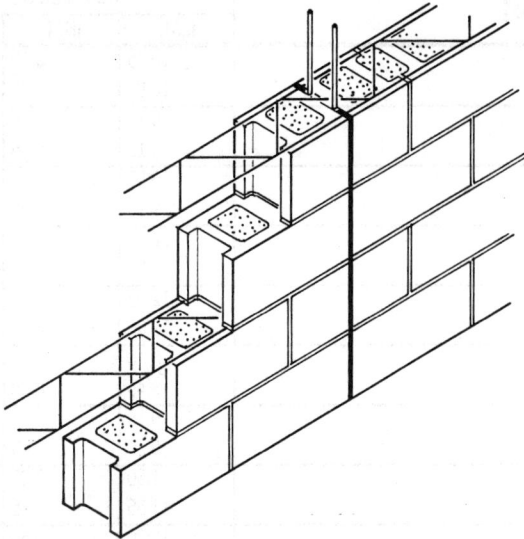

Exterior concrete block walls are defined in the following terms: structural reinforcement, weight, percent solid, size, strength and insulation. Within each of these categories, two to four variations are shown. No costs are included for brick shelf or relieving angles.

System Components	QUANTITY	UNIT	COST PER S.F.		
			MAT.	INST.	TOTAL
SYSTEM B2010 109 1400					
UNREINFORCED CONCRETE BLOCK WALL, 8″ X 8″ X 16″, PERLITE CORE FILL					
Concrete block wall, 8″ thick	1.000	S.F.	2.36	6	8.36
Perlite insulation	1.000	S.F.	.50	.37	.87
Horizontal joint reinforcing, alternate courses	.800	S.F.	.12	.16	.28
Control joint	.050	L.F.	.07	.06	.13
TOTAL			3.05	6.59	9.64

B2010 109		Concrete Block Wall - Regular Weight				COST PER S.F.		
	TYPE	SIZE (IN.)	STRENGTH (P.S.I.)	CORE FILL		MAT.	INST.	TOTAL
1200	Hollow	4x8x16	2,000	none		1.75	5.45	7.20
1250			4,500	none		1.53	5.45	6.98
1300	RB2010 -200	6x8x16	2,000	perlite		2.76	6.15	8.91
1310				styrofoam		4.41	5.85	10.26
1340				none		2.42	5.85	8.27
1350			4,500	perlite		2.28	6.15	8.43
1360				styrofoam		3.93	5.85	9.78
1390				none		1.94	5.85	7.79
1400		8x8x16	2,000	perlite		3.05	6.60	9.65
1410				styrofoam		4.54	6.20	10.74
1440				none		2.55	6.20	8.75
1450			4,500	perlite		2.84	6.60	9.44
1460				styrofoam		4.33	6.20	10.53
1490				none		2.34	6.20	8.54
1500		12x8x16	2,000	perlite		4.70	8.80	13.50
1510				styrofoam		6.35	8.05	14.40
1540				none		3.89	8.05	11.94
1550			4,500	perlite		3.89	8.80	12.69
1560				styrofoam		5.55	8.05	13.60
1590				none		3.08	8.05	11.13
2000	75% solid	4x8x16	2,000	none		1.90	5.50	7.40
2050			4,500	none		2.08	5.50	7.58

B20 Exterior Enclosure

B2010 Exterior Walls

B2010 109			Concrete Block Wall - Regular Weight					

	TYPE	SIZE (IN.)	STRENGTH (P.S.I.)	CORE FILL		COST PER S.F.		
						MAT.	INST.	TOTAL
2100	75% solid	6x8x16	2,000	perlite		2.23	6.05	8.28
2140				none		2.06	5.90	7.96
2150			4,500	perlite		2.72	6.05	8.77
2190				none		2.55	5.90	8.45
2200		8x8x16	2,000	perlite		3.32	6.50	9.82
2240				none		3.07	6.30	9.37
2250			4,500	perlite		3.71	6.50	10.21
2290				none		3.46	6.30	9.76
2300		12x8x16	2,000	perlite		4.08	8.55	12.63
2340				none		3.67	8.15	11.82
2350			4,500	perlite		5.55	8.55	14.10
2390				none		5.10	8.15	13.25
2500	Solid	4x8x16	2,000	none		2.15	5.65	7.80
2550			4,500	none		2.08	5.55	7.63
2600		6x8x16	2,000	none		2.40	6.05	8.45
2650			4,500	none		2.55	6	8.55
2700		8x8x16	2,000	none		3.70	6.45	10.15
2750			4,500	none		3.46	6.40	9.86
2800		12x8x16	2,000	none		5.55	8.40	13.95
2850			4,500	none		5.10	8.25	13.35

B2010 110			Concrete Block Wall - Lightweight					

	TYPE	SIZE (IN.)	WEIGHT (P.C.F.)	CORE FILL		COST PER S.F.		
						MAT.	INST.	TOTAL
3100	Hollow	8x4x16	105	perlite		2.03	6.25	8.28
3110				styrofoam		3.52	5.85	9.37
3140				none		1.53	5.85	7.38
3150			85	perlite		4.88	6.10	10.98
3160				styrofoam		6.35	5.70	12.05
3190				none		4.38	5.70	10.08
3200		4x8x16	105	none		2.06	5.30	7.36
3250			85	none		1.90	5.20	7.10
3300		6x8x16	105	perlite		3.08	6	9.08
3310				styrofoam		4.73	5.70	10.43
3340				none		2.74	5.70	8.44
3350			85	perlite		2.80	5.85	8.65
3360				styrofoam		4.45	5.55	10
3390				none		2.46	5.55	8.01
3400		8x8x16	105	perlite		3.83	6.45	10.28
3410				styrofoam		5.30	6.05	11.35
3440				none		3.33	6.05	9.38
3450			85	perlite		3.20	6.30	9.50
3460				styrofoam		4.69	5.90	10.59
3490				none		2.70	5.90	8.60
3500		12x8x16	105	perlite		5.70	8.55	14.25
3510				styrofoam		7.35	7.80	15.15
3540				none		4.88	7.80	12.68
3550			85	perlite		4.48	8.35	12.83
3560				styrofoam		6.15	7.60	13.75
3590				none		3.67	7.60	11.27
3600		4x8x24	105	none		1.55	5.95	7.50
3650			85	none		2.99	4.99	7.98
3690	For stacked bond add						.40	.40

B2010 Exterior Walls

B2010 110		Concrete Block Wall - Lightweight						

	TYPE	SIZE (IN.)	WEIGHT (P.C.F.)	CORE FILL		COST PER S.F.		
						MAT.	INST.	TOTAL
3700	Hollow	6x8x24	105	perlite		2.37	6.50	8.87
3710				styrofoam		4.02	6.20	10.22
3740				none		2.03	6.20	8.23
3750			85	perlite		4.41	5.50	9.91
3760				styrofoam		6.05	5.20	11.25
3790				none		4.07	5.20	9.27
3800		8x8x24	105	perlite		2.99	6.95	9.94
3810				styrofoam		4.48	6.55	11.03
3840				none		2.49	6.55	9.04
3850			85	perlite		5.30	5.90	11.20
3860				styrofoam		6.80	5.50	12.30
3890				none		4.82	5.50	10.32
3900		12x8x24	105	perlite		4.43	8.25	12.68
3910				styrofoam		6.10	7.50	13.60
3940				none		3.62	7.50	11.12
3950			85	perlite		7.15	8.05	15.20
3960				styrofoam		8.80	7.30	16.10
3990				none		6.35	7.30	13.65
4000	75% solid	4x8x16	105	none		2.08	5.40	7.48
4050			85	none		3.28	5.25	8.53
4100		6x8x16	105	perlite		2.48	5.90	8.38
4140				none		2.31	5.75	8.06
4150			85	perlite		3.95	5.80	9.75
4190				none		3.78	5.65	9.43
4200		8x8x16	105	perlite		3.33	6.35	9.68
4240				none		3.08	6.15	9.23
4250			85	perlite		5.40	6.20	11.60
4290				none		5.15	6	11.15
4300		12x8x16	105	perlite		5.85	8.30	14.15
4340				none		5.45	7.90	13.35
4350			85	perlite		7.35	8.10	15.45
4390				none		6.95	7.70	14.65
4500	Solid	4x8x16	105	none		2.70	5.55	8.25
4550			85	none		2.61	5.30	7.91
4600		6x8x16	105	none		3.10	6	9.10
4650			85	none		4.10	5.70	9.80
4700		8x8x16	105	none		4.11	6.40	10.51
4750			85	none		5.50	6.05	11.55
4800		12x8x16	105	none		7.05	8.25	15.30
4850			85	none		8.35	7.80	16.15
4900	For stacked bond, add						.40	.40

B2010 111		Reinforced Concrete Block Wall - Regular Weight						

	TYPE	SIZE (IN.)	STRENGTH (P.S.I.)	VERT. REINF & GROUT SPACING		COST PER S.F.		
						MAT.	INST.	TOTAL
5200	Hollow	4x8x16	2,000	#4 @ 48"		1.95	6.05	8
5250			4,500	#4 @ 48"		1.73	6.05	7.78
5300		6x8x16	2,000	#4 @ 48"		2.72	6.45	9.17
5330				#5 @ 32"		3.02	6.70	9.72
5340				#5 @ 16"		3.59	7.55	11.14
5350			4,500	#4 @ 28"		2.24	6.45	8.69
5380				#5 @ 32"		2.54	6.70	9.24
5390				#5 @ 16"		3.11	7.55	10.66

B20 Exterior Enclosure

B2010 Exterior Walls

B2010 111 — Reinforced Concrete Block Wall - Regular Weight

	TYPE	SIZE (IN.)	STRENGTH (P.S.I.)	VERT. REINF & GROUT SPACING		COST PER S.F. MAT.	INST.	TOTAL
5400	Hollow	8x8x16	2,000	#4 @ 48"		2.86	6.85	9.71
5430				#5 @ 32"		3.22	7.25	10.47
5440				#5 @ 16"		3.88	8.25	12.13
5450		8x8x16	4,500	#4 @ 48"		2.69	6.95	9.64
5480				#5 @ 32"		3.01	7.25	10.26
5490				#5 @ 16"		3.67	8.25	11.92
5500		12x8x16	2,000	#4 @ 48"		4.37	8.80	13.17
5530				#5 @ 32"		4.75	9.10	13.85
5540				#5 @ 16"		5.60	10.15	15.75
5550			4,500	#4 @ 48"		3.56	8.80	12.36
5580				#5 @ 32"		3.94	9.10	13.04
5590				#5 @ 16"		4.79	10.15	14.94
6100	75% solid	6x8x16	2,000	#4 @ 48"		2.25	6.40	8.65
6130				#5 @ 32"		2.49	6.60	9.09
6140				#5 @ 16"		2.90	7.30	10.20
6150			4,500	#4 @ 48"		2.74	6.40	9.14
6180				#5 @ 32"		2.98	6.60	9.58
6190				#5 @ 16"		3.39	7.30	10.69
6200		8x8x16	2,000	#4 @ 48"		3.27	6.90	10.17
6230				#5 @ 32"		3.52	7.15	10.67
6240				#5 @ 16"		3.95	8	11.95
6250			4,500	#4 @ 48"		3.66	6.90	10.56
6280				#5 @ 32"		3.91	7.15	11.06
6290				#5 @ 16"		4.34	8	12.34
6300		12x8x16	2,000	#4 @ 48"		3.97	8.80	12.77
6330				#5 @ 32"		4.27	9	13.27
6340				#5 @ 16"		4.84	9.85	14.69
6350			4,500	#4 @ 48"		5.40	8.80	14.20
6380				#4 @ 32"		5.70	9	14.70
6390				#5 @ 16"		6.30	9.85	16.15
6500	Solid-double	2-4x8x16	2,000	#4 @ 48" E.W.		5.30	12.75	18.05
6530	Wythe			#5 @ 16" E.W.		6.40	13.55	19.95
6550			4,500	#4 @ 48" E.W.		5.15	12.55	17.70
6580				#5 @ 16" E.W.		6.30	13.35	19.65
6600		2-6x8x16	2,000	#4 @ 48" E.W.		5.80	13.60	19.40
6630				#5 @ 16" E.W.		6.90	14.35	21.25
6650			4,000	#4 @ 48" E.W.		6.10	13.50	19.60
6680				#5 @ 16" E.W.		7.20	14.25	21.45

B2010 112 — Reinforced Concrete Block Wall - Lightweight

	TYPE	SIZE (IN.)	WEIGHT (P.C.F.)	VERT REINF. & GROUT SPACING		COST PER S.F. MAT.	INST.	TOTAL
7100	Hollow	8x4x16	105	#4 @ 48"		1.88	6.60	8.48
7130				#5 @ 32"		2.20	6.90	9.10
7140				#5 @ 16"		2.86	7.90	10.76
7150			85	#4 @ 48"		4.73	6.45	11.18
7180				#5 @ 32"		5.05	6.75	11.80
7190				#5 @ 16"		5.70	7.75	13.45
7200		4x8x16	105	#4 @ 48"		2.26	5.90	8.16
7250			85	#4 @ 48"		2.10	5.80	7.90

B20 Exterior Enclosure

B2010 Exterior Walls

B2010 112 — Reinforced Concrete Block Wall - Lightweight

	TYPE	SIZE (IN.)	WEIGHT (P.C.F.)	VERT REINF. & GROUT SPACING		COST PER S.F. MAT.	INST.	TOTAL
7300	Hollow	6x8x16	105	#4 @ 48"		3.04	6.30	9.34
7330				#5 @ 32"		3.34	6.55	9.89
7340				#5 @ 16"		3.91	7.40	11.31
7350			85	#4 @ 48"		2.76	6.15	8.91
7380				#5 @ 32"		3.06	6.40	9.46
7390				#5 @ 16"		3.63	7.25	10.88
7400		8x8x16	105	#4 @ 48"		3.68	6.80	10.48
7430				#5 @ 32"		4	7.10	11.10
7440				#5 @ 16"		4.66	8.10	12.76
7450		8x8x16	85	#4 @ 48"		3.05	6.65	9.70
7480				#5 @ 32"		3.37	6.95	10.32
7490				#5 @ 16"		4.03	7.95	11.98
7510		12x8x16	105	#4 @ 48"		5.35	8.55	13.90
7530				#5 @ 32"		5.75	8.85	14.60
7540				#5 @ 16"		6.60	9.90	16.50
7550			85	#4 @ 48"		4.15	8.35	12.50
7580				#5 @ 32"		4.53	8.65	13.18
7590				#5 @ 16"		5.40	9.70	15.10
7600		4x8x24	105	#4 @ 48"		1.75	6.55	8.30
7650			85	#4 @ 48"		3.19	5.60	8.79
7700		6x8x24	105	#4 @ 48"		2.33	6.80	9.13
7730				#5 @ 32"		2.63	7.05	9.68
7740				#5 @ 16"		3.20	7.90	11.10
7750			85	#4 @ 48"		4.37	5.80	10.17
7780				#5 @ 32"		4.67	6.05	10.72
7790				#5 @ 16"		5.25	6.90	12.15
7800		8x8x24	105	#4 @ 48"		2.84	7.30	10.14
7840				#5 @ 16"		3.82	8.60	12.42
7850			85	#4 @ 48"		5.15	6.25	11.40
7880				#5 @ 32"		5.50	6.55	12.05
7890				#5 @ 16"		6.15	7.55	13.70
7900		12x8x24	105	#4 @ 48"		4.10	8.25	12.35
7930				#5 @ 32"		4.48	8.55	13.03
7940				#5 @ 16"		5.35	9.60	14.95
7950			85	#4 @ 48"		6.85	8.05	14.90
7980				#5 @ 32"		7.20	8.35	15.55
7990				#5 @ 16"		8.05	9.40	17.45
8100	75% solid	6x8x16	105	#4 @ 48"		2.50	6.25	8.75
8130				#5 @ 32"		2.74	6.45	9.19
8140				#5 @ 16"		3.15	7.15	10.30
8150			85	#4 @ 48"		3.97	6.15	10.12
8180				#5 @ 32"		4.21	6.35	10.56
8190				#5 @ 16"		4.62	7.05	11.67
8200		8x8x16	105	#4 @ 48"		3.28	6.75	10.03
8230				#5 @ 32"		3.53	7	10.53
8240				#5 @ 16"		3.96	7.85	11.81
8250			85	#4 @ 48"		5.35	6.60	11.95
8280				#5 @ 32"		5.60	6.85	12.45
8290				#5 @ 16"		6.05	7.70	13.75

138

B20 Exterior Enclosure

B2010 Exterior Walls

| B2010 112 | | Reinforced Concrete Block Wall - Lightweight | | | | | | |

	TYPE	SIZE (IN.)	WEIGHT (P.C.F.)	VERT REINF. & GROUT SPACING		COST PER S.F.		
						MAT.	INST.	TOTAL
8300		12x8x16	105	#4 @ 48"		5.75	8.55	14.30
8330				#5 @ 32"		6.05	8.75	14.80
8340				#5 @ 16"		6.60	9.60	16.20
8350			85	#4 @ 48"		7.25	8.35	15.60
8380				#5 @ 32"		7.55	8.55	16.10
8390				#5 @ 16"		8.10	9.40	17.50
8500	Solid-double	2-4x8x16	105	#4 @ 48"		6.40	12.55	18.95
8530	Wythe		105	#5 @ 16"		7.50	13.35	20.85
8550			85	#4 @ 48"		6.20	12.05	18.25
8580				#5 @ 16"		7.35	12.85	20.20
8600		2-6x8x16	105	#4 @ 48"		7.20	13.50	20.70
8630				#5 @ 16"		8.30	14.25	22.55
8650			85	#4 @ 48"		9.20	12.90	22.10
8680				#5 @ 16"		10.30	13.65	23.95
8900	For stacked bond add						.40	.40
8910								

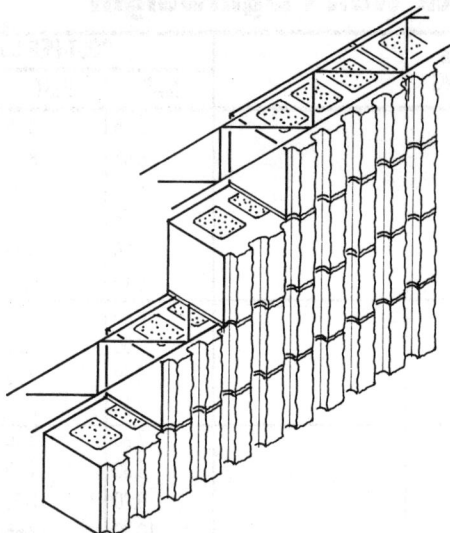

Exterior split ribbed block walls are defined in the following terms: structural reinforcement, weight, percent solid, size, number of ribs and insulation. Within each of these categories two to four variations are shown. No costs are included for brick shelf or relieving angles. Costs include control joints every 20' and horizontal reinforcing.

System Components	QUANTITY	UNIT	COST PER S.F.		
			MAT.	INST.	TOTAL
SYSTEM B2010 113 1430					
UNREINFORCED SPLIT RIB BLOCK WALL, 8"X8"X16", 8 RIBS(HEX), PERLITE FILL					
Split ribbed block wall, 8" thick	1.000	S.F.	5.55	7.40	12.95
Perlite insulation	1.000	S.F.	.50	.37	.87
Horizontal joint reinforcing, alternate courses	.800	L.F.	.12	.16	.28
Control joint	.050	L.F.	.07	.06	.13
TOTAL			6.24	7.99	14.23

B2010 113		Split Ribbed Block Wall - Regular Weight				COST PER S.F.		
	TYPE	SIZE (IN.)	RIBS	CORE FILL		MAT.	INST.	TOTAL
1220	Hollow	4x8x16	4	none		3.15	6.75	9.90
1250			8	none		3.67	6.75	10.42
1280			16	none		2.99	6.85	9.84
1330		6x8x16	4	none		4.88	7.45	12.33
1340	RB2010 -200			styrofoam		6.55	7.15	13.70
1350				none		4.54	7.15	11.69
1360			16	perlite		4.29	7.55	11.84
1370				styrofoam		5.95	7.25	13.20
1380				none		3.95	7.25	11.20
1430		8x8x16	8	perlite		6.25	8	14.25
1440				styrofoam		7.75	7.60	15.35
1450				none		5.75	7.60	13.35
1460			16	perlite		5.10	8.10	13.20
1470				styrofoam		6.60	7.70	14.30
1480				none		4.62	7.70	12.32
1530		12x8x16	8	perlite		7.75	10.60	18.35
1540				styrofoam		9.40	9.85	19.25
1550				none		6.95	9.85	16.80
1560			16	perlite		6.45	10.80	17.25
1570				styrofoam		8.10	10.05	18.15
1580				none		5.65	10.05	15.70

B20 Exterior Enclosure

B2010 Exterior Walls

B2010 113 — Split Ribbed Block Wall - Regular Weight

	TYPE	SIZE (IN.)	RIBS	CORE FILL		COST PER S.F. MAT.	INST.	TOTAL
2120	75% solid	4x8x16	4	none		3.94	6.85	10.79
2150			8	none		4.67	6.85	11.52
2180			16	none		3.79	6.95	10.74
2230		6x8x16	8	perlite		6	7.40	13.40
2250				none		5.85	7.25	13.10
2260			16	perlite		5.20	7.50	12.70
2280				none		5.05	7.35	12.40
2330		8x8x16	8	perlite		7.60	7.90	15.50
2350				none		7.35	7.70	15.05
2360			16	perlite		6.10	8.05	14.15
2380				none		5.85	7.85	13.70
2430		12x8x16	8	perlite		9.15	10.45	19.60
2450				none		8.75	10.05	18.80
2460			16	perlite		7.60	10.65	18.25
2480				none		7.20	10.25	17.45
2520	Solid	4x8x16	4	none		4.81	6.95	11.76
2550			8	none		5.60	6.95	12.55
2580			16	none		4.55	7.05	11.60
2620		6x8x16	8	none		7.05	7.35	14.40
2650			16	none		6.10	7.45	13.55
2680		8x8x16	8	none		8.90	7.85	16.75
2720			16	none		7.05	8	15.05
2750		12x8x16	8	none		10.60	10.25	20.85
2780			16	none		8.70	10.40	19.10

B2010 114 — Split Ribbed Block Wall - Lightweight

	TYPE	SIZE (IN.)	RIBS	CORE FILL		COST PER S.F. MAT.	INST.	TOTAL
3250	Hollow	4x8x16	8	none		3.91	6.55	10.46
3280			16	none		3.27	6.65	9.92
3330		6x8x16	8	perlite		5.20	7.25	12.45
3340				styrofoam		6.85	6.95	13.80
3350				none		4.84	6.95	11.79
3360			16	perlite		4.62	7.35	11.97
3370				styrofoam		6.25	7.05	13.30
3380				none		4.28	7.05	11.33
3430		8x8x16	8	perlite		7.25	7.75	15
3440				styrofoam		8.75	7.35	16.10
3450				none		6.75	7.35	14.10
3460			16	perlite		5.50	7.85	13.35
3470				styrofoam		7	7.45	14.45
3480				none		5	7.45	12.45
3530		12x8x16	8	perlite		8.50	10.30	18.80
3540				styrofoam		10.15	9.55	19.70
3550				none		7.70	9.55	17.25
3560			16	perlite		6.90	10.45	17.35
3570				styrofoam		8.55	9.70	18.25
3580				none		6.10	9.70	15.80
4150	75% solid	4x8x16	8	none		5.05	6.65	11.70
4180			16	none		4.09	6.75	10.84

B2010 Exterior Walls

B2010 114	Split Ribbed Block Wall - Lightweight

	TYPE	SIZE (IN.)	RIBS	CORE FILL		COST PER S.F. MAT.	INST.	TOTAL
4230	75% solid	6x8x16	8	perlite		6.30	7.20	13.50
4250				none		6.15	7.05	13.20
4260			16	perlite		5.60	7.30	12.90
4280				none		5.45	7.15	12.60
4330		8x8x16	8	perlite		8.90	7.65	16.55
4350				none		8.65	7.45	16.10
4360			16	perlite		6.60	7.80	14.40
4380		8x8x16	16	none		6.35	7.60	13.95
4430		12x8x16	8	perlite		10.30	10.10	20.40
4450				none		9.90	9.70	19.60
4460			16	perlite		8.15	10.25	18.40
4480				none		7.75	9.85	17.60
4550	Solid	4x8x16	8	none		6.05	6.75	12.80
4580			16	none		4.98	6.85	11.83
4650		6x8x16	8	none		7.45	7.15	14.60
4680			16	none		6.60	7.25	13.85
4750		8x8x16	8	none		10.45	7.60	18.05
4780			16	none		7.70	7.70	15.40
4850		12x8x16	8	none		11.95	9.85	21.80
4880			16	none		9.40	10.05	19.45

B2010 115	Reinforced Split Ribbed Block Wall - Regular Weight

	TYPE	SIZE (IN.)	RIBS	VERT. REINF. & GROUT SPACING		COST PER S.F. MAT.	INST.	TOTAL
5200	Hollow	4x8x16	4	#4 @ 48"		3.35	7.35	10.70
5230			8	#4 @ 48"		3.87	7.35	11.22
5260			16	#4 @ 48"		3.19	7.45	10.64
5330		6x8x16	8	#4 @ 48"		4.84	7.75	12.59
5340				#5 @ 32"		5.15	8	13.15
5350				#5 @ 16"		5.70	8.85	14.55
5360			16	#4 @ 48"		4.25	7.85	12.10
5370				#5 @ 32"		4.55	8.10	12.65
5380				#5 @ 16"		5.10	8.95	14.05
5430		8x8x16	8	#4 @ 48"		6.10	8.35	14.45
5440				#5 @ 32"		6.40	8.65	15.05
5450				#5 @ 16"		7.05	9.65	16.70
5460			16	#4 @ 48"		4.97	8.45	13.42
5470				#5 @ 32"		5.30	8.75	14.05
5480				#5 @ 16"		5.95	9.75	15.70
5530		12x8x16	8	#4 @ 48"		7.45	10.60	18.05
5540				#5 @ 32"		7.80	10.90	18.70
5550				#5 @ 16"		8.65	11.95	20.60
5560			16	#4 @ 48"		6.15	10.80	16.95
5570				#5 @ 32"		6.50	11.10	17.60
5580				#5 @ 16"		7.35	12.15	19.50
6230	75% solid	6x8x16	8	#4 @ 48"		6	7.75	13.75
6240				#5 @ 32"		6.25	7.95	14.20
6250				#5 @ 16"		6.65	8.65	15.30
6260			16	#4 @ 48"		5.25	7.85	13.10
6270				#5 @ 32"		5.45	8.05	13.50
6280				#5 @ 16"		5.90	8.75	14.65

B20 Exterior Enclosure

B2010 Exterior Walls

| B2010 115 | | Reinforced Split Ribbed Block Wall - Regular Weight | | | | | | |

	TYPE	SIZE (IN.)	RIBS	VERT. REINF. & GROUT SPACING		COST PER S.F.		
						MAT.	INST.	TOTAL
6330	75% solid	8x8x16	8	#4 @ 48"		7.55	8.30	15.85
6340				#5 @ 32"		7.80	8.55	16.35
6350				#5 @ 16"		8.20	9.40	17.60
6360			16	#4 @ 48"		6.05	8.45	14.50
6370				#5 @ 32"		6.30	8.70	15
6380				#5 @ 16"		6.70	9.55	16.25
6430		12x8x16	8	#4 @ 48"		9.05	10.70	19.75
6440				#5 @ 32"		9.35	10.90	20.25
6450				#5 @ 16"		9.90	11.75	21.65
6460			16	#4 @ 48"		7.50	10.90	18.40
6470				#5 @ 32"		7.80	11.10	18.90
6480				#5 @ 16"		8.35	11.95	20.30
6500	Solid-double	2-4x8x16	4	#4 @ 48"		5.80	8.35	14.15
6520	Wythe			#5 @ 16"		6.35	8.75	15.10
6530			8	#4 @ 48"		6.60	8.35	14.95
6550				#5 @ 16"		7.20	8.75	15.95
6560			16	#4 @ 48"		5.55	8.45	14
6580				#5 @ 16"		6.10	8.85	14.95
6630		2-6x8x16	8	#4 @ 48"		8.05	8.80	16.85
6650				#5 @ 16"		8.60	9.20	17.80
6660			16	#4 @ 48"		7.10	8.90	16
6680				#5 @ 16"		7.65	9.30	16.95

| B2010 116 | | Reinforced Split Ribbed Block Wall - Lightweight | | | | | | |

	TYPE	SIZE (IN.)	RIBS	VERT. REINF. & GROUT SPACING		COST PER S.F.		
						MAT.	INST.	TOTAL
7230	Hollow	4x8x16	8	#4 @ 48"		4.11	7.15	11.26
7260			16	#4 @ 48"		3.47	7.25	10.72
7330		6x8x16	8	#4 @ 48"		5.15	7.55	12.70
7340				#5 @ 32"		5.45	7.80	13.25
7350				#5 @ 16"		6	8.65	14.65
7360			16	#4 @ 48"		4.58	7.65	12.23
7370				#5 @ 32"		4.88	7.90	12.78
7380				#5 @ 16"		5.45	8.75	14.20
7430		8x8x16	8	#4 @ 48"		7.10	8.10	15.20
7440				#5 @ 32"		7.40	8.40	15.80
7450				#5 @ 16"		8.05	9.40	17.45
7460			16	#4 @ 48"		5.35	8.20	13.55
7470				#5 @ 32"		5.65	8.50	14.15
7480				#5 @ 16"		6.35	9.50	15.85
7530		12x8x16	8	#4 @ 48"		8.20	10.30	18.50
7540				#5 @ 32"		8.55	10.60	19.15
7550				#5 @ 16"		9.40	11.65	21.05
7560			16	#4 @ 48"		6.60	10.45	17.05
7570				#5 @ 32"		6.95	10.75	17.70
7580				#5 @ 16"		7.80	11.80	19.60
8230	75% solid	6x8x16	8	#4 @ 48"		6.30	7.55	13.85
8240				#5 @ 32"		6.55	7.75	14.30
8250				#5 @ 16"		6.95	8.45	15.40
8260			16	#4 @ 48"		5.60	7.65	13.25
8270				#5 @ 32"		5.85	7.85	13.70
8280				#5 @ 16"		6.25	8.55	14.80

B2010 Exterior Walls

| B2010 116 | | Reinforced Split Ribbed Block Wall - Lightweight |

	TYPE	SIZE (IN.)	RIBS	VERT. REINF. & GROUT SPACING		COST PER S.F.		
						MAT.	INST.	TOTAL
8330	75% solid	8x8x16	8	#4 @ 48"		8.85	8.05	16.90
8340				#5 @ 32"		9.10	8.30	17.40
8350				#5 @ 16"		9.50	9.15	18.65
8360			16	#4 @ 48"		6.55	8.20	14.75
8370				#5 @ 32"		6.80	8.45	15.25
8380				#5 @ 16"		7.20	9.30	16.50
8430		12x8x16	8	#4 @ 48"		10.20	10.35	20.55
8440				#5 @ 32"		10.50	10.55	21.05
8450				#5 @ 16"		11.05	11.40	22.45
8460			16	#4 @ 48"		8.05	10.50	18.55
8470				#5 @ 32"		8.35	10.70	19.05
8480				#5 @ 16"		8.90	11.55	20.45
8530	Solid double	2-4x8x16	8	#4 @ 48" E.W		7.20	8.40	15.60
8550	Wythe			#5 @ 16" E.W.		8.35	9.20	17.55
8560			16	#4 @ 48" E.W.		6.15	8.50	14.65
8580				#5 @ 16" E.W.		7.25	9.30	16.55
8630	Solid double	2-6x8x16	8	#4 @ 48" E.W.		8.60	8.85	17.45
8650	Wythe			#5 @ 16" E.W.		9.75	9.60	19.35
8660			16	#4 @ 48" E.W.		7.75	8.95	16.70
8680				#5 @ 16" E.W.		8.90	9.70	18.60
8900	For stacked bond, add						.40	.40

B2010 Exterior Walls

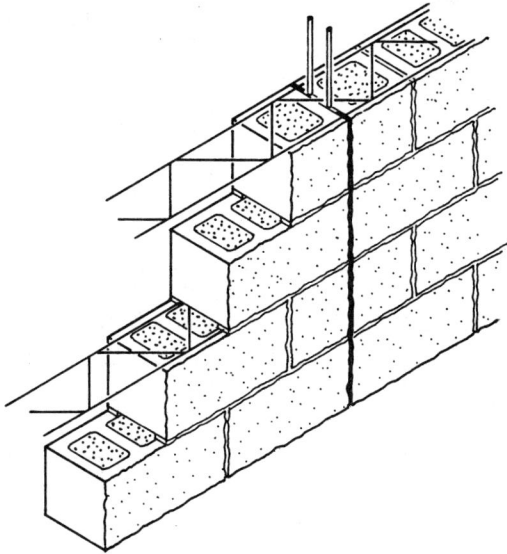

Exterior split face block walls are defined in the following terms: structural reinforcement, weight, percent solid, size, scores and insulation. Within each of these categories two to four variations are shown. No costs are included for brick shelf or relieving angles. Costs include control joints every 20′ and horizontal reinforcing.

System Components	QUANTITY	UNIT	COST PER S.F.		
			MAT.	INST.	TOTAL
SYSTEM B2010 117 1600					
UNREINFORCED SPLIT FACE BLOCK WALL, 8″X8″X16″, 0 SCORES, PERLITE FILL					
Split face block wall, 8″ thick	1.000	S.F.	4.66	7.65	12.31
Perlite insulation	1.000	S.F.	.50	.37	.87
Horizontal joint reinforcing, alternate course	.800	L.F.	.12	.16	.28
Control joint	.050	L.F.	.07	.06	.13
TOTAL			5.35	8.24	13.59

B2010 117		Split Face Block Wall - Regular Weight						
	TYPE	SIZE (IN.)	SCORES	CORE FILL		COST PER S.F.		
						MAT.	INST.	TOTAL
1200	Hollow	8x4x16	0	perlite		9.20	8.80	18
1210				styrofoam		10.70	8.40	19.10
1240				none		8.70	8.40	17.10
1250	RB2010-200		1	perlite		9.85	8.80	18.65
1260				styrofoam		11.35	8.40	19.75
1290				none		9.35	8.40	17.75
1300		12x4x16	0	perlite		11.90	10	21.90
1310				styrofoam		13.55	9.25	22.80
1340				none		11.10	9.25	20.35
1350			1	perlite		12.55	10	22.55
1360				styrofoam		14.20	9.25	23.45
1390				none		11.75	9.25	21
1400		4x8x16	0	none		3.15	6.65	9.80
1450			1	none		3.55	6.75	10.30
1500		6x8x16	0	perlite		4.41	7.65	12.06
1510				styrofoam		6.05	7.35	13.40
1540				none		4.07	7.35	11.42
1550			1	perlite		4.83	7.75	12.58
1560				styrofoam		6.50	7.45	13.95
1590				none		4.49	7.45	11.94

B20 Exterior Enclosure

B2010 Exterior Walls

B2010 117 — Split Face Block Wall - Regular Weight

	TYPE	SIZE (IN.)	SCORES	CORE FILL		MAT.	INST.	TOTAL
1600	Hollow	8x8x16	0	perlite		5.35	8.25	13.60
1610				styrofoam		6.85	7.85	14.70
1640				none		4.85	7.85	12.70
1650		8x8x16	1	perlite		5.65	8.40	14.05
1660				styrofoam		7.15	8	15.15
1690				none		5.15	8	13.15
1700		12x8x16	0	perlite		6.80	10.80	17.60
1710				styrofoam		8.45	10.05	18.50
1740				none		6	10.05	16.05
1750			1	perlite		7.15	11	18.15
1760				styrofoam		8.80	10.25	19.05
1790				none		6.35	10.25	16.60
1800	75% solid	8x4x16	0	perlite		11.20	8.75	19.95
1840				none		10.95	8.55	19.50
1850			1	perlite		11.90	8.75	20.65
1890				none		11.65	8.55	20.20
1900		12x4x16	0	perlite		14.45	9.85	24.30
1940				none		14.05	9.45	23.50
1950			1	perlite		15	9.85	24.85
1990				none		14.60	9.45	24.05
2000		4x8x16	0	none		3.94	6.75	10.69
2050			1	none		4.28	6.85	11.13
2100		6x8x16	0	perlite		5.40	8.90	14.30
2140				none		5.25	8.75	14
2150			1	perlite		5.75	7.75	13.50
2190				none		5.60	7.60	13.20
2200		8x8x16	0	perlite		6.40	8.20	14.60
2240				none		6.15	8	14.15
2250			1	perlite		6.70	8.30	15
2290				none		6.45	8.10	14.55
2300		12x8x16	0	perlite		8	10.65	18.65
2340				none		7.60	10.25	17.85
2350			1	perlite		8.30	10.80	19.10
2390				none		7.90	10.40	18.30
2400	Solid	8x4x16	0	none		13.40	8.70	22.10
2450			1	none		14.05	8.70	22.75
2500		12x4x16	0	none		17	9.65	26.65
2550			1	none		17.60	11.25	28.85
2600		4x8x16	0	none		4.81	6.85	11.66
2650			1	none		5.15	6.95	12.10
2700		6x8x16	0	none		6.45	7.60	14.05
2750			1	none		6.75	7.70	14.45
2800		8x8x16	0	none		7.35	8.10	15.45
2850			1	none		7.70	8.25	15.95
2900		12x8x16	0	none		9.15	10.40	19.55
2950			1	none		9.45	10.60	20.05

B2010 118 — Split Face Block Wall - Lightweight

	TYPE	SIZE (IN.)	SCORES	CORE FILL		MAT.	INST.	TOTAL
3200	Hollow	8x4x16	0	perlite		9.50	9.90	19.40
3210				styrofoam		11	9.50	20.50
3240				none		9	9.50	18.50

146

B2010 118		Split Face Block Wall - Lightweight								

	TYPE	SIZE (IN.)	SCORES	CORE FILL		COST PER S.F.		
						MAT.	INST.	TOTAL
3250	Hollow		1	perlite		10.20	8.50	18.70
3260				styrofoam		11.70	8.10	19.80
3290				none		9.70	8.10	17.80
3300		12x4x16	0	perlite		12.35	9.65	22
3310				styrofoam		14	8.90	22.90
3340				none		11.55	8.90	20.45
3350		12x4x16	1	perlite		13.05	11.15	24.20
3360				styrofoam		14.70	10.40	25.10
3390				none		12.25	10.40	22.65
3400		4x8x16	0	none		3.43	6.45	9.88
3450			1	none		3.71	6.55	10.26
3500		6x8x16	0	perlite		4.92	7.45	12.37
3510				styrofoam		6.55	7.15	13.70
3540				none		4.58	7.15	11.73
3550			1	perlite		5.25	7.55	12.80
3560				styrofoam		6.90	7.25	14.15
3590				none		4.89	7.25	12.14
3600		8x8x16	0	perlite		5.75	8	13.75
3610				styrofoam		7.25	7.60	14.85
3640				none		5.25	7.60	12.85
3650			1	perlite		6.10	8.10	14.20
3660				styrofoam		7.60	7.70	15.30
3690				none		5.60	7.70	13.30
3700		12x8x16	0	perlite		7.25	10.45	17.70
3710				styrofoam		8.90	9.70	18.60
3740				none		6.45	9.70	16.15
3750			1	perlite		7.55	10.60	18.15
3760				styrofoam		9.20	9.85	19.05
3790				none		6.75	9.85	16.60
3800	75% solid	8x4x16	0	perlite		11.90	8.75	20.65
3840				none		11.65	8.55	20.20
3850			1	perlite		12.40	8.45	20.85
3890				none		12.15	8.25	20.40
3900		12x4x16	0	perlite		15	9.85	24.85
3940				none		14.60	9.45	24.05
3950			1	perlite		15.75	11	26.75
3990				none		15.35	10.60	25.95
4000		4x8x16	0	none		4.30	6.55	10.85
4050			1	none		4.66	6.65	11.31
4100		6x8x16	0	perlite		5.95	7.40	13.35
4140				none		5.80	7.25	13.05
4150			1	perlite		6.30	7.50	13.80
4190				none		6.15	7.35	13.50
4200		8x8x16	0	perlite		6.95	7.90	14.85
4240				none		6.70	7.70	14.40
4250			1	perlite		7.25	8.05	15.30
4290				none		7	7.85	14.85
4300		12x8x16	0	perlite		8.55	10.25	18.80
4340				none		8.15	9.85	18
4350			1	perlite		8.85	10.45	19.30
4390				none		8.45	10.05	18.50

B20 Exterior Enclosure

B2010 Exterior Walls

B2010 118 — Split Face Block Wall - Lightweight

	TYPE	SIZE (IN.)	SCORES	CORE FILL		COST PER S.F. MAT.	INST.	TOTAL
4400	Solid	8x4x16	0	none		13.85	8.40	22.25
4450			1	none		14.55	8.40	22.95
4500		12x4x16	0	none		17.80	9.25	27.05
4550			1	none		18.50	10.85	29.35
4600		4x8x16	0	none		5.20	6.65	11.85
4650			1	none		5.55	6.75	12.30
4700		6x8x16	0	none		7.05	7.35	14.40
4750			1	none		7.35	7.45	14.80
4800		8x8x16	0	none		8.15	7.85	16
4850			1	none		8.45	8	16.45
4900		12x8x16	0	none		9.75	10.05	19.80
4950			1	none		10.10	10.25	20.35

B2010 119 — Reinforced Split Face Block Wall - Regular Weight

	TYPE	SIZE (IN.)	SCORES	VERT. REINF. & GROUT SPACING		COST PER S.F. MAT.	INST.	TOTAL
5200	Hollow	8x4x16	0	#4 @ 48"		9.05	9.15	18.20
5210				#5 @ 32"		9.35	9.45	18.80
5240				#5 @ 16"		10	10.45	20.45
5250			1	#4 @ 48"		9.70	9.15	18.85
5260				#5 @ 32"		10	9.45	19.45
5290				#5 @ 16"		10.65	10.45	21.10
5300		12x4x16	0	#4 @ 48"		11.60	10	21.60
5310				#5 @ 32"		11.95	10.30	22.25
5340				#5 @ 16"		12.80	11.35	24.15
5350			1	#4 @ 48"		12.25	10	22.25
5360				#5 @ 32"		12.60	10.30	22.90
5390				#5 @ 16"		13.45	11.35	24.80
5400		4x8x16	0	#4 @ 48"		3.35	7.25	10.60
5450			1	#4 @ 48"		3.75	7.35	11.10
5500		6x8x16	0	#4 @ 48"		4.37	7.95	12.32
5510				#5 @ 32"		4.67	8.20	12.87
5540				#5 @ 16"		5.25	9.05	14.30
5550			1	#4 @ 48"		4.79	8.05	12.84
5560				#5 @ 32"		5.10	8.30	13.40
5590				#5 @ 16"		5.65	9.15	14.80
5600		8x8x16	0	#4 @ 48"		5.20	8.60	13.80
5610				#5 @ 32"		5.50	8.90	14.40
5640				#5 @ 16"		5.60	9.40	15
5650			1	#4 @ 48"		7.85	11.90	19.75
5660				#5 @ 32"		5.85	9.05	14.90
5690				#5 @ 16"		6.50	10.05	16.55
5700		12x8x16	0	#4 @ 48"		6.50	10.80	17.30
5710				#5 @ 32"		6.85	11.10	17.95
5740				#5 @ 16"		7.70	12.15	19.85
5750			1	#4 @ 48"		6.85	11	17.85
5760				#5 @ 32"		7.20	11.30	18.50
5790				#5 @ 16"		8.05	12.35	20.40

B20 Exterior Enclosure

B2010 Exterior Walls

B2010 119 — Reinforced Split Face Block Wall - Regular Weight

	TYPE	SIZE (IN.)	SCORES	VERT. REINF. & GROUT SPACING		COST PER S.F.		
						MAT.	INST.	TOTAL
6000	75% solid	8x4x16	0	#4 @ 48"		11.15	9.15	20.30
6010				#5 @ 32"		11.40	9.40	20.80
6040				#5 @ 16"		11.80	10.25	22.05
6050			1	#4 @ 48"		11.85	9.15	21
6060				#5 @ 32"		12.10	9.40	21.50
6090				#5 @ 16"		12.50	10.25	22.75
6100		12x4x16	0	#4 @ 48"		14.35	10.10	24.45
6110				#5 @ 32"		14.65	10.30	24.95
6140				#5 @ 16"		15.15	11	26.15
6150			1	#4 @ 48"		14.90	10.10	25
6160				#5 @ 32"		15.20	10.30	25.50
6190				#5 @ 16"		15.75	11.15	26.90
6200		6x8x16	0	#4 @ 48"		5.40	9.25	14.65
6210				#5 @ 32"		5.65	9.45	15.10
6240				#5 @ 16"		6.05	10.15	16.20
6250			1	#4 @ 48"		5.75	8.10	13.85
6260				#5 @ 32"		6	8.30	14.30
6290				#5 @ 16"		6.40	9	15.40
6300		8x8x16	0	#4 @ 48"		6.35	8.60	14.95
6310				#5 @ 32"		6.60	8.85	15.45
6340				#5 @ 16"		7	9.70	16.70
6350			1	#4 @ 48"		6.65	8.70	15.35
6360				#5 @ 32"		6.90	8.95	15.85
6390				#5 @ 16"		7.30	9.80	17.10
6400	75% solid	12x8x16	0	#4 @ 48"		7.90	10.90	18.80
6410				#5 @ 32"		8.20	11.10	19.30
6440				#5 @ 16"		8.75	11.95	20.70
6450			1	#4 @ 48"		8.20	11.05	19.25
6460				#5 @ 32"		8.50	11.25	19.75
6490				#5 @ 16"		9.05	12.10	21.15
6700	Solid-double Wythe	2-4x8x16	0	#4 @ 48" E.W.		10.60	15.15	25.75
6710				#5 @ 32" E.W.		11	15.30	26.30
6740				#5 @ 16" E.W.		11.75	15.95	27.70
6750			1	#4 @ 48" E.W.		11.25	15.35	26.60
6760				#5 @ 32" E.W.		11.65	15.50	27.15
6790				#5 @ 16" E.W.		12.40	16.15	28.55
6800		2-6x8x16	0	#4 @ 48" E.W.		13.85	16.70	30.55
6810				#5 @ 32" E.W.		14.25	16.85	31.10
6840				#5 @ 16" E.W.		15	17.45	32.45
6850			1	#4 @ 48" E.W.		14.45	16.90	31.35
6860				#5 @ 32" E.W.		14.85	17.05	31.90
6890				#5 @ 16" E.W.		15.60	17.65	33.25

B2010 120 — Reinforced Split Face Block Wall - Lightweight

	TYPE	SIZE (IN.)	SCORES	VERT. REINF. & GROUT SPACING		COST PER S.F.		
						MAT.	INST.	TOTAL
7200	Hollow	8x4x16	0	#4 @ 48"		9.35	10.25	19.60
7210				#5 @ 32"		9.65	10.55	20.20
7240				#5 @ 16"		10.30	11.55	21.85
7250			1	#4 @ 48"		10.05	8.85	18.90
7260				#5 @ 32"		10.35	9.15	19.50
7290				#5 @ 16"		11	10.15	21.15

B2010 Exterior Walls

B2010 120	Reinforced Split Face Block Wall - Lightweight

	TYPE	SIZE (IN.)	SCORES	VERT. REINF. & GROUT SPACING		COST PER S.F.		
						MAT.	INST.	TOTAL
7300	Hollow	12x4x16	0	#4 @ 48"		12.05	9.65	21.70
7340				#5 @ 16"		13.25	11	24.25
7350			1	#4 @ 48"		12.75	11.15	23.90
7360				#5 @ 32"		13.10	11.45	24.55
7390				#5 @ 16"		13.95	12.50	26.45
7400		4x8x16	0	#4 @ 48"		3.63	7.05	10.68
7450			1	#4 @ 48"		3.91	7.15	11.06
7500		6x8x16	0	#4 @ 48"		4.88	7.75	12.63
7510				#5 @ 32"		5.20	8	13.20
7540				#5 @ 16"		5.75	8.85	14.60
7550			1	#4 @ 48"		5.20	7.85	13.05
7560				#5 @ 32"		5.50	8.10	13.60
7590				#5 @ 16"		6.05	8.95	15
7600		8x8x16	0	#4 @ 48"		5.60	8.35	13.95
7610				#5 @ 32"		5.90	8.65	14.55
7640				#5 @ 16"		6.55	9.65	16.20
7650			1	#4 @ 48"		5.95	8.45	14.40
7660				#5 @ 32"		6.25	8.75	15
7690				#5 @ 16"		6.90	9.75	16.65
7700		12x8x16	0	#4 @ 48"		6.95	10.45	17.40
7710				#5 @ 32"		7.30	10.75	18.05
7740		12x8x16	0	#5 @ 16"		8.15	11.80	19.95
7750			1	#4 @ 48"		7.25	10.60	17.85
7760				#5 @ 16"		7.60	10.90	18.50
7790				#5 @ 16"		8.45	11.95	20.40
8000	75% solid	8x4x16	0	#4 @ 48"		11.85	9.15	21
8010				#5 @ 32"		12.10	9.40	21.50
8040				#5 @ 16"		12.50	10.25	22.75
8050			1	#4 @ 48"		12.35	8.85	21.20
8060				#5 @ 32"		12.60	9.10	21.70
8090				#5 @ 16"		13	9.95	22.95
8100		12x4x16	0	#4 @ 48"		14.90	10.10	25
8110				#5 @ 32"		15.20	10.30	25.50
8140				#5 @ 16"		15.75	11.15	26.90
8150			1	#4 @ 48"		15.65	11.25	26.90
8160				#5 @ 32"		15.95	11.45	27.40
8190				#5 @ 16"		16.50	12.30	28.80
8200		6x8x16	0	#4 @ 48"		5.95	7.75	13.70
8210				#5 @ 32"		6.20	7.95	14.15
8240				#5 @ 16"		6.60	8.65	15.25
8250			1	#4 @ 48"		6.30	7.85	14.15
8260				#5 @ 32"		6.55	8.05	14.60
8290				#5 @ 16"		6.95	8.75	15.70
8300		8x8x16	0	#4 @ 48"		6.90	8.30	15.20
8310				#5 @ 32"		7.15	8.55	15.70
8340				#5 @ 16"		7.55	9.40	16.95
8350			1	#4 @ 48"		7.20	8.45	15.65
8360				#5 @ 32"		7.45	8.70	16.15
8390				#5 @ 16"		7.85	9.55	17.40

B2010 Exterior Walls

B2010 120	Reinforced Split Face Block Wall - Lightweight

	TYPE	SIZE (IN.)	SCORES	VERT. REINF. & GROUT SPACING		COST PER S.F.		
						MAT.	INST.	TOTAL
8400	75% solid	12x8x16	0	#4 @ 48"		8.45	10.50	18.95
8410				#5 @ 32"		8.75	10.70	19.45
8440				#5 @ 16"		9.30	11.55	20.85
8450			1	#4 @ 48"		8.75	10.70	19.45
8460				#5 @ 32"		9.05	10.90	19.95
8490				#5 @ 16"		9.60	11.75	21.35
8700	Solid-double Wythe	2-4x8x16	0	#4 @ 48" E.W.		11.40	14.75	26.15
8710				#5 @ 32" E.W.		11.85	14.90	26.75
8740				#5 @ 16" E.W.		12.55	15.55	28.10
8750			1	#4 @ 48" E.W.		12.10	14.95	27.05
8760				#5 @ 32" E.W.		12.55	15.10	27.65
8790				#5 @ 16" E.W.		13.25	15.75	29
8800		2-6x8x16	0	#4 @ 48" E.W.		15.05	16.20	31.25
8810				#5 @ 32" E.W.		15.45	16.35	31.80
8840				#5 @ 16" E.W.		16.20	16.95	33.15
8850			1	#4 @ 48" E.W.		15.65	16.40	32.05
8860				#5 @ 32" E.W.		16.05	16.55	32.60
8890				#5 @ 16" E.W.		16.80	17.15	33.95

B20 Exterior Enclosure

B2010 Exterior Walls

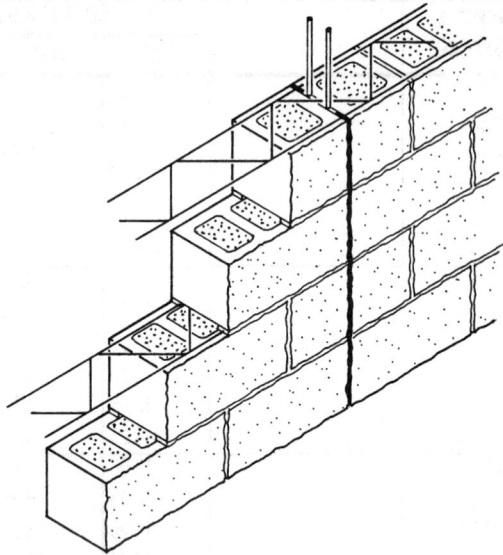

Exterior ground face block walls are defined in the following terms: structural reinforcement, weight, percent solid, size, scores and insulation. Within each of these categories two to four variations are shown. No costs are included for brick shelf or relieving angles. Costs include control joints every 20' and horizontal reinforcing.

System Components	QUANTITY	UNIT	COST PER S.F.		
			MAT.	INST.	TOTAL
SYSTEM B2010 121 1600					
UNREINF. GROUND FACE BLOCK WALL, 8"X8"X16", 0 SCORES, PERLITE FILL					
Ground face block wall, 8" thick	1.000	S.F.	8	7.80	15.80
Perlite insulation	1.000	S.F.	.50	.37	.87
Horizontal joint reinforcing, alternate course	.800	L.F.	.12	.16	.28
Control joint	.050	L.F.	.07	.06	.13
TOTAL			8.69	8.39	17.08

B2010 121 — Ground Face Block Wall

	TYPE	SIZE (IN.)	SCORES	CORE FILL		MAT.	INST.	TOTAL
1200	Hollow	4x8x16	0	none		5.65	6.75	12.40
1250			1	none		5.75	6.85	12.60
1300			2 to 5	none		5.90	6.95	12.85
1400	RB2010 -200	6x8x16	0	perlite		7.20	7.75	14.95
1410			.	styrofoam		8.85	7.45	16.30
1440				none		6.90	7.45	14.35
1450			1	perlite		7.30	7.90	15.20
1460				styrofoam		8.95	7.60	16.55
1490				none		7	7.60	14.60
1500			2 to 5	perlite		7.50	8	15.50
1510				styrofoam		9.15	7.70	16.85
1540				none		7.20	7.70	14.90
1600		8x8x16	0	perlite		8.70	8.40	17.10
1610				styrofoam		10.20	8	18.20
1640				none		8.20	8	16.20
1650			1	perlite		8.80	8.50	17.30
1660				styrofoam		10.30	8.10	18.40
1690				none		8.30	8.10	16.40
1700			2 to 5	perlite		8.95	8.65	17.60
1710				styrofoam		10.45	8.25	18.70
1740				none		8.45	8.25	16.70

152

B20 Exterior Enclosure

B2010 Exterior Walls

B2010 121 — Ground Face Block Wall

	TYPE	SIZE (IN.)	SCORES	CORE FILL		MAT.	INST.	TOTAL
1800	Hollow	12x8x16	0	perlite		11.40	11	22.40
1810				styrofoam		13.05	10.25	23.30
1840				none		10.60	10.25	20.85
1850			1	perlite		11.50	11.15	22.65
1860				styrofoam		13.15	10.40	23.55
1890				none		10.70	10.40	21.10
1900		12x8x16	2 to 5	perlite		11.55	11.35	22.90
1910				styrofoam		13.20	10.60	23.80
1940				none		10.75	10.60	21.35
2200	75% solid	4x8x16	0	none		7.25	6.85	14.10
2250			1	none		7.35	6.95	14.30
2300			2 to 5	none		7.45	7.05	14.50
2400		6x8x16	0	perlite		9	7.75	16.75
2440				none		8.85	7.60	16.45
2450			1	perlite		9.15	7.85	17
2490				none		9	7.70	16.70
2500			2 to 5	perlite		9.25	8	17.25
2540				none		9.10	7.85	16.95
2600		8x8x16	0	perlite		10.75	8.30	19.05
2640				none		10.50	8.10	18.60
2650			1	perlite		10.85	8.45	19.30
2690				none		10.60	8.25	18.85
2700			2 to 5	perlite		10.95	8.60	19.55
2740				none		10.70	8.40	19.10
2800		12x8x16	0	perlite		14	10.80	24.80
2840				none		13.60	10.40	24
2850			1	perlite		14.10	11	25.10
2890				none		13.70	10.60	24.30
2900			2 to 5	perlite		14.15	11.25	25.40
2940				none		13.75	10.85	24.60
3200	Solid	4x8x16	0	none		8.85	6.95	15.80
3250			1	none		8.95	7.05	16
3300			2 to 5	none		9.05	7.15	16.20
3400		6x8x16	0	none		10.75	7.70	18.45
3450			1	none		10.85	7.85	18.70
3500			2 to 5	none		10.95	8	18.95
3600		8x8x16	0	none		12.80	8.25	21.05
3650			1	none		12.90	8.40	21.30
3700			2 to 5	none		13	8.55	21.55
3800		12x8x16	0	none		16.55	10.60	27.15
3850			1	none		16.65	10.85	27.50
3900			2 to 5	none		16.75	11.05	27.80

B2010 122 — Reinforced Ground Face Block Wall

	TYPE	SIZE (IN.)	SCORES	VERT. REINF. & GROUT SPACING		MAT.	INST.	TOTAL
5200	Hollow	4x8x16	0	#4 @ 48"		5.85	7.35	13.20
5250			1	#4 @ 48"		5.95	7.45	13.40
5300			2 to 5	#4 @ 48"		6.10	7.55	13.65
5400		6x8x16	0	#4 @ 48"		7.20	8.05	15.25
5420				#5 @ 32"		7.50	8.30	15.80
5430				#5 @ 16"		8.05	9.15	17.20

B2010 Exterior Walls

B2010 122	Reinforced Ground Face Block Wall

	TYPE	SIZE (IN.)	SCORES	VERT. REINF. & GROUT SPACING		COST PER S.F.		
						MAT.	INST.	TOTAL
5450	Hollow	6x8x16	1	#4 @ 48"		7.30	8.20	15.50
5470				#5 @ 32"		7.60	8.45	16.05
5480				#5 @ 16"		8.15	9.30	17.45
5500			2 to 5	#4 @ 48"		7.50	8.30	15.80
5520				#5 @ 32"		7.80	8.55	16.35
5530				#5 @ 16"		8.35	9.40	17.75
5600		8x8x16	0	#4 @ 48"		8.55	8.75	17.30
5620				#5 @ 32"		8.85	9.05	17.90
5630				#5 @ 16"		9.50	10.05	19.55
5650			1	#4 @ 48"		8.65	8.85	17.50
5670				#5 @ 32"		8.95	9.15	18.10
5680				#5 @ 16"		9.60	10.15	19.75
5700			2 to 5	#4 @ 48"		8.80	9	17.80
5720				#5 @ 32"		9.10	9.30	18.40
5730				#5 @ 16"		9.75	10.30	20.05
5800		12x8x16	0	#4 @ 48"		11.10	11	22.10
5820				#5 @ 32"		11.45	11.30	22.75
5830				#5 @ 16"		12.30	12.35	24.65
5850			1	#4 @ 48"		11.20	11.15	22.35
5870				#5 @ 32"		11.55	11.45	23
5880				#5 @ 16"		12.40	12.50	24.90
5900			2 to 5	#4 @ 48"		11.25	11.35	22.60
5920				#5 @ 32"		11.60	11.65	23.25
5930				#5 @ 16"		12.45	12.70	25.15
6200	75% solid	6x8x16	0	#4 @ 48"		9.05	8	17.05
6220				#5 @ 32"		9.30	8.15	17.45
6230				#5 @ 16"		9.80	8.75	18.55
6250			1	#4 @ 48"		9.20	8.10	17.30
6270				#5 @ 32"		9.45	8.25	17.70
6280				#5 @ 16"		9.95	8.85	18.80
6300			2 to 5	#4 @ 48"		9.30	8.25	17.55
6330				#5 @ 16"		10.05	9	19.05
6400		8x8x16	0	#4 @ 48"		10.75	8.60	19.35
6420				#5 @ 32"		11	8.80	19.80
6430				#5 @ 16"		11.50	9.45	20.95
6450			1	#4 @ 48"		10.85	8.75	19.60
6470				#5 @ 32"		11.10	8.95	20.05
6480				#5 @ 16"		11.60	9.60	21.20
6500			2 to 5	#4 @ 48"		10.95	8.90	19.85
6520				#5 @ 32"		11.20	9.10	20.30
6530				#5 @ 16"		11.70	9.75	21.45
6600		12x8x16	0	#4 @ 48"		13.90	10.90	24.80
6620				#5 @ 32"		14.20	11.15	25.35
6630				#5 @ 16"		14.80	11.80	26.60
6650			1	#4 @ 48"		14	11.10	25.10
6670				#5 @ 32"		14.30	11.35	25.65
6680				#5 @ 16"		14.90	12	26.90
6700			2 to 5	#4 @ 48"		14.05	11.35	25.40
6720				#5 @ 32"		14.35	11.60	25.95
6730				#5 @ 16"		14.95	12.25	27.20

B20 Exterior Enclosure

B2010 Exterior Walls

B2010 122					Reinforced Ground Face Block Wall				

	TYPE	SIZE (IN.)	SCORES	VERT. REINF. & GROUT SPACING		COST PER S.F.		
						MAT.	INST.	TOTAL
6800	Solid-double	2-4x8x16	0	#4 @ 48" E.W.		18.60	15.20	33.80
6830	Wythe			#5 @ 16" E.W.		19.70	15.85	35.55
6850			1	#4 @ 48" E.W.		18.80	15.40	34.20
6880				#5 @ 16" E.W.		19.90	16.05	35.95
6900			2 to 5	#4 @ 48" E.W.		19	15.60	34.60
6930				#5 @ 16" E.W.		20	16.25	36.25
7000		2-6x8x16	0	#4 @ 48" E.W.		22.50	16.70	39.20
7030				#5 @ 16" E.W.		23.50	17.35	40.85
7050			1	#4 @ 48" E.W.		22.50	17	39.50
7080				#5 @ 16" E.W.		23.50	17.65	41.15
7100			2 to 5	#4 @ 48" E.W.		22.50	17.30	39.80
7130				#5 @ 16" E.W.		24	17.95	41.95

B2010 Exterior Walls

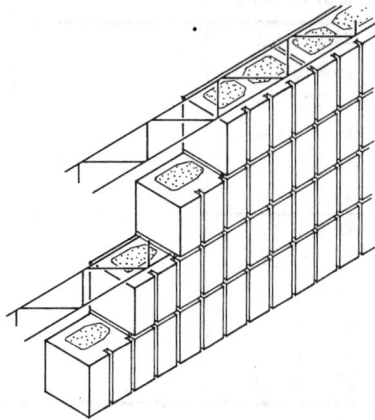

Exterior miscellaneous block walls are defined in the following terms: structural reinforcement, finish texture, percent solid, size, weight and insulation. Within each of these categories two to four variations are shown. No costs are included for brick shelf or relieving angles. Costs include control joints every 20′ and horizontal reinforcing.

System Components		QUANTITY	UNIT	COST PER S.F.		
				MAT.	INST.	TOTAL
SYSTEM B2010 123 2700						
UNREINFORCED HEXAGONAL BLOCK WALL, 8″X8″X16″, 125PCF, PERLITE FILL						
Hexagonal block partition, 8″ thick		1.000	S.F.	4.88	7.80	12.68
Perlite insulation		1.000	S.F.	.50	.37	.87
Horizontal joint reinf., alternate courses		.800	L.F.	.12	.16	.28
Control joint		.050	L.F.	.07	.06	.13
	TOTAL			5.57	8.39	13.96

B2010 123				Miscellaneous Block Wall			

	TYPE	SIZE (IN.)	WEIGHT (P.C.F.)	CORE FILL	COST PER S.F.		
					MAT.	INST.	TOTAL
1100	Deep groove-hollow	4x8x16	125	none	3.77	6.75	10.52
1150			105	none	3.94	6.55	10.49
1200	RB2010 -200	6x8x16	125	perlite	5.20	7.45	12.65
1210				styrofoam	6.85	7.15	14
1240				none	4.84	7.15	11.99
1250			105	perlite	5.40	6.85	12.25
1260				styrofoam	7.05	6.55	13.60
1290				none	5.05	6.55	11.60
1300		8x8x16	125	perlite	6.70	8.10	14.80
1310				styrofoam	8.20	7.70	15.90
1340				none	6.20	7.70	13.90
1350			105	perlite	7.35	7.85	15.20
1360				styrofoam	8.85	7.45	16.30
1390				none	6.85	7.45	14.30
1400	Deep groove-75% solid	4x8x16	125	none	4.83	6.85	11.68
1450			105	none	5.05	6.65	11.70
1500		6x8x16	125	perlite	6.30	7.40	13.70
1540				none	6.15	7.25	13.40
1550			105	perlite	6.60	7.20	13.80
1590				none	6.45	7.05	13.50

B2010 123	Miscellaneous Block Wall

	TYPE	SIZE (IN.)	WEIGHT (P.C.F.)	CORE FILL		COST PER S.F.		
						MAT.	INST.	TOTAL
1600	Deep groove	8x8x16	125	perlite		8.15	8.05	16.20
1640	75% solid			none		7.90	7.85	15.75
1650			105	perlite		8.95	7.80	16.75
1690				none		8.70	7.60	16.30
1700	Deep groove-solid	4x8x16	125	none		5.80	6.95	12.75
1750			105	none		6.10	6.75	12.85
1800		6x8x16	125	none		7.45	7.35	14.80
1850			105	none		7.85	7.15	15
1900		8x8x16	125	none		9.65	8	17.65
1950			105	none		10.60	7.70	18.30
2100	Fluted	4x8x16	125	none		3.79	6.75	10.54
2150			105	none		3.90	6.55	10.45
2200		6x8x16	125	perlite		4.96	7.45	12.41
2210				styrofoam		6.60	7.15	13.75
2240				none		4.62	7.15	11.77
2300		8x8x16	125	perlite		6.05	8.10	14.15
2310				styrofoam		7.55	7.70	15.25
2340				none		5.55	7.70	13.25
2500	Hex-hollow	4x8x16	125	none		2.73	6.75	9.48
2550			105	none		2.92	6.55	9.47
2600		6x8x16	125	perlite		4.16	7.75	11.91
2610				styrofoam		5.80	7.45	13.25
2640				none		3.82	7.45	11.27
2650			105	perlite		4.47	7.55	12.02
2660				styrofoam		6.10	7.25	13.35
2690				none		4.13	7.25	11.38
2700		8x8x16	125	perlite		5.55	8.40	13.95
2710				styrofoam		7.05	8	15.05
2740				none		5.05	8	13.05
2750			105	perlite		5.95	8.10	14.05
2760				styrofoam		7.45	7.70	15.15
2790				none		5.45	7.70	13.15
2800	Hex-solid	4x8x16	125	none		3.70	6.95	10.65
2850			105	none		4.02	6.75	10.77
3100	Slump block	4x4x16	125	none		4.51	5.55	10.06
3200		6x4x16	125	perlite		6.45	6	12.45
3210				styrofoam		8.10	5.70	13.80
3240				none		6.15	5.70	11.85
3300		8x4x16	125	perlite		7.35	6.30	13.65
3310				styrofoam		8.85	5.90	14.75
3340				none		6.85	5.90	12.75
3400		12x4x16	125	perlite		13.55	7.80	21.35
3410				styrofoam		15.10	7.05	22.15
3440				none		12.75	7.05	19.80
3600		6x6x16	125	perlite		5.70	6.20	11.90
3610				styrofoam		7.35	5.90	13.25
3640				none		5.40	5.90	11.30
3700		8x6x16	125	perlite		7.85	6.50	14.35
3710				styrofoam		9.35	6.10	15.45
3740				none		7.35	6.10	13.45
3800		12x6x16	125	perlite		12.40	8.35	20.75
3810				styrofoam		13.95	7.60	21.55
3840				none		11.60	7.60	19.20

B2010 Exterior Walls

| B2010 124 | | | | Reinforced Misc. Block Walls | | | | |

	TYPE	SIZE (IN.)	WEIGHT (P.C.F.)	VERT. REINF. & GROUT SPACING		COST PER S.F.		
						MAT.	INST.	TOTAL
5100	Deep groove-hollow	4x8x16	125	#4 @ 48"		3.97	7.35	11.32
5150			105	#4 @ 48"		4.14	7.15	11.29
5200		6x8x16	125	#4 @ 48"		5.15	7.75	12.90
5220				#5 @ 32"		5.45	8	13.45
5230				#5 @ 16"		6	8.85	14.85
5250			105	#4 @ 48"		5.35	7.15	12.50
5270				#5 @ 32"		5.65	7.40	13.05
5280				#5 @ 16"		6.20	8.25	14.45
5300		8x8x16	125	#4 @ 48"		6.55	8.45	15
5320				#5 @ 32"		6.85	8.75	15.60
5330				#5 @ 16"		7.50	9.75	17.25
5350			105	#4 @ 48"		7.20	8.20	15.40
5370				#5 @ 32"		7.50	8.50	16
5380				#5 @ 16"		8.15	9.50	17.65
5500	Deep groove-75% solid	6x8x16	125	#4 @ 48"		6.35	7.65	14
5520				#5 @ 32"		6.60	7.80	14.40
5530				#5 @ 16"		7.10	8.40	15.50
5550			105	#4 @ 48"		6.65	7.45	14.10
5570				#5 @ 32"		6.95	7.75	14.70
5580				#5 @ 16"		7.40	8.20	15.60
5600		8x8x16	125	#4 @ 48"		8.15	8.35	16.50
5620				#5 @ 32"		8.40	8.55	16.95
5630				#5 @ 16"		8.90	9.20	18.10
5650			105	#4 @ 48"		8.95	8.10	17.05
5670				#5 @ 32"		9.20	8.30	17.50
5680				#5 @ 16"		9.70	8.95	18.65
5700	Deep groove-solid	2-4x8x16	125	#4 @ 48" E.W.		12.50	15.20	27.70
5730	Double wythe			#5 @ 16" E.W.		13.65	15.90	29.55
5750			105	#4 @ 48" E.W.		13.10	14.80	27.90
5780				#5 @ 16" E.W.		14.25	15.50	29.75
5800		2-6x8x16	125	#4 @ 48" E.W.		15.70	16	31.70
5830				#5 @ 16" E.W.		16.85	16.70	33.55
5850			105	#4 @ 48" E.W.		16.50	15.60	32.10
5880				#5 @ 16" E.W.		17.65	16.30	33.95
6100	Fluted	4x8x16	125	#4 @ 48"		3.99	7.35	11.34
6150			105	#4 @ 48"		4.10	7.15	11.25
6200		6x8x16	125	#4 @ 48"		4.92	7.75	12.67
6220				#5 @ 32"		5.20	8	13.20
6230				#5 @ 16"		5.80	8.85	14.65
6300		8x8x16	125	#4 @ 48"		5.90	8.45	14.35
6320				#5 @ 32"		6.20	8.75	14.95
6330				#5 @ 16"		6.85	9.75	16.60
6500	Hex-hollow	4x8x16	125	#4 @ 48"		2.93	7.35	10.28
6550			105	#4 @ 48"		3.12	7.15	10.27
6620				#5 @ 32"		4.42	8.30	12.72
6630				#5 @ 16"		4.99	9.15	14.14
6650		6x8x16	105	#4 @ 48"		4.43	7.85	12.28
6670				#5 @ 32"		4.73	8.10	12.83
6680				#5 @ 16"		5.30	8.95	14.25
6700		8x8x16	125	#4 @ 48"		5.40	8.75	14.15
6720				#5 @ 32"		5.75	9.05	14.80
6730				#5 @ 16"		6.40	10.05	16.45

B20 Exterior Enclosure

B2010 Exterior Walls

| B2010 124 | Reinforced Misc. Block Walls |

	TYPE	SIZE (IN.)	WEIGHT (P.C.F.)	VERT. REINF. & GROUT SPACING		COST PER S.F.		
						MAT.	INST.	TOTAL
6750	Hex-solid	8x8x16	105	#4 @ 48"		5.80	8.45	14.25
6770				#5 @ 32"		6.10	8.75	14.85
6800	Hex-solid	2-4x8x16	125	#4 @ 48" E.W.		8.20	15.15	23.35
6830	Double wythe			#5 @ 16" E.W.		9.40	15.90	25.30
6850			105	#4 @ 48" E.W.		8.85	14.75	23.60
6880				#5 @ 16" E.W.		10.05	15.50	25.55
7100	Slump block	4x4x16	125	#4 @ 48"		4.71	6.15	10.86
7200		6x4x16	125	#4 @ 48"		6.45	6.30	12.75
7220				#5 @ 32"		6.75	6.80	13.55
7230				#5 @ 16"		7.30	7.40	14.70
7300	Slump block	8x4x16	125	#4 @ 48"		7.20	6.65	13.85
7320				#5 @ 32"		7.50	6.95	14.45
7330				#5 @ 16"		8.15	7.95	16.10
7400		12x4x16	125	#4 @ 48"		13.25	7.80	21.05
7420				#5 @ 32"		13.60	8.10	21.70
7430				#5 @ 16"		14.45	9.15	23.60
7600		6x6x16	125	#4 @ 48"		5.70	6.50	12.20
7620				#5 @ 32"		6	6.75	12.75
7630				#5 @ 16"		6.55	7.60	14.15
7700		8x6x16	125	#4 @ 48"		7.70	6.85	14.55
7720				#5 @ 32"		8	7.15	15.15
7730				#5 @ 16"		8.65	8.15	16.80
7800		12x6x16	125	#4 @ 48"		12.10	8.35	20.45
7820				#5 @ 32"		12.45	8.65	21.10
7830				#5 @ 16"		13.30	9.70	23

B2010 Exterior Walls

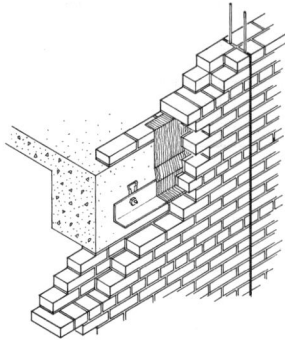

Exterior solid brick walls are defined in the following terms: structural reinforcement, type of brick, thickness, and bond. Sixteen different types of face bricks are presented, with single wythes shown in four different bonds. Six types of reinforced single wythe walls are also given, and twice that many for reinforced double wythe. Shelf angles are included in system components. These walls do not include ties and as such are not tied to a backup wall.

System Components	QUANTITY	UNIT	COST PER S.F.		
			MAT.	INST.	TOTAL
SYSTEM B2010 125 1000					
BRICK WALL, COMMON, SINGLE WYTHE, 4″ THICK, RUNNING BOND					
Common brick wall, running bond	1.000	S.F.	3.11	9.80	12.91
Wash brick	1.000	S.F.	.06	.88	.94
Control joint, backer rod	.100	L.F.		.11	.11
Control joint, sealant	.100	L.F.	.02	.27	.29
Shelf angle	1.000	Lb.	1.27	.89	2.16
Flashing	.100	S.F.	.11	.32	.43
TOTAL			4.57	12.27	16.84

B2010 125			Solid Brick Walls - Single Wythe			COST PER S.F.		
	TYPE	THICKNESS (IN.)	BOND			MAT.	INST.	TOTAL
1000	Common	4	running			4.57	12.25	16.82
1010			common			5.45	14	19.45
1050			Flemish			6	17	23
1100			English			6.65	18	24.65
1150	Standard	4	running			7.60	12.70	20.30
1160			common			9	14.65	23.65
1200			Flemish			9.95	17.50	27.45
1250			English			11.15	18.55	29.70
1300	Glazed	4	running			15.65	13.20	28.85
1310			common			18.65	15.35	34
1350			Flemish			20.50	18.55	39.05
1400			English			23	19.80	42.80
1450	Engineer	4	running			5.55	11.10	16.65
1460			common			6.45	12.70	19.15
1500			Flemish			7.10	15.35	22.45
1550			English			7.95	16.10	24.05
1600	Economy	4	running			5.35	9.70	15.05
1610			common			6.25	11.10	17.35
1650			Flemish			6.85	13.20	20.05
1700			English			16.40	14	30.40
1750	Double	4	running			10.25	7.95	18.20
1760			common			12.10	9	21.10
1800			Flemish			13.35	10.65	24
1850			English			14.95	11.30	26.25

B20 Exterior Enclosure

B2010 Exterior Walls

B2010 125 — Solid Brick Walls - Single Wythe

	TYPE	THICKNESS (IN.)	BOND			COST PER S.F. MAT.	INST.	TOTAL
1900	Fire	4-1/2	running			11.45	11.10	22.55
1910			common			13.55	12.70	26.25
1950			Flemish			15	15.35	30.35
2000			English			16.85	16.10	32.95
2050	King	3-1/2	running			4.73	9.95	14.68
2060			common			5.50	11.45	16.95
2100			Flemish			6.55	13.70	20.25
2150			English			7.25	14.30	21.55
2200	Roman	4	running			7.50	11.45	18.95
2210			common			8.90	13.20	22.10
2250			Flemish			9.80	15.70	25.50
2300			English			11.05	17	28.05
2350	Norman	4	running			6.50	9.50	16
2360			common			7.65	10.80	18.45
2400			Flemish			16.95	12.95	29.90
2450			English			9.45	13.70	23.15
2500	Norwegian	4	running			5.40	8.45	13.85
2510			common			6.30	9.60	15.90
2550			Flemish			6.90	11.45	18.35
2600			English			7.70	12.05	19.75
2650	Utility	4	running			5.05	7.45	12.50
2660			common			5.85	8.40	14.25
2700			Flemish			6.40	9.95	16.35
2750			English			7.15	10.50	17.65
2800	Triple	4	running			4.06	6.75	10.81
2810			common			4.62	7.55	12.17
2850			Flemish			5	8.90	13.90
2900			English			5.50	9.30	14.80
2950	SCR	6	running			7.25	9.70	16.95
2960			common			8.60	11.10	19.70
3000			Flemish			9.55	13.20	22.75
3050			English			10.75	14	24.75
3100	Norwegian	6	running			6.30	8.65	14.95
3110			common			7.35	9.85	17.20
3150			Flemish			8.10	11.65	19.75
3200			English			9.05	12.25	21.30
3250	Jumbo	6	running			6.80	7.65	14.45
3260			common			8	8.65	16.65
3300			Flemish			8.85	10.25	19.10
3350			English			9.95	10.80	20.75

B2010 126 — Solid Brick Walls - Double Wythe

	TYPE	THICKNESS (IN.)	COLLAR JOINT THICKNESS (IN.)			COST PER S.F. MAT.	INST.	TOTAL
4100	Common	8	3/4			8.45	20.50	28.95
4150	Standard	8	1/2			14.45	21.50	35.95
4200	Glazed	8	1/2			30.50	22	52.50
4250	Engineer	8	1/2			10.25	18.80	29.05
4300	Economy	8	3/4			9.90	16.15	26.05
4350	Double	8	3/4			12.50	13.05	25.55
4400	Fire	8	3/4			22	18.80	40.80
4450	King	7	3/4			8.65	16.50	25.15

B2010 Exterior Walls

B2010 126	Solid Brick Walls - Double Wythe							

	TYPE	THICKNESS (IN.)	COLLAR JOINT THICKNESS (IN.)			COST PER S.F.		
						MAT.	INST.	TOTAL
4500	Roman	8	1			14.20	19.35	33.55
4550	Norman	8	3/4			12.25	15.80	28.05
4600	Norwegian	8	3/4			10	14	24
4650	Utility	8	3/4			9.30	12.10	21.40
4700	Triple	8	3/4			7.30	10.90	18.20
4750	SCR	12	3/4			13.70	16.20	29.90
4800	Norwegian	12	3/4			11.75	14.30	26.05
4850	Jumbo	12	3/4			12.80	12.50	25.30

B2010 127	Solid Brick Walls, Reinforced							

	TYPE	THICKNESS (IN.)	WYTHE	REINF. & SPACING		COST PER S.F.		
						MAT.	INST.	TOTAL
7200	Common	4"	1	#4 @ 48"vert		4.82	12.75	17.57
7220				#5 @ 32"vert		5.05	12.90	17.95
7230				#5 @ 16"vert		5.45	13.50	18.95
7300	Utility	4"	1	#4 @ 48"vert		5.30	7.95	13.25
7320				#5 @ 32"vert		5.55	8.10	13.65
7330				#5 @ 16"vert		5.90	8.70	14.60
7400	SCR	6"	1	#4 @ 48"vert		7.55	10.20	17.75
7420				#5 @ 32"vert		7.80	10.35	18.15
7430				#5 @ 16"vert		8.30	10.95	19.25
7500	Jumbo	6"	1	#4 @ 48"vert		7.05	8.10	15.15
7520				#5 @ 32"vert		7.30	8.25	15.55
7530				#5 @ 16"vert		7.65	8.75	16.40
7600	Jumbo	8"	1	#4 @ 48"vert		8.35	8.30	16.65
7620				#5 @ 32"vert		8.60	8.55	17.15
7630				#5 @ 16"vert		9.10	9.15	18.25
7700	King	9"	2	#4 @ 48" E.W.		9.15	17.30	26.45
7720				#5 @ 32" E.W.		9.60	17.40	27
7730				#5 @ 16" E.W.		10.30	17.95	28.25
7800	Common	10"	2	#4 @ 48" E.W.		8.95	21.50	30.45
7820				#5 @ 32" E.W.		9.40	21.50	30.90
7830				#5 @ 16" E.W.		10.10	22	32.10
7900	Standard	10"	2	#4 @ 48" E.W.		14.95	22	36.95
7920				#5 @ 32" E.W.		15.40	22	37.40
7930				#5 @ 16" E.W.		16.10	22.50	38.60
8000	Engineer	10"	2	#4 @ 48" E.W.		10.75	19.60	30.35
8020				#5 @ 32" E.W.		11.20	19.70	30.90
8030				#5 @ 16" E.W.		11.90	20	31.90
8100	Economy	10"	2	#4 @ 48" E.W.		10.40	16.95	27.35
8120				#5 @ 32" E.W.		10.85	17.05	27.90
8130				#5 @ 16" E.W.		11.55	17.60	29.15
8200	Double	10"	2	#4 @ 48" E.W.		13	13.85	26.85
8220				#5 @ 32" E.W.		13.45	13.95	27.40
8230				#5 @ 16" E.W.		14.15	14.50	28.65
8300	Fire	10"	2	#4 @ 48" E.W.		22.50	19.60	42.10
8320				#5 @ 32" E.W.		23	19.70	42.70
8330				#5 @ 16" E.W.		24	20	44
8400	Roman	10"	2	#4 @ 48" E.W.		14.70	20	34.70
8420				#5 @ 32" E.W.		15.15	20	35.15
8430				#5 @ 16" E.W.		15.85	20.50	36.35

B20 Exterior Enclosure

B2010 Exterior Walls

B2010 127	Solid Brick Walls, Reinforced

	TYPE	THICKNESS (IN.)	WYTHE	REINF. & SPACING		COST PER S.F.		
						MAT.	INST.	TOTAL
8500	Norman	10"	2	#4 @ 48" E.W.		12.75	16.60	29.35
8520				#5 @ 32" E.W.		13.20	16.70	29.90
8530				#5 @ 16" E.W.		13.90	17.25	31.15
8600	Norwegian	10"	2	#4 @ 48" E.W.		10.50	14.80	25.30
8620				#5 @ 32" E.W.		10.95	14.90	25.85
8630				#5 @ 16" E.W.		11.65	15.45	27.10
8700	Utility	10"	2	#4 @ 48" E.W.		9.80	12.90	22.70
8720				#5 @ 32" E.W.		10.25	13	23.25
8730				#5 @ 16" E.W.		10.95	13.55	24.50
8800	Triple	10"	2	#4 @ 48" E.W.		7.80	11.70	19.50
8820				#5 @ 32" E.W.		8.25	11.80	20.05
8830				#5 @ 16" E.W.		8.95	12.35	21.30
8900	SCR	14"	2	#4 @ 48" E.W.		14.20	16.95	31.15
8920				#5 @ 32" E.W.		14.65	17.05	31.70
8930				#5 @ 16" E.W.		15.35	17.60	32.95
9000	Jumbo	14"	2	#4 @ 48" E.W.		13.30	13.25	26.55
9030				#5 @ 16" E.W.		14.45	13.90	28.35

B2010 Exterior Walls

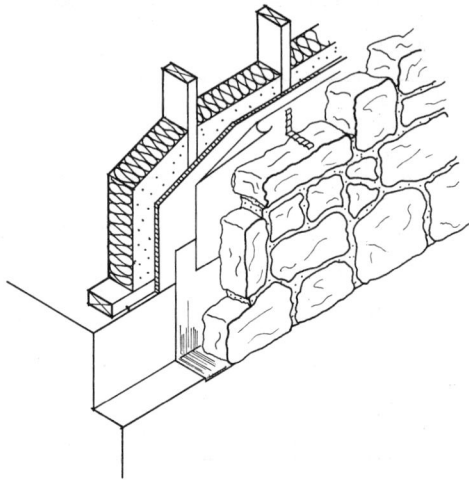

The table below lists costs per S.F. for stone veneer walls on various backup using different stone. Typical components for a system are shown in the component block below.

System Components	QUANTITY	UNIT	COST PER S.F.		
			MAT.	INST.	TOTAL
SYSTEM B2010 128 2000					
ASHLAR STONE VENEER 4″, 2″X4″ STUD 16″ O.C. BACK UP, 8′ HIGH					
Ashlar veneer, 4″ thick, sawn face, split joints, low priced stone	1.000	S.F.	10.20	13.15	23.35
Framing, 2x4 studs 8′ high 16″ O.C.	.920	B.F.	.36	1.24	1.60
Wall ties for stone veneer, galv. corrg 7/8″ x 7″, 22 gauge	.700	Ea.	.08	.33	.41
Asphalt felt sheathing paper, 15 lb.	1.000	S.F.	.05	.13	.18
Sheathing plywood on wall CDX 1/2″	1.000	S.F.	.51	.88	1.39
Fiberglass insulation, 3-1/2″, cr-11	1.000	S.F.	.63	.37	1
Flashing, copper, paperback 1 side, 3 oz	.125	S.F.	.26	.18	.44
TOTAL			12.09	16.28	28.37

B2010 128	Stone Veneer	COST PER S.F.		
		MAT.	INST.	TOTAL
2000	Ashlar veneer, 4″, 2″x4″ stud backup, 16″ O.C., 8′ high, low priced stone	12.10	16.30	28.40
2100	Metal stud backup, 8′ high, 16″ O.C.	13.40	16.55	29.95
2150	24″ O.C.	12.95	16.10	29.05
2200	Conc. block backup, 4″ thick	13.40	19.95	33.35
2300	6″ thick	14.05	20	34.05
2350	8″ thick	14.20	20.50	34.70
2400	10″ thick	15.05	22	37.05
2500	12″ thick	15.50	23.50	39
3100	High priced stone, wood stud backup, 10′ high, 16″ O.C.	18.40	18.45	36.85
3200	Metal stud backup, 10′ high, 16″ O.C.	19.65	18.70	38.35
3250	24″ O.C.	19.20	18.25	37.45
3300	Conc. block backup, 10′ high, 4″ thick	19.65	22	41.65
3350	6″ thick	20.50	22	42.50
3400	8″ thick	20.50	23	43.50
3450	10″ thick	21.50	24	45.50
3500	12″ thick	22	25.50	47.50
4000	Indiana limestone 2″ thick, sawn finish, wood stud backup, 10′ high, 16″ O.C.	40.50	12.50	53
4100	Metal stud backup, 10′ high, 16″ O.C.	41.50	12.75	54.25
4150	24″ O.C.	41	12.30	53.30
4200	Conc. block backup, 4″ thick	41.50	16.15	57.65
4250	6″ thick	42.50	16.35	58.85
4300	8″ thick	42.50	16.90	59.40

B2010 Exterior Walls

B2010 128	Stone Veneer	COST PER S.F.		
		MAT.	INST.	TOTAL
4350	10" thick	43.50	18.10	61.60
4400	12" thick	44	19.80	63.80
4450	2" thick, smooth finish, wood stud backup, 8' high, 16" O.C.	40.50	12.50	53
4550	Metal stud backup, 8' high, 16" O.C.	41.50	12.55	54.05
4600	24" O.C.	41	12.15	53.15
4650	Conc. block backup, 4" thick	41.50	15.95	57.45
4700	6" thick	42.50	16.15	58.65
4750	8" thick	42.50	16.70	59.20
4800	10" thick	43.50	18.10	61.60
4850	12" thick	44	19.80	63.80
5350	4" thick, smooth finish, wood stud backup, 8' high, 16" O.C.	47	12.50	59.50
5450	Metal stud backup, 8' high, 16" O.C.	48	12.75	60.75
5500	24" O.C.	47.50	12.30	59.80
5550	Conc. block backup, 4" thick	48	16.15	64.15
5600	6" thick	49	16.35	65.35
5650	8" thick	49	16.90	65.90
5700	10" thick	50	18.10	68.10
5750	12" thick	50.50	19.80	70.30
6000	Granite, gray or pink, 2" thick, wood stud backup, 8' high, 16" O.C.	24	23.50	47.50
6100	Metal studs, 8' high, 16" O.C.	25	24	49
6150	24" O.C.	24.50	23.50	48
6200	Conc. block backup, 4" thick	25	27.50	52.50
6250	6" thick	26	27.50	53.50
6300	8" thick	26	28	54
6350	10" thick	27	29	56
6400	12" thick	27.50	31	58.50
6900	4" thick, wood stud backup, 8' high, 16" O.C.	38.50	27	65.50
7000	Metal studs, 8' high, 16" O.C.	39.50	27.50	67
7050	24" O.C.	39	27	66
7100	Conc. block backup, 4" thick	39.50	31	70.50
7150	6" thick	40.50	31	71.50
7200	8" thick	40.50	31.50	72
7250	10" thick	41.50	32.50	74
7300	12" thick	42	34.50	76.50

B2010 Exterior Walls

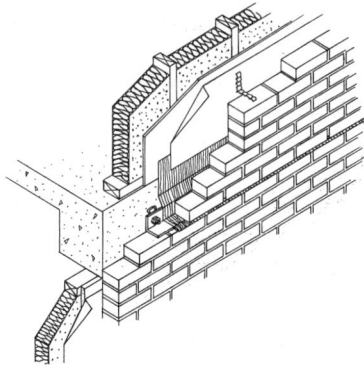

Exterior brick veneer/stud backup walls are defined in the following terms: type of brick and studs, stud spacing and bond. All systems include a back-up wall, a control joint every 20', a brick shelf every 12' of height, ties to the backup and the necessary dampproofing, flashing and insulation.

System Components	QUANTITY	UNIT	COST PER S.F.		
			MAT.	INST.	TOTAL
SYSTEM B2010 129 1100					
STANDARD BRICK VENEER, 2"X4" STUD BACKUP @ 16"O.C., RUNNING BOND					
Standard brick wall, 4" thick, running bond	1.000	S.F.	6.15	10.25	16.40
Wash smooth brick	1.000	S.F.	.06	.88	.94
Joint backer rod	.100	L.F.		.11	.11
Sealant	.100	L.F.	.02	.27	.29
Wall ties, corrugated, 7/8" x 7", 22 gauge	.003	Ea.	.03	.14	.17
Shelf angle	1.000	Lb.	1.27	.89	2.16
Wood stud partition, backup, 2" x 4" @ 16" O.C.	1.000	S.F.	.34	.99	1.33
Sheathing, plywood, CDX, 1/2"	1.000	S.F.	.51	.71	1.22
Building paper, asphalt felt, 15 lb.	1.000	S.F.	.05	.13	.18
Fiberglass insulation, batts, 3-1/2" thick paper backing	1.000	S.F.	.63	.37	1
Flashing, copper, paperbacked	.100	S.F.	.11	.32	.43
TOTAL			9.17	15.06	24.23

B2010 129				Brick Veneer/Wood Stud Backup				
	FACE BRICK	STUD BACKUP	STUD SPACING (IN.)	BOND		COST PER S.F.		
						MAT.	INST.	TOTAL
1100	Standard	2x4-wood	16	running		9.15	15.05	24.20
1120				common		10.55	17	27.55
1140				Flemish		11.50	19.85	31.35
1160				English		12.70	21	33.70
1400		2x6-wood	16	running		9.45	15.15	24.60
1420				common		10.85	17.10	27.95
1440				Flemish		11.80	19.95	31.75
1460				English		13	21	34
1500			24	running		9.30	14.95	24.25
1520				common		10.70	16.90	27.60
1540				Flemish		11.65	19.75	31.40
1560				English		12.85	21	33.85
1700	Glazed	2x4-wood	16	running		17.20	15.55	32.75
1720				common		20	17.70	37.70
1740				Flemish		22	21	43
1760				English		24.50	22	46.50

B2010 Exterior Walls

B2010 129	Brick Veneer/Wood Stud Backup

	FACE BRICK	STUD BACKUP	STUD SPACING (IN.)	BOND		COST PER S.F.		
						MAT.	INST.	TOTAL
2000	Glazed	2x6-wood	16	running		17.50	15.65	33.15
2020				common		20.50	17.80	38.30
2040				Flemish		22.50	21	43.50
2060				English		25	22.50	47.50
2100			24	running		17.35	15.45	32.80
2120				common		20.50	17.60	38.10
2140				Flemish		22.50	21	43.50
2160				English		24.50	22	46.50
2300	Engineer	2x4-wood	16	running		7.10	13.45	20.55
2320				common		8	15.05	23.05
2340				Flemish		8.65	17.70	26.35
2360				English		9.50	18.45	27.95
2600		2x6-wood	16	running		7.35	13.55	20.90
2620				common		8.30	15.15	23.45
2640				Flemish		8.95	17.80	26.75
2660				English		9.80	18.55	28.35
2700			24	running		7.20	13.35	20.55
2720				common		8.15	14.95	23.10
2740				Flemish		8.80	17.60	26.40
2760				English		9.65	18.35	28
2900	Roman	2x4-wood	16	running		9.05	13.80	22.85
2920				common		10.45	15.55	26
2940				Flemish		11.35	18.05	29.40
2960				English		12.60	19.35	31.95
3200		2x6-wood	16	running		9.35	13.90	23.25
3220				common		10.75	15.65	26.40
3240				Flemish		11.65	18.05	29.70
3260				English		12.90	19.45	32.35
3300			24	running		9.20	13.70	22.90
3320				common		10.60	15.45	26.05
3340				Flemish		11.50	17.95	29.45
3360				English		12.75	19.25	32
3500	Norman	2x4-wood	16	running		8.05	11.85	19.90
3520				common		9.20	13.15	22.35
3540				Flemish		18.50	15.30	33.80
3560				English		11	16.05	27.05
3800		2x6-wood	16	running		8.35	11.95	20.30
3820				common		9.50	13.25	22.75
3840				Flemish		18.30	14.80	33.10
3860				English		11.30	16.15	27.45
3900			24	running		8.20	11.75	19.95
3920				common		9.35	13.05	22.40
3940				Flemish		18.65	15.20	33.85
3960				English		11.15	15.95	27.10
4100	Norwegian	2x4-wood	16	running		6.95	10.80	17.75
4120				common		7.85	11.95	19.80
4140				Flemish		8.45	13.80	22.25
4160				English		9.25	14.40	23.65
4400		2x6-wood	16	running		7.25	10.90	18.15
4420				common		8.15	12.05	20.20
4440				Flemish		8.75	13.90	22.65
4460				English		9.55	14.50	24.05

B20 Exterior Enclosure

B2010 Exterior Walls

B2010 129 — Brick Veneer/Wood Stud Backup

	FACE BRICK	STUD BACKUP	STUD SPACING (IN.)	BOND		COST PER S.F.		
						MAT.	INST.	TOTAL
4500	Norwegian	2x6-wood	24	running		7.10	10.70	17.80
4520				common		8	11.85	19.85
4540				Flemish		8.60	13.70	22.30
4560				English		9.40	14.30	23.70

B2010 130 — Brick Veneer/Metal Stud Backup

	FACE BRICK	STUD BACKUP	STUD SPACING (IN.)	BOND		COST PER S.F.		
						MAT.	INST.	TOTAL
5100	Standard	25ga.x6"NLB	24	running		9.15	14.75	23.90
5120				common		10.55	16.70	27.25
5140				Flemish		11	18.95	29.95
5160				English		12.70	20.50	33.20
5200		20ga.x3-5/8"NLB	16	running		9.30	15.35	24.65
5220				common		10.70	17.30	28
5240				Flemish		11.65	20	31.65
5260				English		12.85	21	33.85
5300			24	running		9.10	14.90	24
5320				common		10.50	16.85	27.35
5340				Flemish		11.45	19.70	31.15
5360				English		12.65	21	33.65
5400		16ga.x3-5/8"LB	16	running		10.25	15.55	25.80
5420				common		11.65	17.50	29.15
5440				Flemish		12.60	20.50	33.10
5460				English		13.80	21.50	35.30
5500			24	running		9.80	15.15	24.95
5520				common		11.20	17.10	28.30
5540				Flemish		12.15	19.95	32.10
5560				English		13.35	21	34.35
5700	Glazed	25ga.x6"NLB	24	running		17.20	15.25	32.45
5720				common		20	17.40	37.40
5740				Flemish		22	20.50	42.50
5760				English		24.50	22	46.50
5800		20ga.x3-5/8"NLB	24	running		17.15	15.40	32.55
5820				common		20	17.55	37.55
5840				Flemish		22	21	43
5860				English		24.50	22	46.50
6000		16ga.x3-5/8"LB	16	running		18.30	16.05	34.35
6020				common		21.50	18.20	39.70
6040				Flemish		23.50	21.50	45
6060				English		25.50	22.50	48
6100			24	running		17.85	15.65	33.50
6120				common		21	17.80	38.80
6140				Flemish		23	21	44
6160				English		25	22	47
6300	Engineer	25ga.x6"NLB	24	running		7.05	13.15	20.20
6320				common		8	14.75	22.75
6340				Flemish		8.65	17.40	26.05
6360				English		9.50	18.15	27.65
6400		20ga.x3-5/8"NLB	16	running		7.20	13.75	20.95
6420				common		8.15	15.35	23.50
6440				Flemish		8.80	18	26.80
6460				English		9.65	18.75	28.40

B2010 Exterior Walls

B2010 130 — Brick Veneer/Metal Stud Backup

	FACE BRICK	STUD BACKUP	STUD SPACING (IN.)	BOND		COST PER S.F.		
						MAT.	INST.	TOTAL
6500	Engineer	20ga.x3-5/8"NLB	24	running		6.55	12.55	19.10
6520				common		7.95	14.90	22.85
6540				Flemish		8.60	17.55	26.15
6560				English		9.45	18.30	27.75
6600		16ga.x3-5/8"LB	16	running		8.15	13.95	22.10
6620				common		9.10	15.55	24.65
6640				Flemish		9.75	18.20	27.95
6660				English		10.60	18.95	29.55
6700			24	running		7.70	13.55	21.25
6720				common		8.65	15.15	23.80
6740				Flemish		9.30	17.80	27.10
6760				English		10.15	18.55	28.70
6900	Roman	25ga.x6"NLB	24	running		9.05	13.50	22.55
6920				common		10.45	15.20	25.65
6940				Flemish		11.35	17.75	29.10
6960				English		12.60	19.05	31.65
7000		20ga.x3-5/8"NLB	16	running		9.20	14.10	23.30
7020				common		10.60	15.85	26.45
7040				Flemish		11.50	18.35	29.85
7060				English		12.75	19.65	32.40
7100			24	running		9	13.65	22.65
7120				common		10.40	15.40	25.80
7140				Flemish		11.30	17.90	29.20
7160				English		12.55	19.20	31.75
7200		16ga.x3-5/8"LB	16	running		10.15	14.30	24.45
7220				common		11.55	16.05	27.60
7240				Flemish		12.45	18.55	31
7260				English		13.70	19.85	33.55
7300			24	running		9.70	13.90	23.60
7320				common		11.10	15.65	26.75
7340				Flemish		12	18.15	30.15
7360				English		13.25	19.45	32.70
7500	Norman	25ga.x6"NLB	24	running		8.05	11.55	19.60
7520				common		9.20	12.85	22.05
7540				Flemish		18.50	15	33.50
7560				English		11	15.75	26.75
7600		20ga.x3-5/8"NLB	24	running		8	11.70	19.70
7620				common		9.15	13	22.15
7640				Flemish		18.45	15.15	33.60
7660				English		10.95	15.90	26.85
7800		16ga.x3-5/8"LB	16	running		9.15	12.35	21.50
7820				common		10.30	13.65	23.95
7840				Flemish		19.60	15.80	35.40
7860				English		12.10	16.55	28.65
7900			24	running		8.70	11.95	20.65
7920				common		9.85	13.25	23.10
7940				Flemish		19.15	15.40	34.55
7960				English		11.65	16.15	27.80
8100	Norwegian	25ga.x6"NLB	24	running		6.95	10.50	17.45
8120				common		7.85	11.65	19.50
8140				Flemish		8.45	13.50	21.95
8160				English		9.25	14.10	23.35

B20 Exterior Enclosure

B2010 Exterior Walls

B2010 130	Brick Veneer/Metal Stud Backup

	FACE BRICK	STUD BACKUP	STUD SPACING (IN.)	BOND		COST PER S.F.		
						MAT.	INST.	TOTAL
8200	Norwegian	20ga.x3-5/8"NLB	16	running		7.10	11.10	18.20
8220				common		8	12.25	20.25
8240				Flemish		8.60	14.10	22.70
8260				English		9.40	14.70	24.10
8300			24	running		6.90	10.65	17.55
8320				common		7.80	11.80	19.60
8340				Flemish		8.40	13.65	22.05
8360				English		9.20	14.25	23.45
8400		16ga.x3-5/8"LB	16	running		8.05	11.30	19.35
8420				common		8.95	12.45	21.40
8440				Flemish		9.55	14.30	23.85
8460				English		10.35	14.90	25.25
8500			24	running		7.60	10.90	18.50
8520				common		8.50	12.05	20.55
8540				Flemish		9.10	13.90	23
8560				English		9.90	14.50	24.40

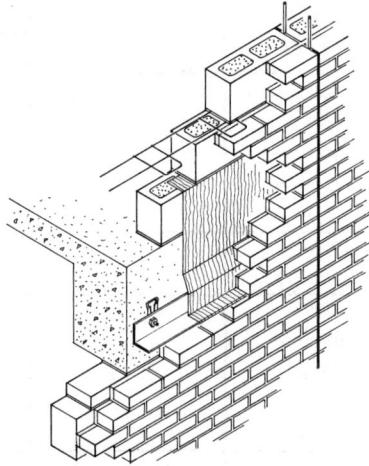

Exterior brick face composite walls are defined in the following terms: type of face brick and backup masonry, thickness of backup masonry and insulation. A special section is included on triple wythe construction at the back. Seven types of face brick are shown with various thicknesses of seven types of backup. All systems include a brick shelf, ties to the backup and necessary dampproofing, flashing, and control joints every 20′.

System Components	QUANTITY	UNIT	COST PER S.F.		
			MAT.	INST.	TOTAL
SYSTEM B2010 132 1120					
COMPOSITE WALL, STANDARD BRICK FACE, 6″ C.M.U. BACKUP, PERLITE FILL					
Face brick veneer, standard, running bond	1.000	S.F.	6.15	10.25	16.40
Wash brick	1.000	S.F.	.06	.88	.94
Concrete block backup, 6″ thick	1.000	S.F.	2.24	5.65	7.89
Wall ties	.300	Ea.	.08	.14	.22
Perlite insulation, poured	1.000	S.F.	.34	.29	.63
Flashing, aluminum	.100	S.F.	.11	.32	.43
Shelf angle	1.000	Lb.	1.27	.89	2.16
Control joint	.050	L.F.	.07	.06	.13
Backer rod	.100	L.F.		.11	.11
Sealant	.100	L.F.	.02	.27	.29
Collar joint	1.000	S.F.	.39	.55	.94
TOTAL			10.73	19.41	30.14

B2010 132		Brick Face Composite Wall - Double Wythe						
	FACE BRICK	BACKUP MASONRY	BACKUP THICKNESS (IN.)	BACKUP CORE FILL		COST PER S.F.		
						MAT.	INST.	TOTAL
1000	Standard	common brick	4	none		11.25	23	34.25
1040		SCR brick	6	none		13.95	20.50	34.45
1080		conc. block	4	none		9.75	18.75	28.50
1120			6	perlite		10.75	19.40	30.15
1160				styrofoam		12.40	19.15	31.55
1200			8	perlite		11	19.85	30.85
1240				styrofoam		12.50	19.50	32
1280		L.W. block	4	none		10.05	18.60	28.65
1320			6	perlite		11.05	19.25	30.30
1360				styrofoam		12.70	19	31.70
1400			8	perlite		11.80	19.70	31.50
1440				styrofoam		13.30	19.35	32.65
1520		glazed block	4	none		17.50	20	37.50
1560			6	perlite		18.55	20.50	39.05
1600				styrofoam		20	20.50	40.50
1640			8	perlite		18.80	21	39.80
1680				styrofoam		20.50	20.50	41

B2010 Exterior Walls

| B2010 132 | Brick Face Composite Wall - Double Wythe |

	FACE BRICK	BACKUP MASONRY	BACKUP THICKNESS (IN.)	BACKUP CORE FILL		COST PER S.F.		
						MAT.	INST.	TOTAL
1720	Standard	clay tile	4	none		15.80	18	33.80
1760			6	none		17.70	18.50	36.20
1800			8	none		20.50	19.15	39.65
1840		glazed tile	4	none		20.50	23.50	44
1880								
2000	Glazed	common brick	4	none		19.30	23.50	42.80
2040		SCR brick	6	none		22	21	43
2080		conc. block	4	none		17.80	19.25	37.05
2120			6	perlite		18.80	19.90	38.70
2160				styrofoam		20.50	19.65	40.15
2200			8	perlite		19.05	20.50	39.55
2240				styrofoam		20.50	20	40.50
2280		L.W. block	4	none		18.10	19.10	37.20
2320			6	perlite		19.10	19.75	38.85
2360				styrofoam		21	19.50	40.50
2400			8	perlite		19.85	20	39.85
2440				styrofoam		21.50	19.85	41.35
2520		glazed block	4	none		25.50	20.50	46
2560			6	perlite		26.50	21	47.50
2600				styrofoam		28	21	49
2640			8	perlite		27	21.50	48.50
2680				styrofoam		28.50	21	49.50
2720		clay tile	4	none		24	18.50	42.50
2760			6	none		26	19	45
2800			8	none		28.50	19.65	48.15
2840		glazed tile	4	none		28.50	24	52.50
2880								
3000	Engineer	common brick	4	none		9.20	21.50	30.70
3040		SCR brick	6	none		11.85	19.15	31
3080		conc. block	4	none		7.65	17.15	24.80
3120			6	perlite		8.65	17.80	26.45
3160				styrofoam		10.30	17.55	27.85
3200			8	perlite		8.95	18.25	27.20
3240				styrofoam		10.40	17.90	28.30
3280		L.W. block	4	none		6.15	12.15	18.30
3320			6	perlite		8.95	17.65	26.60
3360				styrofoam		10.60	17.40	28
3400			8	perlite		9.70	18.10	27.80
3440				styrofoam		11.20	17.75	28.95
3520		glazed block	4	none		15.40	18.45	33.85
3560			6	perlite		16.45	19	35.45
3600				styrofoam		18.10	18.75	36.85
3640			8	perlite		16.70	19.50	36.20
3680				styrofoam		18.20	19.15	37.35
3720		clay tile	4	none		13.70	16.40	30.10
3760			6	none		15.60	16.90	32.50
3800			8	none		18.25	17.55	35.80
3840		glazed tile	4	none		18.20	22	40.20
4000	Roman	common brick	4	none		11.15	22	33.15
4040		SCR brick	6	none		13.85	19.50	33.35

B2010 Exterior Walls

B2010 132 — Brick Face Composite Wall - Double Wythe

	FACE BRICK	BACKUP MASONRY	BACKUP THICKNESS (IN.)	BACKUP CORE FILL		COST PER S.F.		
						MAT.	INST.	TOTAL
4080	Roman	conc. block	4	none		9.65	17.50	27.15
4120			6	perlite		10.65	18.15	28.80
4160				styrofoam		12.30	17.90	30.20
4200			8	perlite		10.90	18.60	29.50
4240				styrofoam		12.40	18.25	30.65
4280		L.W. block	4	none		9.95	17.35	27.30
4320			6	perlite		10.95	18	28.95
4360				styrofoam		12.60	17.75	30.35
4400			8	perlite		11.70	18.45	30.15
4440				styrofoam		13.20	18.10	31.30
4520		glazed block	4	none		17.40	18.80	36.20
4560			6	perlite		18.45	19.35	37.80
4600				styrofoam		20	19.10	39.10
4640			8	perlite		18.70	19.85	38.55
4680				styrofoam		20	19.50	39.50
4720		claytile	4	none		15.70	16.75	32.45
4760			6	none		17.60	17.25	34.85
4800			8	none		20.50	17.90	38.40
4840		glazed tile	4	none		20	22.50	42.50
5000	Norman	common brick	4	none		10.15	20	30.15
5040		SCR brick	6	none		12.85	17.55	30.40
5080		conc. block	4	none		8.65	15.55	24.20
5120			6	perlite		9.65	16.20	25.85
5160				styrofoam		11.30	15.95	27.25
5200			8	perlite		9.90	16.65	26.55
5240				styrofoam		11.40	16.30	27.70
5280		L.W. block	4	none		8.95	15.40	24.35
5320			6	perlite		10	16.25	26.25
5360				styrofoam		11.60	15.80	27.40
5400			8	perlite		10.70	16.50	27.20
5440				styrofoam		12.20	16.15	28.35
5520		glazed block	4	none		16.40	16.85	33.25
5560			6	perlite		17.45	17.40	34.85
5600				styrofoam		19.10	17.15	36.25
5640			8	perlite		17.70	17.90	35.60
5680				styrofoam		19.20	17.55	36.75
5720		clay tile	4	none		14.70	14.80	29.50
5760			6	none		16.60	15.30	31.90
5800			8	none		19.25	15.95	35.20
5840		glazed tile	4	none		19.20	20.50	39.70
6000	Norwegian	common brick	4	none		9.05	19.05	28.10
6040		SCR brick	6	none		11.75	16.50	28.25
6080		conc. block	4	none		7.55	14.50	22.05
6120			6	perlite		8.55	15.15	23.70
6160				styrofoam		10.20	14.90	25.10
6200			8	perlite		8.80	15.60	24.40
6240				styrofoam		10.30	15.25	25.55
6280		L.W. block	4	none		7.85	14.35	22.20
6320			6	perlite		8.85	15	23.85
6360				styrofoam		10.50	14.75	25.25
6400			8	perlite		9.60	15.45	25.05
6440				styrofoam		11.10	15.10	26.20

B2010 Exterior Walls

B2010 132	Brick Face Composite Wall - Double Wythe

	FACE BRICK	BACKUP MASONRY	BACKUP THICKNESS (IN.)	BACKUP CORE FILL		COST PER S.F.		
						MAT.	INST.	TOTAL
6520	Norwegian	glazed block	4	none		15.30	15.80	31.10
6560			6	perlite		16.35	16.35	32.70
6600				styrofoam		18	16.10	34.10
6640			8	perlite		16.60	16.85	33.45
6680				styrofoam		18.10	16.50	34.60
6720		clay tile	4	none		13.60	13.75	27.35
6760			6	none		15.50	14.25	29.75
6800			8	none		18.15	14.90	33.05
6840		glazed tile	4	none		18.10	19.50	37.60
6880								
7000	Utility	common brick	4	none		8.70	18.05	26.75
7040		SCR brick	6	none		11.40	15.50	26.90
7080		conc. block	4	none		7.15	13.50	20.65
7120			6	perlite		8.15	14.15	22.30
7160				styrofoam		9.80	13.90	23.70
7200			8	perlite		8.45	14.60	23.05
7240				styrofoam		9.95	14.25	24.20
7280		L.W. block	4	none		7.50	13.35	20.85
7320			6	perlite		8.50	14	22.50
7360				styrofoam		10.15	13.75	23.90
7400			8	perlite		9.25	14.45	23.70
7440				styrofoam		10.70	14.10	24.80
7520		glazed block	4	none		14.95	14.80	29.75
7560			6	perlite		16	15.35	31.35
7600				styrofoam		17.65	15.10	32.75
7640			8	perlite		16.25	15.85	32.10
7680				styrofoam		17.75	15.50	33.25
7720		clay tile	4	none		13.25	12.75	26
7760			6	none		15.15	13.25	28.40
7800			8	none		17.80	13.90	31.70
7840		glazed tile	4	none		17.75	18.50	36.25
7880								

B2010 133	Brick Face Composite Wall - Triple Wythe

	FACE BRICK	MIDDLE WYTHE	INSIDE MASONRY	TOTAL THICKNESS (IN.)		COST PER S.F.		
						MAT.	INST.	TOTAL
8000	Standard	common brick	standard brick	12		17.90	33	50.90
8100		4" conc. brick	standard brick	12		17.30	34	51.30
8120		4" conc. brick	common brick	12		14.25	34	48.25
8200	Glazed	common brick	standard brick	12		26	33.50	59.50
8300		4" conc. brick	standard brick	12		25.50	34.50	60
8320		4" conc. brick	glazed brick	12		22.50	34.50	57
8400	Engineer	common brick	standard brick	12		15.80	31.50	47.30
8500		4" conc. brick	standard brick	12		15.25	32.50	47.75
8520		4" conc. brick	engineer brick	12		12.20	32	44.20
8600	Roman	common brick	standard brick	12		17.80	32	49.80
8700		4" conc. brick	standard brick	12		17.20	33	50.20
8720		4" conc. brick	Roman brick	12		14.15	32.50	46.65
8800	Norman	common brick	standard brick	12		16.81	30.10	46.90
8900		4" conc. brick	standard brick	12		16.20	31	47.20
8920		4" conc. brick	Norman brick	12		13.15	30.50	43.65

B2010 Exterior Walls

| B2010 133 | Brick Face Composite Wall - Triple Wythe | | | | | | | |

	FACE BRICK	MIDDLE WYTHE	INSIDE MASONRY	TOTAL THICKNESS (IN.)		COST PER S.F.		
						MAT.	INST.	TOTAL
9000	Norwegian	common brick	standard brick	12		15.70	29	44.70
9100		4" conc. brick	standard brick	12		15.10	30	45.10
9120		4" conc. brick	Norwegian brick	12		12.05	29.50	41.55
9200	Utility	common brick	standard brick	12		15.35	28	43.35
9300		4" conc. brick	standard brick	12		14.60	29	43.60
9320		4" conc. brick	utility brick	12		11.70	28.50	40.20

B2010 Exterior Walls

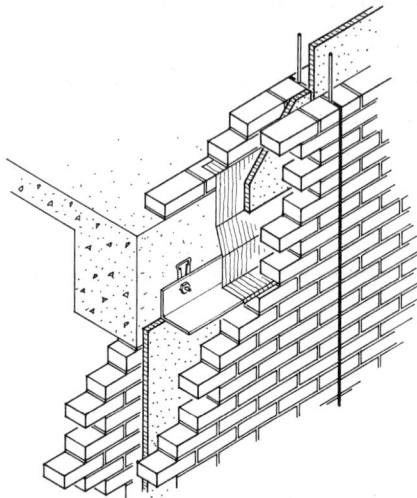

Exterior brick face cavity walls are defined in the following terms: cavity treatment, type of face brick, backup masonry, total thickness and insulation. Seven types of face brick are shown with fourteen types of backup. All systems include a brick shelf, ties to the backups and necessary dampproofing, flashing, and control joints every 20′.

System Components	QUANTITY	UNIT	COST PER S.F.		
			MAT.	INST.	TOTAL
SYSTEM B2010 134 1000					
CAVITY WALL, STANDARD BRICK FACE, COMMON BRICK BACKUP, POLYSTYRENE					
Face brick veneer, standard, running bond	1.000	S.F.	6.15	10.25	16.40
Wash brick	1.000	S.F.	.06	.88	.94
Common brick wall backup, 4″ thick	1.000	S.F.	3.11	9.80	12.91
Wall ties	.300	L.F.	.13	.14	.27
Polystyrene insulation board, 1″ thick	1.000	S.F.	.26	.62	.88
Flashing, aluminum	.100	S.F.	.11	.32	.43
Shelf angle	1.000	Lb.	1.27	.89	2.16
Control joint	.050	L.F.	.07	.06	.13
Backer rod	.100	L.F.		.11	.11
Sealant	.100	L.F.	.02	.27	.29
TOTAL			11.18	23.34	34.52

B2010 134			**Brick Face Cavity Wall**					
	FACE BRICK	BACKUP MASONRY	TOTAL THICKNESS (IN.)	CAVITY INSULATION		COST PER S.F.		
						MAT.	INST.	TOTAL
1000	Standard	4″ common brick	10	polystyrene		11.20	23.50	34.70
1020				none		10.90	22.50	33.40
1040		6″ SCR brick	12	polystyrene		13.90	21	34.90
1060				none		13.65	20	33.65
1080		4″ conc. block	10	polystyrene		9.65	18.80	28.45
1100				none		9.40	18.15	27.55
1120		6″ conc. block	12	polystyrene		10.35	19.20	29.55
1140				none		10.05	18.55	28.60
1160		4″ L.W. block	10	polystyrene		9.95	18.65	28.60
1180				none		9.70	18	27.70
1200		6″ L.W. block	12	polystyrene		10.65	19.05	29.70
1220				none		10.40	18.40	28.80
1240		4″ glazed block	10	polystyrene		17.40	20	37.40
1260				none		17.15	19.45	36.60
1280		6″ glazed block	12	polystyrene		17.45	20	37.45
1300				none		17.20	19.45	36.65

B2010 Exterior Walls

B2010 134	Brick Face Cavity Wall

	FACE BRICK	BACKUP MASONRY	TOTAL THICKNESS (IN.)	CAVITY INSULATION		COST PER S.F.		
						MAT.	INST.	TOTAL
1320	Standard	4" clay tile	10	polystyrene		15.70	18.05	33.75
1340				none		15.45	17.45	32.90
1360		4" glazed tile	10	polystyrene		20	24	44
1380				none		19.95	23	42.95
1500	Glazed	4" common brick	10	polystyrene		19.25	24	43.25
1520				none		18.95	23	41.95
1540		6" SCR brick	12	polystyrene		22	21.50	43.50
1560				none		21.50	20.50	42
1580		4" conc. block	10	polystyrene		17.70	19.30	37
1600				none		17.45	18.65	36.10
1620		6" conc. block	12	polystyrene		18.40	19.70	38.10
1640				none		18.10	19.05	37.15
1660		4" L.W. block	10	polystyrene		18	19.15	37.15
1680				none		17.75	18.50	36.25
1700		6" L.W. block	12	polystyrene		18.70	19.55	38.25
1720				none		18.45	18.90	37.35
1740		4" glazed block	10	polystyrene		25.50	20.50	46
1760				none		25	19.95	44.95
1780		6" glazed block	12	polystyrene		26	21	47
1800				none		26	20.50	46.50
1820		4" clay tile	10	polystyrene		24	18.55	42.55
1840				none		23.50	17.95	41.45
1860		4" glazed tile	10	polystyrene		25.50	19.05	44.55
1880				none		25.50	18.40	43.90
2000	Engineer	4" common brick	10	polystyrene		9.10	21.50	30.60
2020				none		8.85	21	29.85
2040		6" SCR brick	12	polystyrene		11.80	19.20	31
2060				none		11.55	18.55	30.10
2080		4" conc. block	10	polystyrene		7.55	17.20	24.75
2100				none		7.30	16.55	23.85
2120		6" conc. block	12	polystyrene		8.25	17.60	25.85
2140				none		8	16.95	24.95
2160		4" L.W. block	10	polystyrene		7.90	17.05	24.95
2180				none		7.60	16.40	24
2200		6" L.W. block	12	polystyrene		8.55	17.45	26
2220				none		8.30	16.80	25.10
2240		4" glazed block	10	polystyrene		15.35	18.50	33.85
2260				none		15.10	17.85	32.95
2280		6" glazed block	12	polystyrene		16.05	18.80	34.85
2300				none		15.80	18.15	33.95
2320		4" clay tile	10	polystyrene		13.65	16.45	30.10
2340				none		13.40	15.85	29.25
2360		4" glazed tile	10	polystyrene		18.15	22	40.15
2380				none		17.90	21.50	39.40
2500	Roman	4" common brick	10	polystyrene		11.10	22	33.10
2520				none		10.80	21.50	32.30
2540		6" SCR brick	12	polystyrene		13.80	19.55	33.35
2560				none		13.55	18.90	32.45

B2010 134						Brick Face Cavity Wall			
	FACE BRICK	BACKUP MASONRY	TOTAL THICKNESS (IN.)	CAVITY INSULATION			COST PER S.F.		
							MAT.	INST.	TOTAL
2580	Roman	4″ conc. block	10	polystyrene			9.55	17.55	27.10
2600				none			9.30	16.90	26.20
2620		6″ conc. block	12	polystyrene			10.25	17.95	28.20
2640				none			9.95	17.30	27.25
2660		4″ L.W. block	10	polystyrene			9.85	17.40	27.25
2680				none			9.60	16.75	26.35
2700		6″ L.W. block	12	polystyrene			10.55	17.80	28.35
2720				none			10.30	17.15	27.45
2740		4″ glazed block	10	polystyrene			17.30	18.85	36.15
2760				none			17.05	18.20	35.25
2780		6″ glazed block	12	polystyrene			18.05	19.15	37.20
2800				none			17.80	18.50	36.30
2820		4″ clay tile	10	polystyrene			15.60	16.80	32.40
2840				none			15.35	16.20	31.55
2860		4″ glazed tile	10	polystyrene			20	22.50	42.50
2880				none			19.85	22	41.85
3000	Norman	4″ common brick	10	polystyrene			10.10	20	30.10
3020				none			9.80	19.50	29.30
3040		6″ SCR brick	12	polystyrene			12.80	17.60	30.40
3060				none			12.55	16.95	29.50
3080		4″ conc. block	10	polystyrene			8.55	15.60	24.15
3100				none			8.30	14.95	23.25
3120		6″ conc. block	12	polystyrene			9.25	16	25.25
3140				none			8.95	15.35	24.30
3160		4″ L.W. block	10	polystyrene			8.85	15.45	24.30
3180				none			8.60	14.80	23.40
3200		6″ L.W. block	12	polystyrene			9.55	15.85	25.40
3220				none			9.30	15.20	24.50
3240		4″ glazed block	10	polystyrene			16.30	16.90	33.20
3260				none			16.05	16.25	32.30
3320		4″ clay tile	10	polystyrene			14.60	14.85	29.45
3340				none			14.35	14.25	28.60
3360		4″ glazed tile	10	polystyrene			19.10	20.50	39.60
3380				none			18.85	19.95	38.80
3500	Norwegian	4″ common brick	10	polystyrene			9	19.10	28.10
3520				none			8.70	18.45	27.15
3540		6″ SCR brick	12	polystyrene			11.70	16.55	28.25
3560				none			11.45	15.90	27.35
3580		4″ conc. block	10	polystyrene			7.45	14.55	22
3600				none			7.20	13.90	21.10
3620		6″ conc. block	12	polystyrene			8.15	14.95	23.10
3640				none			7.85	14.30	22.15

B20 Exterior Enclosure

B2010 Exterior Walls

B2010 134	Brick Face Cavity Wall							

B2010 134 — Brick Face Cavity Wall

	FACE BRICK	BACKUP MASONRY	TOTAL THICKNESS (IN.)	CAVITY INSULATION		COST PER S.F.		
						MAT.	INST.	TOTAL
3660	Norwegian	4" L.W. block	10	polystyrene		7.75	14.40	22.15
3680				none		7.50	13.75	21.25
3700		6" L.W. block	12	polystyrene		8.45	14.80	23.25
3720				none		8.20	14.15	22.35
3740		4" glazed block	10	polystyrene		15.20	15.85	31.05
3760				none		14.95	15.20	30.15
3780		6" glazed block	12	polystyrene		15.95	16.15	32.10
3800				none		15.70	15.50	31.20
3820		4" clay tile	10	polystyrene		13.50	13.80	27.30
3840				none		13.25	13.20	26.45
3860		4" glazed tile	10	polystyrene		18	19.55	37.55
3880				none		17.75	18.90	36.65
4000	Utility	4" common brick	10	polystyrene		8.60	18.10	26.70
4020				none		8.35	17.45	25.80
4040		6" SCR brick	12	polystyrene		11.35	15.55	26.90
4060				none		11.05	14.90	25.95
4080		4" conc. block	10	polystyrene		7.10	13.55	20.65
4100				none		6.85	12.90	19.75
4120		6" conc. block	12	polystyrene		7.75	13.95	21.70
4140				none		7.50	13.30	20.80
4160		4" L.W. block	10	polystyrene		7.40	13.40	20.80
4180				none		7.15	12.75	19.90
4200		6" L.W. block	12	polystyrene		8.10	13.80	21.90
4220				none		7.85	13.15	21
4240		4" glazed block	10	polystyrene		14.85	14.85	29.70
4260				none		14.60	14.20	28.80
4280		6" glazed block	12	polystyrene		15.60	15.15	30.75
4300				none		15.30	14.50	29.80
4320		4" clay tile	10	polystyrene		13.15	12.80	25.95
4340				none		12.90	12.20	25.10
4360		4" glazed tile	10	polystyrene		17.65	18.55	36.20
4380				none		17.40	17.90	35.30

B2010 135 — Brick Face Cavity Wall - Insulated Backup

	FACE BRICK	BACKUP MASONRY	TOTAL THICKNESS (IN.)	BACKUP CORE FILL		COST PER S.F.		
						MAT.	INST.	TOTAL
5100	Standard	6" conc. block	10	perlite		10.40	18.85	29.25
5120				styrofoam		12.05	18.55	30.60
5140		8" conc. block	12	perlite		10.70	19.30	30
5160				styrofoam		12.20	18.90	31.10
5180		6" L.W. block	10	perlite		10.70	18.70	29.40
5200				styrofoam		12.35	18.40	30.75
5220		8" L.W. block	12	perlite		11.45	19.15	30.60
5240				styrofoam		12.95	18.75	31.70
5260		6" glazed block	10	perlite		18.20	20	38.20
5280				styrofoam		19.85	19.75	39.60
5300		8" glazed block	12	perlite		18.50	20.50	39
5320				styrofoam		19.95	20	39.95
5340		6" clay tile	10	none		17.35	17.90	35.25
5360		8" clay tile	12	none		20	18.55	38.55
5600	Glazed	6" conc. block	10	perlite		18.45	19.35	37.80
5620				styrofoam		20	19.05	39.05

B2010 Exterior Walls

B2010 135				Brick Face Cavity Wall - Insulated Backup			

	FACE BRICK	BACKUP MASONRY	TOTAL THICKNESS (IN.)	BACKUP CORE FILL		COST PER S.F.	
					MAT.	INST.	TOTAL
5640	Glazed	8" conc. block	12	perlite	18.75	19.80	38.55
5660				styrofoam	20	19.40	39.40
5680		6" L.W. block	10	perlite	18.75	19.20	37.95
5700				styrofoam	20.50	18.90	39.40
5720		8" L.W. block	12	perlite	19.50	19.65	39.15
5740				styrofoam	21	19.25	40.25
5760		6" glazed block	10	perlite	26.50	20.50	47
5780				styrofoam	28	20.50	48.50
5800		8" glazed block	12	perlite	26.50	21	47.50
5820				styrofoam	28	20.50	48.50
5840		6" clay tile	10	none	25.50	18.40	43.90
5860		8" clay tile	8	none	28	19.05	47.05
6100	Engineer	6" conc. block	10	perlite	8.30	17.25	25.55
6120				styrofoam	9.95	16.95	26.90
6140		8" conc. block	12	perlite	8.60	17.70	26.30
6160				styrofoam	10.10	17.30	27.40
6180		6" L.W. block	10	perlite	8.65	17.10	25.75
6200				styrofoam	10.30	16.80	27.10
6220		8" L.W. block	12	perlite	9.40	17.55	26.95
6240				styrofoam	10.90	17.15	28.05
6260		6" glazed block	10	perlite	16.10	18.45	34.55
6280				styrofoam	17.75	18.15	35.90
6300		8" glazed block	12	perlite	16.40	18.95	35.35
6320				styrofoam	17.90	18.55	36.45
6340		6" clay tile	10	none	15.30	16.30	31.60
6360		8" clay tile	12	none	17.95	16.95	34.90
6600	Roman	6" conc. block	10	perlite	10.30	17.60	27.90
6620				styrofoam	11.95	17.30	29.25
6640		8" conc. block	12	perlite	10.60	18.05	28.65
6660				styrofoam	12.10	17.65	29.75
6680		6" L.W. block	10	perlite	10.60	17.45	28.05
6700				styrofoam	12.25	17.15	29.40
6720		8" L.W. block	12	perlite	11.35	17.90	29.25
6740				styrofoam	12.85	17.50	30.35
6760		6" glazed block	10	perlite	18.10	18.80	36.90
6780				styrofoam	19.75	18.50	38.25
6800		8" glazed block	12	perlite	18.40	19.30	37.70
6820				styrofoam	19.85	18.90	38.75
6840	Roman	6" clay tile	10	none	17.25	16.65	33.90
6860		8" clay tile	12	none	19.95	17.30	37.25
7100	Norman	6" conc. block	10	perlite	9.30	15.65	24.95
7120				styrofoam	10.95	15.35	26.30
7140		8" conc. block	12	perlite	9.60	16.10	25.70
7160				styrofoam	11.10	15.70	26.80
7180		6" L.W. block	10	perlite	9.60	15.50	25.10
7200				styrofoam	11.25	15.20	26.45
7220		8" L.W. block	12	perlite	10.35	15.95	26.30
7240				styrofoam	11.85	15.55	27.40
7260		6" glazed block	10	perlite	17.10	16.85	33.95
7280				styrofoam	18.75	16.55	35.30
7300		8" glazed block	12	perlite	17.40	17.35	34.75
7320				styrofoam	18.85	16.95	35.80

B2010 Exterior Walls

B2010 135	Brick Face Cavity Wall - Insulated Backup

	FACE BRICK	BACKUP MASONRY	TOTAL THICKNESS (IN.)	BACKUP CORE FILL		COST PER S.F.		
						MAT.	INST.	TOTAL
7340	Norwegian	6" clay tile	10	none		16.25	14.70	30.95
7360		8" clay tile	12	none		18.95	15.35	34.30
7600		6" conc. block	10	perlite		8.20	14.60	22.80
7620				styrofoam		9.85	14.30	24.15
7640		8" conc. block	12	perlite		8.50	15.05	23.55
7660				styrofoam		10	14.65	24.65
7680		6" L.W. block	10	perlite		8.50	14.45	22.95
7700				styrofoam		10.15	14.15	24.30
7720		8" L.W. block	12	perlite		9.25	14.90	24.15
7740				styrofoam		10.75	14.50	25.25
7760		6" glazed block	10	perlite		16	15.80	31.80
7780				styrofoam		17.65	15.50	33.15
7800		8" glazed block	12	perlite		16.30	16.30	32.60
7820				styrofoam		17.75	15.90	33.65
7840		6" clay tile	10	none		15.15	13.65	28.80
7860		8" clay tile	12	none		17.85	14.30	32.15
8100	Utility	6" conc. block	10	perlite		7.85	13.60	21.45
8120				styrofoam		9.50	13.30	22.80
8140		8" conc. block	12	perlite		8.15	14.05	22.20
8160				styrofoam		9.60	13.65	23.25
8180		6" L.W. block	10	perlite		8.15	13.45	21.60
8200				styrofoam		9.80	13.15	22.95
8220		8" L.W. block	12	perlite		8.90	13.90	22.80
8240				styrofoam		10.40	13.50	23.90
8260		6" glazed block	10	perlite		15.65	14.80	30.45
8280				styrofoam		17.30	14.50	31.80
8300		8" glazed block	12	perlite		15.90	15.30	31.20
8320				styrofoam		17.40	14.90	32.30
8340		6" clay tile	10	none		14.80	12.65	27.45
8360		8" clay tile	12	none		17.45	13.30	30.75

B2010 Exterior Walls

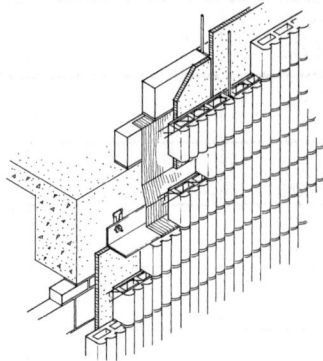

Exterior block face cavity walls are defined in the following terms: cavity treatment, type of face block, backup masonry, total thickness and insulation. Twelve types of face block are shown with four types of backup. All systems include a brick shelf and necessary dampproofing, flashing, and control joints every 20'.

System Components	QUANTITY	UNIT	COST PER S.F.		
			MAT.	INST.	TOTAL
SYSTEM B2010 137 1600					
CAVITY WALL, FLUTED BLOCK FACE, 4″ C.M.U. BACKUP, POLYSTYRENE BOARD					
Fluted block partition, 4″ thick	1.000	S.F.	5.65	6.75	12.40
Conc. block wall backup, 4″ thick	1.000	S.F.	1.58	5.25	6.83
Horizontal joint reinforcing	.800	L.F.	.12	.20	.32
Polystyrene insulation board, 1″ thick	1.000	S.F.	.26	.62	.88
Flashing, aluminum	.100	S.F.	.11	.32	.43
Shelf angle	1.000	Lb.	1.27	.89	2.16
Control joint	.050	L.F.	.07	.06	.13
Backer rod	.100	L.F.		.11	.11
Sealant	.100	L.F.	.02	.27	.29
TOTAL			9.08	14.47	23.55

B2010 137		Block Face Cavity Wall						
	FACE BLOCK	BACKUP MASONRY	TOTAL THICKNESS (IN.)	CAVITY INSULATION		COST PER S.F.		
						MAT.	INST.	TOTAL
1000	Deep groove	4″ conc. block	10	polystyrene		9.10	14.45	23.55
1050	Reg. wt.			none		8.80	13.85	22.65
1060		6″ conc. block	12	polystyrene		9.75	14.85	24.60
1110				none		9.50	14.25	23.75
1120		4″ L.W. block	10	polystyrene		9.40	14.30	23.70
1170				none		9.15	13.70	22.85
1180		6″ L.W. block	12	polystyrene		10.05	14.70	24.75
1230				none		9.80	14.10	23.90
1300	Deep groove	4″ conc. block	10	polystyrene		9.40	14.25	23.65
1350	L.W.			none		9.10	13.65	22.75
1360		6″ conc. block	12	polystyrene		10.05	14.65	24.70
1410				none		9.80	14.05	23.85
1420		4″ L.W. block	10	polystyrene		9.70	14.10	23.80
1470				none		9.45	13.50	22.95
1480		6″ L.W. block	12	polystyrene		10.35	14.50	24.85
1530				none		10.10	13.90	24

B20 Exterior Enclosure

B2010 Exterior Walls

| B2010 137 | | | | Block Face Cavity Wall | | | | |

	FACE BLOCK	BACKUP MASONRY	TOTAL THICKNESS (IN.)	CAVITY INSULATION		COST PER S.F.		
						MAT.	INST.	TOTAL
1600	Fluted	4″ conc. block	10	polystyrene		9.10	14.45	23.55
1650	Reg. wt.			none		6.80	13.65	20.45
1660		6″ conc. block	12	polystyrene		7.70	14.65	22.35
1710				none		7.45	14.05	21.50
1720		4″ L.W. block	10	polystyrene		7.35	14.10	21.45
1770				none		7.10	13.50	20.60
1780		6″ L.W. block	12	polystyrene		8.05	14.50	22.55
1830				none		7.80	13.90	21.70
1900	Fluted	4″ conc. block	10	polystyrene		7.15	14.05	21.20
1950	L.W.			none		6.90	13.45	20.35
1960		6″ conc. block	12	polystyrene		7.85	14.45	22.30
2010				none		7.55	13.85	21.40
2020		4″ L.W. block	10	polystyrene		7.45	13.90	21.35
2070				none		7.20	13.30	20.50
2080		6″ L.W. block	12	polystyrene		8.15	14.30	22.45
2130				none		7.90	13.70	21.60
2200	Ground face	4″ conc. block	10	polystyrene		12.25	14.55	26.80
2250	1 Score			none		11.95	13.95	25.90
2260		6″ conc. block	12	polystyrene		12.90	14.95	27.85
2310				none		12.65	14.35	27
2320		4″ L.W. block	10	polystyrene		12.55	14.40	26.95
2370				none		12.30	13.80	26.10
2380		6″ L.W. block	12	polystyrene		13.20	14.80	28
2430				none		12.95	14.20	27.15
2500	Hexagonal	4″ conc. block	10	polystyrene		6	14.25	20.25
2550	Reg. wt.			none		5.75	13.65	19.40
2560		6″ conc. block	12	polystyrene		6.65	14.65	21.30
2610				none		6.40	14.05	20.45
2620		4″ L.W. block	10	polystyrene		6.30	14.10	20.40
2670				none		6.05	13.50	19.55
2680		6″ L.W. block	12	polystyrene		7	14.50	21.50
2730				none		6.70	13.90	20.60
2800	Hexagonal	4″ conc. block	10	polystyrene		6.20	14.05	20.25
2850	L.W.			none		5.90	13.45	19.35
2860		6″ conc. block	12	polystyrene		6.85	14.45	21.30
2910				none		6.60	13.85	20.45
2920		4″ L.W. block	10	polystyrene		6.50	13.90	20.40
2970				none		6.25	13.30	19.55
2980		6″ L.W. block	12	polystyrene		7.15	14.30	21.45
3030				none		6.90	13.70	20.60
3100	Slump block	4″ conc. block	10	polystyrene		7.75	13.05	20.80
3150	4x16			none		7.50	12.45	19.95
3160		6″ conc. block	12	polystyrene		8.45	13.45	21.90
3210				none		8.20	12.85	21.05
3220		4″ L.W. block	10	polystyrene		8.10	12.90	21
3270				none		7.80	12.30	20.10
3280		6″ L.W. block	12	polystyrene		8.75	13.30	22.05
3330				none		8.50	12.70	21.20

B2010 Exterior Walls

| B2010 137 | | Block Face Cavity Wall | | | | | | |

	FACE BLOCK	BACKUP MASONRY	TOTAL THICKNESS (IN.)	CAVITY INSULATION		COST PER S.F.		
						MAT.	INST.	TOTAL
3400	Split face	4" conc. block	10	polystyrene		8.40	14.45	22.85
3450	1 Score			none		8.15	13.85	22
3460	Reg. wt.	6" conc. block	12	polystyrene		9.05	14.85	23.90
3510				none		8.80	14.25	23.05
3520		4" L.W. block	10	polystyrene		8.70	14.30	23
3570				none		8.45	13.70	22.15
3580		6" L.W. block	12	polystyrene		9.40	14.70	24.10
3630				none		9.15	14.10	23.25
3700	Split face	4" conc. block	10	polystyrene		8.85	14.25	23.10
3750	1 Score			none		8.55	13.65	22.20
3760	L.W.	6" conc. block	12	polystyrene		9.50	14.65	24.15
3810				none		9.25	14.05	23.30
3820		4" L.W. block	10	polystyrene		9.15	14.10	23.25
3870				none		8.90	13.50	22.40
3880		6" L.W. block	12	polystyrene		9.80	14.50	24.30
3930				none		9.55	13.90	23.45
4000	Split rib	4" conc. block	10	polystyrene		8.90	14.45	23.35
4050	8 Rib			none		8.60	13.85	22.45
4060	Reg. wt.	6" conc. block	12	polystyrene		9.55	14.85	24.40
4110				none		9.30	14.25	23.55
4120		4" L.W. block	10	polystyrene		9.20	14.30	23.50
4170				none		8.95	13.70	22.65
4180		6" L.W. block	12	polystyrene		9.85	14.70	24.55
4230				none		9.60	14.10	23.70
4300	Split rib	4" conc. block	10	polystyrene		9.35	14.25	23.60
4350	8 Rib			none		9.05	13.65	22.70
4360	L.W.	6" conc. block	12	polystyrene		10	14.65	24.65
4410				none		9.75	14.05	23.80
4420		4" L.W. block	10	polystyrene		9.65	14.10	23.75
4470				none		9.40	13.50	22.90
4480		6" L.W. block	12	polystyrene		10.30	14.50	24.80
4530				none		10.05	13.90	23.95

| B2010 138 | | Block Face Cavity Wall - Insulated Backup | | | | | | |

	FACE BLOCK	BACKUP MASONRY	TOTAL THICKNESS (IN.)	BACKUP CORE INSULATION		COST PER S.F.		
						MAT.	INST.	TOTAL
5010	Deep groove	6" conc. block	10	perlite		9.80	14.55	24.35
5050	Reg. wt.			styrofoam		11.45	14.25	25.70
5060		8" conc. block	12	perlite		10.10	14.95	25.05
5110				styrofoam		11.60	14.60	26.20
5120		6" L.W. block	10	perlite		10.15	14.40	24.55
5170				styrofoam		11.80	14.10	25.90
5180		8" L.W. block	12	perlite		10.90	14.80	25.70
5230				styrofoam		12.40	14.45	26.85
5300	Deep groove	6" conc. block	10	perlite		10.10	14.35	24.45
5350	L.W.			styrofoam		11.75	14.05	25.80
5360		8" conc. block	12	perlite		10.40	14.75	25.15
5410				styrofoam		11.90	14.40	26.30
5420		6" L.W. block	10	perlite		10.45	14.20	24.65
5470				styrofoam		12.10	13.90	26
5480		8" L.W. block	12	perlite		11.20	14.60	25.80
5530				styrofoam		12.70	14.25	26.95

B20 Exterior Enclosure

B2010 Exterior Walls

B2010 138			Block Face Cavity Wall - Insulated Backup				
	FACE BLOCK	BACKUP MASONRY	TOTAL THICKNESS (IN.)	BACKUP CORE INSULATION		COST PER S.F.	
						MAT.	INST.	TOTAL

	FACE BLOCK	BACKUP MASONRY	TOTAL THICKNESS (IN.)	BACKUP CORE INSULATION		MAT.	INST.	TOTAL
5600	Fluted	6" conc. block	10	perlite		7.80	14.35	22.15
5650	Reg.			styrofoam		9.45	14.05	23.50
5660		8" conc. block	12	perlite		8.10	14.75	22.85
5710				styrofoam		9.55	14.40	23.95
5720		6" L.W. block	10	perlite		8.10	14.20	22.30
5770				styrofoam		9.75	13.90	23.65
5780		8" L.W. block	12	perlite		8.85	14.60	23.45
5830				styrofoam		10.35	14.25	24.60
5900	Fluted	6 conc. block	10	perlite		7.90	14.15	22.05
5950	L.W.			styrofoam		9.55	13.85	23.40
5960		8" conc. block	12	perlite		8.20	14.55	22.75
6010				styrofoam		9.70	14.20	23.90
6020		6" L.W. block	10	perlite		8.20	14	22.20
6070				styrofoam		9.85	13.70	23.55
6080		8" L.W. block	12	perlite		8.95	14.40	23.35
6130				styrofoam		10.45	14.05	24.50
6200	Ground face	6" conc. block	10	perlite		12.95	14.65	27.60
6250	1 score			styrofoam		14.60	14.35	28.95
6260		8" conc. block	12	perlite		13.25	15.05	28.30
6310				styrofoam		14.75	14.70	29.45
6320	Ground face	6" L.W. block	10	perlite		13.30	14.50	27.80
6370	1 Score			styrofoam		14.95	14.20	29.15
6380		8" L.W. block	12	perlite		14.05	14.90	28.95
6430				styrofoam		16.55	17.40	33.95
6500	Hexagonal	6" conc. block	10	perlite		6.75	14.35	21.10
6550	Reg. wt.			styrofoam		8.40	14.05	22.45
6560		8" conc. block	12	perlite		7	14.75	21.75
6610				styrofoam		8.50	14.40	22.90
6620		6" L.W. block	10	perlite		7.05	14.20	21.25
6670				styrofoam		8.70	13.90	22.60
6680		8" L.W. block	12	perlite		7.80	14.60	22.40
6730				styrofoam		9.30	14.25	23.55
6800	Hexagonal	6" conc. block	10	perlite		6.90	14.15	21.05
6850	L.W.			styrofoam		8.55	13.85	22.40
6860		8" conc. block	12	perlite		7.20	14.55	21.75
6910				styrofoam		8.70	14.20	22.90
6920		6" L.W. block	10	perlite		7.25	14	21.25
6970				styrofoam		8.90	13.70	22.60
6980		8" L.W. block	12	perlite		8	14.40	22.40
7030				styrofoam		9.50	14.05	23.55
7100	Slump block	6" conc. block	10	perlite		8.50	13.15	21.65
7150	4x16			styrofoam		10.15	12.85	23
7160		8" conc. block	12	perlite		8.80	13.55	22.35
7210				styrofoam		10.30	13.20	23.50
7220		6" L.W. block	10	perlite		8.85	13	21.85
7270				styrofoam		10.50	12.70	23.20
7280		8" L.W. block	12	perlite		9.60	13.40	23
7330				styrofoam		11.05	13.05	24.10

185

| B2010 138 | | | | Block Face Cavity Wall - Insulated Backup | | | | |

	FACE BLOCK	BACKUP MASONRY	TOTAL THICKNESS (IN.)	BACKUP CORE INSULATION		COST PER S.F.		
						MAT.	INST.	TOTAL
7400	Split face	6" conc. block	10	perlite		9.15	14.55	23.70
7450	1 Score			styrofoam		10.80	14.25	25.05
7460	Reg. wt.	8" conc. block	12	perlite		9.45	14.95	24.40
7510				styrofoam		10.90	14.60	25.50
7520		6" L.W. block	10	perlite		9.45	14.40	23.85
7570				styrofoam		11.10	14.10	25.20
7580		8" L.W. block	12	perlite		10.20	14.80	25
7630				styrofoam		11.70	14.45	26.15
7700	Split rib	6" conc. block	10	perlite		9.60	14.55	24.15
7750	8 Rib			styrofoam		11.25	14.25	25.50
7760	Reg. wt.	8" conc. block	12	perlite		9.90	14.95	24.85
7810				styrofoam		11.40	14.60	26
7820		6" L.W. block	10	perlite		9.95	14.40	24.35
7870				styrofoam		11.60	14.10	25.70
7880		8" L.W. block	12	perlite		10.70	14.80	25.50
7930				styrofoam		12.20	14.45	26.65
8000	Split rib	6" conc. block	10	perlite		10.05	14.35	24.40
8050	8 Rib			styrofoam		11.80	14.10	25.90
8060	L.W.	8" conc. block	12	perlite		10.35	14.75	25.10
8110				styrofoam		11.85	14.40	26.25
8120		6" L.W. block	10	perlite		10.40	14.20	24.60
8170				styrofoam		12.05	13.90	25.95
8180		8" L.W. block	12	perlite		11.15	14.60	25.75
8230				styrofoam		12.65	14.25	26.90

B2010 Exterior Walls

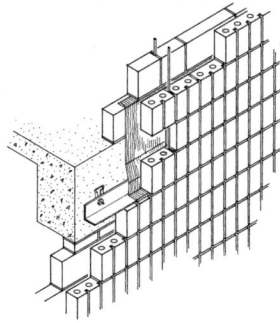

Exterior block face composite walls are defined in the following terms: type of face block and backup masonry, total thickness and insulation. All systems include shelf angles and necessary dampproofing, flashing, and control joints every 20'.

System Components	QUANTITY	UNIT	COST PER S.F.		
			MAT.	INST.	TOTAL
SYSTEM B2010 139 1000					
COMPOSITE WALL, GROOVED BLOCK FACE, 4" C.M.U. BACKUP, PERLITE FILL					
Deep groove block veneer, 4" thick	1.000	S.F.	5.65	6.75	12.40
Concrete block wall, backup, 4" thick	1.000	S.F.	1.58	5.25	6.83
Horizontal joint reinforcing	.800	L.F.	.16	.16	.32
Perlite insulation, poured	1.000	S.F.	.22	.18	.40
Flashing, aluminum	.100	S.F.	.11	.32	.43
Shelf angle	1.000	Lb.	1.27	.89	2.16
Control joint	.050	L.F.	.07	.06	.13
Backer rod	.100	L.F.		.11	.11
Sealant	.100	L.F.	.02	.27	.29
Collar joint	1.000	S.F.	.39	.55	.94
TOTAL			9.47	14.54	24.01

B2010 139		Block Face Composite Wall						
	FACE BLOCK	BACKUP MASONRY	TOTAL THICKNESS (IN.)	BACKUP CORE INSULATION		COST PER S.F.		
						MAT.	INST.	TOTAL
1000	Deep groove	4" conc. block	8	perlite		9.45	14.55	24
1050	Reg. wt.			none		9.25	14.35	23.60
1060		6" conc. block	10	perlite		10.25	15.10	25.35
1110				styrofoam		11.90	14.80	26.70
1120		8" conc. block	12	perlite		10.55	15.55	26.10
1170				styrofoam		12.05	15.15	27.20
1200		4" L.W. block	8	perlite		9.80	14.40	24.20
1250				none		9.55	14.20	23.75
1260		6" L.W. block	10	perlite		10.60	14.95	25.55
1310				styrofoam		12.25	14.65	26.90
1320		8" L.W. block	12	perlite		11.35	15.40	26.75
1370				styrofoam		12.80	15	27.80

B20 Exterior Enclosure

B2010 Exterior Walls

B2010 139	Block Face Composite Wall

	FACE BLOCK	BACKUP MASONRY	TOTAL THICKNESS (IN.)	BACKUP CORE INSULATION		COST PER S.F.		
						MAT.	INST.	TOTAL
1600	Deep groove	4″ conc. block	8	perlite		9.75	14.35	24.10
1650	L.W.			none		9.55	14.15	23.70
1660		6″ conc. block	10	perlite		10.55	14.90	25.45
1710				styrofoam		12.20	14.60	26.80
1720		8″ conc. block	12	perlite		10.85	15.35	26.20
1770				styrofoam		12.35	14.95	27.30
1800		4″ L.W. block	8	perlite		10.10	14.20	24.30
1850				none		9.85	14	23.85
1860		6″ L.W. block	10	perlite		10.90	14.75	25.65
1910				styrofoam		12.55	14.45	27
1920		8″ L.W. block	12	perlite		11.65	15.20	26.85
1970				styrofoam		13.10	14.80	27.90
2200	Fluted	4″ conc. block	8	perlite		7.45	14.35	21.80
2250	Reg. wt.			none		7.20	14.15	21.35
2260		6″ conc. block	10	perlite		8.25	14.90	23.15
2310				styrofoam		9.90	14.60	24.50
2320		8″ conc. block	12	perlite		8.50	15.35	23.85
2370				styrofoam		10	14.95	24.95
2400		4″ L.W. block	8	perlite		7.75	14.20	21.95
2450				styrofoam		7.55	14	21.55
2460		6″ L.W. block	10	perlite		8.55	14.75	23.30
2510				styrofoam		10.20	14.45	24.65
2520		8″ L.W. block	12	perlite		9.30	15.20	24.50
2570				styrofoam		10.80	14.80	25.60
2860	Fluted	6″ conc. block	10	perlite		8.45	15	23.45
2910	L.W.			styrofoam		10	14.40	24.40
2920		8″ conc. block	12	perlite		8.65	15.15	23.80
2970				styrofoam		10.10	14.75	24.85
3060		6″ L.W. block	10	perlite		8.65	14.55	23.20
3110				styrofoam		10.30	14.25	24.55
3120		8″ L.W. block	12	perlite		9.40	15	24.40
3170				styrofoam		10.90	14.60	25.50
3400	Ground face	4″ conc. block	8	perlite		12.60	14.65	27.25
3450	1 Score			none		12.40	14.45	26.85
3460		6″ conc. block	10	perlite		13.40	15.20	28.60
3510				styrofoam		15.05	14.90	29.95
3520		8″ conc. block	12	perlite		13.70	15.75	29.45
3570				styrofoam		15.20	15.25	30.45
3600		4″ L.W. block	8	perlite		12.95	14.50	27.45
3650				none		12.70	14.30	27
3660		6″ L.W. block	10	perlite		13.75	15.05	28.80
3710				styrofoam		15.40	14.75	30.15
3720		8″ L.W. block	12	perlite		14.50	15.50	30
3770				styrofoam		15.95	15.10	31.05

188

B2010 Exterior Walls

B2010 139	Block Face Composite Wall

	FACE BLOCK	BACKUP MASONRY	TOTAL THICKNESS (IN.)	BACKUP CORE INSULATION		COST PER S.F.		
						MAT.	INST.	TOTAL
4000	Hexagonal	4" conc. block	8	perlite		6.40	14.35	20.75
4050	Reg. wt.			none		6.15	14.15	20.30
4060		6" conc. block	10	perlite		7.15	14.90	22.05
4110				styrofoam		8.80	14.60	23.40
4120		8" conc. block	12	perlite		7.50	15.40	22.90
4170				styrofoam		8.95	14.95	23.90
4200		4" L.W. block	8	perlite		6.70	14.20	20.90
4250				none		6.45	14	20.45
4260		6" L.W. block	10	perlite		7.50	14.75	22.25
4310				styrofoam		9.15	14.45	23.60
4320		8" L.W. block	12	perlite		8.25	15.20	23.45
4370				styrofoam		9.75	14.80	24.55
4600	Hexagonal	4" conc. block	8	perlite		6.55	14.15	20.70
4650	L.W.			none		6.35	13.95	20.30
4660		6" conc. block	10	perlite		7.25	14.60	21.85
4710				styrofoam		9	14.40	23.40
4720		8" conc. block	12	perlite		7.65	15.15	22.80
4770				styrofoam		9.15	14.75	23.90
4800		4" L.W. block	8	perlite		6.90	14	20.90
4850				none		6.65	13.80	20.45
4860		6" L.W. block	10	perlite		7.70	14.55	22.25
4910				styrofoam		9.35	14.25	23.60
4920		8" L.W. block	12	perlite		8.45	15	23.45
4970				styrofoam		9.90	14.60	24.50
5200	Slump block	4" conc. block	8	perlite		8.15	13.15	21.30
5250	4x16			none		7.95	12.95	20.90
5260		6" conc. block	10	perlite		8.95	13.70	22.65
5310				styrofoam		10.60	13.40	24
5320		8" conc. block	12	perlite		9.25	14.15	23.40
5370				styrofoam		10.75	13.75	24.50
5400		4" L.W. block	8	perlite		8.45	13	21.45
5450				none		8.25	12.80	21.05
5460		6" L.W. block	10	perlite		9.25	13.55	22.80
5510				styrofoam		10.90	13.25	24.15
5520		8" L.W. block	12	perlite		10	14	24
5570				styrofoam		11.50	13.60	25.10
5800	Split face	4" conc. block	8	perlite		8.80	14.55	23.35
5850	1 Score			none		8.55	14.35	22.90
5860	Reg. wt.	6" conc. block	10	perlite		9.60	15.10	24.70
5910				styrofoam		11.25	14.80	26.05
5920		8" conc. block	12	perlite		9.85	15.55	25.40
5970				styrofoam		11.35	15.15	26.50
6000		4" L.W. block	8	perlite		9.10	14.40	23.50
6050				none		8.90	14.20	23.10
6060		6" L.W. block	10	perlite		9.90	14.95	24.85
6110				styrofoam		11.55	14.65	26.20
6120		8" L.W. block	12	perlite		10.65	15.40	26.05
6170				styrofoam		12.15	15	27.15

B20 Exterior Enclosure

B2010 Exterior Walls

B2010 139	Block Face Composite Wall

	FACE BLOCK	BACKUP MASONRY	TOTAL THICKNESS (IN.)	BACKUP CORE INSULATION		COST PER S.F.		
						MAT.	INST.	TOTAL
6460	Split face	6" conc. block	10	perlite		10	14.90	24.90
6510	1 Score			styrofoam		11.65	14.60	26.25
6520	L.W.	8" conc. block	12	perlite		10.30	15.35	25.65
6570				styrofoam		11.80	14.95	26.75
6660		6" L.W. block	10	perlite		10.35	14.75	25.10
6710				styrofoam		12.05	14.65	26.70
6720		8" L.W. block	12	perlite		11.10	15.20	26.30
6770				styrofoam		12.60	15	27.60
7000	Split rib	4" conc. block	8	perlite		9.25	14.55	23.80
7050	8 Rib			none		9.05	14.35	23.40
7060	Reg. wt.	6" conc. block	10	perlite		10.05	15.10	25.15
7110				styrofoam		11.70	14.80	26.50
7120		8" conc. block	12	perlite		10.35	15.55	25.90
7170				styrofoam		11.85	15.15	27
7200		4" L.W. block	8	perlite		9.60	14.40	24
7250				none		9.35	14.20	23.55
7260		6" L.W. block	10	perlite		10.40	14.95	25.35
7310				styrofoam		12.05	14.65	26.70
7320		8" L.W. block	12	perlite		11.15	15.40	26.55
7370				styrofoam		12.60	15	27.60
7600	Split rib	4" conc. block	8	perlite		9.70	14.35	24.05
7650	8 Rib			none		9.50	14.15	23.65
7660	L.W.	6" conc. block	10	perlite		10.50	14.90	25.40
7710				styrofoam		12.15	14.60	26.75
7720		8" conc. block	12	perlite		10.80	15.35	26.15
7770				styrofoam		12.30	14.95	27.25
7800		4" conc. block	8	perlite		10.05	14.20	24.25
7850				none		9.80	14	23.80
7860		6" conc. block	10	perlite		10.85	14.75	25.60
7910				styrofoam		12.50	14.45	26.95
7920		8" conc. block	12	perlite		11.60	15.20	26.80
7970				styrofoam		13.05	14.80	27.85

B20 Exterior Enclosure

B2010 Exterior Walls

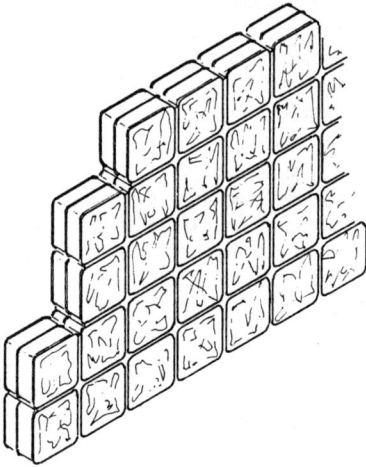

The table below lists costs per S.F. for glass block walls. Included in the costs are the following special accessories required for glass block walls.

Glass block accessories required for proper installation.

Wall ties: Galvanized double steel mesh full length of joint.

Fiberglass expansion joint at sides and top.

Silicone caulking: One gallon does 95 L.F.

Oakum: One lb. does 30 L.F.

Asphalt emulsion: One gallon does 600 L.F.

If block are not set in wall chase, use 2'-0" long wall anchors at 2'-0" O.C.

System Components	QUANTITY	UNIT	COST PER S.F. MAT.	COST PER S.F. INST.	COST PER S.F. TOTAL
SYSTEM B2010 140 2300					
GLASS BLOCK, 4" THICK, 6" X 6" PLAIN, UNDER 1,000 S.F.					
Glass block, 4" thick, 6" x 6" plain, under 1000 S.F.	4.100	Ea.	22.50	19.60	42.10
Glass block, cleaning blocks after installation, both sides add	2.000	S.F.	.12	2.25	2.37
TOTAL			22.62	21.85	44.47

B2010 140	Glass Block	MAT.	INST.	TOTAL
2300	Glass block 4" thick, 6"x6" plain, under 1,000 S.F.	22.50	22	44.50
2400	1,000 to 5,000 S.F.	22	18.95	40.95
2500	Over 5,000 S.F.	21.50	17.80	39.30
2600	Solar reflective, under 1,000 S.F.	31.50	29.50	61
2700	1,000 to 5,000 S.F.	31	25.50	56.50
2800	Over 5,000 S.F.	30	24	54
3500	8"x8" plain, under 1,000 S.F.	15.55	16.35	31.90
3600	1,000 to 5,000 S.F.	15.25	14.10	29.35
3700	Over 5,000 S.F.	14.75	12.75	27.50
3800	Solar reflective, under 1,000 S.F.	22	22	44
3900	1,000 to 5,000 S.F.	21.50	18.85	40.35
4000	Over 5,000 S.F.	20.50	16.95	37.45
5000	12"x12" plain, under 1,000 S.F.	17.70	15.15	32.85
5100	1,000 to 5,000 S.F.	17.35	12.75	30.10
5200	Over 5,000 S.F.	16.85	11.65	28.50
5300	Solar reflective, under 1,000 S.F.	25	20.50	45.50
5400	1,000 to 5,000 S.F.	24.50	16.95	41.45
5600	Over 5,000 S.F.	23.50	15.40	38.90
5800	3" thinline, 6"x6" plain, under 1,000 S.F.	16.65	22	38.65
5900	Over 5,000 S.F.	16.65	22	38.65
6000	Solar reflective, under 1,000 S.F.	23.50	29.50	53
6100	Over 5,000 S.F.	25	24	49
6200	8"x8" plain, under 1,000 S.F.	10.40	16.35	26.75
6300	Over 5,000 S.F.	10.05	12.75	22.80
6400	Solar reflective, under 1,000 S.F.	14.55	22	36.55
6500	Over 5,000 S.F.	14.05	16.95	31

B2010 Exterior Walls

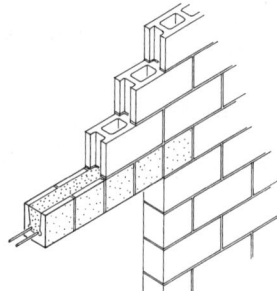

Concrete Block Lintel

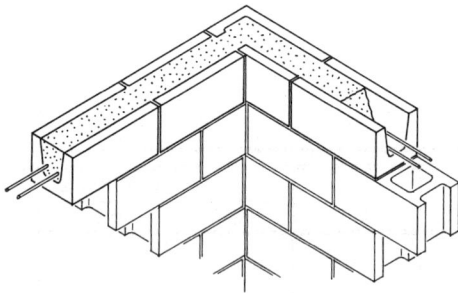

Bond Beam

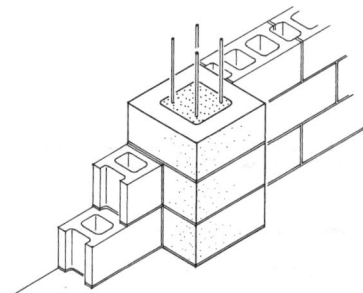

Pilaster

Concrete block specialties are divided into lintels, pilasters, and bond beams. Lintels are defined by span, thickness, height and wall loading. Span refers to the clear opening but the cost includes 8″ of bearing at both ends.

Bond beams and pilasters are defined by height thickness and weight of the masonry unit itself. Components for bond beams also include grout and reinforcing. For pilasters, components include four #5 reinforcing bars and type N mortar.

System Components			COST PER LINTEL		
	QUANTITY	UNIT	MAT.	INST.	TOTAL
SYSTEM B2010 144 3100					
CONCRETE BLOCK LINTEL, 6″X8″, LOAD 300 LB/L.F. 3′-4″ SPAN					
Lintel blocks, 6″ x 8″ x 8″	7.000	Ea.	10.78	20.58	31.36
Joint reinforcing, #3 & #4 steel bars, horizontal	3.120	Lb.	2.65	3.40	6.05
Grouting, 8″ deep, 6″ thick, .15 C.F./L.F.	.700	C.F.	3.78	4.69	8.47
Temporary shoring, lintel forms	1.000	Set	.80	11.26	12.06
Temporary shoring, wood joists	1.000	Set		19.80	19.80
Temporary shoring, vertical members	1.000	Set		36	36
Mortar, masonry cement, 1:3 mix, type N	.270	C.F.	1.29	.09	1.38
TOTAL			19.30	95.82	115.12

B2010 144		Concrete Block Lintel						
	SPAN	THICKNESS (IN.)	HEIGHT (IN.)	WALL LOADING P.L.F.		COST PER LINTEL		
						MAT.	INST.	TOTAL
3100	3′-4″	6	8	300		19.30	96	115.30
3150		6	16	1,000		30	103	133
3200		8	8	300		21.50	103	124.50
3300		8	8	1,000		24	106	130
3400		8	16	1,000		34.50	112	146.50
3450	4′-0″	6	8	300		22.50	106	128.50
3500		6	16	1,000		39	119	158
3600		8	8	300		25	112	137
3700		8	16	1,000		39.50	121	160.50

B2010 Exterior Walls

B2010 144	Concrete Block Lintel							
	SPAN	THICKNESS (IN.)	HEIGHT (IN.)	WALL LOADING P.L.F.		COST PER LINTEL		
						MAT.	INST.	TOTAL
3800	4'-8"	6	8	300		25	107	132
3900		6	16	1,000		39.50	118	157.50
4000		8	8	300		28.50	119	147.50
4100		8	16	1,000		45	130	175
4200	5'-4"	6	8	300		31.50	120	151.50
4300		6	16	1,000		45.50	124	169.50
4400		8	8	300		35	131	166
4500		8	16	1,000		52	141	193
4600	6'-0"	6	8	300		39.50	145	184.50
4700		6	16	300		48	151	199
4800		6	16	1,000		52	157	209
4900		8	8	300		38.50	159	197.50
5000		8	16	1,000		58.50	172	230.50
5100	6'-8"	6	16	300		47	152	199
5200		6	16	1,000		51.50	158	209.50
5300		8	8	300		46.50	165	211.50
5400		8	16	1,000		64.50	182	246.50
5500	7'-4"	6	16	300		52	157	209
5600		6	16	1,000		68.50	169	237.50
5700		8	8	300		59.50	178	237.50
5800		8	16	300		78	190	268
5900		8	16	1,000		78	190	268
6000	8'-0"	6	16	300		56	164	220
6100		6	16	1,000		74	176	250
6200		8	16	300		86.50	199	285.50
6500		8	16	1,000		86.50	199	285.50

B2010 145	Concrete Block Specialties							
	TYPE	HEIGHT (IN.)	THICKNESS (IN.)	WEIGHT (P.C.F.)		COST PER L.F.		
						MAT.	INST.	TOTAL
7100	Bond beam	8	8	125		5.65	6.60	12.25
7200				105		6.25	6.55	12.80
7300			12	125		6.95	8.30	15.25
7400				105		7.70	8.20	15.90
7500	Pilaster	8	16	125		13.05	22	35.05
7600			20	125		15.45	26	41.45

B2010 Exterior Walls

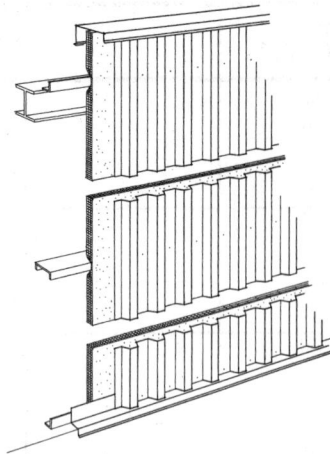

The table below lists costs for metal siding of various descriptions, not including the steel frame, or the structural steel, of a building. Costs are per S.F. including all accessories and insulation.

For steel frame support see System B2010 154.

System Components	QUANTITY	UNIT	COST PER S.F.		
			MAT.	INST.	TOTAL
SYSTEM B2010 146 1400					
METAL SIDING ALUMINUM PANEL, CORRUGATED, .024" THICK, NATURAL					
Corrugated aluminum siding, industrial type, .024" thick	1.000	S.F.	2.08	2.51	4.59
Flashing alum mill finish .032" thick	.111	S.F.	.18	.35	.53
Tapes, sealant , p.v.c. foam adheasive, 1/16"x1"	.004	C.L.F.	.06		.06
Closure strips for aluminum siding, corrugated, .032" thick	.111	L.F.	.09	.27	.36
Pre-engineered steel building insulation, vinyl faced, 1-1/2" thick, R5	1.000	S.F.	.30	.43	.73
TOTAL			2.71	3.56	6.27

B2010 146	Metal Siding Panel	COST PER S.F.		
		MAT.	INST.	TOTAL
1400	Metal siding aluminum panel, corrugated, .024" thick, natural	2.71	3.56	6.27
1450	Painted	2.87	3.56	6.43
1500	.032" thick, natural	3.02	3.56	6.58
1550	Painted	3.55	3.56	7.11
1600	Ribbed 4" pitch, .032" thick, natural	2.96	3.56	6.52
1650	Painted	3.59	3.56	7.15
1700	.040" thick, natural	3.38	3.56	6.94
1750	Painted	3.88	3.56	7.44
1800	.050" thick, natural	3.83	3.56	7.39
1850	Painted	4.33	3.56	7.89
1900	8" pitch panel, .032" thick, natural	2.82	3.41	6.23
1950	Painted	3.44	3.43	6.87
2000	.040" thick, natural	3.25	3.43	6.68
2050	Painted	3.71	3.47	7.18
2100	.050" thick, natural	3.70	3.45	7.15
2150	Painted	4.22	3.48	7.70
3000	Steel, corrugated or ribbed, 29 Ga. .0135" thick, galvanized	2.11	3.25	5.36
3050	Colored	2.97	3.30	6.27
3100	26 Ga. .0179" thick, galvanized	2.18	3.27	5.45
3150	Colored	3.06	3.32	6.38
3200	24 Ga. .0239" thick, galvanized	2.74	3.28	6.02
3250	Colored	3.42	3.33	6.75
3300	22 Ga. .0299" thick, galvanized	2.75	3.30	6.05
3350	Colored	4.09	3.35	7.44
3400	20 Ga. .0359" thick, galvanized	2.75	3.30	6.05
3450	Colored	4.34	3.54	7.88

B20 Exterior Enclosure

B2010 Exterior Walls

B2010 146	Metal Siding Panel	COST PER S.F.		
		MAT.	INST.	TOTAL
4100	Sandwich panels, factory fab., 1" polystyrene, steel core, 26 Ga., galv.	5.30	5.10	10.40
4200	Colored, 1 side	6.50	5.10	11.60
4300	2 sides	8.35	5.10	13.45
4400	2" polystyrene, steel core, 26 Ga., galvanized	6.25	5.10	11.35
4500	Colored, 1 side	7.45	5.10	12.55
4600	2 sides	9.30	5.10	14.40
4700	22 Ga., baked enamel exterior	12.30	5.40	17.70
4800	Polyvinyl chloride exterior	12.95	5.40	18.35
5100	Textured aluminum, 4' x 8' x 5/16" plywood backing, single face	4.10	3.10	7.20
5200	Double face	4.88	3.10	7.98
5300	4' x 10' x 5/16" plywood backing, single face	4.58	3.10	7.68

B20 Exterior Enclosure

B2010 Exterior Walls

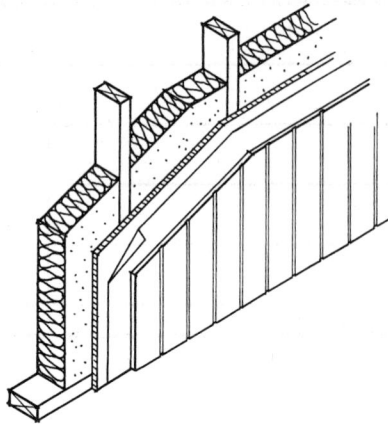

The table below lists costs per S.F. for exterior walls with wood siding. A variety of systems are presented using both wood and metal studs at 16″ and 24″ O.C.

System Components	QUANTITY	UNIT	COST PER S.F.		
			MAT.	INST.	TOTAL
SYSTEM B2010 148 1400					
2″X4″ STUDS, 16″ O.C., INSUL. WALL, W/ 5/8″ TEXTURE 1-11 FIR PLYWOOD					
Partitions, 2″ x 4″ studs 8′ high 16″ O.C.	1.000	B.F.	.36	1.24	1.60
Sheathing plywood on wall CDX 1/2″	1.000	S.F.	.51	.88	1.39
Building paper, asphalt felt sheathing paper 15 lb	1.000	S.F.	.05	.13	.18
Fiberglass insulation batts, paper or foil back, 3-1/2″, R11	1.000	S.F.	.63	.37	1
Siding plywood texture 1-11 fir 5/8″, natural	1.000	S.F.	1.24	1.47	2.71
Exterior wood stain on other than shingles, 2 coats & sealer	1.000	S.F.	.18	.99	1.17
TOTAL			2.97	5.08	8.05

B2010 148	Panel, Shingle & Lap Siding	COST PER S.F.		
		MAT.	INST.	TOTAL
1400	Wood siding w/2″x4″studs, 16″O.C., insul. wall, 5/8″text 1-11 fir plywood	2.97	5.10	8.07
1450	5/8″ text 1-11 cedar plywood	4.43	4.91	9.34
1500	1″ x 4″ vert T.&G. redwood	5.15	5.85	11
1600	1″ x 8″ vert T.&G. redwood	5.30	5.10	10.40
1650	1″ x 5″ rabbetted cedar bev. siding	6.85	5.30	12.15
1700	1″ x 6″ cedar drop siding	6.95	5.30	12.25
1750	1″ x 12″ rough sawn cedar	3.10	5.10	8.20
1800	1″ x 12″ sawn cedar, 1″ x 4″ battens	6.05	4.79	10.84
1850	1″ x 10″ redwood shiplap siding	5.40	4.93	10.33
1900	18″ no. 1 red cedar shingles, 5-1/2″ exposed	5.45	6.70	12.15
1950	6″ exposed	5.20	6.45	11.65
2000	6-1/2″ exposed	4.92	6.25	11.17
2100	7″ exposed	4.66	6.05	10.71
2150	7-1/2″ exposed	4.39	5.80	10.19
3000	8″ wide aluminum siding	3.28	4.54	7.82
3150	8″ plain vinyl siding	2.61	4.62	7.23
3250	8″ insulated vinyl siding	2.91	5.10	8.01
3400	2″ x 6″ studs, 16″ O.C., insul. wall, w/ 5/8″ text 1-11 fir plywood	3.47	5.20	8.67
3500	5/8″ text 1-11 cedar plywood	4.93	5.20	10.13
3600	1″ x 4″ vert T.&G. redwood	5.65	6	11.65
3700	1″ x 8″ vert T.&G. redwood	5.80	5.25	11.05
3800	1″ x 5″ rabbetted cedar bev siding	7.35	5.45	12.80
3900	1″ x 6″ cedar drop siding	7.45	5.45	12.90
4000	1″ x 12″ rough sawn cedar	3.60	5.25	8.85
4200	1″ x 12″ sawn cedar, 1″ x 4″ battens	6.55	4.93	11.48
4500	1″ x 10″ redwood shiplap siding	5.90	5.05	10.95

B2010 Exterior Walls

B2010 148	Panel, Shingle & Lap Siding	COST PER S.F.		
		MAT.	INST.	TOTAL
4550	18" no. 1 red cedar shingles, 5-1/2" exposed	5.95	6.85	12.80
4600	6" exposed	5.70	6.60	12.30
4650	6-1/2" exposed	5.40	6.40	11.80
4700	7" exposed	5.15	6.15	11.30
4750	7-1/2" exposed	4.89	5.95	10.84
4800	8" wide aluminum siding	3.78	4.68	8.46
4850	8" plain vinyl siding	3.11	4.76	7.87
4900	8" insulated vinyl siding	3.41	5.25	8.66
5000	2" x 6" studs, 24" O.C., insul. wall, 5/8" text 1-11, fir plywood	3.33	4.92	8.25
5050	5/8" text 1-11 cedar plywood	4.79	4.92	9.71
5100	1" x 4" vert T.&G. redwood	5.50	5.70	11.20
5150	1" x 8" vert T.&G. redwood	5.65	4.95	10.60
5200	1" x 5" rabbetted cedar bev siding	7.20	5.15	12.35
5250	1" x 6" cedar drop siding	7.30	5.15	12.45
5300	1" x 12" rough sawn cedar	3.46	4.96	8.42
5400	1" x 12" sawn cedar, 1" x 4" battens	6.40	4.63	11.03
5450	1" x 10" redwood shiplap siding	5.75	4.77	10.52
5500	18" no. 1 red cedar shingles, 5-1/2" exposed	5.80	6.55	12.35
5550	6" exposed	5.55	6.30	11.85
5650	7" exposed	5	5.85	10.85
5700	7-1/2" exposed	4.75	5.65	10.40
5750	8" wide aluminum siding	3.64	4.38	8.02
5800	8" plain vinyl siding	2.97	4.46	7.43
5850	8" insulated vinyl siding	3.27	4.96	8.23
5900	3-5/8" metal studs, 16 Ga., 16" OC insul.wall, 5/8"text 1-11 fir plywood	4.27	5.35	9.62
5950	5/8" text 1-11 cedar plywood	5.75	5.35	11.10
6000	1" x 4" vert T.&G. redwood	6.45	6.10	12.55
6050	1" x 8" vert T.&G. redwood	6.60	5.35	11.95
6100	1" x 5" rabbetted cedar bev siding	8.15	5.55	13.70
6150	1" x 6" cedar drop siding	8.25	5.60	13.85
6200	1" x 12" rough sawn cedar	4.40	5.40	9.80
6250	1" x 12" sawn cedar, 1" x 4" battens	7.35	5.05	12.40
6300	1" x 10" redwood shiplap siding	6.70	5.20	11.90
6350	18" no. 1 red cedar shingles, 5-1/2" exposed	6.75	6.95	13.70
6500	6" exposed	6.50	6.75	13.25
6550	6-1/2" exposed	6.20	6.50	12.70
6600	7" exposed	5.95	6.30	12.25
6650	7-1/2" exposed	5.95	6.10	12.05
6700	8" wide aluminum siding	4.58	4.80	9.38
6750	8" plain vinyl siding	3.91	4.71	8.62
6800	8" insulated vinyl siding	4.21	5.20	9.41
7000	3-5/8" metal studs, 16 Ga. 24" OC insul wall, 5/8" text 1-11 fir plywood	3.81	4.75	8.56
7050	5/8" text 1-11 cedar plywood	5.25	4.75	10
7100	1" x 4" vert T.&G. redwood	6	5.70	11.70
7150	1" x 8" vert T.&G. redwood	6.15	4.95	11.10
7200	1" x 5" rabbetted cedar bev siding	7.65	5.15	12.80
7250	1" x 6" cedar drop siding	7.75	5.15	12.90
7300	1" x 12" rough sawn cedar	3.94	4.96	8.90
7350	1" x 12" sawn cedar 1" x 4" battens	6.90	4.63	11.53
7400	1" x 10" redwood shiplap siding	6.20	4.77	10.97
7450	18" no. 1 red cedar shingles, 5-1/2" exposed	6.55	6.75	13.30
7500	6" exposed	6.30	6.55	12.85
7550	6-1/2" exposed	6.05	6.30	12.35
7600	7" exposed	5.75	6.10	11.85
7650	7-1/2" exposed	5.25	5.65	10.90
7700	8" wide aluminum siding	4.12	4.38	8.50
7750	8" plain vinyl siding	3.45	4.46	7.91
7800	8" insul. vinyl siding	3.75	4.96	8.71

B2010 Exterior Walls

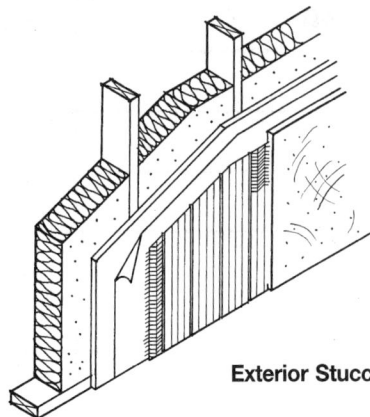

Exterior Stucco Wall

The table below lists costs for some typical stucco walls including all the components as demonstrated in the component block below. Prices are presented for backup walls using wood studs, metal studs and CMU.

System Components			COST PER S.F.		
	QUANTITY	UNIT	MAT.	INST.	TOTAL
SYSTEM B2010 151 2100					
7/8″ CEMENT STUCCO, PLYWOOD SHEATHING, STUD WALL, 2″ X 4″, 16″ O.C.					
Framing 2x4 studs 8′ high 16″ O.C.	1.000	S.F.	.36	1.24	1.60
Plywood sheathing on ext. stud wall, 5/8″ thick	1.000	S.F.	.74	.94	1.68
Building paper, asphalt felt sheathing paper 15 lb	1.000	S.F.	.05	.13	.18
Stucco, 3 coats 7/8″ thk, float finish, on frame construction	1.000	S.F.	.82	4.69	5.51
Fiberglass insulation batts, paper or foil back, 3-1/2″, R-11	1.000	S.F.	.63	.37	1
Paint exterior stucco, brushwork, primer & 2 coats	1.000	S.F.	.20	.84	1.04
TOTAL			2.80	8.21	11.01

B2010 151	Stucco Wall	COST PER S.F.		
		MAT.	INST.	TOTAL
2100	Cement stucco, 7/8″ th., plywood sheathing, stud wall, 2″ x 4″, 16″ O.C.	2.80	8.20	11
2200	24″ O.C.	2.72	7.95	10.67
2300	2″ x 6″, 16″ O.C.	3.30	8.35	11.65
2400	24″ O.C.	3.16	8.05	11.21
2500	No sheathing, metal lath on stud wall, 2″ x 4″, 16″ O.C.	2.01	7.15	9.16
2600	24″ O.C.	1.93	6.90	8.83
2700	2″ x 6″, 16″ O.C.	2.51	7.30	9.81
2800	24″ O.C.	2.37	7	9.37
2900	1/2″ gypsum sheathing, 3-5/8″ metal studs, 16″ O.C.	3.78	7.45	11.23
2950	24″ O.C.	3.32	7	10.32
3000	Cement stucco, 5/8″ th., 2 coats on std. CMU block, 8″x 16″, 8″ thick	2.66	10.05	12.71
3100	10″ thick	3.81	10.35	14.16
3200	12″ thick	3.91	11.90	15.81
3300	Std. light Wt. block 8″ x 16″, 8″ thick	3.70	9.90	13.60
3400	10″ thick	4.66	10.15	14.81
3500	12″ thick	5.25	11.60	16.85
3600	3 coat stucco, self furring metal lath 3.4 Lb/SY, on 8″ x 16″, 8″ thick	2.70	10.65	13.35
3700	10″ thick	3.85	10.95	14.80
3800	12″ thick	3.95	12.50	16.45
3900	Lt. Wt. block, 8″ thick	3.74	10.50	14.24
4000	10″ thick	4.70	10.75	15.45
4100	12″ thick	5.30	12.20	17.50

B20 Exterior Enclosure

B2010 Exterior Walls

B2010 152	E.I.F.S.	COST PER S.F.		
		MAT.	INST.	TOTAL
5100	E.I.F.S., plywood sheathing, stud wall, 2" x 4", 16" O.C., 1" EPS	4.13	10.10	14.23
5110	2" EPS	4.54	10.10	14.64
5120	3" EPS	4.72	10.10	14.82
5130	4" EPS	5.20	10.10	15.30
5140	2" x 6", 16" O.C., 1" EPS	4.63	10.25	14.88
5150	2" EPS	5.05	10.25	15.30
5160	3" EPS	5.20	10.25	15.45
5170	4" EPS	5.70	10.25	15.95
5180	Cement board sheathing, 3-5/8" metal studs, 16" O.C., 1" EPS	5.50	12	17.50
5190	2" EPS	5.90	12	17.90
5200	3" EPS	6.10	12	18.10
5210	4" EPS	7.35	12	19.35
5220	6" metal studs, 16" O.C., 1" EPS	6.25	12.05	18.30
5230	2" EPS	6.65	12.05	18.70
5240	3" EPS	6.85	12.05	18.90
5250	4" EPS	8.10	12.05	20.15
5260	CMU block, 8" x 8" x 16", 1" EPS	4.61	13.65	18.26
5270	2" EPS	5	13.65	18.65
5280	3" EPS	5.20	13.65	18.85
5290	4" EPS	5.70	13.65	19.35
5300	8" x 10" x 16", 1" EPS	5.75	13.95	19.70
5310	2" EPS	6.15	13.95	20.10
5320	3" EPS	6.35	13.95	20.30
5330	4" EPS	6.85	13.95	20.80
5340	8" x 12" x 16", 1" EPS	5.85	15.50	21.35
5350	2" EPS	6.25	15.50	21.75
5360	3" EPS	6.45	15.50	21.95
5370	4" EPS	6.95	15.50	22.45

B2010 Exterior Walls

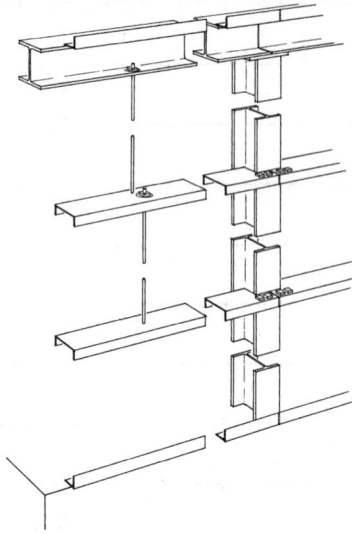

Description: The table below lists costs, $/S.F., for channel girts with sag rods and connector angles top and bottom for various column spacings, building heights and wind loads. Additive costs are shown for wind columns.

How to Use this Table: Add the cost of girts, sag rods, and angles to the framing costs of steel buildings clad in metal or composition siding. If the column spacing is in excess of the column spacing shown, use intermediate wind columns. Additive costs are shown under "Wind Columns."

Column Spacing 20'–0" Girts	$5.23
Wind Columns	1.22
Total Costs for Girts	$6.45/S.F.

Design and Pricing Assumptions: Structural steel is A36.

System Components		QUANTITY	UNIT	COST PER S.F.		
				MAT.	INST.	TOTAL
SYSTEM B2010 154 3000						
METAL SIDING SUPPORT, 18' HIGH, 20 PSF WIND LOAD, 20' SPACING						
Structural steel		1.417	Lb.	2.34	.45	2.79
Lightgage framing, angles less than 4"		.400	Lb.	.40	3.14	3.54
	TOTAL			2.74	3.59	6.33
SYSTEM B2010 154 3100						
WIND COLUMNS						
Structural steel		.856	S.F.	1.41	.28	1.69

B2010 154 — Metal Siding Support

	BLDG. HEIGHT (FT.)	WIND LOAD (P.S.F.)	COL. SPACING (FT.)		INTERMEDIATE COLUMNS	COST PER S.F.		
						MAT.	INST.	TOTAL
3000	18	20	20			2.74	3.59	6.33
3100					wind cols.	1.41	.28	1.69
3200	RB2010 -430	20	25			2.99	3.65	6.64
3300					wind cols.	1.13	.21	1.34
3400		20	30			3.30	3.70	7
3500					wind cols.	.94	.19	1.13
3600		20	35			3.64	3.77	7.41
3700					wind cols.	.81	.16	.97
3800		30	20			3.03	3.65	6.68
3900					wind cols.	1.41	.28	1.69
4000		30	25			3.30	3.70	7
4100					wind cols.	1.13	.21	1.34
4200		30	30			3.65	3.77	7.42
4300					wind cols.	1.28	.25	1.53
4400		30	35			4.86	4.01	8.87
4500					wind cols.	1.27	.24	1.51
4600	30	20	20			2.65	2.55	5.20
4700					wind cols.	2	.39	2.39
4800		20	25			2.98	2.61	5.59
4900					wind cols.	1.60	.31	1.91

B20 Exterior Enclosure

B2010 Exterior Walls

B2010 154	Metal Siding Support

	BLDG. HEIGHT (FT.)	WIND LOAD (P.S.F.)	COL. SPACING (FT.)		INTERMEDIATE COLUMNS	COST PER S.F.		
						MAT.	INST.	TOTAL
5000	30	20	30			3.36	2.68	6.04
5100					wind cols.	1.57	.31	1.88
5200		20	35			3.75	2.76	6.51
5300					wind cols.	1.56	.31	1.87
5400		30	20			3.02	2.61	5.63
5500					wind cols.	2.36	.45	2.81
5600		30	25			3.35	2.68	6.03
5700					wind cols.	2.25	.44	2.69
5800		30	30			3.77	2.76	6.53
5900					wind cols.	2.12	.41	2.53
6000		30	35			5.20	3.04	8.24
6100					wind cols.	2.07	.40	2.47

B2020 Exterior Windows

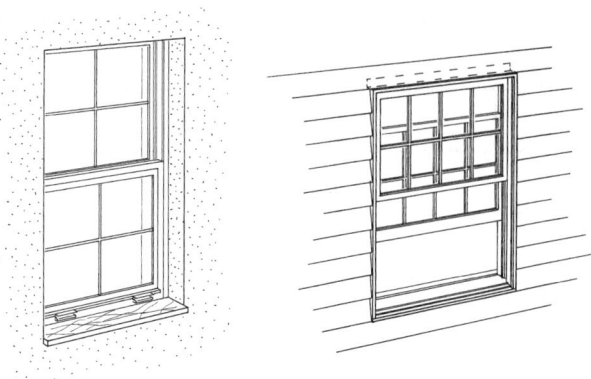

The table below lists window systems by material, type and size. Prices between sizes listed can be interpolated with reasonable accuracy. Prices include frame, hardware, and casing as illustrated in the component block below.

System Components	QUANTITY	UNIT	COST PER UNIT		
			MAT.	INST.	TOTAL
SYSTEM B2020 102 3000					
WOOD, DOUBLE HUNG, STD. GLASS, 2'-8" X 4'-6"					
Framing rough opening, header w/jacks	14.680	B.F.	8.51	25.10	33.61
Casing, stock pine 11/16" x 2-1/2"	16.000	L.F.	23.36	32.96	56.32
Stool cap, pine, 11/16" x 3-1/2"	5.000	L.F.	12.50	12.40	24.90
Residential wood window, double hung, 2'-8" x 4'-6" standard glazed	1.000	Ea.	194	146	340
TOTAL			238.37	216.46	454.83

B2020 102	Wood Windows							
	MATERIAL	TYPE	GLAZING	SIZE	DETAIL	COST PER UNIT		
						MAT.	INST.	TOTAL
3000	Wood	double hung	std. glass	2'-8" x 4'-6"		238	216	454
3050				3'-0" x 5'-6"		315	248	563
3100			insul. glass	2'-8" x 4'-6"		255	216	471
3150				3'-0" x 5'-6"		340	248	588
3200		sliding	std. glass	3'-4" x 2'-7"		288	180	468
3250				4'-4" x 3'-3"		330	197	527
3300				5'-4" x 6'-0"		420	237	657
3350			insul. glass	3'-4" x 2'-7"		350	210	560
3400				4'-4" x 3'-3"		400	229	629
3450				5'-4" x 6'-0"		505	270	775
3500		awning	std. glass	2'-10" x 1'-9"		215	108	323
3600				4'-4" x 2'-8"		350	105	455
3700			insul. glass	2'-10" x 1'-9"		263	125	388
3800				4'-4" x 2'-8"		430	115	545
3900		casement	std. glass	1'-10" x 3'-2"	1 lite	320	142	462
3950				4'-2" x 4'-2"	2 lite	575	189	764
4000				5'-11" x 5'-2"	3 lite	870	244	1,114
4050				7'-11" x 6'-3"	4 lite	1,225	293	1,518
4100				9'-11" x 6'-3"	5 lite	1,600	340	1,940
4150			insul. glass	1'-10" x 3'-2"	1 lite	320	142	462
4200				4'-2" x 4'-2"	2 lite	590	189	779
4250				5'-11" x 5'-2"	3 lite	920	244	1,164
4300				7'-11" x 6'-3"	4 lite	1,300	293	1,593
4350				9'-11" x 6'-3"	5 lite	1,675	340	2,015

B20 Exterior Enclosure

B2020 Exterior Windows

B2020 102 — Wood Windows

	MATERIAL	TYPE	GLAZING	SIZE	DETAIL	COST PER UNIT		
						MAT.	INST.	TOTAL
4400	Wood	picture	std. glass	4'-6" x 4'-6"		455	244	699
4450				5'-8" x 4'-6"		515	271	786
4500		picture	insul. glass	4'-6" x 4'-6"		560	284	844
4550				5'-8" x 4'-6"		630	315	945
4600		fixed bay	std. glass	8' x 5'		1,600	445	2,045
4650				9'-9" x 5'-4"		1,125	630	1,755
4700			insul. glass	8' x 5'		2,150	445	2,595
4750				9'-9" x 5'-4"		1,225	630	1,855
4800		casement bay	std. glass	8' x 5'		1,600	510	2,110
4850			insul. glass	8' x 5'		1,800	510	2,310
4900		vert. bay	std. glass	8' x 5'		1,800	510	2,310
4950			insul. glass	8' x 5'		1,875	510	2,385

B2020 104 — Steel Windows

	MATERIAL	TYPE	GLAZING	SIZE	DETAIL	COST PER UNIT		
						MAT.	INST.	TOTAL
5000	Steel	double hung	1/4" tempered	2'-8" x 4'-6"		835	170	1,005
5050				3'-4" x 5'-6"		1,275	259	1,534
5100			insul. glass	2'-8" x 4'-6"		855	194	1,049
5150				3'-4" x 5'-6"		1,300	296	1,596
5200		horz. pivoted	std. glass	2' x 2'		251	56.50	307.50
5250				3' x 3'		565	127	692
5300				4' x 4'		1,000	226	1,226
5350				6' x 4'		1,500	340	1,840
5400			insul. glass	2' x 2'		257	65	322
5450				3' x 3'		580	146	726
5500				4' x 4'		1,025	259	1,284
5550				6' x 4'		1,550	390	1,940
5600		picture window	std. glass	3' x 3'		355	127	482
5650				6' x 4'		940	340	1,280
5700			insul. glass	3' x 3'		370	146	516
5750				6' x 4'		980	390	1,370
5800		industrial security	std. glass	2'-9" x 4'-1"		755	159	914
5850				4'-1" x 5'-5"		1,475	315	1,790
5900			insul. glass	2'-9" x 4'-1"		775	182	957
5950				4'-1" x 5'-5"		1,525	360	1,885
6000		comm. projected	std. glass	3'-9" x 5'-5"		1,250	288	1,538
6050				6'-9" x 4'-1"		1,675	390	2,065
6100			insul. glass	3'-9" x 5'-5"		1,275	330	1,605
6150				6'-9" x 4'-1"		1,725	445	2,170
6200		casement	std. glass	4'-2" x 4'-2"	2 lite	985	246	1,231
6250			insul. glass	4'-2" x 4'-2"		1,000	281	1,281
6300			std. glass	5'-11" x 5'-2"	3 lite	1,850	435	2,285
6350			insul. glass	5'-11" x 5'-2"		1,900	495	2,395

B2020 106 — Aluminum Windows

	MATERIAL	TYPE	GLAZING	SIZE	DETAIL	COST PER UNIT		
						MAT.	INST.	TOTAL
6400	Aluminum	awning	std. glass	3'-1" x 3'-2"		370	127	497
6450				4'-5" x 5'-3"		420	159	579
6500			insul. glass	3'-1" x 3'-2"		445	152	597
6550				4'-5" x 5'-3"		505	191	696

B20 Exterior Enclosure

B2020 Exterior Windows

B2020 106			Aluminum Windows					

	MATERIAL	TYPE	GLAZING	SIZE	DETAIL	COST PER UNIT		
						MAT.	INST.	TOTAL
6600	Aluminum	sliding	std. glass	3' x 2'		230	127	357
6650				5' x 3'		350	142	492
6700				8' x 4'		370	212	582
6750				9' x 5'		560	320	880
6800			insul. glass	3' x 2'		246	127	373
6850				5' x 3'		410	142	552
6900				8' x 4'		595	212	807
6950				9' x 5'		900	320	1,220
7000		single hung	std. glass	2' x 3'		218	127	345
7050				2'-8" x 6'-8"		385	159	544
7100				3'-4" x 5'-0"		315	142	455
7150			insul. glass	2' x 3'		264	127	391
7200				2'-8" x 6'-8"		495	159	654
7250				3'-4" x 5'		350	142	492
7300		double hung	std. glass	2' x 3'		289	85	374
7350				2'-8" x 6'-8"		860	252	1,112
7400				3'-4" x 5'		805	236	1,041
7450			insul. glass	2' x 3'		299	97	396
7500				2'-8" x 6'-8"		885	288	1,173
7550				3'-4" x 5'-0"		830	270	1,100
7600		casement	std. glass	3'-1" x 3'-2"		247.65	137.96	385.13
7650				4'-5" x 5'-3"		600	335	935
7700			insul. glass	3'-1" x 3'-2"		264	158	422
7750				4'-5" x 5'-3"		640	385	1,025
7800		hinged swing	std. glass	3' x 4'		530	170	700
7850				4' x 5'		885	283	1,168
7900			insul. glass	3' x 4'		550	194	744
7950				4' x 5'		915	325	1,240
8200		picture unit	std. glass	2'-0" x 3'-0"		159	85	244
8250				2'-8" x 6'-8"		470	252	722
8300				3'-4" x 5'-0"		440	236	676
8350			insul. glass	2'-0" x 3'-0"		169	97	266
8400				2'-8" x 6'-8"		500	288	788
8450				3'-4" x 5'-0"		470	270	740
8500		awning type	std. glass	3'-0" x 3'-0"	2 lite	445	91	536
8550				3'-0" x 4'-0"	3 lite	520	127	647
8600				3'-0" x 5'-4"	4 lite	625	127	752
8650				4'-0" x 5'-4"	4 lite	685	142	827
8700			insul. glass	3'-0" x 3'-0"	2 lite	475	91	566
8750				3'-0" x 4'-0"	3 lite	595	127	722
8800				3'-0" x 5'-4"	4 lite	735	127	862
8850				4'-0" x 5'-4"	4 lite	825	142	967

B2020 Exterior Windows

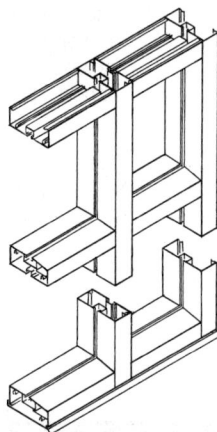

The table below lists costs per S.F of opening for framing with 1-3/4" x 4-1/2" clear anodized tubular aluminum framing. This is the type often used for 1/4" plate glass flush glazing.

For bronze finish, add 18% to material cost. For black finish, add 27% to material cost. For stainless steel, add 75% to material cost. For monumental grade, add 50% to material cost. This tube framing is usually installed by a glazing contractor.

Note: The costs below do not include the glass. For glazing, use Assembly B2020 220.

System Components	QUANTITY	UNIT	COST/S.F. OPNG.		
			MAT.	INST.	TOTAL
SYSTEM B2020 210 1250					
ALUM FLUSH TUBE, FOR 1/4" GLASS, 5'X20' OPENING, 3 INTER. HORIZONTALS					
Flush tube frame, alum mill fin, 1-3/4"x4" open headr for 1/4" glass	.450	L.F.	5.67	5.24	10.91
Flush tube frame, alum mill fin, 1-3/4"x4" open sill for 1/4" glass	.050	L.F.	.50	.57	1.07
Flush tube frame, alum mill fin, 1-3/4"x4" closed back sill, 1/4" glass	.150	L.F.	2.51	1.65	4.16
Aluminum structural shapes, 1" to 10" members, under 1 ton	.040	Lb.	.15	.23	.38
Joints for tube frame, 90° clip type	.100	Ea.	2.50		2.50
Caulking/sealants, polysulfide, 1 or 2 part,1/2x1/4"bead 154 lf/gal	.500	L.F.	.19	1.82	2.01
TOTAL			11.52	9.51	21.03

B2020 210	Tubular Aluminum Framing	COST/S.F. OPNG.		
		MAT.	INST.	TOTAL
1100	Alum flush tube frame, for 1/4" glass, 1-3/4"x4", 5'x6' opng, no inter hori	12.55	11.45	24
1150	One intermediate horizontal	17.10	13.45	30.55
1200	Two intermediate horizontals	21.50	15.45	36.95
1250	5' x 20' opening, three intermediate horizontals	11.50	9.50	21
1400	1-3/4" x 4-1/2", 5' x 6' opening, no intermediate horizontals	14.60	11.45	26.05
1450	One intermediate horizontal	19.35	13.45	32.80
1500	Two intermediate horizontals	24	15.45	39.45
1550	5' x 20' opening, three intermediate horizontals	13.05	9.50	22.55
1700	For insulating glass, 2"x4-1/2", 5'x6' opening, no intermediate horizontals	15.05	12.05	27.10
1750	One intermediate horizontal	19.50	14.15	33.65
1800	Two intermediate horizontals	24	16.25	40.25
1850	5' x 20' opening, three intermediate horizontals	13.10	10	23.10
2000	Thermal break frame, 2-1/4"x4-1/2", 5'x6'opng, no intermediate horizontals	15.70	12.20	27.90
2050	One intermediate horizontal	21	14.65	35.65
2100	Two intermediate horizontals	26	17.15	43.15
2150	5' x 20' opening, three intermediate horizontals	14.20	10.40	24.60

B2020 Exterior Windows

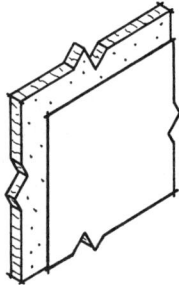

The table below lists costs of curtain wall and spandrel panels per S.F. Costs do not include structural framing used to hang the panels from.

Spandrel Glass Panel   Sandwich Panel

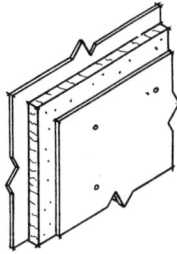

B2020 220	Curtain Wall Panels	COST PER S.F.		
		MAT.	INST.	TOTAL
1000	Glazing panel, insulating, 1/2" thick, 2 lites 1/8" float, clear	9.85	9.85	19.70
1100	Tinted	13.80	9.85	23.65
1200	5/8" thick units, 2 lites 3/16" float, clear	14.25	10.35	24.60
1400	1" thick units, 2 lites, 1/4" float, clear	16.70	12.45	29.15
1700	Light and heat reflective glass, tinted	32.50	11	43.50
2000	Plate glass, 1/4" thick, clear	5.80	7.80	13.60
2050	Tempered	8.20	7.80	16
2100	Tinted	8.15	7.80	15.95
2200	3/8" thick, clear	10.05	12.45	22.50
2250	Tempered	16.60	12.45	29.05
2300	Tinted	16.30	12.45	28.75
2400	1/2" thick, clear	19.95	16.95	36.90
2450	Tempered	24.50	16.95	41.45
2500	Tinted	28.50	16.95	45.45
2600	3/4" thick, clear	36.50	26.50	63
2650	Tempered	43	26.50	69.50
3000	Spandrel glass, panels, 1/4" plate glass insul w/fiberglass, 1" thick	17.40	7.80	25.20
3100	2" thick	20.50	7.80	28.30
3200	Galvanized steel backing, add	6.05		6.05
3300	3/8" plate glass, 1" thick	29	7.80	36.80
3400	2" thick	32	7.80	39.80
4000	Polycarbonate, masked, clear or colored, 1/8" thick	10.15	5.50	15.65
4100	3/16" thick	11.60	5.65	17.25
4200	1/4" thick	13.05	6	19.05
4300	3/8" thick	23	6.20	29.20
5000	Facing panel, textured al, 4' x 8' x 5/16" plywood backing, sgl face	4.10	3.10	7.20
5100	Double face	4.88	3.10	7.98
5200	4' x 10' x 5/16" plywood backing, single face	4.58	3.10	7.68
5300	Double face	6.45	3.10	9.55
5400	4' x 12' x 5/16" plywood backing, single face	6.50	3.10	9.60
5500	Sandwich panel, 22 Ga. galv., both sides 2" insulation, enamel exterior	12.30	5.40	17.70
5600	Polyvinylidene fluoride exterior finish	12.95	5.40	18.35
5700	26 Ga., galv. both sides, 1" insulation, colored 1 side	6.50	5.10	11.60
5800	Colored 2 sides	8.35	5.10	13.45

B2030 Exterior Doors

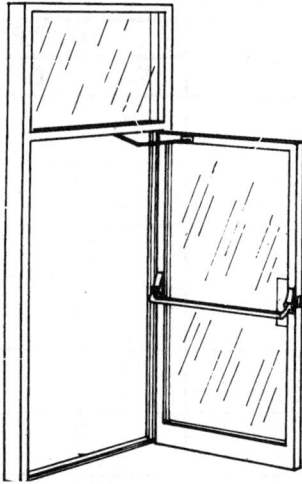

Costs are listed for exterior door systems by material, type and size. Prices between sizes listed can be interpolated with reasonable accuracy. Prices are per opening for a complete door system including frame.

B2030 110		Glazed Doors, Steel or Aluminum						
	MATERIAL	TYPE	DOORS	SPECIFICATION	OPENING	COST PER OPNG.		
						MAT.	INST.	TOTAL
5600	St. Stl. & glass	revolving	stock unit	manual oper.	6'-0" x 7'-0"	41,500	8,450	49,950
5650				auto Cntrls.	6'-10" x 7'-0"	57,000	9,000	66,000
5700	Bronze	revolving	stock unit	manual oper.	6'-10" x 7'-0"	48,400	17,000	65,400
5750				auto Cntrls.	6'-10" x 7'-0"	64,000	17,600	81,600
5800	St. Stl. & glass	balanced	standard	economy	3'-0" x 7'-0"	8,900	1,425	10,325
5850				premium	3'-0" x 7'-0"	15,500	1,875	17,375
6300	Alum. & glass	w/o transom	narrow stile	w/panic Hrdwre.	3'-0" x 7'-0"	1,500	925	2,425
6350				dbl. door, Hrdwre.	6'-0" x 7'-0"	2,600	1,525	4,125
6400			wide stile	hdwre.	3'-0" x 7'-0"	1,825	910	2,735
6450				dbl. door, Hdwre.	6'-0" x 7'-0"	3,550	1,825	5,375
6500			full vision	hdwre.	3'-0" x 7'-0"	2,225	1,450	3,675
6550				dbl. door, Hdwre.	6'-0" x 7'-0"	3,100	2,075	5,175
6600			non-standard	hdwre.	3'-0" x 7'-0"	2,000	910	2,910
6650				dbl. door, Hdwre.	6'-0" x 7'-0"	4,000	1,825	5,825
6700			bronze fin.	hdwre.	3'-0" x 7'-0"	1,475	910	2,385
6750				dbl. door, Hrdwre.	6'-0" x 7'-0"	2,925	1,825	4,750
6800			black fin.	hdwre.	3'-0" x 7'-0"	2,125	910	3,035
6850				dbl. door, Hdwre.	6'-0" x 7'-0"	4,250	1,825	6,075
6900		w/transom	narrow stile	hdwre.	3'-0" x 10'-0"	2,125	1,050	3,175
6950				dbl. door, Hdwre.	6'-0" x 10'-0"	3,025	1,825	4,850
7000			wide stile	hdwre.	3'-0" x 10'-0"	2,450	1,275	3,725
7050				dbl. door, Hdwre.	6'-0" x 10'-0"	3,275	2,175	5,450
7100			full vision	hdwre.	3'-0" x 10'-0"	2,700	1,400	4,100
7150				dbl. door, Hdwre.	6'-0" x 10'-0"	3,550	2,400	5,950

B20 Exterior Enclosure

B2030 Exterior Doors

| B2030 110 | Glazed Doors, Steel or Aluminum | | | | | | |

	MATERIAL	TYPE	DOORS	SPECIFICATION	OPENING	COST PER OPNG.		
						MAT.	INST.	TOTAL
7200			non-standard	hdwre.	3'-0" x 10'-0"	2,075	980	3,055
7250				dbl. door, Hdwre.	6'-0" x 10'-0"	4,150	1,950	6,100
7300			bronze fin.	hdwre.	3'-0" x 10'-0"	1,550	980	2,530
7350				dbl. door, Hdwre.	6'-0" x 10'-0"	3,075	1,950	5,025
7400			black fin.	hdwre.	3'-0" x 10'-0"	2,200	980	3,180
7450				dbl. door, Hdwre.	6'-0" x 10'-0"	4,400	1,950	6,350
7500		revolving	stock design	minimum	6'-10" x 7'-0"	20,600	3,400	24,000
7550				average	6'-0" x 7'-0"	24,900	4,250	29,150
7600				maximum	6'-10" x 7'-0"	43,000	5,675	48,675
7650				min., automatic	6'-10" x 7'-0"	36,300	3,950	40,250
7700				avg., automatic	6'-10" x 7'-0"	40,600	4,800	45,400
7750				max., automatic	6'-10" x 7'-0"	58,500	6,225	64,725
7800		balanced	standard	economy	3'-0" x 7'-0"	6,375	1,425	7,800
7850				premium	3'-0" x 7'-0"	8,000	1,825	9,825
7900		mall front	sliding panels	alum. fin.	16'-0" x 9'-0"	3,325	720	4,045
7950					24'-0" x 9'-0"	4,825	1,325	6,150
8000				bronze fin.	16'-0" x 9'-0"	3,875	840	4,715
8050					24'-0" x 9'-0"	5,625	1,550	7,175
8100			fixed panels	alum. fin.	48'-0" x 9'-0"	9,000	1,025	10,025
8150				bronze fin.	48'-0" x 9'-0"	10,500	1,200	11,700
8200		sliding entrance	5' x 7' door	electric oper.	12'-0" x 7'-6"	8,475	1,325	9,800
8250		sliding patio	temp. glass	economy	6'-0" x 7'-0"	1,475	248	1,723
8300			temp. glass	economy	12'-0" x 7'-0"	3,500	330	3,830
8350				premium	6'-0" x 7'-0"	2,225	370	2,595
8400					12'-0" x 7'-0"	5,250	495	5,745

B2030 Exterior Doors

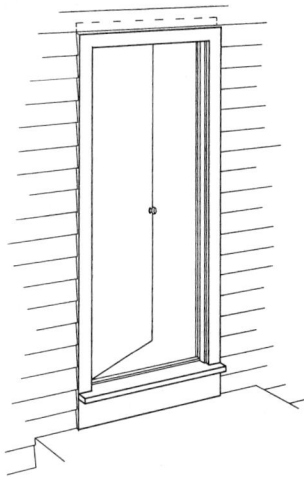

Costs are listed for exterior door systems by material, type and size. Prices between sizes listed can be interpolated with reasonable accuracy. Prices are per opening for a complete door system including frame as illustrated in the component block.

System Components	QUANTITY	UNIT	COST PER OPNG.		
			MAT.	INST.	TOTAL
SYSTEM B2030 210 2500					
WOOD DOOR, SOLID CORE, SINGLE, HINGED, 3'-0" X 7'-0"					
Commercial door, exterior flush solid core birch, 1-3/4" x 7' x 3' wide	1.000	Ea.	204	71	275
Wood door exterior frame, pine, w/trim, 5/4 x 5-3/16" deep	1.000	Set	211.65	44.88	256.53
Hinges, full mortise-high freq, brass base, 4-1/2" x 4-1/2", US10	1.500	Pr.	129.75		129.75
Panic device for mortise locks, single door exit only	1.000	Ea.	625	124	749
Paint exterior door & frame one side 3' x 7', primer & 2 coats	1.000	Ea.	1.98	25	26.98
Sill, 8/4 x 8" deep, oak, 2" horns	1.000	Ea.	22	11	33
TOTAL			1,194.38	275.88	1,470.26

B2030 210 — Wood Doors

	MATERIAL	TYPE	DOORS	SPECIFICATION	OPENING	COST PER OPNG.		
						MAT.	INST.	TOTAL
2350	Birch	solid core	single door	hinged	2'-6" x 6'-8"	1,175	269	1,444
2400					2'-6" x 7'-0"	1,175	271	1,446
2450					2'-8" x 7'-0"	1,175	271	1,446
2500					3'-0" x 7'-0"	1,200	276	1,476
2550			double door	hinged	2'-6" x 6'-8"	2,150	485	2,635
2600					2'-6" x 7'-0"	2,150	495	2,645
2650					2'-8" x 7'-0"	2,150	495	2,645
2700					3'-0" x 7'-0"	2,200	505	2,705
2750	Wood	combination	storm & screen	hinged	3'-0" x 6'-8"	325	66	391
2800					3'-0" x 7'-0"	360	72.50	432.50
2850		overhead	panels, H.D.	manual oper.	8'-0" x 8'-0"	830	495	1,325
2900					10'-0" x 10'-0"	1,250	550	1,800
2950					12'-0" x 12'-0"	1,800	660	2,460
3000					14'-0" x 14'-0"	2,800	760	3,560
3050					20'-0" x 16'-0"	6,550	1,525	8,075
3100				electric oper.	8'-0" x 8'-0"	1,850	745	2,595
3150					10'-0" x 10'-0"	2,275	800	3,075
3200					12'-0" x 12'-0"	2,825	910	3,735
3250					14'-0" x 14'-0"	3,825	1,000	4,825
3300					20'-0" x 16'-0"	7,775	2,025	9,800

B20 Exterior Enclosure

B2030 Exterior Doors

| B2030 220 | | | | Steel Doors | | | |

	MATERIAL	TYPE	DOORS	SPECIFICATION	OPENING	COST PER OPNG.		
						MAT.	INST.	TOTAL
3350	Steel 18 Ga.	hollow metal	1 door w/frame	no label	2'-6" x 7'-0"	1,300	280	1,580
3400					2'-8" x 7'-0"	1,300	280	1,580
3450					3'-0" x 7'-0"	1,300	280	1,580
3500					3'-6" x 7'-0"	1,400	294	1,694
3550		hollow metal	1 door w/frame	no label	4'-0" x 8'-0"	1,625	295	1,920
3600			2 doors w/frame	no label	5'-0" x 7'-0"	2,500	520	3,020
3650					5'-4" x 7'-0"	2,500	520	3,020
3700					6'-0" x 7'-0"	2,500	520	3,020
3750					7'-0" x 7'-0"	2,700	545	3,245
3800					8'-0" x 8'-0"	3,175	550	3,725
3850			1 door w/frame	"A" label	2'-6" x 7'-0"	1,575	335	1,910
3900					2'-8" x 7'-0"	1,550	340	1,890
3950					3'-0" x 7'-0"	1,550	340	1,890
4000					3'-6" x 7'-0"	1,675	350	2,025
4050					4'-0" x 8'-0"	1,800	365	2,165
4100			2 doors w/frame	"A" label	5'-0" x 7'-0"	3,000	625	3,625
4150					5'-4" x 7'-0"	3,050	630	3,680
4200					6'-0" x 7'-0"	3,050	630	3,680
4250					7'-0" x 7'-0"	3,275	645	3,920
4300					8'-0" x 8'-0"	3,300	650	3,950
4350	Steel 24 Ga.	overhead	sectional	manual oper.	8'-0" x 8'-0"	805	495	1,300
4400					10'-0" x 10'-0"	1,025	550	1,575
4450					12'-0" x 12'-0"	1,300	660	1,960
4500					20'-0" x 14'-0"	3,725	1,425	5,150
4550				electric oper.	8'-0" x 8'-0"	1,825	745	2,570
4600					10'-0" x 10'-0"	2,050	800	2,850
4650					12'-0" x 12'-0"	2,325	910	3,235
4700					20'-0" x 14'-0"	4,950	1,925	6,875
4750	Steel	overhead	rolling	manual oper.	8'-0" x 8'-0"	1,125	795	1,920
4800					10'-0" x 10'-0"	1,925	910	2,835
4850					12'-0" x 12'-0"	2,025	1,050	3,075
4900					14'-0" x 14'-0"	3,100	1,600	4,700
4950					20'-0" x 12'-0"	2,175	1,425	3,600
5000					20'-0" x 16'-0"	3,550	2,125	5,675
5050				electric oper.	8'-0" x 8'-0"	2,300	1,050	3,350
5100					10'-0" x 10'-0"	3,100	1,175	4,275
5150					12'-0" x 12'-0"	3,200	1,300	4,500
5200					14'-0" x 14'-0"	4,275	1,850	6,125
5250					20'-0" x 12'-0"	3,325	1,675	5,000
5300					20'-0" x 16'-0"	4,700	2,375	7,075
5350				fire rated	10'-0" x 10'-0"	2,075	1,150	3,225
5400			rolling grille	manual oper.	10'-0" x 10'-0"	2,700	1,275	3,975
5450					15'-0" x 8'-0"	3,125	1,600	4,725
5500		vertical lift	1 door w/frame	motor operator	16'-0" x 16'-0"	23,100	5,325	28,425
5550					32'-0" x 24'-0"	49,200	3,550	52,750

B20 Exterior Enclosure

B2030 Exterior Doors

B2030 230				Aluminum Doors				
	MATERIAL	TYPE	DOORS	SPECIFICATION	OPENING	COST PER OPNG.		
						MAT.	INST.	TOTAL
6000	Aluminum	combination	storm & screen	hinged	3'-0" x 6'-8"	335	71	406
6050					3'-0" x 7'-0"	370	78	448
6100		overhead	rolling grille	manual oper.	12'-0" x 12'-0"	4,325	2,250	6,575
6150				motor oper.	12'-0" x 12'-0"	5,700	2,500	8,200
6200	Alum. & Fbrgls.	overhead	heavy duty	manual oper.	12'-0" x 12'-0"	2,800	660	3,460
6250				electric oper.	12'-0" x 12'-0"	3,825	910	4,735

B3010 Roof Coverings

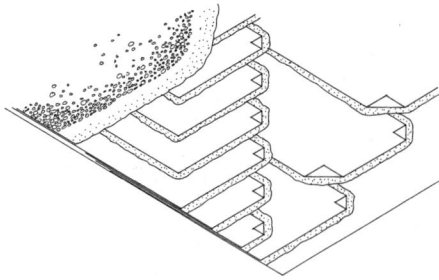

Multiple ply roofing is the most popular covering for minimum pitch roofs. Lines 1200 through 6300 list the costs of the various types, plies and weights per S.F.

System Components	QUANTITY	UNIT	COST PER S.F.		
			MAT.	INST.	TOTAL
SYSTEM B3010 105 2500					
ASPHALT FLOOD COAT, W/GRAVEL, 4 PLY ORGANIC FELT					
Organic #30 base felt	1.000	S.F.	.11	.08	.19
Organic #15 felt, 3 plies	3.000	S.F.	.16	.24	.40
Asphalt mopping of felts	4.000	S.F.	.40	.73	1.13
Asphalt flood coat	1.000	S.F.	.25	.59	.84
Gravel aggregate, washed river stone	4.000	Lb.	.07	.13	.20
TOTAL			.99	1.77	2.76

B3010 105	Built-Up	COST PER S.F.		
		MAT.	INST.	TOTAL
1200	Asphalt flood coat w/gravel; not incl. insul, flash., nailers			
1300				
1400	Asphalt base sheets & 3 plies #15 asphalt felt, mopped	.93	1.59	2.52
1500	On nailable deck	.98	1.67	2.65
1600	4 plies #15 asphalt felt, mopped	1.28	1.75	3.03
1700	On nailable deck	1.14	1.85	2.99
1800	Coated glass base sheet, 2 plies glass (type IV), mopped	.94	1.59	2.53
1900	For 3 plies	1.12	1.75	2.87
2000	On nailable deck	1.05	1.85	2.90
2300	4 plies glass fiber felt (type IV), mopped	1.37	1.75	3.12
2400	On nailable deck	1.23	1.85	3.08
2500	Organic base sheet & 3 plies #15 organic felt, mopped	.99	1.77	2.76
2600	On nailable deck	.96	1.85	2.81
2700	4 plies #15 organic felt, mopped	1.20	1.59	2.79
2750				
2800	Asphalt flood coat, smooth surface, not incl. insul, flash., nailers			
2850				
2900	Asphalt base sheet & 3 plies #15 asphalt felt, mopped	.99	1.46	2.45
3000	On nailable deck	.92	1.52	2.44
3100	Coated glass fiber base sheet & 2 plies glass fiber felt, mopped	.88	1.40	2.28
3200	On nailable deck	.83	1.46	2.29
3300	For 3 plies, mopped	1.06	1.52	2.58
3400	On nailable deck	.99	1.59	2.58
3700	4 plies glass fiber felt (type IV), mopped	1.24	1.52	2.76
3800	On nailable deck	1.17	1.59	2.76
3900	Organic base sheet & 3 plies #15 organic felt, mopped	.97	1.46	2.43
4000	On nailable decks	.90	1.52	2.42
4100	4 plies #15 organic felt, mopped	1.14	1.59	2.73
4200	Coal tar pitch with gravel surfacing			
4300	4 plies #15 tarred felt, mopped	1.72	1.67	3.39
4400	3 plies glass fiber felt (type IV), mopped	1.40	1.85	3.25
4500	Coated glass fiber base sheets 2 plies glass fiber felt, mopped	1.40	1.85	3.25

B30 Roofing

B3010 Roof Coverings

B3010 105	Built-Up	COST PER S.F.		
		MAT.	**INST.**	**TOTAL**
4600	On nailable decks	1.24	1.95	3.19
4800	3 plies glass fiber felt (type IV), mopped	1.95	1.67	3.62
4900	On nailable decks	1.79	1.75	3.54
5300	Asphalt mineral surface, roll roofing			
5400	1 ply #15 organic felt, 1 ply mineral surfaced			
5500	Selvage roofing, lap 19", nailed and mopped	.75	1.30	2.05
5600	3 plies glass fiber felt (type IV), 1 ply mineral surfaced			
5700	Selvage roofing, lapped 19", mopped	1.14	1.40	2.54
5800	Coated glass fiber base sheet			
5900	2 plies glass fiber, felt (type IV), 1 ply mineral surfaced			
6000	Selvage, roofing, lapped 19", mopped	1.23	1.40	2.63
6100	On nailable deck	1.13	1.46	2.59
6200	3 plies glass fiber felt (type IV), 1 ply mineral surfaced			
6300	Selvage roofing, lapped 19", mopped	1.14	1.40	2.54

B30 Roofing

B3010 Roof Coverings

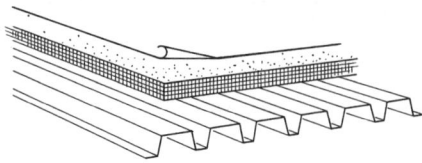

Fully Adhered

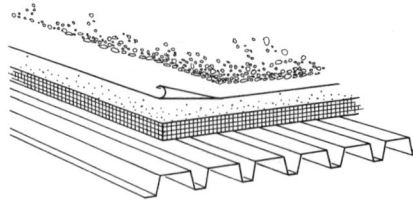

Ballasted

The systems listed below reflect only the cost for the single ply membrane.

For additional components see:

Insulation	B3010 320
Base Flashing	B3010 410
Roof Edge	B3010 420
Roof Openings	B3020 210

B3010 120	Single Ply Membrane	COST PER S.F.		
		MAT.	INST.	TOTAL
1000	CSPE (Chlorosulfonated polyethylene), 35 mils, fully adhered	2.04	.88	2.92
1100	Loosely laid with stone ballast	2.12	.45	2.57
2000	EPDM (Ethylene propylene diene monomer), 45 mils, fully adhered	1.02	.88	1.90
2100	Loosely laid with stone ballast	.81	.45	1.26
2200	Mechanically fastened with batten strips	.71	.65	1.36
3300	60 mils, fully adhered	1.22	.88	2.10
3400	Loosely laid with stone ballast	1.02	.45	1.47
3500	Mechanically fastened with batten strips	.91	.65	1.56
4000	Modified bit., SBS modified, granule surface cap sheet, mopped, 150 mils	.58	1.75	2.33
4100	Smooth surface cap sheet, mopped, 145 mils	.54	1.67	2.21
4500	APP modified, granule surface cap sheet, torched, 180 mils	.65	1.15	1.80
4600	Smooth surface cap sheet, torched, 170 mils	.61	1.09	1.70
5000	PIB (Polyisobutylene), 100 mils, fully adhered with contact cement	2.58	.88	3.46
5100	Loosely laid with stone ballast	1.99	.45	2.44
5200	Partially adhered with adhesive	2.50	.65	3.15
5300	Hot asphalt attachment	2.38	.65	3.03
6000	Reinforced PVC, 48 mils, loose laid and ballasted with stone	1.28	.45	1.73
6100	Partially adhered with mechanical fasteners	1.17	.65	1.82
6200	Fully adhered with adhesive	1.71	.88	2.59
6300	Reinforced PVC, 60 mils, loose laid and ballasted with stone	1.30	.45	1.75
6400	Partially adhered with mechanical fasteners	1.19	.65	1.84
6500	Fully adhered with adhesive	1.73	.88	2.61

B3010 Roof Coverings

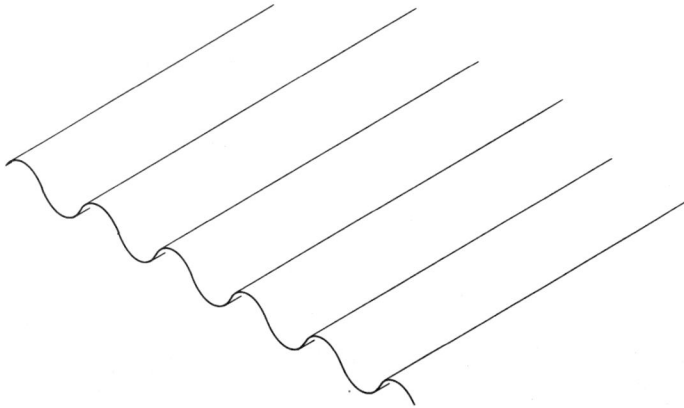

Listed below are installed prices for preformed roofing materials with applicable specifications for each material.

B3010 130	Preformed Metal Roofing	COST PER S.F.		
		MAT.	INST.	TOTAL
0200	Corrugated roofing, aluminum, mill finish, .0175" thick, .272 P.S.F.	.96	1.62	2.58
0250	.0215" thick, .334 P.S.F.	1.27	1.62	2.89
0300	.0240" thick, .412 P.S.F.	1.80	1.62	3.42
0350	.0320" thick, .552 P.S.F.	2.32	1.62	3.94
0400	Painted, .0175" thick, .280 P.S.F.	1.40	1.62	3.02
0450	.0215" thick, .344 P.S.F.	1.55	1.62	3.17
0500	.0240" thick, .426 P.S.F.	2.19	1.62	3.81
0550	.0320" thick, .569 P.S.F.	2.97	1.62	4.59
0700	Fiberglass, 6 oz., .375 P.S.F.	1.74	1.95	3.69
0750	8 oz., .5 P.S.F.	3.77	1.95	5.72
0900	Steel, galvanized, 29 ga., .72 P.S.F.	1.85	1.77	3.62
0930	26 ga., .91 P.S.F.	1.97	1.85	3.82
0950	24 ga., 1.26 P.S.F.	2.04	1.95	3.99
0970	22 ga., 1.45 P.S.F.	3.20	2.05	5.25
1000	Colored, 26 ga., 1.08 P.S.F.	1.85	1.85	3.70
1050	24 ga., 1.43 P.S.F.	2.19	1.95	4.14

B3010 Roof Coverings

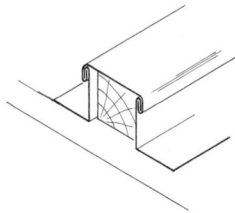

Batten Seam

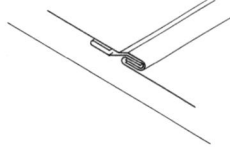

Flat Seam

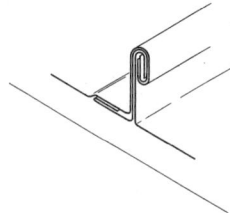

Standing Seam

Formed metal roofing is practical on all sloped roofs from 1/4″ per foot of rise to vertical. Its use is more aesthetic than economical. Table below lists the various materials used and the weight per S.F.

System Components	QUANTITY	UNIT	COST PER S.F.		
			MAT.	INST.	TOTAL
SYSTEM B3010 135 1000					
BATTEN SEAM, FORMED COPPER ROOFING 3″ MIN. SLOPE, 16 OZ., 1.2 PSF.					
Copper roof, batten seam, 16 oz. over 10 squares	1.000	S.F.	11.75	5.30	17.05
Asphalt impregnated felt, 30 lb, 2 sq. per roll, not mopped	1.000	S.F.	.11	.08	.19
TOTAL			11.86	5.38	17.24

B3010 135	Formed Metal	COST PER S.F.		
		MAT.	INST.	TOTAL
1000	Batten seam, formed copper roofing, 3″ min slope, 16 oz., 1.2 P.S.F.	11.85	5.40	17.25
1100	18 oz., 1.35 P.S.F.	13.10	5.90	19
1200	20 oz., 1.50 P.S.F.	14.60	5.90	20.50
1500	Lead, 3″ min slope, 3 Lb., 3.6 P.S.F.	6.35	4.93	11.28
2000	Zinc copper alloy, 3″ min slope, .020″ thick, .88 P.S.F.	11.85	4.93	16.78
2050	.027″ thick, 1.18 P.S.F.	16.85	5.15	22
2100	.032″ thick, 1.41 P.S.F.	19.85	5.40	25.25
2150	.040″ thick, 1.77 P.S.F.	24	5.65	29.65
3000	Flat seam, copper, 1/4″ min. slope, 16 oz., 1.2 P.S.F.	11.85	4.93	16.78
3100	18 oz., 1.35 P.S.F.	13.10	5.15	18.25
3200	20 oz., 1.50 P.S.F.	14.60	5.40	20
3500	Lead, 1/4″ min. slope, 3 Lb. 3.6 P.S.F.	6.35	4.53	10.88
4000	Zinc copper alloy, 1/4″ min. slope, .020″ thick, .84 P.S.F.	11.85	4.93	16.78
4050	.027″ thick, 1.13 P.S.F.	16.85	5.15	22
4100	.032″ thick, 1.33 P.S.F.	19.85	5.30	25.15
4150	.040″ thick, 1.67 P.S.F.	24	5.65	29.65
5000	Standing seam, copper, 2-1/2″ min. slope, 16 oz., 1.25 P.S.F.	11.85	4.53	16.38
5100	18 oz., 1.40 P.S.F.	13.10	4.93	18.03
5200	20 oz., 1.56 P.S.F.	14.60	5.40	20
5500	Lead, 2-1/2″ min. slope, 3 Lb., 3.6 P.S.F.	6.35	4.53	10.88
6000	Zinc copper alloy, 2-1/2″ min. slope, .020″ thick, .87 P.S.F.	11.85	4.93	16.78
6050	.027″ thick, 1.18 P.S.F.	16.85	5.15	22
6100	.032″ thick, 1.39 P.S.F.	19.85	5.40	25.25
6150	.040″ thick, 1.74 P.S.F.	24	5.65	29.65

B3010 Roof Coverings

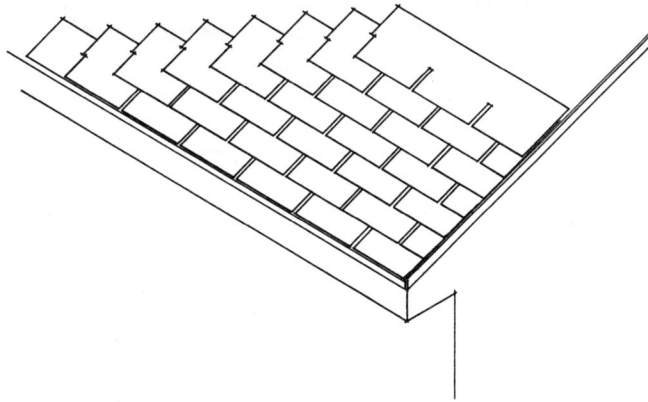

Shingles and tiles are practical in applications where the roof slope is more than 3-1/2" per foot of rise. Table below lists the various materials and the weight per S.F.

System Components	QUANTITY	UNIT	COST PER S.F.		
			MAT.	INST.	TOTAL
SYSTEM B3010 140 1200					
STRIP SHINGLES, 4" SLOPE, MULTI-LAYERED, CLASS A					
Class A, 260-300 lb./sq., 4 bundles/sq.	1.000	S.F.	.88	1.32	2.20
Building paper, asphalt felt sheathing paper 15 lb	1.000	S.F.	.05	.13	.18
TOTAL			.93	1.45	2.38

B3010 140	Shingle & Tile	COST PER S.F.		
		MAT.	INST.	TOTAL
1095	Asphalt roofing			
1100	Strip shingles, 4" slope, inorganic class A 210-235 lb/Sq.	.60	.97	1.57
1150	Organic, class C, 235-240 lb./sq.	.61	1.06	1.67
1200	Premium laminated multi-layered, class A, 260-300 lb./Sq.	.93	1.45	2.38
1250	Class C, 300-385 lb/sq	1.22	1.67	2.89
1545	Metal roofing			
1550	Alum., shingles, colors, 3" min slope, .019" thick, 0.4 PSF	2.64	1.07	3.71
1600	.020" thick, 0.6 PSF	2.57	1.07	3.64
1850	Steel, colors, 3" min slope, 26 gauge, 1.0 PSF	3.28	2.17	5.45
2000	24 gauge, 1.7 PSF	3.41	2.17	5.58
2295	Mineral fiber roofing			
2300	3" min slope, class A, shingles, 14" x 30" 3.25 psf	3.81	2.18	5.99
2400	Shakes, 9" x 16", 5.0 PSF	3.46	2.18	5.64
2500	16" x 16", 5.4 PSF	2.35	1.62	3.97
2795	Slate roofing			
2800	4" min. slope, shingles, 3/16" thick, 8.0 PSF	5.35	2.71	8.06
2900	1/4" thick, 9.0 PSF	5.35	2.71	8.06
3495	Wood roofing			
3500	4" min slope, cedar shingles, 16" x 5", 5" exposure 1.6 PSF	3.76	2.14	5.90
4000	Shakes, 18", 8-1/2" exposure, 2.8 PSF	2.74	2.64	5.38
4100	24", 10" exposure, 3.5 PSF	2.89	2.14	5.03
5095	Tile roofing			
5100	Aluminum, mission, 3" min slope, .019" thick, 0.65 PSF	9.05	2.06	11.11
5200	S type, 0.48 PSF	6.25	1.73	7.98
6000	Clay, Americana, 3" minimum slope, 8 PSF	7.40	2.87	10.27
6500	Mission, 4" minimum slope, 12.2 PSF	8.25	4.08	12.33

B3010 Roof Coverings

B3010 320	Roof Deck Rigid Insulation	COST PER S.F.		
		MAT.	INST.	TOTAL
0100	Fiberboard low density, 1/2" thick, R1.39			
0150	1" thick R2.78 RB3010 -010	.46	.58	1.04
0300	1 1/2" thick R4.17	.70	.58	1.28
0350	2" thick R5.56	.94	.58	1.52
0370	Fiberboard high density, 1/2" thick R1.3	.24	.46	.70
0380	1" thick R2.5	.48	.58	1.06
0390	1 1/2" thick R3.8	.73	.58	1.31
0410	Fiberglass, 3/4" thick R2.78	.62	.46	1.08
0450	15/16" thick R3.70	.81	.46	1.27
0500	1-1/16" thick R4.17	1.02	.46	1.48
0550	1-5/16" thick R5.26	1.39	.46	1.85
0600	2-1/16" thick R8.33	1.49	.58	2.07
0650	2 7/16" thick R10	1.71	.58	2.29
1000	Foamglass, 1 1/2" thick R4.55	1.49	.58	2.07
1100	3" thick R9.09	3.03	.66	3.69
1200	Tapered for drainage	1.28	.77	2.05
1260	Perlite, 1/2" thick R1.32	.37	.44	.81
1300	3/4" thick R2.08	.41	.58	.99
1350	1" thick R2.78	.51	.58	1.09
1400	1 1/2" thick R4.17	.53	.58	1.11
1450	2" thick R5.56	.88	.66	1.54
1510	Polyisocyanurate 2#/CF density, 1" thick R7.14	.55	.33	.88
1550	1 1/2" thick R10.87	.69	.37	1.06
1600	2" thick R14.29	.89	.42	1.31
1650	2 1/2" thick R16.67	1.11	.44	1.55
1700	3" thick R21.74	1.39	.46	1.85
1750	3 1/2" thick R25	2.12	.46	2.58
1800	Tapered for drainage	2.12	.33	2.45
1810	Expanded polystyrene, 1#/CF density, 3/4" thick R2.89	.34	.31	.65
1820	2" thick R7.69	.68	.37	1.05
1825	Extruded Polystyrene			
1830	Extruded polystyrene, 15 PSI compressive strength, 1" thick, R5	.54	.31	.85
1835	2" thick R10	.68	.37	1.05
1840	3" thick R15	1.40	.46	1.86
1900	25 PSI compressive strength, 1" thick R5	.74	.31	1.05
1950	2" thick R10	1.40	.37	1.77
2000	3" thick R15	2.12	.46	2.58
2050	4" thick R20	2.99	.46	3.45
2150	Tapered for drainage	.62	.31	.93
2550	40 PSI compressive strength, 1" thick R5	.56	.31	.87
2600	2" thick R10	1.06	.37	1.43
2650	3" thick R15	1.54	.46	2
2700	4" thick R20	2.07	.46	2.53
2750	Tapered for drainage	.78	.33	1.11
2810	60 PSI compressive strength, 1" thick R5	.65	.32	.97
2850	2" thick R10	1.16	.38	1.54
2900	Tapered for drainage	.94	.33	1.27
4000	Composites with 1-1/2" polyisocyanurate			
4010	1" fiberboard	1.61	.58	2.19
4020	1" perlite	1.68	.54	2.22
4030	7/16" oriented strand board	1.95	.58	2.53

B3010 Roof Coverings

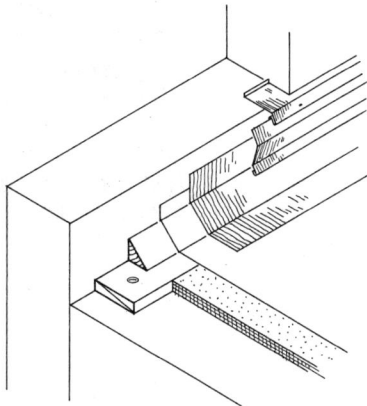

The table below lists the costs for various types of base flashing per L.F.

All the systems below are also based on using a 4″ x 4″ diagonally cut treated wood cant and a 2″ x 8″ treated wood nailer fastened at 4′-0″ O.C.

System Components	QUANTITY	UNIT	COST PER L.F.		
			MAT.	INST.	TOTAL
SYSTEM B3010 410 1000					
ALUMINUM BASE FLASHING MILL FINISH, .019″ THICK					
Treated 2″ x 8″ wood blocking	1.333	B.F.	1.15	1.53	2.68
Anchor bolts, 4′ O.C.	1.333	B.F.	.61	2.17	2.78
Treated wood cant, 4″ x 4″ split	1.000	L.F.	1.67	1.42	3.09
Roof covering, splice sheet	1.000	S.F.	2.12	.45	2.57
Base flashing, .019″ thick, aluminum	1.000	L.F.	1.12	3.18	4.30
Reglet .025″ thick counter flashing .032″ thick	1.000	L.F.	1.40	2.20	3.60
Counter flashing, aluminum	1.000	L.F.	1.65	3.87	5.52
TOTAL			9.72	14.82	24.54

B3010 410 — Base Flashing

	TYPE	DESCRIPTION	SPECIFICATION	REGLET	COUNTER FLASHING	COST PER L.F.		
						MAT.	INST.	TOTAL
1000	Aluminum	mill finish	.019″ thick	alum. .025	alum. .032	9.70	14.85	24.55
1100			.050″ thick	.025	.032	11.35	14.85	26.20
1200		fabric, 2 sides	.004″ thick	.025	.032	10.15	13.05	23.20
1300			.016″ thick	.025	.032	10.45	13.05	23.50
1400		mastic, 2 sides	.004″ thick	.025	.032	10.15	13.05	23.20
1500			.016″ thick	.025	.032	10.60	13.05	23.65
1700	Copper	sheets, plain	16 oz.	copper 10 oz.	copper 10 oz.	20.50	15.65	36.15
1800			20 oz.	10 oz.	10 oz.	22	15.85	37.85
1900			24 oz.	10 oz.	10 oz.	27	16.05	43.05
2000			32 oz.	10 oz.	10 oz.	30.50	16.25	46.75
2200	Sheets, Metal	galvanized	20 Ga.	galv. 24 Ga.	galv. 24 Ga.	11.05	33.50	44.55
2300			24 Ga.	24 Ga.	24 Ga.	9.95	27	36.95
2400			30 Ga.	24 Ga.	24 Ga.	8.85	20.50	29.35
2600	Rubber	butyl	1/32″ thick	galv. 24 Ga.	galv. 24 Ga.	8.50	13.25	21.75
2700			1/16″ thick	24 Ga.	24 Ga.	9	13.25	22.25
2800		neoprene	1/16″ thick	24 Ga.	24 Ga.	9.65	13.25	22.90
2900			1/8″ thick	24 Ga.	24 Ga.	11.95	13.25	25.20

B30 Roofing

B3010 Roof Coverings

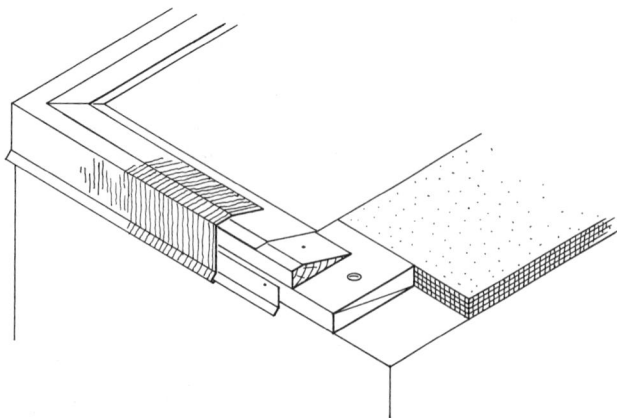

The table below lists the costs for various types of roof perimeter edge treatments.

All systems below include the cost per L.F. for a 2″ x 8″ treated wood nailer fastened at 4′-0″ O.C. and a diagonally cut 4″ x 6″ treated wood cant.

Roof edge and base flashing are assumed to be made from the same material.

System Components	QUANTITY	UNIT	COST PER L.F.		
			MAT.	INST.	TOTAL
SYSTEM B3010 420 1000					
ALUMINUM ROOF EDGE, MILL FINISH, .050″ THICK, 4″ HIGH					
Treated 2″ x 8″ wood blocking	1.333	B.F.	1.15	1.53	2.68
Anchor bolts, 4′ O.C.	1.333	B.F.	.61	2.17	2.78
Treated wood cant	1.000	L.F.	1.67	1.42	3.09
Roof covering, splice sheet	1.000	S.F.	2.12	.45	2.57
Aluminum roof edge, .050″ thick, 4″ high	1.000	L.F.	6.55	4.01	10.56
TOTAL			12.10	9.58	21.68

B3010 420				Roof Edges				
	EDGE TYPE	DESCRIPTION	SPECIFICATION	FACE HEIGHT		COST PER L.F.		
						MAT.	INST.	TOTAL
1000	Aluminum	mill finish	.050″ thick	4″		12.10	9.60	21.70
1100				6″		12.55	9.90	22.45
1200				8″		14.35	10.25	24.60
1300		duranodic	.050″ thick	4″		11.85	9.60	21.45
1400				6″		13	9.90	22.90
1500				8″		14.10	10.25	24.35
1600		painted	.050″ thick	4″		12.85	9.60	22.45
1700				6″		14.20	9.90	24.10
1800				8″		14.25	10.25	24.50
2000	Copper	plain	16 oz.	4″		8.10	9.60	17.70
2100				6″		9.45	9.90	19.35
2200				8″		10.75	11.35	22.10
2300			20 oz.	4″		9	10.95	19.95
2400				6″		10.05	10.55	20.60
2500				8″		11.55	12.20	23.75
2700	Sheet Metal	galvanized	20 Ga.	4″		9.25	11.35	20.60
2800				6″		11.10	11.35	22.45
2900				8″		12.95	13.25	26.20
3000			24 Ga.	4″		8.10	9.60	17.70
3100				6″		9.45	9.60	19.05
3200				8″		10.70	10.95	21.65

B30 Roofing

B3010 Roof Coverings

B3010 430			Flashing				
	MATERIAL	BACKING	SIDES	SPECIFICATION	QUANTITY	MAT.	INST.	TOTAL

	MATERIAL	BACKING	SIDES	SPECIFICATION	QUANTITY	COST PER S.F. MAT.	INST.	TOTAL
0040	Aluminum	none		.019″		1.12	3.18	4.30
0050				.032″		1.63	3.18	4.81
0100				.040″		2.43	3.18	5.61
0150				.050″		2.77	3.18	5.95
0300		fabric	2	.004″		1.55	1.40	2.95
0350				.016″		1.84	1.40	3.24
0400		mastic		.004″		1.55	1.40	2.95
0450				.005″		1.84	1.40	3.24
0500				.016″		2.02	1.40	3.42
0510								
0700	Copper	none		16 oz.	<500 lbs.	7.45	4.01	11.46
0710					>2000 lbs.	8.80	2.98	11.78
0750				20 oz.	<500 lbs.	8.70	4.20	12.90
0760					>2000 lbs.	8.80	3.18	11.98
0800				24 oz.	<500 lbs.	10.60	4.40	15
0810					>2000 lbs.	10.70	3.42	14.12
0850				32 oz.	<500 lbs.	14.10	4.62	18.72
0860					>2000 lbs.	22	3.55	25.55
1000		paper backed	1	2 oz.		1.56	1.40	2.96
1100				3 oz.		2.04	1.40	3.44
1200			2	2 oz.		1.57	1.40	2.97
1300				5 oz.		3.04	1.40	4.44
2000	Copper lead	fabric	1	2 oz.		2.66	1.40	4.06
2100	Coated			5 oz.		3.06	1.40	4.46
2200		mastic	2	2 oz.		2.08	1.40	3.48
2300				5 oz.		2.57	1.40	3.97
2400		paper	1	2 oz.		1.56	1.40	2.96
2500				3 oz.		2.04	1.40	3.44
2600			2	2 oz.		1.57	1.40	2.97
2700				5 oz.		3.04	1.40	4.44
3000	Lead	none		2.5 lb.	to 12″ wide	4.55	3.42	7.97
3100					over 12″	4.55	3.42	7.97
3500	PVC black	none		.010″		.29	1.62	1.91
3600				.020″		.40	1.62	2.02
3700				.030″		.52	1.62	2.14
3800				.056″		1.22	1.62	2.84
4000	Rubber butyl	none		1/32″		1.06	1.62	2.68
4100				1/16″		1.57	1.62	3.19
4200	Neoprene			1/16″		2.23	1.62	3.85
4300				1/8″		4.50	1.62	6.12
4500	Stainless steel	none		.015″	<500 lbs.	5.75	4.01	9.76
4600	Copper clad				>2000 lbs.	5.55	2.98	8.53
4700				.018″	<500 lbs.	7.60	4.62	12.22
4800					>2000 lbs.	5.55	3.18	8.73
5000	Plain			32 ga.		4.17	2.98	7.15
5100				28 ga.		5.15	2.98	8.13
5200				26 ga.		6.25	2.98	9.23
5300				24 ga.		8.15	2.98	11.13
5400	Terne coated			28 ga.		8.10	2.98	11.08
5500				26 ga.		9.10	2.98	12.08

B3010 Roof Coverings

B3010 610		Gutters				COST PER L.F.		
	SECTION	MATERIAL	THICKNESS	SIZE	FINISH	MAT.	INST.	TOTAL
0050	Box	aluminum	.027"	5"	enameled	2.70	4.84	7.54
0100					mill	2.55	4.84	7.39
0200			.032"	5"	enameled	3.21	4.84	8.05
0300					mill	3.60	4.84	8.44
0500		copper	16 Oz.	4"	lead coated	12.50	4.84	17.34
0600					mill	6.75	4.84	11.59
0700				5"	lead coated	15.65	4.84	20.49
0800					mill	8.05	4.84	12.89
1000		steel galv.	28 Ga.	5"	enameled	1.66	4.84	6.50
1100					mill	1.61	4.84	6.45
1200			26 Ga.	5"	mill	1.75	4.84	6.59
1300				6"	mill	2.53	4.84	7.37
1500		stainless steel		4"	mill	6.45	4.84	11.29
1600		stainless		5"	mill	6.95	4.84	11.79
1800		vinyl		4"	colors	1.25	4.51	5.76
1900				5"	colors	1.49	4.51	6
2200		wood clear,		3"x4"	treated	7.70	4.96	12.66
2300		hemlock or fir		4"x5"	treated	8.80	4.96	13.76
3000	Half round	copper	16 Oz.	4"	lead coated	11.55	4.84	16.39
3100					mill	6.90	4.84	11.74
3200				5"	lead coated	13.30	5.55	18.85
3300					mill	6.95	4.84	11.79
3600		steel galv.	28 Ga.	5"	enameled	1.66	4.84	6.50
3700					mill	1.61	4.84	6.45
3800			26 Ga.	5"	mill	1.75	4.84	6.59
4000		stainless steel		4"	mill	6.45	4.84	11.29
4100				5"	mill	6.95	4.84	11.79
5000		vinyl		4"	white	1	4.51	5.51

B3010 620		Downspouts				COST PER V.L.F.		
	MATERIALS	SECTION	SIZE	FINISH	THICKNESS	MAT.	INST.	TOTAL
0100	Aluminum	rectangular	2"x3"	embossed mill	.020"	1.14	3.06	4.20
0150				enameled	.020"	1.75	3.06	4.81
0200				enameled	.024"	2.02	3.23	5.25
0250			3"x4"	enameled	.024"	2.81	4.15	6.96
0300		round corrugated	3"	enameled	.020"	1.56	3.06	4.62
0350			4"	enameled	.025"	2.63	4.15	6.78
0500	Copper	rectangular corr.	2"x3"	mill	16 Oz.	7.90	3.06	10.96
0550				lead coated	16 Oz.	14.05	3.06	17.11
0600		smooth		mill	16 Oz.	7.70	3.06	10.76
0650				lead coated	16 Oz.	10.75	3.06	13.81
0700		rectangular corr.	3"x4"	mill	16 Oz.	10.20	4.01	14.21
0750				lead coated	16 Oz.	9.55	3.47	13.02
0800		smooth		mill	16 Oz.	10	4.01	14.01
0850				lead coated	16 Oz.	11	4.01	15.01
0900		round corrugated	2"	mill	16 Oz.	10.20	3.15	13.35
0950		smooth		lead coated	16 Oz.	12.60	3.06	15.66
1000		corrugated	3"	mill	16 Oz.	9.80	3.17	12.97
1050		smooth		lead coated	16 Oz.	13.85	3.06	16.91

B30 Roofing

B3010 Roof Coverings

B3010 620 — Downspouts

	MATERIALS	SECTION	SIZE	FINISH	THICKNESS	COST PER V.L.F.		
						MAT.	INST.	TOTAL
1100		corrugated	4″	mill	16 Oz.	10.30	4.17	14.47
1150		smooth		lead coated	16 Oz.	16.20	4.01	20.21
1200		corrugated	5″	mill	16 Oz.	17	4.67	21.67
1250				lead coated	16 Oz.	25.50	4.47	29.97
1300	Steel	rectangular corr.	2″x3″	galvanized	28 Ga.	1.76	3.06	4.82
1350				epoxy coated	24 Ga.	2.09	3.06	5.15
1400		smooth		galvanized	28 Ga.	1.91	3.06	4.97
1450		rectangular corr.	3″x4″	galvanized	28 Ga.	1.76	4.01	5.77
1500				epoxy coated	24 Ga.	2.33	4.01	6.34
1550		smooth		galvanized	28 Ga.	2.76	4.01	6.77
1600		round corrugated	2″	galvanized	28 Ga.	1.91	3.06	4.97
1650			3″	galvanized	28 Ga.	1.91	3.06	4.97
1700			4″	galvanized	28 Ga.	2.73	4.01	6.74
1750			5″	galvanized	28 Ga.	4.05	4.47	8.52
1800				galvanized	26 Ga.	3.27	4.47	7.74
1850			6″	galvanized	28 Ga.	5.30	5.55	10.85
1900				galvanized	26 Ga.	6.75	5.55	12.30
2500	Stainless steel	rectangular	2″x3″	mill		54	3.06	57.06
2550	Tubing sch.5		3″x4″	mill		69	4.01	73.01
2600			4″x5″	mill		142	4.31	146.31
2650		round	3″	mill		54	3.06	57.06
2700			4″	mill		69	4.01	73.01
2750			5″	mill		142	4.31	146.31

B3010 630 — Gravel Stop

	MATERIALS	SECTION	SIZE	FINISH	THICKNESS	COST PER L.F.		
						MAT.	INST.	TOTAL
5100	Aluminum	extruded	4″	mill	.050″	6.55	4.01	10.56
5200			4″	duranodic	.050″	6.30	4.01	10.31
5300			8″	mill	.050″	8.80	4.65	13.45
5400			8″	duranodic	.050″	8.55	4.65	13.20
5500			12″-2 pc.	mill	.050″	11.70	5.80	17.50
6000			12″-2 pc.	duranodic	.050″	10.75	5.80	16.55
6100	Stainless	formed	6″	mill	24 Ga.	12.25	4.31	16.56
6200			12″	mill	24 Ga.	25.50	5.80	31.30

B3020 Roof Openings

Roof Hatch

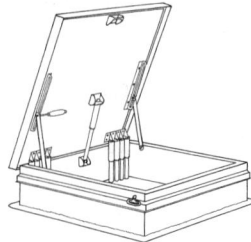

Smoke Hatch

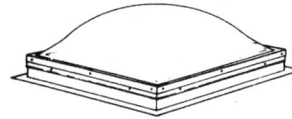

Skylight

B3020 110	Skylights	COST PER S.F.		
		MAT.	INST.	TOTAL
5100	Skylights, plastic domes, insul curbs, nom. size to 10 S.F., single glaze	25.50	12.15	37.65
5200	Double glazing	29	14.95	43.95
5300	10 S.F. to 20 S.F., single glazing	25	4.93	29.93
5400	Double glazing	24.50	6.20	30.70
5500	20 S.F. to 30 S.F., single glazing	21	4.19	25.19
5600	Double glazing	23	4.93	27.93
5700	30 S.F. to 65 S.F., single glazing	18.15	3.19	21.34
5800	Double glazing	23	4.19	27.19
6000	Sandwich panels fiberglass, 1-9/16″ thick, 2 S.F. to 10 S.F.	17.60	9.75	27.35
6100	10 S.F. to 18 S.F.	15.75	7.35	23.10
6200	2-3/4″ thick, 25 S.F. to 40 S.F.	25.50	6.60	32.10
6300	40 S.F. to 70 S.F.	21	5.90	26.90

B3020 210	Hatches	COST PER OPNG.		
		MAT.	INST.	TOTAL
0200	Roof hatches with curb and 1″ fiberglass insulation, 2′-6″x3′-0″, aluminum	1,075	195	1,270
0300	Galvanized steel 165 lbs.	655	195	850
0400	Primed steel 164 lbs.	550	195	745
0500	2′-6″x4′-6″ aluminum curb and cover, 150 lbs.	935	216	1,151
0600	Galvanized steel 220 lbs.	660	216	876
0650	Primed steel 218 lbs.	910	216	1,126
0800	2′x6″x8′-0″ aluminum curb and cover, 260 lbs.	1,825	295	2,120
0900	Galvanized steel, 360 lbs.	1,750	295	2,045
0950	Primed steel 358 lbs.	1,325	295	1,620
1200	For plexiglass panels, add to the above	440		440
2100	Smoke hatches, unlabeled not incl. hand winch operator, 2′-6″x3′, galv	795	236	1,031
2200	Plain steel, 160 lbs.	665	236	901
2400	2′-6″x8′-0″,galvanized steel, 360 lbs.	1,925	325	2,250
2500	Plain steel, 350 lbs.	1,450	325	1,775
3000	4′-0″x8′-0″, double leaf low profile, aluminum cover, 359 lb.	2,850	243	3,093
3100	Galvanized steel 475 lbs.	2,425	243	2,668
3200	High profile, aluminum cover, galvanized curb, 361 lbs.	2,750	243	2,993

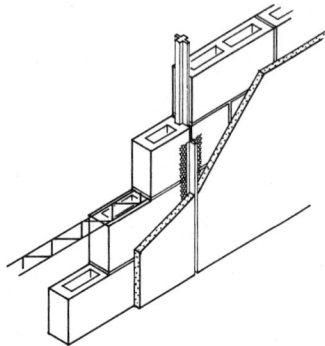

The Concrete Block Partition Systems are defined by weight and type of block, thickness, type of finish and number of sides finished. System components include joint reinforcing on alternate courses and vertical control joints.

System Components	QUANTITY	UNIT	COST PER S.F.		
			MAT.	INST.	TOTAL
SYSTEM C1010 102 1020					
CONC. BLOCK PARTITION, 8" X 16", 4" TK., 2 CT. GYP. PLASTER 2 SIDES					
Conc. block partition, 4" thick	1.000	S.F.	1.58	5.25	6.83
Control joint	.050	L.F.	.10	.13	.23
Horizontal joint reinforcing	.800	L.F.	.07	.06	.13
Gypsum plaster, 2 coat, on masonry	2.000	S.F.	.76	4.97	5.73
TOTAL			2.51	10.41	12.92

C1010 102 — Concrete Block Partitions - Regular Weight

	TYPE	THICKNESS (IN.)	TYPE FINISH	SIDES FINISHED		COST PER S.F.		
						MAT.	INST.	TOTAL
1000	Hollow	4	none	0		1.75	5.45	7.20
1010			gyp. plaster 2 coat	1		2.13	7.90	10.03
1020				2		2.51	10.40	12.91
1100			lime plaster - 2 coat	1		1.98	7.90	9.88
1150			lime portland - 2 coat	1		2	7.90	9.90
1200			portland - 3 coat	1		2.02	8.30	10.32
1400			5/8" drywall	1		2.38	7.15	9.53
1410				2		3.01	8.85	11.86
1500		6	none	0		2.42	5.85	8.27
1510			gyp. plaster 2 coat	1		2.80	8.30	11.10
1520				2		3.18	10.80	13.98
1600			lime plaster - 2 coat	1		2.65	8.30	10.95
1650			lime portland - 2 coat	1		2.67	8.30	10.97
1700			portland - 3 coat	1		2.69	8.70	11.39
1900			5/8" drywall	1		3.05	7.55	10.60
1910				2		3.68	9.25	12.93
2000		8	none	0		2.55	6.20	8.75
2010			gyp. plaster 2 coat	1		2.93	8.70	11.63
2020			gyp. plaster 2 coat	2		3.31	11.20	14.51
2100			lime plaster - 2 coat	1		2.78	8.70	11.48
2150			lime portland - 2 coat	1		2.80	8.70	11.50
2200			portland - 3 coat	1		2.82	9.10	11.92
2400			5/8" drywall	1		3.18	7.90	11.08
2410				2		3.81	9.60	13.41

C10 Interior Construction

C1010 Partitions

C1010 102 — Concrete Block Partitions - Regular Weight

	TYPE	THICKNESS (IN.)	TYPE FINISH	SIDES FINISHED		COST PER S.F.		
						MAT.	INST.	TOTAL
2500	Hollow	10	none	0		3.43	6.50	9.93
2510			gyp. plaster 2 coat	1		3.81	9	12.81
2520				2		4.19	11.45	15.64
2600			lime plaster - 2 coat	1		3.66	9	12.66
2650			lime portland - 2 coat	1		3.68	9	12.68
2700			portland - 3 coat	1		3.70	9.35	13.05
2900			5/8" drywall	1		4.06	8.20	12.26
2910				2		4.69	9.90	14.59
3000	Solid	2	none	0		1.61	5.40	7.01
3010			gyp. plaster	1		1.99	7.85	9.84
3020				2		2.37	10.35	12.72
3100			lime plaster - 2 coat	1		1.84	7.85	9.69
3150			lime portland - 2 coat	1		1.86	7.85	9.71
3200			portland - 3 coat	1		1.88	8.25	10.13
3400			5/8" drywall	1		2.24	7.10	9.34
3410				2		2.87	8.80	11.67
3500		4	none	0		2.15	5.65	7.80
3510			gyp. plaster	1		2.62	8.15	10.77
3520				2		2.91	10.60	13.51
3600			lime plaster - 2 coat	1		2.38	8.10	10.48
3650			lime portland - 2 coat	1		2.40	8.10	10.50
3700			portland - 3 coat	1		2.42	8.50	10.92
3900			5/8" drywall	1		2.78	7.35	10.13
3910				2		3.41	9.05	12.46
4000		6	none	0		2.40	6.05	8.45
4010			gyp. plaster	1		2.78	8.50	11.28
4020				2		3.16	11	14.16
4100			lime plaster - 2 coat	1		2.63	8.50	11.13
4150			lime portland - 2 coat	1		2.65	8.50	11.15
4200			portland - 3 coat	1		2.67	8.90	11.57
4400			5/8" drywall	1		3.03	7.75	10.78
4410				2		3.66	9.45	13.11

C1010 104 — Concrete Block Partitions - Lightweight

	TYPE	THICKNESS (IN.)	TYPE FINISH	SIDES FINISHED		COST PER S.F.		
						MAT.	INST.	TOTAL
5000	Hollow	4	none	0		2.06	5.30	7.36
5010			gyp. plaster	1		2.44	7.75	10.19
5020				2		2.82	10.25	13.07
5100			lime plaster - 2 coat	1		2.29	7.75	10.04
5150			lime portland - 2 coat	1		2.31	7.75	10.06
5200			portland - 3 coat	1		2.33	8.15	10.48
5400			5/8" drywall	1		2.69	7	9.69
5410				2		3.32	8.70	12.02
5500		6	none	0		2.74	5.70	8.44
5510			gyp. plaster	1		3.12	8.15	11.27
5520			gyp. plaster	2		3.50	10.65	14.15
5600			lime plaster - 2 coat	1		2.97	8.15	11.12
5650			lime portland - 2 coat	1		2.99	8.15	11.14
5700			portland - 3 coat	1		3.01	8.55	11.56
5900			5/8" drywall	1		3.37	7.40	10.77
5910				2		4	9.10	13.10

C10 Interior Construction

C1010 Partitions

C1010 104	Concrete Block Partitions - Lightweight

	TYPE	THICKNESS (IN.)	TYPE FINISH	SIDES FINISHED		COST PER S.F.		
						MAT.	INST.	TOTAL
6000	Hollow	8	none	0		3.33	6.05	9.38
6010			gyp. plaster	1		3.71	8.55	12.26
6020				2		4.09	11.05	15.14
6100			lime plaster - 2 coat	1		3.56	8.55	12.11
6150			lime portland - 2 coat	1		3.58	8.55	12.13
6200			portland - 3 coat	1		3.60	8.95	12.55
6400			5/8" drywall	1		3.96	7.75	11.71
6410				2		4.59	9.45	14.04
6500		10	none	0		4.29	6.35	10.64
6510			gyp. plaster	1		4.67	8.85	13.52
6520				2		5.05	11.30	16.35
6700			portland - 3 coat	1		4.56	9.20	13.76
6900			5/8" drywall	1		4.92	8.05	12.97
6910				2		5.55	9.75	15.30
7000	Solid	4	none	0		2.70	5.55	8.25
7010			gyp. plaster	1		3.08	8	11.08
7020				2		3.46	10.50	13.96
7100			lime plaster - 2 coat	1		2.93	8	10.93
7150			lime portland - 2 coat	1		2.95	8	10.95
7200			portland - 3 coat	1		2.97	8.40	11.37
7400			5/8" drywall	1		3.33	7.25	10.58
7410				2		3.96	8.95	12.91
7500		6	none	0		3.10	6	9.10
7510			gyp. plaster	1		3.57	8.55	12.12
7520				2		3.86	10.95	14.81
7600			lime plaster - 2 coat	1		3.33	8.45	11.78
7650			lime portland - 2 coat	1		3.35	8.45	11.80
7700			portland - 3 coat	1		3.37	8.85	12.22
7900			5/8" drywall	1		3.73	7.70	11.43
7910				2		4.36	9.40	13.76
8000		8	none	0		4.11	6.40	10.51
8010			gyp. plaster	1		4.49	8.90	13.39
8020				2		4.87	11.40	16.27
8100			lime plaster - 2 coat	1		4.34	8.90	13.24
8150			lime portland - 2 coat	1		4.36	8.90	13.26
8200			portland - 3 coat	1		4.38	9.30	13.68
8400			5/8" drywall	1		4.74	8.10	12.84
8410				2		5.35	9.80	15.15

C1010 Partitions

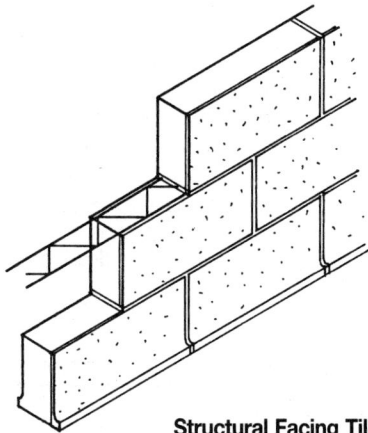

Structural Facing Tile
8W Series
8" x 16"

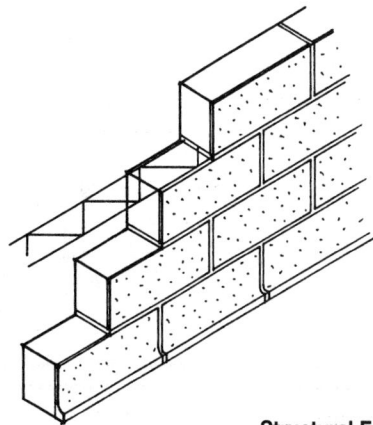

Structural Facing Tile
6T Series
5-1/3" x 12"

C1010 120	Tile Partitions	COST PER S.F.		
		MAT.	INST.	TOTAL
1000	8W series 8"x16", 4" thick wall, reinf every 2 courses, glazed 1 side	13.25	6.55	19.80
1100	Glazed 2 sides	14.80	6.95	21.75
1200	Glazed 2 sides, using 2 wythes of 2" thick tile	20.50	12.50	33
1300	6" thick wall, horizontal reinf every 2 courses, glazed 1 side	23.50	6.85	30.35
1400	Glazed 2 sides, each face different color, 2" and 4" tile	23.50	12.80	36.30
1500	8" thick wall, glazed 2 sides using 2 wythes of 4" thick tile	26.50	13.10	39.60
1600	10" thick wall, glazed 2 sides using 1 wythe of 4" tile & 1 wythe of 6" tile	37	13.40	50.40
1700	Glazed 2 sides cavity wall, using 2 wythes of 4" thick tile	26.50	13.10	39.60
1800	12" thick wall, glazed 2 sides using 2 wythes of 6" thick tile	47	13.70	60.70
1900	Glazed 2 sides cavity wall, using 2 wythes of 4" thick tile	26.50	13.10	39.60
2100	6T series 5-1/3"x12" tile, 4" thick, non load bearing glazed one side,	12.15	10.25	22.40
2200	Glazed two sides	15.60	11.55	27.15
2300	Glazed two sides, using two wythes of 2" thick tile	18.60	20	38.60
2400	6" thick, glazed one side	18.25	10.75	29
2500	Glazed two sides	21.50	12.20	33.70
2600	Glazed two sides using 2" thick tile and 4" thick tile	21.50	20.50	42
2700	8" thick, glazed one side	24	12.55	36.55
2800	Glazed two sides using two wythes of 4" thick tile	24.50	20.50	45
2900	Glazed two sides using 6" thick tile and 2" thick tile	27.50	21	48.50
3000	10" thick cavity wall, glazed two sides using two wythes of 4" tile	24.50	20.50	45
3100	12" thick, glazed two sides using 4" thick tile and 8" thick tile	36	23	59
3200	2" thick facing tile, glazed one side, on 6" concrete block	12.50	11.90	24.40
3300	On 8" concrete block	12.60	12.25	24.85
3400	On 10" concrete block	13.50	12.50	26

C1010 Partitions

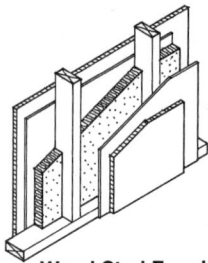
Wood Stud Framing

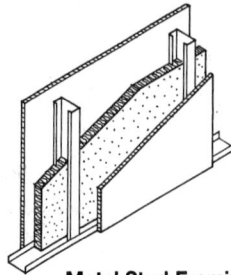
Metal Stud Framing

The Drywall Partitions/Stud Framing Systems are defined by type of drywall and number of layers, type and spacing of stud framing, and treatment on the opposite face. Components include taping and finishing.

Cost differences between regular and fire resistant drywall are negligible, and terminology is interchangeable. In some cases fiberglass insulation is included for additional sound deadening.

System Components	QUANTITY	UNIT	COST PER S.F.		
			MAT.	INST.	TOTAL
SYSTEM C1010 124 1250					
DRYWALL PARTITION,5/8″ F.R.1 SIDE,5/8″ REG.1 SIDE,2″X4″STUDS,16″ O.C.					
Gypsum plasterboard, nailed/screwed to studs, 5/8″F.R. fire resistant	1.000	S.F.	.40	.50	.90
Gypsum plasterboard, nailed/screwed to studs, 5/8″ regular	1.000	S.F.	.37	.50	.87
Taping and finishing joints	2.000	S.F.	.10	1	1.10
Framing, 2 x 4 studs @ 16″ O.C., 10′ high	1.000	S.F.	.34	.99	1.33
TOTAL			1.21	2.99	4.20

C1010 124		Drywall Partitions/Wood Stud Framing						
	FACE LAYER	BASE LAYER	FRAMING	OPPOSITE FACE	INSULATION	COST PER S.F.		
						MAT.	INST.	TOTAL
1200	5/8″ FR drywall	none	2 x 4, @ 16″ O.C.	same	0	1.24	2.99	4.23
1250				5/8″ reg. drywall	0	1.21	2.99	4.20
1300				nothing	0	.79	1.99	2.78
1400		1/4″ SD gypsum	2 x 4 @ 16″ O.C.	same	1-1/2″ fiberglass	2.76	4.59	7.35
1450				5/8″ FR drywall	1-1/2″ fiberglass	2.35	4.04	6.39
1500				nothing	1-1/2″ fiberglass	1.90	3.04	4.94
1600		resil. channels	2 x 4 @ 16″, O.C.	same	1-1/2″ fiberglass	2.29	5.80	8.09
1650				5/8″ FR drywall	1-1/2″ fiberglass	2.12	4.65	6.77
1700				nothing	1-1/2″ fiberglass	1.67	3.65	5.32
1800		5/8″ FR drywall	2 x 4 @ 24″ O.C.	same	0	1.96	3.80	5.76
1850				5/8″ FR drywall	0	1.56	3.30	4.86
1900				nothing	0	1.11	2.30	3.41
1950		5/8″ FR drywall	2 x 4, 16″ O.C.	same	0	2.04	3.99	6.03
1955				5/8″ FR drywall	0	1.64	3.49	5.13
2000				nothing	0	1.19	2.49	3.68
2010		5/8″ FR drywall	staggered, 6″ plate	same	0	2.38	5	7.38
2015				5/8″ FR drywall	0	1.98	4.50	6.48
2020				nothing	0	1.53	3.50	5.03
2200		5/8″ FR drywall	2 rows-2 x 4	same	2″ fiberglass	3.13	5.50	8.63
2250			16″O.C.	5/8″ FR drywall	2″ fiberglass	2.73	4.98	7.71
2300				nothing	2″ fiberglass	2.28	3.98	6.26
2400	5/8″ WR drywall	none	2 x 4, @ 16″ O.C.	same	0	1.48	2.99	4.47
2450				5/8″ FR drywall	0	1.36	2.99	4.35
2500				nothing	0	.91	1.99	2.90
2600		5/8″ FR drywall	2 x 4, @ 24″ O.C.	same	0	2.20	3.80	6
2650				5/8″ FR drywall	0	1.68	3.30	4.98
2700				nothing	0	1.23	2.30	3.53

C1010 Partitions

C1010 124	Drywall Partitions/Wood Stud Framing

	FACE LAYER	BASE LAYER	FRAMING	OPPOSITE FACE	INSULATION	COST PER S.F.		
						MAT.	INST.	TOTAL
2800	5/8 VF drywall	none	2 x 4, @ 16" O.C.	same	0	2.58	3.19	5.77
2850				5/8" FR drywall	0	1.91	3.09	5
2900				nothing	0	1.46	2.09	3.55
3000	5/8 VF drywall	5/8" FR drywall	2 x 4 , 24" O.C.	same	0	3.30	4	7.30
3050				5/8" FR drywall	0	2.23	3.40	5.63
3100				nothing	0	1.78	2.40	4.18
3200	1/2" reg drywall	3/8" reg drywall	2 x 4, @ 16" O.C.	same	0	1.80	3.99	5.79
3250				5/8" FR drywall	0	1.52	3.49	5.01
3300				nothing	0	1.07	2.49	3.56

C1010 126	Drywall Partitions/Metal Stud Framing

	FACE LAYER	BASE LAYER	FRAMING	OPPOSITE FACE	INSULATION	COST PER S.F.		
						MAT.	INST.	TOTAL
5200	5/8" FR drywall	none	1-5/8" @ 24" O.C.	same	0	1.15	2.65	3.80
5250				5/8" reg. drywall	0	1.12	2.65	3.77
5300				nothing	0	.70	1.65	2.35
5400			3-5/8" @ 24" O.C.	same	0	1.26	2.67	3.93
5450				5/8" reg. drywall	0	1.23	2.67	3.90
5500				nothing	0	.81	1.67	2.48
5600		1/4" SD gypsum	1-5/8" @ 24" O.C.	same	0	1.97	3.75	5.72
5650				5/8" FR drywall	0	1.56	3.20	4.76
5700				nothing	0	1.11	2.20	3.31
5800			2-1/2" @ 24" O.C.	same	0	2.04	3.76	5.80
5850				5/8" FR drywall	0	1.63	3.21	4.84
5900				nothing	0	1.18	2.21	3.39
6000		5/8" FR drywall	2-1/2" @ 16" O.C.	same	0	2.35	4.27	6.62
6050				5/8" FR drywall	0	1.95	3.77	5.72
6100				nothing	0	1.50	2.77	4.27
6200			3-5/8" @ 24" O.C.	same	0	2.06	3.67	5.73
6250				5/8"FR drywall	3-1/2" fiberglass	2.29	3.54	5.83
6300				nothing	0	1.21	2.17	3.38
6400	5/8" WR drywall	none	1-5/8" @ 24" O.C.	same	0	1.39	2.65	4.04
6450				5/8" FR drywall	0	1.27	2.65	3.92
6500				nothing	0	.82	1.65	2.47
6600			3-5/8" @ 24" O.C.	same	0	1.50	2.67	4.17
6650				5/8" FR drywall	0	1.38	2.67	4.05
6700				nothing	0	.93	1.67	2.60
6800		5/8" FR drywall	2-1/2" @ 16" O.C.	same	0	2.59	4.27	6.86
6850				5/8" FR drywall	0	2.07	3.77	5.84
6900				nothing	0	1.62	2.77	4.39
7000			3-5/8" @ 24" O.C.	same	0	2.30	3.67	5.97
7050				5/8"FR drywall	3-1/2" fiberglass	2.41	3.54	5.95
7100				nothing	0	1.33	2.17	3.50
7200	5/8" VF drywall	none	1-5/8" @ 24" O.C.	same	0	2.49	2.85	5.34
7250				5/8" FR drywall	0	1.82	2.75	4.57
7300				nothing	0	1.37	1.75	3.12
7400			3-5/8" @ 24" O.C.	same	0	2.60	2.87	5.47
7450				5/8" FR drywall	0	1.93	2.77	4.70
7500				nothing	0	1.48	1.77	3.25

C10 Interior Construction

C1010 Partitions

C1010 126	Drywall Partitions/Metal Stud Framing

	FACE LAYER	BASE LAYER	FRAMING	OPPOSITE FACE	INSULATION	COST PER S.F.		
						MAT.	INST.	TOTAL
7600	5/8" VF drywall	5/8" FR drywall	2-1/2" @ 16" O.C.	same	0	3.69	4.47	8.16
7650				5/8" FR drywall	0	2.62	3.87	6.49
7700				nothing	0	2.17	2.87	5.04
7800			3-5/8" @ 24" O.C.	same	0	3.40	3.87	7.27
7850				5/8"FR drywall	3-1/2" fiberglass	2.96	3.64	6.60
7900				nothing	0	1.88	2.27	4.15

C1010 Partitions

C1010 128	Drywall Components	COST PER S.F.		
		MAT.	INST.	TOTAL
0140	Metal studs, 24" O.C. including track, load bearing, 18 gage, 2-1/2"	.87	.93	1.80
0160	3-5/8"	1.03	.95	1.98
0180	4"	.93	.96	1.89
0200	6"	1.38	.98	2.36
0220	16 gage, 2-1/2"	1	1.06	2.06
0240	3-5/8"	1.20	1.08	2.28
0260	4"	1.27	1.10	2.37
0280	6"	1.60	1.13	2.73
0300	Non load bearing, 25 gage, 1-5/8"	.25	.65	.90
0340	3-5/8"	.36	.67	1.03
0360	4"	.39	.67	1.06
0380	6"	.55	.68	1.23
0400	20 gage, 2-1/2"	.48	.83	1.31
0420	3-5/8"	.52	.84	1.36
0440	4"	.61	.84	1.45
0460	6"	.72	.85	1.57
0540	Wood studs including blocking, shoe and double top plate, 2"x4", 12"O.C.	.42	1.24	1.66
0560	16" O.C.	.34	.99	1.33
0580	24" O.C.	.26	.80	1.06
0600	2"x6", 12" O.C.	.75	1.42	2.17
0620	16" O.C.	.60	1.10	1.70
0640	24" O.C.	.46	.86	1.32
0642	Furring one side only, steel channels, 3/4", 12" O.C.	.38	1.79	2.17
0644	16" O.C.	.34	1.59	1.93
0646	24" O.C.	.23	1.20	1.43
0647	1-1/2" , 12" O.C.	.51	2	2.51
0648	16" O.C.	.46	1.75	2.21
0649	24" O.C.	.30	1.38	1.68
0650	Wood strips, 1" x 3", on wood, 12" O.C.	.38	.90	1.28
0651	16" O.C.	.29	.68	.97
0652	On masonry, 12" O.C.	.38	1	1.38
0653	16" O.C.	.29	.75	1.04
0654	On concrete, 12" O.C.	.38	1.91	2.29
0655	16" O.C.	.29	1.43	1.72
0665	Gypsum board, one face only, exterior sheathing, 1/2"	.67	.88	1.55
0680	Interior, fire resistant, 1/2"	.35	.50	.85
0700	5/8"	.40	.50	.90
0720	Sound deadening board 1/4"	.41	.55	.96
0740	Standard drywall 3/8"	.34	.50	.84
0760	1/2"	.34	.50	.84
0780	5/8"	.37	.50	.87
0800	Tongue & groove coreboard 1"	.97	2.06	3.03
0820	Water resistant, 1/2"	.53	.50	1.03
0840	5/8"	.52	.50	1.02
0860	Add for the following:, foil backing	.15		.15
0880	Fiberglass insulation, 3-1/2"	.63	.37	1
0900	6"	.85	.37	1.22
0920	Rigid insulation 1"	.54	.50	1.04
0940	Resilient furring @ 16" O.C.	.24	1.55	1.79
0960	Taping and finishing	.05	.50	.55
0980	Texture spray	.09	.53	.62
1000	Thin coat plaster	.09	.62	.71
1040	2"x4" staggered studs 2"x6" plates & blocking	.48	1.08	1.56

C1010 Partitions

Plaster Partitions are defined as follows: type of plaster, type and spacing of framing, type of lath and treatment of opposite face.

Included in the system components are expansion joints. Metal studs are assumed to be non-loadbearing.

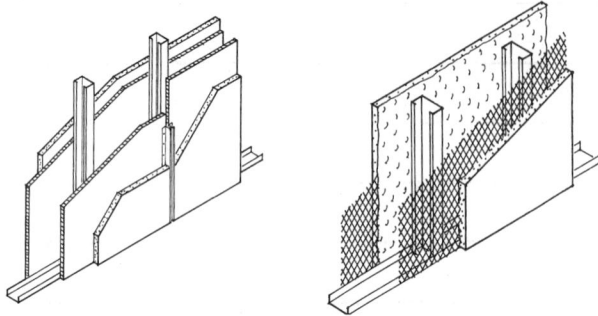

System Components			COST PER S.F.		
	QUANTITY	UNIT	MAT.	INST.	TOTAL
SYSTEM C1010 140 1000					
GYP PLASTER PART'N, 2 COATS,2-1/2″ MTL. STUD,16″O.C.,3/8″GYP. LATH					
Gypsum plaster, 2 coats, on walls	2.000	S.F.	1.04	4.72	5.76
Gypsum lath, 3/8″ thick, on metal studs	2.000	S.F.	1.34	1.24	2.58
Metal studs, 25 ga., 2-1/2″ @ 16″ O.C.	1.000	S.F.	.43	1.01	1.44
Expansion joint	.100	L.F.	.12	.16	.28
TOTAL			2.93	7.13	10.06

C1010 140	Plaster Partitions/Metal Stud Framing							
	TYPE	FRAMING	LATH	OPPOSITE FACE		COST PER S.F.		
						MAT.	INST.	TOTAL
1000	2 coat gypsum	2-1/2″ @ 16″O.C.	3/8″ gypsum	same		2.93	7.15	10.08
1010				nothing		1.74	4.15	5.89
1100		3-1/4″ @ 24″O.C.	1/2″ gypsum	same		2.95	6.90	9.85
1110				nothing		1.72	3.86	5.58
1500	2 coat vermiculite	2-1/2″ @ 16″O.C.	3/8″ gypsum	same		2.88	7.75	10.63
1510				nothing		1.72	4.46	6.18
1600		3-1/4″ @ 24″O.C.	1/2″ gypsum	same		2.90	7.50	10.40
1610				nothing		1.70	4.17	5.87
2000	3 coat gypsum	2-1/2″ @ 16″O.C.	3/8″ gypsum	same		2.81	8.10	10.91
2010				nothing		1.68	4.63	6.31
2020			3.4lb. diamond	same		2.59	8.10	10.69
2030				nothing		1.57	4.63	6.20
2040			2.75lb. ribbed	same		2.35	8.10	10.45
2050				nothing		1.45	4.63	6.08
2100		3-1/4″ @ 24″O.C.	1/2″ gypsum	same		2.83	7.85	10.68
2110				nothing		1.66	4.34	6
2120			3.4lb. ribbed	same		2.77	7.85	10.62
2130				nothing		1.62	4.34	5.96
2500	3 coat lime	2-1/2″ @ 16″O.C.	3.4lb. diamond	same		2.25	8.10	10.35
2510				nothing		1.40	4.63	6.03
2520			2.75lb. ribbed	same		2.01	8.10	10.11
2530				nothing		1.28	4.63	5.91
2600		3-1/4″ @ 24″O.C.	3.4lb. ribbed	same		2.43	7.85	10.28
2610				nothing		1.45	4.34	5.79

C10 Interior Construction

C1010 Partitions

C1010 140 — Plaster Partitions/Metal Stud Framing

	TYPE	FRAMING	LATH	OPPOSITE FACE		COST PER S.F. MAT.	INST.	TOTAL
3000	3 coat Portland	2-1/2" @ 16"O.C.	3.4lb. diamond	same		2.21	8.10	10.31
3010				nothing		1.38	4.63	6.01
3020			2.75lb. ribbed	same		1.97	8.10	10.07
3030				nothing		1.26	4.63	5.89
3100		3-1/4" @ 24"O.C.	3.4lb. ribbed	same		2.39	7.85	10.24
3110				nothing		1.43	4.34	5.77
3500	3 coat gypsum	2-1/2" @ 16"O.C.	3/8" gypsum	same		3.42	10.40	13.82
3510	W/med. Keenes			nothing		1.99	5.80	7.79
3520			3.4lb. diamond	same		3.20	10.40	13.60
3530				nothing		1.88	5.80	7.68
3540			2.75lb. ribbed	same		2.96	10.40	13.36
3550				nothing		1.76	5.80	7.56
3600		3-1/4" @ 24"O.C.	1/2" gypsum	same		3.44	10.15	13.59
3610				nothing		1.97	5.50	7.47
3620			3.4lb. ribbed	same		3.38	10.15	13.53
3630				nothing		1.93	5.50	7.43
4000	3 coat gypsum	2-1/2" @ 16"O.C.	3/8" gypsum	same		3.43	11.55	14.98
4010	W/hard Keenes			nothing		1.99	6.35	8.34
4020			3.4lb. diamond	same		3.21	11.55	14.76
4030				nothing		1.88	6.35	8.23
4040			2.75lb. ribbed	same		2.97	11.55	14.52
4050				nothing		1.76	6.35	8.11
4100		3-1/4" @ 24"O.C.	1/2" gypsum	same		3.45	11.30	14.75
4110				nothing		1.97	6.05	8.02
4120			3.4lb. ribbed	same		3.39	11.30	14.69
4130				nothing		1.93	6.05	7.98
4500	3 coat lime	2-1/2" @ 16"O.C.	3.4lb. diamond	same		2.19	8.10	10.29
4510	Portland			nothing		1.37	4.63	6
4520			2.75lb. ribbed	same		1.95	8.10	10.05
4530				nothing		1.25	4.63	5.88
4600		3-1/4" @ 24"O.C.	3.4lb. ribbed	same		2.37	7.85	10.22
4610				nothing		1.42	4.34	5.76

C1010 142 — Plaster Partitions/Wood Stud Framing

	TYPE	FRAMING	LATH	OPPOSITE FACE		COST PER S.F. MAT.	INST.	TOTAL
5000	2 coat gypsum	2"x4" @ 16"O.C.	3/8" gypsum	same		2.84	6.95	9.79
5010				nothing		1.65	4.06	5.71
5100		2"x4" @ 24"O.C.	1/2" gypsum	same		2.85	6.85	9.70
5110				nothing		1.62	3.90	5.52
5500	2 coat vermiculite	2"x4" @ 16"O.C.	3/8" gypsum	same		2.79	7.60	10.39
5510				nothing		1.63	4.37	6
5600		2"x4" @ 24"O.C.	1/2" gypsum	same		2.80	7.50	10.30
5610				nothing		1.60	4.21	5.81

C1010 Partitions

C1010 142	Plaster Partitions/Wood Stud Framing

	TYPE	FRAMING	LATH	OPPOSITE FACE		COST PER S.F.		
						MAT.	INST.	TOTAL
6000	3 coat gypsum	2"x4" @ 16"O.C.	3/8" gypsum	same		2.72	7.95	10.67
6010				nothing		1.59	4.54	6.13
6020			3.4lb. diamond	same		2.50	8	10.50
6030				nothing		1.48	4.57	6.05
6040			2.75lb. ribbed	same		2.24	8.05	10.29
6050				nothing		1.35	4.59	5.94
6100		2"x4" @ 24"O.C.	1/2" gypsum	same		2.73	7.80	10.53
6110				nothing		1.56	4.38	5.94
6120			3.4lb. ribbed	same		2.18	7.90	10.08
6130				nothing		1.28	4.42	5.70
6500	3 coat lime	2"x4" @ 16"O.C.	3.4lb. diamond	same		2.16	8	10.16
6510				nothing		1.31	4.57	5.88
6520			2.75lb. ribbed	same		1.90	8.05	9.95
6530				nothing		1.18	4.59	5.77
6600		2"x4" @ 24"O.C.	3.4lb. ribbed	same		1.84	7.90	9.74
6610				nothing		1.11	4.42	5.53
7000	3 coat Portland	2"x4" @ 16"O.C.	3.4lb. diamond	same		2.12	8	10.12
7010				nothing		1.29	4.57	5.86
7020			2.75lb. ribbed	same		1.86	8.05	9.91
7030				nothing		1.16	4.59	5.75
7100		2"x4" @ 24"O.C.	3.4lb. ribbed	same		2.29	7.95	10.24
7110				nothing		1.33	4.47	5.80
7500	3 coat gypsum W/med Keenes	2"x4" @ 16"O.C.	3/8" gypsum	same		3.33	10.25	13.58
7510				nothing		1.90	5.70	7.60
7520			3.4lb. diamond	same		3.11	10.30	13.41
7530				nothing		1.79	5.70	7.49
7540			2.75lb. ribbed	same		2.85	10.35	13.20
7550				nothing		1.66	5.75	7.41
7600		2"x4" @ 24"O.C.	1/2" gypsum	same		3.34	10.10	13.44
7610				nothing		1.87	5.55	7.42
7620			3.4lb. ribbed	same		3.28	10.25	13.53
7630				nothing		1.83	5.60	7.43
8000	3 coat gypsum W/hard Keenes	2"x4" @ 16"O.C.	3/8" gypsum	same		3.34	11.40	14.74
8010				nothing		1.90	6.25	8.15
8020			3.4lb. diamond	same		3.12	11.45	14.57
8030				nothing		1.79	6.30	8.09
8040			2.75lb. ribbed	same		2.86	11.50	14.36
8050				nothing		1.66	6.30	7.96
8100		2"x4" @ 24"O.C.	1/2" gypsum	same		3.35	11.30	14.65
8110				nothing		1.87	6.10	7.97
8120			3.4lb. ribbed	same		3.29	11.45	14.74
8130				nothing		1.83	6.20	8.03
8500	3 coat lime Portland	2"x4" @ 16"O.C.	3.4lb. diamond	same		2.10	8	10.10
8510				nothing		1.28	4.57	5.85
8520			2.75lb. ribbed	same		1.84	8.05	9.89
8530				nothing		1.15	4.59	5.74
8600		2"x4" @ 24"O.C.	3.4lb. ribbed	same		2.27	7.95	10.22
8610				nothing		1.32	4.47	5.79

C10 Interior Construction

C1010 Partitions

C1010 144	Plaster Partition Components	COST PER S.F.		
		MAT.	INST.	TOTAL
0060	Metal studs, 16" O.C., including track, non load bearing, 25 gage, 1-5/8"	.43	1.01	1.44
0080	2-1/2"	.43	1.01	1.44
0100	3-1/4"	.50	1.03	1.53
0120	3-5/8"	.50	1.03	1.53
0140	4"	.53	1.04	1.57
0160	6"	.75	1.05	1.80
0180	Load bearing, 20 gage, 2-1/2"	1.18	1.29	2.47
0200	3-5/8"	1.40	1.31	2.71
0220	4"	1.47	1.34	2.81
0240	6"	1.87	1.36	3.23
0260	16 gage 2-1/2"	1.38	1.46	2.84
0280	3-5/8"	1.66	1.50	3.16
0300	4"	1.75	1.53	3.28
0320	6"	2.20	1.55	3.75
0340	Wood studs, including blocking, shoe and double plate, 2"x4", 12" O.C.	.42	1.24	1.66
0360	16" O.C.	.34	.99	1.33
0380	24" O.C.	.26	.80	1.06
0400	2"x6", 12" O.C.	.75	1.42	2.17
0420	16" O.C.	.60	1.10	1.70
0440	24" O.C.	.46	.86	1.32
0460	Furring one face only, steel channels, 3/4", 12" O.C.	.38	1.79	2.17
0480	16" O.C.	.34	1.59	1.93
0500	24" O.C.	.23	1.20	1.43
0520	1-1/2", 12" O.C.	.51	2	2.51
0540	16" O.C.	.46	1.75	2.21
0560	24"O.C.	.30	1.38	1.68
0580	Wood strips 1"x3", on wood., 12" O.C.	.38	.90	1.28
0600	16"O.C.	.29	.68	.97
0620	On masonry, 12" O.C.	.38	1	1.38
0640	16" O.C.	.29	.75	1.04
0660	On concrete, 12" O.C.	.38	1.91	2.29
0680	16" O.C.	.29	1.43	1.72
0700	Gypsum lath. plain or perforated, nailed to studs, 3/8" thick	.67	.55	1.22
0720	1/2" thick	.72	.58	1.30
0740	Clipped to studs, 3/8" thick	.67	.62	1.29
0760	1/2" thick	.72	.67	1.39
0780	Metal lath, diamond painted, nailed to wood studs, 2.5 lb.	.41	.55	.96
0800	3.4 lb.	.56	.58	1.14
0820	Screwed to steel studs, 2.5 lb.	.41	.58	.99
0840	3.4 lb.	.49	.62	1.11
0860	Rib painted, wired to steel, 2.75 lb	.44	.62	1.06
0880	3.4 lb	.68	.67	1.35
0900	4.0 lb	.70	.72	1.42
0910				
0920	Gypsum plaster, 2 coats	.46	2.36	2.82
0940	3 coats	.66	2.84	3.50
0960	Perlite or vermiculite plaster, 2 coats	.50	2.67	3.17
0980	3 coats	.77	3.32	4.09
1000	Stucco, 3 coats, 1" thick, on wood framing	.82	4.69	5.51
1020	On masonry	.29	3.67	3.96
1100	Metal base galvanized and painted 2-1/2" high	.76	1.75	2.51

C1010 Partitions

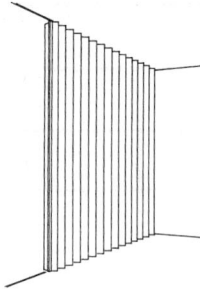

Folding Accordion

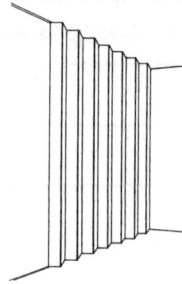

Folding Leaf

Movable and Borrow Lites

C1010 205	Partitions	COST PER S.F.		
		MAT.	INST.	TOTAL
0360	Folding accordion, vinyl covered, acoustical, 3 lb. S.F., 17 ft max. hgt	28	9.90	37.90
0380	5 lb. per S.F. 27 ft max height	39	10.45	49.45
0400	5.5 lb. per S.F., 17 ft. max height	46	11	57
0420	Commercial, 1.75 lb per S.F., 8 ft. max height	24.50	4.41	28.91
0440	2.0 Lb per S.F., 17 ft. max height	25	6.60	31.60
0460	Industrial, 4.0 lb. per S.F. 27 ft max height	36	13.20	49.20
0480	Vinyl clad wood or steel, electric operation 6 psf	53	6.20	59.20
0500	Wood, non acoustic, birch or mahogany	28.50	3.30	31.80
0560	Folding leaf, aluminum framed acoustical 12 ft.high.,5.5 lb per S.F.,min.	39	16.50	55.50
0580	Maximum	47	33	80
0600	6.5 lb. per S.F., minimum	40.50	16.50	57
0620	Maximum	50.50	33	83.50
0640	Steel acoustical, 7.5 per S.F., vinyl faced, minimum	57.50	16.50	74
0660	Maximum	70	33	103
0665	Steel with vinyl face, economy	57.50	16.50	74
0670	Deluxe	70	33	103
0680	Wood acoustic type, vinyl faced to 18' high 6 psf, minimum	54	16.50	70.50
0700	Average	64.50	22	86.50
0720	Maximum	83.50	33	116.50
0740	Formica or hardwood faced, minimum	55.50	16.50	72
0760	Maximum	59.50	33	92.50
0780	Wood, low acoustical type to 12 ft. high 4.5 psf	40.50	19.80	60.30
0840	Demountable, trackless wall, cork finish, semi acous, 1-5/8"th, min	42.50	3.05	45.55
0860	Maximum	40	5.20	45.20
0880	Acoustic, 2" thick, minimum	34	3.25	37.25
0900	Maximum	59	4.41	63.41
0920	In-plant modular office system, w/prehung steel door			
0940	3" thick honeycomb core panels			
0960	12' x 12', 2 wall	10.35	.62	10.97
0970	4 wall	9.65	.83	10.48
0980	16' x 16', 2 wall	10.10	.44	10.54
0990	4 wall	6.80	.44	7.24
1000	Gypsum, demountable, 3" to 3-3/4" thick x 9' high, vinyl clad	6.95	2.28	9.23
1020	Fabric clad	17.20	2.50	19.70
1040	1.75 system, vinyl clad hardboard, paper honeycomb core panel			
1060	1-3/4" to 2-1/2" thick x 9' high	11.65	2.28	13.93
1080	Unitized gypsum panel system, 2" to 2-1/2" thick x 9' high			
1100	Vinyl clad gypsum	15	2.28	17.28
1120	Fabric clad gypsum	24.50	2.50	27
1140	Movable steel walls, modular system			
1160	Unitized panels, 48" wide x 9' high			
1180	Baked enamel, pre-finished	16.85	1.83	18.68
1200	Fabric clad	24.50	1.96	26.46

C10 Interior Construction

C1010 Partitions

C1010 210	Partitions	COST PER S.F.		
		MAT.	INST.	TOTAL
0100	Partitions, folding accordion, 1.25 lb. vinyl (residential), to 8' high	21.50	3.30	24.80
0110	4 lb. vinyl (industrial), to 27' high	36	13.20	49.20
0120	Acoustical, 3 lb. vinyl, to 17' high	28	9.90	37.90
0130	4.5 lb. vinyl, fire rated 40, 20' high	46	6.20	52.20
0485	Operable acoustic, no track, 1-5/8" thick, economy	33	2.64	35.64
0490	With track, 3" thick, deluxe	100	3.30	103.30

C1010 230	Demountable Partitions, L.F.	COST PER L.F.		
		MAT.	INST.	TOTAL
0100	Office-demountable gypsum, 3" to 3-3/4" thick x 9' high, vinyl clad	62.50	20.50	83
0110	Fabric clad	155	22.50	177.50
0120	1.75 system, vinyl clad hardboard, paper honeycomb			
0130	core panel, 1-3/4" to 2-1/2" thick x 9' high	105	20.50	125.50
0140	Steel, modular, unitized panels, 48" wide x 9' high			
0150	Baked enamel, pre-finished	152	16.50	168.50
0160	Portable, fiber core, 4' high, fabric face	84.50	5.65	90.15

C1010 Partitions

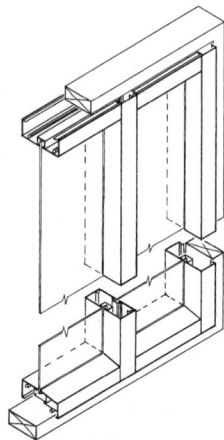

Alum. Tube Frame

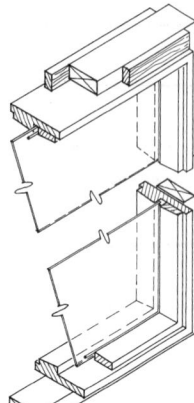

Oakwood Frame

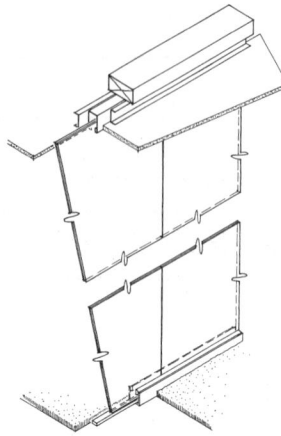

Concealed Frame Butt Glazed

Interior Glazed Openings are defined as follows: framing material, glass, size and intermediate framing members. Components for each system include gasket setting or glazing bead and typical wood blocking.

System Components	QUANTITY	UNIT	COST PER OPNG.		
			MAT.	INST.	TOTAL
SYSTEM C1010 710 1000					
GLAZED OPENING, ALUMINUM TUBE FRAME, FLUSH, 1/4″ FLOAT GLASS, 6′ X 4′					
Aluminum tube frame, flush, anodized bronze, head & jamb	16.500	L.F.	207.90	192.23	400.13
Aluminum tube frame, flush, anodized bronze, open sill	7.000	L.F.	69.30	79.80	149.10
Joints for tube frame, clip type	4.000	Ea.	100		100
Gasket setting, add	20.000	L.F.	122		122
Wood blocking	8.000	B.F.	3.84	22	25.84
Float glass, 1/4″ thick, clear, plain	24.000	S.F.	139.20	187.20	326.40
TOTAL			642.24	481.23	1,123.47

C1010 710		Interior Glazed Opening					
	FRAME	GLASS	OPENING-SIZE W X H	INTERMEDIATE MULLION	INTERMEDIATE HORIZONTAL	COST PER OPNG.		
						MAT.	INST.	TOTAL
1000	Aluminum flush	1/4″ float	6′x4′	0	0	640	480	1,120
1040	Tube		12′x4′	3	0	1,575	1,025	2,600
1080			4′x5′	0	0	585	415	1,000
1120			8′x5′	1	0	1,075	770	1,845
1160			12′x5′	2	0	1,575	1,125	2,700
1240		3/8″ float	9′x6′	2	0	1,750	1,300	3,050
1280			4′x8′-6″	0	1	1,125	830	1,955
1320			16′x10′	3	1	4,575	3,400	7,975
1400		1/4″ tempered	6′x4′	0	0	700	480	1,180
1440			12′x4′	3	0	1,700	1,025	2,725
1480			4′x5′	0	0	630	415	1,045
1520			8′x5′	1	0	1,175	770	1,945
1560			12′x5′	2	0	1,725	1,125	2,850
1640		3/8″ tempered	9′x6′	2	0	2,125	1,300	3,425
1680			4′x8′-6″	0	0	1,175	780	1,955
1720			16′x10′	3	0	4,975	3,200	8,175
1800		1/4″ one way mirror	6′x4′	0	0	970	475	1,445
1840			12′x4′	3	0	2,225	1,025	3,250

C1010 Partitions

C1010 710		Interior Glazed Opening						

	FRAME	GLASS	OPENING-SIZE W X H	INTERMEDIATE MULLION	INTERMEDIATE HORIZONTAL	COST PER OPNG.		
						MAT.	INST.	TOTAL
1880	Aluminum flush	1/4" one way mirror	4'x5'	0	0	855	410	1,265
1920	Tube		8'x5'	1	0	1,625	755	2,380
1960			12'x5'	2	0	2,400	1,100	3,500
2040		3/8" laminated safety	9'x6'	2	0	2,425	1,275	3,700
2080			4'x8'-6"	0	0	1,375	765	2,140
2120			16'x10'	3	0	5,925	3,125	9,050
2200		1-3/16" bullet proof	6'x4'	0	0	2,900	1,725	4,625
2240			12'x4'	3	0	6,075	3,500	9,575
2280			4'x5'	0	0	2,475	1,450	3,925
2320			8'x5'	1	0	4,825	2,825	7,650
2360			12'x5'	2	0	7,175	4,200	11,375
2440		1" acoustical	9'x6'	2	0	3,250	1,150	4,400
2480			4'x8'-6"	0	0	1,925	695	2,620
2520			16'x10'	3	0	8,225	2,750	10,975
3000	Oakwood	1/4" float	6'x4'	0	0	288	360	648
3040			12'x4'	3	0	700	870	1,570
3080			4'x5'	0	0	247	293	540
3120			8'x5'	1	0	500	620	1,120
3160			12'x5'	2	0	755	945	1,700
3240		3/8" float	9'x6'	2	0	945	1,125	2,070
3280			4'x8'-6"	0	1	580	660	1,240
3320			16'x10'	3	1	2,700	3,250	5,950
3400		1/4" tempered	6'x4'	0	0	345	360	705
3440			12'x4'	3	0	815	870	1,685
3480			4'x5'	0	0	295	293	588
3520			8'x5'	1	0	595	620	1,215
3560			12'x5'	2	0	900	945	1,845
3640		3/8" tempered	9'x6'	2	0	1,300	1,125	2,425
3680			4'x8'-6"	0	0	740	595	1,335
3720			16'x10'	3	0	3,500	2,925	6,425
3800		1/4" one way mirror	6'x4'	0	0	615	350	965
3840			12'x4'	3	0	1,350	855	2,205
3880			4'x5'	0	0	520	286	806
3920			8'x5'	1	0	1,050	605	1,655
3960			12'x5'	2	0	1,575	920	2,495
4040		3/8" laminated safety	9'x6'	2	0	1,625	1,100	2,725
4080			4'x8'-6"	0	0	945	580	1,525
4120			16'x10'	3	0	4,450	2,850	7,300
4200		1-3/16" bullet proof	6'x4'	0	0	2,475	1,575	4,050
4240			12'x4'	3	0	5,075	3,300	8,375
4280			4'x5'	0	0	2,075	1,300	3,375
4360			12'x5'	2	0	6,225	3,975	10,200
4440		1" acoustical	9'x6'	2	0	2,325	955	3,280
4520			16'x10'	3	0	6,525	2,425	8,950

C1010 Partitions

C1010 710	Interior Glazed Opening

	FRAME	GLASS	OPENING-SIZE W X H	INTERMEDIATE MULLION	INTERMEDIATE HORIZONTAL	COST PER OPNG.		
						MAT.	INST.	TOTAL
5000	Concealed frame	1/2" float	6'x4'	1		680	870	1,550
5040	Butt glazed		9'x4'	2		980	1,250	2,230
5080			8'x5'	1		1,050	1,275	2,325
5120			16'x5'	3		2,025	2,400	4,425
5200		3/4" float	6'x8'	1		2,025	1,950	3,975
5240			9'x8'	2		2,975	2,800	5,775
5280			8'x10'	1		3,275	3,000	6,275
5320			16'x10'	3		6,375	5,650	12,025
5400		1/4" tempered	6'x4'	1		395	650	1,045
5440			9'x4'	2		555	915	1,470
5480			8'x5'	1		585	915	1,500
5520			16'x5'	3		1,075	1,675	2,750
5600		1/2" tempered	6'x8'	1		1,450	1,500	2,950
5640			9'x8'	2		2,100	2,100	4,200
5680			8'x10'	1		2,325	2,225	4,550
5720			16'x10'	3		4,450	4,125	8,575
5800		1/4" laminated safety	6'x4'	1		490	715	1,205
5840			9'x4'	2		700	1,000	1,700
5880			8'x5'	1		745	1,025	1,770
5920			16'x5'	3		1,400	1,875	3,275
6000		1/2" laminated safety	6'x8'	1		1,525	1,375	2,900
6040			9'x8'	2		2,225	1,925	4,150
6080			8'x10'	1		2,450	2,025	4,475
6120			16'x10'	3		4,700	3,700	8,400
6200		1-3/16" bullet proof	6'x4'	1		2,525	1,875	4,400
6240			9'x4'	2		3,750	2,750	6,500
6280			8'x5'	1		4,150	2,950	7,100
6320			16'x5'	3		8,175	5,725	13,900
6400		2" bullet proof	6'x8'	1		4,550	4,425	8,975
6440			9'x8'	2		6,750	6,500	13,250
6480			8'x10'	1		7,475	7,100	14,575
6520			16'x10'	3		14,800	13,900	28,700

C10 Interior Construction

C1020 Interior Doors

Sliding Panel-Mall Front

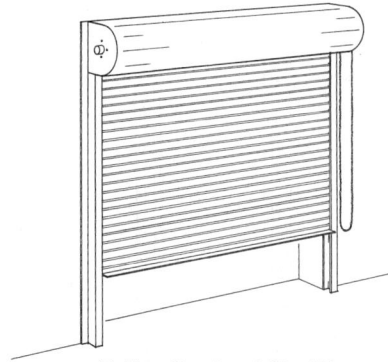

Rolling Overhead Steel Door

C1020 102	Special Doors	COST PER OPNG.		
		MAT.	INST.	TOTAL
3800	Sliding entrance door and system mill finish	9,625	2,350	11,975
3900	Bronze finish	10,500	2,550	13,050
4000	Black finish	11,000	2,650	13,650
4100	Sliding panel mall front, 16'x9' opening, mill finish	3,325	720	4,045
4200	Bronze finish	4,325	935	5,260
4300	Black finish	5,325	1,150	6,475
4400	24'x9' opening mill finish	4,825	1,325	6,150
4500	Bronze finish	6,275	1,725	8,000
4600	Black finish	7,725	2,125	9,850
4700	48'x9' opening mill finish	9,000	1,025	10,025
4800	Bronze finish	11,700	1,325	13,025
4900	Black finish	14,400	1,650	16,050
5000	Rolling overhead steel door, manual, 8' x 8' high	1,125	795	1,920
5100	10' x 10' high	1,925	910	2,835
5200	20' x 10' high	3,225	1,275	4,500
5300	12' x 12' high	2,025	1,050	3,075
5400	Motor operated, 8' x 8' high	2,300	1,050	3,350
5500	10' x 10' high	3,100	1,175	4,275
5600	20' x 10' high	4,400	1,525	5,925
5700	12' x 12' high	3,200	1,300	4,500
5800	Roll up grille, aluminum, manual, 10' x 10' high, mill finish	3,000	1,550	4,550
5900	Bronze anodized	4,800	1,550	6,350
6000	Motor operated, 10' x 10' high, mill finish	4,375	1,800	6,175
6100	Bronze anodized	6,175	1,800	7,975
6200	Steel, manual, 10' x 10' high	2,700	1,275	3,975
6300	15' x 8' high	3,125	1,600	4,725
6400	Motor operated, 10' x 10' high	4,075	1,525	5,600
6500	15' x 8' high	4,500	1,850	6,350

C10 Interior Construction

C1020   Interior Doors

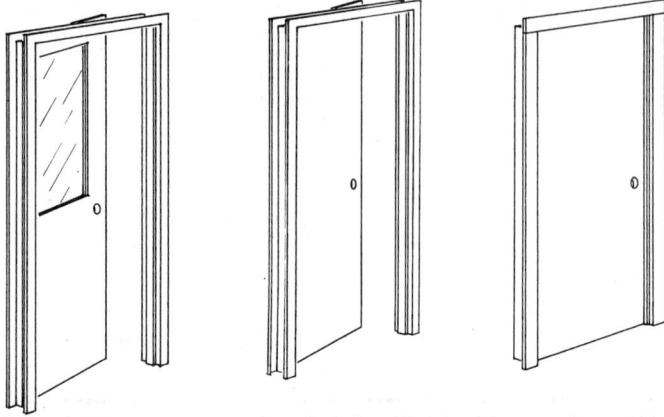

The Metal Door/Metal Frame Systems are defined as follows: door type, design and size, frame type and depth. Included in the components for each system is painting the door and frame. No hardware has been included in the systems.

Steel Door, Half Glass Steel Frame

Steel Door, Flush Steel Frame

Wood Door, Flush Wood Frame

System Components	QUANTITY	UNIT	COST EACH		
			MAT.	INST.	TOTAL
SYSTEM C1020 114 1200					
STEEL DOOR, HOLLOW, 20 GA., HALF GLASS, 2'-8"X6'-8", D.W.FRAME, 4-7/8" DP					
Steel door, flush, hollow core, 1-3/4" thick,half glass, 20 ga.,2'-8"x6'-8"	1.000	Ea.	475	49.50	524.50
Steel frame, KD, 16 ga., drywall, 4-7/8" deep, 2'-8"x6'-8", single	1.000	Ea.	122	66	188
Float glass, 3/16" thick, clear, tempered	5.000	S.F.	33.25	36	69.25
Paint door and frame each side, primer	1.000	Ea.	5.16	84	89.16
Paint door and frame each side, 2 coats	1.000	Ea.	8.14	141	149.14
TOTAL			643.55	376.50	1,020.05

C1020 114 — Metal Door/Metal Frame

	TYPE	DESIGN	SIZE	FRAME	DEPTH	COST EACH		
						MAT.	INST.	TOTAL
1000	Flush-hollow	20 ga. full panel	2'-8" x 6'-8"	drywall K.D.	4-7/8"	470	345	815
1020				butt welded	8-3/4"	580	390	970
1160			6'-0" x 7'-0"	drywall K.D.	4-7/8"	920	560	1,480
1180				butt welded	8-3/4"	1,050	605	1,655
1200		20 ga. half glass	2'-8" x 6'-8"	drywall K.D.	4-7/8"	645	375	1,020
1220				butt welded	8-3/4"	755	420	1,175
1360			6'-0" x 7'-0"	drywall K.D.	4-7/8"	1,200	625	1,825
1380				butt welded	8-3/4"	1,325	670	1,995
1800		18 ga. full panel	2'-8" x 6'-8"	drywall K.D.	4-7/8"	475	350	825
1820				butt welded	8-3/4"	585	395	980
1960			6'-0" x 7'-0"	drywall K.D.	4-7/8"	1,000	560	1,560
1980				butt welded	8-3/4"	1,150	605	1,755
2000		18 ga. half glass	2'-8" x 6'-8"	drywall K.D.	4-7/8"	675	380	1,055
2020				butt welded	8-3/4"	785	425	1,210
2160			6'-0" x 7'-0"	drywall K.D.	4-7/8"	1,300	630	1,930
2180				butt welded	8-3/4"	1,425	675	2,100

C10 Interior Construction

C1020 Interior Doors

C1020 116 — Labeled Metal Door/Metal Frames

	TYPE	DESIGN	SIZE	FRAME	DEPTH	COST EACH MAT.	COST EACH INST.	COST EACH TOTAL
6000	Hollow-1/2 hour	20 ga. full panel	2'-8" x 6'-8"	drywall K.D.	4-7/8"	580	345	925
6020				butt welded	8-3/4"	690	390	1,080
6160			6'-0" x 7'-0"	drywall K.D.	4-7/8"	1,050	560	1,610
6180				butt welded	8-3/4"	1,125	605	1,730
6200		20 ga. vision	2'-8" x 6'-8"	drywall K.D.	4-7/8"	710	355	1,065
6220				butt welded	8-3/4"	820	400	1,220
6360			6'-0" x 7'-0"	drywall K.D.	4-7/8"	1,300	575	1,875
6380				butt welded	8-3/4"	1,400	620	2,020
6600		18 ga. full panel	2'-8" x 7'-0"	drywall K.D.	4-7/8"	655	350	1,005
6620				butt welded	8-3/4"	765	395	1,160
6760			6'-0" x 7'-0"	drywall K.D.	4-7/8"	1,125	565	1,690
6780				butt welded	8-3/4"	1,225	610	1,835
6800		18 ga. vision	2'-8" x 7'-0"	drywall K.D.	4-7/8"	785	355	1,140
6820				butt welded	8-3/4"	895	400	1,295
6960			6'-0" x 7'-0"	drywall K.D.	4-7/8"	1,375	580	1,955
6980				butt welded	8-3/4"	1,475	625	2,100
7200	Hollow-3 hour	18 ga. full panel	2'-8" x 7'-0"	butt K.D.	5-3/4"	655	350	1,005
7220				butt welded	8-3/4"	765	395	1,160
7360			6'-0" x 7'-0"	butt K.D.	5-3/4"	1,125	565	1,690
7380				butt welded	8-3/4"	1,225	610	1,835
8000	Composite-1-1/2 hour	20 ga. full panel	2'-8" x 6'-8"	drywall K.D.	4-3/8"	680	350	1,030
8020				butt welded	8-3/4"	790	395	1,185
8160			6'-0" x 7'-0"	drywall K.D.	4-7/8"	1,250	565	1,815
8180				butt welded	8-3/4"	1,325	610	1,935
8200		20 ga. vision	2'-8" x 6'-8"	drywall K.D.	4-7/8"	810	355	1,165
8220				butt welded	8-3/4"	920	400	1,320
8360			6'-0" x 7'-0"	drywall K.D.	4-7/8"	1,500	580	2,080
8400	Composite-3 hour	18 ga. full panel	2'-8" x 7'-0"	drywall K.D.	5-3/4"	770	355	1,125
8420				butt welded	8-3/4"	880	400	1,280
8560			6'-0" x 7'-0"	drywall K.D.	5-3/4"	1,350	575	1,925

C1020 120 — Wood Door/Wood Frame

	TYPE	FACE	SIZE	FRAME	DEPTH	COST EACH MAT.	COST EACH INST.	COST EACH TOTAL
1600	Hollow core/flush	lauan	2'-8" x 6'-8"	pine	3-5/8"	167	256	423
1620					5-3/16"	157	274	431
1760			6'-0" x 6'-8"	pine	3-5/8"	271	400	671
1780					5-3/16"	258	425	683
1800		birch	2'-8" x 6'-8"	pine	3-5/8"	216	256	472
1820					5-3/16"	206	274	480
1960			6'-0" x 6'-8"	pine	3-5/8"	360	400	760
1980					5-3/16"	345	425	770
2000		oak	2'-8" x 6'-8"	oak	3-5/8"	137	218	355
2020					5-3/16"	221	277	498
2160			6'-0" x 6'-8"	oak	3-5/8"	350	405	755
2180					5-3/16"	365	430	795
3000	Particle core/flush	lauan	2'-8" x 6'-8"	pine	3-5/8"	218	263	481
3020					5-3/16"	207	281	488
3160			6'-0" x 7'-0"	pine	3-5/8"	365	440	805
3180					5-3/16"	350	460	810

245

C10 Interior Construction

C1020 Interior Doors

C1020 120	Wood Door/Wood Frame

	TYPE	FACE	SIZE	FRAME	DEPTH	COST EACH MAT.	COST EACH INST.	COST EACH TOTAL
3200	Particle core/flush	birch	2'-8" x 6'-8"	pine	3-5/8"	230	263	493
3220					5-3/16"	220	281	501
3360			6'-0" x 7'-0"	pine	3-5/8"	390	440	830
3380					5-3/16"	380	460	840
3398								
3400		oak	2'-8" x 6'-8"	oak	3-5/8"	221	266	487
3420					5-3/16"	236	284	520
3560			6'-0" x 7'-0"	oak	3-5/8"	390	440	830
3580					5-3/16"	410	465	875
5000	Solid core/flush	birch	2'-8" x 6'-8"	pine	3-5/8"	268	263	531
5020					5-3/16"	258	281	539
5160			6'-0" x 7'-0"	pine	3-5/8"	465	440	905
5180					5-3/16"	455	460	915
5200		oak	2'-8" x 6'-8"	oak	3-5/8"	259	266	525
5220					5-3/16"	274	284	558
5360			6'-0" x 7'-0"	oak	3-5/8"	470	440	910
5380					5-3/16"	485	465	950
5400		M.D. overlay	2'-8" x 6'-8"	pine	3-5/8"	293	420	713

C1020 122	Wood Door/Metal Frame

	TYPE	FACE	SIZE	FRAME	DEPTH	COST EACH MAT.	COST EACH INST.	COST EACH TOTAL
1600	Hollow core/flush	lauan	2'-8" x 6'-8"	drywall K.D.	4-7/8"	168	196	364
1620				butt welded	8-3/4"	278	204	482
1760			6'-0" x 7'-0"	drywall K.D.	4-7/8"	300	330	630
1780				butt welded	8-3/4"	430	340	770
1800		birch	2'-8" x 6'-8"	drywall K.D.	4-7/8"	217	196	413
1820				butt welded	8-3/4"	325	204	529
1960			6'-0" x 7'-0"	drywall K.D.	4-7/8"	390	330	720
1980				butt welded	8-3/4"	515	340	855
2000		oak	2'-8" x 6'-8"	drywall K.D.	4-7/8"	227	196	423
2020				butt welded	8-3/4"	335	204	539
2160			6'-0" x 7'-0"	drywall K.D.	4-7/8"	405	330	735
2180				butt welded	8-3/4"	530	340	870
3000	Particle core/flush	lauan	2'-8" x 6'-8"	drywall K.D.	4-7/8"	219	203	422
3020				butt welded	8-3/4"	330	212	542
3160			6'-0" x 7'-0"	drywall K.D.	4-7/8"	395	365	760
3180				butt welded	8-3/4"	525	380	905
3200		birch	2'-8" x 6'-8"	drywall K.D.	4-7/8"	231	203	434
3220				butt welded	8-3/4"	340	212	552
3360			6'-0" x 7'-0"	drywall K.D.	4-7/8"	420	365	785
3380				butt welded	8-3/4"	550	380	930
3400		oak	2'-8" x 6'-8"	drywall K.D.	4-7/8"	242	203	445
3420				butt welded	8-3/4"	350	212	562
3560			6'-0" x 7'-0"	drywall K.D.	4-7/8"	445	365	810
3580				butt welded	8-3/4"	575	380	955
5000	Solid core/flush	birch	2'-8" x 6'-8"	drywall K.D.	4-7/8"	269	203	472
5020				butt welded	8-3/4"	380	212	592
5160			6'-0" x 7'-0"	drywall K.D.	4-7/8"	495	365	860
5180				butt welded	8-3/4"	625	380	1,005

C10 Interior Construction

C1020 Interior Doors

C1020 122	Wood Door/Metal Frame

	TYPE	FACE	SIZE	FRAME	DEPTH	COST EACH		
						MAT.	INST.	TOTAL
5200	Solid core/flush	oak	2'-8" x 6'-8"	drywall K.D.	4-7/8"	280	203	483
5220				butt welded	8-3/4"	390	212	602
5360			6'-0" x 7'-0"	drywall K.D.	4-7/8"	525	365	890
5380				butt welded	8-3/4"	650	380	1,030
6000	1 hr/flush	birch	2'-8" x 6'-8"	drywall K.D.	4-7/8"	545	208	753
6020				butt welded	8-3/4"	650	217	867
6160			6'-0" x 7'-0"	drywall K.D.	4-7/8"	1,000	380	1,380
6180				butt welded	8-3/4"	1,100	390	1,490
6200		oak	2'-8" x 6'-8"	drywall K.D.	4-7/8"	540	208	748
6220				butt welded	8-3/4"	645	217	862
6360			6'-0" x 7'-0"	drywall K.D.	4-7/8"	980	380	1,360
6380				butt welded	8-3/4"	1,075	390	1,465
6400		walnut	2'-8" x 6'-8"	drywall K.D.	4-7/8"	645	208	853
6420				butt welded	8-3/4"	750	217	967
6560			6'-0" x 7'-0"	drywall K.D.	4-7/8"	1,200	380	1,580
6580				butt welded	8-3/4"	1,300	390	1,690
7600	1-1/2 hr/flush	birch	2'-8" x 6'-8"	drywall K.D.	4-7/8"	535	208	743
7620				butt welded	8-3/4"	640	217	857
7760			6'-0" x 7'-0"	drywall K.D.	4-7/8"	990	380	1,370
7780				butt welded	8-3/4"	1,075	390	1,465
7800		oak	2'-8" x 6'-8"	drywall K.D.	4-7/8"	505	208	713
7820				butt welded	8-3/4"	610	217	827
7960			6'-0" x 7'-0"	drywall K.D.	4-7/8"	910	380	1,290
7980				butt welded	8-3/4"	1,000	390	1,390
8000		walnut	2'-8" x 6'-8"	drywall K.D.	4-7/8"	620	208	828
8020				butt welded	8-3/4"	725	217	942
8160			6'-0" x 7'-0"	drywall K. D.	4 7/8"	1,150	380	1,530
8180				butt welded	8-3/4"	1,250	390	1,640

C1020 124	Wood Fire Doors/Metal Frames

	FIRE RATING (HOURS)	CORE MATERIAL	THICKNESS (IN.)	FACE MATERIAL	SIZE W(FT)XH(FT)	COST EACH		
						MAT.	INST.	TOTAL
0840	1	mineral	1-3/4"	birch	2/8 6/8	545	208	753
0860		3 ply stile			3/0 7/0	575	225	800
0880				oak	2/8 6/8	540	208	748
0900					3/0 7/0	565	225	790
0920				walnut	2/8 6/8	645	208	853
0940					3/0 7/0	675	225	900
0960				m.d. overlay	2/8 6/8	500	355	855
0980					3/0 7/0	530	370	900
1000	1-1/2	mineral	1-3/4"	birch	2/8 6/8	535	208	743
1020		3 ply stile			3/0 7/0	570	225	795
1040				oak	2/8 6/8	505	208	713
1060					3/0 7/0	530	225	755
1080				walnut	2/8 6/8	620	208	828
1100					3/0 7/0	650	225	875
1120				m.d. overlay	2/8 6/8	535	355	890
1140					3/0 7/0	560	370	930

C10 Interior Construction

C1020 Interior Doors

C1020 310	Hardware	COST EACH		
		MAT.	INST.	TOTAL
0060	**HINGES**			
0080				
0100	Full mortise, low frequency, steel base, 4-1/2" x 4-1/2", USP	6.65		6.65
0120	5" x 5" USP	16.50		16.50
0140	6" x 6" USP	33		33
0160	Average frequency, steel base, 4-1/2" x 4-1/2" USP	13.50		13.50
0180	5" x 5" USP	22.50		22.50
0200	6" x 6" USP	48		48
0220	High frequency, steel base, 4-1/2" x 4-1/2", USP	32.50		32.50
0240	5" x 5" USP	30		30
0260	6" x 6" USP	73.50		73.50
0280				
0300	**LOCKSETS**			
0320				
0340	Heavy duty cylindrical, passage door			
0360	Non-keyed, passage	53.50	41.50	95
0380	Privacy	67	41.50	108.50
0400	Keyed, single cylinder function	92	49.50	141.50
0420	Hotel	132	62	194
0440				
0460	For re-core cylinder, add	37.50		37.50
0480				
0500	**CLOSERS**			
0520				
0540	Rack & pinion			
0560	Adjustable backcheck, 3 way mount, all sizes, regular arm	172	82.50	254.50
0580	Hold open arm	172	82.50	254.50
0600	Fusible link	143	76.50	219.50
0620	Non-sized, regular arm	162	82.50	244.50
0640	4 way mount, non-sized, regular arm	229	82.50	311.50
0660	Hold open arm	235	82.50	317.50
0680				
0700	**PUSH, PULL**			
0720				
0740	Push plate, aluminum	9.50	41.50	51
0760	Bronze	22.50	41.50	64
0780	Pull handle, push bar, aluminum	139	45	184
0800	Bronze	180	49.50	229.50
0810				
0820	**PANIC DEVICES**			
0840				
0860	Narrow stile, rim mounted, bar, exit only	680	82.50	762.50
0880	Outside key & pull	740	99	839
0900	Bar and vertical rod, exit only	700	99	799
0920	Outside key & pull	700	124	824
0940	Bar and concealed rod, exit only	815	165	980
0960	Mortise, bar, exit only	625	124	749
0980	Touch bar, exit only	910	124	1,034
1000				
1020	**WEATHERSTRIPPING**			
1040				
1060	Interlocking, 3' x 7', zinc	17.05	165	182.05
1080	Spring type, 3' x 7', bronze	39	165	204

C10 Interior Construction

C1030 Fittings

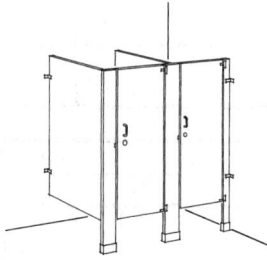

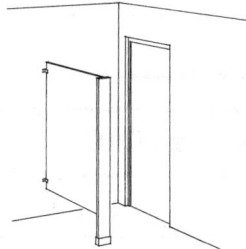

Toilet Units   **Entrance Screens**   **Urinal Screens**

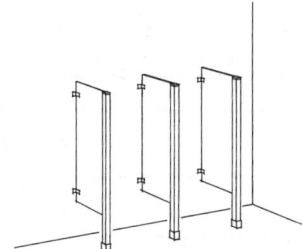

C1030 110	Toilet Partitions	MAT.	INST.	TOTAL
		COST PER UNIT		
0380	Toilet partitions, cubicles, ceiling hung, marble	1,825	470	2,295
0400	Painted metal	515	248	763
0420	Plastic laminate	665	248	913
0460	Stainless steel	1,950	248	2,198
0480	Handicap addition	430		430
0520	Floor and ceiling anchored, marble	2,000	375	2,375
0540	Painted metal	535	198	733
0560	Plastic laminate	675	198	873
0600	Stainless steel	1,850	198	2,048
0620	Handicap addition	305		305
0660	Floor mounted marble	1,175	315	1,490
0680	Painted metal	595	142	737
0700	Plastic laminate	600	142	742
0740	Stainless steel	1,675	142	1,817
0760	Handicap addition	300		300
0780	Juvenile deduction	43.50		43.50
0820	Floor mounted with handrail marble	1,125	315	1,440
0840	Painted metal	485	165	650
0860	Plastic laminate	635	165	800
0900	Stainless steel	1,250	165	1,415
0920	Handicap addition	365		365
0960	Wall hung, painted metal	655	142	797
1020	Stainless steel	1,725	142	1,867
1040	Handicap addition	365		365
1080	Entrance screens, floor mounted, 54" high, marble	745	104	849
1100	Painted metal	256	66	322
1140	Stainless steel	935	66	1,001
1300	Urinal screens, floor mounted, 24" wide, laminated plastic	400	124	524
1320	Marble	630	147	777
1340	Painted metal	219	124	343
1380	Stainless steel	615	124	739
1428	Wall mounted wedge type, painted metal	141	99	240
1430	Urinal screens, wall hung, plastic laminate/particle board	202	124	326
1460	Stainless steel	595	99	694
1500	Partitions, shower stall, single wall, painted steel, 2'-8" x 2'-8"	1,000	232	1,232
1510	Fiberglass, 2'-8" x 2'-8"	775	258	1,033
1520	Double wall, enameled steel, 2'-8" x 2'-8"	1,075	232	1,307
1530	Stainless steel, 2'-8" x 2'-8"	1,875	232	2,107
1540	Doors, plastic, 2' wide	140	64.50	204.50
1550	Tempered glass, chrome/brass frame-deluxe	970	580	1,550
1560	Tub enclosure, sliding panels, tempered glass, aluminum frame	400	291	691
1570	Chrome/brass frame-deluxe	1,175	385	1,560

C10 Interior Construction

C1030 Fittings

C1030 210	Fabricated Compact Units and Cubicles, EACH	COST EACH		
		MAT.	INST.	TOTAL
0500	Telephone enclosure, shelf type, wall hung, economy	1,425	198	1,623
0510	Deluxe	2,975	198	3,173
0520	Booth type, painted steel, economy	3,825	660	4,485
0530	Stainless steel, deluxe	12,700	660	13,360

C1030 310	Storage Specialties, EACH	COST EACH		
		MAT.	INST.	TOTAL
0200	Lockers, steel, single tier, 5' to 6' high, per opening, min.	149	41.50	190.50
0210	Maximum	261	48.50	309.50
0220	Two tier, minimum	84.50	22.50	107
0230	Maximum	112	29	141
0440	Parts bins, 3' wide x 7' high, 14 bins, 18" x 12" x 12"	300	78.50	378.50
0450	84 bins, 6" x 6" x 12"	715	98	813
0600	Shelving, metal industrial, braced, 3' wide, 1' deep	21.50	10.90	32.40
0610	2' deep	31	11.60	42.60
0620	Enclosed, 3' wide, 1' deep	38	12.75	50.75
0630	2' deep	50.50	15	65.50

C1030 510	Identifying/Visual Aid Specialties, EACH	COST EACH		
		MAT.	INST.	TOTAL
0100	Control boards, magnetic, porcelain finish, framed, 24" x 18"	219	124	343
0110	96" x 48"	930	198	1,128
0120	Directory boards, outdoor, black plastic, 36" x 24"	810	495	1,305
0130	36" x 36"	935	660	1,595
0140	Indoor, economy, open faced, 18" x 24"	146	142	288
0150	36" x 48"	270	165	435
0160	Building, aluminum, black felt panel, 24" x 18"	350	248	598
0170	48" x 72"	1,025	990	2,015
0500	Signs, interior electric exit sign, wall mounted, 6"	117	70	187
0510	Street, reflective alum., dbl. face, 4 way, w/bracket	184	33	217
0520	Letters, cast aluminum, 1/2" deep, 4" high	24.50	27.50	52
0530	1" deep, 10" high	54.50	27.50	82
0540	Plaques, cast aluminum, 20" x 30"	1,150	248	1,398
0550	Cast bronze, 36" x 48"	4,450	495	4,945

C1030 520	Identifying/Visual Aid Specialties, S.F.	COST PER S.F.		
		MAT.	INST.	TOTAL
0100	Bulletin board, cork sheets, no frame, 1/4" thick	3.96	3.42	7.38
0110	1/2" thick	6.95	3.42	10.37
0120	Aluminum frame, 1/4" thick, 3' x 5'	8.05	4.15	12.20
0130	4' x 8'	5.50	2.20	7.70
0200	Chalkboards, wall hung, alum, frame & chalktrough	11.80	2.20	14
0210	Wood frame & chalktrough	11.45	2.37	13.82
0220	Sliding board, one board with back panel	45	2.11	47.11
0230	Two boards with back panel	69	2.11	71.11
0240	Liquid chalk type, alum. frame & chalktrough	10.70	2.20	12.90
0250	Wood frame & chalktrough	21.50	2.20	23.70

C1030 710	Bath and Toilet Accessories, EACH	COST EACH		
		MAT.	INST.	TOTAL
0100	Specialties, bathroom accessories, st. steel, curtain rod, 5' long, 1" diam	45.50	38	83.50
0110	1-1/2" diam.	40	38	78
0120	Dispenser, towel, surface mounted	46	31	77
0130	Flush mounted with waste receptacle	435	49.50	484.50
0140	Grab bar, 1-1/4" diam., 12" long	32.50	20.50	53
0150	1-1/2" diam. 36" long	48	25	73

C1030 Fittings

C1030 710	Bath and Toilet Accessories, EACH	COST EACH		
		MAT.	INST.	TOTAL
0160	Mirror, framed with shelf, 18" x 24"	249	25	274
0170	72" x 24"	415	82.50	497.50
0180	Toilet tissue dispenser, surface mounted, single roll	21.50	16.50	38
0190	Double roll	26	20.50	46.50
0200	Towel bar, 18" long	43.50	21.50	65
0210	30" long	111	23.50	134.50
0300	Medicine cabinets, sliding mirror doors, 20" x 16" x 4-3/4", unlighted	111	71	182
0310	24" x 19" x 8-1/2", lighted	175	99	274
0320	Triple door, 30" x 32", unlighted, plywood body	263	71	334
0330	Steel body	345	71	416
0340	Oak door, wood body, beveled mirror, single door	169	71	240
0350	Double door	410	82.50	492.50

C1030 730	Bath and Toilet Accessories, L.F.	COST PER L.F.		
		MAT.	INST.	TOTAL
0100	Partitions, hospital curtain, ceiling hung, polyester oxford cloth	19.10	4.84	23.94
0110	Designer oxford cloth	35	6.15	41.15

C1030 830	Fabricated Cabinets & Counters, L.F.	COST PER L.F.		
		MAT.	INST.	TOTAL
0110	Household, base, hardwood, one top drawer & one door below x 12" wide	250	40	290
0120	Four drawer x 24" wide	450	44.50	494.50
0130	Wall, hardwood, 30" high with one door x 12" wide	211	45	256
0140	Two doors x 48" wide	440	54	494
0150	Counter top-laminated plastic, stock, economy	9.55	16.50	26.05
0160	Custom-square edge, 7/8" thick	13.80	37	50.80
0170	School, counter, wood, 32" high	225	49.50	274.50
0180	Metal, 84" high	405	66	471

C1030 910	Other Fittings, EACH	COST EACH		
		MAT.	INST.	TOTAL
0500	Mail boxes, horizontal, rear loaded, aluminum, 5" x 6" x 15" deep	38.50	14.60	53.10
0510	Front loaded, aluminum, 10" x 12" x 15" deep	119	25	144
0520	Vertical, front loaded, aluminum, 15" x 5" x 6" deep	33	14.60	47.60
0530	Bronze, duranodic finish	44	14.60	58.60
0540	Letter slot, post office	121	62	183
0550	Mail counter, window, post office, with grille	605	248	853
0600	Partition, woven wire, wall panel, 4' x 7' high	145	39.50	184.50
0610	Wall panel with window & shelf, 5' x 8' high	360	66	426
0620	Doors, swinging w/ no transom, 3' x 7' high	320	165	485
0630	Sliding, 6' x 10' high	825	248	1,073
0700	Turnstiles, one way, 4' arm, 46" diam., manual	855	198	1,053
0710	Electric	1,400	825	2,225
0720	3 arm, 5'-5" diam. & 7' high, manual	4,250	990	5,240
0730	Electric	4,800	1,650	6,450

C20 Stairs

C2010 Stair Construction

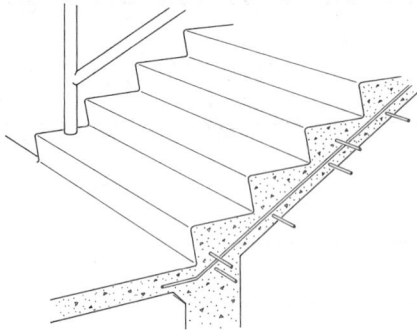

General Design: See reference section for code requirements. Maximum height between landings is 12'; usual stair angle is 20° to 50° with 30° to 35° best. Usual relation of riser to treads is:

Riser + tread = 17.5.
2x (Riser) + tread = 25.
Riser x tread = 70 or 75.

Maximum riser height is 7" for commercial, 8-1/4" for residential.
Usual riser height is 6-1/2" to 7-1/4".

Minimum tread width is 11" for commercial and 9" for residential.

For additional information please see reference section.

Cost Per Flight: Table below lists the cost per flight for 4'-0" wide stairs.
Side walls are not included.
Railings are included.

System Components	QUANTITY	UNIT	COST PER FLIGHT		
			MAT.	INST.	TOTAL
SYSTEM C2010 110 0560					
STAIRS, C.I.P. CONCRETE WITH LANDING, 12 RISERS					
Concrete in place, free standing stairs not incl. safety treads	48.000	L.F.	328.80	1,720.80	2,049.60
Concrete in place, free standing stair landing	32.000	S.F.	184	473.28	657.28
Stair tread C.I. abrasive 4" wide	48.000	L.F.	748.80	297.60	1,046.40
Industrial railing, welded, 2 rail 3'-6" high 1-1/2" pipe	18.000	L.F.	612	192.24	804.24
Wall railing with returns, steel pipe	17.000	L.F.	326.40	181.56	507.96
TOTAL			2,200	2,865.48	5,065.48

C2010 110	Stairs		COST PER FLIGHT		
			MAT.	INST.	TOTAL
0470	Stairs, C.I.P. concrete, w/o landing, 12 risers, w/o nosing	RC2010 -100	1,275	2,075	3,350
0480	With nosing		2,025	2,375	4,400
0550	W/landing, 12 risers, w/o nosing		1,450	2,575	4,025
0560	With nosing		2,200	2,875	5,075
0570	16 risers, w/o nosing		1,775	3,225	5,000
0580	With nosing		2,775	3,625	6,400
0590	20 risers, w/o nosing		2,100	3,875	5,975
0600	With nosing		3,350	4,375	7,725
0610	24 risers, w/o nosing		2,425	4,550	6,975
0620	With nosing		3,925	5,150	9,075
0630	Steel, grate type w/nosing & rails, 12 risers, w/o landing		6,425	1,075	7,500
0640	With landing		8,400	1,500	9,900
0660	16 risers, with landing		10,500	1,850	12,350
0680	20 risers, with landing		12,700	2,225	14,925
0700	24 risers, with landing		14,800	2,600	17,400
0710	Cement fill metal pan & picket rail, 12 risers, w/o landing		8,400	1,075	9,475
0720	With landing		11,000	1,650	12,650
0740	16 risers, with landing		13,800	2,000	15,800
0760	20 risers, with landing		16,600	2,350	18,950
0780	24 risers, with landing		19,400	2,725	22,125
0790	Cast iron tread & pipe rail, 12 risers, w/o landing		8,400	1,075	9,475
0800	With landing		11,000	1,650	12,650
0820	16 risers, with landing		13,800	2,000	15,800
0840	20 risers, with landing		16,600	2,350	18,950
0860	24 risers, with landing		19,400	2,725	22,125
0870	Pan tread & flat bar rail, pre-assembled, 12 risers, w/o landing		6,900	840	7,740
0880	With landing		12,500	1,300	13,800
0900	16 risers, with landing		13,800	1,475	15,275
0920	20 risers, with landing		16,100	1,725	17,825
0940	24 risers, with landing		18,400	2,025	20,425

C20 Stairs

C2010 Stair Construction

C2010 110	Stairs	COST PER FLIGHT		
		MAT.	INST.	TOTAL
0950	Spiral steel, industrial checkered plate 4'-6' dia., 12 risers	7,975	730	8,705
0960	16 risers	10,600	975	11,575
0970	20 risers	13,300	1,225	14,525
0980	24 risers	16,000	1,450	17,450
0990	Spiral steel, industrial checkered plate 6'-0" dia., 12 risers	9,725	820	10,545
1000	16 risers	13,000	1,075	14,075
1010	20 risers	16,200	1,375	17,575
1020	24 risers	19,400	1,650	21,050
1030	Aluminum, spiral, stock units, 5'-0" dia., 12 risers	9,000	730	9,730
1040	16 risers	12,000	975	12,975
1050	20 risers	15,000	1,225	16,225
1060	24 risers	18,000	1,450	19,450
1070	Custom 5'-0" dia., 12 risers	17,100	730	17,830
1080	16 risers	22,800	975	23,775
1090	20 risers	28,500	1,225	29,725
1100	24 risers	34,200	1,450	35,650
1120	Wood, prefab box type, oak treads, wood rails 3'-6" wide, 14 risers	2,100	415	2,515
1150	Prefab basement type, oak treads, wood rails 3'-0" wide, 14 risers	1,050	102	1,152

C30 Interior Finishes

C3010 Wall Finishes

C3010 230	Paint & Covering	COST PER S.F.		
		MAT.	INST.	TOTAL
0060	Painting, interior on plaster and drywall, brushwork, primer & 1 coat	.11	.65	.76
0080	Primer & 2 coats	.17	.86	1.03
0100	Primer & 3 coats	.24	1.06	1.30
0120	Walls & ceilings, roller work, primer & 1 coat	.11	.43	.54
0140	Primer & 2 coats	.17	.56	.73
0160	Woodwork incl. puttying, brushwork, primer & 1 coat	.11	.94	1.05
0180	Primer & 2 coats	.17	1.24	1.41
0200	Primer & 3 coats	.24	1.69	1.93
0260	Cabinets and casework, enamel, primer & 1 coat	.12	1.06	1.18
0280	Primer & 2 coats	.18	1.30	1.48
0300	Masonry or concrete, latex, brushwork, primer & 1 coat	.23	.88	1.11
0320	Primer & 2 coats	.30	1.26	1.56
0340	Addition for block filler	.15	1.08	1.23
0380	Fireproof paints, intumescent, 1/8" thick 3/4 hour	2.14	.86	3
0420	7/16" thick 2 hour	6.15	3.01	9.16
0440	1-1/16" thick 3 hour	10.05	6.05	16.10
0480	Miscellaneous metal brushwork, exposed metal, primer & 1 coat	.10	1.06	1.16
0500	Gratings, primer & 1 coat	.26	1.32	1.58
0600	Pipes over 12" diameter	.49	4.22	4.71
0700	Structural steel, brushwork, light framing 300-500 S.F./Ton	.08	1.63	1.71
0720	Heavy framing 50-100 S.F./Ton	.08	.82	.90
0740	Spraywork, light framing 300-500 S.F./Ton	.09	.36	.45
0800	Varnish, interior wood trim, no sanding sealer & 1 coat	.08	1.06	1.14
0820	Hardwood floor, no sanding 2 coats	.17	.22	.39
0840	Wall coatings, acrylic glazed coatings, minimum	.31	.80	1.11
0860	Maximum	.66	1.38	2.04
0880	Epoxy coatings, minimum	.41	.80	1.21
1080	Maximum	1.06	2.91	3.97
1100	High build epoxy 50 mil, minimum	.68	1.08	1.76
1120	Maximum	1.17	4.44	5.61
1140	Laminated epoxy with fiberglass minimum	.74	1.43	2.17
1160	Maximum	1.31	2.91	4.22
1180	Sprayed perlite or vermiculite 1/16" thick, minimum	.26	.14	.40
1200	Maximum	.75	.66	1.41
1260	Wall coatings, vinyl plastic, minimum	.33	.57	.90
1280	Maximum	.83	1.76	2.59
1300	Urethane on smooth surface, 2 coats, minimum	.26	.37	.63
1320	Maximum	.58	.63	1.21
1340	3 coats, minimum	.35	.50	.85
1360	Maximum	.79	.90	1.69
1380	Ceramic-like glazed coating, cementitious, minimum	.48	.96	1.44
1420	Resin base, minimum	.33	.66	.99
1440	Maximum	.54	1.28	1.82
1460	Wall coverings, aluminum foil	1.07	1.54	2.61
1500	Vinyl backing	5.70	1.77	7.47
1520	Cork tiles, 12"x12", light or dark, 3/16" thick	4.68	1.77	6.45
1540	5/16" thick	3.98	1.81	5.79
1580	Natural, non-directional, 1/2" thick	7.35	1.77	9.12
1600	12"x36", granular, 3/16" thick	1.32	1.10	2.42
1620	1" thick	1.71	1.15	2.86
1660	5/16" thick	5.90	1.81	7.71
1661	Paneling, prefinished plywood, birch	1.49	2.36	3.85
1662	Mahogany, African	2.76	2.48	5.24
1664	Oak or cherry	3.54	2.48	6.02
1665	Rosewood	5.05	3.10	8.15
1666	Teak	3.54	2.48	6.02
1720	Gypsum based, fabric backed, minimum	.88	.53	1.41
1740	Average	1.29	.59	1.88

254

C30 Interior Finishes

C3010 Wall Finishes

C3010 230	Paint & Covering		COST PER S.F.		
			MAT.	INST.	TOTAL
1760	Maximum		1.43	.66	2.09
1780	Vinyl wall covering, fabric back, light weight		.72	.66	1.38
1800	Medium weight		.85	.89	1.74
1820	Heavy weight		1.75	.98	2.73
1840	Wall paper, double roll, solid pattern, avg. workmanship		.39	.66	1.05
1860	Basic pattern, avg. workmanship		.69	.79	1.48
1880	Basic pattern, quality workmanship		2.59	.98	3.57
1900	Grass cloths with lining paper, minimum		.81	1.06	1.87
1920	Maximum		2.61	1.21	3.82
1940	Ceramic tile, thin set, 4-1/4" x 4-1/4"		2.74	4.20	6.94
1960	12" x 12"		3.98	2.66	6.64

C3010 235	Paint Trim		COST PER L.F.		
			MAT.	INST.	TOTAL
2040	Painting, wood trim, to 6" wide, enamel, primer & 1 coat		.12	.53	.65
2060	Primer & 2 coats		.18	.67	.85
2080	Misc. metal brushwork, ladders		.52	5.30	5.82
2120	6" to 8" dia.		.13	2.22	2.35
2140	10" to 12" dia.		.38	3.32	3.70
2160	Railings, 2" pipe		.18	2.64	2.82
2180	Handrail, single		.10	1.06	1.16
2185	Caulking & Sealants, Polyureathane, In place, 1 or 2 component, 1/2" X 1/4"		.39	3.39	3.78

C3020 410	Tile & Covering	COST PER S.F.		
		MAT.	INST.	TOTAL
0060	Carpet tile, nylon, fusion bonded, 18" x 18" or 24" x 24", 24 oz.	3.55	.62	4.17
0080	35 oz.	4.16	.62	4.78
0100	42 oz.	5.55	.62	6.17
0140	Carpet, tufted, nylon, roll goods, 12' wide, 26 oz.	5.10	.71	5.81
0160	36 oz.	7.85	.71	8.56
0180	Woven, wool, 36 oz.	13.30	.71	14.01
0200	42 oz.	13.65	.71	14.36
0220	Padding, add to above, minimum	.56	.33	.89
0240	Maximum	.95	.33	1.28
0260	Composition flooring, acrylic, 1/4" thick	1.60	4.79	6.39
0280	3/8" thick	2.04	5.55	7.59
0300	Epoxy, minimum	2.74	3.68	6.42
0320	Maximum	3.32	5.10	8.42
0340	Epoxy terrazzo, minimum	5.80	5.40	11.20
0360	Maximum	8.50	7.20	15.70
0380	Mastic, hot laid, 1-1/2" thick, minimum	4.02	3.61	7.63
0400	Maximum	5.15	4.79	9.94
0420	Neoprene 1/4" thick, minimum	3.95	4.56	8.51
0440	Maximum	5.45	5.80	11.25
0460	Polyacrylate with ground granite 1/4", minimum	3.28	3.39	6.67
0480	Maximum	5.65	5.15	10.80
0500	Polyester with colored quart 2 chips 1/16", minimum	2.94	2.33	5.27
0520	Maximum	4.64	3.68	8.32
0540	Polyurethane with vinyl chips, minimum	7.70	2.33	10.03
0560	Maximum	11.10	2.90	14
0600	Concrete topping, granolithic concrete, 1/2" thick	.32	3.96	4.28
0620	1" thick	.63	4.07	4.70
0640	2" thick	1.26	4.68	5.94
0660	Heavy duty 3/4" thick, minimum	.43	6.15	6.58
0680	Maximum	.80	7.35	8.15
0700	For colors, add to above, minimum	.47	1.42	1.89
0740	Exposed aggregate finish, minimum	.25	.72	.97
0760	Maximum	.73	.96	1.69
0780	Abrasives, .25 P.S.F. add to above, minimum	.48	.53	1.01
0800	Maximum	.67	.53	1.20
0820	Dust on coloring, add, minimum	.47	.34	.81
0860	Floor coloring using 0.6 psf powdered color, 1/2" integral, minimum	5.35	3.96	9.31
0880	Maximum	5.65	3.96	9.61
1020	Integral topping and finish, 1:1:2 mix, 3/16" thick	.11	2.33	2.44
1040	1/2" thick	.28	2.46	2.74
1060	3/4" thick	.43	2.75	3.18
1080	1" thick	.57	3.11	3.68
1100	Terrazzo, minimum	3.16	14.40	17.56
1120	Maximum	5.90	18.05	23.95
1280	Resilient, asphalt tile, 1/8" thick on concrete, minimum	1.24	1.12	2.36
1300	Maximum	1.36	1.12	2.48
1320	On wood, add for felt underlay	.22		.22
1340	Cork tile, minimum	6.35	1.42	7.77
1360	Maximum	13.65	1.42	15.07
1380	Polyethylene, in rolls, minimum	3.36	1.62	4.98
1400	Maximum	6.35	1.62	7.97
1420	Polyurethane, thermoset, minimum	5.05	4.46	9.51
1440	Maximum	6.05	8.95	15
1460	Rubber, sheet goods, minimum	6.90	3.72	10.62
1480	Maximum	11	4.96	15.96
1500	Tile, minimum	6.80	1.12	7.92
1520	Maximum	12.65	1.62	14.27
1540	Synthetic turf, minimum	4.95	2.13	7.08

C3020 Floor Finishes

C3020 410	Tile & Covering	COST PER S.F.		
		MAT.	INST.	TOTAL
1560	Maximum	12.40	2.35	14.75
1580	Vinyl, composition tile, minimum	.96	.89	1.85
1600	Maximum	2.35	.89	3.24
1620	Vinyl tile, minimum	3.25	.89	4.14
1640	Maximum	6.45	.89	7.34
1660	Sheet goods, minimum	3.85	1.79	5.64
1680	Maximum	6.55	2.23	8.78
1720	Tile, ceramic natural clay	5.15	4.36	9.51
1730	Marble, synthetic 12"x12"x5/8"	6.65	13.30	19.95
1740	Porcelain type, minimum	5.20	4.36	9.56
1800	Quarry tile, mud set, minimum	7.75	5.70	13.45
1820	Maximum	10	7.25	17.25
1840	Thin set, deduct		1.14	1.14
1860	Terrazzo precast, minimum	4.57	5.95	10.52
1880	Maximum	9.85	5.95	15.80
1900	Non-slip, minimum	19.45	14.70	34.15
1920	Maximum	21.50	19.60	41.10
2020	Wood block, end grain factory type, natural finish, 2" thick	5.10	3.96	9.06
2040	Fir, vertical grain, 1"x4", no finish, minimum	2.84	1.94	4.78
2060	Maximum	3.01	1.94	4.95
2080	Prefinished white oak, prime grade, 2-1/4" wide	7.05	2.92	9.97
2100	3-1/4" wide	9	2.68	11.68
2120	Maple strip, sanded and finished, minimum	4.51	4.25	8.76
2140	Maximum	5.45	4.25	9.70
2160	Oak strip, sanded and finished, minimum	4.36	4.25	8.61
2180	Maximum	5.25	4.25	9.50
2200	Parquetry, sanded and finished, minimum	5.70	4.43	10.13
2220	Maximum	6.30	6.30	12.60
2260	Add for sleepers on concrete, treated, 24" O.C., 1"x2"	2.50	2.55	5.05
2280	1"x3"	2.28	1.98	4.26
2300	2"x4"	.73	1	1.73
2320	2"x6"	.76	.76	1.52
2340	Underlayment, plywood, 3/8" thick	1.13	.66	1.79
2350	1/2" thick	1.10	.68	1.78
2360	5/8" thick	1.42	.71	2.13
2370	3/4" thick	1.46	.76	2.22
2380	Particle board, 3/8" thick	.45	.66	1.11
2390	1/2" thick	.50	.68	1.18
2400	5/8" thick	.59	.71	1.30
2410	3/4" thick	.72	.76	1.48
2420	Hardboard, 4' x 4', .215" thick	.55	.66	1.21

C3030 Ceiling Finishes

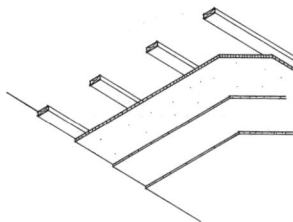

2 Coats of Plaster on Gypsum
Lath on Wood Furring

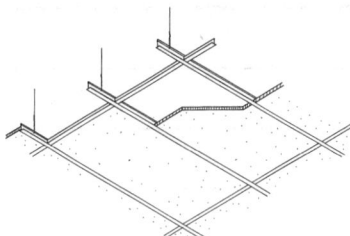

Fiberglass Board on
Exposed Suspended Grid System

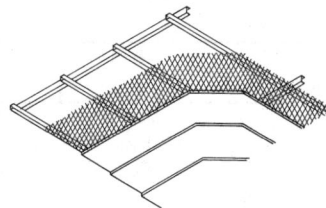

Plaster and Metal Lath
on Metal Furring

System Components			COST PER S.F.		
	QUANTITY	UNIT	MAT.	INST.	TOTAL
SYSTEM C3030 105 2400					
GYPSUM PLASTER, 2 COATS, 3/8″GYP. LATH, WOOD FURRING, FRAMING					
Gypsum plaster 2 coats no lath, on ceilings	.110	S.Y.	.46	2.67	3.13
Gypsum lath plain/perforated nailed, 3/8″ thick	.110	S.Y.	.67	.55	1.22
Add for ceiling installation	.110	S.Y.		.22	.22
Furring, 1″ x 3″ wood strips on wood joists	.750	L.F.	.29	1.07	1.36
Paint, primer and one coat	1.000	S.F.	.11	.43	.54
TOTAL			1.53	4.94	6.47

C3030 105 — Plaster Ceilings

	TYPE	LATH	FURRING	SUPPORT		COST PER S.F. MAT.	INST.	TOTAL
2400	2 coat gypsum	3/8″ gypsum	1″x3″ wood, 16″ O.C.	wood		1.53	4.94	6.47
2500	Painted			masonry		1.53	5.05	6.58
2600				concrete		1.53	5.60	7.13
2700	3 coat gypsum	3.4# metal	1″x3″ wood, 16″ O.C.	wood		1.62	5.35	6.97
2800	Painted			masonry		1.62	5.45	7.07
2900				concrete		1.62	6.05	7.67
3000	2 coat perlite	3/8″ gypsum	1″x3″ wood, 16″ O.C.	wood		1.57	5.20	6.77
3100	Painted			masonry		1.57	5.30	6.87
3200				concrete		1.57	5.90	7.47
3300	3 coat perlite	3.4# metal	1″x3″ wood, 16″ O.C.	wood		1.46	5.30	6.76
3400	Painted			masonry		1.46	5.40	6.86
3500				concrete		1.46	6	7.46
3600	2 coat gypsum	3/8″ gypsum	3/4″ CRC, 12″ O.C.	1-1/2″ CRC, 48″O.C.		1.62	5.85	7.47
3700	Painted		3/4″ CRC, 16″ O.C.	1-1/2″ CRC, 48″O.C.		1.58	5.30	6.88
3800			3/4″ CRC, 24″ O.C.	1-1/2″ CRC, 48″O.C.		1.47	4.87	6.34
3900	2 coat perlite	3/8″ gypsum	3/4″ CRC, 12″ O.C.	1-1/2″ CRC, 48″O.C		1.66	6.35	8.01
4000	Painted		3/4″ CRC, 16″ O.C.	1-1/2″ CRC, 48″O.C.		1.62	5.80	7.42
4100			3/4″ CRC, 24″ O.C.	1-1/2″ CRC, 48″O.C.		1.51	5.35	6.86
4200	3 coat gypsum	3.4# metal	3/4″ CRC, 12″ O.C.	1-1/2″ CRC, 36″ O.C.		2.02	7.80	9.82
4300	Painted		3/4″ CRC, 16″ O.C.	1-1/2″ CRC, 36″ O.C.		1.98	7.25	9.23
4400			3/4″ CRC, 24″ O.C.	1-1/2″ CRC, 36″ O.C.		1.87	6.80	8.67
4500	3 coat perlite	3.4# metal	3/4″ CRC, 12″ O.C.	1-1/2″ CRC,36″ O.C.		2.13	8.50	10.63
4600	Painted		3/4″ CRC, 16″ O.C.	1-1/2″ CRC, 36″ O.C.		2.09	7.95	10.04
4700			3/4″ CRC, 24″ O.C.	1-1/2″ CRC, 36″ O.C.		1.98	7.50	9.48

C30 Interior Finishes

C3030 Ceiling Finishes

| C3030 110 | Drywall Ceilings |

	TYPE	FINISH	FURRING	SUPPORT		COST PER S.F.		
						MAT.	INST.	TOTAL
4800	1/2" F.R. drywall	painted and textured	1"x3" wood, 16" O.C.	wood		.89	3.06	3.95
4900				masonry		.89	3.15	4.04
5000				concrete		.89	3.76	4.65
5100	5/8" F.R. drywall	painted and textured	1"x3" wood, 16" O.C.	wood		.93	3.06	3.99
5200				masonry		.93	3.15	4.08
5300				concrete		.93	3.76	4.69
5400	1/2" F.R. drywall	painted and textured	7/8"resil. channels	24" O.C.		.75	2.96	3.71
5500			1"x2" wood	stud clips		.93	2.85	3.78
5600			1-5/8"metal studs	24" O.C.		.85	2.64	3.49
5700	5/8" F.R. drywall	painted and textured	1-5/8"metal studs	24" O.C.		.89	2.64	3.53
5702								

C3030 Ceiling Finishes

C3030 140	Plaster Ceiling Components	COST PER S.F.		
		MAT.	INST.	TOTAL
0060	Plaster, gypsum incl. finish, 2 coats			
0080	3 coats	.66	3.20	3.86
0100	Perlite, incl. finish, 2 coats	.50	3.14	3.64
0120	3 coats	.77	3.91	4.68
0140	Thin coat on drywall	.09	.62	.71
0200	Lath, gypsum, 3/8" thick	.67	.77	1.44
0220	1/2" thick	.72	.80	1.52
0240	5/8" thick	.40	.94	1.34
0260	Metal, diamond, 2.5 lb.	.41	.62	1.03
0280	3.4 lb.	.56	.67	1.23
0300	Flat rib, 2.75 lb.	.44	.62	1.06
0320	3.4 lb.	.68	.67	1.35
0440	Furring, steel channels, 3/4" galvanized , 12" O.C.	.38	2	2.38
0460	16" O.C.	.34	1.45	1.79
0480	24" O.C.	.23	1	1.23
0500	1-1/2" galvanized , 12" O.C.	.51	2.22	2.73
0520	16" O.C.	.46	1.62	2.08
0540	24" O.C.	.30	1.08	1.38
0560	Wood strips, 1"x3", on wood, 12" O.C.	.38	1.42	1.80
0580	16" O.C.	.29	1.07	1.36
0600	24" O.C.	.19	.71	.90
0620	On masonry, 12" O.C.	.38	1.55	1.93
0640	16" O.C.	.29	1.16	1.45
0660	24" O.C.	.19	.78	.97
0680	On concrete, 12" O.C.	.38	2.36	2.74
0700	16" O.C.	.29	1.77	2.06
0720	24" O.C.	.19	1.18	1.37

C3030 210	Acoustical Ceilings					COST PER S.F.		
	TYPE	TILE	GRID	SUPPORT		MAT.	INST.	TOTAL
5800	5/8" fiberglass board	24" x 48"	tee	suspended		1.72	1.48	3.20
5900		24" x 24"	tee	suspended		1.89	1.62	3.51
6000	3/4" fiberglass board	24" x 48"	tee	suspended		2.78	1.52	4.30
6100		24" x 24"	tee	suspended		2.95	1.66	4.61
6500	5/8" mineral fiber	12" x 12"	1"x3" wood, 12" O.C.	wood		2.42	3.07	5.49
6600				masonry		2.42	3.20	5.62
6700				concrete		2.42	4.01	6.43
6800	3/4" mineral fiber	12" x 12"	1"x3" wood, 12" O.C.	wood		2.44	3.07	5.51
6900				masonry		2.44	3.07	5.51
7000				concrete		2.44	3.07	5.51
7100	3/4"mineral fiber on	12" x 12"	25 ga. channels	runners		2.98	3.77	6.75
7102	5/8" F.R. drywall							
7200	5/8" plastic coated	12" x 12"		adhesive backed		2.45	1.65	4.10
7201	Mineral fiber							
7202								
7300	3/4" plastic coated	12" x 12"		adhesive backed		2.47	1.65	4.12
7301	Mineral fiber							
7302								
7400	3/4" mineral fiber	12" x 12"	conceal 2" bar & channels	suspended		2.66	3.72	6.38
7401								
7402								

C30 Interior Finishes

C3030 Ceiling Finishes

C3030 240	Acoustical Ceiling Components	COST PER S.F.		
		MAT.	INST.	TOTAL
2480	Ceiling boards, eggcrate, acrylic, 1/2" x 1/2" x 1/2" cubes	2	.99	2.99
2500	Polystyrene, 3/8" x 3/8" x 1/2" cubes	1.68	.97	2.65
2520	1/2" x 1/2" x 1/2" cubes	2.24	.99	3.23
2540	Fiberglass boards, plain, 5/8" thick	.80	.79	1.59
2560	3/4" thick	1.86	.83	2.69
2580	Grass cloth faced, 3/4" thick	3.34	.99	4.33
2600	1" thick	2.38	1.02	3.40
2620	Luminous panels, prismatic, acrylic	2.44	1.24	3.68
2640	Polystyrene	1.25	1.24	2.49
2660	Flat or ribbed, acrylic	4.25	1.24	5.49
2680	Polystyrene	2.92	1.24	4.16
2700	Drop pan, white, acrylic	6.25	1.24	7.49
2720	Polystyrene	5.20	1.24	6.44
2740	Mineral fiber boards, 5/8" thick, standard	.86	.73	1.59
2760	Plastic faced	1.35	1.24	2.59
2780	2 hour rating	1.29	.73	2.02
2800	Perforated aluminum sheets, .024 thick, corrugated painted	2.48	1.01	3.49
2820	Plain	4.14	.99	5.13
3080	Mineral fiber, plastic coated, 12" x 12" or 12" x 24", 5/8" thick	2.04	1.65	3.69
3100	3/4" thick	2.06	1.65	3.71
3120	Fire rated, 3/4" thick, plain faced	1.54	1.65	3.19
3140	Mylar faced	1.30	1.65	2.95
3160	Add for ceiling primer	.14		.14
3180	Add for ceiling cement	.41		.41
3240	Suspension system, furring, 1" x 3" wood 12" O.C.	.38	1.42	1.80
3260	T bar suspension system, 2' x 4' grid	.69	.62	1.31
3280	2' x 2' grid	.86	.76	1.62
3300	Concealed Z bar suspension system 12" module	.72	.95	1.67
3320	Add to above for 1-1/2" carrier channels 4' O.C.	.17	1.05	1.22
3340	Add to above for carrier channels for recessed lighting	.30	1.08	1.38

261

D1010 Elevators and Lifts

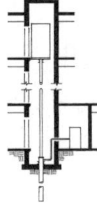

The hydraulic elevator obtains its motion from the movement of liquid under pressure in the piston connected to the car bottom. These pistons can provide travel to a maximum rise of 70′ and are sized for the intended load. As the rise reaches the upper limits the cost tends to exceed that of a geared electric unit.

System Components	QUANTITY	UNIT	COST EACH		
			MAT.	INST.	TOTAL
SYSTEM D1010 110 2000					
PASS. ELEV. HYDRAULIC 2500 LB. 5 FLOORS, 100 FPM					
Passenger elevator, hydraulic, 1500 lb capacity, 2 stop, 100 FPM	1.000	Ea.	43,800	13,500	57,300
Over 10′ travel height, passenger elevator, hydraulic, add	40.000	V.L.F.	39,200	7,440	46,640
Passenger elevator, hydraulic, 2500 lb. capacity over standard, add	1.000	Ea.	3,550		3,550
Over 2 stops, passenger elevator, hydraulic, add	3.000	Stop	8,475	15,000	23,475
Hall lantern	5.000	Ea.	2,975	845	3,820
Maintenance agreement for pass. elev. 9 months	1.000	Ea.	4,000		4,000
Position indicator at lobby	1.000	Ea.	685	42	727
TOTAL			102,685	36,827	139,512

D1010 110	Hydraulic		COST EACH		
			MAT.	INST.	TOTAL
1300	Pass. elev., 1500 lb., 2 Floors, 100 FPM	RD1010 -010	49,000	13,800	62,800
1400	5 Floors, 100 FPM		99,000	36,800	135,800
1600	2000 lb., 2 Floors, 100 FPM		51,000	13,800	64,800
1700	5 floors, 100 FPM		101,500	36,800	138,300
1900	2500 lb., 2 Floors, 100 FPM		52,500	13,800	66,300
2000	5 floors, 100 FPM		102,500	36,800	139,300
2200	3000 lb., 2 Floors, 100 FPM		53,500	13,800	67,300
2300	5 floors, 100 FPM		104,000	36,800	140,800
2500	3500 lb., 2 Floors, 100 FPM		56,000	13,800	69,800
2600	5 floors, 100 FPM		106,000	36,800	142,800
2800	4000 lb., 2 Floors, 100 FPM		57,500	13,800	71,300
2900	5 floors, 100 FPM		107,500	36,800	144,300
3100	4500 lb., 2 Floors, 100 FPM		59,000	13,800	72,800
3200	5 floors, 100 FPM		109,000	36,800	145,800
4000	Hospital elevators, 3500 lb., 2 Floors, 100 FPM		82,500	13,800	96,300
4100	5 floors, 100 FPM		143,000	36,800	179,800
4300	4000 lb., 2 Floors, 100 FPM		82,500	13,800	96,300
4400	5 floors, 100 FPM		143,000	36,800	179,800
4600	4500 lb., 2 Floors, 100 FPM		90,000	13,800	103,800
4800	5 floors, 100 FPM		150,500	36,800	187,300
4900	5000 lb., 2 Floors, 100 FPM		92,500	13,800	106,300
5000	5 floors, 100 FPM		153,000	36,800	189,800
6700	Freight elevators (Class "B"), 3000 lb., 2 Floors, 50 FPM		132,000	17,600	149,600
6800	5 floors, 100 FPM		257,000	46,200	303,200
7000	4000 lb., 2 Floors, 50 FPM		141,000	17,600	158,600
7100	5 floors, 100 FPM		266,500	46,200	312,700
7500	10,000 lb., 2 Floors, 50 FPM		180,000	17,600	197,600
7600	5 floors, 100 FPM		305,000	46,200	351,200
8100	20,000 lb., 2 Floors, 50 FPM		248,000	17,600	265,600
8200	5 Floors, 100 FPM		373,000	46,200	419,200

D1010   Elevators and Lifts

Geared traction elevators are the intermediate group both in cost and in operating areas. These are in buildings of four to fifteen floors and speed ranges from 200′ to 500′ per minute.

System Components	QUANTITY	UNIT	COST EACH		
			MAT.	INST.	TOTAL
SYSTEM D1010 140 1600					
PASSENGER 2500 LB., 5 FLOORS, 200 FPM					
Passenger elevator, geared, 2000 lb. capacity, 4 stop, 200 FPM	1.000	Ea.	100,500	27,000	127,500
Over 40′ travel height, passenger elevator electric, add	10.000	V.L.F.	7,150	1,860	9,010
Passenger elevator, electric, 2500 lb. cap., add	1.000	Ea.	4,200		4,200
Over 4 stops, passenger elevator, electric, add	1.000	Stop	8,600	5,000	13,600
Hall lantern	5.000	Ea.	2,975	845	3,820
Maintenance agreement for passenger elevator, 9 months	1.000	Ea.	4,000		4,000
Position indicator at lobby	1.000	Ea.	685	42	727
TOTAL			128,110	34,747	162,857

D1010 140	Traction Geared Elevators		COST EACH		
			MAT.	INST.	TOTAL
1300	Passenger, 2000 Lb., 5 floors, 200 FPM		124,000	34,700	158,700
1500	15 floors, 350 FPM		300,500	105,000	405,500
1600	2500 Lb., 5 floors, 200 FPM	RD1010 -010	128,000	34,700	162,700
1800	15 floors, 350 FPM		305,000	105,000	410,000
1900	3000 Lb., 5 floors, 200 FPM		129,000	34,700	163,700
2100	15 floors, 350 FPM		305,500	105,000	410,500
2200	3500 Lb., 5 floors, 200 FPM		132,500	34,700	167,200
2400	15 floors, 350 FPM		309,000	105,000	414,000
2500	4000 Lb., 5 floors, 200 FPM		131,500	34,700	166,200
2700	15 floors, 350 FPM		308,500	105,000	413,500
2800	4500 Lb., 5 floors, 200 FPM		134,000	34,700	168,700
3000	15 floors, 350 FPM		310,500	105,000	415,500
3100	5000 Lb., 5 floors, 200 FPM		136,000	34,700	170,700
3300	15 floors, 350 FPM		312,500	105,000	417,500
4000	Hospital, 3500 Lb., 5 floors, 200 FPM		127,000	34,700	161,700
4200	15 floors, 350 FPM		301,500	105,000	406,500
4300	4000 Lb., 5 floors, 200 FPM		127,000	34,700	161,700
4500	15 floors, 350 FPM		301,500	105,000	406,500
4600	4500 Lb., 5 floors, 200 FPM		133,000	34,700	167,700
4800	15 floors, 350 FPM		307,500	105,000	412,500
4900	5000 Lb., 5 floors, 200 FPM		134,500	34,700	169,200
5100	15 floors, 350 FPM		309,000	105,000	414,000
6000	Freight, 4000 Lb., 5 floors, 50 FPM class 'B'		142,000	36,000	178,000
6200	15 floors, 200 FPM class 'B'		357,500	125,000	482,500
6300	8000 Lb., 5 floors, 50 FPM class 'B'		174,500	36,000	210,500
6500	15 floors, 200 FPM class 'B'		390,000	125,000	515,000
7000	10,000 Lb., 5 floors, 50 FPM class 'B'		192,500	36,000	228,500
7200	15 floors, 200 FPM class 'B'		556,000	125,000	681,000
8000	20,000 Lb., 5 floors, 50 FPM class 'B'		222,500	36,000	258,500
8200	15 floors, 200 FPM class 'B'		585,500	125,000	710,500

D1010   Elevators and Lifts

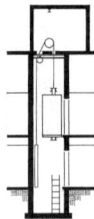

Gearless traction elevators are used in high rise situations where speeds to over 1000 FPM are required to move passengers efficiently. This type of installation is also the most costly.

System Components	QUANTITY	UNIT	COST EACH		
			MAT.	INST.	TOTAL
SYSTEM D1010 150 1700					
PASSENGER, 2500 LB., 10 FLOORS, 200 FPM					
Passenger elevator, geared, 2000 lb capacity, 4 stop 100 FPM	1.000	Ea.	100,500	27,000	127,500
Over 10' travel height, passenger elevator electric, add	60.000	V.L.F.	42,900	11,160	54,060
Passenger elevator, electric, 2500 lb. cap., add	1.000	Ea.	4,200		4,200
Over 2 stops, passenger elevator, electric, add	6.000	Stop	51,600	30,000	81,600
Gearless variable voltage machinery, overhead	1.000	Ea.	116,000	19,300	135,300
Hall lantern	9.000	Ea.	5,950	1,690	7,640
Position indicator at lobby	1.000	Ea.	685	42	727
Fireman services control	1.000	Ea.	12,600	1,125	13,725
Emergency power manual switching, add	1.000	Ea.	550	169	719
Maintenance agreement for passenger elevator, 9 months	1.000	Ea.	4,000		4,000
TOTAL			338,985	90,486	429,471

D1010 150	Traction Gearless Elevators		COST EACH		
			MAT.	INST.	TOTAL
1700	Passenger, 2500 Lb., 10 floors, 200 FPM		339,000	90,500	429,500
1900	30 floors, 600 FPM		705,500	231,000	936,500
2000	3000 Lb., 10 floors, 200 FPM	RD1010 -010	340,000	90,500	430,500
2200	30 floors, 600 FPM		706,500	231,000	937,500
2300	3500 Lb., 10 floors, 200 FPM		343,000	90,500	433,500
2500	30 floors, 600 FPM		710,000	231,000	941,000
2700	50 floors, 800 FPM		1,048,500	371,500	1,420,000
2800	4000 lb., 10 floors, 200 FPM		342,500	90,500	433,000
3000	30 floors, 600 FPM		709,000	231,000	940,000
3200	50 floors, 800 FPM		1,048,000	371,500	1,419,500
3300	4500 lb., 10 floors, 200 FPM		344,500	90,500	435,000
3500	30 floors, 600 FPM		711,500	231,000	942,500
3700	50 floors, 800 FPM		1,050,000	371,500	1,421,500
3800	5000 lb., 10 floors, 200 FPM		346,500	90,500	437,000
4000	30 floors, 600 FPM		713,500	231,000	944,500
4200	50 floors, 800 FPM		1,052,000	371,500	1,423,500
6000	Hospital, 3500 Lb., 10 floors, 200 FPM		336,500	90,500	427,000
6200	30 floors, 600 FPM		699,000	231,000	930,000
6400	4000 Lb., 10 floors, 200 FPM		336,500	90,500	427,000
6600	30 floors, 600 FPM		699,000	231,000	930,000
6800	4500 Lb., 10 floors, 200 FPM		343,000	90,500	433,500
7000	30 floors, 600 FPM		705,000	231,000	936,000
7200	5000 Lb., 10 floors, 200 FPM		344,000	90,500	434,500
7400	30 floors, 600 FPM		706,000	231,000	937,000

D10 Conveying

D1020 Escalators and Moving Walks

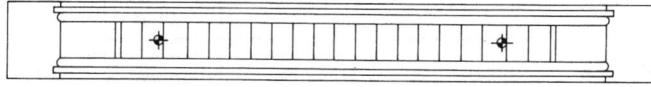

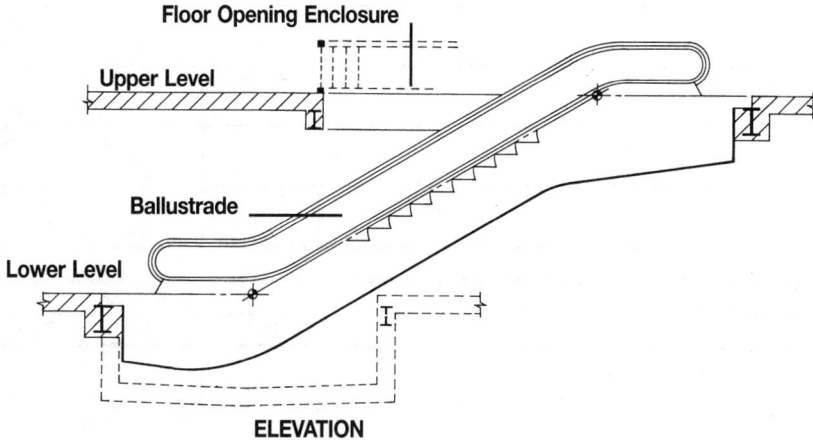

PLAN

Floor Opening Enclosure

Upper Level

Ballustrade

Lower Level

ELEVATION

Moving stairs can be used for buildings where 600 or more people are to be carried to the second floor or beyond. Freight cannot be carried on escalators and at least one elevator must be available for this function.

Carrying capacity is 5000 to 8000 people per hour. Power requirement is 2 KW to 3 KW per hour and incline angle is 30°.

D1020 110				Moving Stairs				
	TYPE	HEIGHT	WIDTH	BALUSTRADE MATERIAL		COST EACH		
						MAT.	INST.	TOTAL
0100	Escalator	10ft	32″	glass		99,500	37,500	137,000
0150				metal		105,500	37,500	143,000
0200			48″	glass		10,800	37,500	48,300
0250				metal		114,500	37,500	152,000
0300		15ft	32″	glass		105,500	43,800	149,300
0350				metal		114,500	43,800	158,300
0400			48″	glass		112,000	43,800	155,800
0450				metal		114,500	43,800	158,300
0500		20ft	32″	glass		111,000	54,000	165,000
0550				metal		118,000	54,000	172,000
0600			48″	glass		121,000	54,000	175,000
0650				metal		126,500	54,000	180,500
0700		25ft	32″	glass		117,500	65,500	183,000
0750				metal		126,500	65,500	192,000
0800			48″	glass		136,500	65,500	202,000
0850				metal		126,500	65,500	192,000

D1020 210				Moving Walks				
	TYPE	STORY HEIGHT	DEGREE SLOPE	WIDTH	COST RANGE	COST PER L.F.		
						MAT.	INST.	TOTAL
1500	Ramp	10′-23′	12	3′-4″	minimum	2,525	740	3,265
1600					maximum	3,200	900	4,100
2000	Walk	0′	0	2′-0″	minimum	940	405	1,345
2500					maximum	1,300	595	1,895
3000				3′-4″		2,125	595	2,720
3500					maximum	2,475	685	3,160

D1090 Other Conveying Systems

D1090 410	Miscellaneous Types	COST EACH		
		MAT.	INST.	TOTAL
1050	Conveyors, Horiz. belt, ctr drive, 60 FPM, cap 40 lb/LF, belt x 26.5'L	3,475	1,950	5,425
1100	24" belt, 42' length	5,150	2,450	7,600
1200	24" belt, 62' length	6,625	3,250	9,875
1500	Inclined belt, horiz. loading and end idler assembly, 16" belt x 27.5'L	10,100	3,250	13,350
1600	24" belt, 27.5' length	11,400	6,500	17,900
3500	Pneumatic tube system, single, 2 station-100', stock economy model, 3" dia.	7,250	9,875	17,125
3600	4" diameter	8,200	13,200	21,400
4000	Twin tube system base cost, minimum	7,025	1,575	8,600
4100	Maximum	13,900	4,750	18,650
4200	Add per LF, of system, minimum	9.05	12.70	21.75
4300	Maximum	26.50	31.50	58
5000	Completely automatic system, 15 to 50 stations, 4" dia., cost/station	22,100	4,075	26,175
5100	50 to 144 stations, 4" diameter, cost per station	17,200	3,700	20,900
7100	Corresp. lift, Self cont., 45 lb. cap., 2 stop, elec. enam. steel doors	9,200	5,600	14,800

D1090 510	Chutes	COST EACH		
		MAT.	INST.	TOTAL
0100	Chutes, linen or refuse incl. sprinklers, galv. steel, 18" diam., per floor	1,225	330	1,555
0110	Aluminized steel, 36" diam., per floor	1,650	415	2,065
0120	Mail, 8-3/4" x 3-1/2", aluminum & glass, per floor	715	232	947
0130	Bronze or stainless, per floor	1,100	258	1,358

D20 Plumbing

D2010 Plumbing Fixtures

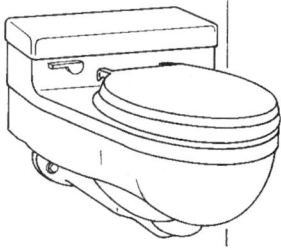

Systems are complete with trim seat and rough-in (supply, waste and vent) for connection to supply branches and waste mains.

One Piece Wall Hung   **Supply**   **Waste/Vent**   **Floor Mount**

System Components	QUANTITY	UNIT	COST EACH		
			MAT.	INST.	TOTAL
SYSTEM D2010 110 1880					
WATER CLOSET, VITREOUS CHINA, ELONGATED					
TANK TYPE, WALL HUNG, TWO PIECE		Ea.			
Water closet, tank type vit china wall hung 2 pc. w/seat supply & stop	1.000	Ea.	610	199	809
Pipe Steel galvanized, schedule 40, threaded, 2" diam.	4.000	L.F.	45.60	65.80	111.40
Pipe, CI soil, no hub, cplg 10' OC, hanger 5' OC, 4" diam.	2.000	L.F.	27.90	36.30	64.20
Pipe, coupling, standard coupling, CI soil, no hub, 4" diam.	2.000	Ea.	45	64	109
Copper tubing type L solder joint, hangar 10' O.C., 1/2" diam.	6.000	L.F.	25.26	43.50	68.76
Wrought copper 90° elbow for solder joints 1/2" diam.	2.000	Ea.	2.70	59	61.70
Wrought copper Tee for solder joints 1/2" diam.	1.000	Ea.	2.32	45	47.32
Supports/carrier, water closet, siphon jet, horiz, single, 4" waste	1.000	Ea.	550	110	660
TOTAL			1,308.78	622.60	1,931.38

D2010 110		Water Closet Systems	COST EACH		
			MAT.	INST.	TOTAL
1800	Water closet, vitreous china, elongated				
1840	Tank type, wall hung				
1880	Close coupled two piece	1,300	625	1,925
1920	Floor mount, one piece	915	660	1,575
1960	One piece low profile	775	660	1,435
2000	Two piece close coupled	535	660	1,195
2040	Bowl only with flush valve				
2080	Wall hung	1,175	705	1,880
2120	Floor mount	720	670	1,390
2160	Floor mount, ADA compliant with 18" high bowl	685	690	1,375

Note: Row 1880 / 1920 include reference box "RD2010 -400".

D2010   Plumbing Fixtures

Systems are complete with trim, seat, flush valve and rough-in (supply, waste and vent) for connection to supply branches and waste mains.

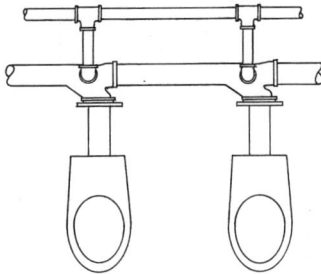

Side by Side

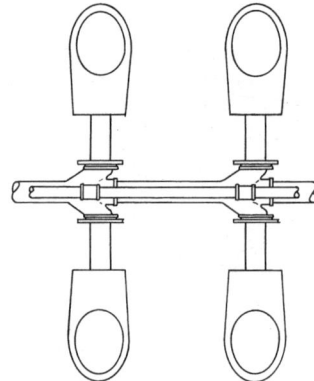

Back to Back

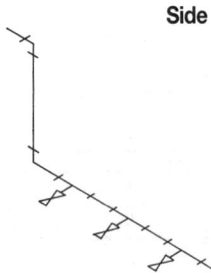

Supply

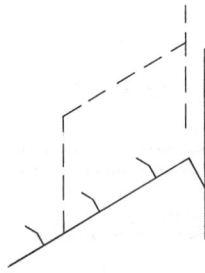

Waste/Vent

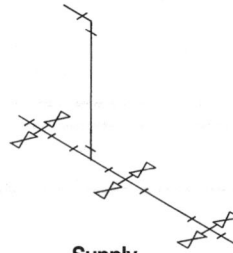

Supply

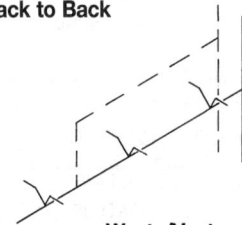

Waste/Vent

System Components	QUANTITY	UNIT	COST EACH		
			MAT.	INST.	TOTAL
SYSTEM D2010 120 1760					
WATER CLOSETS, BATTERY MOUNT, WALL HUNG, SIDE BY SIDE, FIRST CLOSET					
Water closet, bowl only w/flush valve, seat, wall hung	1.000	Ea.	315	182	497
Pipe, CI soil, no hub, cplg 10' OC, hanger 5' OC, 4" diam	3.000	L.F.	41.85	54.45	96.30
Coupling, standard, CI, soil, no hub, 4" diam	2.000	Ea.	45	64	109
Copper tubing type L, solder joints, hangers 10' OC, 1" diam	6.000	L.F.	59.70	51.60	111.30
Copper tubing, type DWV, solder joints, hangers 10'OC, 2" diam	6.000	L.F.	123	79.80	202.80
Wrought copper 90° elbow for solder joints 1" diam	1.000	Ea.	7.45	36.50	43.95
Wrought copper Tee for solder joints, 1" diam	1.000	Ea.	17.25	58.50	75.75
Support/carrier, siphon jet, horiz, adjustable single, 4" pipe	1.000	Ea.	550	110	660
Valve, gate, bronze, 125 lb, NRS, soldered 1" diam	1.000	Ea.	55.50	31	86.50
Wrought copper, DWV, 90° elbow, 2" diam	1.000	Ea.	25.50	58.50	84
TOTAL			1,240.25	726.35	1,966.60

D2010 120	Water Closets, Group		COST EACH		
			MAT.	INST.	TOTAL
1760	Water closets, battery mount, wall hung, side by side, first closet		1,250	725	1,975
1800	Each additional water closet, add		1,150	685	1,835
3000	Back to back, first pair of closets	RD2010 -400	2,075	965	3,040
3100	Each additional pair of closets, back to back		2,025	950	2,975

D2010 Plumbing Fixtures

Systems are complete with trim, flush valve and rough-in (supply, waste and vent) for connection to supply branches and waste mains.

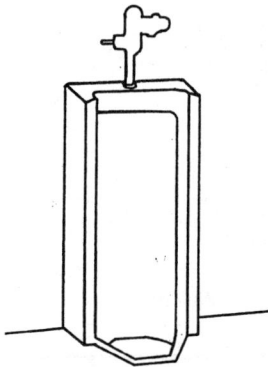

Stall Type

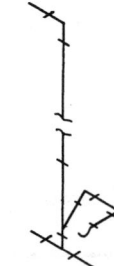

Supply   **Waste/Vent**

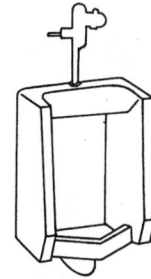

Wall Hung

System Components	QUANTITY	UNIT	COST EACH		
			MAT.	INST.	TOTAL
SYSTEM D2010 210 2000					
URINAL, VITREOUS CHINA, WALL HUNG					
Urinal, wall hung, vitreous china, incl. hanger	1.000	Ea.	273	350	623
Pipe, steel, galvanized, schedule 40, threaded, 1-1/2" diam.	5.000	L.F.	42.75	65.75	108.50
Copper tubing type DWV, solder joint, hangers 10' OC, 2" diam.	3.000	L.F.	61.50	39.90	101.40
Combination Y & 1/8 bend for CI soil pipe, no hub, 3" diam.	1.000	Ea.	15.65		15.65
Pipe, CI, no hub, cplg. 10' OC, hanger 5' OC, 3" diam.	4.000	L.F.	43.20	65.80	109
Pipe coupling standard, CI soil, no hub, 3" diam.	3.000	Ea.	38	55	93
Copper tubing type L, solder joint, hanger 10' OC 3/4" diam.	5.000	L.F.	33	38.50	71.50
Wrought copper 90° elbow for solder joints 3/4" diam.	1.000	Ea.	3.04	31	34.04
Wrought copper Tee for solder joints, 3/4" diam.	1.000	Ea.	5.60	49	54.60
TOTAL			515.74	694.95	1,210.69

D2010 210	Urinal Systems	COST EACH		
		MAT.	INST.	TOTAL
2000	Urinal, vitreous china, wall hung	515	695	1,210
2040	Stall type	1,025	830	1,855

D20 Plumbing

D2010 Plumbing Fixtures

Systems are complete with trim and rough-in (supply, waste and vent) to connect to supply branches and waste mains.

Vanity Top

Supply   Waste/Vent

Wall Hung

System Components	QUANTITY	UNIT	COST EACH MAT.	COST EACH INST.	COST EACH TOTAL
SYSTEM D2010 310 1560					
LAVATORY W/TRIM, VANITY TOP, P.E. ON C.I., 20" X 18"					
Lavatory w/trim, PE on CI, white, vanity top, 20" x 18" oval	1.000	Ea.	254	165	419
Pipe, steel, galvanized, schedule 40, threaded, 1-1/4" diam.	4.000	L.F.	28.80	47.40	76.20
Copper tubing type DWV, solder joint, hanger 10' OC 1-1/4" diam.	4.000	L.F.	48	39	87
Wrought copper DWV, Tee, sanitary, 1-1/4" diam.	1.000	Ea.	26	65	91
P trap w/cleanout, 20 ga., 1-1/4" diam.	1.000	Ea.	79	32.50	111.50
Copper tubing type L, solder joint, hanger 10' OC 1/2" diam.	10.000	L.F.	42.10	72.50	114.60
Wrought copper 90° elbow for solder joints 1/2" diam.	2.000	Ea.	2.70	59	61.70
Wrought copper Tee for solder joints, 1/2" diam.	2.000	Ea.	4.64	90	94.64
Stop, chrome, angle supply, 1/2" diam.	2.000	Ea.	16.20	53	69.20
TOTAL			501.44	623.40	1,124.84

D2010 310	Lavatory Systems		COST EACH MAT.	COST EACH INST.	COST EACH TOTAL
1560	Lavatory w/trim, vanity top, PE on CI, 20" x 18", Vanity top by others.		500	625	1,125
1600	19" x 16" oval		420	625	1,045
1640	18" round	RD2010	485	625	1,110
1680	Cultured marble, 19" x 17"	-400	435	625	1,060
1720	25" x 19"		445	625	1,070
1760	Stainless, self-rimming, 25" x 22"		605	625	1,230
1800	17" x 22"		595	625	1,220
1840	Steel enameled, 20" x 17"		420	640	1,060
1880	19" round		420	640	1,060
1920	Vitreous china, 20" x 16"		530	655	1,185
1960	19" x 16"		535	655	1,190
2000	22" x 13"		540	655	1,195
2040	Wall hung, PE on CI, 18" x 15"		760	690	1,450
2080	19" x 17"		775	690	1,465
2120	20" x 18"		725	690	1,415
2160	Vitreous china, 18" x 15"		655	705	1,360
2200	19" x 17"		590	705	1,295
2240	24" x 20"		925	705	1,630

D2010 Plumbing Fixtures

Systems are complete with trim and rough-in (supply, waste and vent) to connect to supply branches and waste mains.

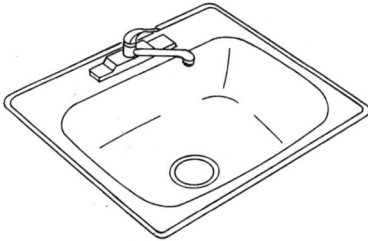

Countertop Single Bowl

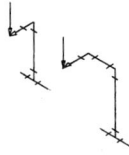

Supply

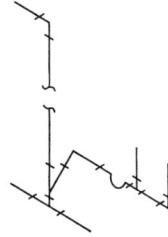

Waste/Vent

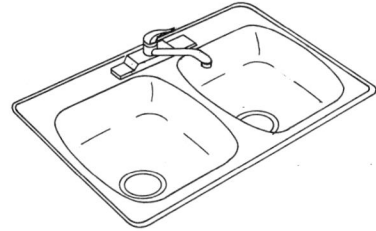

Countertop Double Bowl

System Components	QUANTITY	UNIT	COST EACH		
			MAT.	INST.	TOTAL
SYSTEM D2010 410 1720					
KITCHEN SINK W/TRIM, COUNTERTOP, P.E. ON C.I., 24" X 21", SINGLE BOWL					
Kitchen sink, countertop, PE on CI, 1 bowl, 24" x 21" OD	1.000	Ea.	299	188	487
Pipe, steel, galvanized, schedule 40, threaded, 1-1/4" diam.	4.000	L.F.	28.80	47.40	76.20
Copper tubing, type DWV, solder, hangers 10' OC 1-1/2" diam.	6.000	L.F.	90.60	65.10	155.70
Wrought copper, DWV, Tee, sanitary, 1-1/2" diam.	1.000	Ea.	32	73	105
P trap, standard, copper, 1-1/2" diam.	1.000	Ea.	76.50	34.50	111
Copper tubing, type L, solder joints, hangers 10' OC 1/2" diam.	10.000	L.F.	42.10	72.50	114.60
Wrought copper 90° elbow for solder joints 1/2" diam.	2.000	Ea.	2.70	59	61.70
Wrought copper Tee for solder joints, 1/2" diam.	2.000	Ea.	4.64	90	94.64
Stop, angle supply, chrome, 1/2" CTS	2.000	Ea.	16.20	53	69.20
TOTAL			592.54	682.50	1,275.04

D2010 410	Kitchen Sink Systems	COST EACH		
		MAT.	INST.	TOTAL
1720	Kitchen sink w/trim, countertop, PE on CI, 24"x21", single bowl	595	685	1,280
1760	30" x 21" single bowl	600	685	1,285
1800	32" x 21" double bowl	735	735	1,470
1840	42" x 21" double bowl	890	745	1,635
1880	Stainless steel, 19" x 18" single bowl	855	685	1,540
1920	25" x 22" single bowl	920	685	1,605
1960	33" x 22" double bowl	1,225	735	1,960
2000	43" x 22" double bowl	1,400	745	2,145
2040	44" x 22" triple bowl	1,425	775	2,200
2080	44" x 24" corner double bowl	1,075	745	1,820
2120	Steel, enameled, 24" x 21" single bowl	455	685	1,140
2160	32" x 21" double bowl	505	735	1,240
2240	Raised deck, PE on CI, 32" x 21", dual level, double bowl	660	930	1,590
2280	42" x 21" dual level, triple bowl	1,250	1,025	2,275

D20 Plumbing

D2010 Plumbing Fixtures

Systems are complete with trim and rough-in (supply, waste and vent) to connect to supply branches and waste mains.

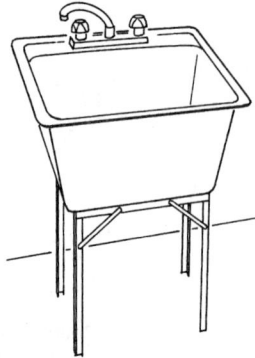

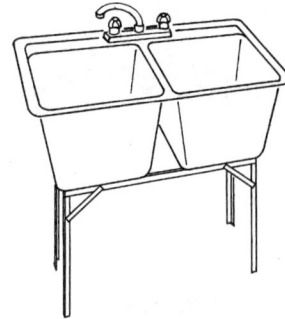

Single Compartment Sink   **Supply**   **Waste/Vent**   **Double Compartment Sink**

System Components	QUANTITY	UNIT	MAT.	INST.	TOTAL
SYSTEM D2010 420 1760					
LAUNDRY SINK W/TRIM, PE ON CI, BLACK IRON FRAME					
24″ X 20″ OD, SINGLE COMPARTMENT					
Laundry sink PE on CI w/trim & frame, 24″ x 21″ OD, 1 compartment	1.000	Ea.	455	176	631
Pipe, steel, galvanized, schedule 40, threaded, 1-1/4″ diam	4.000	L.F.	28.80	47.40	76.20
Copper tubing, type DWV, solder joint, hanger 10′ OC 1-1/2″diam	6.000	L.F.	90.60	65.10	155.70
Wrought copper, DWV, Tee, sanitary, 1-1/2″ diam	1.000	Ea.	32	73	105
P trap, standard, copper, 1-1/2″ diam	1.000	Ea.	76.50	34.50	111
Copper tubing type L, solder joints, hangers 10′ OC, 1/2″ diam	10.000	L.F.	42.10	72.50	114.60
Wrought copper 90° elbow for solder joints 1/2″ diam	2.000	Ea.	2.70	59	61.70
Wrought copper Tee for solder joints, 1/2″ diam	2.000	Ea.	4.64	90	94.64
Stop, angle supply, 1/2″ diam	2.000	Ea.	16.20	53	69.20
TOTAL			748.54	670.50	1,419.04

D2010 420	Laundry Sink Systems	MAT.	INST.	TOTAL
1740	Laundry sink w/trim, PE on CI, black iron frame			
1760	24″ x 20″, single compartment	750	670	1,420
1800	24″ x 23″ single compartment	740	670	1,410
1840	48″ x 21″ double compartment	1,225	725	1,950
1920	Molded stone, on wall, 22″ x 21″ single compartment	450	670	1,120
1960	45″x 21″ double compartment	560	725	1,285
2040	Plastic, on wall or legs, 18″ x 23″ single compartment	405	655	1,060
2080	20″ x 24″ single compartment	440	655	1,095
2120	36″ x 23″ double compartment	500	710	1,210
2160	40″ x 24″ double compartment	585	710	1,295

RD2010-400

D20 Plumbing

D2010 Plumbing Fixtures

Corrosion resistant laboratory sink systems are complete with trim and rough-in (supply, waste and vent) to connect to supply branches and waste mains.

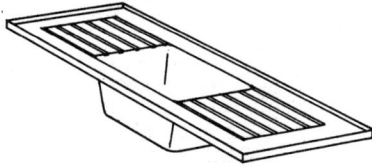

Laboratory Sink

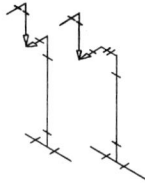

Supply

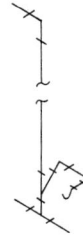

Waste/Vent

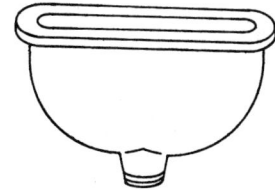

Polypropylene Cup Sink

System Components	QUANTITY	UNIT	COST EACH MAT.	COST EACH INST.	COST EACH TOTAL
SYSTEM D2010 430 1600					
LABORATORY SINK W/TRIM, POLYETHYLENE, SINGLE BOWL					
DOUBLE DRAINBOARD, 54″ X 24″ OD					
Sink w/trim, polyethylene, 1 bowl, 2 drainboards 54″ x 24″ OD	1.000	Ea.	1,150	350	1,500
Pipe, polypropylene, schedule 40, acid resistant 1-1/2″ diam	10.000	L.F.	55	155	210
Tee, sanitary, polypropylene, acid resistant, 1-1/2″ diam	1.000	Ea.	13.20	58.50	71.70
P trap, polypropylene, acid resistant, 1-1/2″ diam	1.000	Ea.	43	34.50	77.50
Copper tubing type L, solder joint, hanger 10′ OC 1/2″ diam	10.000	L.F.	42.10	72.50	114.60
Wrought copper 90° elbow for solder joints 1/2″ diam	2.000	Ea.	2.70	59	61.70
Wrought copper Tee for solder joints, 1/2″ diam	2.000	Ea.	4.64	90	94.64
Stop, angle supply, chrome, 1/2″ diam	2.000	Ea.	16.20	53	69.20
TOTAL			1,326.84	872.50	2,199.34

D2010 430	Laboratory Sink Systems		COST EACH MAT.	COST EACH INST.	COST EACH TOTAL
1580	Laboratory sink w/trim, polyethylene, single bowl,				
1600	Double drainboard, 54″ x 24″ O.D.		1,325	875	2,200
1640	Single drainboard, 47″ x 24″O.D.	RD2010 -400	1,525	875	2,400
1680	70″ x 24″ O.D.		1,400	875	2,275
1760	Flanged, 14-1/2″ x 14-1/2″ O.D.		400	785	1,185
1800	18-1/2″ x 18-1/2″ O.D.		490	785	1,275
1840	23-1/2″ x 20-1/2″ O.D.		530	785	1,315
1920	Polypropylene, cup sink, oval, 7″ x 4″ O.D.		282	695	977
1960	10″ x 4-1/2″ O.D.		266	695	961

D2010 Plumbing Fixtures

Corrosion resistant laboratory sink systems are complete with trim and rough–in (supply, waste and vent) to connect to supply branches and waste mains.

Wall Hung

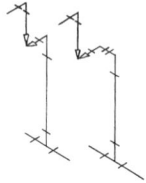

Supply

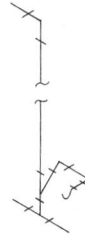

Waste/Vent

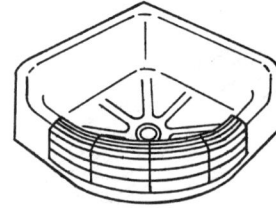

Corner, Floor

System Components	QUANTITY	UNIT	COST EACH		
			MAT.	INST.	TOTAL
SYSTEM D2010 440 4260					
SERVICE SINK, PE ON CI, CORNER FLOOR, 28"X28", W/RIM GUARD & TRIM					
Service sink, corner floor, PE on CI, 28" x 28", w/rim guard & trim	1.000	Ea.	690	239	929
Copper tubing type DWV, solder joint, hanger 10'OC 3" diam	6.000	L.F.	216	108.90	324.90
Copper tubing type DWV, solder joint, hanger 10'OC 2" diam	4.000	L.F.	82	53.20	135.20
Wrought copper DWV, Tee, sanitary, 3" diam.	1.000	Ea.	139	151	290
P trap with cleanout & slip joint, copper 3" diam	1.000	Ea.	284	53	337
Copper tubing, type L, solder joints, hangers 10' OC, 1/2" diam	10.000	L.F.	42.10	72.50	114.60
Wrought copper 90° elbow for solder joints 1/2" diam	2.000	Ea.	2.70	59	61.70
Wrought copper Tee for solder joints, 1/2" diam	2.000	Ea.	4.64	90	94.64
Stop, angle supply, chrome, 1/2" diam	2.000	Ea.	16.20	53	69.20
TOTAL			1,476.64	879.60	2,356.24

D2010 440	Service Sink Systems	COST EACH		
		MAT.	INST.	TOTAL
4260	Service sink w/trim, PE on CI, corner floor, 28" x 28", w/rim guard	1,475	880	2,355
4300	Wall hung w/rim guard, 22" x 18"	1,750	1,025	2,775
4340	24" x 20"	1,800	1,025	2,825
4380	Vitreous china, wall hung 22" x 20"	1,625	1,025	2,650

D20 Plumbing

D2010 Plumbing Fixtures

Systems are complete with trim and rough-in (supply, waste and vent) to connect to supply branches and waste mains.

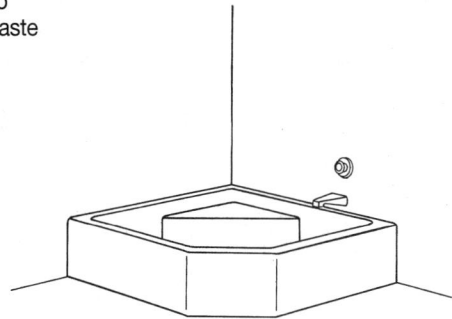

Recessed Bathtub **Supply** **Waste/Vent** **Corner Bathtub**

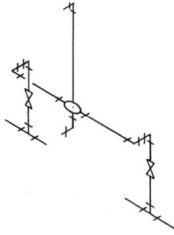

System Components	QUANTITY	UNIT	COST EACH		
			MAT.	INST.	TOTAL
SYSTEM D2010 510 2000					
BATHTUB, RECESSED, PORCELAIN ENAMEL ON CAST IRON,, 48″ x 42″					
Bath tub, porcelain enamel on cast iron, w/fittings, 48″ x 42″	1.000	Ea.	1,975	263	2,238
Pipe, steel, galvanized, schedule 40, threaded, 1-1/4″ diam	4.000	L.F.	28.80	47.40	76.20
Pipe, CI no hub soil w/couplings 10′ OC, hangers 5′ OC, 4″ diam	3.000	L.F.	41.85	54.45	96.30
Combination Y and 1/8 bend for C.I. soil pipe, no hub, 4″ pipe size	1.000	Ea.	37.50		37.50
Drum trap, 3″ x 5″, copper, 1-1/2″ diam	1.000	Ea.	107	36.50	143.50
Copper tubing type L, solder joints, hangers 10′ OC 1/2″ diam	10.000	L.F.	42.10	72.50	114.60
Wrought copper 90° elbow, solder joints, 1/2″ diam	2.000	Ea.	2.70	59	61.70
Wrought copper Tee, solder joints, 1/2″ diam	2.000	Ea.	4.64	90	94.64
Stop, angle supply, 1/2″ diameter	2.000	Ea.	16.20	53	69.20
Copper tubing type DWV, solder joints, hanger 10′ OC 1-1/2″ diam	3.000	L.F.	45.30	32.55	77.85
Pipe coupling, standard, C.I. soil no hub, 4″ pipe size	2.000	Ea.	45	64	109
TOTAL			2,346.09	772.40	3,118.49

D2010 510	Bathtub Systems		COST EACH		
			MAT.	INST.	TOTAL
2000	Bathtub, recessed, P.E. on Cl., 48″ x 42″	RD2010 -400	2,350	770	3,120
2040	72″ x 36″		2,425	860	3,285
2080	Mat bottom, 5′ long		1,250	750	2,000
2120	5′-6″ long		1,750	770	2,520
2160	Corner, 48″ x 42″		2,350	750	3,100
2200	Formed steel, enameled, 4′-6″ long		770	690	1,460

D2010 Plumbing Fixtures

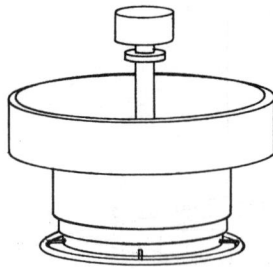

Systems are complete with trim, flush valve and rough-in (supply, waste and vent) for connection to supply branches and waste mains.

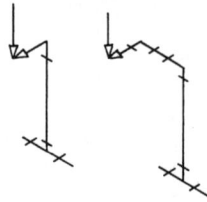

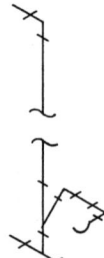

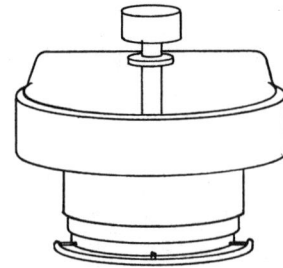

Circular Fountain

Supply

Waste/Vent

Semi-Circular Fountain

System Components	QUANTITY	UNIT	COST EACH		
			MAT.	INST.	TOTAL
SYSTEM D2010 610 1760					
GROUP WASH FOUNTAIN, PRECAST TERRAZZO					
CIRCULAR, 36" DIAMETER					
Wash fountain, group, precast terrazzo, foot control 36" diam	1.000	Ea.	4,475	545	5,020
Copper tubing type DWV, solder joint, hanger 10' OC, 2" diam.	10.000	L.F.	205	133	338
P trap, standard, copper, 2" diam.	1.000	Ea.	118	39	157
Wrought copper, Tee, sanitary, 2" diam.	1.000	Ea.	37.50	83.50	121
Copper tubing type L, solder joint, hanger 10' OC 1/2" diam.	20.000	L.F.	84.20	145	229.20
Wrought copper 90° elbow for solder joints 1/2" diam.	3.000	Ea.	4.05	88.50	92.55
Wrought copper Tee for solder joints, 1/2" diam.	2.000	Ea.	4.64	90	94.64
TOTAL			4,928.39	1,124	6,052.39

D2010 610	Group Wash Fountain Systems	COST EACH		
		MAT.	INST.	TOTAL
1740	Group wash fountain, precast terrazzo			
1760	Circular, 36" diameter	4,925	1,125	6,050
1800	54" diameter	6,025	1,225	7,250
1840	Semi-circular, 36" diameter	4,575	1,125	5,700
1880	54" diameter	5,425	1,225	6,650
1960	Stainless steel, circular, 36" diameter	4,825	1,050	5,875
2000	54" diameter	6,125	1,175	7,300
2040	Semi-circular, 36" diameter	4,275	1,050	5,325
2080	54" diameter	5,350	1,175	6,525
2160	Thermoplastic, circular, 36" diameter	3,725	850	4,575
2200	54" diameter	4,275	990	5,265
2240	Semi-circular, 36" diameter	3,475	850	4,325
2280	54" diameter	4,150	990	5,140

D2010 Plumbing Fixtures

Systems are complete with trim
and rough-in (supply, waste and
vent) for connection to supply branches
and waste mains.

Three Wall

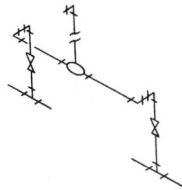

Supply

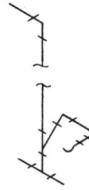

Waste/Vent

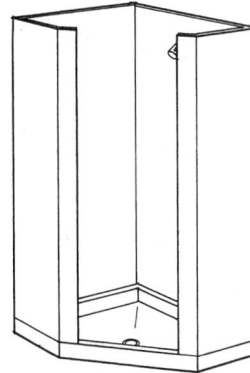

Corner Angle

System Components	QUANTITY	UNIT	COST EACH		
			MAT.	INST.	TOTAL
SYSTEM D2010 710 1560					
SHOWER, STALL, BAKED ENAMEL, MOLDED STONE RECEPTOR, 30″ SQUARE					
Shower stall, enameled steel, molded stone receptor, 30″ square	1.000	Ea.	805	203	1,008
Copper tubing type DWV, solder joints, hangers 10′ OC, 2″ diam.	6.000	L.F.	90.60	65.10	155.70
Wrought copper DWV, Tee, sanitary, 2″ diam.	1.000	Ea.	32	73	105
Trap, standard, copper, 2″ diam.	1.000	Ea.	76.50	34.50	111
Copper tubing type L, solder joint, hanger 10′ OC 1/2″ diam.	16.000	L.F.	67.36	116	183.36
Wrought copper 90° elbow for solder joints 1/2″ diam.	3.000	Ea.	4.05	88.50	92.55
Wrought copper Tee for solder joints, 1/2″ diam.	2.000	Ea.	4.64	90	94.64
Stop and waste, straightway, bronze, solder joint 1/2″ diam	2.000	Ea.	6.18	49	55.18
TOTAL			1,086.33	719.10	1,805.43

D2010 710	Shower Systems		COST EACH		
			MAT.	INST.	TOTAL
1560	Shower, stall, baked enamel, molded stone receptor, 30″ square		1,075	720	1,795
1600	32″ square		680	725	1,405
1640	Terrazzo receptor, 32″ square	RD2010 -400	1,100	725	1,825
1680	36″ square		1,550	735	2,285
1720	36″ corner angle		1,675	735	2,410
1800	Fiberglass one piece, three walls, 32″ square		790	710	1,500
1840	36″ square		855	710	1,565
1880	Polypropylene, molded stone receptor, 30″ square		830	1,050	1,880
1920	32″ square		845	1,050	1,895
1960	Built-in head, arm, bypass, stops and handles		254	271	525

D2010 Plumbing Fixtures

Systems are complete with trim and rough-in (supply, waste and vent) to connect to supply branches and waste mains.

Wall Mounted, No Back

Supply   **Waste/Vent**

Wall Mounted, Low Back

System Components	QUANTITY	UNIT	COST EACH		
			MAT.	INST.	TOTAL
SYSTEM D2010 810 1800					
DRINKING FOUNTAIN, ONE BUBBLER, WALL MOUNTED					
NON RECESSED, BRONZE, NO BACK					
Drinking fountain, wall mount, bronze, 1 bubbler	1.000	Ea.	1,025	146	1,171
Copper tubing, type L, solder joint, hanger 10' OC 3/8" diam	5.000	L.F.	17.90	34.75	52.65
Stop, supply, straight, chrome, 3/8" diam	1.000	Ea.	6.75	24.50	31.25
Wrought copper 90° elbow for solder joints 3/8" diam.	1.000	Ea.	4.06	26.50	30.56
Wrought copper Tee for solder joints, 3/8" diam.	1.000	Ea.	6.85	42	48.85
Copper tubing, type DWV, solder joint, hanger 10' OC 1-1/4" diam	4.000	L.F.	48	39	87
P trap, standard, copper drainage, 1-1/4" diam	1.000	Ea.	79	32.50	111.50
Wrought copper, DWV, Tee, sanitary, 1-1/4" diam	1.000	Ea.	26	65	91
TOTAL			1,213.56	410.25	1,623.81

D2010 810	Drinking Fountain Systems		COST EACH		
			MAT.	INST.	TOTAL
1740	Drinking fountain, one bubbler, wall mounted				
1760	Non recessed				
1800	Bronze, no back	RD2010 -400	1,225	410	1,635
1840	Cast iron, enameled, low back		830	410	1,240
1880	Fiberglass, 12" back		1,475	410	1,885
1920	Stainless steel, no back		1,525	410	1,935
1960	Semi-recessed, poly marble		1,050	410	1,460
2040	Stainless steel		1,175	410	1,585
2080	Vitreous china		860	410	1,270
2120	Full recessed, poly marble		1,175	410	1,585
2200	Stainless steel		1,150	410	1,560
2240	Floor mounted, pedestal type, aluminum		770	555	1,325
2320	Bronze		1,700	555	2,255
2360	Stainless steel		1,700	555	2,255

D2010 Plumbing Fixtures

Systems are complete with trim and rough-in (supply, waste and vent) for connection to supply branches and waste mains.

Wall Hung

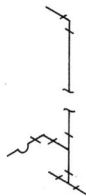

Supply

Waste/Vent

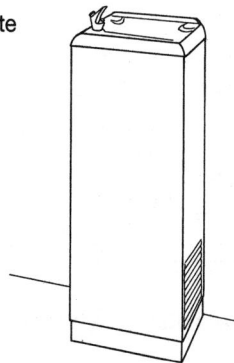

Floor Mounted

System Components	QUANTITY	UNIT	COST EACH		
			MAT.	INST.	TOTAL
SYSTEM D2010 820 1840					
WATER COOLER, ELECTRIC, SELF CONTAINED, WALL HUNG, 8.2 GPH					
Water cooler, wall mounted, 8.2 GPH	1.000	Ea.	710	263	973
Copper tubing type DWV, solder joint, hanger 10' OC 1-1/4" diam.	4.000	L.F.	48	39	87
Wrought copper DWV, Tee, sanitary 1-1/4" diam.	1.000	Ea.	26	65	91
P trap, copper drainage, 1-1/4" diam.	1.000	Ea.	79	32.50	111.50
Copper tubing type L, solder joint, hanger 10' OC 3/8" diam.	5.000	L.F.	17.90	34.75	52.65
Wrought copper 90° elbow for solder joints 3/8" diam.	1.000	Ea.	4.06	26.50	30.56
Wrought copper Tee for solder joints, 3/8" diam.	1.000	Ea.	6.85	42	48.85
Stop and waste, straightway, bronze, solder, 3/8" diam.	1.000	Ea.	6.75	24.50	31.25
TOTAL			898.56	527.25	1,425.81

D2010 820	Water Cooler Systems		COST EACH		
			MAT.	INST.	TOTAL
1840	Water cooler, electric, wall hung, 8.2 GPH		900	525	1,425
1880	Dual height, 14.3 GPH		1,275	540	1,815
1920	Wheelchair type, 7.5 G.P.H.	RD2010 -400	2,000	525	2,525
1960	Semi recessed, 8.1 G.P.H.		1,250	525	1,775
2000	Full recessed, 8 G.P.H.		1,875	565	2,440
2040	Floor mounted, 14.3 G.P.H.		975	460	1,435
2080	Dual height, 14.3 G.P.H.		1,350	555	1,905
2120	Refrigerated compartment type, 1.5 G.P.H.		1,600	460	2,060

D2010 Plumbing Fixtures

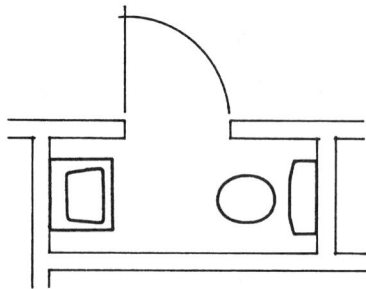

Two Fixture Bathroom Systems consisting of a lavatory, water closet, and rough-in service piping.

• Prices for plumbing and fixtures only.

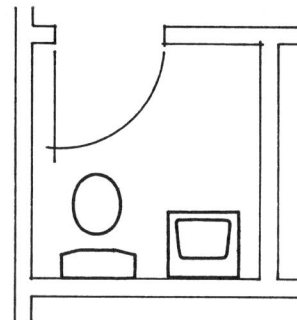

*Common wall is with an adjacent bathroom

System Components	QUANTITY	UNIT	COST EACH		
			MAT.	INST.	TOTAL
SYSTEM D2010 920 1180					
BATHROOM, LAVATORY & WATER CLOSET, 2 WALL PLUMBING, STAND ALONE					
Water closet, 2 Pc. close cpld. vit .china flr. mntd. w/seat, supply & stop	1.000	Ea.	203	199	402
Water closet, rough-in waste & vent	1.000	Set	300	345	645
Lavatory w/ftngs., wall hung, white, PE on CI, 20" x 18"	1.000	Ea.	325	132	457
Lavatory, rough-in waste & vent	1.000	Set	400	635	1,035
Copper tubing type L, solder joint, hanger 10' OC 1/2" diam.	10.000	L.F.	42.10	72.50	114.60
Pipe, steel, galvanized, schedule 40, threaded, 2" diam.	12.000	L.F.	136.80	197.40	334.20
Pipe, CI soil, no hub, coupling 10' OC, hanger 5' OC, 4" diam.	7.000	L.F.	88.55	134.05	222.60
TOTAL			1,495.45	1,714.95	3,210.40

D2010 920	Two Fixture Bathroom, Two Wall Plumbing	COST EACH		
		MAT.	INST.	TOTAL
1180	Bathroom, lavatory & water closet, 2 wall plumbing, stand alone	1,500	1,725	3,225
1200	Share common plumbing wall*	1,350	1,475	2,825

D2010 922	Two Fixture Bathroom, One Wall Plumbing	COST EACH		
		MAT.	INST.	TOTAL
2220	Bathroom, lavatory & water closet, one wall plumbing, stand alone	1,375	1,550	2,925
2240	Share common plumbing wall*	1,225	1,300	2,525

D2010 Plumbing Fixtures

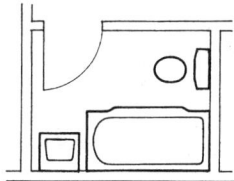

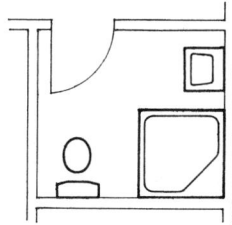

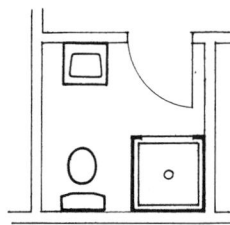

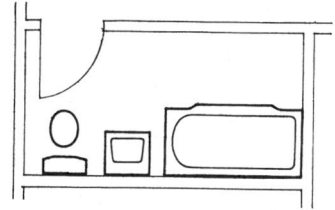

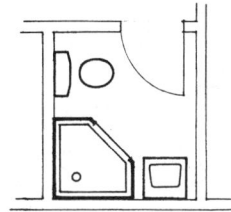

Three Fixture Bathroom Systems consisting of a lavatory, water closet, bathtub or shower and rough-in service piping.

- Prices for plumbing and fixtures only.

*Common wall is with an adjacent bathroom

System Components			COST EACH		
	QUANTITY	UNIT	MAT.	INST.	TOTAL
SYSTEM D2010 924 1170					
BATHROOM, LAVATORY, WATER CLOSET & BATHTUB					
ONE WALL PLUMBING, STAND ALONE					
Wtr closet, 2 pc close cpld vit china flr mntd w/seat supply & stop	1.000	Ea.	203	199	402
Water closet, rough-in waste & vent	1.000	Set	300	345	645
Lavatory w/ftngs, wall hung, white, PE on CI, 20" x 18"	1.000	Ea.	325	132	457
Lavatory, rough-in waste & vent	1.000	Set	400	635	1,035
Bathtub, white PE on CI, w/ftgs, mat bottom, recessed, 5' long	1.000	Ea.	890	239	1,129
Baths, rough-in waste and vent	1.000	Set	297	459	756
TOTAL			2,415	2,009	4,424

D2010 924	Three Fixture Bathroom, One Wall Plumbing	COST EACH		
		MAT.	INST.	TOTAL
1150	Bathroom, three fixture, one wall plumbing			
1160	Lavatory, water closet & bathtub			
1170	Stand alone	2,425	2,000	4,425
1180	Share common plumbing wall *	2,075	1,450	3,525

D2010 926	Three Fixture Bathroom, Two Wall Plumbing	COST EACH		
		MAT.	INST.	TOTAL
2130	Bathroom, three fixture, two wall plumbing			
2140	Lavatory, water closet & bathtub			
2160	Stand alone	2,425	2,025	4,450
2180	Long plumbing wall common *	2,175	1,625	3,800
3610	Lavatory, bathtub & water closet			
3620	Stand alone	2,650	2,325	4,975
3640	Long plumbing wall common *	2,475	2,100	4,575
4660	Water closet, corner bathtub & lavatory			
4680	Stand alone	3,525	2,050	5,575
4700	Long plumbing wall common *	3,200	1,550	4,750
6100	Water closet, stall shower & lavatory			
6120	Stand alone	2,275	2,300	4,575
6140	Long plumbing wall common *	2,125	2,125	4,250
7060	Lavatory, corner stall shower & water closet			
7080	Stand alone	3,075	2,050	5,125
7100	Short plumbing wall common *	2,525	1,375	3,900

D2010 Plumbing Fixtures

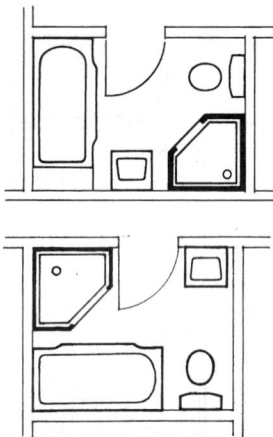

Four Fixture Bathroom Systems consisting of a lavatory, water closet, bathtub, shower and rough-in service piping.

- Prices for plumbing and fixtures only.

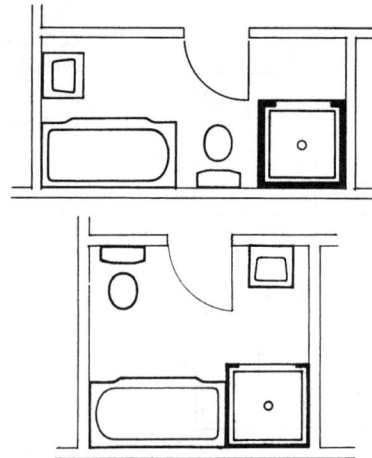

*Common wall is with an adjacent bathroom

System Components	QUANTITY	UNIT	COST EACH		
			MAT.	INST.	TOTAL
SYSTEM D2010 928 1160					
BATHROOM, BATHTUB, WATER CLOSET, STALL SHOWER & LAVATORY					
TWO WALL PLUMBING, STAND ALONE					
Wtr closet, 2 pc close cpld vit china flr mntd w/seat supply & stop	1.000	Ea.	203	199	402
Water closet, rough-in waste & vent	1.000	Set	300	345	645
Lavatory w/ftngs, wall hung, white PE on CI, 20" x 18"	1.000	Ea.	325	132	457
Lavatory, rough-in waste & vent	1.000	Set	40	63.50	103.50
Bathtub, white PE on CI, w/ftgs, mat bottom, recessed, 5' long	1.000	Ea.	890	239	1,129
Baths, rough-in waste and vent	1.000	Set	330	510	840
Shower stall, bkd enam, molded stone receptor, door & trim 32" sq.	1.000	Ea.	400	211	611
Shower stall, rough-in supply, waste & vent	1.000	Set	435	515	950
TOTAL			2,923	2,214.50	5,137.50

D2010 928	Four Fixture Bathroom, Two Wall Plumbing	COST EACH		
		MAT.	INST.	TOTAL
1140	Bathroom, four fixture, two wall plumbing			
1150	Bathtub, water closet, stall shower & lavatory			
1160	Stand alone	2,925	2,225	5,150
1180	Long plumbing wall common *	2,600	1,700	4,300
2260	Bathtub, lavatory, corner stall shower & water closet			
2280	Stand alone	3,925	2,225	6,150
2320	Long plumbing wall common *	3,600	1,725	5,325
3620	Bathtub, stall shower, lavatory & water closet			
3640	Stand alone	3,275	2,775	6,050
3660	Long plumbing wall (opp. door) common *	2,950	2,275	5,225

D2010 930	Four Fixture Bathroom, Three Wall Plumbing	COST EACH		
		MAT.	INST.	TOTAL
4680	Bathroom, four fixture, three wall plumbing			
4700	Bathtub, stall shower, lavatory & water closet			
4720	Stand alone	4,450	3,075	7,525
4760	Long plumbing wall (opposite door) common *	4,300	2,825	7,125

D2010   Plumbing Fixtures

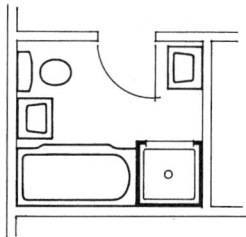

Five Fixture Bathroom Systems consisting of two lavatories, a water closet, bathtub, shower and rough-in service piping.

- Prices for plumbing and fixtures only.

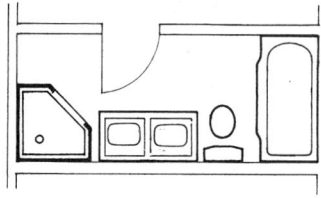

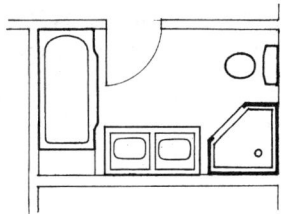

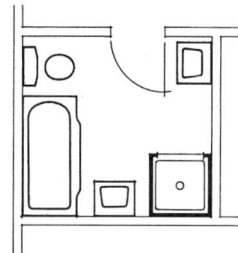

*Common wall is with an adjacent bathroom.

System Components	QUANTITY	UNIT	COST EACH		
			MAT.	INST.	TOTAL
SYSTEM D2010 932 1360					
BATHROOM, BATHTUB, WATER CLOSET, STALL SHOWER & TWO LAVATORIES					
TWO WALL PLUMBING, STAND ALONE					
Wtr closet, 2 pc close cpld vit china flr mntd incl seat,supply & stop	1.000	Ea.	203	199	402
Water closet, rough-in waste & vent	1.000	Set	300	345	645
Lavatory w/ftngs, wall hung, white PE on CI, 20″ x 18″	2.000	Ea.	650	264	914
Lavatory, rough-in waste & vent	2.000	Set	800	1,270	2,070
Bathtub, white PE on CI, w/ftgs, mat bottom, recessed, 5′ long	1.000	Ea.	890	239	1,129
Baths, rough-in waste and vent	1.000	Set	330	510	840
Shower stall, bkd enam molded stone receptor, door & ftng, 32″ sq.	1.000	Ea.	400	211	611
Shower stall, rough-in supply, waste & vent,	1.000	Set	435	515	950
TOTAL			4,008	3,553	7,561

D2010 932	Five Fixture Bathroom, Two Wall Plumbing	COST EACH		
		MAT.	INST.	TOTAL
1320	Bathroom, five fixture, two wall plumbing			
1340	Bathtub, water closet, stall shower & two lavatories			
1360	Stand alone	4,000	3,550	7,550
1400	One short plumbing wall common *	3,675	3,050	6,725
1500	Bathtub, two lavatories, corner stall shower & water closet			
1520	Stand alone	5,000	3,550	8,550
1540	Long plumbing wall common*	4,500	2,775	7,275

D2010 934	Five Fixture Bathroom, Three Wall Plumbing	COST EACH		
		MAT.	INST.	TOTAL
2360	Bathroom, five fixture, three wall plumbing			
2380	Water closet, bathtub, two lavatories & stall shower			
2400	Stand alone	5,000	3,550	8,550
2440	One short plumbing wall common *	4,675	3,050	7,725

D2010 936	Five Fixture Bathroom, One Wall Plumbing	COST EACH		
		MAT.	INST.	TOTAL
4080	Bathroom, five fixture, one wall plumbing			
4100	Bathtub, two lavatories, corner stall shower & water closet			
4120	Stand alone	4,775	3,175	7,950
4160	Share common wall *	4,000	2,050	6,050

D20 Plumbing

D2010 Plumbing Fixtures

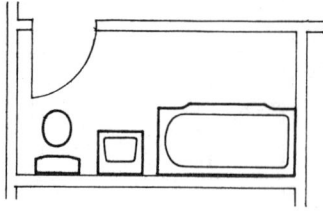

Example of Plumbing Cost Calculations: The bathroom system includes the individual fixtures such as bathtub, lavatory, shower and water closet. These fixtures are listed below as separate items merely as a checklist.

D2010 951	Plumbing Systems 20 Unit, 2 Story Apartment Building							
							COST EACH	
	FIXTURE	SYSTEM	LINE	QUANTITY	UNIT	MAT.	INST.	TOTAL
0440	Bathroom	D2010 926	3640	20	Ea.	49,400	41,900	91,300
0480	Bathtub							
0520	Booster pump[1]	not req'd.						
0560	Drinking fountain							
0600	Garbage disposal[1]	not incl.						
0660								
0680	Grease interceptor							
0720	Water heater	D2020 250	2140	1	Ea.	9,875	2,850	12,725
0760	Kitchen sink	D2010 410	1960	20	Ea.	24,700	14,700	39,400
0800	Laundry sink	D2010 420	1840	4	Ea.	4,925	2,900	7,825
0840	Lavatory							
0900								
0920	Roof drain, 1 floor	D2040 210	4200	2	Ea.	1,325	1,525	2,850
0960	Roof drain, add'l floor	D2040 210	4240	20	L.F.	253	385	638
1000	Service sink	D2010 440	4300	1	Ea.	1,750	1,025	2,775
1040	Sewage ejector[1]	not req'd.						
1080	Shower							
1100								
1160	Sump pump							
1200	Urinal							
1240	Water closet							
1320								
1360	**SUB TOTAL**					92,000	65,500	157,500
1480	Water controls	RD2010-031		10%[2]		9,225	6,525	15,750
1520	Pipe & fittings[3]	RD2010-031		30%[2]		27,700	19,600	47,300
1560	Other							
1600	Quality/complexity	RD2010-031		15%[2]		13,800	9,800	23,600
1680								
1720	**TOTAL**					143,000	101,500	244,500
1740								

[1]**Note:** Cost for items such as booster pumps, backflow preventers, sewage ejectors, water meters, etc., may be obtained from RSMeans "Plumbing Cost Data" book, or be added as an increase in the Water Controls percentage. Water Controls, Pipe and Fittings, and the Quality/Complexity factors come from Table RD2010-031.

[2]Percentage of subtotal.

[3]Long easily discernable runs of pipe would be more accurately priced from Table D2090 810. If this is done, reduce the miscellaneous percentage in proportion.

D2020 Domestic Water Distribution

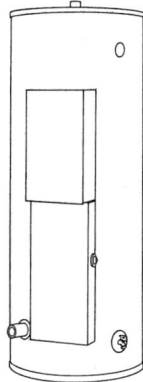

Systems below include piping and fittings within 10' of heater. Electric water heaters do not require venting.

System Components	QUANTITY	UNIT	COST EACH		
			MAT.	INST.	TOTAL
SYSTEM D2020 240 1820					
ELECTRIC WATER HEATER, COMMERCIAL, 100°F RISE					
50 GALLON TANK, 9 KW, 37 GPH					
Water heater, commercial, electric, 50 Gal, 9 KW, 37 GPH	1.000	Ea.	4,475	325	4,800
Copper tubing, type L, solder joint, hanger 10' OC, 3/4" diam	34.000	L.F.	224.40	261.80	486.20
Wrought copper 90° elbow for solder joints 3/4" diam	5.000	Ea.	15.20	155	170.20
Wrought copper Tee for solder joints, 3/4" diam	2.000	Ea.	11.20	98	109.20
Wrought copper union for soldered joints, 3/4" diam	2.000	Ea.	35.60	65	100.60
Valve, gate, bronze, 125 lb, NRS, soldered 3/4" diam	2.000	Ea.	78	59	137
Relief valve, bronze, press & temp, self-close, 3/4" IPS	1.000	Ea.	130	21	151
Wrought copper adapter, copper tubing to male, 3/4" IPS	1.000	Ea.	4.76	34.50	39.26
TOTAL			4,974.16	1,019.30	5,993.46

D2020 240	Electric Water Heaters - Commercial Systems	COST EACH		
		MAT.	INST.	TOTAL
1800	Electric water heater, commercial, 100°F rise			
1820	50 gallon tank, 9 KW 37 GPH RD2020 -100	4,975	1,025	6,000
1860	80 gal, 12 KW 49 GPH	6,475	1,250	7,725
1900	36 KW 147 GPH	8,925	1,350	10,275
1940	120 gal, 36 KW 147 GPH	9,575	1,475	11,050
1980	150 gal, 120 KW 490 GPH	29,000	1,575	30,575
2020	200 gal, 120 KW 490 GPH	30,700	1,600	32,300
2060	250 gal, 150 KW 615 GPH	34,300	1,875	36,175
2100	300 gal, 180 KW 738 GPH	38,500	1,975	40,475
2140	350 gal, 30 KW 123 GPH	28,300	2,150	30,450
2180	180 KW 738 GPH	37,800	2,150	39,950
2220	500 gal, 30 KW 123 GPH	36,200	2,525	38,725
2260	240 KW 984 GPH	54,000	2,525	56,525
2300	700 gal, 30 KW 123 GPH	45,100	2,875	47,975
2340	300 KW 1230 GPH	63,000	2,875	65,875
2380	1000 gal, 60 KW 245 GPH	52,500	4,025	56,525
2420	480 KW 1970 GPH	85,000	4,025	89,025
2460	1500 gal, 60 KW 245 GPH	76,500	4,950	81,450
2500	480 KW 1970 GPH	121,000	4,950	125,950

D20 Plumbing

D2020 Domestic Water Distribution

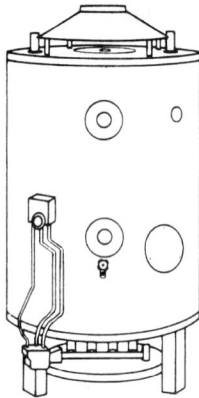

Units may be installed in multiples for increased capacity.

Included below is the heater with self-energizing gas controls, safety pilots, insulated jacket, hi-limit aquastat and pressure relief valve.

Installation includes piping and fittings within 10' of heater. Gas heaters require vent piping (not included in these prices).

System Components	QUANTITY	UNIT	COST EACH		
			MAT.	INST.	TOTAL
SYSTEM D2020 250 1780					
GAS FIRED WATER HEATER, COMMERCIAL, 100°F RISE					
75.5 MBH INPUT, 63 GPH					
Water heater, commercial, gas, 75.5 MBH, 63 GPH	1.000	Ea.	1,900	420	2,320
Copper tubing, type L, solder joint, hanger 10' OC, 1-1/4" diam	30.000	L.F.	412.50	303	715.50
Wrought copper 90° elbow for solder joints 1-1/4" diam	4.000	Ea.	45.20	156	201.20
Wrought copper Tee for solder joints, 1-1/4" diam	2.000	Ea.	47	130	177
Wrought copper union for soldered joints, 1-1/4" diam	2.000	Ea.	106	84	190
Valve, gate, bronze, 125 lb, NRS, soldered 1-1/4" diam	2.000	Ea.	177	78	255
Relief valve, bronze, press & temp, self-close, 3/4" IPS	1.000	Ea.	130	21	151
Copper tubing, type L, solder joints, 3/4" diam	8.000	L.F.	52.80	61.60	114.40
Wrought copper 90° elbow for solder joints 3/4" diam	1.000	Ea.	3.04	31	34.04
Wrought copper, adapter, CTS to MPT, 3/4" IPS	1.000	Ea.	4.76	34.50	39.26
Pipe steel black, schedule 40, threaded, 3/4" diam	10.000	L.F.	29	96	125
Pipe, 90° elbow, malleable iron black, 150 lb threaded, 3/4" diam	2.000	Ea.	7.42	84	91.42
Pipe, union with brass seat, malleable iron black, 3/4" diam	1.000	Ea.	11.35	45	56.35
Valve, gas stop w/o check, brass, 3/4" IPS	1.000	Ea.	15.05	26.50	41.55
TOTAL			2,941.12	1,570.60	4,511.72

D2020 250	Gas Fired Water Heaters - Commercial Systems		COST EACH		
			MAT.	INST.	TOTAL
1760	Gas fired water heater, commercial, 100°F rise				
1780	75.5 MBH input, 63 GPH		2,950	1,575	4,525
1860	100 MBH input, 91 GPH	RD2020 -100	6,075	1,650	7,725
1980	155 MBH input, 150 GPH		7,525	1,900	9,425
2060	200 MBH input, 192 GPH		8,325	2,300	10,625
2140	300 MBH input, 278 GPH		9,875	2,850	12,725
2180	390 MBH input, 374 GPH		11,500	2,875	14,375
2220	500 MBH input, 480 GPH		15,300	3,100	18,400
2260	600 MBH input, 576 GPH		21,700	3,350	25,050

D2020 Domestic Water Distribution

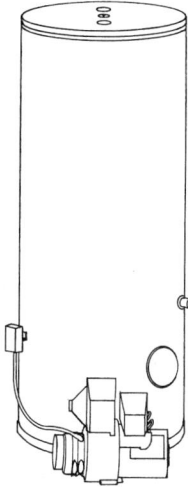

Units may be installed in multiples for increased capacity.

Included below is the heater, wired-in flame retention burners, cadmium cell primary controls, hi-limit controls, ASME pressure relief valves, draft controls, and insulated jacket.

Oil fired water heater systems include piping and fittings within 10′ of heater. Oil fired heaters require vent piping (not included in these systems).

System Components	QUANTITY	UNIT	COST EACH		
			MAT.	INST.	TOTAL
SYSTEM D2020 260 1820					
OIL FIRED WATER HEATER, COMMERCIAL, 100°F RISE					
140 GAL., 140 MBH INPUT, 134 GPH					
Water heater, commercial, oil, 140 gal., 140 MBH input, 134 GPH	1.000	Ea.	15,600	495	16,095
Copper tubing, type L, solder joint, hanger 10′ OC, 3/4″ diam	34.000	L.F.	224.40	261.80	486.20
Wrought copper 90° elbow for solder joints 3/4″ diam	5.000	Ea.	15.20	155	170.20
Wrought copper Tee for solder joints, 3/4″ diam	2.000	Ea.	11.20	98	109.20
Wrought copper union for soldered joints, 3/4″ diam	2.000	Ea.	35.60	65	100.60
Valve, bronze, 125 lb, NRS, soldered 3/4″ diam	2.000	Ea.	78	59	137
Relief valve, bronze, press & temp, self-close, 3/4″ IPS	1.000	Ea.	130	21	151
Wrought copper adapter, copper tubing to male, 3/4″ IPS	1.000	Ea.	4.76	34.50	39.26
Copper tubing, type L, solder joint, hanger 10′ OC, 3/8″ diam	10.000	L.F.	35.80	69.50	105.30
Wrought copper 90° elbow for solder joints 3/8″ diam.	2.000	Ea.	8.12	53	61.12
Valve, globe, fusible, 3/8″ IPS	1.000	Ea.	12.55	24.50	37.05
TOTAL			16,155.63	1,336.30	17,491.93

D2020 260	Oil Fired Water Heaters - Commercial Systems		COST EACH		
			MAT.	INST.	TOTAL
1800	Oil fired water heater, commercial, 100°F rise				
1820	140 gal., 140 MBH input, 134 GPH		16,200	1,325	17,525
1900	140 gal., 255 MBH input, 247 GPH	RD2020	17,500	1,675	19,175
1940	140 gal., 270 MBH input, 259 GPH	-100	21,400	1,900	23,300
1980	140 gal., 400 MBH input, 384 GPH		22,200	2,200	24,400
2060	140 gal., 720 MBH input, 691 GPH		23,600	2,300	25,900
2100	221 gal., 300 MBH input, 288 GPH		31,400	2,525	33,925
2140	221 gal., 600 MBH input, 576 GPH		34,800	2,550	37,350
2180	221 gal., 800 MBH input, 768 GPH		35,300	2,650	37,950
2220	201 gal., 1000 MBH input, 960 GPH		36,100	2,675	38,775
2260	201 gal., 1250 MBH input, 1200 GPH		36,800	2,750	39,550
2300	201 gal., 1500 MBH input, 1441 GPH		39,800	2,825	42,625
2340	411 gal., 600 MBH input, 576 GPH		40,200	2,875	43,075
2380	411 gal., 800 MBH input, 768 GPH		40,500	2,950	43,450
2420	411 gal., 1000 MBH input, 960 GPH		43,900	3,400	47,300
2460	411 gal., 1250 MBH input, 1200 GPH		44,800	3,500	48,300
2500	397 gal., 1500 MBH input, 1441 GPH		45,200	3,600	48,800
2540	397 gal., 1750 MBH input, 1681 GPH		48,600	3,725	52,325

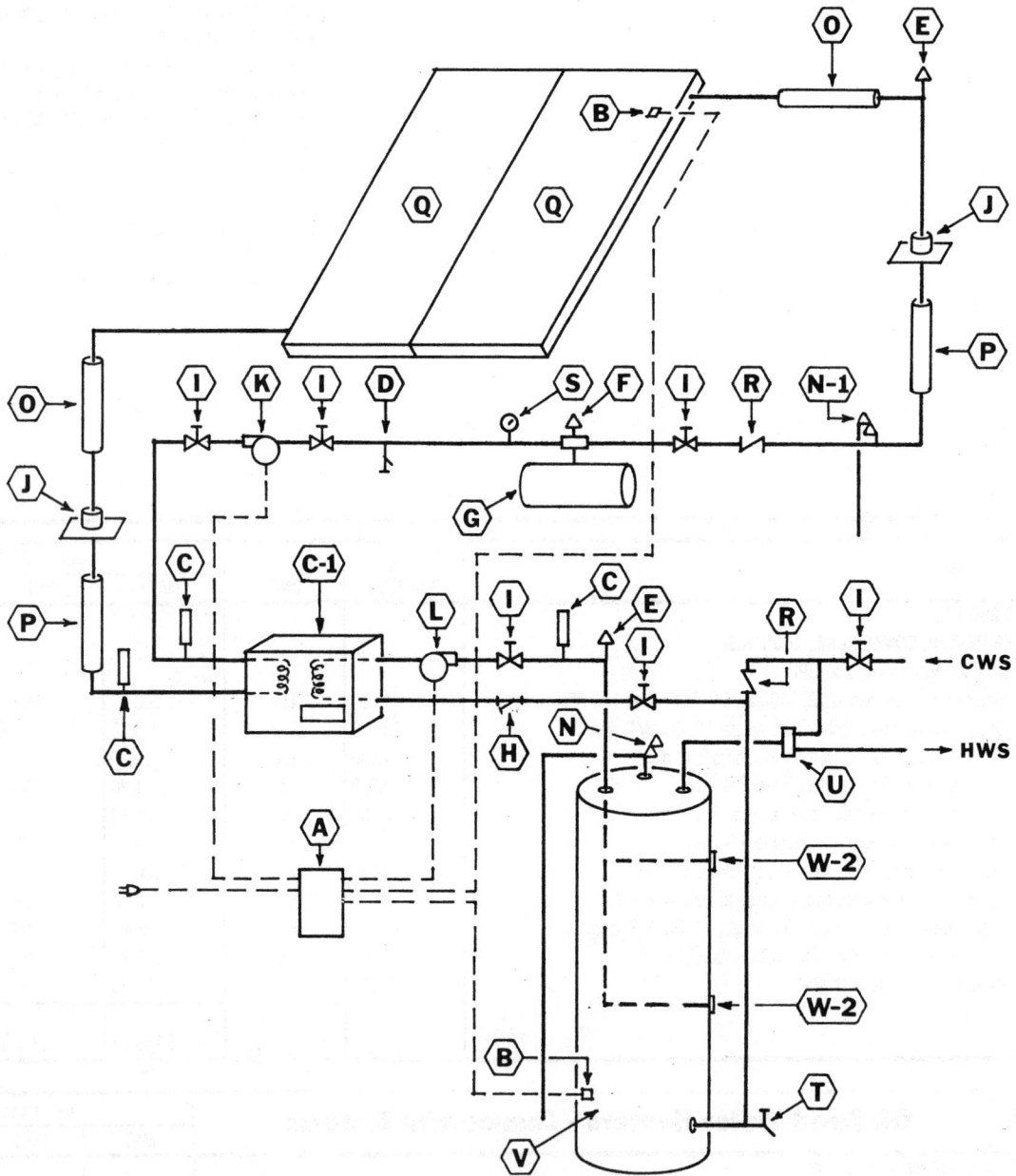

In this closed-loop indirect collection system, fluid with a low freezing temperature, such as propylene glycol, transports heat from the collectors to water storage. The transfer fluid is contained in a closed-loop consisting of collectors, supply and return piping, and a remote heat exchanger. The heat exchanger transfers heat energy from the fluid in the collector loop to potable water circulated in a storage loop. A typical two-or-three panel system contains 5 to 6 gallons of heat transfer fluid.

When the collectors become approximately 20°F warmer than the storage temperature, a controller activates the circulator on the collector and storage loops. The circulators will move the fluid and potable water through the heat exchanger until heat collection no longer occurs. At that point, the system shuts down. Since the heat transfer medium is a fluid with a very low freezing temperature, there is no need for it to be drained from the system between periods of collection.

D2020 Domestic Water Distribution

System Components	QUANTITY	UNIT	COST EACH		
			MAT.	INST.	TOTAL
SYSTEM D2020 265 2760					
SOLAR, CLOSED LOOP, ADD-ON HOT WATER SYS., EXTERNAL HEAT EXCHANGER					
3/4" TUBING, TWO 3'X7' BLACK CHROME COLLECTORS					
A,B,G,L,K,M Heat exchanger fluid-fluid pkg incl 2 circulators, expansion tank,					
Check valve, relief valve, controller, hi temp cutoff, & 2 sensors	1.000	Ea.	765	420	1,185
C Thermometer, 2" dial	3.000	Ea.	66	109.50	175.50
D, T Fill & drain valve, brass, 3/4" connection	1.000	Ea.	8.85	24.50	33.35
E Air vent, manual, 1/8" fitting	2.000	Ea.	6.12	36.60	42.72
F Air purger	1.000	Ea.	47.50	49	96.50
H Strainer, Y type, bronze body, 3/4" IPS	1.000	Ea.	24	31	55
I Valve, gate, bronze, NRS, soldered 3/4" diam	6.000	Ea.	234	177	411
J Neoprene vent flashing	2.000	Ea.	23.80	59	82.80
N-1, N Relief valve temp & press, 150 psi 210°F self-closing 3/4" IPS	1.000	Ea.	16.70	19.50	36.20
O Pipe covering, urethane, ultraviolet cover, 1" wall 3/4" diam	20.000	L.F.	52.40	108	160.40
P Pipe covering, fiberglass, all service jacket, 1" wall, 3/4" diam	50.000	L.F.	53.50	216.50	270
Q Collector panel solar energy blk chrome on copper, 1/8" temp glass 3'x7'	2.000	Ea.	1,460	222	1,682
Roof clamps for solar energy collector panels	2.000	Set	5.18	30.10	35.28
R Valve, swing check, bronze, regrinding disc, 3/4" diam	2.000	Ea.	140	59	199
S Pressure gauge, 60 psi, 2" dial	1.000	Ea.	24.50	18.30	42.80
U Valve, water tempering, bronze, sweat connections, 3/4" diam	1.000	Ea.	95.50	29.50	125
W-2, V Tank water storage w/heating element, drain, relief valve, existing	1.000	Ea.			
Copper tubing type L, solder joint, hanger 10' OC 3/4" diam	20.000	L.F.	132	154	286
Copper tubing, type M, solder joint, hanger 10' OC 3/4" diam	70.000	L.F.	339.50	525	864.50
Sensor wire, #22-2 conductor multistranded	.500	C.L.F.	8.45	28	36.45
Solar energy heat transfer fluid, propylene glycol anti-freeze	6.000	Gal.	91.50	126	217.50
Wrought copper fittings & solder, 3/4" diam	76.000	Ea.	231.04	2,356	2,587.04
TOTAL			3,825.54	4,798.50	8,624.04

D2020 265	Solar, Closed Loop, Add-On Hot Water Systems		COST EACH		
			MAT.	INST.	TOTAL
2550	Solar, closed loop, add-on hot water system, external heat exchanger				
2700	3/4" tubing, 3 ea 3'x7' black chrome collectors		4,550	4,925	9,475
2720	3 ea 3'x7' flat black absorber plate collectors	RD3010 -600	4,225	4,950	9,175
2740	2 ea 4'x9' flat black w/plastic glazing collectors		3,875	4,950	8,825
2760	2 ea 3'x7' black chrome collectors		3,825	4,800	8,625
2780	1" tubing,4 ea 2'x9' plastic absorber & glazing collectors		5,325	5,550	10,875
2800	4 ea 3'x7' black chrome absorber collectors		6,050	5,575	11,625
2820	4 ea 3'x7' flat black absorber collectors		5,625	5,600	11,225

In the drainback indirect-collection system, the heat transfer fluid is distilled water contained in a loop consisting of collectors, supply and return piping, and an unpressurized holding tank. A large heat exchanger containing incoming potable water is immersed in the holding tank. When a controller activates solar collection, the distilled water is pumped through the collectors and heated and pumped back down to the holding tank. When the temperature differential between the water in the collectors and water in storage is such that collection no longer occurs, the pump turns off and gravity causes the distilled water in the collector loop to drain back to the holding tank. All the loop piping is pitched so that the water can drain out of the collectors and piping and not freeze there. As hot water is needed in the home, incoming water first flows through the holding tank with the immersed heat exchanger and is warmed and then flows through a conventional heater for any supplemental heating that is necessary.

D20 Plumbing

D2020 Domestic Water Distribution

System Components	QUANTITY	UNIT	COST EACH MAT.	COST EACH INST.	COST EACH TOTAL
SYSTEM D2020 270 2760					
SOLAR, DRAINBACK, ADD ON, HOT WATER, IMMERSED HEAT EXCHANGER					
3/4″ TUBING, THREE EA 3′X7′ BLACK CHROME COLLECTOR					
A, B Differential controller 2 sensors, thermostat, solar energy system	1.000	Ea.	380	49	429
C Thermometer 2″ dial	3.000	Ea.	66	109.50	175.50
D, T Fill & drain valve, brass, 3/4″ connection	1.000	Ea.	8.85	24.50	33.35
E-1 Automatic air vent 1/8″ fitting	1.000	Ea.	13.95	18.30	32.25
H Strainer, Y type, bronze body, 3/4″ IPS	1.000	Ea.	24	31	55
I Valve, gate, bronze, NRS, soldered 3/4″ diam	2.000	Ea.	78	59	137
J Neoprene vent flashing	2.000	Ea.	23.80	59	82.80
L Circulator, solar heated liquid, 1/20 HP	1.000	Ea.	170	88	258
N Relief valve temp. & press. 150 psi 210°F self-closing 3/4″ IPS	1.000	Ea.	16.70	19.50	36.20
O Pipe covering, urethane, ultraviolet cover, 1″ wall, 3/4″ diam	20.000	L.F.	52.40	108	160.40
P Pipe covering, fiberglass, all service jacket, 1″ wall, 3/4″ diam	50.000	L.F.	53.50	216.50	270
Q Collector panel solar energy blk chrome on copper, 1/8″ temp glas 3′x7′	3.000	Ea.	2,190	333	2,523
Roof clamps for solar energy collector panels	3.000	Set	7.77	45.15	52.92
R Valve, swing check, bronze, regrinding disc, 3/4″ diam	1.000	Ea.	70	29.50	99.50
U Valve, water tempering, bronze sweat connections, 3/4″ diam	1.000	Ea.	95.50	29.50	125
V Tank, water storage w/heating element, drain, relief valve, existing	1.000	Ea.			
W Tank, water storage immersed heat exchr elec 2″x1/2# insul 120 gal	1.000	Ea.	1,275	420	1,695
X Valve, globe, bronze, rising stem, 3/4″ diam, soldered	3.000	Ea.	297	88.50	385.50
Y Flow control valve	1.000	Ea.	112	26.50	138.50
Z Valve, ball, bronze, solder 3/4″ diam, solar loop flow control	1.000	Ea.	18.65	29.50	48.15
Copper tubing, type L, solder joint, hanger 10′ OC 3/4″ diam	20.000	L.F.	132	154	286
Copper tubing, type M, solder joint, hanger 10′ OC 3/4″ diam	70.000	L.F.	339.50	525	864.50
Sensor wire, #22-2 conductor, multistranded	.500	C.L.F.	8.45	28	36.45
Wrought copper fittings & solder, 3/4″ diam	76.000	Ea.	231.04	2,356	2,587.04
TOTAL			5,664.11	4,846.95	10,511.06

D2020 270	Solar, Drainback, Hot Water Systems		MAT.	INST.	TOTAL
2550	Solar, drainback, hot water, immersed heat exchanger				
2560	3/8″ tubing, 3 ea. 4′ x 4′-4″ vacuum tube collectors		5,900	4,350	10,250
2760	3/4″ tubing, 3 ea 3′x7′ black chrome collectors, 120 gal tank	RD3010-600	5,675	4,850	10,525
2780	3 ea. 3′x7′ flat black absorber collectors, 120 gal tank		5,325	4,875	10,200
2800	2 ea. 4′x9′ flat blk w/plastic glazing collectors 120 gal tank		4,975	4,875	9,850
2840	1″ tubing, 4 ea. 2′x9′ plastic absorber & glazing collectors, 120 gal tank		6,500	5,475	11,975
2860	4 ea. 3′x7′ black chrome absorber collectors, 120 gal tank		7,250	5,500	12,750
2880	4 ea. 3′x7′ flat black absorber collectors, 120 gal tank		6,800	5,525	12,325

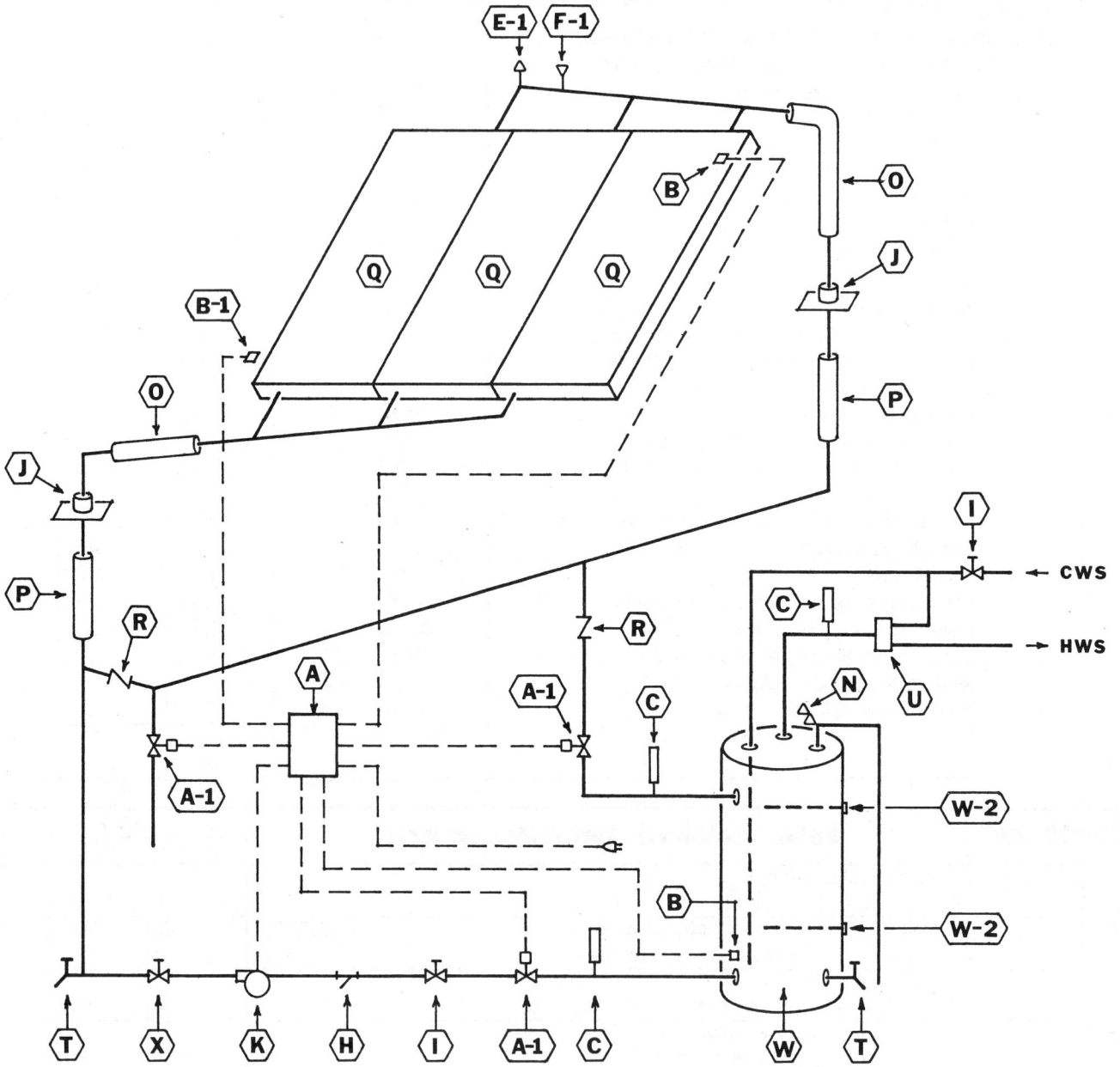

In the draindown direct-collection system, incoming domestic water is heated in the collectors. When the controller activates solar collection, domestic water is first heated as it flows through the collectors and is then pumped to storage. When conditions are no longer suitable for heat collection, the pump shuts off and the water in the loop drains down and out of the system by means of solenoid valves and properly pitched piping.

D2020 Domestic Water Distribution

System Components	QUANTITY	UNIT	COST EACH		
			MAT.	INST.	TOTAL
SYSTEM D2020 275 2760					
SOLAR, DRAINDOWN, HOT WATER, DIRECT COLLECTION					
3/4" TUBING, THREE 3'X7' BLACK CHROME COLLECTORS					
A, B Differential controller, 2 sensors, thermostat, solar energy system	1.000	Ea.	380	49	429
A-1 Solenoid valve, solar heating loop, brass, 3/4" diam, 24 volts	3.000	Ea.	339	195	534
B-1 Solar energy sensor, freeze prevention	1.000	Ea.	24	18.30	42.30
C Thermometer, 2" dial	3.000	Ea.	66	109.50	175.50
E-1 Vacuum relief valve, 3/4" diam	1.000	Ea.	29	18.30	47.30
F-1 Air vent, automatic, 1/8" fitting	1.000	Ea.	13.95	18.30	32.25
H Strainer, Y type, bronze body, 3/4" IPS	1.000	Ea.	24	31	55
I Valve, gate, bronze, NRS, soldered, 3/4" diam	2.000	Ea.	78	59	137
J Vent flashing neoprene	2.000	Ea.	23.80	59	82.80
K Circulator, solar heated liquid, 1/25 HP	1.000	Ea.	250	75.50	325.50
N Relief valve temp & press 150 psi 210°F self-closing 3/4" IPS	1.000	Ea.	16.70	19.50	36.20
O Pipe covering, urethane, ultraviolet cover, 1" wall, 3/4" diam	20.000	L.F.	52.40	108	160.40
P Pipe covering, fiberglass, all service jacket, 1" wall, 3/4" diam	50.000	L.F.	53.50	216.50	270
Roof clamps for solar energy collector panels	3.000	Set	7.77	45.15	52.92
Q Collector panel solar energy blk chrome on copper, 1/8" temp glass 3'x7'	3.000	Ea.	2,190	333	2,523
R Valve, swing check, bronze, regrinding disc, 3/4" diam, soldered	2.000	Ea.	140	59	199
T Drain valve, brass, 3/4" connection	2.000	Ea.	17.70	49	66.70
U Valve, water tempering, bronze, sweat connections, 3/4" diam	1.000	Ea.	95.50	29.50	125
W-2, W Tank, water storage elec elem 2"x1/2# insul 120 gal	1.000	Ea.	1,275	420	1,695
X Valve, globe, bronze, rising stem, 3/4" diam, soldered	1.000	Ea.	99	29.50	128.50
Copper tubing, type L, solder joints, hangers 10' OC 3/4" diam	20.000	L.F.	132	154	286
Copper tubing, type M, solder joints, hangers 10' OC 3/4" diam	70.000	L.F.	339.50	525	864.50
Sensor wire, #22-2 conductor, multistranded	.500	C.L.F.	8.45	28	36.45
Wrought copper fittings & solder, 3/4" diam	76.000	Ea.	231.04	2,356	2,587.04
TOTAL			5,886.31	5,005.05	10,891.36

D2020 275	Solar, Draindown, Hot Water Systems		COST EACH		
			MAT.	INST.	TOTAL
2550	Solar, draindown, hot water				
2580	1/2" tubing, 4 ea. 4' x 4'-4" vacuum tube collectors, 80 gal tank		6,900	4,900	11,800
2760	3/4" tubing, 3 ea 3'x7' black chrome collectors, 120 gal tank	RD3010 -600	5,875	5,000	10,875
2780	3 ea 3'x7' flat collectors, 120 gal tank		5,550	5,025	10,575
2800	2 ea. 4'x9' flat black & plastic glazing collectors, 120 gal tank		5,200	5,050	10,250
2840	1" tubing, 4 ea. 2'x9' plastic absorber & glazing collectors, 120 gal tank		8,975	5,625	14,600
2860	4 ea 3'x7' black chrome absorber collectors, 120 gal tank		9,725	5,650	15,375
2880	4 ea 3'x7' flat black absorber collectors, 120 gal tank		9,275	5,675	14,950

D2020 Domestic Water Distribution

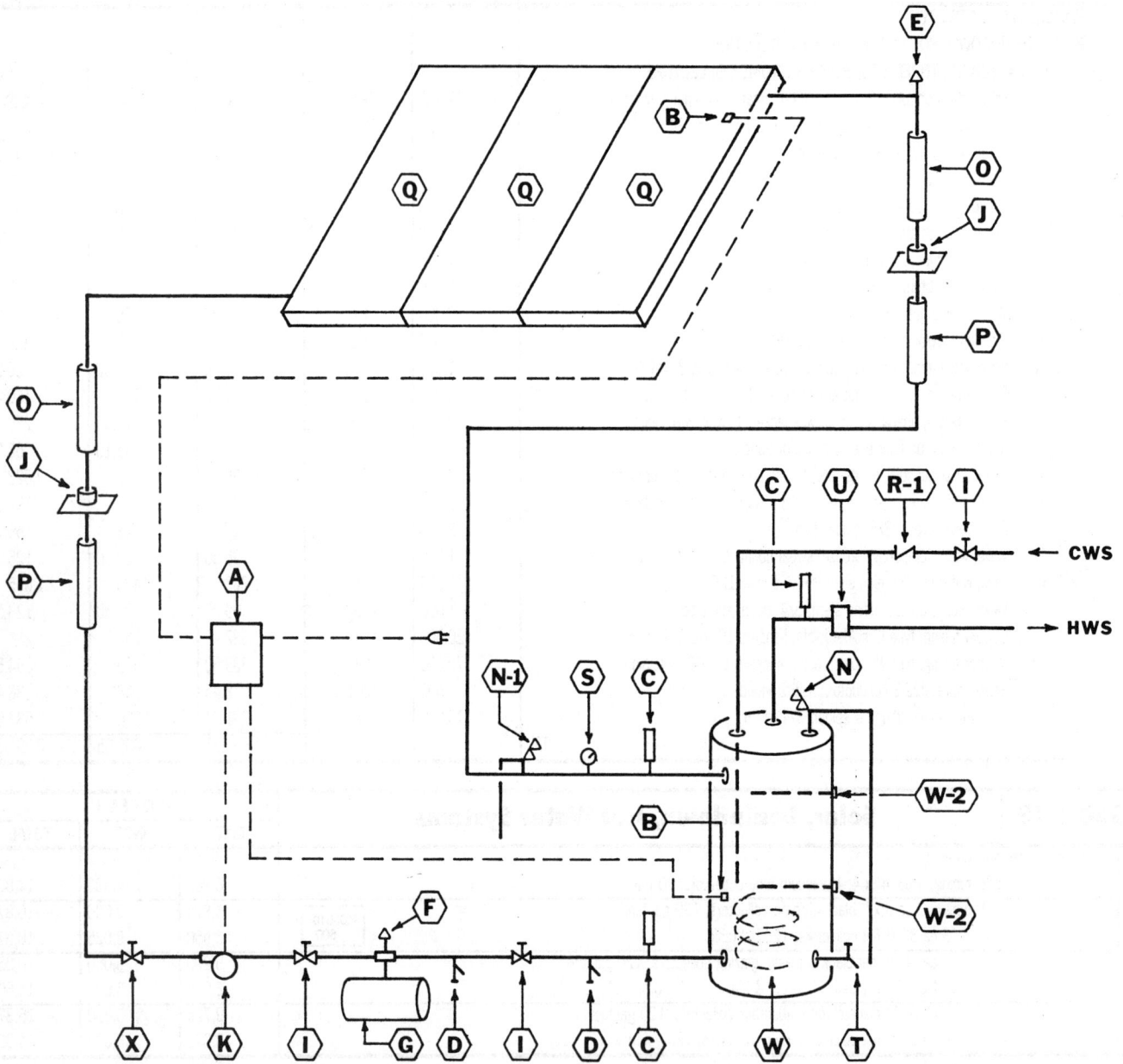

In this closed-loop indirect collection system, fluid with a low freezing temperature, such as propylene glycol, transports heat from the collectors to water storage. The transfer fluid is contained in a closed-loop consisting of collectors, supply and return piping, and a heat exchanger immersed in the storage tank. A typical two-or-three panel system contains 5 to 6 gallons of heat transfer fluid.

When the collectors become approximately 20°F warmer than the storage temperature, a controller activates the circulator. The circulator moves the fluid continuously through the collectors until the temperature difference between the collectors and storage is such that heat collection no longer occurs; at that point, the circulator shuts off. Since the heat transfer fluid has a very low freezing temperature, there is no need for it to be drained from the collectors between periods of collection.

D20 Plumbing

D2020 Domestic Water Distribution

System Components	QUANTITY	UNIT	COST EACH MAT.	COST EACH INST.	COST EACH TOTAL
SYSTEM D2020 295 2760					
SOLAR, CLOSED LOOP, HOT WATER SYSTEM, IMMERSED HEAT EXCHANGER					
3/4″ TUBING, THREE 3′ X 7′ BLACK CHROME COLLECTORS					
A, B Differential controller, 2 sensors, thermostat, solar energy system	1.000	Ea.	380	49	429
C Thermometer 2″ dial	3.000	Ea.	66	109.50	175.50
D, T Fill & drain valves, brass, 3/4″ connection	3.000	Ea.	26.55	73.50	100.05
E Air vent, manual, 1/8″ fitting	1.000	Ea.	3.06	18.30	21.36
F Air purger	1.000	Ea.	47.50	49	96.50
G Expansion tank	1.000	Ea.	67.50	18.30	85.80
I Valve, gate, bronze, NRS, soldered 3/4″ diam	3.000	Ea.	117	88.50	205.50
J Neoprene vent flashing	2.000	Ea.	23.80	59	82.80
K Circulator, solar heated liquid, 1/25 HP	1.000	Ea.	250	75.50	325.50
N-1, N Relief valve, temp & press 150 psi 210°F self-closing 3/4″ IPS	2.000	Ea.	33.40	39	72.40
O Pipe covering, urethane ultraviolet cover, 1″ wall, 3/4″ diam	20.000	L.F.	52.40	108	160.40
P Pipe covering, fiberglass, all service jacket, 1″ wall, 3/4″ diam	50.000	L.F.	53.50	216.50	270
Roof clamps for solar energy collector panel	3.000	Set	7.77	45.15	52.92
Q Collector panel solar blk chrome on copper, 1/8″ temp glass, 3′x7′	3.000	Ea.	2,190	333	2,523
R-1 Valve, swing check, bronze, regrinding disc, 3/4″ diam, soldered	1.000	Ea.	70	29.50	99.50
S Pressure gauge, 60 psi, 2-1/2″ dial	1.000	Ea.	24.50	18.30	42.80
U Valve, water tempering, bronze, sweat connections, 3/4″ diam	1.000	Ea.	95.50	29.50	125
W-2, W Tank, water storage immersed heat exchr elec elem 2″x2# insul 120 Gal	1.000	Ea.	1,275	420	1,695
X Valve, globe, bronze, rising stem, 3/4″ diam, soldered	1.000	Ea.	99	29.50	128.50
Copper tubing type L, solder joint, hanger 10′ OC 3/4″ diam	20.000	L.F.	132	154	286
Copper tubing, type M, solder joint, hanger 10′ OC 3/4″ diam	70.000	L.F.	339.50	525	864.50
Sensor wire, #22-2 conductor multistranded	.500	C.L.F.	8.45	28	36.45
Solar energy heat transfer fluid, propylene glycol, anti-freeze	6.000	Gal.	91.50	126	217.50
Wrought copper fittings & solder, 3/4″ diam	76.000	Ea.	231.04	2,356	2,587.04
TOTAL			5,684.97	4,998.05	10,683.02

D2020 295	Solar, Closed Loop, Hot Water Systems		COST EACH MAT.	COST EACH INST.	COST EACH TOTAL
2550	Solar, closed loop, hot water system, immersed heat exchanger				
2560	3/8″ tubing, 3 ea. 4′ x 4′-4″ vacuum tube collectors, 80 gal. tank		5,975	4,500	10,475
2580	1/2″ tubing, 4 ea. 4′ x 4′-4″ vacuum tube collectors, 80 gal. tank	RD3010 -600	6,725	4,900	11,625
2600	120 gal. tank		6,850	4,950	11,800
2640	2 ea. 3′x7′ black chrome collectors, 80 gal. tank		4,500	4,675	9,175
2660	120 gal. tank		4,550	4,675	9,225
2760	3/4″ tubing, 3 ea. 3′x7′ black chrome collectors, 120 gal. tank		5,675	5,000	10,675
2780	3 ea. 3′x7′ flat black collectors, 120 gal. tank		5,350	5,025	10,375
2840	1″ tubing, 4 ea. 2′x9′ plastic absorber & glazing collectors 120 gal. tank		6,350	5,600	11,950
2860	4 ea. 3′x7′ black chrome collectors, 120 gal. tank		7,100	5,625	12,725

D2040 Rain Water Drainage

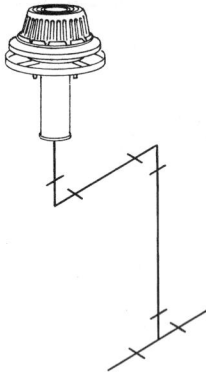

Design Assumptions: Vertical conductor size is based on a maximum rate of rainfall of 4" per hour. To convert roof area to other rates multiply "Max. S.F. Roof Area" shown by four and divide the result by desired local rate. The answer is the local roof area that may be handled by the indicated pipe diameter.

Basic cost is for roof drain, 10' of vertical leader and 10' of horizontal, plus connection to the main.

Pipe Dia.	Max. S.F. Roof Area	Gallons per Min.
2"	544	23
3"	1610	67
4"	3460	144
5"	6280	261
6"	10,200	424
8"	22,000	913

System Components	QUANTITY	UNIT	COST EACH MAT.	COST EACH INST.	COST EACH TOTAL
SYSTEM D2040 210 1880					
ROOF DRAIN, DWV PVC PIPE, 2" DIAM., 10' HIGH					
Drain, roof, main, PVC, dome type 2" pipe size	1.000	Ea.	127	75.50	202.50
Clamp, roof drain, underdeck	1.000	Ea.	27.50	44	71.50
Pipe, Tee, PVC DWV, schedule 40, 2" pipe size	1.000	Ea.	4.16	52.50	56.66
Pipe, PVC, DWV, schedule 40, 2" diam.	20.000	L.F.	48.40	357	405.40
Pipe, elbow, PVC schedule 40, 2" diam.	2.000	Ea.	5.32	58	63.32
TOTAL			**212.38**	**587**	**799.38**

D2040 210	Roof Drain Systems	COST EACH MAT.	COST EACH INST.	COST EACH TOTAL
1880	Roof drain, DWV PVC, 2" diam., piping, 10' high	212	585	797
1920	For each additional foot add	2.42	17.85	20.27
1960	3" diam., 10' high	277	685	962
2000	For each additional foot add	4.63	19.90	24.53
2040	4" diam., 10' high	335	770	1,105
2080	For each additional foot add	6.35	22	28.35
2120	5" diam., 10' high	985	890	1,875
2160	For each additional foot add	14.15	24.50	38.65
2200	6" diam., 10' high	970	985	1,955
2240	For each additional foot add	11.60	27	38.60
2280	8" diam., 10' high	2,775	1,675	4,450
2320	For each additional foot add	27	34	61
3940	C.I., soil, single hub, service wt., 2" diam. piping, 10' high	410	640	1,050
3980	For each additional foot add	7	16.70	23.70
4120	3" diam., 10' high	580	695	1,275
4160	For each additional foot add	9.80	17.55	27.35
4200	4" diam., 10' high	660	760	1,420
4240	For each additional foot add	12.65	19.15	31.80
4280	5" diam., 10' high	1,050	835	1,885
4320	For each additional foot add	17.40	21.50	38.90
4360	6" diam., 10' high	1,175	895	2,070
4400	For each additional foot add	21.50	22.50	44
4440	8" diam., 10' high	2,975	1,825	4,800
4480	For each additional foot add	33.50	38	71.50
6040	Steel galv. sch 40 threaded, 2" diam. piping, 10' high	550	620	1,170
6080	For each additional foot add	11.40	16.45	27.85
6120	3" diam., 10' high	1,075	895	1,970
6160	For each additional foot add	24	24.50	48.50

D20 Plumbing

D2040 Rain Water Drainage

D2040 210	Roof Drain Systems	COST EACH		
		MAT.	INST.	TOTAL
6200	4" diam., 10' high	1,525	1,150	2,675
6240	For each additional foot add	35.50	29.50	65
6280	5" diam., 10' high	1,650	955	2,605
6320	For each additional foot add	38	28.50	66.50
6360	6" diam., 10' high	1,875	1,225	3,100
6400	For each additional foot add	44.50	39	83.50
6440	8" diam., 10' high	3,675	1,800	5,475
6480	For each additional foot add	53	44.50	97.50

D20 Plumbing

D2090 Other Plumbing Systems

Pipe Material Considerations:

1. Malleable iron fittings should be used for gas service.

2. Malleable fittings are used where there are stresses/strains due to expansion and vibration.

3. Cast iron fittings may be broken as an aid to disassembling of piping joints frozen by long use, temperature and mineral deposits.

4. Cast iron pipe is extensively used for underground and submerged service.

5. Type M (light wall) copper tubing is available in hard temper only and is used for nonpressure and less severe applications than K and L.

6. Type L (medium wall) copper tubing, available hard or soft for interior service.

7. Type K (heavy wall) copper tubing, available in hard or soft temper for use where conditions are severe. For underground and interior service.

8. Hard drawn tubing requires fewer hangers or supports but should not be bent. Silver brazed fittings should be used.

Piping costs include hangers spaced for the material type and size, and the couplings required.

D2090 810	Piping - Installed - Unit Costs		MAT.	INST.	TOTAL
			COST PER L.F.		
0840	Cast iron, soil, B & S, service weight, 2" diameter	RD2010 -030	7	16.70	23.70
0860	3" diameter		9.80	17.55	27.35
0880	4" diameter		12.65	19.15	31.80
0900	5" diameter		17.40	21.50	38.90
0920	6" diameter		21.50	22.50	44
0940	8" diameter		33.50	38	71.50
0960	10" diameter		55.50	41.50	97
0980	12" diameter		79	46.50	125.50
1040	No hub, 1-1/2" diameter		7.55	14.85	22.40
1060	2" diameter		7.85	15.75	23.60
1080	3" diameter		10.80	16.45	27.25
1100	4" diameter		13.95	18.15	32.10
1120	5" diameter		19.85	19.75	39.60
1140	6" diameter		24	20.50	44.50
1160	8" diameter		44	32.50	76.50
1180	10" diameter		73.50	36.50	110
1220	Copper tubing, hard temper, solder, type K, 1/2" diameter		5.50	7.50	13
1260	3/4" diameter		10.10	7.90	18
1280	1" diameter		13.30	8.85	22.15
1300	1-1/4" diameter		16.60	10.45	27.05
1320	1-1/2" diameter		21.50	11.70	33.20
1340	2" diameter		33.50	14.65	48.15
1360	2-1/2" diameter		50	17.55	67.55
1380	3" diameter		70	19.50	89.50
1400	4" diameter		118	27.50	145.50
1480	5" diameter		262	33	295
1500	6" diameter		380	43	423
1520	8" diameter		675	48	723
1560	Type L, 1/2" diameter		4.21	7.25	11.46
1600	3/4" diameter		6.60	7.70	14.30
1620	1" diameter		9.95	8.60	18.55
1640	1-1/4" diameter		13.75	10.10	23.85
1660	1-1/2" diameter		17.70	11.25	28.95
1680	2" diameter		27.50	13.95	41.45
1700	2-1/2" diameter		42	17	59
1720	3" diameter		57	18.80	75.80
1740	4" diameter		96	27	123
1760	5" diameter		204	31	235
1780	6" diameter		282	41	323
1800	8" diameter		495	45.50	540.50
1840	Type M, 1/2" diameter		3.07	6.95	10.02
1880	3/4" diameter		4.85	7.50	12.35
1900	1" diameter		7.60	8.35	15.95
1920	1-1/4" diameter		11.25	9.75	21

D2090 Other Plumbing Systems

D2090 810	Piping - Installed - Unit Costs	COST PER L.F.		
		MAT.	INST.	TOTAL
1940	1-1/2" diameter	15.40	10.85	26.25
1960	2" diameter	24	13.30	37.30
1980	2-1/2" diameter	36.50	16.45	52.95
2000	3" diameter	48.50	18.15	66.65
2020	4" diameter	87	26.50	113.50
2040	5" diameter	204	29.50	233.50
2060	6" diameter	282	39	321
2080	8" diameter	495	43	538
2120	Type DWV, 1-1/4" diameter	12	9.75	21.75
2160	1-1/2" diameter	15.10	10.85	25.95
2180	2" diameter	20.50	13.30	33.80
2200	3" diameter	36	18.15	54.15
2220	4" diameter	63.50	26.50	90
2240	5" diameter	181	29.50	210.50
2260	6" diameter	259	39	298
2280	8" diameter	630	43	673
2800	Plastic, PVC, DWV, schedule 40, 1-1/4" diameter	2.03	13.95	15.98
2820	1-1/2" diameter	2	16.25	18.25
2830	2" diameter	2.42	17.85	20.27
2840	3" diameter	4.63	19.90	24.53
2850	4" diameter	6.35	22	28.35
2890	6" diameter	11.60	27	38.60
3010	Pressure pipe 200 PSI, 1/2" diameter	1.33	10.85	12.18
3030	3/4" diameter	1.63	11.45	13.08
3040	1" diameter	1.80	12.70	14.50
3050	1-1/4" diameter	2.36	13.95	16.31
3060	1-1/2" diameter	2.62	16.25	18.87
3070	2" diameter	3.19	17.85	21.04
3080	2-1/2" diameter	5.65	18.80	24.45
3090	3" diameter	6.60	19.90	26.50
3100	4" diameter	10.90	22	32.90
3110	6" diameter	19.60	27	46.60
3120	8" diameter	30.50	34	64.50
4000	Steel, schedule 40, threaded, black, 1/2" diameter	2.46	9.30	11.76
4020	3/4" diameter	2.90	9.60	12.50
4030	1" diameter	4.26	11.05	15.31
4040	1-1/4" diameter	5.40	11.85	17.25
4050	1-1/2" diameter	6.35	13.15	19.50
4060	2" diameter	8.45	16.45	24.90
4070	2-1/2" diameter	13.25	21	34.25
4080	3" diameter	17.05	24.50	41.55
4090	4" diameter	25	29.50	54.50
4100	Grooved threaded 5" diameter	29	28.50	57.50
4110	6" diameter	37	39	76
4120	8" diameter	57	44.50	101.50
4130	10" diameter	85.50	53	138.50
4140	12" diameter	91	60.50	151.50
4200	Galvanized, 1/2" diameter	3.42	9.30	12.72
4220	3/4" diameter	4.14	9.60	13.74
4230	1" diameter	5.65	11.05	16.70
4240	1-1/4" diameter	7.20	11.85	19.05
4250	1-1/2" diameter	8.55	13.15	21.70
4260	2" diameter	11.40	16.45	27.85
4270	2-1/2" diameter	19.35	21	40.35
4280	3" diameter	24	24.50	48.50
4290	4" diameter	35.50	29.50	65
4300	Grooved 5" diameter	38	28.50	66.50
4310	6" diameter	44.50	39	83.50

D20 Plumbing

D2090 Other Plumbing Systems

D2090 810	Piping - Installed - Unit Costs	COST PER L.F.		
		MAT.	INST.	TOTAL
4320	8" diameter	53	44.50	97.50
4330	10" diameter	79	53	132
4340	12" diameter	102	60.50	162.50
5010	Flanged, black, 1" diameter	9.30	15.95	25.25
5020	1-1/4" diameter	10.70	17.40	28.10
5030	1-1/2" diameter	11.45	19.20	30.65
5040	2" diameter	13.90	25	38.90
5050	2-1/2" diameter	16.40	31	47.40
5060	3" diameter	23	35	58
5070	4" diameter	28.50	43	71.50
5080	5" diameter	39.50	53	92.50
5090	6" diameter	48.50	68	116.50
5100	8" diameter	73.50	89.50	163
5110	10" diameter	120	106	226
5120	12" diameter	142	121	263
5720	Grooved joints, black, 3/4" diameter	3.89	8.25	12.14
5730	1" diameter	4.65	9.30	13.95
5740	1-1/4" diameter	6	10.10	16.10
5750	1-1/2" diameter	6.95	11.45	18.40
5760	2" diameter	8.85	14.65	23.50
5770	2-1/2" diameter	11.10	18.50	29.60
5900	3" diameter	17.70	21	38.70
5910	4" diameter	21	23.50	44.50
5920	5" diameter	29	28.50	57.50
5930	6" diameter	37	39	76
5940	8" diameter	57	44.50	101.50
5950	10" diameter	85.50	53	138.50
5960	12" diameter	91	60.50	151.50

D2090 820	Standard Pipe Fittings - Installed - Unit Costs	COST EACH		
		MAT.	INST.	TOTAL
0100	Cast iron, soil, B&S service weight, 1/8 Bend, 2" diameter	10.80	66	76.80
0110	3" diameter	16.85	75.50	92.35
0120	4" diameter	24.50	81	105.50
0130	5" diameter	34	91	125
0140	6" diameter	42	96.50	138.50
0150	8" diameter	125	203	328
0160	10" diameter	180	223	403
0170	12" diameter	340	248	588
0220	1/4 Bend, 2" diameter	15.10	66	81.10
0230	3" diameter	20.50	75.50	96
0240	4" diameter	31.50	81	112.50
0250	5" diameter	44.50	91	135.50
0260	6" diameter	55	96.50	151.50
0270	8" diameter	166	203	369
0280	10" diameter	242	223	465
0290	12" diameter	330	248	578
0360	Tee, 2" diameter	28	105	133
0370	3" diameter	45.50	117	162.50
0380	4" diameter	58.50	132	190.50
0390	5" diameter	89.50	137	226.50
0400	6" diameter	122	149	271
0410	8" diameter	380	320	700
0810	No hub, 1/8 Bend, 1-1/2" diameter	7.10		7.10
0820	2" diameter	7.10		7.10
0830	3" diameter	10.70		10.70
0840	4" diameter	13.55		13.55

D20 Plumbing

D2090 Other Plumbing Systems

D2090 820	Standard Pipe Fittings - Installed - Unit Costs	COST EACH		
		MAT.	INST.	TOTAL
0850	5" diameter	29		29
0860	6" diameter	31.50		31.50
0870	8" diameter	90.50		90.50
0880	10" diameter	168		168
0960	1/4 Bend, 1-1/2" diameter	8.45		8.45
0970	2" diameter	9.30		9.30
0980	3" diameter	12.80		12.80
0990	4" diameter	18.50		18.50
1000	5" diameter	42.50		42.50
1010	6" diameter	46.50		46.50
1020	8" diameter	129		129
1080	Sanitary tee, 1-1/2" diameter	11.80		11.80
1090	2" diameter	12.80		12.80
1110	3" diameter	15.65		15.65
1120	4" diameter	24		24
1130	5" diameter	69		69
1140	6" diameter	70.50		70.50
1150	8" diameter	285		285
1300	Coupling clamp & gasket, 1-1/2" diameter	7.85	22	29.85
1310	2" diameter	8.25	24	32.25
1320	3" diameter	9.90	27.50	37.40
1330	4" diameter	13.55	32	45.55
1340	5" diameter	22	37	59
1350	6" diameter	23	41	64
1360	8" diameter	105	67.50	172.50
1370	10" diameter	183	86	269
1510	Steel pipe fitting, CI, threaded, 125 Lb. black, 45° Elbow, 1/2" diameter	6.20	39	45.20
1520	3/4" diameter	6.25	42	48.25
1530	1" diameter	7.30	45	52.30
1540	1-1/4" diameter	9.85	48	57.85
1550	1-1/2" diameter	16.30	52.50	68.80
1560	2" diameter	18.80	58.50	77.30
1570	2-1/2" diameter	49	75.50	124.50
1580	3" diameter	77.50	105	182.50
1590	4" diameter	161	176	337
1600	5" diameter	179	106	285
1610	6" diameter	222	132	354
1620	8" diameter	410	156	566
1700	90° Elbow, 1/2" diameter	4.06	39	43.06
1710	3/4" diameter	4.22	42	46.22
1720	1" diameter	4.99	45	49.99
1730	1-1/4" diameter	7.10	48	55.10
1740	1-1/2" diameter	9.80	52.50	62.30
1750	2" diameter	15.30	58.50	73.80
1760	2-1/2" diameter	37	75.50	112.50
1770	3" diameter	60	105	165
1780	4" diameter	112	176	288
1790	5" diameter	179	106	285
1800	6" diameter	222	132	354
1810	8" diameter	410	156	566
1900	Tee, 1/2" diameter	6.30	65	71.30
1910	3/4" diameter	7.35	65	72.35
1920	1" diameter	6.55	73	79.55
1930	1-1/4" diameter	11.95	75.50	87.45
1940	1-1/2" diameter	15.55	81	96.55
1950	2" diameter	21.50	96	117.50
1960	2-1/2" diameter	56	117	173
1970	3" diameter	87	176	263

D2090 Other Plumbing Systems

D2090 820	Standard Pipe Fittings - Installed - Unit Costs	COST EACH		
		MAT.	INST.	TOTAL
1980	4" diameter	168	263	431
1990	5" diameter	283	161	444
2000	6" diameter	345	196	541
2010	8" diameter	665	234	899
2300	Copper, wrought, solder joints, 45° Elbow, 1/2" diameter	2.49	29.50	31.99
2310	3/4" diameter	4.36	31	35.36
2320	1" diameter	10.95	36.50	47.45
2330	1-1/4" diameter	14.80	39	53.80
2340	1-1/2" diameter	17.80	45	62.80
2350	2" diameter	30	53	83
2360	2-1/2" diameter	63.50	81	144.50
2370	3" diameter	94	81	175
2380	4" diameter	200	117	317
2390	5" diameter	775	176	951
2400	6" diameter	1,225	182	1,407
2500	90° Elbow, 1/2" diameter	1.35	29.50	30.85
2510	3/4" diameter	3.04	31	34.04
2520	1" diameter	7.45	36.50	43.95
2530	1-1/4" diameter	11.30	39	50.30
2540	1-1/2" diameter	17.75	45	62.75
2550	2" diameter	32	53	85
2560	2-1/2" diameter	64	81	145
2610	3" diameter	86	96	182
2620	4" diameter	219	117	336
2630	5" diameter	920	176	1,096
2640	6" diameter	1,225	182	1,407
2700	Tee, 1/2" diameter	2.32	45	47.32
2710	3/4" diameter	5.60	49	54.60
2720	1" diameter	17.25	58.50	75.75
2730	1-1/4" diameter	23.50	65	88.50
2740	1-1/2" diameter	36.50	73	109.50
2750	2" diameter	56.50	83.50	140
2760	2-1/2" diameter	114	132	246
2770	3" diameter	175	151	326
2780	4" diameter	420	211	631
2790	5" diameter	1,375	263	1,638
2800	6" diameter	1,900	273	2,173
2880	Coupling, 1/2" diameter	1.03	26.50	27.53
2890	3/4" diameter	2.06	28	30.06
2900	1" diameter	4.10	32.50	36.60
2910	1-1/4" diameter	7.30	34.50	41.80
2920	1-1/2" diameter	9.65	39	48.65
2930	2" diameter	16.10	45	61.10
2940	2-1/2" diameter	34	70	104
2950	3" diameter	51.50	81	132.50
2960	4" diameter	108	151	259
2970	5" diameter	266	176	442
2980	6" diameter	440	205	645
3200	Malleable iron, 150 Lb. threaded, black, 45° elbow, 1/2" diameter	3.70	39	42.70
3210	3/4" diameter	4.58	42	46.58
3220	1" diameter	5.75	45	50.75
3230	1-1/4" diameter	10.20	48	58.20
3240	1-1/2" diameter	12.60	52.50	65.10
3250	2" diameter	19.05	58.50	77.55
3260	2-1/2" diameter	55	75.50	130.50
3270	3" diameter	71.50	105	176.50
3280	4" diameter	141	176	317
3290	5" diameter	179	106	285

D2090 Other Plumbing Systems

D2090 820	Standard Pipe Fittings - Installed - Unit Costs	COST EACH		
		MAT.	INST.	TOTAL
3300	6" diameter	222	132	354
3400	90° Elbow, 1/2" diameter	3.23	39	42.23
3500	3/4" diameter	3.71	42	45.71
3510	1" diameter	4.70	45	49.70
3520	1-1/4" diameter	7.75	48	55.75
3530	1-1/2" diameter	10.20	52.50	62.70
3540	2" diameter	18.35	58.50	76.85
3550	2-1/2" diameter	39	75.50	114.50
3560	3" diameter	57.50	105	162.50
3570	4" diameter	123	176	299
3580	5" diameter	179	106	285
3590	6" diameter	222	132	354
3700	Tee, 1/2" diameter	2.99	65	67.99
3800	3/4" diameter	4.30	65	69.30
3810	1" diameter	7.35	73	80.35
3820	1-1/4" diameter	11.90	75.50	87.40
3830	1-1/2" diameter	14.85	81	95.85
3910	2" diameter	25.50	96	121.50
3920	2-1/2" diameter	54.50	117	171.50
3930	3" diameter	80.50	176	256.50
3940	4" diameter	194	263	457
3950	5" diameter	283	161	444
3960	6" diameter	169	98.50	267.50
4000	Coupling, 1/2" diameter	3.09	31	34.09
4010	3/4" diameter	3.63	32.50	36.13
4020	1" diameter	5.40	39	44.40
4030	1-1/4" diameter	7.05	40.50	47.55
4040	1-1/2" diameter	9.50	44	53.50
4050	2" diameter	14.10	50	64.10
4060	2-1/2" diameter	39	58.50	97.50
4070	3" diameter	52.50	75.50	128
4080	4" diameter	106	105	211
4090	5" diameter	40	26.50	66.50
4100	6" diameter	53	33	86
5100	Plastic, PVC, high impact/pressure sch 40, 45° Elbow, 1/2" diameter	.74	17.55	18.29
5110	3/4" diameter	1.14	20.50	21.64
5120	1" diameter	1.38	23.50	24.88
5130	1-1/4" diameter	1.90	26.50	28.40
5140	1-1/2" diameter	2.40	29.50	31.90
5150	2" diameter	3.12	29	32.12
5160	3" diameter	12.65	46	58.65
5170	4" diameter	22.50	58	80.50
5180	6" diameter	56	95	151
5190	8" diameter	135	159	294
5260	90° Elbow, 1/2" diameter	.45	17.55	18
5270	3/4" diameter	.50	20.50	21
5280	1" diameter	.90	23.50	24.40
5290	1-1/4" diameter	1.58	26.50	28.08
5300	1-1/2" diameter	1.71	29.50	31.21
5400	2" diameter	2.66	29	31.66
5410	3" diameter	9.70	46	55.70
5420	4" diameter	17.40	58	75.40
5430	6" diameter	55.50	95	150.50
5440	8" diameter	142	159	301
5500	Tee, 1/2" diameter	.55	26.50	27.05
5510	3/4" diameter	.64	31	31.64
5520	1" diameter	1.19	35	36.19
5530	1-1/4" diameter	1.85	39.50	41.35

D2090 Other Plumbing Systems

D2090 820	Standard Pipe Fittings - Installed - Unit Costs	COST EACH		
		MAT.	INST.	TOTAL
5540	1-1/2" diameter	2.27	44	46.27
5550	2" diameter	3.29	43.50	46.79
5560	3" diameter	14.25	69.50	83.75
5570	4" diameter	26	87	113
5580	6" diameter	87	142	229
5590	8" diameter	201	241	442
5680	Coupling, 1/2" diameter	.30	17.55	17.85
5690	3/4" diameter	.41	20.50	20.91
5700	1" diameter	.73	23.50	24.23
5710	1-1/4" diameter	.98	26.50	27.48
5720	1-1/2" diameter	.98	26.50	27.48
5730	2" diameter	1.60	29	30.60
5740	3" diameter	5.50	46	51.50
5750	4" diameter	8	58	66
6010	6" diameter	25	95	120
6020	8" diameter	47	159	206
7810	DWV, socket joints, sch. 40, 1/8 Bend, 1-1/4" diameter	2.95	29	31.95
7820	1-1/2" diameter	1.50	32	33.50
7830	2" diameter	2.31	32	34.31
7840	3" diameter	6.35	50.50	56.85
7850	4" diameter	10.90	64	74.90
7860	6" diameter	46	104	150
7960	1/4 bend, 1-1/4" diameter	6.30	29	35.30
8000	1-1/2" diameter	2.09	32	34.09
8010	2" diameter	3.18	32	35.18
8020	3" diameter	8.05	50.50	58.55
8030	4" diameter	15.15	64	79.15
8210	6" diameter	70	104	174
8300	Sanitary tee, 1-1/4" diameter	4.51	43.50	48.01
8310	1-1/2" diameter	2.83	48.50	51.33
8320	2" diameter	4.16	52.50	56.66
8330	3" diameter	10.65	76	86.65
8340	4" diameter	18.55	96	114.55
8400	Coupling, 1-1/4" diameter	23	29	52
8410	1-1/2" diameter	24.50	32	56.50
8420	2" diameter	29	32	61
8430	3" diameter	60.50	50.50	111
8440	4" diameter	123	64	187

D3010 Energy Supply

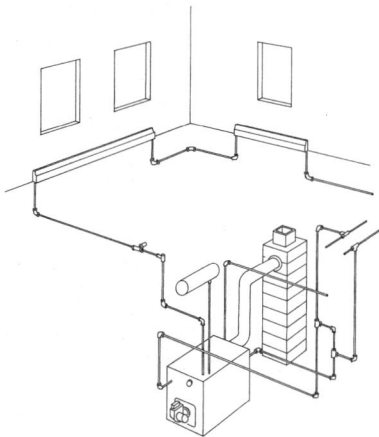

Basis for Heat Loss Estimate, Apartment Type Structures:

1. Masonry walls and flat roof are insulated. U factor is assumed at .08.
2. Window glass area taken as BOCA minimum, 1/10th of floor area. Double insulating glass with 1/4″ air space, U = .65.
3. Infiltration = 0.3 C.F. per hour per S.F. of net wall.
4. Concrete floor loss is 2 BTUH per S.F.
5. Temperature difference taken as 70°F.
6. Ventilating or makeup air has not been included and must be added if desired. Air shafts are not used.

System Components	QUANTITY	UNIT	COST EACH		
			MAT.	INST.	TOTAL
SYSTEM D3010 510 1760					
HEATING SYSTEM, FIN TUBE RADIATION, FORCED HOT WATER					
1,000 S.F. AREA, 10,000 C.F. VOLUME					
Boiler, oil fired, CI, burner, ctrls/insul/breech/pipe/ftng/valves, 109 MBH	1.000	Ea.	3,631.25	3,281.25	6,912.50
Circulating pump, CI flange connection, 1/12 HP	1.000	Ea.	315	176	491
Expansion tank, painted steel, ASME 18 Gal capacity	1.000	Ea.	520	76	596
Storage tank, steel, above ground, 275 Gal capacity w/supports	1.000	Ea.	420	213	633
Copper tubing type L, solder joint, hanger 10′ OC, 3/4″ diam	100.000	L.F.	660	770	1,430
Radiation, 3/4″ copper tube w/alum fin baseboard pkg, 7″ high	30.000	L.F.	252	552	804
Pipe covering, calcium silicate w/cover, 1″ wall, 3/4″ diam	100.000	L.F.	282	585	867
TOTAL			6,080.25	5,653.25	11,733.50
COST PER S.F.			6.08	5.65	11.73

D3010 510	Apartment Building Heating - Fin Tube Radiation		COST PER S.F.		
			MAT.	INST.	TOTAL
1740	Heating systems, fin tube radiation, forced hot water				
1760	1,000 S.F. area, 10,000 C.F. volume	RD3020 -010	6.08	5.65	11.73
1800	10,000 S.F. area, 100,000 C.F. volume		2.84	3.42	6.26
1840	20,000 S.F. area, 200,000 C.F. volume	RD3020 -020	3.31	3.83	7.14
1880	30,000 S.F. area, 300,000 C.F. volume		3.19	3.72	6.91
1890					

D3010 Energy Supply

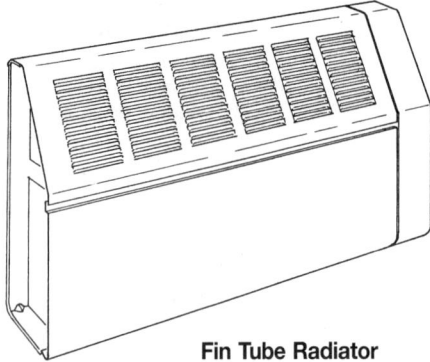

Fin Tube Radiator

Basis for Heat Loss Estimate, Factory or Commercial Type Structures:

1. Walls and flat roof are of lightly insulated concrete block or metal. U factor is assumed at .17.
2. Windows and doors are figured at 1-1/2 S.F. per linear foot of building perimeter. The U factor for flat single glass is 1.13.
3. Infiltration is approximately 5 C.F. per hour per S.F. of wall.
4. Concrete floor loss is 2 BTUH per S.F. of floor.
5. Temperature difference is assumed at 70°F.
6. Ventilation or makeup air has not been included and must be added if desired.

System Components	QUANTITY	UNIT	COST EACH		
			MAT.	INST.	TOTAL
SYSTEM D3010 520 1960					
HEATING SYSTEM, FIN TUBE RADIATION, FORCED HOT WATER					
1,000 S.F. BLDG., ONE FLOOR					
Boiler, oil fired, CI, burner/ctrls/insul/breech/pipe/ftngs/valves, 109 MBH	1.000	Ea.	3,631.25	3,281.25	6,912.50
Expansion tank, painted steel, ASME 18 Gal capacity	1.000	Ea.	2,125	89	2,214
Storage tank, steel, above ground, 550 Gal capacity w/supports	1.000	Ea.	2,525	395	2,920
Circulating pump, CI flanged, 1/8 HP	1.000	Ea.	525	176	701
Pipe, steel, black, sch. 40, threaded, cplg & hngr 10'OC, 1-1/2" diam.	260.000	L.F.	1,651	3,419	5,070
Pipe covering, calcium silicate w/cover , 1" wall, 1-1/2" diam	260.000	L.F.	743.60	1,573	2,316.60
Radiation, steel 1-1/4" tube & 4-1/4" fin w/cover & damper, wall hung	42.000	L.F.	1,701	1,239	2,940
Rough in, steel fin tube radiation w/supply & balance valves	4.000	Set	900	2,600	3,500
TOTAL			13,801.85	12,772.25	26,574.10
COST PER S.F.			13.80	12.77	26.57

D3010 520	Commercial Building Heating - Fin Tube Radiation		COST PER S.F.		
			MAT.	INST.	TOTAL
1940	Heating systems, fin tube radiation, forced hot water				
1960	1,000 S.F. bldg, one floor		13.80	12.80	26.60
2000	10,000 S.F., 100,000 C.F., total two floors		3.87	4.86	8.73
2040	100,000 S.F., 1,000,000 C.F., total three floors	RD3020 -010	1.61	2.16	3.77
2080	1,000,000 S.F., 10,000,000 C.F., total five floors		.82	1.15	1.97
2090		RD3020 -020			

D30 HVAC

D3010 Energy Supply

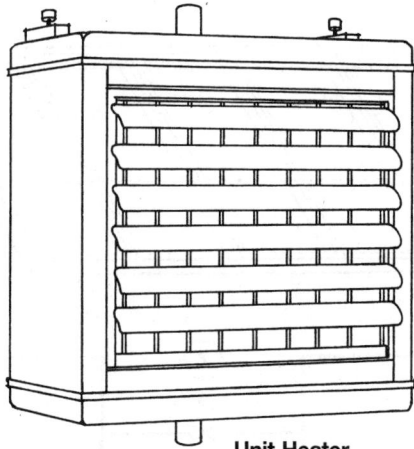

Unit Heater

Basis for Heat Loss Estimate, Factory or Commercial Type Structures:

1. Walls and flat roof are of lightly insulated concrete block or metal. U factor is assumed at .17.
2. Windows and doors are figured at 1-1/2 S.F. per linear foot of building perimeter. The U factor for flat single glass is 1.13.
3. Infiltration is approximately 5 C.F. per hour per S.F. of wall.
4. Concrete floor loss is 2 BTUH per S.F. of floor.
5. Temperature difference is assumed at 70°F.
6. Ventilation or makeup air has not been included and must be added if desired.

System Components	QUANTITY	UNIT	COST EACH		
			MAT.	INST.	TOTAL
SYSTEM D3010 530 1880					
HEATING SYSTEM, TERMINAL UNIT HEATERS, FORCED HOT WATER					
1,000 S.F. BLDG., ONE FLOOR					
Boiler oil fired, CI, burner/ctrls/insul/breech/pipe/ftngs/valves, 109 MBH	1.000	Ea.	3,631.25	3,281.25	6,912.50
Expansion tank, painted steel, ASME 18 Gal capacity	1.000	Ea.	2,125	89	2,214
Storage tank, steel, above ground, 550 Gal capacity w/supports	1.000	Ea.	2,525	395	2,920
Circulating pump, CI, flanged, 1/8 HP	1.000	Ea.	525	176	701
Pipe, steel, black, schedule 40, threaded, cplg & hngr 10'OC 1-1/2" diam	260.000	L.F.	1,651	3,419	5,070
Pipe covering, calcium silicate w/cover, 1" wall, 1-1/2" diam	260.000	L.F.	743.60	1,573	2,316.60
Unit heater, 1 speed propeller, horizontal, 200° EWT, 26.9 MBH	2.000	Ea.	1,040	266	1,306
Unit heater piping hookup with controls	2.000	Set	1,080	2,500	3,580
TOTAL			13,320.85	11,699.25	25,020.10
COST PER S.F.			13.32	11.70	25.02

D3010 530	Commercial Bldg. Heating - Terminal Unit Heaters		COST PER S.F.		
			MAT.	INST.	TOTAL
1860	Heating systems, terminal unit heaters, forced hot water				
1880	1,000 S.F. bldg., one floor		13.30	11.70	25
1920	10,000 S.F. bldg., 100,000 C.F. total two floors	RD3020 -010	3.44	4.02	7.46
1960	100,000 S.F. bldg., 1,000,000 C.F. total three floors		1.60	1.98	3.58
2000	1,000,000 S.F. bldg., 10,000,000 C.F. total five floors	RD3020 -020	1.03	1.28	2.31
2010					

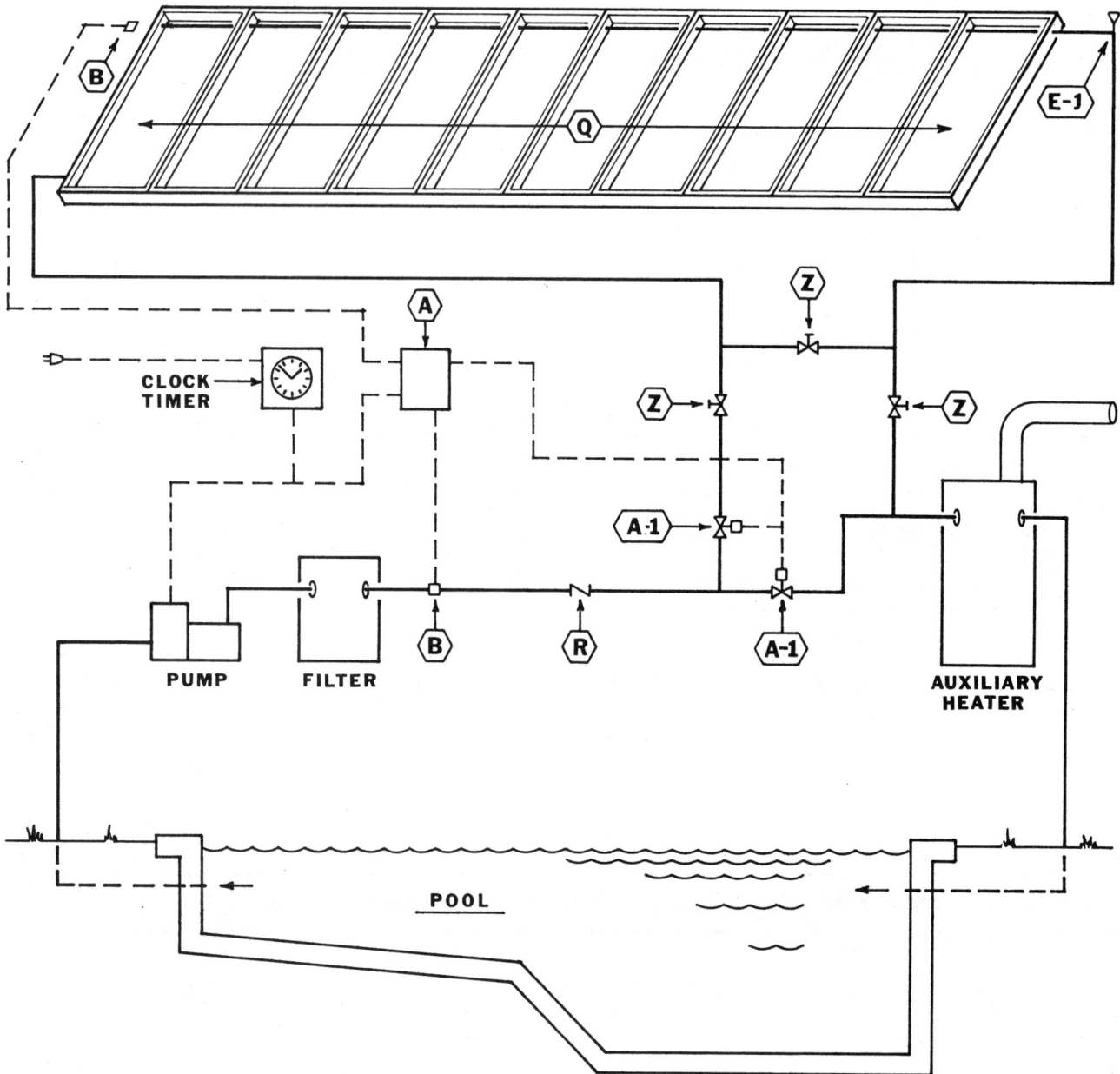

B

Q

E-1

A

Z

CLOCK TIMER

Z

Z

A·1

B

R

A·1

PUMP

FILTER

AUXILIARY HEATER

POOL

This draindown pool system uses a differential thermostat similar to those used in solar domestic hot water and space heating applications. To heat the pool, the pool water passes through the conventional pump-filter loop and then flows through the collectors. When collection is not possible, or when the pool temperature is reached, all water drains from the solar loop back to the pool through the existing piping. The modes are controlled by solenoid valves or other automatic valves in conjunction with a vacuum breaker relief valve, which facilitates draindown.

D30 HVAC

D3010 Energy Supply

System Components	QUANTITY	UNIT	COST EACH MAT.	INST.	TOTAL
SYSTEM D3010 660 2640					
SOLAR SWIMMING POOL HEATER, ROOF MOUNTED COLLECTORS					
TEN 4′ X 10′ FULLY WETTED UNGLAZED PLASTIC ABSORBERS					
A Differential thermostat/controller, 110V, adj pool pump system	1.000	Ea.	305	293	598
A-1 Solenoid valve, PVC, normally 1 open 1 closed (included)	2.000	Ea.			
B Sensor, thermistor type (included)	2.000	Ea.			
E-1 Valve, vacuum relief	1.000	Ea.	29	18.30	47.30
Q Collector panel, solar energy, plastic, liquid full wetted, 4′ x 10′	10.000	Ea.	2,570	2,110	4,680
R Valve, ball check, PVC, socket, 1-1/2″ diam	1.000	Ea.	83	29.50	112.50
Z Valve, ball, PVC, socket, 1-1/2″ diam	3.000	Ea.	156	88.50	244.50
Pipe, PVC, sch 40, 1-1/2″ diam	80.000	L.F.	240	1,300	1,540
Pipe fittings, PVC sch 40, socket joint, 1-1/2″ diam	10.000	Ea.	17.10	295	312.10
Sensor wire, #22-2 conductor, multistranded	.500	C.L.F.	8.45	28	36.45
Roof clamps for solar energy collector panels	10.000	Set	25.90	150.50	176.40
Roof strap, teflon for solar energy collector panels	26.000	L.F.	572	74.10	646.10
TOTAL			4,006.45	4,386.90	8,393.35

D3010 660	Solar Swimming Pool Heater Systems		COST EACH MAT.	INST.	TOTAL
2530	Solar swimming pool heater systems, roof mounted collectors				
2540	10 ea. 3′x7′ black chrome absorber, 1/8″ temp. glass		8,725	3,375	12,100
2560	10 ea. 4′x8′ black chrome absorber, 3/16″ temp. glass	RD3010 -600	10,300	4,025	14,325
2580	10 ea. 3′8″x6′ flat black absorber, 3/16″ temp. glass		7,625	3,450	11,075
2600	10 ea. 4′x9′ flat black absorber, plastic glazing		8,975	4,200	13,175
2620	10 ea. 2′x9′ rubber absorber, plastic glazing		12,400	4,525	16,925
2640	10 ea. 4′x10′ fully wetted unglazed plastic absorber		4,000	4,375	8,375
2660	Ground mounted collectors				
2680	10 ea. 3′x7′ black chrome absorber, 1/8″ temp. glass		8,825	3,900	12,725
2700	10 ea. 4′x8′ black chrome absorber, 3/16″ temp glass		10,400	4,550	14,950
2720	10 ea. 3′8″x6′ flat blk absorber, 3/16″ temp. glass		7,725	3,950	11,675
2740	10 ea. 4′x9′ flat blk absorber, plastic glazing		9,075	4,700	13,775
2760	10 ea. 2′x9′ rubber absorber, plastic glazing		12,400	4,875	17,275
2780	10 ea. 4′x10′ fully wetted unglazed plastic absorber		4,100	4,900	9,000

D3010 Energy Supply

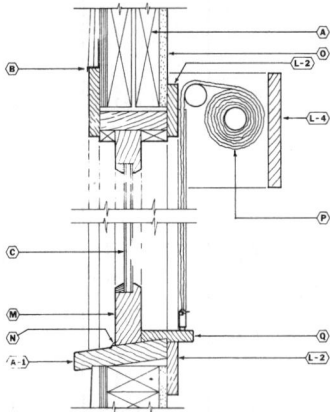

Solar Direct Gain Glazing — Door Panels: In a Direct Gain System the collection, absorption, and storage of solar energy occur directly within the building's living space. In this system, a section of the building's south wall is removed and replaced with glazing. The glazing in the System Component example consists of three standard insulating glass door panels preassembled in wood frames. A direct gain system also includes a thermal storage mass and night insulation.

During the heating season, solar radiation enters the living space directly through the glazing and is converted to heat, which either warms the air or is stored in the mass. The storage mass is typically masonry or water. Masonry often serves as a basic structural component of the building (e.g., a wall or floor). In this system the cost of adding mass to the building is not included. To reduce heat loss through the south facing glass, thereby increasing overall thermal performance, night insulation is included in this system.

System Components		QUANTITY	UNIT	COST EACH		
				MAT.	INST.	TOTAL
SYSTEM D3010 682 2600						
SOLAR PASSIVE HEATING, DIRECT GAIN, 3'X6'8" DOUBLE GLAZED DOOR PANEL						
3 PANELS WIDE						
A	Framing headers over openings, fir 2" x 12"	.025	M.B.F.	22.80	42.75	65.55
A-1	Treated wood sleeper, 1" x 2"	.002	M.B.F.	5	5.10	10.10
B	Aluminum flashing, mill finish, .013 thick	5.000	S.F.	5.92	25.44	31.36
	Studs, 8' high wall, 2" x 4"	.020	M.B.F.	9.60	21.50	31.10
L-2	Interior casing, stock pine, 11/16" x 2-1/2"	25.000	L.F.	36.50	51.50	88
L-4	Valance 1" x 8"	12.000	L.F.	17.64	29.76	47.40
M	Glass slid, Vinyl clad, insulated glass, 6'-0" x 6'-10" high	1.000	Ea.	1,888	317.44	2,205.44
N	Caulking/sealants, silicone rubber, cartridges	.250	Gal.	11.25		11.25
O	Drywall, gypsum plasterboard, nailed to studs, 5/8" standard	39.000	S.F.	14.43	19.50	33.93
O	For taping and finishing, add	39.000	S.F.	1.95	19.50	21.45
P	Mylar shade, dbl. layer heat reflective	49.640	S.F.	496.40	35.74	532.14
Q	Stool cap, pine, 11/16" x 3-1/2"	12.000	L.F.	30	29.76	59.76
	Paint wood trim	40.000	S.F.	4.80	21.20	26
	Demolition, framing	84.000	S.F.		16.38	16.38
	Demolition, shingles	84.000	S.F.		94.08	94.08
	Demolition, drywall	84.000	S.F.		32.76	32.76
	TOTAL			2,544.29	762.41	3,306.70

D3010 682	Solar Direct Gain Glazing, Door Panels	COST EACH		
		MAT.	INST.	TOTAL
2550	Solar passive heating, direct gain, 3'x6'8", double glazed door			
2560	One panel wide	1,250	410	1,660
2580	Two panels wide	1,925	595	2,520
2600	Three panels wide	2,550	760	3,310

D3010 Energy Supply

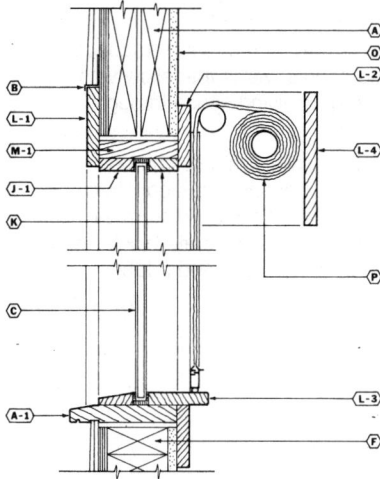

Solar Direct Gain Glazing — Window Units: In a Direct Gain System, collection, absorption, and storage of solar energy occur directly within the building's living space. In this system, a section of the building's south wall is removed and replaced with glazing. The glazing consists of standard insulating replacement glass units in wood site-built frames. A direct gain system also includes a thermal storage mass and night insulation.

During the heating season, solar radiation enters the living space directly through the glazing and is converted to heat, which either warms the air or is stored in the mass. The thermal storage mass is typically masonry or water. Masonry often serves as a basic structural component of the building (e.g., a wall or floor). In this system the cost of adding mass to the building is not included. To reduce heat loss through the south facing glass, thereby increasing overall thermal performance, night insulation is included in this system.

System Components		QUANTITY	UNIT	COST EACH MAT.	COST EACH INST.	COST EACH TOTAL
SYSTEM D3010 684 2600						
SOLAR PASSIVE HEATING, DIRECT GAIN, 2'-6"X5' DBL. GLAZED WINDOW PANEL						
3 PANELS WIDE						
A	Framing headers over openings, fir 2" x 8"	.022	M.B.F.	12.87	48.40	61.27
A-1	Wood exterior sill, oak, no horns, 8/4 x 8" deep	8.000	L.F.	168	79.20	247.20
B	Aluminum flashing, mill finish, .013 thick	4.160	S.F.	3.08	13.23	16.31
C	Glazing, insul glass 2 lite 3/16" float 5/8" thick unit, 15-30 S.F.	38.000	S.F.	541.50	393.30	934.80
F	2"x4" sill blocking	.013	M.B.F.	6.24	38.03	44.27
	Mullion	.027	M.B.F.	12.96	29.03	41.99
J-1	Exterior window stop 1" x 2"	45.000	L.F.	13.50	67.50	81
	Exterior sill stock water repellent 1" x 3"	8.000	L.F.	20	19.84	39.84
K	Interior window stop 1" x 3"	45.000	L.F.	34.20	92.70	126.90
L-1	Exterior casing	28.000	L.F.	27.84	59.52	87.36
L-2	Interior casing 1" x 4"	28.000	L.F.	58.24	64.68	122.92
L-3	Apron	8.000	M.B.F.	16.56	18	34.56
L-4	Valance 1" x 8"	10.000	L.F.	14.70	24.80	39.50
M-1	Window frame 2" x 6"	52.000	L.F.	113.36	156	269.36
O	Drywall, gypsum plasterboard, nailed/screw to studs, 5/8" standard	39.000	S.F.	14.43	19.50	33.93
O	For taping and finishing, add	39.000	S.F.	1.95	19.50	21.45
	Demolition, wall framing	60.000	S.F.		13.26	13.26
	Demolition, shingles	60.000	S.F.		67.20	67.20
P	Mylar triple layer heat reflective shade (see desc. sheet)	37.500	S.F.	436.88	27	463.88
	Demolition, drywall	60.000	S.F.		23.40	23.40
	Paint interior & exterior trim	200.000	L.F.	24	106	130
	TOTAL			1,520.31	1,380.09	2,900.40

D3010 684	Solar Direct Gain Glazing-Window Panels	COST EACH MAT.	COST EACH INST.	COST EACH TOTAL
2550	Solar passive heating, direct gain, 2'-6"x5', double glazed window			
2560	One panel wide	555	500	1,055
2580	Two panels wide	1,125	1,325	2,450
2600	Three panels wide	1,525	1,375	2,900

D3010 Energy Supply

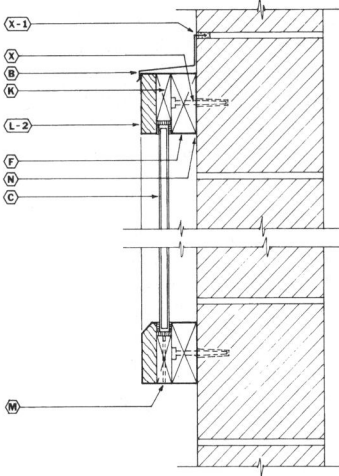

Solar Indirect Gain — Thermal Wall: The thermal storage wall is an Indirect Gain System because collection, absorption, and storage of solar energy occur outside the living space. This system consists of an unvented existing masonry wall located between south facing glazing and the space to be heated. The outside surface of the masonry wall is painted black to enable it to absorb maximum thermal energy. Standard double glazed units in wooden site-built frames are added to the south side of the building.

Thermal energy conducts through the masonry wall and radiates into the living space. A masonry thermal storage wall can also contain vents that allow warm air to flow into the living space through convection.

Indirect gain solar systems can be retrofitted to masonry walls constructed from a wide variety of building materials, including concrete, concrete block (solid and filled), brick, stone and adobe.

System Components		QUANTITY	UNIT	COST EACH		
				MAT.	INST.	TOTAL
SYSTEM D3010 686 2600						
SOLAR PASSIVE HEATING, INDIRECT GAIN, THERMAL STORAGE WALL						
3' X 6'-8" DOUBLE GLAZED PANEL, THREE PANELS WIDE						
B	Aluminum flashing, mill finish, .013 thick	5.500	S.F.	4.07	17.49	21.56
C	Glazing, insul glass 2 lite 3/16" float 5/8" thick unit, 15-30 S.F.	54.000	S.F.	769.50	558.90	1,328.40
F	Framing 2" x 4" lumber	.030	M.B.F.	14.40	59.25	73.65
K	Molding, miscellaneous, 1" x 3"	44.000	L.F.	33.44	90.64	124.08
L-2	Wood batten, 1" x 4", custom pine or cedar	44.000	L.F.	33.44	109.12	142.56
M	Drill holes, 4 per panel 1/2" diameter	30.000	Inch		33	33
N	Caulking/sealants, silicone rubber, cartridges	.250	Gal.	11.25		11.25
X	Lag screws 3/8" x 3" long	20.000	Ea.	5.80	66	71.80
X-1	Reglet	8.000	L.F.		29.60	29.60
	Trim, primer & 1 coat	44.000	L.F.	7.92	29.48	37.40
	Layout and drill holes for lag screws	20.000	Ea.	1.20	157	158.20
	Expansion shields for lag screws	20.000	Ea.	6.20	62	68.20
	TOTAL			887.22	1,212.48	2,099.70

D3010 686		Passive Solar Indirect Gain Wall	COST EACH		
			MAT.	INST.	TOTAL
2550	Solar passive heating, indirect gain				
2560	3'x6'8", double glazed thermal storage wall, one panel wide		310	460	770
2580	Two panels wide		600	820	1,420
2600	Three panels wide		885	1,200	2,085

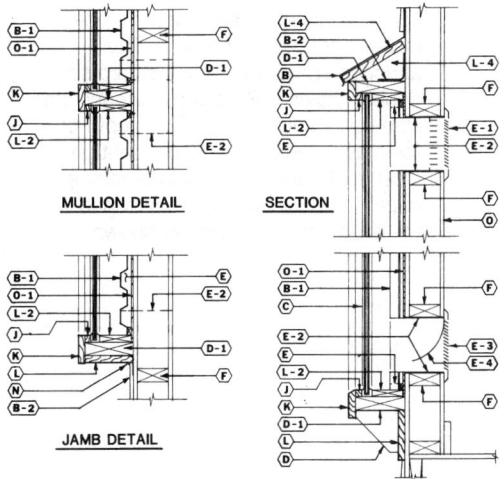

MULLION DETAIL

SECTION

JAMB DETAIL

Solar Indirect Gain — Thermosyphon Panel: The Thermosyphon Air Panel (TAP) System is an indirect gain system used primarily for daytime heating because it does not include thermal mass. This site-built TAP System makes use of double glazing and a ribbed aluminum absorber plate. The panel is framed in wood and attached to the south side of a building. Solar radiation passes through the glass and is absorbed by the absorber plate. As the absorber plate heats, the air between it and the glass also heats. Vents cut through the building's south wall and the absorber plate near the top and bottom of the TAP allow hot air to circulate into the building by means of natural convection.

System Components		QUANTITY	UNIT	COST EACH		
				MAT.	INST.	TOTAL
SYSTEM D3010 688 2600						
SOLAR PASSIVE HEATING, INDIRECT GAIN, THERMOSYPHON PANEL						
3'X6'8" DOUBLE GLAZED PANEL, THREE PANELS WIDE						
B	Aluminum flashing and sleeves, mill finish, .013 thick	8.000	S.F.	17.02	73.14	90.16
B-1	Aluminum ribbed, 4" pitch, on steel frame .032" mill finish	45.000	S.F.	104.85	112.95	217.80
B-2	Angle brackets	9.000	Ea.	6.03	25.47	31.50
C	Glazing, insul glass 2 lite 3/16" float 5/8" thick unit, 15-30 S.F.	45.000	S.F.	641.25	465.75	1,107
D	Support brackets, fir 2"x6"	.003	M.B.F.	1.44	8.78	10.22
D-1	Framing fir 2" x 6"	.040	M.B.F.	30.20	99	129.20
E	Corrugated end closure	15.000	L.F.	12.45	36.45	48.90
E-1	Register, multilouvre operable 6" x 30"	3.000	Ea.	180	97.50	277.50
E-3	Grille 6" x 30"	3.000	Ea.	120	76.50	196.50
E-4	Back draft damper	.038	C.S.F.	.25	.51	.76
F	House wall blocking 2" x 4"	.012	M.B.F.	19.20	48	67.20
J	Absorber supports & glass stop	54.000	L.F.	19.98	99.36	119.34
K	Exterior & bottom trim & interior glazing stop	40.000	L.F.	75.20	186.12	261.32
L	Support ledger & side trim	24.000	L.F.	30.48	47.52	78
L-2	Trim cap	8.000	L.F.	11.76	19.84	31.60
N	Caulking/sealants, silicone rubber, cartridges	.250	Gal.	11.25		11.25
O	Drywall, nailed & taped, 5/8" thick	60.000	S.F.	22.20	30	52.20
O	For taping and finishing, add	60.000	S.F.	3	30	33
O-1	Insulation foil faced exterior wall	45.000	S.F.	21.60	16.65	38.25
	Demolition, framing	60.000	S.F.		16.38	16.38
	Demolition, drywall	60.000	S.F.		23.40	23.40
	Paint absorber plate black	45.000	S.F.	4.50	47.70	52.20
	Finish paint	125.000	S.F.	15	66.25	81.25
	Paint walls & ceiling with roller, primer & 1 coat	60.000	S.F.	6.60	25.80	32.40
	TOTAL			1,354.26	1,653.07	3,007.33

D3010 688		Passive Solar Indirect Gain Panel	COST EACH		
			MAT.	INST.	TOTAL
2550	Solar passive heating, indirect gain				
2560	3'x6'8", double glazed thermosyphon panel, one panel wide		465	585	1,050
2580	Two panels wide		910	1,125	2,035
2600	Three panels wide		1,350	1,650	3,000

D3010 Energy Supply

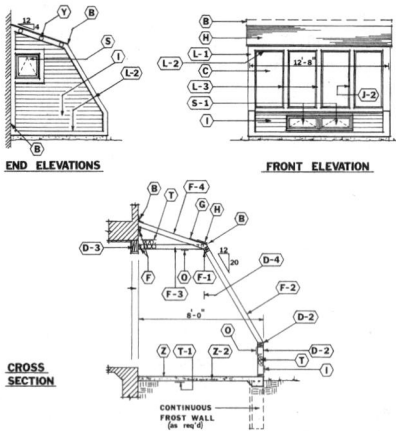

END ELEVATIONS

FRONT ELEVATION

CROSS SECTION

CONTINUOUS
FROST WALL
(as req'd)

Solar Attached Sunspace: The attached sunspace, a room adjacent to the south side of a building, is designed solely for heat collection. During the heating season, solar radiation enters the sunspace through south facing glass. Radiation is absorbed by elements in the space and converted to heat. This sunspace is thermally isolated from the building. A set of sliding glass doors connecting the sunspace to the building space is closed to prevent night heat losses from the living space. This attached sunspace uses standard size, double glazed units and has a concrete slab floor. Since the required depth of footings (if required) is dependent on local frost conditions, costs of footings at depths of 2', 3' and 4' are provided with the base cost of system as shown.

System Components	QUANTITY	UNIT	COST EACH		
			MAT.	INST.	TOTAL
SYSTEM D3010 690 2560					
SOLAR PASSIVE HEATING, DIRECT GAIN, ATTACHED SUNSPACE					
WOOD FRAMED, SLAB ON GRADE					
B Aluminum, flashing, mill finish	21.000	S.F.	15.54	66.78	82.32
C Glazing, double, tempered glass	72.000	S.F.	1,026	745.20	1,771.20
D-2-4 Knee & end wall framing	.188	M.B.F.	108.10	186.12	294.22
F Ledgers, headers and rafters	.028	M.B.F.	86.88	217.20	304.08
Building paper, asphalt felt sheathing paper 15 lb	240.000	S.F.	12.60	32.16	44.76
G Sheathing plywood on roof CDX 1/2"	240.000	S.F.	122.40	170.40	292.80
H Asphalt shingles inorganic class A 210-235 lb/sq	1.000	Sq.	55	84	139
I Wood siding, match existing	164.000	S.F.	836.40	275.52	1,111.92
Window stop, interior or external	96.000	L.F.	57.60	288	345.60
L-1 Exterior trim	44.000	L.F.	51.04	87.12	138.16
L-2, L-3 Exterior trim 1" x 4"	206.000	L.F.	156.56	510.88	667.44
O Drywall, taping & finishing joints	200.000	S.F.	10	100	110
O Drywall, 1/2" on walls, standard, no finish	200.000	S.F.	68	100	168
S, S-1 Windows, wood, awning type, double glazed, with screens	4.000	Ea.	2,300	248	2,548
T Insulation batts, fiberglass, faced	210.000	S.F.	178.50	77.70	256.20
T-1 Polyethylene vapor barrier, standard, .004" thick	2.500	C.S.F.	7.78	26.80	34.58
Y Button vents	6.000	Ea.	10.68	49.50	60.18
Z Concrete floor, reinforced, 10' x 10' x 4"	1.234	C.Y.	172.76	98.17	270.93
Floor finishing, steel trowel	100.000	S.F.		78	78
Z-2 Mesh, welded wire 6 x 6, #10/10	1.000	C.S.F.	19.85	33.50	53.35
Paint interior, exterior and trim	330.000	S.F.	39.60	231	270.60
TOTAL			5,335.29	3,706.05	9,041.34

D3010 690	**Passive Solar Sunspace**	COST EACH		
		MAT.	INST.	TOTAL
2550	Solar passive heating, direct gain, attached sunspace,100 S.F.			
2560	Wood framed, slab on grade	5,325	3,700	9,025
2580	2' deep frost wall	5,825	4,550	10,375
2600	3' deep frost wall	5,975	4,875	10,850
2620	4' deep frost wall	6,150	5,225	11,375

D3020 Heat Generating Systems

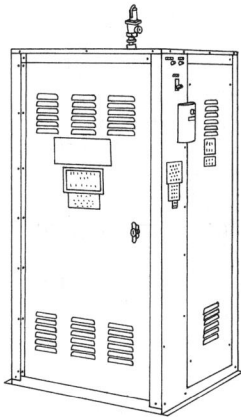

Boiler

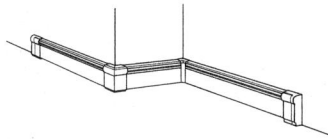

Baseboard Radiation

Small Electric Boiler
System Considerations:
1. Terminal units are fin tube baseboard radiation rated at 720 BTU/hr with 200° water temperature or 820 BTU/hr steam.
2. Primary use being for residential or smaller supplementary areas, the floor levels are based on 7-1/2′ ceiling heights.
3. All distribution piping is copper for boilers through 205 MBH. All piping for larger systems is steel pipe.

System Components	QUANTITY	UNIT	COST EACH		
			MAT.	INST.	TOTAL
SYSTEM D3020 102 1120					
SMALL HEATING SYSTEM, HYDRONIC, ELECTRIC BOILER					
1,480 S.F., 61 MBH, STEAM, 1 FLOOR					
Boiler, electric steam, std cntrls, trim, ftngs and valves, 18 KW, 61.4 MBH	1.000	Ea.	4,372.50	1,485	5,857.50
Copper tubing type L, solder joint, hanger 10′OC, 1-1/4″ diam	160.000	L.F.	2,200	1,616	3,816
Radiation, 3/4″ copper tube w/alum fin baseboard pkg 7″ high	60.000	L.F.	504	1,104	1,608
Rough in baseboard panel or fin tube with valves & traps	10.000	Set	2,290	5,600	7,890
Pipe covering, calcium silicate w/cover 1″ wall 1-1/4″ diam	160.000	L.F.	451.20	968	1,419.20
Low water cut-off, quick hookup, in gage glass tappings	1.000	Ea.	218	37	255
TOTAL			10,035.70	10,810	20,845.70
COST PER S.F.			6.78	7.30	14.08

D3020 102	Small Heating Systems, Hydronic, Electric Boilers		COST PER S.F.		
			MAT.	INST.	TOTAL
1100	Small heating systems, hydronic, electric boilers				
1120	Steam, 1 floor, 1480 S.F., 61 M.B.H.		6.76	7.30	14.06
1160	3,000 S.F., 123 M.B.H.	RD3020 -010	5.25	6.40	11.65
1200	5,000 S.F., 205 M.B.H.		4.60	5.90	10.50
1240	2 floors, 12,400 S.F., 512 M.B.H.	RD3020 -020	3.38	5.85	9.23
1280	3 floors, 24,800 S.F., 1023 M.B.H.		3.53	5.80	9.33
1320	34,750 S.F., 1,433 M.B.H.		3.25	5.65	8.90
1360	Hot water, 1 floor, 1,000 S.F., 41 M.B.H.		9.75	4.07	13.82
1400	2,500 S.F., 103 M.B.H.		7.70	7.30	15
1440	2 floors, 4,850 S.F., 205 M.B.H.		8.10	8.75	16.85
1480	3 floors, 9,700 S.F., 410 M.B.H.		9.15	9.05	18.20

D3020 Heat Generating Systems

Boiler

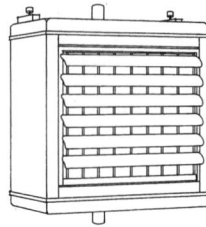

Unit Heater

Large Electric Boiler System Considerations:

1. Terminal units are all unit heaters of the same size. Quantities are varied to accommodate total requirements.
2. All air is circulated through the heaters a minimum of three times per hour.
3. As the capacities are adequate for commercial use, floor levels are based on 10′ ceiling heights.
4. All distribution piping is black steel pipe.

System Components	QUANTITY	UNIT	COST EACH		
			MAT.	INST.	TOTAL
SYSTEM D3020 104 1240					
LARGE HEATING SYSTEM, HYDRONIC, ELECTRIC BOILER					
9,280 S.F., 135 KW, 461 MBH, 1 FLOOR					
Boiler, electric hot water, std ctrls, trim, ftngs, valves, 135 KW, 461 MBH	1.000	Ea.	10,500	3,262.50	13,762.50
Expansion tank, painted steel, 60 Gal capacity ASME	1.000	Ea.	3,325	178	3,503
Circulating pump, CI, close cpld, 50 GPM, 2 HP, 2″ pipe conn	1.000	Ea.	2,100	350	2,450
Unit heater, 1 speed propeller, horizontal, 200° EWT, 72.7 MBH	7.000	Ea.	5,285	1,358	6,643
Unit heater piping hookup with controls	7.000	Set	3,780	8,750	12,530
Pipe, steel, black, schedule 40, welded, 2-1/2″ diam	380.000	L.F.	4,104	9,044	13,148
Pipe covering, calcium silicate w/cover, 1″ wall, 2-1/2″ diam	380.000	L.F.	1,326.20	2,356	3,682.20
TOTAL			30,420.20	25,298.50	55,718.70
COST PER S.F.			3.28	2.73	6.01

D3020 104	Large Heating Systems, Hydronic, Electric Boilers		COST PER S.F.		
			MAT.	INST.	TOTAL
1230	Large heating systems, hydronic, electric boilers				
1240	9,280 S.F., 135 K.W., 461 M.B.H., 1 floor		3.28	2.73	6.01
1280	14,900 S.F., 240 K.W., 820 M.B.H., 2 floors	RD3020 -010	4.11	4.43	8.54
1320	18,600 S.F., 296 K.W., 1,010 M.B.H., 3 floors		4.07	4.81	8.88
1360	26,100 S.F., 420 K.W., 1,432 M.B.H., 4 floors	RD3020 -020	4.02	4.74	8.76
1400	39,100 S.F., 666 K.W., 2,273 M.B.H., 4 floors		3.45	3.95	7.40
1440	57,700 S.F., 900 K.W., 3,071 M.B.H., 5 floors		3.58	3.87	7.45
1480	111,700 S.F., 1,800 K.W., 6,148 M.B.H., 6 floors		3.23	3.33	6.56
1520	149,000 S.F., 2,400 K.W., 8,191 M.B.H., 8 floors		3.22	3.33	6.55
1560	223,300 S.F., 3,600 K.W., 12,283 M.B.H., 14 floors		3.46	3.78	7.24

D3020 Heat Generating Systems

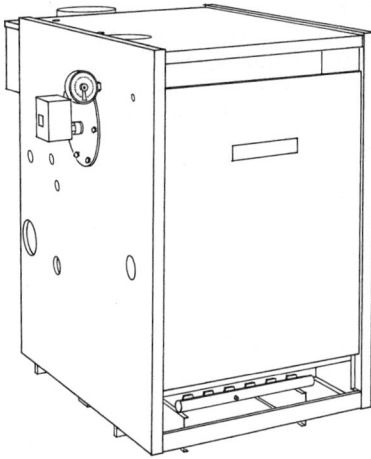

Boiler Selection: The maximum allowable working pressures are limited by ASME "Code for Heating Boilers" to 15 PSI for steam and 160 PSI for hot water heating boilers, with a maximum temperature limitation of 250°F. Hot water boilers are generally rated for a working pressure of 30 PSI. High pressure boilers are governed by the ASME "Code for Power Boilers" which is used almost universally for boilers operating over 15 PSIG. High pressure boilers used for a combination of heating/process loads are usually designed for 150 PSIG.

Boiler ratings are usually indicated as either Gross or Net Output. The Gross Load is equal to the Net Load plus a piping and pickup allowance. When this allowance cannot be determined, divide the gross output rating by 1.25 for a value equal to or greater than the next heat loss requirement of the building.

Table below lists installed cost per boiler and includes insulating jacket, standard controls, burner and safety controls. Costs do not include piping or boiler base pad. Outputs are Gross.

D3020 106	Boilers, Hot Water & Steam		COST EACH		
			MAT.	INST.	TOTAL
0600	Boiler, electric, steel, hot water, 12 K.W., 41 M.B.H.		3,950	1,250	5,200
0620	30 K.W., 103 M.B.H.		4,700	1,350	6,050
0640	60 K.W., 205 M.B.H.	RD3020 -010	5,750	1,475	7,225
0660	120 K.W., 410 M.B.H.		6,450	1,800	8,250
0680	210 K.W., 716 M.B.H.	RD3020 -020	7,550	2,700	10,250
0700	510 K.W., 1,739 M.B.H.		18,900	5,050	23,950
0720	720 K.W., 2,452 M.B.H.		22,800	5,700	28,500
0740	1,200 K.W., 4,095 M.B.H.		29,600	6,525	36,125
0760	2,100 K.W., 7,167 M.B.H.		56,500	8,225	64,725
0780	3,600 K.W., 12,283 M.B.H.		87,000	13,900	100,900
0820	Steam, 6 K.W., 20.5 M.B.H.		3,725	1,350	5,075
0840	24 K.W., 81.8 M.B.H.		4,625	1,475	6,100
0860	60 K.W., 205 M.B.H.		6,375	1,625	8,000
0880	150 K.W., 512 M.B.H.		9,225	2,500	11,725
0900	510 K.W., 1,740 M.B.H.		24,300	6,175	30,475
0920	1,080 K.W., 3,685 M.B.H.		34,100	8,875	42,975
0940	2,340 K.W., 7,984 M.B.H.		69,500	13,900	83,400
0980	Gas, cast iron, hot water, 80 M.B.H.		1,875	1,550	3,425
1000	100 M.B.H.		2,150	1,675	3,825
1020	163 M.B.H.		2,800	2,250	5,050
1040	280 M.B.H.		4,075	2,500	6,575
1060	544 M.B.H.		8,650	4,450	13,100
1080	1,088 M.B.H.		13,100	5,650	18,750
1100	2,000 M.B.H.		23,100	8,825	31,925
1120	2,856 M.B.H.		25,500	11,300	36,800
1140	4,720 M.B.H.		68,500	15,600	84,100
1160	6,970 M.B.H.		98,500	25,400	123,900
1180	For steam systems under 2,856 M.B.H., add 8%				
1520	Oil, cast iron, hot water, 109 M.B.H.		2,075	1,875	3,950
1540	173 M.B.H.		2,600	2,250	4,850
1560	236 M.B.H.		3,375	2,650	6,025
1580	1,084 M.B.H.		9,600	6,000	15,600
1600	1,600 M.B.H.		12,600	8,625	21,225
1620	2,480 M.B.H.		19,500	11,000	30,500
1640	3,550 M.B.H.		25,000	13,200	38,200
1660	Steam systems same price as hot water				

D3020 Heat Generating Systems

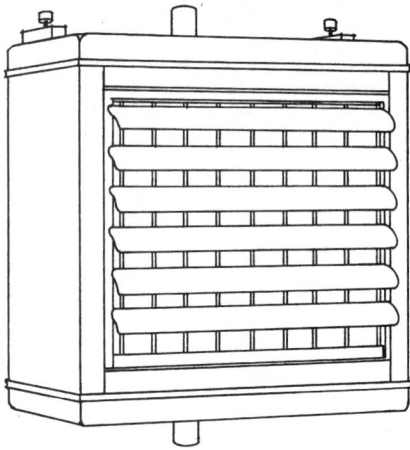

Unit Heater

Fossil Fuel Boiler System Considerations:

1. Terminal units are horizontal unit heaters. Quantities are varied to accommodate total heat loss per building.
2. Unit heater selection was determined by their capacity to circulate the building volume a minimum of three times per hour in addition to the BTU output.
3. Systems shown are forced hot water. Steam boilers cost slightly more than hot water boilers. However, this is compensated for by the smaller size or fewer terminal units required with steam.
4. Floor levels are based on 10' story heights.
5. MBH requirements are gross boiler output.

System Components	QUANTITY	UNIT	COST EACH		
			MAT.	INST.	TOTAL
SYSTEM D3020 108 1280					
HEATING SYSTEM, HYDRONIC, FOSSIL FUEL, TERMINAL UNIT HEATERS					
CAST IRON BOILER, GAS, 80 MBH, 1,070 S.F. BUILDING					
Boiler, gas, hot water, CI, burner, controls, insulation, breeching, 80 MBH	1.000	Ea.	1,968.75	1,627.50	3,596.25
Pipe, steel, black, schedule 40, threaded, cplg & hngr 10'OC, 2" diam	200.000	L.F.	1,690	3,290	4,980
Unit heater, 1 speed propeller, horizontal, 200° EWT, 72.7 MBH	2.000	Ea.	1,510	388	1,898
Unit heater piping hookup with controls	2.000	Set	1,080	2,500	3,580
Expansion tank, painted steel, ASME, 18 Gal capacity	1.000	Ea.	2,125	89	2,214
Circulating pump, CI, flange connection, 1/12 HP	1.000	Ea.	315	176	491
Pipe covering, calcium silicate w/cover, 1" wall, 2" diam	200.000	L.F.	656	1,240	1,896
TOTAL			9,344.75	9,310.50	18,655.25
COST PER S.F.			8.73	8.70	17.43

D3020 108	Heating Systems, Unit Heaters		COST PER S.F.		
			MAT.	INST.	TOTAL
1260	Heating systems, hydronic, fossil fuel, terminal unit heaters,				
1280	Cast iron boiler, gas, 80 M.B.H., 1,070 S.F. bldg.		8.74	8.69	17.43
1320	163 M.B.H., 2,140 S.F. bldg.	RD3020 -010	5.80	5.90	11.70
1360	544 M.B.H., 7,250 S.F. bldg.		4.23	4.18	8.41
1400	1,088 M.B.H., 14,500 S.F. bldg.	RD3020 -020	3.59	3.96	7.55
1440	3,264 M.B.H., 43,500 S.F. bldg.		3	3.01	6.01
1480	5,032 M.B.H., 67,100 S.F. bldg.		3.45	3.06	6.51
1520	Oil, 109 M.B.H., 1,420 S.F. bldg.		9	7.75	16.75
1560	235 M.B.H., 3,150 S.F. bldg.		6	5.50	11.50
1600	940 M.B.H., 12,500 S.F. bldg.		4.36	3.62	7.98
1640	1,600 M.B.H., 21,300 S.F. bldg.		3.97	3.48	7.45
1680	2,480 M.B.H., 33,100 S.F. bldg.		3.95	3.13	7.08
1720	3,350 M.B.H., 44,500 S.F. bldg.		3.63	3.20	6.83
1760	Coal, 148 M.B.H., 1,975 S.F. bldg.		8.10	5.05	13.15
1800	300 M.B.H., 4,000 S.F. bldg.		6.10	3.94	10.04
1840	2,360 M.B.H., 31,500 S.F. bldg.		4.45	3.19	7.64

D30 HVAC

D3020 Heat Generating Systems

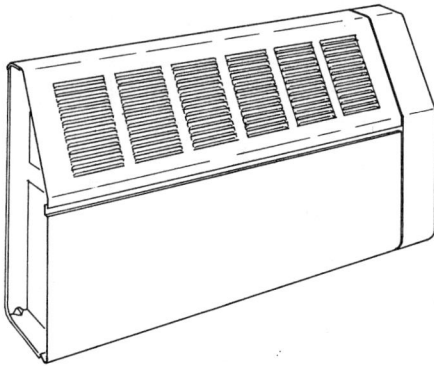

Fin Tube Radiator

Fossil Fuel Boiler System Considerations:

1. Terminal units are commercial steel fin tube radiation. Quantities are varied to accommodate total heat loss per building.
2. Systems shown are forced hot water. Steam boilers cost slightly more than hot water boilers. However, this is compensated for by the smaller size or fewer terminal units required with steam.
3. Floor levels are based on 10' story heights.
4. MBH requirements are gross boiler output.

System Components	QUANTITY	UNIT	COST EACH		
			MAT.	INST.	TOTAL
SYSTEM D3020 110 3240					
HEATING SYSTEM, HYDRONIC, FOSSIL FUEL, FIN TUBE RADIATION					
CAST IRON BOILER, GAS, 80 MBH, 1,070 S.F. BLDG.					
Boiler, gas, hot water, CI, burner, controls, insulation, breeching, 80 MBH	1.000	Ea.	1,968.75	1,627.50	3,596.25
Pipe, steel, black, schedule 40, threaded, 2″ diam	200.000	L.F.	1,690	3,290	4,980
Radiation, steel 1-1/4″ tube & 4-1/4″ fin w/cover & damper, wall hung	80.000	L.F.	3,240	2,360	5,600
Rough-in, wall hung steel fin radiation with supply & balance valves	8.000	Set	1,800	5,200	7,000
Expansion tank, painted steel, ASME, 18 Gal capacity	1.000	Ea.	2,125	89	2,214
Circulating pump, CI flanged, 1/12 HP	1.000	Ea.	315	176	491
Pipe covering, calcium silicate w/cover, 1″ wall, 2″ diam	200.000	L.F.	656	1,240	1,896
TOTAL			11,794.75	13,982.50	25,777.25
COST PER S.F.			11.02	13.07	24.09

D3020 110	Heating System, Fin Tube Radiation	COST PER S.F.		
		MAT.	INST.	TOTAL
3230	Heating systems, hydronic, fossil fuel, fin tube radiation			
3240	Cast iron boiler, gas, 80 MBH, 1,070 S.F. bldg.	11.05	13.05	24.10
3280	169 M.B.H., 2,140 S.F. bldg.	6.90	8.25	15.15
3320	544 M.B.H., 7,250 S.F. bldg.	5.85	7.10	12.95
3360	1,088 M.B.H., 14,500 S.F. bldg.	5.30	6.95	12.25
3400	3,264 M.B.H., 43,500 S.F. bldg.	4.82	6.15	10.97
3440	5,032 M.B.H., 67,100 S.F. bldg.	5.30	6.20	11.50
3480	Oil, 109 M.B.H., 1,420 S.F. bldg.	12.80	14.25	27.05
3520	235 M.B.H., 3,150 S.F. bldg.	7.60	8.35	15.95
3560	940 M.B.H., 12,500 S.F. bldg.	6.10	6.65	12.75
3600	1,600 M.B.H., 21,300 S.F. bldg.	5.80	6.60	12.40
3640	2,480 M.B.H., 33,100 S.F. bldg.	5.80	6.30	12.10
3680	3,350 M.B.H., 44,500 S.F. bldg.	5.45	6.35	11.80
3720	Coal, 148 M.B.H., 1,975 S.F. bldg.	9.65	7.90	17.55
3760	300 M.B.H., 4,000 S.F. bldg.	7.65	6.75	14.40
3800	2,360 M.B.H., 31,500 S.F. bldg.	6.25	6.25	12.50
4080	Steel boiler, oil, 97 M.B.H., 1,300 S.F. bldg.	11.55	11.75	23.30
4120	315 M.B.H., 4,550 S.F. bldg.	5.20	5.95	11.15
4160	525 M.B.H., 7,000 S.F. bldg.	7.15	6.85	14
4200	1,050 M.B.H., 14,000 S.F. bldg.	6.25	6.85	13.10
4240	2,310 M.B.H., 30,800 S.F. bldg.	5.75	6.30	12.05
4280	3,150 M.B.H., 42,000 S.F. bldg.	5.75	6.40	12.15

RD3020 -010

RD3020 -020

D3020 Heat Generating Systems

D3020 310	Gas Vent, Galvanized Steel, Double Wall	COST PER V.L.F.		
		MAT.	INST.	TOTAL
0030	Gas vent, galvanized steel, double wall, round pipe diameter, 3"	5.90	14.55	20.45
0040	4"	7.35	15.40	22.75
0050	5"	8.60	16.35	24.95
0060	6"	10	17.45	27.45
0070	7"	14.65	18.70	33.35
0080	8"	16.35	20	36.35
0090	10"	37	22	59
0100	12"	49.50	24	73.50
0110	14"	83.50	25	108.50
0120	16	113	26	139
0130	18"	146	27.50	173.50
0140	20"	153	45	198
0150	22"	219	48	267
0160	24"	269	51	320
0180	30"	360	58	418
0220	38"	550	68	618
0300	48"	865	85.50	950.50
0400				

D3020 310	Gas Vent Fittings, Galvanized Steel, Double Wall	COST EACH		
		MAT.	INST.	TOTAL
1020	Gas vent, fittings, 45° Elbow, adjustable thru 8" diameter			
1030	Pipe diameter, 3"	14.60	29	43.60
1040	4"	17.30	31	48.30
1050	5"	20.50	32.50	53
1060	6"	24.50	35	59.50
1070	7"	35	37.50	72.50
1080	8"	40.50	40	80.50
1090	10"	108	43.50	151.50
1100	12"	137	47.50	184.50
1110	14"	191	50	241
1120	16"	240	52.50	292.50
1130	18"	310	55	365
1140	20"	365	90.50	455.50
1150	22"	560	95.50	655.50
1170	24"	700	102	802
1200	30"	1,500	116	1,616
1250	38"	2,025	136	2,161
1300	48"	2,675	163	2,838
1410	90° Elbow, adjustable, pipe diameter, 3"	24.50	29	53.50
1420	4"	29.50	31	60.50
1430	5"	34.50	32.50	67
1440	6"	41.50	35	76.50
1450	7"	65.50	37.50	103
1460	8"	69.50	40	109.50
1510	Tees, standard, pipe diameter, 3"	36	38.50	74.50
1520	4"	37.50	40	77.50
1530	5"	41	42	83
1540	6"	44	43.50	87.50
1550	7"	48.50	45.50	94
1560	8"	66.50	47.50	114
1570	10"	194	50	244
1580	12"	233	52.50	285.50
1590	14"	380	58	438
1600	16"	540	65.50	605.50
1610	18"	675	74.50	749.50
1620	20"	915	95.50	1,010.50

D3020 Heat Generating Systems

D3020 310	Gas Vent Fittings, Galvanized Steel, Double Wall	COST EACH		
		MAT.	INST.	TOTAL
1630	22"	1,125	125	1,250
1640	24"	1,325	136	1,461
1660	30"	2,300	163	2,463
1700	38"	3,225	203	3,428
1750	48"	4,900	325	5,225
1810	Tee, access caps, pipe diameter, 3"	2.57	23.50	26.07
1820	4"	2.89	25	27.89
1830	5"	3.83	26	29.83
1840	6"	4.47	28.50	32.97
1850	7"	7.25	30	37.25
1860	8"	8.50	31	39.50
1870	10"	17.55	32.50	50.05
1880	12"	23.50	35	58.50
1890	14"	48	37.50	85.50
1900	16"	56	42	98
1910	18"	63	43.50	106.50
1920	20"	67.50	60.50	128
1930	22"	77.50	74	151.50
1940	24"	87.50	77.50	165
1950	30"	410	90.50	500.50
1970	38"	505	116	621
1990	48"	555	148	703
2110	Tops, bird proof, pipe diameter, 3"	13.50	22.50	36
2120	4"	17.80	24	41.80
2130	5"	19.90	25	44.90
2140	6"	24.50	26	50.50
2150	7"	36.50	27.50	64
2160	8"	47.50	29	76.50
2170	10"	120	31	151
2180	12"	181	32.50	213.50
2190	14"	207	35	242
2200	16"	375	37.50	412.50
2210	18"	655	40	695
2220	20"	730	58	788
2230	22"	995	74	1,069
2240	24"	1,250	81.50	1,331.50
2410	Roof flashing, tall cone/adjustable, pipe diameter, 3"	27	29	56
2420	4"	31	31	62
2430	5"	32.50	32.50	65
2440	6"	34.50	35	69.50
2450	7"	58	37.50	95.50
2460	8"	59.50	40	99.50
2470	10"	91.50	43.50	135
2480	12"	117	47.50	164.50
2490	14"	137	52.50	189.50
2500	16"	169	58	227
2510	18"	229	65.50	294.50
2520	20"	297	90.50	387.50
2530	22"	375	116	491
2540	24"	450	136	586

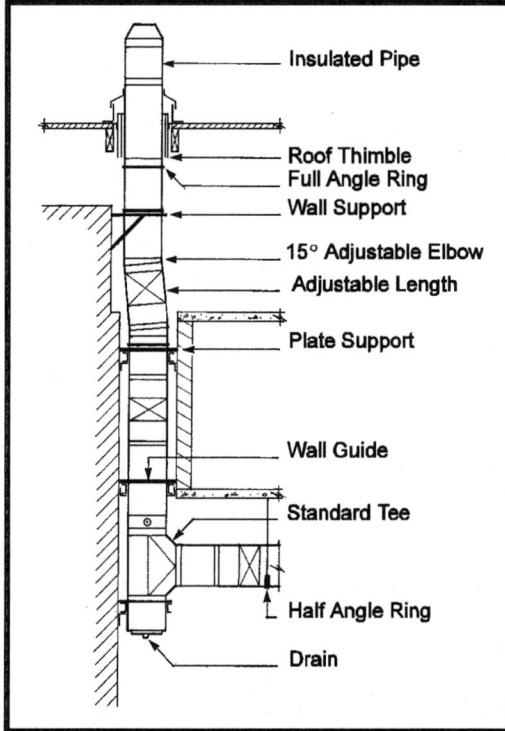

Diagram labels:
- Insulated Pipe
- Roof Thimble
- Full Angle Ring
- Wall Support
- 15° Adjustable Elbow
- Adjustable Length
- Plate Support
- Wall Guide
- Standard Tee
- Half Angle Ring
- Drain

D3020 Heat Generating Systems

D3020 315	Chimney, Stainless Steel, Insulated, Wood & Oil	COST PER V.L.F.		
		MAT.	INST.	TOTAL
0530	Chimney, stainless steel, insulated, diameter, 6"	62	17.45	79.45
0540	7"	80.50	18.70	99.20
0550	8"	95	20	115
0560	10"	137	22	159
0570	12"	184	24	208
0580	14"	241	25	266

D3030 Cooling Generating Systems

Chilled Water Supply & Return Piping

Air Cooled Water Chiller Unit

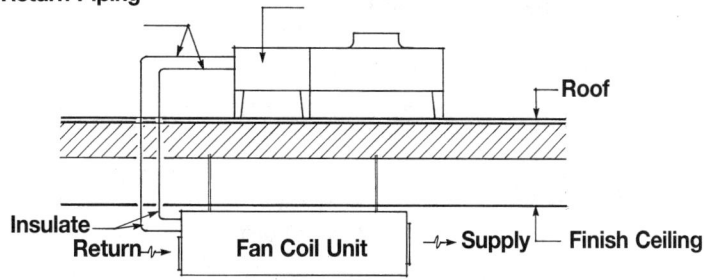

Roof

Insulate

Return

Fan Coil Unit

Supply — Finish Ceiling

Design Assumptions: The chilled water, air cooled systems priced, utilize reciprocating hermetic compressors and propeller-type condenser fans. Piping with pumps and expansion tanks is included based on a two pipe system. No ducting is included and the fan-coil units are cooling only. Water treatment and balancing are not included. Chilled water piping is insulated. Area distribution is through the use of multiple fan coil units. Fewer but larger fan coil units with duct distribution would be approximately the same S.F. cost.

System Components	QUANTITY	UNIT	COST EACH		
			MAT.	INST.	TOTAL
SYSTEM D3030 110 1200					
PACKAGED CHILLER, AIR COOLED, WITH FAN COIL UNIT					
APARTMENT CORRIDORS, 3,000 S.F., 5.50 TON					
Fan coil air conditioning unit, cabinet mounted & filters chilled water	1.000	Ea.	3,345.23	489.41	3,834.64
Water chiller,air conditioning unit, reciprocating, air cooled,	1.000	Ea.	6,572.50	1,828.75	8,401.25
Chilled water unit coil connections	1.000	Ea.	980	1,325	2,305
Chilled water distribution piping	440.000	L.F.	10,340	18,040	28,380
TOTAL			21,237.73	21,683.16	42,920.89
COST PER S.F.			7.08	7.23	14.31

D3030 110	Chilled Water, Air Cooled Condenser Systems		COST PER S.F.		
			MAT.	INST.	TOTAL
1180	Packaged chiller, air cooled, with fan coil unit				
1200	Apartment corridors, 3,000 S.F., 5.50 ton		7.10	7.25	14.35
1360	40,000 S.F., 73.33 ton	RD3030 -010	4.16	3.20	7.36
1440	Banks and libraries, 3,000 S.F., 12.50 ton		10.60	8.50	19.10
1560	20,000 S.F., 83.33 ton		7.30	4.39	11.69
1680	Bars and taverns, 3,000 S.F., 33.25 ton		17.30	10.10	27.40
1760	10,000 S.F., 110.83 ton		12.75	2.50	15.25
1920	Bowling alleys, 3,000 S.F., 17.00 ton		13.15	9.55	22.70
2040	20,000 S.F., 113.33 ton		8.40	4.44	12.84
2160	Department stores, 3,000 S.F., 8.75 ton		9.55	8.05	17.60
2320	40,000 S.F., 116.66 ton		5.45	3.31	8.76
2400	Drug stores, 3,000 S.F., 20.00 ton		14.85	9.90	24.75
2520	20,000 S.F., 133.33 ton		10.40	4.68	15.08
2640	Factories, 2,000 S.F., 10.00 ton		9.30	8.15	17.45
2800	40,000 S.F., 133.33 ton		5.95	3.40	9.35
2880	Food supermarkets, 3,000 S.F., 8.50 ton		9.40	8	17.40
3040	40,000 S.F., 113.33 ton		5.20	3.32	8.52
3120	Medical centers, 3,000 S.F., 7.00 ton		8.50	7.80	16.30
3280	40,000 S.F., 93.33 ton		4.68	3.25	7.93
3360	Offices, 3,000 S.F., 9.50 ton		8.65	7.95	16.60
3520	40,000 S.F., 126.66 ton		5.85	3.43	9.28
3600	Restaurants, 3,000 S.F., 15.00 ton		11.85	8.85	20.70
3720	20,000 S.F., 100.00 ton		8.45	4.65	13.10
3840	Schools and colleges, 3,000 S.F., 11.50 ton		10.10	8.35	18.45
3960	20,000 S.F., 76.66 ton		7.10	4.43	11.53

D3030 Cooling Generating Systems

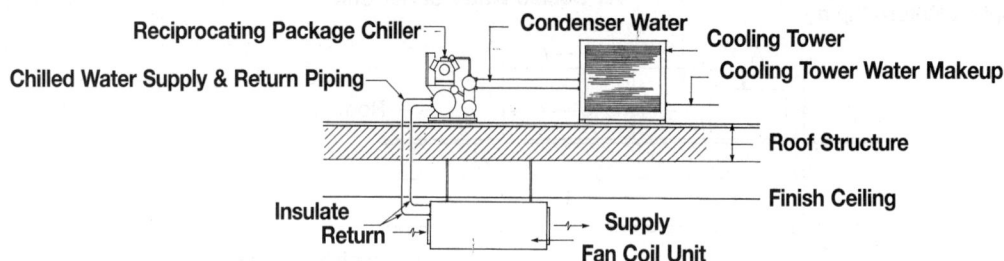

Reciprocating Package Chiller
Condenser Water
Cooling Tower
Cooling Tower Water Makeup
Chilled Water Supply & Return Piping
Roof Structure
Finish Ceiling
Insulate Return
Supply
Fan Coil Unit

General: Water cooled chillers are available in the same sizes as air cooled units. They are also available in larger capacities.

Design Assumptions: The chilled water systems with water cooled condenser include reciprocating hermetic compressors, water cooling tower, pumps, piping and expansion tanks and are based on a two pipe system. Chilled water piping is insulated. No ducts are included and fan-coil units are cooling only. Area distribution is through use of multiple fan coil units. Fewer but larger fan coil units with duct distribution would be approximately the same S.F. cost. Water treatment and balancing are not included.

System Components	QUANTITY	UNIT	COST EACH		
			MAT.	INST.	TOTAL
SYSTEM D3030 115 1320					
PACKAGED CHILLER, WATER COOLED, WITH FAN COIL UNIT					
APARTMENT CORRIDORS, 4,000 S.F., 7.33 TON					
Fan coil air conditioner unit, cabinet mounted & filters, chilled water	2.000	Ea.	4,458.48	652.28	5,110.76
Water chiller, water cooled, 1 compressor, hermetic scroll,	1.000	Ea.	5,358.60	3,183.10	8,541.70
Cooling tower, draw thru single flow, belt drive	1.000	Ea.	1,106.83	135.24	1,242.07
Cooling tower pumps & piping	1.000	System	553.42	318.86	872.28
Chilled water unit coil connections	2.000	Ea.	1,960	2,650	4,610
Chilled water distribution piping	520.000	L.F.	12,220	21,320	33,540
TOTAL			25,657.33	28,259.48	53,916.81
COST PER S.F.			6.41	7.06	13.47

*Cooling requirements would lead to choosing a water cooled unit

D3030 115	Chilled Water, Cooling Tower Systems		COST PER S.F.		
			MAT.	INST.	TOTAL
1300	Packaged chiller, water cooled, with fan coil unit				
1320	Apartment corridors, 4,000 S.F., 7.33 ton		6.45	7.05	13.45
1600	Banks and libraries, 4,000 S.F., 16.66 ton	RD3030 -010	10.05	7.75	17.80
1800	60,000 S.F., 250.00 ton		7.30	6.30	13.60
1880	Bars and taverns, 4,000 S.F., 44.33 ton		16.25	9.80	26.05
2000	20,000 S.F., 221.66 ton		17.80	8.15	25.95
2160	Bowling alleys, 4,000 S.F., 22.66 ton		11.75	8.55	20.30
2320	40,000 S.F., 226.66 ton		10.05	5.90	15.95
2440	Department stores, 4,000 S.F., 11.66 ton		7.25	7.65	14.90
2640	60,000 S.F., 175.00 ton		6.55	5.75	12.30
2720	Drug stores, 4,000 S.F., 26.66 ton		12.55	8.80	21.35
2880	40,000 S.F., 266.67 ton		9.90	6.70	16.60
3000	Factories, 4,000 S.F., 13.33 ton		8.80	7.40	16.20
3200	60,000 S.F., 200.00 ton		6.60	6	12.60
3280	Food supermarkets, 4,000 S.F., 11.33 ton		7.10	7.60	14.70
3480	60,000 S.F., 170.00 ton		6.50	5.70	12.20
3560	Medical centers, 4,000 S.F., 9.33 ton		6.25	7	13.25
3760	60,000 S.F., 140.00 ton		5.40	5.80	11.20
3840	Offices, 4,000 S.F., 12.66 ton		8.55	7.30	15.85
4040	60,000 S.F., 190.00 ton		6.40	5.95	12.35
4120	Restaurants, 4,000 S.F., 20.00 ton		10.60	7.95	18.55
4320	60,000 S.F., 300.00 ton		8.20	6.50	14.70
4400	Schools and colleges, 4,000 S.F., 15.33 ton		9.55	7.60	17.15
4600	60,000 S.F., 230.00 ton		6.80	6.05	12.85

D30 HVAC

D3030 Cooling Generating Systems

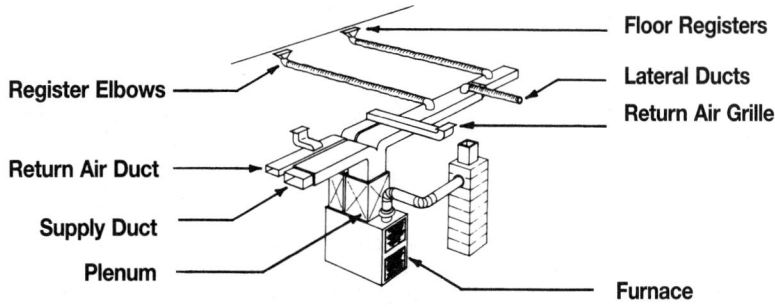

System Components	QUANTITY	UNIT	COST EACH		
			MAT.	INST.	TOTAL
SYSTEM D3030 214 1200					
HEATING/COOLING, GAS FIRED FORCED AIR,					
ONE ZONE, 1200 SF BLDG, SEER 14					
Thermostat manual	1.000	Ea.	29.50	72.50	102
Intermittent pilot	1.000	Ea.	164		164
Furnace, 3 Ton cooling, 115 MBH	1.000	Ea.	1,975	350	2,325
Cooling tubing 25 feet	1.000	Ea.	251		251
Ductwork	158.000	Lb.	194.34	1,090.20	1,284.54
Ductwork connection	12.000	Ea.	354	209.40	563.40
Supply ductwork	176.000	SF Surf	140.80	818.40	959.20
Supply grill	2.000	Ea.	51	53	104
Duct insulation	1.000	L.F.	365.76	578.88	944.64
Return register	1.000	Ea.	312	217.80	529.80
TOTAL			3,837.40	3,390.18	7,227.58

D3030 214	Heating/Cooling System	COST EACH		
		MAT.	INST.	TOTAL
1200	Heating/Cooling system, gas fired, SEER 14, 1200 SF Bldg	3,825	3,400	7,225
1300	2000 SF Bldg	5,050	5,350	10,400
1400	Heating/Cooling system, heat pump 3 ton, SEER 14, 1200 SF Bldg	5,600	4,375	9,975
1500	5 ton, SEER 14, 2000 SF Bldg	7,575	5,000	12,575

D3050 Terminal & Package Units

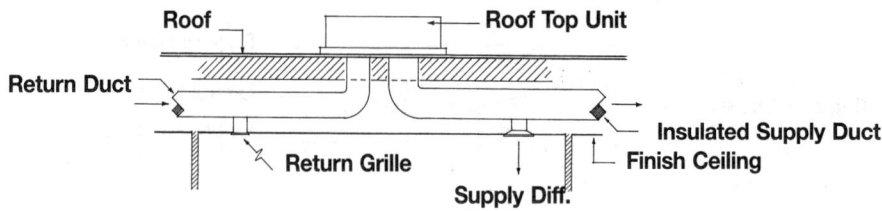

Roof — Roof Top Unit
Return Duct — Insulated Supply Duct — Finish Ceiling
Return Grille — Supply Diff.

System Description: Rooftop single zone units are electric cooling and gas heat. Duct systems are low velocity, galvanized steel supply and return. Price variations between sizes are due to several factors. Jumps in the cost of the rooftop unit occur when the manufacturer shifts from the largest capacity unit on a small frame to the smallest capacity on the next larger frame, or changes from one compressor to two. As the unit capacity increases for larger areas the duct distribution grows in proportion. For most applications there is a tradeoff point where it is less expensive and more efficient to utilize smaller units with short simple distribution systems. Larger units also require larger initial supply and return ducts which can create a space problem. Supplemental heat may be desired in colder locations. The table below is based on one unit supplying the area listed. The 10,000 S.F. unit for bars and taverns is not listed because a nominal 110 ton unit would be required and this is above the normal single zone rooftop capacity.

System Components	QUANTITY	UNIT	COST EACH		
			MAT.	INST.	TOTAL
SYSTEM D3050 150 1280					
ROOFTOP, SINGLE ZONE, AIR CONDITIONER					
APARTMENT CORRIDORS, 500 S.F., .92 TON					
Rooftop air-conditioner, 1 zone, electric cool, standard controls, curb	1.000	Ea.	1,621.50	701.50	2,323
Ductwork package for rooftop single zone units	1.000	System	289.80	1,012	1,301.80
TOTAL			1,911.30	1,713.50	3,624.80
COST PER S.F.			3.82	3.43	7.25

D3050 150	Rooftop Single Zone Unit Systems	COST PER S.F.		
		MAT.	INST.	TOTAL
1260	Rooftop, single zone, air conditioner			
1280	Apartment corridors, 500 S.F., .92 ton	3.82	3.43	7.25
1480	10,000 S.F., 18.33 ton	2.61	2.33	4.94
1560	Banks or libraries, 500 S.F., 2.08 ton	8.65	7.75	16.40
1760	10,000 S.F., 41.67 ton	5.10	5.25	10.35
1840	Bars and taverns, 500 S.F. 5.54 ton	11.90	10.30	22.20
2000	5,000 S.F., 55.42 ton	11.30	7.90	19.20
2080	Bowling alleys, 500 S.F., 2.83 ton	7.60	8.70	16.30
2280	10,000 S.F., 56.67 ton	6.70	7.15	13.85
2360	Department stores, 500 S.F., 1.46 ton	6.05	5.45	11.50
2560	10,000 S.F., 29.17 ton	3.67	3.68	7.35
2640	Drug stores, 500 S.F., 3.33 ton	8.95	10.25	19.20
2840	10,000 S.F., 66.67 ton	7.85	8.40	16.25
2920	Factories, 500 S.F., 1.67 ton	6.90	6.20	13.10
3120	10,000 S.F., 33.33 ton	4.19	4.21	8.40
3200	Food supermarkets, 500 S.F., 1.42 ton	5.90	5.30	11.20
3400	10,000 S.F., 28.33 ton	3.56	3.58	7.14
3480	Medical centers, 500 S.F., 1.17 ton	4.85	4.35	9.20
3680	10,000 S.F., 23.33 ton	3.31	2.96	6.27
3760	Offices, 500 S.F., 1.58 ton	6.60	5.90	12.50
3960	10,000 S.F., 31.67 ton	3.99	3.99	7.98
4000	Restaurants, 500 S.F., 2.50 ton	10.40	9.30	19.70
4200	10,000 S.F., 50.00 ton	5.90	6.30	12.20
4240	Schools and colleges, 500 S.F., 1.92 ton	7.95	7.15	15.10
4440	10,000 S.F., 38.33 ton	4.68	4.84	9.52

Note next to 1480/1560 rows: RD3030 -010

D30 HVAC

D3050 Terminal & Package Units

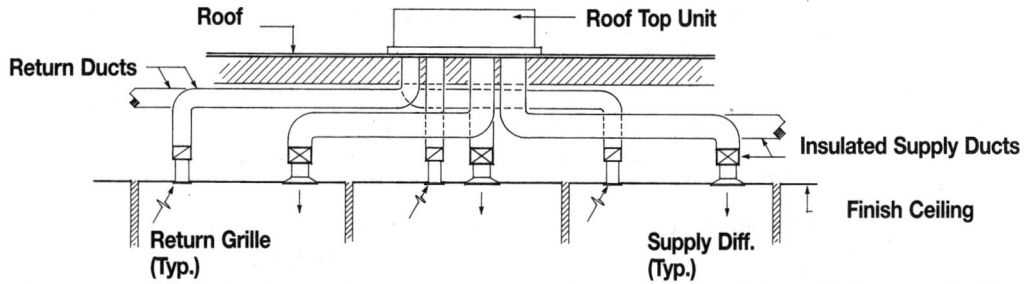

System Description: Rooftop units are multizone with up to 12 zones, and include electric cooling, gas heat, thermostats, filters, supply and return fans complete. Duct systems are low velocity, galvanized steel supply and return with insulated supplies.

Multizone units cost more per ton of cooling than single zone. However, they offer flexibility where load conditions are varied due to heat generating areas or exposure to radiational heating. For example, perimeter offices on the "sunny side" may require cooling at the same

time "shady side" or central offices may require heating. It is possible to accomplish similar results using duct heaters in branches of the single zone unit. However, heater location could be a problem and total system operating energy efficiency could be lower.

System Components	QUANTITY	UNIT	COST EACH		
			MAT.	INST.	TOTAL
SYSTEM D3050 155 1280					
ROOFTOP, MULTIZONE, AIR CONDITIONER					
APARTMENT CORRIDORS, 3,000 S.F., 5.50 TON					
Rooftop multizone unit, standard controls, curb	1.000	Ea.	25,960	1,628	27,588
Ductwork package for rooftop multizone units	1.000	System	2,530	11,550	14,080
TOTAL			28,490	13,178	41,668
COST PER S.F.			9.50	4.39	13.89

Note A: Small single zone unit recommended

D3050 155	Rooftop Multizone Unit Systems		COST PER S.F.		
			MAT.	INST.	TOTAL
1240	Rooftop, multizone, air conditioner				
1260	Apartment corridors, 1,500 S.F., 2.75 ton. See Note A.				
1280	3,000 S.F., 5.50 ton	RD3030 -010	9.50	4.40	13.90
1440	25,000 S.F., 45.80 ton		6.35	4.25	10.60
1520	Banks or libraries, 1,500 S.F., 6.25 ton		21.50	10	31.50
1640	15,000 S.F., 62.50 ton		10.80	9.60	20.40
1720	25,000 S.F., 104.00 ton		9.75	9.55	19.30
1800	Bars and taverns, 1,500 S.F., 16.62 ton		46	14.35	60.35
1840	3,000 S.F., 33.24 ton		38.50	13.85	52.35
1880	10,000 S.F., 110.83 ton		23.50	13.80	37.30
2080	Bowling alleys, 1,500 S.F., 8.50 ton		29.50	13.60	43.10
2160	10,000 S.F., 56.70 ton		19.60	13.15	32.75
2240	20,000 S.F., 113.00 ton		13.25	12.95	26.20
2640	Drug stores, 1,500 S.F., 10.00 ton		34.50	15.95	50.45
2680	3,000 S.F., 20.00 ton		24	15.40	39.40
2760	15,000 S.F., 100.00 ton		15.60	15.30	30.90
3760	Offices, 1,500 S.F., 4.75 ton, See Note A				
3880	15,000 S.F., 47.50 ton		10.95	7.35	18.30
3960	25,000 S.F., 79.16 ton		8.20	7.30	15.50
4000	Restaurants, 1,500 S.F., 7.50 ton		26	12	38
4080	10,000 S.F., 50.00 ton		17.30	11.60	28.90
4160	20,000 S.F., 100.00 ton		11.70	11.50	23.20
4240	Schools and colleges, 1,500 S.F., 5.75 ton		19.85	9.20	29.05
4360	15,000 S.F., 57.50 ton		9.95	8.80	18.75
4441	25,000 S.F., 95.83 ton		9.40	8.80	18.20

D3050 Terminal & Package Units

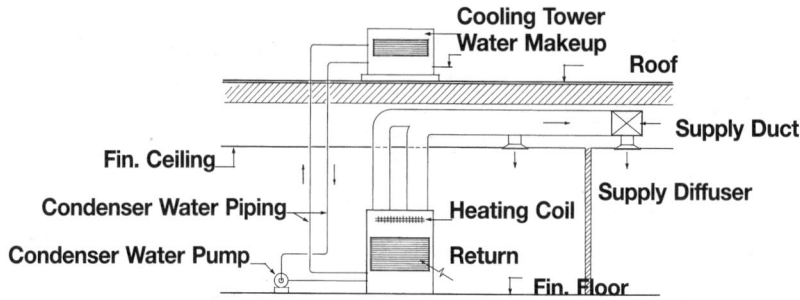

System Description: Self-contained, single package water cooled units include cooling tower, pump, piping allowance. Systems for 1000 S.F. and up include duct and diffusers to provide for even distribution of air. Smaller units distribute air through a supply air plenum, which is integral with the unit.

Returns are not ducted and supplies are not insulated.

Hot water or steam heating coils are included but piping to boiler and the boiler itself is not included.

Where local codes or conditions permit single pass cooling for the smaller units, deduct 10%.

System Components			COST EACH		
	QUANTITY	UNIT	MAT.	INST.	TOTAL
SYSTEM D3050 160 1300					
SELF-CONTAINED, WATER COOLED UNIT					
APARTMENT CORRIDORS, 500 S.F., .92 TON					
Self contained, water cooled, single package air conditioner unit	1.000	Ea.	1,142.40	554.40	1,696.80
Ductwork package for water or air cooled packaged units	1.000	System	108.56	763.60	872.16
Cooling tower, draw thru single flow, belt drive	1.000	Ea.	138.92	16.97	155.89
Cooling tower pumps & piping	1.000	System	69.46	40.02	109.48
TOTAL			1,459.34	1,374.99	2,834.33
COST PER S.F.			2.92	2.75	5.67

D3050 160	Self-contained, Water Cooled Unit Systems	COST PER S.F.		
		MAT.	INST.	TOTAL
1280	Self-contained, water cooled unit	2.92	2.75	5.67
1300	Apartment corridors, 500 S.F., .92 ton	2.90	2.75	5.65
1440	10,000 S.F., 18.33 ton	3.50	1.92	5.42
1520	Banks or libraries, 500 S.F., 2.08 ton	6.10	2.77	8.87
1680	10,000 S.F., 41.66 ton	6.90	4.37	11.27
1760	Bars and taverns, 500 S.F., 5.54 ton	13.20	4.67	17.87
1920	10,000 S.F., 110.00 ton	16.95	6.80	23.75
2000	Bowling alleys, 500 S.F., 2.83 ton	8.35	3.78	12.13
2160	10,000 S.F., 56.66 ton	9.10	5.85	14.95
2200	Department stores, 500 S.F., 1.46 ton	4.30	1.95	6.25
2360	10,000 S.F., 29.17 ton	5	2.94	7.94
2440	Drug stores, 500 S.F., 3.33 ton	9.80	4.44	14.24
2600	10,000 S.F., 66.66 ton	10.75	6.85	17.60
2680	Factories, 500 S.F., 1.66 ton	4.90	2.21	7.11
2840	10,000 S.F., 33.33 ton	5.75	3.36	9.11
2920	Food supermarkets, 500 S.F., 1.42 ton	4.17	1.88	6.05
3080	10,000 S.F., 28.33 ton	4.88	2.86	7.74
3160	Medical centers, 500 S.F., 1.17 ton	3.43	1.55	4.98
3320	10,000 S.F., 23.33 ton	4.44	2.44	6.88
3400	Offices, 500 S.F., 1.58 ton	4.66	2.11	6.77
3560	10,000 S.F., 31.67 ton	5.45	3.20	8.65
3640	Restaurants, 500 S.F., 2.50 ton	7.35	3.33	10.68
3800	10,000 S.F., 50.00 ton	5.85	4.81	10.66
3880	Schools and colleges, 500 S.F., 1.92 ton	5.65	2.56	8.21
4040	10,000 S.F., 38.33 ton	6.35	4.02	10.37

Note: Row 1440/1520 includes a reference box reading **RD3030 -010**.

D3050 Terminal & Package Units

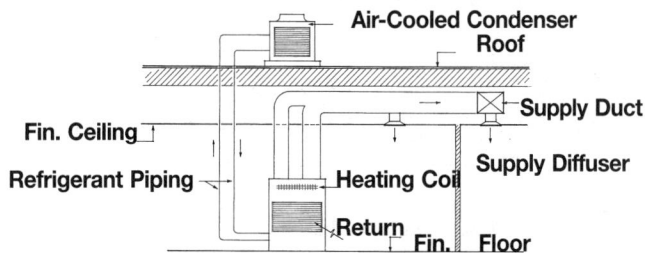

System Description: Self-contained air cooled units with remote air cooled condenser and interconnecting tubing. Systems for 1000 S.F. and up include duct and diffusers. Smaller units distribute air directly.

Returns are not ducted and supplies are not insulated.

Potential savings may be realized by using a single zone rooftop system or through-the-wall unit, especially in the smaller capacities, if the application permits.

Hot water or steam heating coils are included but piping to boiler and the boiler itself is not included.

Condenserless models are available for 15% less where remote refrigerant source is available.

System Components	QUANTITY	UNIT	COST EACH		
			MAT.	INST.	TOTAL
SYSTEM D3050 165 1320					
SELF-CONTAINED, AIR COOLED UNIT					
APARTMENT CORRIDORS, 500 S.F., .92 TON					
Air cooled, package unit	1.000	Ea.	1,198.50	365.50	1,564
Ductwork package for water or air cooled packaged units	1.000	System	108.56	763.60	872.16
Refrigerant piping	1.000	System	506	492.20	998.20
Air cooled condenser, direct drive, propeller fan	1.000	Ea.	348.75	137.95	486.70
TOTAL			2,161.81	1,759.25	3,921.06
COST PER S.F.			4.32	3.52	7.84

D3050 165	Self-contained, Air Cooled Unit Systems		COST PER S.F.		
			MAT.	INST.	TOTAL
1300	Self-contained, air cooled unit				
1320	Apartment corridors, 500 S.F., .92 ton		4.32	3.52	7.82
1480	10,000 S.F., 18.33 ton		3.15	2.84	5.99
1560	Banks or libraries, 500 S.F., 2.08 ton	RD3030 -010	9.20	4.47	13.67
1720	10,000 S.F., 41.66 ton		8.30	6.35	14.65
1800	Bars and taverns, 500 S.F., 5.54 ton		19.30	9.75	29.05
1960	10,000 S.F., 110.00 ton		20.50	12.20	32.70
2040	Bowling alleys, 500 S.F., 2.83 ton		12.55	6.10	18.65
2200	10,000 S.F., 56.66 ton		10.70	8.60	19.30
2240	Department stores, 500 S.F., 1.46 ton		6.45	3.14	9.59
2400	10,000 S.F., 29.17 ton		5.10	4.43	9.53
2480	Drug stores, 500 S.F., 3.33 ton		14.75	7.20	21.95
2640	10,000 S.F., 66.66 ton		12.65	10.15	22.80
2720	Factories, 500 S.F., 1.66 ton		7.45	3.61	11.06
2880	10,000 S.F., 33.33 ton		5.85	5.05	10.90
3200	Medical centers, 500 S.F., 1.17 ton		5.20	2.52	7.72
3360	10,000 S.F., 23.33 ton		4.02	3.63	7.65
3440	Offices, 500 S.F., 1.58 ton		7	3.41	10.41
3600	10,000 S.F., 31.66 ton		5.60	4.81	10.41
3680	Restaurants, 500 S.F., 2.50 ton		11.05	5.40	16.45
3840	10,000 S.F., 50.00 ton		10.15	7.55	17.70
3920	Schools and colleges, 500 S.F., 1.92 ton		8.50	4.13	12.63
4080	10,000 S.F., 38.33 ton		7.65	5.85	13.50

D3050 Terminal & Package Units

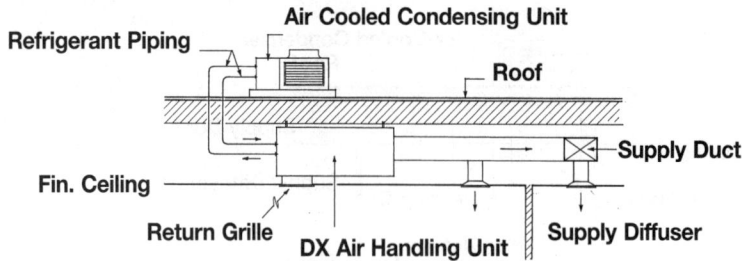

General: Split systems offer several important advantages which should be evaluated when a selection is to be made. They provide a greater degree of flexibility in component selection which permits an accurate match-up of the proper equipment size and type with the particular needs of the building. This allows for maximum use of modern energy saving concepts in heating and cooling. Outdoor installation of the air cooled condensing unit allows space savings in the building and also isolates the equipment operating sounds from building occupants.

Design Assumptions: The systems below are comprised of a direct expansion air handling unit and air cooled condensing unit with interconnecting copper tubing. Ducts and diffusers are also included for distribution of air. Systems are priced for cooling only. Heat can be added as desired either by putting hot water/steam coils into the air unit or into the duct supplying the particular area of need. Gas fired duct furnaces are also available. Refrigerant liquid line is insulated.

System Components	QUANTITY	UNIT	COST EACH		
			MAT.	INST.	TOTAL
SYSTEM D3050 170 1280					
SPLIT SYSTEM, AIR COOLED CONDENSING UNIT					
APARTMENT CORRIDORS, 1,000 S.F., 1.80 TON					
Fan coil AC unit, cabinet mntd. & filters direct expansion air cool	1.000	Ea.	411.75	129.93	541.68
Ductwork package, for split system, remote condensing unit	1.000	System	101.57	722.85	824.42
Refrigeration piping	1.000	System	406.26	814.35	1,220.61
Condensing unit, air cooled, incls. compressor & standard controls	1.000	Ea.	1,525	518.50	2,043.50
TOTAL			2,444.58	2,185.63	4,630.21
COST PER S.F.			2.44	2.19	4.63

D3050 170	Split Systems With Air Cooled Condensing Units		COST PER S.F.		
			MAT.	INST.	TOTAL
1260	Split system, air cooled condensing unit				
1280	Apartment corridors, 1,000 S.F., 1.83 ton		2.44	2.19	4.63
1440	20,000 S.F., 36.66 ton		2.58	3.08	5.66
1520	Banks and libraries, 1,000 S.F., 4.17 ton	RD3030 -010	4.02	5	9.02
1680	20,000 S.F., 83.32 ton		7.05	7.25	14.30
1760	Bars and taverns, 1,000 S.F., 11.08 ton		10.50	9.90	20.40
1880	10,000 S.F., 110.84 ton		16.05	11.55	27.60
2000	Bowling alleys, 1,000 S.F., 5.66 ton		5.70	9.40	15.10
2160	20,000 S.F., 113.32 ton		11.10	10.30	21.40
2320	Department stores, 1,000 S.F., 2.92 ton		2.86	3.46	6.32
2480	20,000 S.F., 58.33 ton		4.10	4.89	8.99
2560	Drug stores, 1,000 S.F., 6.66 ton		6.70	11.05	17.75
2720	20,000 S.F., 133.32 ton*				
2800	Factories, 1,000 S.F., 3.33 ton		3.25	3.94	7.19
2960	20,000 S.F., 66.66 ton		5.10	5.85	10.95
3040	Food supermarkets, 1,000 S.F., 2.83 ton		2.78	3.35	6.13
3200	20,000 S.F., 56.66 ton		3.98	4.76	8.74
3280	Medical centers, 1,000 S.F., 2.33 ton		2.46	2.72	5.18
3440	20,000 S.F., 46.66 ton		3.28	3.91	7.19
3520	Offices, 1,000 S.F., 3.17 ton		3.09	3.75	6.84
3680	20,000 S.F., 63.32 ton		4.85	5.55	10.40
3760	Restaurants, 1,000 S.F., 5.00 ton		5.05	8.30	13.35
3920	20,000 S.F., 100.00 ton		9.80	9.10	18.90
4000	Schools and colleges, 1,000 S.F., 3.83 ton		3.69	4.62	8.31
4160	20,000 S.F., 76.66 ton		5.90	6.70	12.60

D3050 Terminal & Package Units

Computer rooms impose special requirements on air conditioning systems. A prime requirement is reliability, due to the potential monetary loss that could be incurred by a system failure. A second basic requirement is the tolerance of control with which temperature and humidity are regulated, and dust eliminated. As the air conditioning system reliability is so vital, the additional cost of reserve capacity and redundant components is often justified.

System Descriptions: Computer areas may be environmentally controlled by one of three methods as follows:

1. **Self-contained Units**
   These are units built to higher standards of performance and reliability. They usually contain alarms and controls to indicate component operation failure, filter change, etc. It should be remembered that these units in the room will occupy space that is relatively expensive to build and that all alterations and service of the equipment will also have to be accomplished within the computer area.
2. **Decentralized Air Handling Units** In operation these are similar to the self-contained units except that their cooling capability comes from remotely located refrigeration equipment as refrigerant or chilled water. As no compressors or refrigerating equipment are required in the air units, they are smaller and require less service than self-contained units. An added plus for this type of system occurs if some of the computer components themselves also require chilled water for cooling.
3. **Central System Supply** Cooling is obtained from a central source which, since it is not located within the computer room, may have excess capacity and permit greater flexibility without interfering with the computer components. System performance criteria must still be met.

Note: The costs shown below do not include an allowance for ductwork or piping.

D3050 185	Computer Room Cooling Units	COST EACH		
		MAT.	INST.	TOTAL
0560	Computer room unit, air cooled, includes remote condenser			
0580	3 ton	17,200	2,125	19,325
0600	5 ton (RD3030 -010)	18,400	2,375	20,775
0620	8 ton	34,500	3,950	38,450
0640	10 ton	36,100	4,275	40,375
0660	15 ton	39,700	4,850	44,550
0680	20 ton	48,200	6,925	55,125
0700	23 ton	59,500	7,900	67,400
0800	Chilled water, for connection to existing chiller system			
0820	5 ton	13,200	1,450	14,650
0840	8 ton	13,300	2,125	15,425
0860	10 ton	13,400	2,175	15,575
0880	15 ton	14,200	2,225	16,425
0900	20 ton	15,000	2,325	17,325
0920	23 ton	16,000	2,625	18,625
1000	Glycol system, complete except for interconnecting tubing			
1020	3 ton	21,700	2,675	24,375
1040	5 ton	23,700	2,800	26,500
1060	8 ton	38,100	4,650	42,750
1080	10 ton	40,500	5,075	45,575
1100	15 ton	50,000	6,375	56,375
1120	20 ton	55,500	6,925	62,425
1140	23 ton	58,000	7,225	65,225
1240	Water cooled, not including condenser water supply or cooling tower			
1260	3 ton	17,500	1,725	19,225
1280	5 ton	19,100	1,975	21,075
1300	8 ton	31,300	3,225	34,525
1320	15 ton	38,000	3,950	41,950
1340	20 ton	40,900	4,375	45,275
1360	23 ton	42,800	4,750	47,550

D4010 Sprinklers

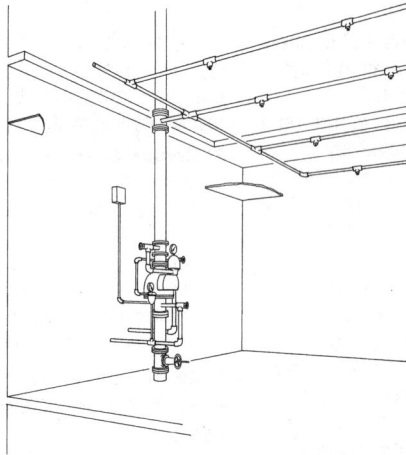

Dry Pipe System: A system employing automatic sprinklers attached to a piping system containing air under pressure, the release of which as from the opening of sprinklers permits the water pressure to open a valve known as a "dry pipe valve". The water then flows into the piping system and out the opened sprinklers.

All areas are assumed to be open.

System Components	QUANTITY	UNIT	COST EACH		
			MAT.	INST.	TOTAL
SYSTEM D4010 310 0580					
DRY PIPE SPRINKLER, STEEL BLACK, SCH. 40 PIPE					
LIGHT HAZARD, ONE FLOOR, 2000 S.F.					
Valve, gate, iron body 125 lb., OS&Y, flanged, 4" pipe size	1.000	Ea.	420	262.50	682.50
Valve, swing check, bronze, 125 lb., regrinding disc, 2-1/2" pipe size	1.000	Ea.	420	52.50	472.50
Valve, angle, bronze, 150 lb., rising stem, threaded, 2" pipe size	1.000	Ea.	393.75	39.75	433.50
*Alarm valve, 2-1/2" pipe size	1.000	Ea.	1,012.50	258.75	1,271.25
Alarm, water motor, complete with gong,	1.000	Ea.	251.25	108.75	360
Fire alarm horn, electric	1.000	Ea.	33.38	62.63	96.01
Valve swing check w/balldrip CI with brass trim, 4" pipe size	1.000	Ea.	210	258.75	468.75
Pipe, steel, black, schedule 40, 4" diam.	10.000	L.F.	149.25	226.20	375.45
Dry pipe valve, trim & gauges, 4" pipe size	1.000	Ea.	1,668.75	787.50	2,456.25
Pipe, steel, black, schedule 40, threaded, cplg & hngr 10'OC 2-1/2" diam.	20.000	L.F.	198.75	315	513.75
Pipe, steel, black, schedule 40, threaded, cplg & hngr 10'OC 2" diam.	12.500	L.F.	79.22	154.22	233.44
Pipe, steel, black, schedule 40, threaded, cplg & hngr 10'OC 1-1/4" diam.	37.500	L.F.	151.88	333.28	485.16
Pipe, steel, black, schedule 40, threaded, cplg & hngr 10'OC 1" diam.	112.000	L.F.	357.84	928.20	1,286.04
Pipe Tee, malleable iron black, 150 lb. threaded, 4" pipe size	2.000	Ea.	291	394.50	685.50
Pipe Tee, malleable iron black, 150 lb. threaded, 2-1/2" pipe size	2.000	Ea.	81.75	175.50	257.25
Pipe Tee, malleable iron black, 150 lb. threaded, 2" pipe size	1.000	Ea.	19.13	72	91.13
Pipe Tee, malleable iron black, 150 lb. threaded, 1-1/4" pipe size	5.000	Ea.	44.63	283.13	327.76
Pipe Tee, malleable iron black, 150 lb. threaded, 1" pipe size	4.000	Ea.	22.05	219	241.05
Pipe 90° elbow malleable iron black, 150 lb. threaded, 1" pipe size	6.000	Ea.	21.15	202.50	223.65
Sprinkler head dry 1/2" orifice 1" NPT, 3" to 4-3/4" length	12.000	Ea.	1,194	498	1,692
Air compressor, 200 Gal sprinkler system capacity, 1/3 HP	1.000	Ea.	573.75	333.75	907.50
*Standpipe connection, wall, flush, brs. w/plug & chain 2-1/2"x2-1/2"	1.000	Ea.	101.25	156	257.25
Valve gate bronze, 300 psi, NRS, class 150, threaded, 1" pipe size	1.000	Ea.	58.88	23.25	82.13
TOTAL			7,754.16	6,145.66	13,899.82
COST PER S.F.			3.88	3.07	6.95

*Not included in systems under 2000 S.F.

D4010 310	Dry Pipe Sprinkler Systems		COST PER S.F.		
			MAT.	INST.	TOTAL
0520	Dry pipe sprinkler systems, steel, black, sch. 40 pipe				
0530	Light hazard, one floor, 500 S.F.		7.25	5.25	12.50
0560	1000 S.F.		4.06	3.03	7.09
0580	2000 S.F.	RD4010 -100	3.88	3.06	6.94
0600	5000 S.F.		2.05	2.08	4.13
0620	10,000 S.F.	RD4020 -300	1.47	1.74	3.21
0640	50,000 S.F.		1.18	1.54	2.01
0660	Each additional floor, 500 S.F.		1.64	2.54	4.18

D40 Fire Protection

D4010 Sprinklers

D4010 310	Dry Pipe Sprinkler Systems	COST PER S.F.		
		MAT.	INST.	TOTAL
0680	1000 S.F.	1.35	2.08	3.43
0700	2000 S.F.	1.32	1.95	3.27
0720	5000 S.F.	1.10	1.65	2.75
0740	10,000 S.F.	1.01	1.52	2.53
0760	50,000 S.F.	.93	1.36	2.30
1000	Ordinary hazard, one floor, 500 S.F.	7.40	5.30	12.70
1020	1000 S.F.	4.18	3.07	7.25
1040	2000 S.F.	3.98	3.21	7.19
1060	5000 S.F.	2.34	2.22	4.56
1080	10,000 S.F.	1.93	2.29	4.22
1100	50,000 S.F.	1.70	2.13	3.84
1140	Each additional floor, 500 S.F.	1.79	2.60	4.39
1160	1000 S.F.	1.58	2.34	3.92
1180	2000 S.F.	1.56	2.13	3.69
1200	5000 S.F.	1.44	1.82	3.26
1220	10,000 S.F.	1.33	1.80	3.13
1240	50,000 S.F.	1.24	1.54	2.79
1500	Extra hazard, one floor, 500 S.F.	10.15	6.60	16.75
1520	1000 S.F.	6.20	4.73	10.93
1540	2000 S.F.	4.47	4.12	8.59
1560	5000 S.F.	2.77	3.10	5.87
1580	10,000 S.F.	2.83	2.95	5.78
1600	50,000 S.F.	2.99	2.83	5.82
1660	Each additional floor, 500 S.F.	2.37	3.23	5.60
1680	1000 S.F.	2.32	3.06	5.38
1700	2000 S.F.	2.17	3.07	5.24
1720	5000 S.F.	1.87	2.67	4.54
1740	10,000 S.F.	2.14	2.44	4.58
1760	50,000 S.F.	2.15	2.33	4.48
2020	Grooved steel, black, sch. 40 pipe, light hazard, one floor, 2000 S.F.	3.73	2.65	6.38
2060	10,000 S.F.	1.49	1.52	3.01
2100	Each additional floor, 2000 S.F.	1.31	1.57	2.88
2150	10,000 S.F.	1.03	1.30	2.33
2200	Ordinary hazard, one floor, 2000 S.F.	3.95	2.79	6.74
2250	10,000 S.F.	1.89	1.92	3.81
2300	Each additional floor, 2000 S.F.	1.53	1.71	3.24
2350	10,000 S.F.	1.44	1.71	3.15
2400	Extra hazard, one floor, 2000 S.F.	4.47	3.54	8.01
2450	10,000 S.F.	2.56	2.48	5.04
2500	Each additional floor, 2000 S.F.	2.18	2.54	4.72
2550	10,000 S.F.	1.98	2.18	4.16
3050	Grooved steel black sch. 10 pipe, light hazard, one floor, 2000 S.F.	3.66	2.62	6.28
3100	10,000 S.F.	1.43	1.49	2.92
3150	Each additional floor, 2000 S.F.	1.24	1.54	2.78
3200	10,000 S.F.	.97	1.27	2.24
3250	Ordinary hazard, one floor, 2000 S.F.	3.88	2.78	6.66
3300	10,000 S.F.	1.75	1.88	3.63
3350	Each additional floor, 2000 S.F.	1.46	1.70	3.16
3400	10,000 S.F.	1.30	1.67	2.97
3450	Extra hazard, one floor, 2000 S.F.	4.41	3.51	7.92
3500	10,000 S.F.	2.42	2.44	4.86
3550	Each additional floor, 2000 S.F.	2.12	2.51	4.63
3600	10,000 S.F.	1.90	2.15	4.05
4050	Copper tubing, type M, light hazard, one floor, 2000 S.F.	4.56	2.61	7.17
4100	10,000 S.F.	2.26	1.52	3.78
4150	Each additional floor, 2000 S.F.	2.16	1.56	3.72
4200	10,000 S.F.	1.80	1.31	3.11
4250	Ordinary hazard, one floor, 2000 S.F.	4.94	2.92	7.86

D4010 Sprinklers

D4010 310	Dry Pipe Sprinkler Systems	COST PER S.F.		
		MAT.	INST.	TOTAL
4300	10,000 S.F.	2.82	1.78	4.60
4350	Each additional floor, 2000 S.F.	2.96	1.82	4.78
4400	10,000 S.F.	2.30	1.54	3.84
4450	Extra hazard, one floor, 2000 S.F.	5.75	3.55	9.30
4500	10,000 S.F.	5.50	2.66	8.16
4550	Each additional floor, 2000 S.F.	3.47	2.55	6.02
4600	10,000 S.F.	3.96	2.34	6.30
5050	Copper tubing, type M, T-drill system, light hazard, one floor			
5060	2000 S.F.	4.58	2.45	7.03
5100	10,000 S.F.	2.14	1.27	3.41
5150	Each additional floor, 2000 S.F.	2.18	1.40	3.58
5200	10,000 S.F.	1.68	1.06	2.74
5250	Ordinary hazard, one floor, 2000 S.F.	4.75	2.51	7.26
5300	10,000 S.F.	2.73	1.60	4.33
5350	Each additional floor, 2000 S.F.	2.33	1.43	3.76
5400	10,000 S.F.	2.15	1.32	3.47
5450	Extra hazard, one floor, 2000 S.F.	5.20	2.94	8.14
5500	10,000 S.F.	4.55	1.95	6.50
5550	Each additional floor, 2000 S.F.	2.91	1.94	4.85
5600	10,000 S.F.	3.03	1.63	4.66

D4010 Sprinklers

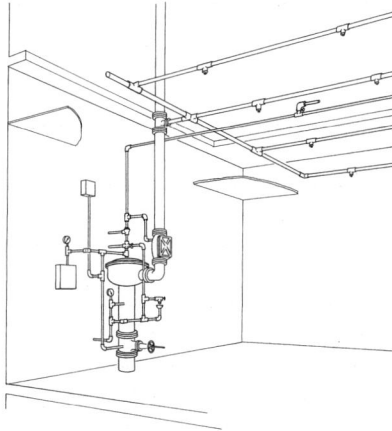

Pre-Action System: A system employing automatic sprinklers attached to a piping system containing air that may or may not be under pressure, with a supplemental heat responsive system of generally more sensitive characteristics than the automatic sprinklers themselves, installed in the same areas as the sprinklers. Actuation of the heat responsive system, as from a fire, opens a valve which permits water to flow into the sprinkler piping system and to be discharged from those sprinklers which were opened by heat from the fire.

All areas are assumed to be open.

System Components	QUANTITY	UNIT	COST EACH MAT.	COST EACH INST.	COST EACH TOTAL
SYSTEM D4010 350 0580					
PREACTION SPRINKLER SYSTEM, STEEL BLACK SCH. 40 PIPE					
LIGHT HAZARD, 1 FLOOR, 2000 S.F.					
Valve, gate, iron body 125 lb., OS&Y, flanged, 4" pipe size	1.000	Ea.	420	262.50	682.50
*Valve, swing check w/ball drip CI with brass trim 4" pipe size	1.000	Ea.	210	258.75	468.75
Valve, swing check, bronze, 125 lb., regrinding disc, 2-1/2" pipe size	1.000	Ea.	420	52.50	472.50
Valve, angle, bronze, 150 lb., rising stem, threaded, 2" pipe size	1.000	Ea.	393.75	39.75	433.50
*Alarm valve, 2-1/2" pipe size	1.000	Ea.	1,012.50	258.75	1,271.25
Alarm, water motor, complete with gong	1.000	Ea.	251.25	108.75	360
Fire alarm horn, electric	1.000	Ea.	33.38	62.63	96.01
Thermostatic release for release line	2.000	Ea.	825	43.50	868.50
Pipe, steel, black, schedule 40, 4" diam	10.000	L.F.	149.25	226.20	375.45
Dry pipe valve, trim & gauges, 4" pipe size	1.000	Ea.	1,668.75	787.50	2,456.25
Pipe, steel, black, schedule 40, threaded, cplg. & hngr. 10'OC 2-1/2" diam.	20.000	L.F.	198.75	315	513.75
Pipe steel black, schedule 40, threaded, cplg. & hngr.10'OC 2" diam.	12.500	L.F.	79.22	154.22	233.44
Pipe, steel, black, schedule 40, threaded, cplg. & hngr. 10'OC 1-1/4" diam.	37.500	L.F.	151.88	333.28	485.16
Pipe, steel, black, schedule 40, threaded, cplg. & hngr. 10'OC 1" diam.	112.000	L.F.	357.84	928.20	1,286.04
Pipe, Tee, malleable iron, black, 150 lb. threaded, 4" diam	2.000	Ea.	291	394.50	685.50
Pipe, Tee, malleable iron, black, 150 lb. threaded, 2-1/2" pipe size	2.000	Ea.	81.75	175.50	257.25
Pipe, Tee, malleable iron, black, 150 lb. threaded, 2" pipe size	1.000	Ea.	19.13	72	91.13
Pipe, Tee, malleable iron, black, 150 lb. threaded, 1-1/4" pipe size	5.000	Ea.	44.63	283.13	327.76
Pipe, Tee, malleable iron, black, 150 lb. threaded, 1" pipe size	4.000	Ea.	22.05	219	241.05
Pipe, 90° elbow, malleable iron, blk., 150 lb. threaded, 1" pipe size	6.000	Ea.	21.15	202.50	223.65
Sprinkler head, std. spray, brass 135°-286° F 1/2" NPT, 3/8" orifice	12.000	Ea.	151.20	432	583.20
Air compressor auto complete 200 Gal sprinkler sys. cap., 1/3 HP	1.000	Ea.	573.75	333.75	907.50
*Standpipe conn.,wall, flush, brass w/plug & chain 2-1/2" x 2-1/2"	1.000	Ea.	101.25	156	257.25
Valve, gate, bronze, 300 psi, NRS, class 150, threaded, 1" pipe size	1.000	Ea.	58.88	23.25	82.13
TOTAL			7,536.36	6,123.16	13,659.52
COST PER S.F.			3.77	3.06	6.83

*Not included in systems under 2000 S.F.

D4010 350	Preaction Sprinkler Systems		COST PER S.F. MAT.	COST PER S.F. INST.	COST PER S.F. TOTAL
0520	Preaction sprinkler systems, steel, black, sch. 40 pipe				
0530	Light hazard, one floor, 500 S.F.	RD4010 -100	7.10	4.17	11.27
0560	1000 S.F.		4.10	3.10	7.20
0580	2000 S.F.	RD4020 -300	3.76	3.05	6.81

D4010 Sprinklers

D4010 350	Preaction Sprinkler Systems	COST PER S.F.		
		MAT.	INST.	TOTAL
0600	5000 S.F.	1.96	2.07	4.03
0620	10,000 S.F.	1.39	1.74	3.13
0640	50,000 S.F.	1.10	1.53	2.63
0660	Each additional floor, 500 S.F.	1.81	2.25	4.06
0680	1000 S.F.	1.37	2.07	3.44
0700	2000 S.F.	1.34	1.94	3.28
0720	5000 S.F.	1.01	1.64	2.65
0740	10,000 S.F.	.93	1.52	2.45
0760	50,000 S.F.	1.94	1.40	2.35
1000	Ordinary hazard, one floor, 500 S.F.	7.40	4.53	11.93
1020	1000 S.F.	4.07	3.06	7.13
1040	2000 S.F.	4.08	3.21	7.29
1060	5000 S.F.	2.15	2.21	4.36
1080	10,000 S.F.	1.70	2.28	3.98
1100	50,000 S.F.	1.45	2.13	3.58
1140	Each additional floor, 500 S.F.	2.09	2.61	4.70
1160	1000 S.F.	1.36	2.11	3.47
1180	2000 S.F.	1.25	2.11	3.36
1200	5000 S.F.	1.36	1.97	3.33
1220	10,000 S.F.	1.25	2.07	3.32
1240	50,000 S.F.	1.13	1.82	2.95
1500	Extra hazard, one floor, 500 S.F.	9.85	5.80	15.65
1520	1000 S.F.	5.65	4.30	9.95
1540	2000 S.F.	4.09	4.09	8.18
1560	5000 S.F.	2.66	3.35	6.01
1580	10,000 S.F.	2.50	3.27	5.77
1600	50,000 S.F.	2.59	3.15	5.74
1660	Each additional floor, 500 S.F.	2.40	3.22	5.62
1680	1000 S.F.	1.94	3.03	4.97
1700	2000 S.F.	1.79	3.04	4.83
1720	5000 S.F.	1.52	2.69	4.21
1740	10,000 S.F.	1.66	2.46	4.12
1760	50,000 S.F.	1.59	2.29	3.88
2020	Grooved steel, black, sch. 40 pipe, light hazard, one floor, 2000 S.F.	3.75	2.64	6.39
2060	10,000 S.F.	1.41	1.52	2.93
2100	Each additional floor of 2000 S.F.	1.33	1.56	2.89
2150	10,000 S.F.	.95	1.30	2.25
2200	Ordinary hazard, one floor, 2000 S.F.	3.84	2.78	6.62
2250	10,000 S.F.	1.66	1.91	3.57
2300	Each additional floor, 2000 S.F.	1.42	1.70	3.12
2350	10,000 S.F.	1.21	1.70	2.91
2400	Extra hazard, one floor, 2000 S.F.	4.09	3.51	7.60
2450	10,000 S.F.	2.09	2.45	4.54
2500	Each additional floor, 2000 S.F.	1.80	2.51	4.31
2550	10,000 S.F.	1.48	2.15	3.63
3050	Grooved steel, black, sch. 10 pipe light hazard, one floor, 2000 S.F.	3.68	2.61	6.29
3100	10,000 S.F.	1.35	1.49	2.84
3150	Each additional floor, 2000 S.F.	1.26	1.53	2.79
3200	10,000 S.F.	.89	1.27	2.16
3250	Ordinary hazard, one floor, 2000 S.F.	3.71	2.59	6.30
3300	10,000 S.F.	1.31	1.85	3.16
3350	Each additional floor, 2000 S.F.	1.35	1.69	3.04
3400	10,000 S.F.	1.07	1.66	2.73
3450	Extra hazard, one floor, 2000 S.F.	4.03	3.48	7.51
3500	10,000 S.F.	1.92	2.41	4.33
3550	Each additional floor, 2000 S.F.	1.74	2.48	4.22
3600	10,000 S.F.	1.40	2.12	3.52
4050	Copper tubing, type M, light hazard, one floor, 2000 S.F.	4.59	2.60	7.19

D40 Fire Protection

D4010 Sprinklers

D4010 350	Preaction Sprinkler Systems	COST PER S.F.		
		MAT.	INST.	TOTAL
4100	10,000 S.F.	2.19	1.52	3.71
4150	Each additional floor, 2000 S.F.	2.19	1.55	3.74
4200	10,000 S.F.	1.51	1.29	2.80
4250	Ordinary hazard, one floor, 2000 S.F.	4.83	2.91	7.74
4300	10,000 S.F.	2.59	1.77	4.36
4350	Each additional floor, 2000 S.F.	2.20	1.61	3.81
4400	10,000 S.F.	1.85	1.41	3.26
4450	Extra hazard, one floor, 2000 S.F.	5.40	3.52	8.92
4500	10,000 S.F.	4.95	2.63	7.58
4550	Each additional floor, 2000 S.F.	3.09	2.52	5.61
4600	10,000 S.F.	3.46	2.31	5.77
5050	Copper tubing, type M, T-drill system, light hazard, one floor			
5060	2000 S.F.	4.60	2.44	7.04
5100	10,000 S.F.	2.06	1.27	3.33
5150	Each additional floor, 2000 S.F.	2.20	1.39	3.59
5200	10,000 S.F.	1.60	1.06	2.66
5250	Ordinary hazard, one floor, 2000 S.F.	4.64	2.50	7.14
5300	10,000 S.F.	2.50	1.59	4.09
5350	Each additional floor, 2000 S.F.	2.22	1.42	3.64
5400	10,000 S.F.	2.05	1.38	3.43
5450	Extra hazard, one floor, 2000 S.F.	4.82	2.91	7.73
5500	10,000 S.F.	4.02	1.92	5.94
5550	Each additional floor, 2000 S.F.	2.53	1.91	4.44
5600	10,000 S.F.	2.53	1.60	4.13

D40 Fire Protection

D4010 Sprinklers

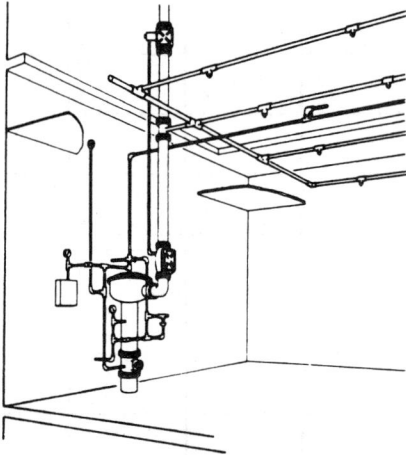

Deluge System: A system employing open sprinklers attached to a piping system connected to a water supply through a valve which is opened by the operation of a heat responsive system installed in the same areas as the sprinklers. When this valve opens, water flows into the piping system and discharges from all sprinklers attached thereto.

All areas are assumed to be open.

System Components	QUANTITY	UNIT	COST EACH		
			MAT.	INST.	TOTAL
SYSTEM D4010 370 0580					
DELUGE SPRINKLER SYSTEM, STEEL BLACK SCH. 40 PIPE					
LIGHT HAZARD, 1 FLOOR, 2000 S.F.					
Valve, gate, iron body 125 lb., OS&Y, flanged, 4" pipe size	1.000	Ea.	420	262.50	682.50
Valve, swing check w/ball drip, CI w/brass ftngs., 4" pipe size	1.000	Ea.	210	258.75	468.75
Valve, swing check, bronze, 125 lb., regrinding disc, 2-1/2" pipe size	1.000	Ea.	420	52.50	472.50
Valve, angle, bronze, 150 lb., rising stem, threaded, 2" pipe size	1.000	Ea.	393.75	39.75	433.50
*Alarm valve, 2-1/2" pipe size	1.000	Ea.	1,012.50	258.75	1,271.25
Alarm, water motor, complete with gong	1.000	Ea.	251.25	108.75	360
Fire alarm horn, electric	1.000	Ea.	33.38	62.63	96.01
Thermostatic release for release line	2.000	Ea.	825	43.50	868.50
Pipe, steel, black, schedule 40, 4" diam.	10.000	L.F.	149.25	226.20	375.45
Deluge valve trim, pressure relief, emergency release, gauge, 4" pipe size	1.000	Ea.	2,268.75	787.50	3,056.25
Deluge system pressured monitoring panel, 120V	1.000	Ea.	6,862.50	24	6,886.50
Pipe, steel, black, schedule 40, threaded, cplg & hngr 10' OC 2-1/2" diam.	20.000	L.F.	198.75	315	513.75
Pipe, steel, black, schedule 40, threaded, cplg & hngr 10' OC 2" diam.	12.500	L.F.	79.22	154.22	233.44
Pipe, steel, black, schedule 40, threaded, cplg & hngr 10' OC 1-1/4" diam.	37.500	L.F.	151.88	333.28	485.16
Pipe, steel, black, schedule 40, threaded, cplg & hngr 10' OC 1" diam.	112.000	L.F.	357.84	928.20	1,286.04
Pipe, Tee, malleable iron, black, 150 lb. threaded, 4" pipe size	2.000	Ea.	291	394.50	685.50
Pipe, Tee, malleable iron, black, 150 lb. threaded, 2-1/2" pipe size	2.000	Ea.	81.75	175.50	257.25
Pipe, Tee, malleable iron, black, 150 lb. threaded, 2" pipe size	1.000	Ea.	19.13	72	91.13
Pipe, Tee, malleable iron, black, 150 lb. threaded, 1-1/4" pipe size	5.000	Ea.	44.63	283.13	327.76
Pipe, Tee, malleable iron, black, 150 lb. threaded, 1" pipe size	4.000	Ea.	22.05	219	241.05
Pipe, 90° elbow, malleable iron, black, 150 lb. threaded 1" pipe size	6.000	Ea.	21.15	202.50	223.65
Sprinkler head, std spray, brass 135°-286° F 1/2" NPT, 3/8" orifice	9.720	Ea.	151.20	432	583.20
Air compressor, auto, complete, 200 Gal sprinkler sys. cap., 1/3 HP	1.000	Ea.	573.75	333.75	907.50
*Standpipe connection, wall, flush w/plug & chain 2-1/2" x 2-1/2"	1.000	Ea.	101.25	156	257.25
Valve, gate, bronze, 300 psi, NRS, class 150, threaded, 1" pipe size	1.000	Ea.	58.88	23.25	82.13
TOTAL			14,998.86	6,147.16	21,146.02
COST PER S.F.			7.50	3.07	10.57

*Not included in systems under 2000 S.F.

D4010 370	Deluge Sprinkler Systems		COST PER S.F.		
			MAT.	INST.	TOTAL
0520	Deluge sprinkler systems, steel, black, sch. 40 pipe	RD4010 -100			
0530	Light hazard, one floor, 500 S.F.		21	4.22	25.22
0560	1000 S.F.	RD4020 -300	11.05	2.97	14.02
0580	2000 S.F.		7.50	3.06	10.56

D40 Fire Protection

D4010 Sprinklers

D4010 370	Deluge Sprinkler Systems	COST PER S.F.		
		MAT.	INST.	TOTAL
0600	5000 S.F.	3.45	2.07	5.52
0620	10,000 S.F.	2.14	1.74	3.88
0640	50,000 S.F.	1.26	1.51	2.77
0660	Each additional floor, 500 S.F.	1.81	2.25	4.06
0680	1000 S.F.	1.37	2.07	3.44
0700	2000 S.F.	1.34	1.94	3.28
0720	5000 S.F.	1.01	1.64	2.65
0740	10,000 S.F.	.93	1.52	2.45
0760	50,000 S.F.	.95	1.38	2.33
1000	Ordinary hazard, one floor, 500 S.F.	22	4.84	26.84
1020	1000 S.F.	11.05	3.08	14.13
1040	2000 S.F.	7.80	3.22	11.02
1060	5000 S.F.	3.64	2.21	5.85
1080	10,000 S.F.	2.45	2.28	4.73
1100	50,000 S.F.	1.64	2.16	3.80
1140	Each additional floor, 500 S.F.	2.09	2.61	4.70
1160	1000 S.F.	1.36	2.11	3.47
1180	2000 S.F.	1.25	2.11	3.36
1200	5000 S.F.	1.25	1.81	3.06
1220	10,000 S.F.	1.19	1.84	3.03
1240	50,000 S.F.	1.06	1.66	2.72
1500	Extra hazard, one floor, 500 S.F.	24	5.85	29.85
1520	1000 S.F.	12.95	4.47	17.42
1540	2000 S.F.	7.80	4.10	11.90
1560	5000 S.F.	3.87	3.08	6.95
1580	10,000 S.F.	3.13	2.99	6.12
1600	50,000 S.F.	3.07	2.88	5.95
1660	Each additional floor, 500 S.F.	2.40	3.22	5.62
1680	1000 S.F.	1.94	3.03	4.97
1700	2000 S.F.	1.79	3.04	4.83
1720	5000 S.F.	1.52	2.69	4.21
1740	10,000 S.F.	1.71	2.56	4.27
1760	50,000 S.F.	1.69	2.44	4.13
2000	Grooved steel, black, sch. 40 pipe, light hazard, one floor			
2020	2000 S.F.	7.50	2.65	10.15
2060	10,000 S.F.	2.18	1.54	3.72
2100	Each additional floor, 2,000 S.F.	1.33	1.56	2.89
2150	10,000 S.F.	.95	1.30	2.25
2200	Ordinary hazard, one floor, 2000 S.F.	3.84	2.78	6.62
2250	10,000 S.F.	2.41	1.91	4.32
2300	Each additional floor, 2000 S.F.	1.42	1.70	3.12
2350	10,000 S.F.	1.21	1.70	2.91
2400	Extra hazard, one floor, 2000 S.F.	7.80	3.52	11.32
2450	10,000 S.F.	2.86	2.46	5.32
2500	Each additional floor, 2000 S.F.	1.80	2.51	4.31
2550	10,000 S.F.	1.48	2.15	3.63
3000	Grooved steel, black, sch. 10 pipe, light hazard, one floor			
3050	2000 S.F.	6.95	2.50	9.45
3100	10,000 S.F.	2.10	1.49	3.59
3150	Each additional floor, 2000 S.F.	1.26	1.53	2.79
3200	10,000 S.F.	.89	1.27	2.16
3250	Ordinary hazard, one floor, 2000 S.F.	7.50	2.78	10.28
3300	10,000 S.F.	2.06	1.85	3.91
3350	Each additional floor, 2000 S.F.	1.35	1.69	3.04
3400	10,000 S.F.	1.07	1.66	2.73
3450	Extra hazard, one floor, 2000 S.F.	7.75	3.49	11.24
3500	10,000 S.F.	2.67	2.41	5.08
3550	Each additional floor, 2000 S.F.	1.74	2.48	4.22

D4010 Sprinklers

D4010 370	Deluge Sprinkler Systems	COST PER S.F.		
		MAT.	INST.	TOTAL
3600	10,000 S.F.	1.40	2.12	3.52
4000	Copper tubing, type M, light hazard, one floor			
4050	2000 S.F.	8.30	2.61	10.91
4100	10,000 S.F.	2.93	1.52	4.45
4150	Each additional floor, 2000 S.F.	2.18	1.55	3.73
4200	10,000 S.F.	1.51	1.29	2.80
4250	Ordinary hazard, one floor, 2000 S.F.	8.55	2.92	11.47
4300	10,000 S.F.	3.34	1.77	5.11
4350	Each additional floor, 2000 S.F.	2.20	1.61	3.81
4400	10,000 S.F.	1.85	1.41	3.26
4450	Extra hazard, one floor, 2000 S.F.	9.10	3.53	12.63
4500	10,000 S.F.	5.75	2.64	8.39
4550	Each additional floor, 2000 S.F.	3.09	2.52	5.61
4600	10,000 S.F.	3.46	2.31	5.77
5000	Copper tubing, type M, T-drill system, light hazard, one floor			
5050	2000 S.F.	8.35	2.45	10.80
5100	10,000 S.F.	2.81	1.27	4.08
5150	Each additional floor, 2000 S.F.	2.11	1.41	3.52
5200	10,000 S.F.	1.60	1.06	2.66
5250	Ordinary hazard, one floor, 2000 S.F.	8.35	2.51	10.86
5300	10,000 S.F.	3.25	1.59	4.84
5350	Each additional floor, 2000 S.F.	2.22	1.42	3.64
5400	10,000 S.F.	2.05	1.38	3.43
5450	Extra hazard, one floor, 2000 S.F.	8.55	2.92	11.47
5500	10,000 S.F.	4.77	1.92	6.69
5550	Each additional floor, 2000 S.F.	2.53	1.91	4.44
5600	10,000 S.F.	2.53	1.60	4.13

D40 Fire Protection

D4010 Sprinklers

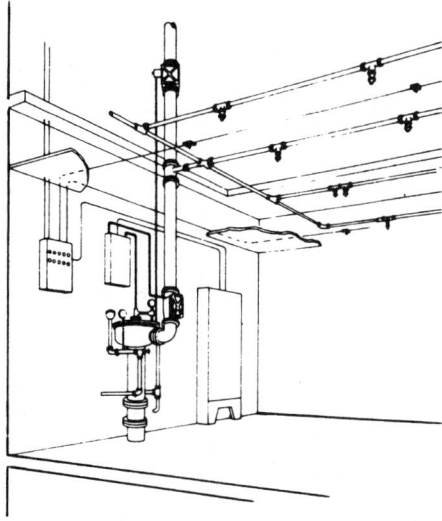

Firecycle is a fixed fire protection sprinkler system utilizing water as its extinguishing agent. It is a time delayed, recycling, preaction type which automatically shuts the water off when heat is reduced below the detector operating temperature and turns the water back on when that temperature is exceeded.

The system senses a fire condition through a closed circuit electrical detector system which controls water flow to the fire automatically. Batteries supply up to 90 hour emergency power supply for system operation. The piping system is dry (until water is required) and is monitored with pressurized air. Should any leak in the system piping occur, an alarm will sound, but water will not enter the system until heat is sensed by a Firecycle detector.

All areas are assumed to be open.

System Components	QUANTITY	UNIT	COST EACH		
			MAT.	INST.	TOTAL
SYSTEM D4010 390 0580					
FIRECYCLE SPRINKLER SYSTEM, STEEL BLACK SCH. 40 PIPE					
LIGHT HAZARD, ONE FLOOR, 2000 S.F.					
Valve, gate, iron body 125 lb., OS&Y, flanged, 4" pipe size	1.000	Ea.	420	262.50	682.50
Valve, angle, bronze, 150 lb., rising stem, threaded, 2" pipe size	1.000	Ea.	393.75	39.75	433.50
Valve, swing check, bronze, 125 lb., regrinding disc, 2-1/2" pipe size	1.000	Ea.	420	52.50	472.50
*Alarm valve, 2-1/2" pipe size	1.000	Ea.	1,012.50	258.75	1,271.25
Alarm, water motor, complete with gong	1.000	Ea.	251.25	108.75	360
Pipe, steel, black, schedule 40, 4" diam.	10.000	L.F.	149.25	226.20	375.45
Fire alarm, horn, electric	1.000	Ea.	33.38	62.63	96.01
Pipe, steel, black, schedule 40, threaded, cplg & hngr 10' OC 2-1/2" diam.	20.000	L.F.	198.75	315	513.75
Pipe, steel, black, schedule 40, threaded, cplg & hngr 10' OC 2" diam.	12.500	L.F.	79.22	154.22	233.44
Pipe, steel, black, schedule 40, threaded, cplg & hngr 10' OC 1-1/4" diam.	37.500	L.F.	151.88	333.28	485.16
Pipe, steel, black, schedule 40, threaded, cplg & hngr 10' OC 1" diam.	112.000	L.F.	357.84	928.20	1,286.04
Pipe, Tee, malleable iron, black, 150 lb. threaded, 4" pipe size	2.000	Ea.	291	394.50	685.50
Pipe, Tee, malleable iron, black, 150 lb. threaded, 2-1/2" pipe size	2.000	Ea.	81.75	175.50	257.25
Pipe, Tee, malleable iron, black, 150 lb. threaded, 2" pipe size	1.000	Ea.	19.13	72	91.13
Pipe, Tee, malleable iron, black, 150 lb. threaded, 1-1/4" pipe size	5.000	Ea.	44.63	283.13	327.76
Pipe, Tee, malleable iron, black, 150 lb. threaded, 1" pipe size	4.000	Ea.	22.05	219	241.05
Pipe, 90° elbow, malleable iron, black, 150 lb. threaded, 1" pipe size	6.000	Ea.	21.15	202.50	223.65
Sprinkler head std spray, brass 135°-286° F 1/2" NPT, 3/8" orifice	12.000	Ea.	151.20	432	583.20
Firecycle controls, incls panel, battery, solenoid valves, press switches	1.000	Ea.	8,925	1,650	10,575
Detector, firecycle system	2.000	Ea.	862.50	54	916.50
Firecycle pkg, swing check & flow control valves w/trim 4" pipe size	1.000	Ea.	4,087.50	787.50	4,875
Air compressor, auto, complete, 200 Gal sprinkler sys. cap., 1/3 HP	1.000	Ea.	573.75	333.75	907.50
*Standpipe connection, wall, flush, brass w/plug & chain 2-1/2"x2-1/2"	1.000	Ea.	101.25	156	257.25
Valve, gate, bronze 300 psi, NRS, class 150, threaded, 1" diam.	1.000	Ea.	58.88	23.25	82.13
TOTAL			18,707.61	7,524.91	26,232.52
COST PER S.F.			9.35	3.76	13.11

*Not included in systems under 2000 S.F.

D4010 Sprinklers

D4010 390	Firecycle Sprinkler Systems	COST PER S.F.		
		MAT.	INST.	TOTAL
0520	Firecycle sprinkler systems, steel black sch. 40 pipe			
0530	Light hazard, one floor, 500 S.F. [RD4010 -100]	30	8.20	38.20
0560	1000 S.F. [RD4020 -300]	15.45	5.10	20.55
0580	2000 S.F.	9.35	3.75	13.10
0600	5000 S.F.	4.20	2.35	6.55
0620	10,000 S.F.	2.56	1.88	4.44
0640	50,000 S.F.	1.37	1.54	2.91
0660	Each additional floor of 500 S.F.	1.85	2.26	4.11
0680	1000 S.F.	1.39	2.08	3.47
0700	2000 S.F.	1.15	1.93	3.08
0720	5000 S.F.	1.02	1.64	2.66
0740	10,000 S.F.	.99	1.52	2.51
0760	50,000 S.F.	.99	1.38	2.37
1000	Ordinary hazard, one floor, 500 S.F.	30	8.55	38.55
1020	1000 S.F.	15.40	5.10	20.50
1040	2000 S.F.	9.45	3.90	13.35
1060	5000 S.F.	4.39	2.49	6.88
1080	10,000 S.F.	2.87	2.42	5.29
1100	50,000 S.F.	1.91	2.41	4.32
1140	Each additional floor, 500 S.F.	2.13	2.62	4.75
1160	1000 S.F.	1.38	2.12	3.50
1180	2000 S.F.	1.40	1.95	3.35
1200	5000 S.F.	1.26	1.81	3.07
1220	10,000 S.F.	1.16	1.79	2.95
1240	50,000 S.F.	1.07	1.59	2.66
1500	Extra hazard, one floor, 500 S.F.	32.50	9.80	42.30
1520	1000 S.F.	16.75	6.30	23.05
1540	2000 S.F.	9.70	4.79	14.49
1560	5000 S.F.	4.62	3.36	7.98
1580	10,000 S.F.	3.62	3.40	7.02
1600	50,000 S.F.	3.10	3.74	6.84
1660	Each additional floor, 500 S.F.	2.44	3.23	5.67
1680	1000 S.F.	1.96	3.04	5
1700	2000 S.F.	1.81	3.05	4.86
1720	5000 S.F.	1.53	2.69	4.22
1740	10,000 S.F.	1.72	2.46	4.18
1760	50,000 S.F.	1.70	2.36	4.06
2020	Grooved steel, black, sch. 40 pipe, light hazard, one floor			
2030	2000 S.F.	9.35	3.34	12.69
2060	10,000 S.F.	2.81	2.30	5.11
2100	Each additional floor, 2000 S.F.	1.35	1.57	2.92
2150	10,000 S.F.	1.01	1.30	2.31
2200	Ordinary hazard, one floor, 2000 S.F.	9.45	3.48	12.93
2250	10,000 S.F.	3.01	2.17	5.18
2300	Each additional floor, 2000 S.F.	1.44	1.71	3.15
2350	10,000 S.F.	1.27	1.70	2.97
2400	Extra hazard, one floor, 2000 S.F.	9.70	4.21	13.91
2450	10,000 S.F.	3.21	2.58	5.79
2500	Each additional floor, 2000 S.F.	1.82	2.52	4.34
2550	10,000 S.F.	1.54	2.15	3.69
3050	Grooved steel, black, sch. 10 pipe light hazard, one floor,			
3060	2000 S.F.	9.25	3.31	12.56
3100	10,000 S.F.	2.52	1.63	4.15
3150	Each additional floor, 2000 S.F.	1.28	1.54	2.82
3200	10,000 S.F.	.95	1.27	2.22
3250	Ordinary hazard, one floor, 2000 S.F.	9.35	3.47	12.82
3300	10,000 S.F.	2.69	2.01	4.70
3350	Each additional floor, 2000 S.F.	1.37	1.70	3.07

D4010 Sprinklers

D4010 390	Firecycle Sprinkler Systems	COST PER S.F.		
		MAT.	INST.	TOTAL
3400	10,000 S.F.	1.13	1.66	2.79
3450	Extra hazard, one floor, 2000 S.F.	9.60	4.18	13.78
3500	10,000 S.F.	3.07	2.54	5.61
3550	Each additional floor, 2000 S.F.	1.76	2.49	4.25
3600	10,000 S.F.	1.46	2.12	3.58
4060	Copper tubing, type M, light hazard, one floor, 2000 S.F.	10.15	3.30	13.45
4100	10,000 S.F.	3.35	1.66	5.01
4150	Each additional floor, 2000 S.F.	2.20	1.56	3.76
4200	10,000 S.F.	1.78	1.31	3.09
4250	Ordinary hazard, one floor, 2000 S.F.	10.40	3.61	14.01
4300	10,000 S.F.	3.76	1.91	5.67
4350	Each additional floor, 2000 S.F.	2.22	1.62	3.84
4400	10,000 S.F.	1.88	1.39	3.27
4450	Extra hazard, one floor, 2000 S.F.	10.95	4.22	15.17
4500	10,000 S.F.	6.20	2.80	9
4550	Each additional floor, 2000 S.F.	3.11	2.53	5.64
4600	10,000 S.F.	3.52	2.31	5.83
5060	Copper tubing, type M, T-drill system, light hazard, one floor 2000 S.F.	10.20	3.14	13.34
5100	10,000 S.F.	3.23	1.41	4.64
5150	Each additional floor, 2000 S.F.	2.35	1.48	3.83
5200	10,000 S.F.	1.66	1.06	2.72
5250	Ordinary hazard, one floor, 2000 S.F.	10.25	3.20	13.45
5300	10,000 S.F.	3.67	1.73	5.40
5350	Each additional floor, 2000 S.F.	2.24	1.43	3.67
5400	10,000 S.F.	2.11	1.38	3.49
5450	Extra hazard, one floor, 2000 S.F.	10.40	3.61	14.01
5500	10,000 S.F.	5.15	2.05	7.20
5550	Each additional floor, 2000 S.F.	2.55	1.92	4.47
5600	10,000 S.F.	2.59	1.60	4.19

D4010 Sprinklers

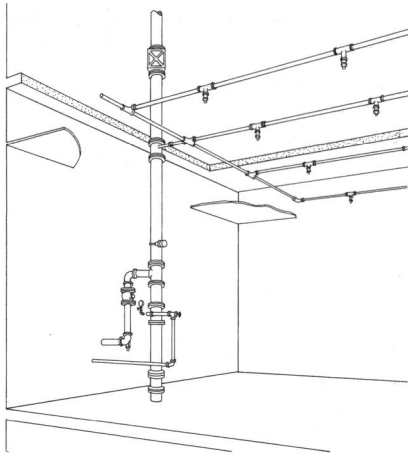

Wet Pipe System. A system employing automatic sprinklers attached to a piping system containing water and connected to a water supply so that water discharges immediately from sprinklers opened by heat from a fire.

All areas are assumed to be open.

System Components	QUANTITY	UNIT	COST EACH		
			MAT.	INST.	TOTAL
SYSTEM D4010 410 0580					
WET PIPE SPRINKLER, STEEL, BLACK, SCH. 40 PIPE					
LIGHT HAZARD, ONE FLOOR, 2000 S.F.					
Valve, gate, iron body, 125 lb., OS&Y, flanged, 4" diam.	1.000	Ea.	420	262.50	682.50
Valve, swing check, bronze, 125 lb., regrinding disc, 2-1/2" pipe size	1.000	Ea.	420	52.50	472.50
Valve, angle, bronze, 150 lb., rising stem, threaded, 2" diam.	1.000	Ea.	393.75	39.75	433.50
*Alarm valve, 2-1/2" pipe size	1.000	Ea.	1,012.50	258.75	1,271.25
Alarm, water motor, complete with gong	1.000	Ea.	251.25	108.75	360
Valve, swing check, w/balldrip CI with brass trim 4" pipe size	1.000	Ea.	210	258.75	468.75
Pipe, steel, black, schedule 40, 4" diam.	10.000	L.F.	149.25	226.20	375.45
*Flow control valve, trim & gauges, 4" pipe size	1.000	Set	2,325	588.75	2,913.75
Fire alarm horn, electric	1.000	Ea.	33.38	62.63	96.01
Pipe, steel, black, schedule 40, threaded, cplg & hngr 10' OC, 2-1/2" diam.	20.000	L.F.	198.75	315	513.75
Pipe, steel, black, schedule 40, threaded, cplg & hngr 10' OC, 2" diam.	12.500	L.F.	79.22	154.22	233.44
Pipe, steel, black, schedule 40, threaded, cplg & hngr 10' OC, 1-1/4" diam.	37.500	L.F.	151.88	333.28	485.16
Pipe steel, black, schedule 40, threaded cplg & hngr 10' OC, 1" diam.	112.000	L.F.	357.84	928.20	1,286.04
Pipe Tee, malleable iron black, 150 lb. threaded, 4" pipe size	2.000	Ea.	291	394.50	685.50
Pipe Tee, malleable iron black, 150 lb. threaded, 2-1/2" pipe size	2.000	Ea.	81.75	175.50	257.25
Pipe Tee, malleable iron black, 150 lb. threaded, 2" pipe size	1.000	Ea.	19.13	72	91.13
Pipe Tee, malleable iron black, 150 lb. threaded, 1-1/4" pipe size	5.000	Ea.	44.63	283.13	327.76
Pipe Tee, malleable iron black, 150 lb. threaded, 1" pipe size	4.000	Ea.	22.05	219	241.05
Pipe 90° elbow, malleable iron black, 150 lb. threaded, 1" pipe size	6.000	Ea.	21.15	202.50	223.65
Sprinkler head, standard spray, brass 135°-286° F 1/2" NPT, 3/8" orifice	12.000	Ea.	151.20	432	583.20
Valve, gate, bronze, NRS, class 150, threaded, 1" pipe size	1.000	Ea.	58.88	23.25	82.13
*Standpipe connection, wall, single, flush w/plug & chain 2-1/2"x2-1/2"	1.000	Ea.	101.25	156	257.25
TOTAL			6,793.86	5,547.16	12,341.02
COST PER S.F.			3.40	2.76	6.16

*Not included in systems under 2000 S.F.

D4010 410	Wet Pipe Sprinkler Systems		COST PER S.F.		
			MAT.	INST.	TOTAL
0520	Wet pipe sprinkler systems, steel, black, sch. 40 pipe				
0530	Light hazard, one floor, 500 S.F.		2.10	2.68	4.78
0560	1000 S.F.	RD4010 -100	3.27	2.75	6.02
0580	2000 S.F.		3.40	2.76	6.16
0600	5000 S.F.	RD4020 -300	1.65	1.94	3.59
0620	10,000 S.F.		1.10	1.67	2.77
0640	50,000 S.F.		.85	1.48	2.33
0660	Each additional floor, 500 S.F.		1.01	2.27	3.28

D40 Fire Protection

D4010 Sprinklers

D4010 410	Wet Pipe Sprinkler Systems	COST PER S.F.		
		MAT.	INST.	TOTAL
0680	1000 S.F.	.98	2.11	3.09
0700	2000 S.F.	.93	1.92	2.85
0720	5000 S.F.	.68	1.62	2.30
0740	10,000 S.F.	.64	1.50	2.14
0760	50,000 S.F.	.57	1.16	1.73
1000	Ordinary hazard, one floor, 500 S.F.	2.32	2.88	5.20
1020	1000 S.F.	3.24	2.71	5.95
1040	2000 S.F.	3.50	2.91	6.41
1060	5000 S.F.	1.84	2.08	3.92
1080	10,000 S.F.	1.41	2.21	3.62
1100	50,000 S.F.	1.17	2.07	3.24
1140	Each additional floor, 500 S.F.	1.27	2.57	3.84
1160	1000 S.F.	.95	2.09	3.04
1180	2000 S.F.	1.04	2.10	3.14
1200	5000 S.F.	1.05	1.98	3.03
1220	10,000 S.F.	.96	2.05	3.01
1240	50,000 S.F.	.87	1.81	2.68
1500	Extra hazard, one floor, 500 S.F.	6.15	4.44	10.59
1520	1000 S.F.	4.09	3.82	7.91
1540	2000 S.F.	3.73	3.91	7.64
1560	5000 S.F.	2.55	3.41	5.96
1580	10,000 S.F.	2.19	3.24	5.43
1600	50,000 S.F.	2.37	3.11	5.48
1660	Each additional floor, 500 S.F.	1.58	3.18	4.76
1680	1000 S.F.	1.53	3.01	4.54
1700	2000 S.F.	1.38	3.02	4.40
1720	5000 S.F.	1.19	2.67	3.86
1740	10,000 S.F.	1.37	2.44	3.81
1760	50,000 S.F.	1.36	2.32	3.68
2020	Grooved steel, black sch. 40 pipe, light hazard, one floor, 2000 S.F.	3.38	2.35	5.73
2060	10,000 S.F.	1.36	1.51	2.87
2100	Each additional floor, 2000 S.F.	.92	1.54	2.46
2150	10,000 S.F.	.66	1.28	1.94
2200	Ordinary hazard, one floor, 2000 S.F.	3.47	2.49	5.96
2250	10,000 S.F.	1.37	1.84	3.21
2300	Each additional floor, 2000 S.F.	1.01	1.68	2.69
2350	10,000 S.F.	.92	1.68	2.60
2400	Extra hazard, one floor, 2000 S.F.	3.72	3.22	6.94
2450	10,000 S.F.	1.77	2.38	4.15
2500	Each additional floor, 2000 S.F.	1.39	2.49	3.88
2550	10,000 S.F.	1.19	2.13	3.32
3050	Grooved steel black sch. 10 pipe, light hazard, one floor, 2000 S.F.	3.31	2.32	5.63
3100	10,000 S.F.	1.06	1.42	2.48
3150	Each additional floor, 2000 S.F.	.85	1.51	2.36
3200	10,000 S.F.	.60	1.25	1.85
3250	Ordinary hazard, one floor, 2000 S.F.	3.40	2.48	5.88
3300	10,000 S.F.	1.23	1.80	3.03
3350	Each additional floor, 2000 S.F.	.94	1.67	2.61
3400	10,000 S.F.	.78	1.64	2.42
3450	Extra hazard, one floor, 2000 S.F.	3.66	3.19	6.85
3500	10,000 S.F.	1.63	2.34	3.97
3550	Each additional floor, 2000 S.F.	1.33	2.46	3.79
3600	10,000 S.F.	1.11	2.10	3.21
4050	Copper tubing, type M, light hazard, one floor, 2000 S.F.	4.21	2.31	6.52
4100	10,000 S.F.	1.89	1.45	3.34
4150	Each additional floor, 2000 S.F.	1.77	1.53	3.30
4200	10,000 S.F.	1.43	1.29	2.72
4250	Ordinary hazard, one floor, 2000 S.F.	4.46	2.62	7.08

D40 Fire Protection

D4010 Sprinklers

D4010 410	Wet Pipe Sprinkler Systems	COST PER S.F.		
		MAT.	INST.	TOTAL
4300	10,000 S.F.	2.30	1.70	4
4350	Each additional floor, 2000 S.F.	2.09	1.73	3.82
4400	10,000 S.F.	1.78	1.51	3.29
4450	Extra hazard, one floor, 2000 S.F.	5	3.23	8.23
4500	10,000 S.F.	4.66	2.56	7.22
4550	Each additional floor, 2000 S.F.	2.68	2.50	5.18
4600	10,000 S.F.	3.17	2.29	5.46
5050	Copper tubing, type M, T-drill system, light hazard, one floor			
5060	2000 S.F.	4.23	2.15	6.38
5100	10,000 S.F.	1.77	1.20	2.97
5150	Each additional floor, 2000 S.F.	1.79	1.37	3.16
5200	10,000 S.F.	1.31	1.04	2.35
5250	Ordinary hazard, one floor, 2000 S.F.	4.27	2.21	6.48
5300	10,000 S.F.	2.21	1.52	3.73
5350	Each additional floor, 2000 S.F.	1.81	1.40	3.21
5400	10,000 S.F.	1.76	1.36	3.12
5450	Extra hazard, one floor, 2000 S.F.	4.45	2.62	7.07
5500	10,000 S.F.	3.73	1.85	5.58
5550	Each additional floor, 2000 S.F.	2.19	1.93	4.12
5600	10,000 S.F.	2.24	1.58	3.82

D4020 Standpipes

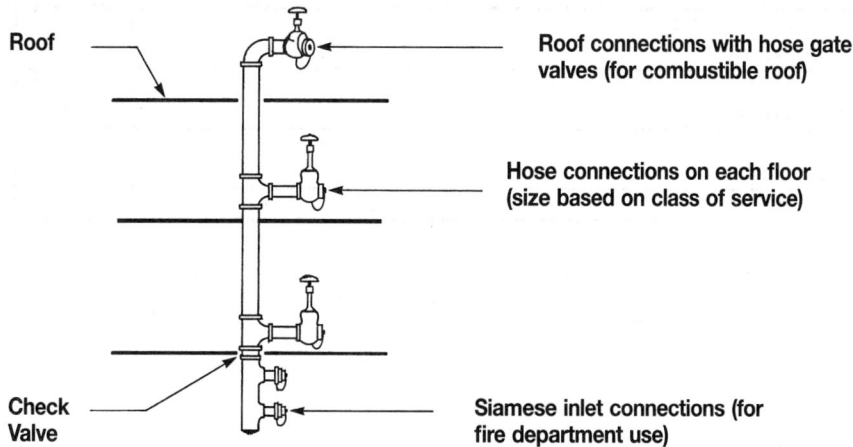

Roof — Roof connections with hose gate valves (for combustible roof)

Hose connections on each floor (size based on class of service)

Check Valve

Siamese inlet connections (for fire department use)

System Components	QUANTITY	UNIT	COST PER FLOOR		
			MAT.	INST.	TOTAL
SYSTEM D4020 310 0560					
WET STANDPIPE RISER, CLASS I, STEEL, BLACK, SCH. 40 PIPE, 10' HEIGHT					
4" DIAMETER PIPE, ONE FLOOR					
Pipe, steel, black, schedule 40, threaded, 4" diam.	20.000	L.F.	500	590	1,090
Pipe, Tee, malleable iron, black, 150 lb. threaded, 4" pipe size	2.000	Ea.	388	526	914
Pipe, 90° elbow, malleable iron, black, 150 lb. threaded 4" pipe size	1.000	Ea.	123	176	299
Pipe, nipple, steel, black, schedule 40, 2-1/2" pipe size x 3" long	2.000	Ea.	24.90	132	156.90
Fire valve, gate, 300 lb., brass w/handwheel, 2-1/2" pipe size	1.000	Ea.	171	82.50	253.50
Fire valve, pressure restricting, adj., rgh. brs., 2-1/2" pipe size	1.000	Ea.	304	165	469
Valve, swing check, w/ball drip, CI w/brs. ftngs., 4" pipe size	1.000	Ea.	280	345	625
Standpipe conn wall dble. flush brs. w/plugs & chains 2-1/2"x2-1/2"x4"	1.000	Ea.	470	208	678
Valve, swing check, bronze, 125 lb., regrinding disc, 2-1/2" pipe size	1.000	Ea.	560	70	630
Roof manifold, fire, w/valves & caps, horiz./vert. brs. 2-1/2"x2-1/2"x4"	1.000	Ea.	152	217	369
Fire, hydrolator, vent & drain, 2-1/2" pipe size	1.000	Ea.	64.50	48	112.50
Valve, gate, iron body 125 lb., OS&Y, threaded, 4" pipe size	1.000	Ea.	560	350	910
TOTAL			**3,597.40**	**2,909.50**	**6,506.90**

D4020 310		Wet Standpipe Risers, Class I	COST PER FLOOR		
			MAT.	INST.	TOTAL
0550	Wet standpipe risers, Class I, steel black sch. 40, 10' height				
0560		4" diameter pipe, one floor	3,600	2,900	6,500
0580		Additional floors	855	905	1,760
0600		6" diameter pipe, one floor	5,925	5,100	11,025
0620		Additional floors	1,450	1,425	2,875
0640		8" diameter pipe, one floor	8,850	6,175	15,025
0660		Additional floors	2,075	1,725	3,800
0680					

RD4020 -300

D4020 310		Wet Standpipe Risers, Class II	COST PER FLOOR		
			MAT.	INST.	TOTAL
1030	Wet standpipe risers, Class II, steel black sch. 40, 10' height				
1040		2" diameter pipe, one floor	1,425	1,050	2,475
1060		Additional floors	465	405	870
1080		2-1/2" diameter pipe, one floor	2,075	1,525	3,600
1100		Additional floors	535	470	1,005
1120					

D40 Fire Protection

D4020 Standpipes

D4020 310	Wet Standpipe Risers, Class III	COST PER FLOOR		
		MAT.	INST.	TOTAL
1530	Wet standpipe risers, Class III, steel black sch. 40, 10' height			
1540	4" diameter pipe, one floor	3,700	2,900	6,600
1560	Additional floors	725	755	1,480
1580	6" diameter pipe, one floor	6,025	5,100	11,125
1600	Additional floors	1,500	1,425	2,925
1620	8" diameter pipe, one floor	8,950	6,175	15,125
1640	Additional floors	2,125	1,725	3,850

D4020 Standpipes

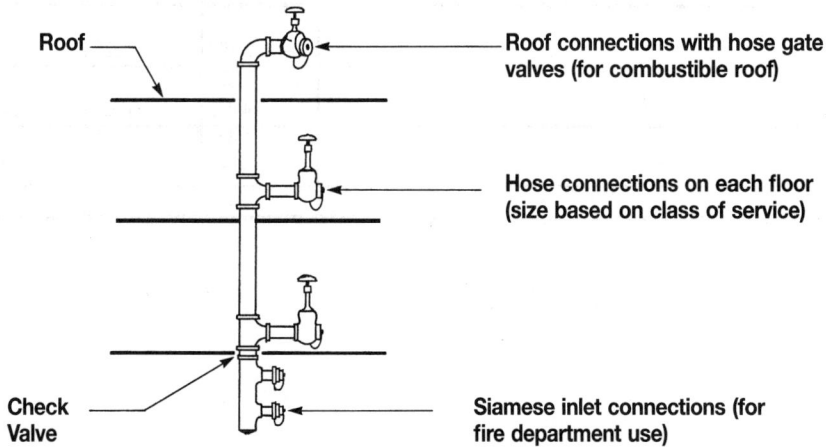

Roof — Roof connections with hose gate valves (for combustible roof)

Hose connections on each floor (size based on class of service)

Check Valve

Siamese inlet connections (for fire department use)

System Components	QUANTITY	UNIT	COST PER FLOOR		
			MAT.	INST.	TOTAL
SYSTEM D4020 330 0540					
DRY STANDPIPE RISER, CLASS I, PIPE, STEEL, BLACK, SCH 40, 10' HEIGHT					
4" DIAMETER PIPE, ONE FLOOR					
Pipe, steel, black, schedule 40, threaded, 4" diam.	20.000	L.F.	500	590	1,090
Pipe, Tee, malleable iron, black, 150 lb. threaded, 4" pipe size	2.000	Ea.	388	526	914
Pipe, 90° elbow, malleable iron, black, 150 lb. threaded 4" pipe size	1.000	Ea.	123	176	299
Pipe, nipple, steel, black, schedule 40, 2-1/2" pipe size x 3" long	2.000	Ea.	24.90	132	156.90
Fire valve gate NRS 300 lb., brass w/handwheel, 2-1/2" pipe size	1.000	Ea.	171	82.50	253.50
Fire valve, pressure restricting, adj., rgh. brs., 2-1/2" pipe size	1.000	Ea.	152	82.50	234.50
Standpipe conn wall dble. flush brs. w/plugs & chains 2-1/2"x2-1/2"x4"	1.000	Ea.	470	208	678
Valve swing check w/ball drip CI w/brs. ftngs., 4"pipe size	1.000	Ea.	280	345	625
Roof manifold, fire, w/valves & caps, horiz./vert. brs. 2-1/2"x2-1/2"x4"	1.000	Ea.	152	217	369
TOTAL			2,260.90	2,359	4,619.90

D4020 330	Dry Standpipe Risers, Class I	COST PER FLOOR		
		MAT.	INST.	TOTAL
0530	Dry standpipe riser, Class I, steel black sch. 40, 10' height			
0540	4" diameter pipe, one floor	2,250	2,350	4,600
0560	Additional floors	790	855	1,645
0580	6" diameter pipe, one floor	4,325	4,050	8,375
0600	Additional floors	1,375	1,375	2,750
0620	8" diameter pipe, one floor	6,500	4,925	11,425
0640	Additional floors	2,000	1,675	3,675
0660				

D4020 330	Dry Standpipe Risers, Class II	COST PER FLOOR		
		MAT.	INST.	TOTAL
1030	Dry standpipe risers, Class II, steel black sch. 40, 10' height			
1040	2" diameter pipe, one floor	1,225	1,100	2,325
1060	Additional floor	400	355	755
1080	2-1/2" diameter pipe, one floor	1,650	1,275	2,925
1100	Additional floors	470	425	895
1120				

D40 Fire Protection

D4020 Standpipes

D4020 330	Dry Standpipe Risers, Class III	COST PER FLOOR		
		MAT.	INST.	TOTAL
1530	Dry standpipe risers, Class III, steel black sch. 40, 10' height			
1540	4" diameter pipe, one floor	2,300	2,325	4,625
1560	Additional floors	670	775	1,445
1580	6" diameter pipe, one floor	4,375	4,050	8,425
1600	Additional floors	1,425	1,375	2,800
1620	8" diameter pipe, one floor	6,550	4,925	11,475
1640	Additional floor	2,050	1,675	3,725

D4020 Standpipes

D4020 410	Fire Hose Equipment	COST EACH		
		MAT.	INST.	TOTAL
0100	Adapters, reducing, 1 piece, FxM, hexagon, cast brass, 2-1/2" x 1-1/2"	50		50
0200	Pin lug, 1-1/2" x 1"	43.50		43.50
0250	3" x 2-1/2"	113		113
0300	For polished chrome, add 75% mat.			
0400	Cabinets, D.S. glass in door, recessed, steel box, not equipped			
0500	Single extinguisher, steel door & frame	95.50	130	225.50
0550	Stainless steel door & frame	204	130	334
0600	Valve, 2-1/2" angle, steel door & frame	123	86.50	209.50
0650	Aluminum door & frame	149	86.50	235.50
0700	Stainless steel door & frame	200	86.50	286.50
0750	Hose rack assy, 2-1/2" x 1-1/2" valve & 100' hose, steel door & frame	229	173	402
0800	Aluminum door & frame	335	173	508
0850	Stainless steel door & frame	450	173	623
0900	Hose rack assy,& extinguisher,2-1/2"x1-1/2" valve & hose,steel door & frame	228	208	436
0950	Aluminum	430	208	638
1000	Stainless steel	785	208	993
1550	Compressor, air, dry pipe system, automatic, 200 gal., 3/4 H.P.	765	445	1,210
1600	520 gal., 1 H.P.	805	445	1,250
1650	Alarm, electric pressure switch (circuit closer)	265	22	287
2500	Couplings, hose, rocker lug, cast brass, 1-1/2"	48		48
2550	2-1/2"	63		63
3000	Escutcheon plate, for angle valves, polished brass, 1-1/2"	12.15		12.15
3050	2-1/2"	29		29
3500	Fire pump, electric, w/controller, fittings, relief valve			
3550	4" pump, 30 H.P., 500 G.P.M.	21,800	3,250	25,050
3600	5" pump, 40 H.P., 1000 G.P.M.	31,800	3,675	35,475
3650	5" pump, 100 H.P., 1000 G.P.M.	35,700	4,075	39,775
3700	For jockey pump system, add	3,900	520	4,420
5000	Hose, per linear foot, synthetic jacket, lined,			
5100	300 lb. test, 1-1/2" diameter	2.63	.40	3.03
5150	2-1/2" diameter	4.38	.47	4.85
5200	500 lb. test, 1-1/2" diameter	2.72	.40	3.12
5250	2-1/2" diameter	4.72	.47	5.19
5500	Nozzle, plain stream, polished brass, 1-1/2" x 10"	43.50		43.50
5550	2-1/2" x 15" x 13/16" or 1-1/2"	78.50		78.50
5600	Heavy duty combination adjustable fog and straight stream w/handle 1-1/2"	405		405
5650	2-1/2" direct connection	505		505
6000	Rack, for 1-1/2" diameter hose 100 ft. long, steel	54	52	106
6050	Brass	66.50	52	118.50
6500	Reel, steel, for 50 ft. long 1-1/2" diameter hose	118	74.50	192.50
6550	For 75 ft. long 2-1/2" diameter hose	192	74.50	266.50
7050	Siamese, w/plugs & chains, polished brass, sidewalk, 4" x 2-1/2" x 2-1/2"	520	415	935
7100	6" x 2-1/2" x 2-1/2"	710	520	1,230
7200	Wall type, flush, 4" x 2-1/2" x 2-1/2"	470	208	678
7250	6" x 2-1/2" x 2-1/2"	655	226	881
7300	Projecting, 4" x 2-1/2" x 2-1/2"	430	208	638
7350	6" x 2-1/2" x 2-1/2"	725	226	951
7400	For chrome plate, add 15% mat.			
8000	Valves, angle, wheel handle, 300 Lb., rough brass, 1-1/2"	63.50	48	111.50
8050	2-1/2"	66	82.50	148.50
8100	Combination pressure restricting, 1-1/2"	77	48	125
8150	2-1/2"	153	82.50	235.50
8200	Pressure restricting, adjustable, satin brass, 1-1/2"	109	48	157
8250	2-1/2"	152	82.50	234.50
8300	Hydrolator, vent and drain, rough brass, 1-1/2"	64.50	48	112.50
8350	2-1/2"	64.50	48	112.50
8400	Cabinet assy, incls. adapter, rack, hose, and nozzle	675	315	990

D4090 Other Fire Protection Systems

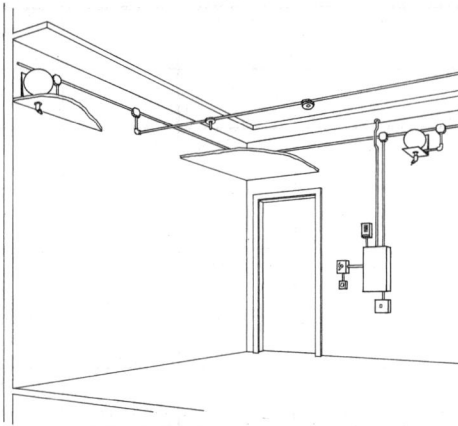

General: Automatic fire protection (suppression) systems other than water sprinklers may be desired for special environments, high risk areas, isolated locations or unusual hazards. Some typical applications would include:

Paint dip tanks
Securities vaults
Electronic data processing
Tape and data storage
Transformer rooms
Spray booths
Petroleum storage
High rack storage

Piping and wiring costs are dependent on the individual application and must be added to the component costs shown below.

All areas are assumed to be open.

D4090 910	Unit Components	COST EACH		
		MAT.	INST.	TOTAL
0020	Detectors with brackets			
0040	Fixed temperature heat detector	36	70	106
0060	Rate of temperature rise detector	42.50	70	112.50
0080	Ion detector (smoke) detector	96	90.50	186.50
0200	Extinguisher agent			
0240	200 lb FM200, container	6,700	263	6,963
0280	75 lb carbon dioxide cylinder	1,100	176	1,276
0320	Dispersion nozzle			
0340	FM200 1-1/2" dispersion nozzle	58	42	100
0380	Carbon dioxide 3" x 5" dispersion nozzle	58	32.50	90.50
0420	Control station			
0440	Single zone control station with batteries	1,475	560	2,035
0470	Multizone (4) control station with batteries	2,825	1,125	3,950
0490				
0500	Electric mechanical release	144	286	430
0520				
0550	Manual pull station	52	97.50	149.50
0570				
0640	Battery standby power 10" x 10" x 17"	885	140	1,025
0700				
0740	Bell signalling device	63.50	70	133.50

D4090 920	FM200 Systems	COST PER C.F.		
		MAT.	INST.	TOTAL
0820	Average FM200 system, minimum			1.52
0840	Maximum			3.03

D5010 Electrical Service/Distribution

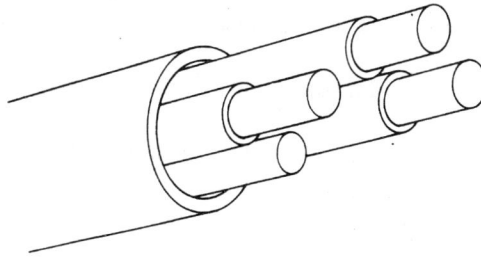

System Components	QUANTITY	UNIT	COST PER L.F.		
			MAT.	INST.	TOTAL
SYSTEM D5010 110 0200					
HIGH VOLTAGE CABLE, NEUTRAL AND CONDUIT INCLUDED, COPPER #2, 5 kV					
Shielded cable, no splice/termn, copper, XLP shielding, 5 kV, #2	.030	C.L.F.	8.52	8.40	16.92
Wire 600 volt, type THW, copper, stranded, #4	.010	C.L.F.	1.17	1.06	2.23
Rigid galv steel conduit to 15' H, 2" diam, w/ term, ftng & support	1.000	L.F.	10.20	12.45	22.65
TOTAL			19.89	21.91	41.80

D5010 110	High Voltage Shielded Conductors	COST PER L.F.		
		MAT.	INST.	TOTAL
0200	High voltage cable, neutral & conduit included, copper #2, 5 kV	19.90	22	41.90
0240	Copper #1, 5 kV	22	22	44
0280	15 kV	37	32	69
0320	Copper 1/0, 5 kV	23	22.50	45.50
0360	15 kV	39.50	32.50	72
0400	25 kV	45	33	78
0440	35 kV	51.50	36.50	88
0480	Copper 2/0, 5 kV	35	26.50	61.50
0520	15 kV	42	33	75
0560	25 kV	53	37	90
0600	35 kV	61	40	101
0640	Copper 4/0, 5 kV	44.50	35	79.50
0680	15 kV	54	38	92
0720	25 kV	59.50	38.50	98
0760	35 kV	68.50	42	110.50
0800	Copper 250 kcmil, 5 kV	50	36	86
0840	15 kV	57	39	96
0880	25 kV	72.50	42	114.50
0920	35 kV	108	52.50	160.50
0960	Copper 350 kcmil, 5 kV	58.50	38	96.50
1000	15 kV	70.50	43.50	114
1040	25 kV	80.50	45	125.50
1080	35 kV	117	55.50	172.50
1120	Copper 500 kcmil, 5 kV	73	42.50	115.50
1160	15 kV	113	54.50	167.50
1200	25 kV	123	56	179
1240	35 kV	127	57.50	184.50

D5010   Electrical Service/Distribution

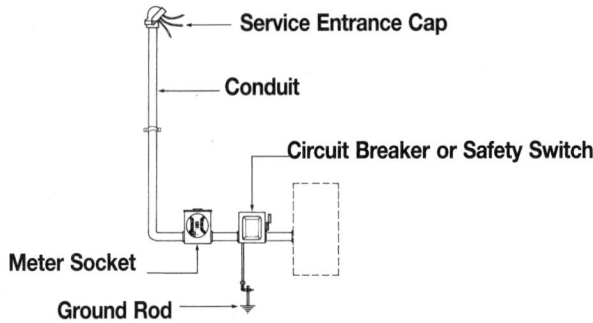

Service Entrance Cap

Conduit

Circuit Breaker or Safety Switch

Meter Socket

Ground Rod

System Components	QUANTITY	UNIT	COST EACH		
			MAT.	INST.	TOTAL
SYSTEM D5010 120 0220					
SERVICE INSTALLATION, INCLUDES BREAKERS, METERING, 20' CONDUIT & WIRE					
3 PHASE, 4 WIRE, 60 A					
Circuit breaker, enclosed (NEMA 1), 600 volt, 3 pole, 60 A	1.000	Ea.	670	200	870
Meter socket, single position, 4 terminal, 100 A	1.000	Ea.	47.50	175	222.50
Rigid galvanized steel conduit, 3/4", including fittings	20.000	L.F.	71	140	211
Wire, 600V type XHHW, copper stranded #6	.900	C.L.F.	110.70	77.40	188.10
Service entrance cap 3/4" diameter	1.000	Ea.	11.55	43	54.55
Conduit LB fitting with cover, 3/4" diameter	1.000	Ea.	15.30	43	58.30
Ground rod, copper clad, 8' long, 3/4" diameter	1.000	Ea.	30.50	106	136.50
Ground rod clamp, bronze, 3/4" diameter	1.000	Ea.	7.35	17.50	24.85
Ground wire, bare armored, #6-1 conductor	.200	C.L.F.	28.80	62	90.80
TOTAL			992.70	863.90	1,856.60

D5010 120	Electric Service, 3 Phase - 4 Wire		COST EACH		
			MAT.	INST.	TOTAL
0200	Service installation, includes breakers, metering, 20' conduit & wire				
0220	3 phase, 4 wire, 120/208 volts, 60 A		995	865	1,860
0240	100 A		1,225	1,050	2,275
0280	200 A		1,875	1,600	3,475
0320	400 A		4,450	2,925	7,375
0360	600 A	RD5010 -110	8,400	3,975	12,375
0400	800 A		10,800	4,800	15,600
0440	1000 A		13,300	5,500	18,800
0480	1200 A		16,700	5,625	22,325
0520	1600 A		30,200	8,075	38,275
0560	2000 A		33,300	9,200	42,500
0570	Add 25% for 277/480 volt				
0580					
0610	1 phase, 3 wire, 120/240 volts, 100 A		525	940	1,465
0620	200 A		1,100	1,375	2,475

D5010 Electrical Service/Distribution

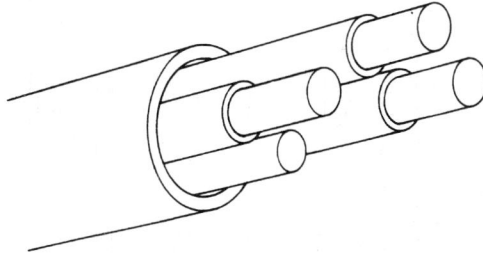

System Components	QUANTITY	UNIT	COST PER L.F.		
			MAT.	INST.	TOTAL
SYSTEM D5010 230 0200					
FEEDERS, INCLUDING STEEL CONDUIT & WIRE, 60 A					
Rigid galvanized steel conduit, 3/4", including fittings	1.000	L.F.	3.55	7	10.55
Wire 600 volt, type XHHW copper stranded #6	.040	C.L.F.	4.92	3.44	8.36
TOTAL			8.47	10.44	18.91

D5010 230	Feeder Installation	COST PER L.F.		
		MAT.	INST.	TOTAL
0200	Feeder installation 600 V, including RGS conduit and XHHW wire, 60 A	8.45	10.45	18.90
0240	100 A	12.65	13.85	26.50
0280	200 A	28	21.50	49.50
0320	400 A	56.50	43	99.50
0360	600 A	122	70	192
0400	800 A	177	83.50	260.50
0440	1000 A	200	107	307
0480	1200 A	263	110	373
0520	1600 A	355	167	522
0560	2000 A	400	214	614
1200	Branch installation 600 V, including EMT conduit and THW wire, 15 A	1.44	5.35	6.79
1240	20 A	1.44	5.35	6.79
1280	30 A	2.26	6.55	8.81
1320	50 A	3.74	7.55	11.29
1360	65 A	4.54	8	12.54
1400	85 A	7.25	9.50	16.75
1440	100 A	9.30	10	19.30
1480	130 A	11.95	11.25	23.20
1520	150 A	14.65	12.95	27.60
1560	200 A	19.60	14.60	34.20

RD5010 -140

D5010 Electrical Service/Distribution

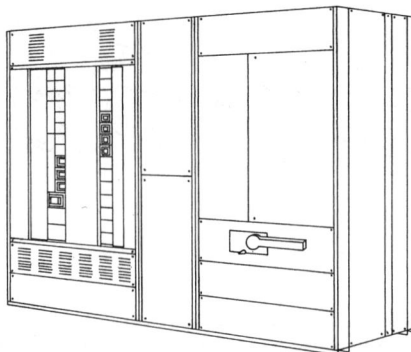

System Components	QUANTITY	UNIT	COST EACH		
			MAT.	INST.	TOTAL
SYSTEM D5010 240 0240					
SWITCHGEAR INSTALLATION, INCL SWBD, PANELS & CIRC BREAKERS, 600 A					
Panelboard, NQOD 225A 4W 120/208V main CB, w/20A bkrs 42 circ	1.000	Ea.	2,350	2,000	4,350
Switchboard, alum. bus bars, 120/208V, 4 wire, 600V	1.000	Ea.	4,675	1,125	5,800
Distribution sect., alum. bus bar, 120/208 or 277/480 V, 4 wire, 600A	1.000	Ea.	2,400	1,125	3,525
Feeder section circuit breakers, KA frame, 70 to 225 A	3.000	Ea.	3,975	525	4,500
TOTAL			13,400	4,775	18,175

D5010 240	Switchgear		COST EACH		
			MAT.	INST.	TOTAL
0200	Switchgear inst., incl. swbd., panels & circ bkr, 400 A, 120/208volt		4,300	3,525	7,825
0240	600 A		13,400	4,775	18,175
0280	800 A	RD5010	17,000	6,800	23,800
0320	1200 A	-110	20,300	10,400	30,700
0360	1600 A		27,400	14,600	42,000
0400	2000 A		34,600	18,600	53,200
0410	Add 20% for 277/480 volt				

D5020 Lighting and Branch Wiring

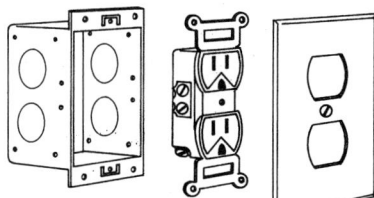

Duplex Receptacle

System Components	QUANTITY	UNIT	COST PER S.F.		
			MAT.	INST.	TOTAL
SYSTEM D5020 110 0200					
RECEPTACLES INCL. PLATE, BOX, CONDUIT, WIRE & TRANS. WHEN REQUIRED					
2.5 PER 1000 S.F., .3 WATTS PER S.F.					
Steel intermediate conduit, (IMC) 1/2" diam	167.000	L.F.	.37	.94	1.31
Wire 600V type THWN-THHN, copper solid #12	3.382	C.L.F.	.06	.17	.23
Wiring device, receptacle, duplex, 120V grounded, 15 amp	2.500	Ea.		.04	.04
Wall plate, 1 gang, brown plastic	2.500	Ea.		.02	.02
Steel outlet box 4" square	2.500	Ea.	.01	.07	.08
Steel outlet box 4" plaster rings	2.500	Ea.	.01	.02	.03
TOTAL			.45	1.26	1.71

D5020 110	Receptacle (by Wattage)		COST PER S.F.		
			MAT.	INST.	TOTAL
0190	Receptacles include plate, box, conduit, wire & transformer when required				
0200	2.5 per 1000 S.F., .3 watts per S.F.		.45	1.26	1.71
0240	With transformer	RD5010 -110	.52	1.32	1.84
0280	4 per 1000 S.F., .5 watts per S.F.		.50	1.47	1.97
0320	With transformer		.60	1.56	2.16
0360	5 per 1000 S.F., .6 watts per S.F.		.60	1.72	2.32
0400	With transformer		.73	1.84	2.57
0440	8 per 1000 S.F., .9 watts per S.F.		.62	1.90	2.52
0480	With transformer		.80	2.07	2.87
0520	10 per 1000 S.F., 1.2 watts per S.F.		.65	2.07	2.72
0560	With transformer		.94	2.35	3.29
0600	16.5 per 1000 S.F., 2.0 watts per S.F.		.76	2.59	3.35
0640	With transformer		1.26	3.07	4.33
0680	20 per 1000 S.F., 2.4 watts per S.F.		.80	2.82	3.62
0720	With transformer		1.39	3.38	4.77

D5020 Lighting and Branch Wiring

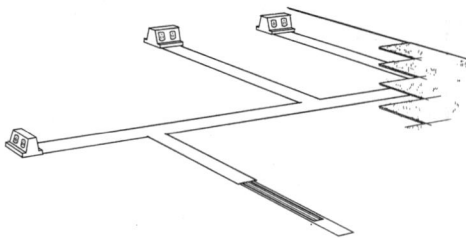

Underfloor Receptacle System

Description: Table D5020 115 includes installed costs of raceways and copper wire from panel to and including receptacle.

National Electrical Code prohibits use of undercarpet system in residential, school or hospital buildings. Can only be used with carpet squares.

Low density = (1) Outlet per 259 S.F. of floor area.

High density = (1) Outlet per 127 S.F. of floor area.

System Components	QUANTITY	UNIT	COST PER S.F.		
			MAT.	INST.	TOTAL
SYSTEM D5020 115 0200					
RECEPTACLE SYSTEMS, UNDERFLOOR DUCT, 5' ON CENTER, LOW DENSITY					
Underfloor duct 3-1/8" x 7/8" w/insert 24" on center	.190	L.F.	3.55	1.52	5.07
Vertical elbow for underfloor duct, 3-1/8", included					
Underfloor duct conduit adapter, 2" x 1-1/4", included					
Underfloor duct junction box, single duct, 3-1/8"	.003	Ea.	1.26	.42	1.68
Underfloor junction box carpet pan	.003	Ea.	1.05	.02	1.07
Underfloor duct outlet, high tension receptacle	.004	Ea.	.39	.28	.67
Wire 600V type THWN-THHN copper solid #12	.010	C.L.F.	.18	.51	.69
TOTAL			6.43	2.75	9.18

D5020 115	Receptacles, Floor		COST PER S.F.		
			MAT.	INST.	TOTAL
0200	Receptacle systems, underfloor duct, 5' on center, low density		6.45	2.75	9.20
0240	High density		7	3.54	10.54
0280	7' on center, low density	RD5010 -110	5.10	2.36	7.46
0320	High density		5.65	3.15	8.80
0400	Poke thru fittings, low density		1.15	1.34	2.49
0440	High density		2.26	2.65	4.91
0520	Telepoles, using Romex, low density		1.14	.82	1.96
0560	High density		2.27	1.66	3.93
0600	Using EMT, low density		1.21	1.08	2.29
0640	High density		2.40	2.19	4.59
0720	Conduit system with floor boxes, low density		1.09	.93	2.02
0760	High density		2.19	1.90	4.09
0840	Undercarpet power system, 3 conductor with 5 conductor feeder, low density		1.46	.33	1.79
0880	High density		2.81	.70	3.51

D5020 Lighting and Branch Wiring

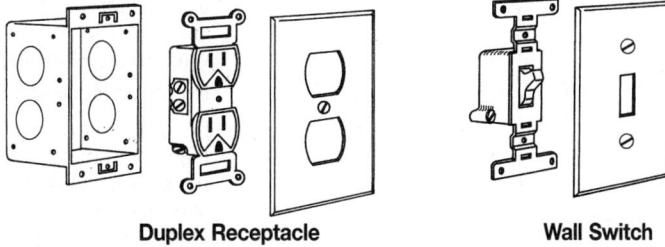

Duplex Receptacle   **Wall Switch**

System Components	QUANTITY	UNIT	COST PER EACH		
			MAT.	INST.	TOTAL
SYSTEM D5020 125 0520					
RECEPTACLES AND WALL SWITCHES, RECEPTICLE DUPLEX 120 V GROUNDED, 15 A					
Electric metallic tubing conduit, (EMT), 3/4″ diam	22.000	L.F.	25.52	94.60	120.12
Wire, 600 volt, type THWN-THHN, copper, solid #12	.630	C.L.F.	11.03	32.13	43.16
Steel outlet box 4″ square	1.000	Ea.	2.62	28	30.62
Steel outlet box, 4″ square, plaster rings	1.000	Ea.	3.12	8.75	11.87
Receptacle, duplex, 120 volt grounded, 15 amp	1.000	Ea.	1.49	14	15.49
Wall plate, 1 gang, brown plastic	1.000	Ea.	.37	7	7.37
TOTAL			44.15	184.48	228.63

D5020 125	Receptacles & Switches by Each	COST PER EACH		
		MAT.	INST.	TOTAL
0460	Receptacles & Switches, with box, plate, 3/4″ EMT conduit & wire			
0520	Receptacle duplex 120 V grounded, 15 A	44	184	228
0560	20 A	53.50	191	244.50
0600	Receptacle duplex ground fault interrupting, 15 A	78.50	191	269.50
0640	20 A	80.50	191	271.50
0680	Toggle switch single, 15 A	48.50	184	232.50
0720	20 A	51	191	242
0760	3 way switch, 15 A	51	195	246
0800	20 A	52.50	201	253.50
0840	4 way switch, 15 A	70	208	278
0880	20 A	86	221	307

D50 Electrical

D5020 Lighting and Branch Wiring

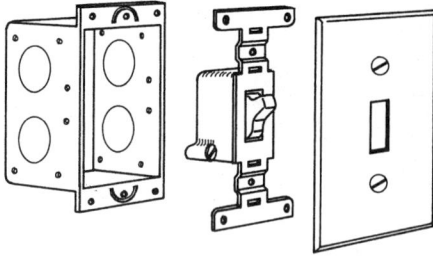

Description: Table D5020 130 includes the cost for switch, plate, box, conduit in slab or EMT exposed and copper wire. Add 20% for exposed conduit.

No power required for switches.

Federal energy guidelines recommend the maximum lighting area controlled per switch shall not exceed 1000 S.F. and that areas over 500 S.F. shall be so controlled that total illumination can be reduced by at least 50%.

System Components	QUANTITY	UNIT	COST PER S.F.		
			MAT.	INST.	TOTAL
SYSTEM D5020 130 0360					
WALL SWITCHES, 5.0 PER 1000 S.F.					
Steel, intermediate conduit (IMC), 1/2" diameter	88.000	L.F.	.20	.49	.69
Wire, 600V type THWN-THHN, copper solid #12	1.710	C.L.F.	.03	.09	.12
Toggle switch, single pole, 15 amp	5.000	Ea.	.03	.07	.10
Wall plate, 1 gang, brown plastic	5.000	Ea.		.04	.04
Steel outlet box 4" plaster rings	5.000	Ea.	.01	.14	.15
Plaster rings	5.000	Ea.	.02	.04	.06
TOTAL			.29	.87	1.16

D5020 130	Wall Switch by Sq. Ft.	COST PER S.F.		
		MAT.	INST.	TOTAL
0200	Wall switches, 1.0 per 1000 S.F.	.07	.20	.27
0240	1.2 per 1000 S.F.	.07	.22	.29
0280	2.0 per 1000 S.F.	.11	.32	.43
0320	2.5 per 1000 S.F.	.13	.41	.54
0360	5.0 per 1000 S.F.	.29	.87	1.16
0400	10.0 per 1000 S.F.	.58	1.76	2.34

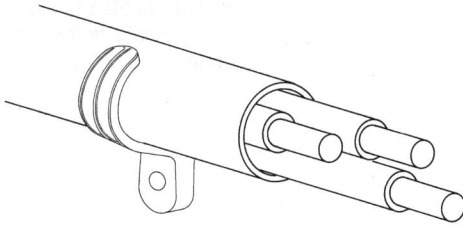

System D5020 135 includes all wiring and connections.

System Components	QUANTITY	UNIT	COST PER S.F.		
			MAT.	INST.	TOTAL
SYSTEM D5020 135 0200					
MISCELLANEOUS POWER, TO .5 WATTS					
Steel intermediate conduit, (IMC) 1/2" diam	15.000	L.F.	.03	.08	.11
Wire 600V type THWN-THHN, copper solid #12	.325	C.L.F.	.01	.02	.03
TOTAL			.04	.10	.14

D5020 135	Miscellaneous Power	COST PER S.F.		
		MAT.	INST.	TOTAL
0200	Miscellaneous power, to .5 watts	.04	.10	.14
0240	.8 watts	.06	.14	.20
0280	1 watt	.07	.19	.26
0320	1.2 watts	.08	.22	.30
0360	1.5 watts	.10	.26	.36
0400	1.8 watts	.12	.30	.42
0440	2 watts	.14	.36	.50
0480	2.5 watts	.17	.45	.62
0520	3 watts	.21	.52	.73

D5020 Lighting and Branch Wiring

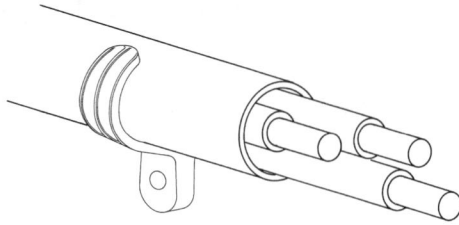

System D5020 140 includes all wiring and connections for central air conditioning units.

System Components	QUANTITY	UNIT	COST PER S.F.		
			MAT.	INST.	TOTAL
SYSTEM D5020 140 0200					
CENTRAL AIR CONDITIONING POWER, 1 WATT					
Steel intermediate conduit, 1/2" diam.	.030	L.F.	.07	.17	.24
Wire 600V type THWN-THHN, copper solid #12	.001	C.L.F.	.02	.05	.07
TOTAL			.09	.22	.31

D5020 140	Central A. C. Power (by Wattage)	COST PER S.F.		
		MAT.	INST.	TOTAL
0200	Central air conditioning power, 1 watt	.09	.22	.31
0220	2 watts	.10	.25	.35
0240	3 watts	.14	.28	.42
0280	4 watts	.20	.38	.58
0320	6 watts	.49	.52	1.01
0360	8 watts	.46	.54	1
0400	10 watts	.64	.63	1.27

D5020 Lighting and Branch Wiring

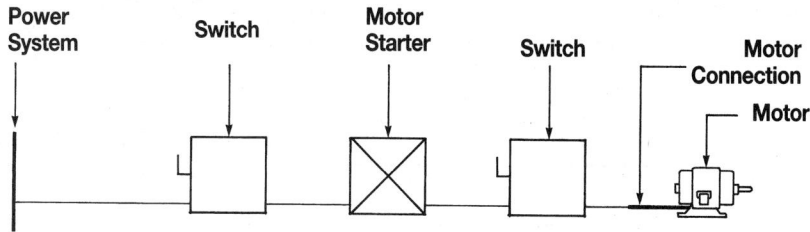

Power System → Switch → Motor Starter → Switch → Motor Connection → Motor

System D5020 145 installed cost of motor wiring as per Table D5010 170 using 50' of rigid conduit and copper wire. **Cost and setting of motor not included.**

System Components	QUANTITY	UNIT	COST EACH		
			MAT.	INST.	TOTAL
SYSTEM D5020 145 0200					
MOTOR INST., SINGLE PHASE, 115V, TO AND INCLUDING 1/3 HP MOTOR SIZE					
Wire 600V type THWN-THHN, copper solid #12	1.250	C.L.F.	21.88	63.75	85.63
Steel intermediate conduit, (IMC) 1/2" diam.	50.000	L.F.	112	280	392
Magnetic FVNR, 115V, 1/3 HP, size 00 starter	1.000	Ea.	205	140	345
Safety switch, fused, heavy duty, 240V 2P 30 amp	1.000	Ea.	151	160	311
Safety switch, non fused, heavy duty, 600V, 3 phase, 30 A	1.000	Ea.	177	175	352
Flexible metallic conduit, Greenfield 1/2" diam.	1.500	L.F.	.81	4.20	5.01
Connectors for flexible metallic conduit Greenfield 1/2" diam.	1.000	Ea.	3.70	7	10.70
Coupling for Greenfield to conduit 1/2" diam flexible metalic conduit	1.000	Ea.	2.20	11.20	13.40
Fuse cartridge nonrenewable, 250V 30 amp	1.000	Ea.	2.18	11.20	13.38
TOTAL			675.77	852.35	1,528.12

D5020 145	Motor Installation	COST EACH		
		MAT.	INST.	TOTAL
0200	Motor installation, single phase, 115V, 1/3 HP motor size	675	850	1,525
0240	1 HP motor size	700	850	1,550
0280	2 HP motor size	750	905	1,655
0320	3 HP motor size	835	925	1,760
0360	230V, 1 HP motor size	680	865	1,545
0400	2 HP motor size	710	865	1,575
0440	3 HP motor size	790	930	1,720
0520	Three phase, 200V, 1-1/2 HP motor size	795	950	1,745
0560	3 HP motor size	845	1,025	1,870
0600	5 HP motor size	905	1,150	2,055
0640	7-1/2 HP motor size	940	1,175	2,115
0680	10 HP motor size	1,475	1,475	2,950
0720	15 HP motor size	1,900	1,625	3,525
0760	20 HP motor size	2,475	1,875	4,350
0800	25 HP motor size	2,525	1,900	4,425
0840	30 HP motor size	3,925	2,225	6,150
0880	40 HP motor size	5,050	2,625	7,675
0920	50 HP motor size	8,375	3,075	11,450
0960	60 HP motor size	8,825	3,250	12,075
1000	75 HP motor size	11,500	3,700	15,200
1040	100 HP motor size	27,100	4,375	31,475
1080	125 HP motor size	28,300	4,800	33,100
1120	150 HP motor size	33,600	5,625	39,225
1160	200 HP motor size	34,100	6,875	40,975
1240	230V, 1-1/2 HP motor size	760	940	1,700
1280	3 HP motor size	810	1,025	1,835
1320	5 HP motor size	865	1,125	1,990
1360	7-1/2 HP motor size	865	1,125	1,990
1400	10 HP motor size	1,325	1,400	2,725
1440	15 HP motor size	1,550	1,525	3,075
1480	20 HP motor size	2,275	1,825	4,100
1520	25 HP motor size	2,475	1,875	4,350

Note: RD5010 -170 (reference shown near rows 0280/0320)

D50 Electrical

D5020 Lighting and Branch Wiring

D5020 145	Motor Installation	COST EACH		
		MAT.	INST.	TOTAL
1560	30 HP motor size	2,525	1,900	4,425
1600	40 HP motor size	4,625	2,575	7,200
1640	50 HP motor size	4,900	2,725	7,625
1680	60 HP motor size	8,375	3,075	11,450
1720	75 HP motor size	10,500	3,500	14,000
1760	100 HP motor size	11,800	3,900	15,700
1800	125 HP motor size	27,500	4,500	32,000
1840	150 HP motor size	30,200	5,125	35,325
1880	200 HP motor size	31,800	5,700	37,500
1960	460V, 2 HP motor size	930	945	1,875
2000	5 HP motor size	975	1,025	2,000
2040	10 HP motor size	1,000	1,125	2,125
2080	15 HP motor size	1,300	1,300	2,600
2120	20 HP motor size	1,350	1,400	2,750
2160	25 HP motor size	1,500	1,475	2,975
2200	30 HP motor size	1,875	1,575	3,450
2240	40 HP motor size	2,425	1,700	4,125
2280	50 HP motor size	2,750	1,875	4,625
2320	60 HP motor size	3,925	2,200	6,125
2360	75 HP motor size	4,650	2,450	7,100
2400	100 HP motor size	5,200	2,725	7,925
2440	125 HP motor size	8,675	3,100	11,775
2480	150 HP motor size	10,500	3,450	13,950
2520	200 HP motor size	12,600	3,900	16,500
2600	575V, 2 HP motor size	930	945	1,875
2640	5 HP motor size	975	1,025	2,000
2680	10 HP motor size	1,000	1,125	2,125
2720	20 HP motor size	1,300	1,300	2,600
2760	25 HP motor size	1,350	1,400	2,750
2800	30 HP motor size	1,875	1,575	3,450
2840	50 HP motor size	2,025	1,625	3,650
2880	60 HP motor size	3,875	2,200	6,075
2920	75 HP motor size	3,925	2,200	6,125
2960	100 HP motor size	4,650	2,450	7,100
3000	125 HP motor size	8,000	3,050	11,050
3040	150 HP motor size	8,675	3,100	11,775
3080	200 HP motor size	11,200	3,500	14,700

D5020 Lighting and Branch Wiring

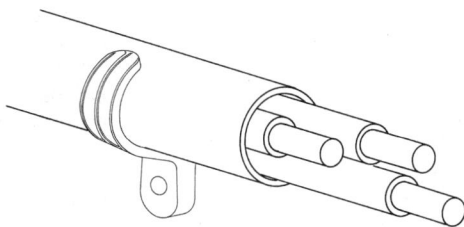

System Components	QUANTITY	UNIT	COST PER L.F.		
			MAT.	INST.	TOTAL
SYSTEM D5020 155 0200					
MOTOR FEEDER SYSTEMS, SINGLE PHASE, UP TO 115V, 1HP OR 230V, 2HP					
Steel intermediate conduit, (IMC) 1/2" diam	1.000	L.F.	2.24	5.60	7.84
Wire 600V type THWN-THHN, copper solid #12	.020	C.L.F.	.35	1.02	1.37
TOTAL			2.59	6.62	9.21

D5020 155	Motor Feeder	COST PER L.F.		
		MAT.	INST.	TOTAL
0200	Motor feeder systems, single phase, feed up to 115V 1HP or 230V 2 HP	2.59	6.60	9.19
0240	115V 2HP, 230V 3HP	2.79	6.70	9.49
0280	115V 3HP	3.20	7	10.20
0360	Three phase, feed to 200V 3HP, 230V 5HP, 460V 10HP, 575V 10HP RD5010 -170	2.77	7.15	9.92
0440	200V 5HP, 230V 7.5HP, 460V 15HP, 575V 20HP	3.07	7.30	10.37
0520	200V 10HP, 230V 10HP, 460V 30HP, 575V 30HP	3.68	7.70	11.38
0600	200V 15HP, 230V 15HP, 460V 40HP, 575V 50HP	4.92	8.80	13.72
0680	200V 20HP, 230V 25HP, 460V 50HP, 575V 60HP	7.25	11.20	18.45
0760	200V 25HP, 230V 30HP, 460V 60HP, 575V 75HP	8.20	11.35	19.55
0840	200V 30HP	12.60	11.70	24.30
0920	230V 40HP, 460V 75HP, 575V 100HP	11.95	12.80	24.75
1000	200V 40HP	19	13.70	32.70
1080	230V 50HP, 460V 100HP, 575V 125HP	16.75	15.15	31.90
1160	200V 50HP, 230V 60HP, 460V 125HP, 575V 150HP	28	16.05	44.05
1240	200V 60HP, 460V 150HP	24.50	18.80	43.30
1320	230V 75HP, 575V 200HP	36.50	19.60	56.10
1400	200V 75HP	37.50	20	57.50
1480	230V 100HP, 460V 200HP	43	23.50	66.50
1560	200V 100HP	58.50	29	87.50
1640	230V 125HP	66	31.50	97.50
1720	200V 125HP, 230V 150HP	79.50	36.50	116
1800	200V 150HP	81	39	120
1880	200V 200HP	117	58.50	175.50
1960	230V 200HP	106	43.50	149.50

D50 Electrical

D5020 Lighting and Branch Wiring

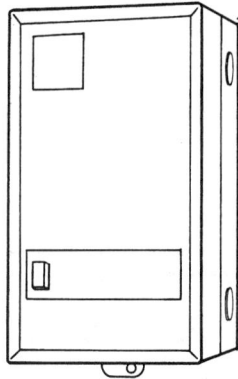

Starters are full voltage, type NEMA 1 for general purpose indoor application with motor overload protection and include mounting and wire connections.

System Components	QUANTITY	UNIT	COST EACH		
			MAT.	INST.	TOTAL
SYSTEM D5020 160 0200					
MAGNETIC STARTER, SIZE 00 TO 1/3 HP, 1 PHASE 115V OR 1 HP 230V					
MAGNETIC STARTER,SIZE 00 TO 1/3 HP, 1 PHASE 115V OR 1 HP 230V	1.000	Ea.	205	140	345
TOTAL			205	140	345

D5020 160	Magnetic Starter	COST EACH		
		MAT.	INST.	TOTAL
0200	Magnetic starter, size 00, to 1/3 HP, 1 phase, 115V or 1 HP 230V	205	140	345
0280	Size 00, to 1-1/2 HP, 3 phase, 200-230V or 2 HP 460-575V	226	160	386
0360	Size 0, to 1 HP, 1 phase, 115V or 2 HP 230V	229	140	369
0440	Size 0, to 3 HP, 3 phase, 200-230V or 5 HP 460-575V	273	243	516
0520	Size 1, to 2 HP, 1 phase, 115V or 3 HP 230V	263	187	450
0600	Size 1, to 7-1/2 HP, 3 phase, 200-230V or 10 HP 460-575V	305	350	655
0680	Size 2, to 10 HP, 3 phase, 200V, 15 HP-230V or 25 HP 460-575V	575	510	1,085
0760	Size 3, to 25 HP, 3 phase, 200V, 30 HP-230V or 50 HP 460-575V	940	620	1,560
0840	Size 4, to 40 HP, 3 phase, 200V, 50 HP-230V or 100 HP 460-575V	2,075	935	3,010
0920	Size 5, to 75 HP, 3 phase, 200V, 100 HP-230V or 200 HP 460-575V	4,875	1,250	6,125
1000	Size 6, to 150 HP, 3 phase, 200V, 200 HP-230V or 400 HP 460-575V	19,100	1,400	20,500

D5020 Lighting and Branch Wiring

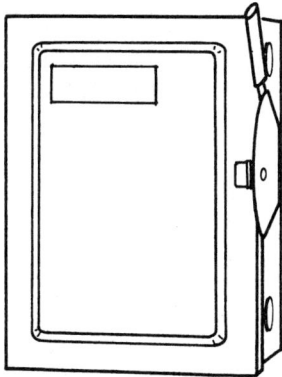

Safety switches are type NEMA 1 for general purpose indoor application, and include time delay fuses, insulation and wire terminations.

System Components	QUANTITY	UNIT	COST EACH		
			MAT.	INST.	TOTAL
SYSTEM D5020 165 0200					
SAFETY SWITCH, 30A FUSED, 1 PHASE, 115V OR 230V.					
Safety switch fused, hvy duty, 240V 2p 30 amp	1.000	Ea.	151	160	311
Fuse, dual element time delay 250V, 30 amp	2.000	Ea.	10.80	22.40	33.20
TOTAL			161.80	182.40	344.20

D5020 165	Safety Switches	COST EACH		
		MAT.	INST.	TOTAL
0200	Safety switch, 30 A fused, 1 phase, 2 HP 115 V or 3 HP, 230 V.	162	182	344
0280	3 phase, 5 HP, 200 V or 7 1/2 HP, 230 V	217	209	426
0360	15 HP, 460 V or 20 HP, 575 V	375	217	592
0440	60 A fused, 3 phase, 15 HP 200 V or 15 HP 230 V	370	277	647
0520	30 HP 460 V or 40 HP 575 V	475	285	760
0600	100 A fused, 3 phase, 20 HP 200 V or 25 HP 230 V	605	335	940
0680	50 HP 460 V or 60 HP 575 V	885	340	1,225
0760	200 A fused, 3 phase, 50 HP 200 V or 60 HP 230 V	1,075	475	1,550
0840	125 HP 460 V or 150 HP 575 V	1,350	485	1,835
0920	400 A fused, 3 phase, 100 HP 200 V or 125 HP 230 V	2,650	675	3,325
1000	250 HP 460 V or 350 HP 575 V	3,375	690	4,065
1020	600 A fused, 3 phase, 150 HP 200 V or 200 HP 230 V	4,500	1,000	5,500
1040	400 HP 460 V	5,525	1,025	6,550

D5020 Lighting and Branch Wiring

Straight
Connector

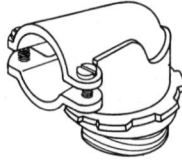

Angle
Connector

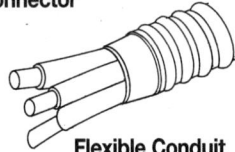

Flexible Conduit

Table below includes costs for the flexible conduit. Not included are wire terminations and testing motor for correct rotation.

System Components	QUANTITY	UNIT	COST EACH		
			MAT.	INST.	TOTAL
SYSTEM D5020 170 0200 **MOTOR CONNECTIONS, SINGLE PHASE, 115V/230V UP TO 1 HP** Motor connection, flexible conduit & fittings, 1 HP motor 115V	1.000	Ea.	9.50	66.50	76
TOTAL			9.50	66.50	76

D5020 170	Motor Connections		COST EACH		
			MAT.	INST.	TOTAL
0200	Motor connections, single phase, 115/230V, up to 1 HP		9.50	66.50	76
0240	Up to 3 HP		9.25	77.50	86.75
0280	Three phase, 200/230/460/575V, up to 3 HP	RD5010 -170	10.30	86	96.30
0320	Up to 5 HP		10.30	86	96.30
0360	Up to 7-1/2 HP		16	102	118
0400	Up to 10 HP		17.20	133	150.20
0440	Up to 15 HP		29.50	170	199.50
0480	Up to 25 HP		34	207	241
0520	Up to 50 HP		87	254	341
0560	Up to 100 HP		214	375	589

D5020 Lighting and Branch Wiring

Manual Starter

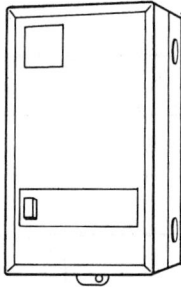

Magnetic Starter

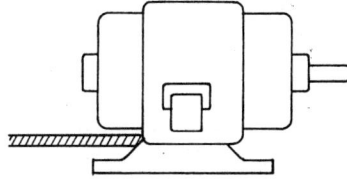

Induction Motor

For 230/460 Volt A.C., 3 phase, 60 cycle ball bearing squirrel cage induction motors, NEMA Class B standard line. Installation included.

No conduit, wire, or terminations included.

System Components	QUANTITY	UNIT	COST EACH		
			MAT.	INST.	TOTAL
SYSTEM D5020 175 0220					
MOTOR, DRIPPROOF CLASS B INSULATION, 1.15 SERVICE FACTOR, WITH STARTER					
1 H.P., 1200 RPM WITH MANUAL STARTER					
Motor, dripproof, class B insul, 1.15 serv fact, 1200 RPM, 1 HP	1.000	Ea.	290	124	414
Motor starter, manual, 3 phase, 1 HP motor	1.000	Ea.	223	160	383
TOTAL			513	284	797

D5020 175	Motor & Starter		COST EACH		
			MAT.	INST.	TOTAL
0190	Motor, dripproof, premium efficient, 1.15 service factor				
0200	1 HP, 1200 RPM, motor only		290	124	414
0220	With manual starter	RD5010 -170	515	284	799
0240	With magnetic starter		515	284	799
0260	1800 RPM, motor only		290	124	414
0280	With manual starter		515	284	799
0300	With magnetic starter		515	284	799
0320	2 HP, 1200 RPM, motor only		400	124	524
0340	With manual starter		625	284	909
0360	With magnetic starter		675	365	1,040
0380	1800 RPM, motor only		294	124	418
0400	With manual starter		515	284	799
0420	With magnetic starter		565	365	930
0440	3600 RPM, motor only		340	124	464
0460	With manual starter		565	284	849
0480	With magnetic starter		615	365	980
0500	3 HP, 1200 RPM, motor only		565	124	689
0520	With manual starter		790	284	1,074
0540	With magnetic starter		840	365	1,205
0560	1800 RPM, motor only		296	124	420
0580	With manual starter		520	284	804
0600	With magnetic starter		570	365	935
0620	3600 RPM, motor only		370	124	494
0640	With manual starter		595	284	879
0660	With magnetic starter		645	365	1,010
0680	5 HP, 1200 RPM, motor only		640	124	764
0700	With manual starter		905	405	1,310
0720	With magnetic starter		945	475	1,420
0740	1800 RPM, motor only		410	124	534
0760	With manual starter		675	405	1,080
0780	With magnetic starter		715	475	1,190
0800	3600 RPM, motor only		400	124	524

D5020 Lighting and Branch Wiring

D5020 175	Motor & Starter	COST EACH		
		MAT.	INST.	TOTAL
0820	With manual starter	665	405	1,070
0840	With magnetic starter	705	475	1,180
0860	7.5 HP, 1800 RPM, motor only	560	133	693
0880	With manual starter	825	415	1,240
0900	With magnetic starter	1,125	645	1,770
0920	10 HP, 1800 RPM, motor only	700	140	840
0940	With manual starter	965	420	1,385
0960	With magnetic starter	1,275	650	1,925
0980	15 HP, 1800 RPM, motor only	865	175	1,040
1000	With magnetic starter	1,450	685	2,135
1040	20 HP, 1800 RPM, motor only	1,175	215	1,390
1060	With magnetic starter	2,125	835	2,960
1100	25 HP, 1800 RPM, motor only	1,350	224	1,574
1120	With magnetic starter	2,300	845	3,145
1160	30 HP, 1800 RPM, motor only	1,575	233	1,808
1180	With magnetic starter	2,525	855	3,380
1220	40 HP, 1800 RPM, motor only	1,950	280	2,230
1240	With magnetic starter	4,025	1,225	5,250
1280	50 HP, 1800 RPM, motor only	2,275	350	2,625
1300	With magnetic starter	4,350	1,275	5,625
1340	60 HP, 1800 RPM, motor only	2,800	400	3,200
1360	With magnetic starter	7,675	1,650	9,325
1400	75 HP, 1800 RPM, motor only	3,400	465	3,865
1420	With magnetic starter	8,275	1,725	10,000
1460	100 HP, 1800 RPM, motor only	4,325	620	4,945
1480	With magnetic starter	9,200	1,875	11,075
1520	125 HP, 1800 RPM, motor only	5,175	800	5,975
1540	With magnetic starter	24,300	2,200	26,500
1580	150 HP, 1800 RPM, motor only	7,475	935	8,410
1600	With magnetic starter	26,600	2,325	28,925
1640	200 HP, 1800 RPM, motor only	9,450	1,125	10,575
1660	With magnetic starter	28,600	2,525	31,125
1680	Totally encl, premium efficient, 1.0 ser. fac., 1HP, 1200RPM, motor only	310	124	434
1700	With manual starter	535	284	819
1720	With magnetic starter	535	284	819
1740	1800 RPM, motor only	310	124	434
1760	With manual starter	535	284	819
1780	With magnetic starter	535	284	819
1800	2 HP, 1200 RPM, motor only	355	124	479
1820	With manual starter	580	284	864
1840	With magnetic starter	630	365	995
1860	1800 RPM, motor only	375	124	499
1880	With manual starter	600	284	884
1900	With magnetic starter	650	365	1,015
1920	3600 RPM, motor only	281	124	405
1940	With manual starter	505	284	789
1960	With magnetic starter	555	365	920
1980	3 HP, 1200 RPM, motor only	480	124	604
2000	With manual starter	705	284	989
2020	With magnetic starter	755	365	1,120
2040	1800 RPM, motor only	450	124	574
2060	With manual starter	675	284	959
2080	With magnetic starter	725	365	1,090
2100	3600 RPM, motor only	340	124	464
2120	With manual starter	565	284	849
2140	With magnetic starter	615	365	980
2160	5 HP, 1200 RPM, motor only	665	124	789
2180	With manual starter	930	405	1,335

D5020 Lighting and Branch Wiring

D5020 175	Motor & Starter	COST EACH		
		MAT.	INST.	TOTAL
2200	With magnetic starter	970	475	1,445
2220	1800 RPM, motor only	515	124	639
2240	With manual starter	780	405	1,185
2260	With magnetic starter	820	475	1,295
2280	3600 RPM, motor only	430	124	554
2300	With manual starter	695	405	1,100
2320	With magnetic starter	735	475	1,210
2340	7.5 HP, 1800 RPM, motor only	740	133	873
2360	With manual starter	1,000	415	1,415
2380	With magnetic starter	1,325	645	1,970
2400	10 HP, 1800 RPM, motor only	890	140	1,030
2420	With manual starter	1,150	420	1,570
2440	With magnetic starter	1,475	650	2,125
2460	15 HP, 1800 RPM, motor only	1,250	175	1,425
2480	With magnetic starter	1,825	685	2,510
2500	20 HP, 1800 RPM, motor only	1,450	215	1,665
2520	With magnetic starter	2,400	835	3,235
2540	25 HP, 1800 RPM, motor only	1,675	224	1,899
2560	With magnetic starter	2,625	845	3,470
2580	30 HP, 1800 RPM, motor only	2,000	233	2,233
2600	With magnetic starter	2,950	855	3,805
2620	40 HP, 1800 RPM, motor only	2,550	280	2,830
2640	With magnetic starter	4,625	1,225	5,850
2660	50 HP, 1800 RPM, motor only	3,150	350	3,500
2680	With magnetic starter	5,225	1,275	6,500
2700	60 HP, 1800 RPM, motor only	4,675	400	5,075
2720	With magnetic starter	9,550	1,650	11,200
2740	75 HP, 1800 RPM, motor only	6,025	465	6,490
2760	With magnetic starter	10,900	1,725	12,625
2780	100 HP, 1800 RPM, motor only	9,225	620	9,845
2800	With magnetic starter	14,100	1,875	15,975
2820	125 HP, 1800 RPM, motor only	12,100	800	12,900
2840	With magnetic starter	31,200	2,200	33,400
2860	150 HP, 1800 RPM, motor only	14,100	935	15,035
2880	With magnetic starter	33,200	2,325	35,525
2900	200 HP, 1800 RPM, motor only	17,200	1,125	18,325
2920	With magnetic starter	36,300	2,525	38,825

D5020 Lighting and Branch Wiring

Design Assumptions:

1. A 100 footcandle average maintained level of illumination.
2. Ceiling heights range from 9' to 11'.
3. Average reflectance values are assumed for ceilings, walls and floors.
4. Cool white (CW) fluorescent lamps with 3150 lumens for 40 watt lamps and 6300 lumens for 8' slimline lamps.
5. Four 40 watt lamps per 4' fixture and two 8' lamps per 8' fixture.
6. Average fixture efficiency values and spacing to mounting height ratios.
7. Installation labor is average U.S. rate as of January 1.

A. Strip Fixture   B. Surface Mounted

C. Recessed   D. Pendent Mounted

System Components	QUANTITY	UNIT	COST PER S.F.		
			MAT.	INST.	TOTAL
SYSTEM D5020 208 0520					
FLUORESCENT FIXTURES MOUNTED 9'-11" ABOVE FLOOR, 100 FC					
TYPE A, 8 FIXTURES PER 400 S.F.					
Steel intermediate conduit, (IMC) 1/2" diam.	.404	L.F.	.90	2.26	3.16
Wire, 600V, type THWN-THHN, copper, solid, #12	.008	C.L.F.	.14	.41	.55
Fluorescent strip fixture 8' long, surface mounted, two 75W SL	.020	Ea.	1.23	1.81	3.04
Steel outlet box 4" concrete	.020	Ea.	.33	.56	.89
Steel outlet box plate with stud, 4" concrete	.020	Ea.	.13	.14	.27
TOTAL			2.73	5.18	7.91

D5020 208	Fluorescent Fixtures (by Type)	COST PER S.F.		
		MAT.	INST.	TOTAL
0520	Fluorescent fixtures, type A, 8 fixtures per 400 S.F.	2.73	5.20	7.93
0560	11 fixtures per 600 S.F.	2.60	5	7.60
0600	17 fixtures per 1000 S.F.	2.54	4.95	7.49
0640	23 fixtures per 1600 S.F.	2.33	4.68	7.01
0680	28 fixtures per 2000 S.F.	2.33	4.68	7.01
0720	41 fixtures per 3000 S.F.	2.29	4.67	6.96
0800	53 fixtures per 4000 S.F.	2.25	4.55	6.80
0840	64 fixtures per 5000 S.F.	2.25	4.55	6.80
0880	Type B, 11 fixtures per 400 S.F.	4.94	7.60	12.54
0920	15 fixtures per 600 S.F.	4.63	7.30	11.93
0960	24 fixtures per 1000 S.F.	4.53	7.25	11.78
1000	35 fixtures per 1600 S.F.	4.27	6.90	11.17
1040	42 fixtures per 2000 S.F.	4.19	6.95	11.14
1080	61 fixtures per 3000 S.F.	4.18	6.70	10.88
1160	80 fixtures per 4000 S.F.	4.09	6.85	10.94
1200	98 fixtures per 5000 S.F.	4.07	6.80	10.87
1240	Type C, 11 fixtures per 400 S.F.	4.05	8	12.05
1280	14 fixtures per 600 S.F.	3.67	7.45	11.12
1320	23 fixtures per 1000 S.F.	3.65	7.40	11.05
1360	34 fixtures per 1600 S.F.	3.56	7.30	10.86
1400	43 fixtures per 2000 S.F.	3.57	7.25	10.82
1440	63 fixtures per 3000 S.F.	3.49	7.15	10.64
1520	81 fixtures per 4000 S.F.	3.42	7.05	10.47
1560	101 fixtures per 5000 S.F.	3.42	7.05	10.47
1600	Type D, 8 fixtures per 400 S.F.	3.73	6.20	9.93
1640	12 fixtures per 600 S.F.	3.73	6.15	9.88
1680	19 fixtures per 1000 S.F.	3.60	6	9.60
1720	27 fixtures per 1600 S.F.	3.40	5.85	9.25
1760	34 fixtures per 2000 S.F.	3.38	5.80	9.18
1800	48 fixtures per 3000 S.F.	3.26	5.65	8.91
1880	64 fixtures per 4000 S.F.	3.26	5.65	8.91
1920	79 fixtures per 5000 S.F.	3.26	5.65	8.91

RD5020 -200

D5020 Lighting and Branch Wiring

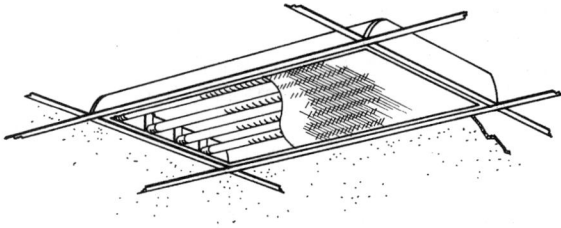

Type C. Recessed, mounted on grid ceiling suspension system, 2' x 4', four 40 watt lamps, acrylic prismatic diffusers.

5.3 watts per S.F. for 100 footcandles.

3 watts per S.F. for 57 footcandles.

System Components	QUANTITY	UNIT	COST PER S.F.		
			MAT.	INST.	TOTAL
SYSTEM D5020 210 0200					
FLUORESCENT FIXTURES RECESS MOUNTED IN CEILING					
1 WATT PER S.F., 20 FC, 5 FIXTURES PER 1000 S.F.					
Steel intermediate conduit, (IMC) 1/2" diam.	.128	L.F.	.29	.72	1.01
Wire, 600 volt, type THW, copper, solid, #12	.003	C.L.F.	.05	.15	.20
Fluorescent fixture, recessed, 2'x 4', four 40W, w/ lens, for grid ceiling	.005	Ea.	.35	.60	.95
Steel outlet box 4" square	.005	Ea.	.08	.14	.22
Fixture whip, Greenfield w/#12 THHN wire	.005	Ea.	.03	.04	.07
TOTAL			.80	1.65	2.45

D5020 210	Fluorescent Fixtures (by Wattage)		COST PER S.F.		
			MAT.	INST.	TOTAL
0190	Fluorescent fixtures recess mounted in ceiling				
0195	T-12, standard 40 watt lamps				
0200	1 watt per S.F., 20 FC, 5 fixtures @40 watts per 1000 S.F.	RD5020 -200	.80	1.65	2.45
0240	2 watt per S.F., 40 FC, 10 fixtures @40 watt per 1000 S.F.		1.60	3.23	4.83
0280	3 watt per S.F., 60 FC, 15 fixtures @40 watt per 1000 S.F		2.40	4.88	7.28
0320	4 watt per S.F., 80 FC, 20 fixtures @40 watt per 1000 S.F.		3.20	6.45	9.65
0400	5 watt per S.F., 100 FC, 25 fixtures @40 watt per 1000 S.F.		4	8.10	12.10
0450	T-8, energy saver 32 watt lamps				
0500	0.8 watt per S.F., 20 FC, 5 fixtures @32 watt per 1000 S.F.		.88	1.65	2.53
0520	1.6 watt per S.F., 40 FC, 10 fixtures @32 watt per 1000 S.F.		1.76	3.23	4.99
0540	2.4 watt per S.F., 60 FC, 15 fixtures @ 32 watt per 1000 S.F		2.64	4.88	7.52
0560	3.2 watt per S.F., 80 FC, 20 fixtures @32 watt per 1000 S.F.		3.52	6.45	9.97
0580	4 watt per S.F., 100 FC, 25 fixtures @32 watt per 1000 S.F.		4.40	8.10	12.50

D5020 Lighting and Branch Wiring

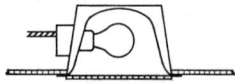

Type A. Recessed wide distribution reflector with flat glass lens 150 W.

Maximum spacing = 1.2 x mounting height.

13 watts per S.F. for 100 footcandles.

Type B

Type B. Recessed reflector down light with baffles 150 W.

Maximum spacing = 0.8 x mounting height.

18 watts per S.F. for 100 footcandles.

Type C. Recessed PAR–38 flood lamp with concentric louver 150 W.

Maximum spacing = 0.5 x mounting height.

19 watts per S.F. for 100 footcandles.

Type D. Recessed R–40 flood lamp with reflector skirt.

Maximum spacing = 0.7 x mounting height.

15 watts per S.F. for 100 footcandles.

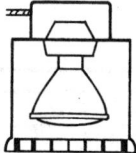

Type C

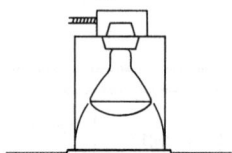

Type D

System Components	QUANTITY	UNIT	COST PER S.F.		
			MAT.	INST.	TOTAL
SYSTEM D5020 214 0400					
INCANDESCENT FIXTURE RECESS MOUNTED, 100 FC					
TYPE A, 34 FIXTURES PER 400 S.F.					
Steel intermediate conduit, (IMC) 1/2" diam	1.060	L.F.	2.37	5.94	8.31
Wire, 600V, type THWN-THHN, copper, solid, #12	.033	C.L.F.	.58	1.68	2.26
Steel outlet box 4" square	.085	Ea.	1.39	2.38	3.77
Fixture whip, Greenfield w/#12 THHN wire	.085	Ea.	.56	.60	1.16
Incandescent fixture, recessed, w/lens, prewired, square trim, 200W	.085	Ea.	8.93	7.10	16.03
TOTAL			13.83	17.70	31.53

D5020 214	Incandescent Fixture (by Type)	COST PER S.F.		
		MAT.	INST.	TOTAL
0380	Incandescent fixture recess mounted, 100 FC			
0400	Type A, 34 fixtures per 400 S.F.	13.85	17.70	31.55
0440	49 fixtures per 600 S.F.	13.45	17.40	30.85
0480	63 fixtures per 800 S.F.	13.15	17.20	30.35
0520	90 fixtures per 1200 S.F.	12.70	16.85	29.55
0560	116 fixtures per 1600 S.F.	12.45	16.70	29.15
0600	143 fixtures per 2000 S.F.	12.35	16.65	29
0640	Type B, 47 fixtures per 400 S.F.	12.95	22.50	35.45
0680	66 fixtures per 600 S.F.	12.45	22	34.45
0720	88 fixtures per 800 S.F.	12.45	22	34.45
0760	127 fixtures per 1200 S.F.	12.20	21.50	33.70
0800	160 fixtures per 1600 S.F.	12.05	21.50	33.55
0840	206 fixtures per 2000 S.F.	12	21.50	33.50
0880	Type C, 51 fixtures per 400 S.F.	17.60	23.50	41.10
0920	74 fixtures per 600 S.F.	17.10	23	40.10
0960	97 fixtures per 800 S.F.	16.90	23	39.90
1000	142 fixtures per 1200 S.F.	16.60	22.50	39.10
1040	186 fixtures per 1600 S.F.	16.45	22.50	38.95
1080	230 fixtures per 2000 S.F.	16.35	22.50	38.85
1120	Type D, 39 fixtures per 400 S.F.	16.40	18.40	34.80
1160	57 fixtures per 600 S.F.	16	18.15	34.15
1200	75 fixtures per 800 S.F.	15.90	18.15	34.05
1240	109 fixtures per 1200 S.F.	15.50	17.95	33.45
1280	143 fixtures per 1600 S.F.	15.25	17.75	33
1320	176 fixtures per 2000 S.F.	15.15	17.70	32.85

RD5020 -200

D5020 Lighting and Branch Wiring

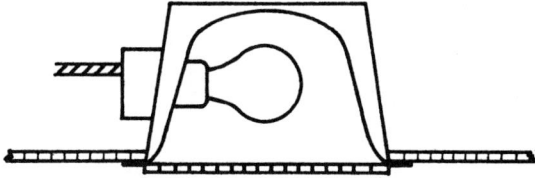

Type A. Recessed, wide distribution reflector with flat glass lens.

150 watt inside frost—2500 lumens per lamp.

PS–25 extended service lamp.

Maximum spacing = 1.2 x mounting height.

13 watts per S.F. for 100 footcandles.

System Components	QUANTITY	UNIT	COST PER S.F.		
			MAT.	INST.	TOTAL
SYSTEM D5020 216 0200					
INCANDESCENT FIXTURE RECESS MOUNTED, TYPE A					
1 WATT PER S.F., 8 FC, 6 FIXT PER 1000 S.F.					
Steel intermediate conduit, (IMC) 1/2" diam	.091	L.F.	.20	.51	.71
Wire, 600V, type THWN-THHN, copper, solid, #12	.002	C.L.F.	.04	.10	.14
Incandescent fixture, recessed, w/lens, prewired, square trim, 200W	.006	Ea.	.63	.50	1.13
Steel outlet box 4" square	.006	Ea.	.10	.17	.27
Fixture whip, Greenfield w/#12 THHN wire	.006	Ea.	.04	.04	.08
TOTAL			1.01	1.32	2.33

D5020 216	Incandescent Fixture (by Wattage)		COST PER S.F.		
			MAT.	INST.	TOTAL
0190	Incandescent fixture recess mounted, type A				
0200	1 watt per S.F., 8 FC, 6 fixtures per 1000 S.F.		1.01	1.32	2.33
0240	2 watt per S.F., 16 FC, 12 fixtures per 1000 S.F.	RD5020 -200	2.02	2.64	4.66
0280	3 watt per S.F., 24 FC, 18 fixtures, per 1000 S.F.		3	3.92	6.92
0320	4 watt per S.F., 32 FC, 24 fixtures per 1000 S.F.		4.01	5.25	9.26
0400	5 watt per S.F., 40 FC, 30 fixtures per 1000 S.F.		5	6.55	11.55

D5020 Lighting and Branch Wiring

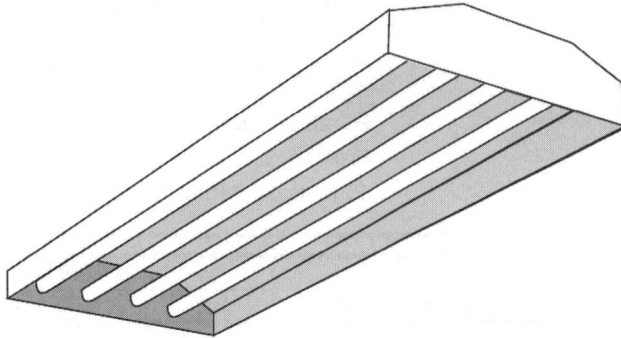

High Bay Flourescent Fixtures
Four T5 HO/54 watt lamps

System Components	QUANTITY	UNIT	COST PER S.F.		
			MAT.	INST.	TOTAL
SYSTEM D5020 218 0200					
FLUORESCENT HIGH BAY FIXTURE, 4 LAMP, 8'-10' ABOVE WORK PLANE					
0.5 WATT/S.F., 29 FC, 2 FIXTURES/1000 S.F.					
Steel intermediate conduit, (IMC) 1/2" diam	.100	L.F.	.22	.56	.78
Wire, 600V, type THWN-THHN, copper, solid, #10	.002	C.L.F.	.06	.11	.17
Steel outlet box 4" concrete	.002	Ea.	.03	.06	.09
Steel outlet box plate with stud, 4" concrete	.002	Ea.	.01	.01	.02
Flourescent , hi bay, 4 lamp, T-5 fixture	.002	Ea.	.44	.25	.69
TOTAL			.76	.99	1.75

D5020 218	Fluorescent Fixture, High Bay, 8'-10' (by Wattage)	COST PER S.F.		
		MAT.	INST.	TOTAL
0190	Fluorescent high bay-4 lamp fixture, 8'-10' above work plane			
0200	.5 watt/SF, 29 FC, 2 fixtures per 1000 SF	.76	.99	1.75
0400	1 watt/SF, 59 FC, 4 fixtures per 1000 SF	1.50	1.90	3.40
0600	1.5 watt/SF, 103 FC, 7 fixtures per 1000 SF	2.46	2.98	5.44
0800	2 watt/SF, 133 FC, 9 fixtures per 1000 SF	3.19	3.91	7.10
1000	2.5 watt/SF, 162 FC, 11 fixtures per 1000 SF	3.94	4.88	8.82

D5020   Lighting and Branch Wiring

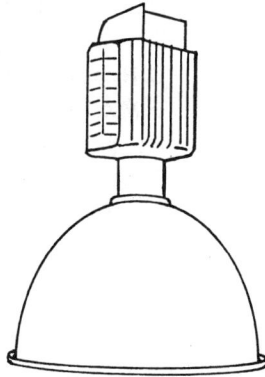

HIGH BAY FIXTURES

B. Metal halide 400 watt

C. High pressure sodium 400 watt

E. Metal halide 1000 watt

F. High pressure sodium 1000 watt

G. Metal halide 1000 watt
   125,000 lumen lamp

System Components	QUANTITY	UNIT	COST PER S.F.		
			MAT.	INST.	TOTAL
SYSTEM D5020 220 0880					
HIGH INTENSITY DISCHARGE FIXTURE, 8'-10' ABOVE WORK PLANE, 100 FC					
TYPE B, 8 FIXTURES PER 900 S.F.					
Steel intermediate conduit, (IMC) 1/2" diam	.460	L.F.	1.03	2.58	3.61
Wire, 600V, type THWN-THHN, copper, solid, #10	.009	C.L.F.	.25	.50	.75
Steel outlet box 4" concrete	.009	Ea.	.15	.25	.40
Steel outlet box plate with stud 4" concrete	.009	Ea.	.06	.06	.12
Metal halide, hi bay, aluminum reflector, 400 W lamp	.009	Ea.	4.01	2.19	6.20
TOTAL			5.50	5.58	11.08

D5020 220	H.I.D. Fixture, High Bay, 8'-10' (by Type)		COST PER S.F.		
			MAT.	INST.	TOTAL
0500	High intensity discharge fixture, 8'-10' above work plane, 100 FC				
0880	Type B, 8 fixtures per 900 S.F.		5.50	5.60	11.10
0920	15 fixtures per 1800 S.F.		5.05	5.35	10.40
0960	24 fixtures per 3000 S.F.	RD5020 -200	5.05	5.35	10.40
1000	31 fixtures per 4000 S.F.		5.05	5.35	10.40
1040	38 fixtures per 5000 S.F.	RD5020 -240	5.05	5.35	10.40
1080	60 fixtures per 8000 S.F.		5.05	5.35	10.40
1120	72 fixtures per 10000 S.F.		4.54	4.97	9.51
1160	115 fixtures per 16000 S.F.		4.54	4.97	9.51
1200	230 fixtures per 32000 S.F.		4.54	4.97	9.51
1240	Type C, 4 fixtures per 900 S.F.		2.71	3.40	6.11
1280	8 fixtures per 1800 S.F.		2.71	3.40	6.11
1320	13 fixtures per 3000 S.F.		2.71	3.40	6.11
1360	17 fixtures per 4000 S.F.		2.71	3.40	6.11
1400	21 fixtures per 5000 S.F.		2.58	3.13	5.71
1440	33 fixtures per 8000 S.F.		2.55	3.07	5.62
1480	40 fixtures per 10000 S.F.		2.48	2.90	5.38
1520	63 fixtures per 16000 S.F.		2.48	2.90	5.38
1560	126 fixtures per 32000 S.F.		2.48	2.90	5.38

D50 Electrical

D5020 Lighting and Branch Wiring

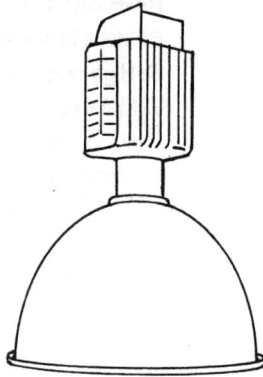

HIGH BAY FIXTURES

B. Metal halide 400 watt

C. High pressure sodium 400 watt

E. Metal halide 1000 watt

F. High pressure sodium 1000 watt

G. Metal halide 1000 watt
   125,000 lumen lamp

System Components	QUANTITY	UNIT	COST PER S.F.		
			MAT.	INST.	TOTAL
SYSTEM D5020 222 0240					
HIGH INTENSITY DISCHARGE FIXTURE, 8'-10' ABOVE WORK PLANE					
1 WATT/S.F., TYPE B, 29 FC, 2 FIXTURES/1000 S.F.					
Steel intermediate conduit, (IMC) 1/2" diam	.100	L.F.	.22	.56	.78
Wire, 600V, type THWN-THHN, copper, solid, #10	.002	C.L.F.	.06	.11	.17
Steel outlet box 4" concrete	.002	Ea.	.03	.06	.09
Steel outlet box plate with stud, 4" concrete	.002	Ea.	.01	.01	.02
Metal halide, hi bay, aluminum reflector, 400 W lamp	.002	Ea.	.89	.49	1.38
TOTAL			1.21	1.23	2.44

D5020 222	H.I.D. Fixture, High Bay, 8'-10' (by Wattage)		COST PER S.F.		
			MAT.	INST.	TOTAL
0190	High intensity discharge fixture, 8'-10' above work plane				
0240	1 watt/S.F., type B, 29 FC, 2 fixtures/1000 S.F.		1.21	1.23	2.44
0280	Type C, 54 FC, 2 fixtures/1000 S.F.	RD5020 -200	1.02	.94	1.96
0400	2 watt/S.F., type B, 59 FC, 4 fixtures/1000 S.F.		2.41	2.37	4.78
0440	Type C, 108 FC, 4 fixtures/1000 S.F.	RD5020 -240	2.04	1.81	3.85
0560	3 watt/S.F., type B, 103 FC, 7 fixtures/1000 S.F.		4.05	3.80	7.85
0600	Type C, 189 FC, 6 fixtures/1000 S.F.		3.20	2.34	5.54
0720	4 watt/S.F., type B, 133 FC, 9 fixtures/1000 S.F.		5.25	4.97	10.22
0760	Type C, 243 FC, 9 fixtures/1000 S.F.		4.54	4.03	8.57
0880	5 watt/S.F., type B, 162 FC, 11 fixtures/1000 S.F.		6.45	6.15	12.60
0920	Type C, 297 FC, 11 fixtures/1000 S.F.		5.55	5	10.55

D5020 Lighting and Branch Wiring

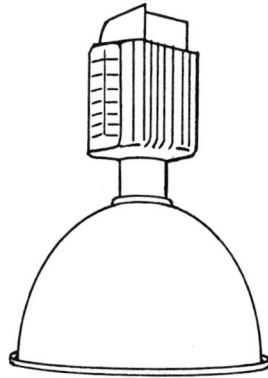

HIGH BAY FIXTURES

B. Metal halide 400 watt

C. High pressure sodium 400 watt

E. Metal halide 1000 watt

F. High pressure sodium 1000 watt

G. Metal halide 1000 watt
125,000 lumen lamp

System Components	QUANTITY	UNIT	COST PER S.F.		
			MAT.	INST.	TOTAL
SYSTEM D5020 224 1240					
HIGH INTENSITY DISCHARGE FIXTURE, 16' ABOVE WORK PLANE, 100 FC					
TYPE C, 5 FIXTURES PER 900 S.F.					
Steel intermediate conduit, (IMC) 1/2" diam	.260	L.F.	.58	1.46	2.04
Wire, 600V, type THWN-THHN, copper, solid, #10	.007	C.L.F.	.19	.39	.58
Steel outlet box 4" concrete	.006	Ea.	.10	.17	.27
Steel outlet box plate with stud, 4" concrete	.006	Ea.	.04	.04	.08
High pressure sodium, hi bay, aluminum reflector, 400 W lamp	.006	Ea.	2.46	1.46	3.92
TOTAL			3.37	3.52	6.89

D5020 224	H.I.D. Fixture, High Bay, 16' (by Type)		COST PER S.F.		
			MAT.	INST.	TOTAL
0510	High intensity discharge fixture, 16' above work plane, 100 FC				
1240	Type C, 5 fixtures per 900 S.F.		3.37	3.52	6.89
1280	9 fixtures per 1800 S.F.		3.12	3.70	6.82
1320	15 fixtures per 3000 S.F.	RD5020 -200	3.12	3.70	6.82
1360	18 fixtures per 4000 S.F.		2.96	3.31	6.27
1400	22 fixtures per 5000 S.F.	RD5020 -240	2.96	3.31	6.27
1440	36 fixtures per 8000 S.F.		2.96	3.31	6.27
1480	42 fixtures per 10,000 S.F.		2.70	3.41	6.11
1520	65 fixtures per 16,000 S.F.		2.70	3.41	6.11
1600	Type G, 4 fixtures per 900 S.F.		4.46	5.55	10.01
1640	6 fixtures per 1800 S.F.		3.77	5.20	8.97
1720	9 fixtures per 4000 S.F.		3.16	5.05	8.21
1760	11 fixtures per 5000 S.F.		3.16	5.05	8.21
1840	21 fixtures per 10,000 S.F.		2.96	4.55	7.51
1880	33 fixtures per 16,000 S.F.		2.96	4.55	7.51

D5020 Lighting and Branch Wiring

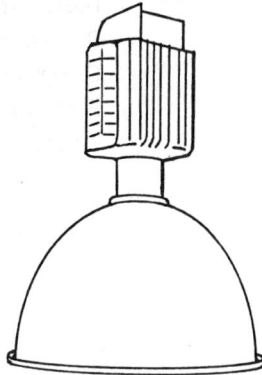

HIGH BAY FIXTURES

B. Metal halide 400 watt

C. High pressure sodium 400 watt

E. Metal halide 1000 watt

F. High pressure sodium 1000 watt

G. Metal halide 1000 watt
125,000 lumen lamp

System Components	QUANTITY	UNIT	COST PER S.F.		
			MAT.	INST.	TOTAL
SYSTEM D5020 226 0240					
HIGH INTENSITY DISCHARGE FIXTURE, 16′ ABOVE WORK PLANE					
1 WATT/S.F., TYPE E, 42 FC, 1 FIXTURE/1000 S.F.					
Steel intermediate conduit, (IMC) 1/2″ diam	.160	L.F.	.36	.90	1.26
Wire, 600V, type THWN-THHN, copper, solid, #10	.003	C.L.F.	.08	.17	.25
Steel outlet box 4″ concrete	.001	Ea.	.02	.03	.05
Steel outlet box plate with stud, 4″ concrete	.001	Ea.	.01	.01	.02
Metal halide, hi bay, aluminum reflector, 1000 W lamp	.001	Ea.	.64	.28	.92
TOTAL			1.11	1.39	2.50

D5020 226	H.I.D. Fixture, High Bay, 16′ (by Wattage)		COST PER S.F.		
			MAT.	INST.	TOTAL
0190	High intensity discharge fixture, 16′ above work plane				
0240	1 watt/S.F., type E, 42 FC, 1 fixture/1000 S.F.		1.11	1.39	2.50
0280	Type G, 52 FC, 1 fixture/1000 S.F.		1.11	1.39	2.50
0320	Type C, 54 FC, 2 fixture/1000 S.F.	RD5020 -200	1.27	1.55	2.82
0440	2 watt/S.F., type E, 84 FC, 2 fixture/1000 S.F.	RD5020 -240	2.23	2.81	5.04
0480	Type G, 105 FC, 2 fixture/1000 S.F.		2.23	2.81	5.04
0520	Type C, 108 FC, 4 fixture/1000 S.F.		2.56	3.09	5.65
0640	3 watt/S.F., type E, 126 FC, 3 fixture/1000 S.F.		3.35	4.19	7.54
0680	Type G, 157 FC, 3 fixture/1000 S.F.		3.35	4.19	7.54
0720	Type C, 162 FC, 6 fixture/1000 S.F.		3.83	4.62	8.45
0840	4 watt/S.F., type E, 168 FC, 4 fixture/1000 S.F.		4.48	5.60	10.08
0880	Type G, 210 FC, 4 fixture/1000 S.F.		4.48	5.60	10.08
0920	Type C, 243 FC, 9 fixture/1000 S.F.		5.55	6.45	12
1040	5 watt/S.F., type E, 210 FC, 5 fixture/1000 S.F.		5.55	7	12.55
1080	Type G, 262 FC, 5 fixture/1000 S.F.		5.55	7	12.55
1120	Type C, 297 FC, 11 fixture/1000 S.F.		6.80	8	14.80

D5020 Lighting and Branch Wiring

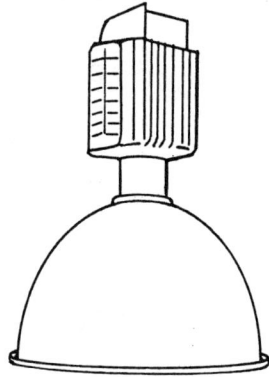

HIGH BAY FIXTURES

B. Metal halide 400 watt

C. High pressure sodium 400 watt

E. Metal halide 1000 watt

F. High pressure sodium 1000 watt

G. Metal halide 1000 watt
   125,000 lumen lamp

System Components	QUANTITY	UNIT	COST PER S.F.		
			MAT.	INST.	TOTAL
SYSTEM D5020 228 1240					
HIGH INTENSITY DISCHARGE FIXTURE, 20' ABOVE WORK PLANE, 100 FC					
TYPE C, 6 FIXTURES PER 900 S.F.					
Steel intermediate conduit, (IMC) 1/2" diam	.350	L.F.	.78	1.96	2.74
Wire, 600V, type THWN-THHN, copper, solid, #10.	.011	C.L.F.	.30	.62	.92
Steel outlet box 4" concrete	.007	Ea.	.11	.20	.31
Steel outlet box plate with stud, 4" concrete	.007	Ea.	.05	.05	.10
High pressure sodium, hi bay, aluminum reflector, 400 W lamp	.007	Ea.	2.87	1.70	4.57
TOTAL			4.11	4.53	8.64

D5020 228	H.I.D. Fixture, High Bay, 20' (by Type)		COST PER S.F.		
			MAT.	INST.	TOTAL
0510	High intensity discharge fixture 20' above work plane, 100 FC				
1240	Type C, 6 fixtures per 900 S.F.		4.11	4.53	8.64
1280	10 fixtures per 1800 S.F.	RD5020 -200	3.68	4.25	7.93
1320	16 fixtures per 3000 S.F.		3.30	4.09	7.39
1360	20 fixtures per 4000 S.F.	RD5020 -240	3.30	4.09	7.39
1400	24 fixtures per 5000 S.F.		3.30	4.09	7.39
1440	38 fixtures per 8000 S.F.		3.23	3.92	7.15
1520	68 fixtures per 16000 S.F.		3.07	4.31	7.38
1560	132 fixtures per 32000 S.F.		3.07	4.31	7.38
1600	Type G, 4 fixtures per 900 S.F.		4.43	5.45	9.88
1640	6 fixtures per 1800 S.F.		3.79	5.20	8.99
1680	7 fixtures per 3000 S.F.		3.72	5.05	8.77
1720	10 fixtures per 4000 S.F.		3.69	4.97	8.66
1760	11 fixtures per 5000 S.F.		3.30	5.35	8.65
1800	18 fixtures per 8000 S.F.		3.30	5.35	8.65
1840	22 fixtures per 10000 S.F.		3.30	5.35	8.65
1880	34 fixtures per 16000 S.F.		3.23	5.15	8.38
1920	66 fixtures per 32000 S.F.		3.05	4.72	7.77

D5020 Lighting and Branch Wiring

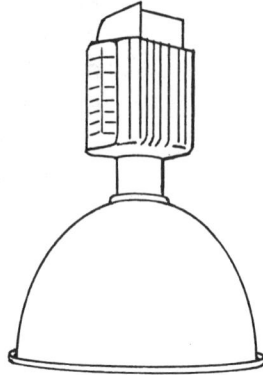

HIGH BAY FIXTURES

B. Metal halide 400 watt

C. High pressure sodium 400 watt

E. Metal halide 1000 watt

F. High pressure sodium 1000 watt

G. Metal halide 1000 watt
125,000 lumen lamp

System Components	QUANTITY	UNIT	COST PER S.F.		
			MAT.	INST.	TOTAL
SYSTEM D5020 230 0240					
HIGH INTENSITY DISCHARGE FIXTURE, 20' ABOVE WORK PLANE					
1 WATT/S.F., TYPE E, 40 FC, 1 FIXTURE 1000 S.F.					
Steel intermediate conduit, (IMC) 1/2" diam	.160	L.F.	.36	.90	1.26
Wire, 600V, type THWN-THHN, copper, solid, #10	.005	C.L.F.	.14	.28	.42
Steel outlet box 4" concrete	.001	Ea.	.02	.03	.05
Steel outlet box plate with stud, 4" concrete	.001	Ea.	.01	.01	.02
Metal halide, hi bay, aluminum reflector, 1000 W lamp	.001	Ea.	.64	.28	.92
TOTAL			1.17	1.50	2.67

D5020 230	H.I.D. Fixture, High Bay, 20' (by Wattage)		COST PER S.F.		
			MAT.	INST.	TOTAL
0190	High intensity discharge fixture, 20' above work plane				
0240	1 watt/S.F., type E, 40 FC, 1 fixture/1000 S.F.		1.17	1.50	2.67
0280	Type G, 50 FC, 1 fixture/1000 S.F.		1.17	1.50	2.67
0320	Type C, 52 FC, 2 fixtures/1000 S.F.	RD5020 -200	1.35	1.72	3.07
0440	2 watt/S.F., type E, 81 FC, 2 fixtures/1000 S.F.	RD5020 -240	2.32	2.98	5.30
0480	Type G, 101 FC, 2 fixtures/1000 S.F.		2.32	2.98	5.30
0520	Type C, 104 FC, 4 fixtures/1000 S.F.		2.70	3.38	6.08
0640	3 watt/S.F., type E, 121 FC, 3 fixtures/1000 S.F.		3.48	4.47	7.95
0680	Type G, 151 FC, 3 fixtures/1000 S.F.		3.48	4.47	7.95
0720	Type C, 155 FC, 6 fixtures/1000 S.F.		4.05	5.10	9.15
0840	4 watt/S.F., type E, 161 FC, 4 fixtures/1000 S.F.		4.64	5.95	10.59
0880	Type G, 202 FC, 4 fixtures/1000 S.F.		4.64	5.95	10.59
0920	Type C, 233 FC, 9 fixtures/1000 S.F.		5.80	7.05	12.85
1040	5 watt/S.F., type E, 202 FC, 5 fixtures/1000 S.F.		5.80	7.45	13.25
1080	Type G, 252 FC, 5 fixtures/1000 S.F.		5.80	7.45	13.25
1120	Type C, 285 FC, 11 fixtures/1000 S.F.		7.15	8.75	15.90

D5020 Lighting and Branch Wiring

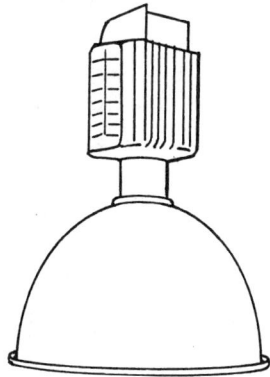

HIGH BAY FIXTURES

B. Metal halide 400 watt

C. High pressure sodium 400 watt

E. Metal halide 1000 watt

F. High pressure sodium 1000 watt

G. Metal halide 1000 watt
   125,000 lumen lamp

System Components	QUANTITY	UNIT	COST PER S.F.		
			MAT.	INST.	TOTAL
SYSTEM D5020 232 1240					
HIGH INTENSITY DISCHARGE FIXTURE, 30' ABOVE WORK PLANE, 100 FC					
TYPE F, 4 FIXTURES PER 900 S.F.					
Steel intermediate conduit, (IMC) 1/2" diam	.580	L.F.	1.30	3.25	4.55
Wire, 600V, type THWN-THHN, copper, solid, #10	.018	C.L.F.	.50	1.01	1.51
Steel outlet box 4" concrete	.004	Ea.	.07	.11	.18
Steel outlet box plate with stud, 4" concrete	.004	Ea.	.03	.03	.06
High pressure sodium, hi bay, aluminum, 1000 W lamp	.004	Ea.	2.36	1.12	3.48
TOTAL			4.26	5.52	9.78

D5020 232	H.I.D. Fixture, High Bay, 30' (by Type)		COST PER S.F.		
			MAT.	INST.	TOTAL
0510	High intensity discharge fixture, 30' above work plane, 100 FC				
1240	Type F, 4 fixtures per 900 S.F.		4.26	5.50	9.76
1280	6 fixtures per 1800 S.F.		3.66	5.25	8.91
1320	8 fixtures per 3000 S.F.	RD5020 -200	3.66	5.25	8.91
1360	9 fixtures per 4000 S.F.		3.06	5	8.06
1400	10 fixtures per 5000 S.F.	RD5020 -240	3.06	5	8.06
1440	17 fixtures per 8000 S.F.		3.06	5	8.06
1480	18 fixtures per 10,000 S.F.		2.91	4.61	7.52
1520	27 fixtures per 16,000 S.F.		2.91	4.61	7.52
1560	52 fixtures per 32000 S.F.		2.88	4.55	7.43
1600	Type G, 4 fixtures per 900 S.F.		4.59	5.85	10.44
1640	6 fixtures per 1800 S.F.		3.83	5.30	9.13
1680	9 fixtures per 3000 S.F.		3.76	5.15	8.91
1720	11 fixtures per 4000 S.F.		3.69	4.97	8.66
1760	13 fixtures per 5000 S.F.		3.69	4.97	8.66
1800	21 fixtures per 8000 S.F.		3.69	4.97	8.66
1840	23 fixtures per 10,000 S.F.		3.32	5.30	8.62
1880	36 fixtures per 16,000 S.F.		3.32	5.30	8.62
1920	70 fixtures per 32,000 S.F.		3.32	5.30	8.62

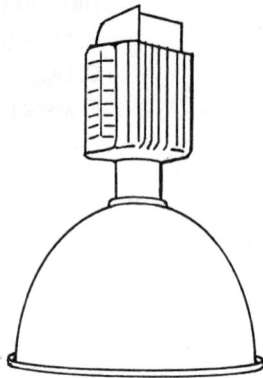

HIGH BAY FIXTURES

B. Metal halide 400 watt

C. High pressure sodium 400 watt

E. Metal halide 1000 watt

F. High pressure sodium 1000 watt

G. Metal halide 1000 watt
125,000 lumen lamp

System Components	QUANTITY	UNIT	COST PER S.F.		
			MAT.	INST.	TOTAL
SYSTEM D5020 234 0240					
HIGH INTENSITY DISCHARGE FIXTURE, 30' ABOVE WORK PLANE					
1 WATT/S.F., TYPE E, 37 FC, 1 FIXTURE/1000 S.F.					
Steel intermediate conduit, (IMC) 1/2" diam	.196	L.F.	.44	1.10	1.54
Wire, 600V type THWN-THHN, copper, solid, #10	.006	C.L.F.	.17	.34	.51
Steel outlet box 4" concrete	.001	Ea.	.02	.03	.05
Steel outlet box plate with stud, 4" concrete	.001	Ea.	.01	.01	.02
Metal halide, hi bay, aluminum reflector, 1000 W lamp	.001	Ea.	.64	.28	.92
TOTAL			1.28	1.76	3.04

D5020 234	H.I.D. Fixture, High Bay, 30' (by Wattage)		COST PER S.F.		
			MAT.	INST.	TOTAL
0190	High intensity discharge fixture, 30' above work plane				
0240	1 watt/S.F., type E, 37 FC, 1 fixture/1000 S.F.		1.28	1.76	3.04
0280	Type G, 45 FC., 1 fixture/1000 S.F.		1.28	1.76	3.04
0320	Type F, 50 FC, 1 fixture/1000 S.F.	RD5020 -200	1.06	1.37	2.43
0440	2 watt/S.F., type E, 74 FC, 2 fixtures/1000 S.F.	RD5020 -240	2.53	3.50	6.03
0480	Type G, 92 FC, 2 fixtures/1000 S.F.		2.53	3.50	6.03
0520	Type F, 100 FC, 2 fixtures/1000 S.F.		2.13	2.79	4.92
0640	3 watt/S.F., type E, 110 FC, 3 fixtures/1000 S.F.		3.83	5.30	9.13
0680	Type G, 138FC, 3 fixtures/1000 S.F.		3.83	5.30	9.13
0720	Type F, 150 FC, 3 fixtures/1000 S.F.		3.19	4.16	7.35
0840	4 watt/S.F., type E, 148 FC, 4 fixtures/1000 S.F.		5.10	7.05	12.15
0880	Type G, 185 FC, 4 fixtures/1000 S.F.		5.10	7.05	12.15
0920	Type F, 200 FC, 4 fixtures/1000 S.F.		4.29	5.60	9.89
1040	5 watt/S.F., type E, 185 FC, 5 fixtures/1000 S.F.		6.35	8.80	15.15
1080	Type G, 230 FC, 5 fixtures/1000 S.F.		6.35	8.80	15.15
1120	Type F, 250 FC, 5 fixtures/1000 S.F.		5.35	6.95	12.30

D5020 Lighting and Branch Wiring

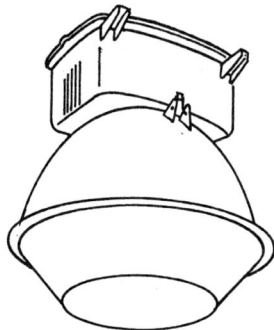

LOW BAY FIXTURES
J. Metal halide 250 watt
K. High pressure sodium 150 watt

System Components	QUANTITY	UNIT	COST PER S.F.		
			MAT.	INST.	TOTAL
SYSTEM D5020 236 0920					
HIGH INTENSITY DISCHARGE FIXTURE, 8'-10' ABOVE WORK PLANE, 50 FC					
TYPE J, 13 FIXTURES PER 1800 S.F.					
Steel intermediate conduit, (IMC) 1/2" diam	.550	L.F.	1.23	3.08	4.31
Wire, 600V, type THWN-THHN, copper, solid, #10	.012	C.L.F.	.33	.67	1
Steel outlet box 4" concrete	.007	Ea.	.11	.20	.31
Steel outlet box plate with stud, 4" concrete	.007	Ea.	.05	.05	.10
Metal halide, lo bay, aluminum reflector, 250 W DX lamp	.007	Ea.	2.73	1.23	3.96
TOTAL			4.45	5.23	9.68

D5020 236	H.I.D. Fixture, Low Bay, 8'-10' (by Type)		COST PER S.F.		
			MAT.	INST.	TOTAL
0510	High intensity discharge fixture, 8'-10' above work plane, 50 FC				
0880	Type J, 7 fixtures per 900 S.F.		4.79	5.25	10.04
0920	13 fixtures per 1800 S.F.	RD5020 -200	4.45	5.25	9.70
0960	21 fixtures per 3000 S.F.		4.45	5.25	9.70
1000	28 fixtures per 4000 S.F.		4.45	5.25	9.70
1040	35 fixtures per 5000 S.F.		4.45	5.25	9.70
1120	62 fixtures per 10,000 S.F.		4.04	5	9.04
1160	99 fixtures per 16,000 S.F.		4.04	5	9.04
1200	199 fixtures per 32,000 S.F.		4.04	5	9.04
1240	Type K, 9 fixtures per 900 S.F.		4.78	4.51	9.29
1280	16 fixtures per 1800 S.F.		4.44	4.36	8.80
1320	26 fixtures per 3000 S.F.		4.41	4.30	8.71
1360	31 fixtures per 4000 S.F.		4.07	4.20	8.27
1400	39 fixtures per 5000 S.F.		4.07	4.20	8.27
1440	62 fixtures per 8000 S.F.		4.07	4.20	8.27
1480	78 fixtures per 10,000 S.F.		4.07	4.20	8.27
1520	124 fixtures per 16,000 S.F.		4.02	4.09	8.11
1560	248 fixtures per 32,000 S.F.		3.93	4.56	8.49

D5020 Lighting and Branch Wiring

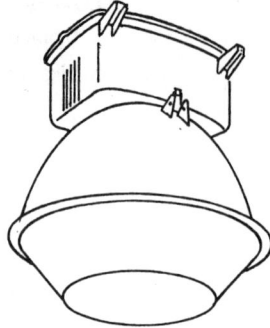

LOW BAY FIXTURES

J. Metal halide 250 watt

K. High pressure sodium 150 watt

System Components	QUANTITY	UNIT	COST PER S.F.		
			MAT.	INST.	TOTAL
SYSTEM D5020 238 0240					
HIGH INTENSITY DISCHARGE FIXTURE, 8'-10' ABOVE WORK PLANE					
1 WATT/S.F., TYPE J, 30 FC, 4 FIXTURES/1000 S.F.					
Steel intermediate conduit, (IMC) 1/2" diam	.280	L.F.	.63	1.57	2.20
Wire, 600V, type THWN-THHN, copper, solid, #10	.008	C.L.F.	.22	.45	.67
Steel outlet box 4" concrete	.004	Ea.	.07	.11	.18
Steel outlet box plate with stud, 4" concrete	.004	Ea.	.03	.03	.06
Metal halide, lo bay, aluminum reflector, 250 W DX lamp	.004	Ea.	1.56	.70	2.26
TOTAL			2.51	2.86	5.37

D5020 238	H.I.D. Fixture, Low Bay, 8'-10' (by Wattage)	COST PER S.F.		
		MAT.	INST.	TOTAL
0190	High intensity discharge fixture, 8'-10' above work plane			
0240	1 watt/S.F., type J, 30 FC, 4 fixtures/1000 S.F.	2.51	2.86	5.37
0280	Type K, 29 FC, 5 fixtures/1000 S.F.	2.53	2.58	5.11
0400	2 watt/S.F., type J, 52 FC, 7 fixtures/1000 S.F.	4.48	5.25	9.73
0440	Type K, 63 FC, 11 fixtures/1000 S.F.	5.45	5.40	10.85
0560	3 watt/S.F., type J, 81 FC, 11 fixtures/1000 S.F.	6.90	7.90	14.80
0600	Type K, 92 FC, 16 fixtures/1000 S.F.	8	7.95	15.95
0720	4 watt/S.F., type J, 103 FC, 14 fixtures/1000 S.F.	8.90	10.40	19.30
0760	Type K, 127 FC, 22 fixtures/1000 S.F.	10.90	10.80	21.70
0880	5 watt/S.F., type J, 133 FC, 18 fixtures/1000 S.F.	11.35	13.15	24.50
0920	Type K, 155 FC, 27 fixtures/1000 S.F.	13.45	13.35	26.80

RD5020 -200

D50 Electrical

D5020 Lighting and Branch Wiring

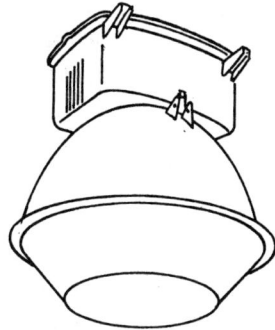

LOW BAY FIXTURES

J. Metal halide 250 watt

K. High pressure sodium 150 watt

System Components	QUANTITY	UNIT	COST PER S.F.		
			MAT.	INST.	TOTAL
SYSTEM D5020 240 0880					
HIGH INTENSITY DISCHARGE FIXTURE, 16' ABOVE WORK PLANE, 50 FC					
TYPE J, 9 FIXTURES PER 900 S.F.					
Steel intermediate conduit, (IMC) 1/2" diam	.630	L.F.	1.41	3.53	4.94
Wire, 600V type, THWN-THHN, copper, solid, #10	.012	C.L.F.	.33	.67	1
Steel outlet box 4" concrete	.010	Ea.	.16	.28	.44
Steel outlet box plate with stud, 4" concrete	.010	Ea.	.07	.07	.14
Metal halide, lo bay, aluminum reflector, 250 W DX lamp	.010	Ea.	3.90	1.75	5.65
TOTAL			5.87	6.30	12.17

D5020 240	H.I.D. Fixture, Low Bay, 16' (by Type)	COST PER S.F.		
		MAT.	INST.	TOTAL
0510	High intensity discharge fixture, 16' above work plane, 50 FC			
0880	Type J, 9 fixtures per 900 S.F.	5.85	6.30	12.15
0920	14 fixtures per 1800 S.F.	5.05	5.95	11
0960	24 fixtures per 3000 S.F.	5.10	6.05	11.15
1000	32 fixtures per 4000 S.F.	5.10	6.05	11.15
1040	35 fixtures per 5000 S.F.	4.73	5.90	10.63
1080	56 fixtures per 8000 S.F.	4.73	5.90	10.63
1120	70 fixtures per 10,000 S.F.	4.73	5.90	10.63
1160	111 fixtures per 16,000 S.F.	4.74	5.90	10.64
1200	222 fixtures per 32,000 S.F.	4.74	5.90	10.64
1240	Type K, 11 fixtures per 900 S.F.	5.95	5.90	11.85
1280	20 fixtures per 1800 S.F.	5.45	5.45	10.90
1320	29 fixtures per 3000 S.F.	5.15	5.35	10.50
1360	39 fixtures per 4000 S.F.	5.10	5.30	10.40
1400	44 fixtures per 5000 S.F.	4.79	5.20	9.99
1440	62 fixtures per 8000 S.F.	4.79	5.20	9.99
1480	87 fixtures per 10,000 S.F.	4.79	5.20	9.99
1520	138 fixtures per 16,000 S.F.	4.79	5.20	9.99

D50 Electrical

D5020 Lighting and Branch Wiring

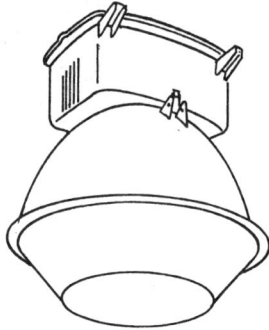

LOW BAY FIXTURES

J. Metal halide 250 watt

K. High pressure sodium 150 watt

System Components			COST PER S.F.		
	QUANTITY	UNIT	MAT.	INST.	TOTAL
SYSTEM D5020 242 0240					
HIGH INTENSITY DISCHARGE FIXTURE, 16' ABOVE WORK PLANE					
1 WATT/S.F., TYPE J, 28 FC, 4 FIXTURES/1000 S.F.					
Steel intermediate conduit, (IMC) 1/2" diam	.328	L.F.	.73	1.84	2.57
Wire, 600V, type THWN-THHN, copper, solid, #10	.010	C.L.F.	.28	.56	.84
Steel outlet box 4" concrete	.004	Ea.	.07	.11	.18
Steel outlet box plate with stud, 4" concrete	.004	Ea.	.03	.03	.06
Metal halide, lo bay, aluminum reflector, 250 W DX lamp	.004	Ea.	1.56	.70	2.26
TOTAL			2.67	3.24	5.91

D5020 242	H.I.D. Fixture, Low Bay, 16' (by Wattage)	COST PER S.F.		
		MAT.	INST.	TOTAL
0190	High intensity discharge fixture, mounted 16' above work plane			
0240	1 watt/S.F., type J, 28 FC, 4 fixt./1000 S.F.	2.67	3.24	5.91
0280	Type K, 27 FC, 5 fixt./1000 S.F.	2.97	3.64	6.61
0400	2 watt/S.F., type J, 48 FC, 7 fixt/1000 S.F.	4.91	6.25	11.16
0440	Type K, 58 FC, 11 fixt/1000 S.F.	6.30	7.40	13.70
0560	3 watt/S.F., type J, 75 FC, 11 fixt/1000 S.F.	7.55	9.50	17.05
0600	Type K, 85 FC, 16 fixt/1000 S.F.	9.30	11.05	20.35
0720	4 watt/S.F., type J, 95 FC, 14 fixt/1000 S.F.	9.80	12.55	22.35
0760	Type K, 117 FC, 22 fixt/1000 S.F.	12.60	14.80	27.40
0880	5 watt/S.F., type J, 122 FC, 18 fixt/1000 S.F.	12.50	15.75	28.25
0920	Type K, 143 FC, 27 fixt/1000 S.F.	15.60	18.45	34.05

RD5020 -200 (note near row 0400)

D50 Electrical

D5030 Communications and Security

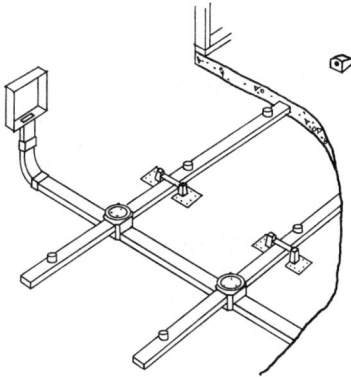

Description: System below includes telephone fitting installed. Does not include cable.

When poke thru fittings and telepoles are used for power, they can also be used for telephones at a negligible additional cost.

System Components	QUANTITY	UNIT	COST PER S.F.		
			MAT.	INST.	TOTAL
SYSTEM D5030 310 0200					
TELEPHONE SYSTEMS, UNDERFLOOR DUCT, 5' ON CENTER, LOW DENSITY					
Underfloor duct 7-1/4" w/insert 2' O.C. 1-3/8" x 7-1/4" super duct	.190	L.F.	6.18	2.13	8.31
Vertical elbow for underfloor superduct, 7-1/4", included					
Underfloor duct conduit adapter, 2" x 1-1/4", included					
Underfloor duct junction box, single duct, 7-1/4" x 3 1/8"	.003	Ea.	1.47	.42	1.89
Underfloor junction box carpet pan	.003	Ea.	1.05	.02	1.07
Underfloor duct outlet, low tension	.004	Ea.	.39	.28	.67
TOTAL			9.09	2.85	11.94

D5030 310	Telephone Systems	COST PER S.F.		
		MAT.	INST.	TOTAL
0200	Telephone systems, underfloor duct, 5' on center, low density	9.10	2.85	11.95
0240	5' on center, high density	9.50	3.13	12.63
0280	7' on center, low density	7.25	2.36	9.61
0320	7' on center, high density	7.65	2.64	10.29
0400	Poke thru fittings, low density	1.06	1.01	2.07
0440	High density	2.12	2.03	4.15
0520	Telepoles, low density	1.08	.60	1.68
0560	High density	2.16	1.21	3.37
0640	Conduit system with floor boxes, low density	1.23	.99	2.22
0680	High density	2.44	1.97	4.41
1020	Telephone wiring for offices & laboratories, 8 jacks/MSF	.56	2.42	2.98

D5030 Communications and Security

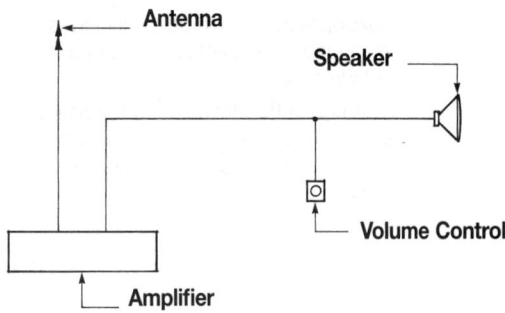

Antenna

Speaker

Volume Control

Amplifier

Sound System Includes AM–FM antenna, outlets, rigid conduit, and copper wire.
Fire Detection System Includes pull stations, signals, smoke and heat detectors, rigid conduit, and copper wire.
Intercom System Includes master and remote stations, rigid conduit, and copper wire.
Master Clock System Includes clocks, bells, rigid conduit, and copper wire.
Master TV Antenna Includes antenna, VHF–UHF reception and distribution, rigid conduit, and copper wire.

System Components	QUANTITY	UNIT	COST EACH		
			MAT.	INST.	TOTAL
SYSTEM D5030 910 0220					
SOUND SYSTEM, INCLUDES OUTLETS, BOXES, CONDUIT & WIRE					
Steel intermediate conduit, (IMC) 1/2″ diam	1200.000	L.F.	2,688	6,720	9,408
Wire sound shielded w/drain, #22-2 conductor	15.500	C.L.F.	465	1,085	1,550
Sound system speakers ceiling or wall	12.000	Ea.	1,452	840	2,292
Sound system volume control	12.000	Ea.	1,080	840	1,920
Sound system amplifier, 250 Watts	1.000	Ea.	1,800	560	2,360
Sound system antenna, AM FM	1.000	Ea.	225	140	365
Sound system monitor panel	1.000	Ea.	400	140	540
Sound system cabinet	1.000	Ea.	875	560	1,435
Steel outlet box 4″ square	12.000	Ea.	31.44	336	367.44
Steel outlet box 4″ plaster rings	12.000	Ea.	37.44	105	142.44
TOTAL			9,053.88	11,326	20,379.88

D5030 910		Communication & Alarm Systems	COST EACH		
			MAT.	INST.	TOTAL
0200	Communication & alarm systems, includes outlets, boxes, conduit & wire				
0210	Sound system, 6 outlets		6,475	7,075	13,550
0220	12 outlets	RD5010 -110	9,050	11,300	20,350
0240	30 outlets		15,700	21,400	37,100
0280	100 outlets		47,800	71,500	119,300
0320	Fire detection systems, non-addressable, 12 detectors		3,133.75	5,873	9,022.50
0360	25 detectors		5,448	9,934.50	15,411
0400	50 detectors		10,719	19,693	30,469
0440	100 detectors		19,850.50	35,804	55,762
0450	Addressable type, 12 detectors		4,221.75	5,923	10,160.50
0452	25 detectors		7,667	10,023.50	17,719
0454	50 detectors		14,936	19,681	34,674
0456	100 detectors		28,638.50	36,132	64,878
0458	Fire alarm control panel, 8 zone		740	1,125	1,865
0459	12 zone		2,525	1,675	4,200
0460	Fire alarm command center, addressable without voice		2,600	980	3,580
0480	Intercom systems, 6 stations		3,625	4,850	8,475
0520	12 stations		7,375	9,650	17,025
0560	25 stations		12,700	18,600	31,300
0600	50 stations		24,500	34,900	59,400
0640	100 stations		48,300	68,000	116,300
0680	Master clock systems, 6 rooms		4,800	7,900	12,700
0720	12 rooms		7,425	13,500	20,925
0760	20 rooms		10,100	19,100	29,200
0800	30 rooms		17,100	35,100	52,200
0840	50 rooms		27,800	59,000	86,800

D50 Electrical

D5030 Communications and Security

D5030 910	Communication & Alarm Systems	COST EACH		
		MAT.	INST.	TOTAL
0880	100 rooms	54,000	117,500	171,500
0920	Master TV antenna systems, 6 outlets	3,250	5,000	8,250
0960	12 outlets	6,050	9,300	15,350
1000	30 outlets	12,400	21,600	34,000
1040	100 outlets	41,100	70,500	111,600

D5030 920	Data Communication	COST PER M.S.F.		
		MAT.	INST.	TOTAL
0100	Data communication system, incl data/voice outlets, boxes, conduit, & cable			
0102	Data and voice system, 2 data/voice outlets per 1000 S.F.	238	415	653
0104	4 data/voice outlets per 1000 S.F.	460	815	1,275
0106	6 data/voice outlets per 1000 S.F.	670	1,200	1,870
0110	8 data/voice outlets per 1000 S.F.	880	1,600	2,480

D5090 Other Electrical Systems

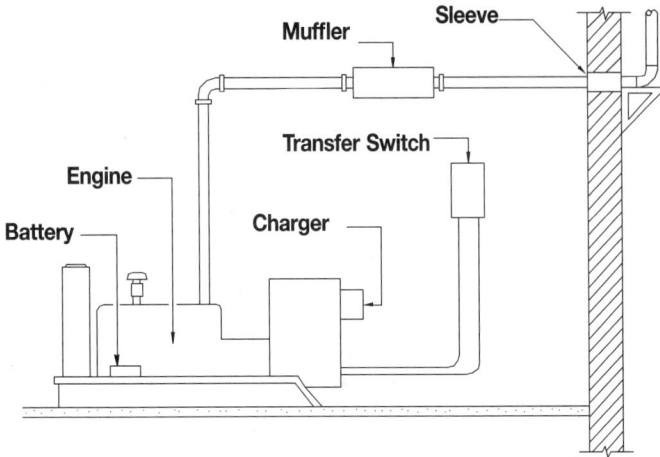

Description: System below tabulates the installed cost for generators by kW. Included in costs are battery, charger, muffler, and transfer switch.

No conduit, wire, or terminations included.

System Components	QUANTITY	UNIT	COST PER kW		
			MAT.	INST.	TOTAL
SYSTEM D5090 210 0200					
GENERATOR SET, INCL. BATTERY, CHARGER, MUFFLER & TRANSFER SWITCH					
GAS/GASOLINE OPER., 3 PHASE, 4 WIRE, 277/480V, 7.5 kW					
Generator set, gas or gasoline operated, 3 ph 4 W, 277/480 V, 7.5 kW	.133	Ea.	1,130	246.13	1,376.13
TOTAL			1,130	246.13	1,376.13

D5090 210	Generators (by kW)		COST PER kW		
			MAT.	INST.	TOTAL
0190	Generator sets, include battery, charger, muffler & transfer switch				
0200	Gas/gasoline operated, 3 phase, 4 wire, 277/480 volt, 7.5 kW		1,125	246	1,371
0240	11.5 kW	RD5010 -110	1,050	187	1,237
0280	20 kW		705	121	826
0320	35 kW		480	79	559
0360	80 kW		345	47.50	392.50
0400	100 kW		305	46.50	351.50
0440	125 kW		495	43.50	538.50
0480	185 kW		445	33	478
0560	Diesel engine with fuel tank, 30 kW		720	92	812
0600	50 kW		530	73	603
0720	125 kW		325	42	367
0760	150 kW		310	39	349
0800	175 kW		289	35	324
0840	200 kW		260	32	292
0880	250 kW		246	26.50	272.50
0920	300 kW		222	23	245
0960	350 kW		214	21.50	235.50
1000	400 kW		231	20	251
1040	500 kW		233	16.95	249.95
1200	750 kW		256	10.65	266.65
1400	1000 kW		237	11.25	248.25

D5090 Special Construction

PV power system, stand alone, AC and DC loads

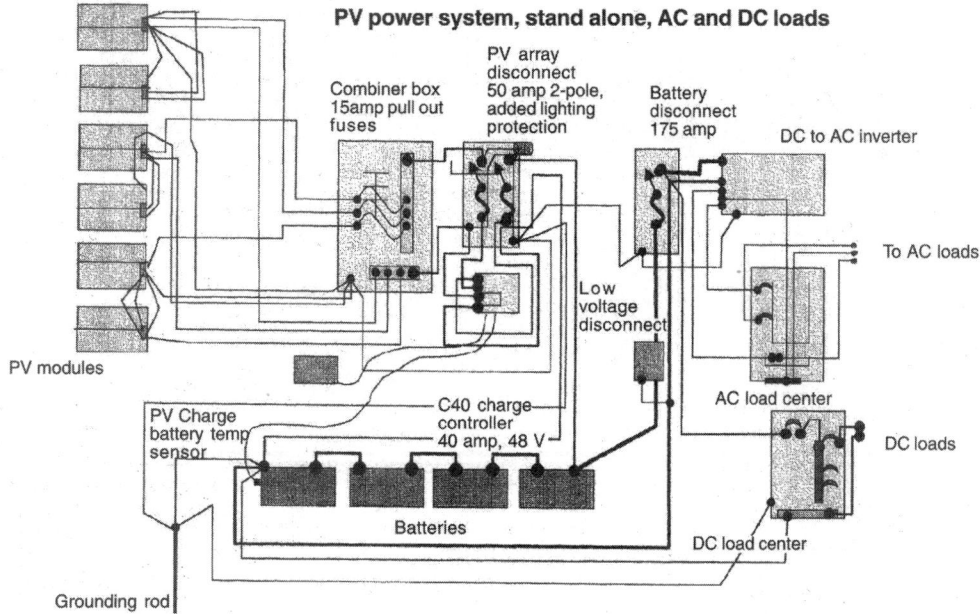

System Components	QUANTITY	UNIT	MAT.	INST.	TOTAL
SYSTEM D5090 420 0100					
PHOTOVOLTAIC POWER SYSTEM, STAND ALONE					
Alternative energy sources, photovoltaic module, 75 watt, 17 volts	12.000	Ea.	6,000	840	6,840
DC to AC inverter for, 48V, 5,500 watt	1.000	Ea.	3,025	280	3,305
PV components, combiner box, 10 lug, NEMA 3R encl	1.000	Ea.	177	140	317
Alternative energy sources, PV components, fuse, 15 A for combiner box	10.000	Ea.	120.50	140	260.50
Battery charger controller w/temp sensor	1.000	Ea.	500	140	640
Digital readout panel, displays hours, volts, amps, etc	1.000	Ea.	88.50	140	228.50
Deep cycle solar battery, 6V, 180 Ah (C/20)	4.000	Ea.	536	280	816
Battery intercon, 15" AWG #2/0, sld w/copper ring lugs	3.000	Ea.	25.65	105	130.65
Batt connec, 24" AWG #2/0, sealed w/copper ring lugs	2.000	Ea.	21.10	70	91.10
Batt connec, 60" AWG #2/0, sealed w/copper ring lugs	2.000	Ea.	27.80	70	97.80
Batt temp computer probe, RJ11 jack, 15' cord	1.000	Ea.	23	35	58
System disconnect, DC 175 Amp circuit breaker	1.000	Ea.	223	70	293
Conduit box for inverter	1.000	Ea.	72.50	70	142.50
Low voltage disconnect	1.000	Ea.	36	70	106
Vented battery enclosure, wood	1.000	Ea.	154	248	402
Grounding, rod, copper clad, 8' long, 5/8" diameter	1.000	Ea.	18	102	120
Grounding, clamp, bronze, 5/8" dia	1.000	Ea.	5.60	17.50	23.10
Bare copper wire, #8 stranded	1.000	C.L.F.	46	51	97
Wire, 600 volt, type THW, copper, stranded, #12	3.600	C.L.F.	64.44	183.60	248.04
Wire, 600 volt, type THW, copper, stranded, #8	1.050	C.L.F.	50.40	73.50	123.90
Rigid galvanized steel conduit, 3/4", including fittings	120.000	L.F.	426	840	1,266
Conduit,to 15' H,incl 2 termn,2 elb&11 bm clp per 100',(EMT), 1" dia	30.000	L.F.	60.60	146.10	206.70
Lightning surge suppressor	1.000	Ea.	55.50	17.50	73
General duty 240 volt, 2 pole NEMA 1, nonfusible, NEMA 3R, 60 amp	1.000	Ea.	283	243	526
Load centers, 3 wire, 120/240V, 100 amp main lugs, indoor, 8 circuits	2.000	Ea.	388	800	1,188
Plug-in panel or load center, 120/240 volt, to 60 amp, 1 pole	5.000	Ea.	67.50	232.50	300
Roof mounting frame, metal, for 6 modules	2.000	Ea.	2,400	230	2,630
Fuses, dual element, time delay, 250 volt, 50 amp	2.000	Ea.	19.80	22.40	42.20
TOTAL			14,914.89	5,657.10	20,571.99

D5090 420		Photovoltaic System, EACH	COST EACH		
			MAT.	INST.	TOTAL
0100		Photovoltaic power system, stand alone, AC and DC loads	14,900	5,650	20,550

D5090 Special Construction

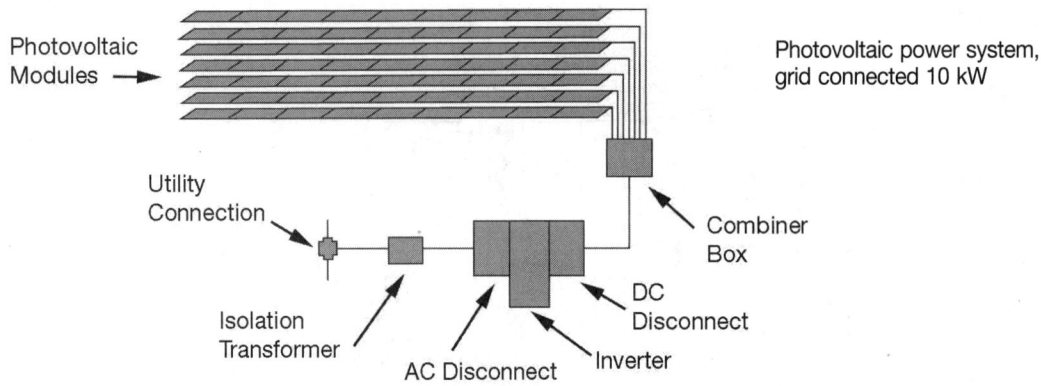

System Components	QUANTITY	UNIT	MAT.	INST.	TOTAL
SYSTEM D5090 430 0100					
PHOTOVOLTAIC POWER SYSTEM, GRID CONNECTED 10 kW					
PV modules, 167 watt, 23.5 V	60.000	Ea.	66,000	4,200	70,200
Mounting frame 6 modules	10.000	Ea.	12,000	1,150	13,150
Steel angle support	400.000	L.F.	220	5,208	5,428
# 12 wire	4.000	C.L.F.	168	204	372
AC Disconnect switch, 60 A	1.000	Ea.	565	254	819
Fuse 75 A	3.000	Ea.	67.50	42	109.50
4/0 wire	2.000	C.L.F.	1,050	392	1,442
Module connection 4/0	4.000	Ea.	40	320	360
Combiner box 10 lug, NEMA 3R	1.000	Ea.	177	140	317
Utility connection, 3 pole breaker	1.000	Ea.	275	90.50	365.50
15 amp fuses	10.000	Ea.	120.50	140	260.50
Enclosure 24" x 24" x 10", NEMA 4	1.000	Ea.	4,000	1,125	5,125
Inverter 12 kW	1.000	Ea.	7,275	840	8,115
Conduit w/fittings & support	200.000	L.F.	404	974	1,378
# 6 ground wire	8.000	C.L.F.	596	688	1,284
Fuse, 60 A	3.000	Ea.	29.85	33.60	63.45
DC disconnect switch, 75 A	1.000	Ea.	1,050	310	1,360
10 kVA isolation tranformer	1.000	Ea.	1,775	700	2,475
# 6 ground wire	.400	C.L.F.	29.80	34.40	64.20
# 6 ground connection	2.000	Ea.	29.60	93	122.60
TOTAL			95,872.25	16,938.50	112,810.75

D5090 430	Photovoltaic System Grid Connected, AC and DC Loads	COST EACH		
		MAT.	INST.	TOTAL
0100	Photovoltaic power system, grid connected 10 kW	96,000	17,000	113,000

D5090 Other Electrical Systems

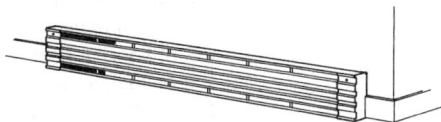

Low Density and Medium Density Baseboard Radiation

The costs shown in Table below are based on the following system considerations:

1. The heat loss per square foot is based on approximately 34 BTU/hr. per S.F. of floor or 10 watts per S.F. of floor.
2. Baseboard radiation is based on the low watt density type rated 187 watts per L.F. and the medium density type rated 250 watts per L.F.
3. Thermostat is not included.
4. Wiring costs include branch circuit wiring.

System Components	QUANTITY	UNIT	COST PER S.F.		
			MAT.	INST.	TOTAL
SYSTEM D5090 510 1000					
ELECTRIC BASEBOARD RADIATION, LOW DENSITY, 900 S.F., 31 MBH, 9 kW					
Electric baseboard radiator, 5' long 935 watt	.011	Ea.	.83	1.08	1.91
Steel intermediate conduit, (IMC) 1/2" diam	.170	L.F.	.38	.95	1.33
Wire 600 volt, type THW, copper, solid, #12	.005	C.L.F.	.09	.26	.35
TOTAL			1.30	2.29	3.59

D5090 510	Electric Baseboard Radiation (Low Density)	COST PER S.F.		
		MAT.	INST.	TOTAL
1000	Electric baseboard radiation, low density, 900 S.F., 31 MBH, 9 kW	1.30	2.29	3.59
1200	1500 S.F., 51 MBH, 15 kW	1.28	2.24	3.52
1400	2100 S.F., 72 MBH, 21 kW	1.14	1.99	3.13
1600	3000 S.F., 102 MBH, 30 kW	1.04	1.81	2.85
2000	Medium density, 900 S.F., 31 MBH, 9 kW	1.15	2.09	3.24
2200	1500 S.F., 51 MBH, 15 kW	1.13	2.04	3.17
2400	2100 S.F., 72 MBH, 21 kW	1.07	1.89	2.96
2600	3000 S.F., 102 MBH, 30 kW	.96	1.71	2.67

D50 Electrical

D5090 Other Electrical Systems

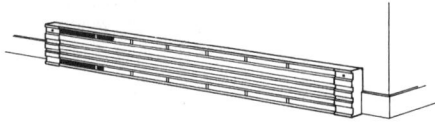

Commercial Duty Baseboard Radiation
The costs shown in Table below are based on the following system considerations:

1. The heat loss per square foot is based on approximately 41 BTU/hr. per S.F. of floor or 12 watts per S.F.
2. The baseboard radiation is of the commercial duty type rated 250 watts per L.F. served by 277 volt, single phase power.
3. Thermostat is not included.
4. Wiring costs include branch circuit wiring.

System Components	QUANTITY	UNIT	COST PER S.F.		
			MAT.	INST.	TOTAL
SYSTEM D5090 520 1000					
ELECTRIC BASEBOARD RADIATION, MEDIUM DENSITY, 1230 S.F., 51 MBH, 15 kW					
Electric baseboard radiator, 5' long	.013	Ea.	.98	1.27	2.25
Steel intermediate conduit, (IMC) 1/2" diam	.154	L.F.	.34	.86	1.20
Wire 600 volt, type THW, copper, solid, #12	.004	C.L.F.	.07	.20	.27
TOTAL			1.39	2.33	3.72

D5090 520	Electric Baseboard Radiation (Medium Density)	COST PER S.F.		
		MAT.	INST.	TOTAL
1000	Electric baseboard radiation, medium density, 1230 SF, 51 MBH, 15 kW	1.39	2.33	3.72
1200	2500 S.F. floor area, 106 MBH, 31 kW	1.29	2.19	3.48
1400	3700 S.F. floor area, 157 MBH, 46 kW	1.27	2.13	3.40
1600	4800 S.F. floor area, 201 MBH, 59 kW	1.18	1.97	3.15
1800	11,300 S.F. floor area, 464 MBH, 136 kW	1.11	1.81	2.92
2000	30,000 S.F. floor area, 1229 MBH, 360 kW	1.09	1.76	2.85

E10 Equipment

E1010 Commercial Equipment

E1010 110	Security/Vault, EACH	COST EACH		
		MAT.	INST.	TOTAL
0100	Bank equipment, drive up window, drawer & mike, no glazing, economy	6,150	1,275	7,425
0110	Deluxe	9,775	2,550	12,325
0120	Night depository, economy	7,975	1,275	9,250
0130	Deluxe	11,300	2,550	13,850
0140	Pneumatic tube systems, 2 station, standard	27,100	4,275	31,375
0150	Teller, automated, 24 hour, single unit	47,000	4,275	51,275
0160	Teller window, bullet proof glazing, 44" x 60"	3,550	780	4,330
0170	Pass through, painted steel, 72" x 40"	3,775	1,600	5,375
0300	Safe, office type, 1 hr. rating, 34" x 20" x 20"	2,400		2,400
0310	4 hr. rating, 62" x 33" x 20"	9,375		9,375
0320	Data storage, 4 hr. rating, 63" x 44" x 16"	14,400		14,400
0330	Jewelers, 63" x 44" x 16"	24,700		24,700
0340	Money, "B" label, 9" x 14" x 14"	580		580
0350	Tool and torch resistive, 24" x 24" x 20"	10,100	196	10,296
0500	Security gates-scissors type, painted steel, single, 6' high, 5-1/2' wide	198	320	518
0510	Double gate, 7-1/2' high, 14' wide	595	635	1,230

E1010 510	Mercantile Equipment, EACH	COST EACH		
		MAT.	INST.	TOTAL
0015	Barber equipment, chair, hydraulic, economy	445	20.50	465.50
0020	Deluxe	3,025	31	3,056
0030	Console, including mirrors, deluxe	390	109	499
0040	Sink, hair washing basin	430	73	503
0100	Checkout counter, single belt	2,725	78.50	2,803.50
0110	Double belt, power take-away	4,600	87	4,687
0200	Display cases, freestanding, glass and aluminum, 3'-6" x 3' x 1'-0" deep	1,325	124	1,449
0210	5'-10" x 4' x 1'-6" deep	2,650	165	2,815
0220	Wall mounted, glass and aluminum, 3' x 4' x 1'-4" deep	2,350	198	2,548
0230	16' x 4' x 1'-4" deep	3,975	660	4,635
0240	Table exhibit, flat, 3' x 4' x 2'-0" wide	1,150	198	1,348
0250	Sloping, 3' x 4' x 3'-0" wide	1,375	330	1,705
0300	Refrigerated food cases, dairy, multi-deck, rear sliding doors, 6 ft. long	12,500	355	12,855
0310	Delicatessen case, service type, 12 ft. long	7,350	274	7,624
0320	Frozen food, chest type, 12 ft. long	7,375	325	7,700
0330	Glass door, reach-in, 5 door	14,100	355	14,455
0340	Meat case, 12 ft. long, single deck	6,075	325	6,400
0350	Multi-deck	10,400	345	10,745
0360	Produce case, 12 ft. long, single deck	8,025	325	8,350
0370	Multi-deck	8,775	345	9,120

E1010 610	Laundry/Dry Cleaning, EACH	COST EACH		
		MAT.	INST.	TOTAL
0100	Laundry equipment, dryers, gas fired, residential, 16 lb. capacity	690	195	885
0110	Commercial, 30 lb. capacity, single	3,400	195	3,595
0120	Dry cleaners, electric, 20 lb. capacity	36,600	5,725	42,325
0130	30 lb. capacity	52,500	7,625	60,125
0140	Ironers, commercial, 120" with canopy, 8 roll	178,500	16,400	194,900
0150	Institutional, 110", single roll	32,700	2,800	35,500
0160	Washers, residential, 4 cycle	880	195	1,075
0170	Commercial, coin operated, deluxe	3,325	195	3,520

E10 Equipment

E1020 Institutional Equipment

E1020 110	Ecclesiastical Equipment, EACH	COST EACH		
		MAT.	INST.	TOTAL
0090	Church equipment, altar, wood, custom, plain	2,100	355	2,455
0100	Granite, custom, deluxe	33,400	4,700	38,100
0110	Baptistry, fiberglass, economy	4,200	1,275	5,475
0120	Bells & carillons, keyboard operation	18,000	8,100	26,100
0130	Confessional, wood, single, economy	2,675	825	3,500
0140	Double, deluxe	16,700	2,475	19,175
0150	Steeples, translucent fiberglas, 30" square, 15' high	6,000	1,675	7,675
0160	Porcelain enamel, custom, 60' high	27,000	11,200	38,200

E1020 130	Ecclesiastical Equipment, L.F.	COST PER L.F.		
		MAT.	INST.	TOTAL
0100	Arch. equip., church equip. pews, bench type, hardwood, economy	81.50	25	106.50
0110	Deluxe	161	33	194

E1020 210	Library Equipment, EACH	COST EACH		
		MAT.	INST.	TOTAL
0110	Library equipment, carrels, metal, economy	265	99	364
0120	Hardwood, deluxe	865	124	989

E1020 230	Library Equipment, L.F.	COST PER L.F.		
		MAT.	INST.	TOTAL
0100	Library equipment, book shelf, metal, single face, 90" high x 10" shelf	124	41.50	165.50
0110	Double face, 90" high x 10" shelf	270	89.50	359.50
0120	Charging desk, built-in, with counter, plastic laminate	286	71	357

E1020 310	Theater and Stage Equipment, EACH	COST EACH		
		MAT.	INST.	TOTAL
0200	Movie equipment, changeover, economy	510		510
0210	Film transport, incl. platters and autowind, economy	5,525		5,525
0220	Lamphouses, incl. rectifiers, xenon, 1000W	7,300	280	7,580
0230	4000W	10,400	375	10,775
0240	Projector mechanisms, 35 mm, economy	12,200		12,200
0250	Deluxe	16,700		16,700
0260	Sound systems, incl. amplifier, single, economy	3,625	620	4,245
0270	Dual, Dolby/super sound	18,200	1,400	19,600
0280	Projection screens, wall hung, manual operation, 50 S.F., economy	325	99	424
0290	Electric operation, 100 S.F., deluxe	2,175	495	2,670
0400	Stage equipment, control boards, incl. dimmers and breakers, economy	12,500	560	13,060
0410	Deluxe	131,000	2,800	133,800
0420	Spotlight, stationary, quartz, 6" lens	144	140	284
0430	Follow, incl. transformer, 2100W	3,375	140	3,515

E1020 320	Theater and Stage Equipment, S.F.	COST PER S.F.		
		MAT.	INST.	TOTAL
0090	Movie equipment, projection screens, rigid in wall, acrylic, 1/4" thick	46	4.79	50.79
0100	1/2" thick	53.50	7.20	60.70
0110	Stage equipment, curtains, velour, medium weight	8.70	1.65	10.35
0120	Silica based yarn, fireproof	16.70	19.80	36.50
0130	Stages, portable with steps, folding legs, 8" high	28.50		28.50
0140	Telescoping platforms, aluminum, deluxe	41.50	26	67.50

E1020 330	Theater and Stage Equipment, L.F.	COST PER L.F.		
		MAT.	INST.	TOTAL
0100	Stage equipment, curtain track, heavy duty	65	55	120
0110	Lights, border, quartz, colored	190	28	218

E1020 Institutional Equipment

E1020 610	Detention Equipment, EACH	COST EACH		
		MAT.	INST.	TOTAL
0110	Detention equipment, cell front rolling door, 7/8" bars, 5' x 7' high	6,425	1,375	7,800
0120	Cells, prefab., including front, 5' x 7' x 7' deep	10,600	1,825	12,425
0130	Doors and frames, 3' x 7', single plate	4,625	680	5,305
0140	Double plate	5,700	680	6,380
0150	Toilet apparatus, incl wash basin	3,700	855	4,555
0160	Visitor cubicle, vision panel, no intercom	3,225	1,375	4,600

E1020 710	Laboratory Equipment, EACH	COST EACH		
		MAT.	INST.	TOTAL
0110	Laboratory equipment, glassware washer, distilled water, economy	6,000	635	6,635
0120	Deluxe	13,400	1,150	14,550
0130	Glove box, fiberglass, bacteriological	18,600		18,600
0140	Radio isotope	18,600		18,600
0150	Titration unit, 4-2000 ml. reservoirs	6,050		6,050

E1020 720	Laboratory Equipment, S.F.	COST PER S.F.		
		MAT.	INST.	TOTAL
0100	Arch. equip., lab equip., counter tops, acid proof, economy	34.50	12.10	46.60
0110	Stainless steel	105	12.10	117.10

E1020 730	Laboratory Equipment, L.F.	COST PER L.F.		
		MAT.	INST.	TOTAL
0110	Laboratory equipment, cabinets, wall, open	142	49.50	191.50
0120	Base, drawer units	450	55	505
0130	Fume hoods, not incl. HVAC, economy	570	184	754
0140	Deluxe incl. fixtures	1,850	415	2,265

E1020 810	Medical Equipment, EACH	COST EACH		
		MAT.	INST.	TOTAL
0100	Dental equipment, central suction system, economy	1,550	490	2,040
0110	Compressor-air, deluxe	7,550	1,000	8,550
0120	Chair, hydraulic, economy	2,925	1,000	3,925
0130	Deluxe	15,100	2,025	17,125
0140	Drill console with accessories, economy	1,650	315	1,965
0150	Deluxe	5,225	315	5,540
0160	X-ray unit, portable	2,400	127	2,527
0170	Panoramic unit	15,300	845	16,145
0300	Medical equipment, autopsy table, standard	9,275	585	9,860
0310	Deluxe	18,400	975	19,375
0320	Incubators, economy	3,575		3,575
0330	Deluxe	21,100		21,100
0700	Station, scrub-surgical, single, economy	4,325	195	4,520
0710	Dietary, medium, with ice	16,800		16,800
0720	Sterilizers, general purpose, single door, 20" x 20" x 28"	16,000		16,000
0730	Floor loading, double door, 28" x 67" x 52"	213,500		213,500
0740	Surgery tables, standard	14,500	910	15,410
0750	Deluxe	32,900	1,275	34,175
0770	Tables, standard, with base cabinets, economy	1,375	330	1,705
0780	Deluxe	4,050	495	4,545
0790	X-ray, mobile, economy	14,000		14,000
0800	Stationary, deluxe	228,000		228,000

E10 Equipment

E1030 Vehicular Equipment

E1030 110	Vehicular Service Equipment, EACH	COST EACH		
		MAT.	INST.	TOTAL
0110	Automotive equipment, compressors, electric, 1-1/2 H.P., std. controls	410	925	1,335
0120	5 H.P., dual controls	2,225	1,375	3,600
0130	Hoists, single post, 4 ton capacity, swivel arms	5,475	3,475	8,950
0140	Dual post, 12 ton capacity, adjustable frame	9,400	9,250	18,650
0150	Lube equipment, 3 reel type, with pumps	8,700	2,775	11,475
0160	Product dispenser, 6 nozzles, w/vapor recovery, not incl. piping, installed	22,900		22,900
0800	Scales, dial type, built in floor, 5 ton capacity, 8' x 6' platform	8,000	2,975	10,975
0810	10 ton capacity, 9' x 7' platform	11,800	4,250	16,050
0820	Truck (including weigh bridge), 20 ton capacity, 24' x 10'	12,300	4,950	17,250
0830	Digital type, truck, 60 ton capacity, 75' x 10'	37,400	12,400	49,800
0840	Concrete foundations, 8' x 6'	1,025	3,750	4,775
0850	70' x 10'	4,400	12,500	16,900

E1030 210	Parking Control Equipment, EACH	COST EACH		
		MAT.	INST.	TOTAL
0110	Parking equipment, automatic gates, 8 ft. arm, one way	4,050	1,025	5,075
0120	Traffic detectors, single treadle	2,600	465	3,065
0130	Booth for attendant, economy	6,400		6,400
0140	Deluxe	27,400		27,400
0150	Ticket printer/dispenser, rate computing	8,875	800	9,675
0160	Key station on pedestal	800	273	1,073

E1030 310	Loading Dock Equipment, EACH	COST EACH		
		MAT.	INST.	TOTAL
0110	Dock bumpers, rubber blocks, 4-1/2" thick, 10" high, 14" long	50	19.05	69.05
0120	6" thick, 20" high, 11" long	147	38	185
0130	Dock boards, H.D., 5' x 5', aluminum, 5000 lb. capacity	1,250		1,250
0140	16,000 lb. capacity	1,650		1,650
0150	Dock levelers, hydraulic, 7' x 8', 10 ton capacity	9,325	1,325	10,650
0160	Dock lifters, platform, 6' x 6', portable, 3000 lb. capacity	9,150		9,150
0170	Dock shelters, truck, scissor arms, economy	1,625	495	2,120
0180	Deluxe	2,100	990	3,090

E1090 Other Equipment

E1090 110	Maintenance Equipment, EACH	MAT.	INST.	TOTAL
0110	Vacuum cleaning, central, residential, 3 valves	1,025	560	1,585
0120	7 valves	1,525	1,275	2,800

E1090 210	Solid Waste Handling Equipment, EACH	COST EACH		
		MAT.	INST.	TOTAL
0110	Waste handling, compactors, single bag, 250 lbs./hr., hand fed	8,775	580	9,355
0120	Heavy duty industrial, 5 C.Y. capacity	20,100	2,775	22,875
0130	Incinerator, electric, 100 lbs./hr., economy	12,800	2,800	15,600
0140	Gas, 2000 lbs./hr., deluxe	554,500	22,300	576,800
0150	Shredder, no baling, 35 tons/hr.	288,500		288,500
0160	Incl. baling, 50 tons/day	576,000		576,000

E1090 350	Food Service Equipment, EACH	COST EACH		
		MAT.	INST.	TOTAL
0110	Kitchen equipment, bake oven, single deck	4,750	132	4,882
0120	Broiler, without oven	3,900	132	4,032
0130	Commercial dish washer, semiautomatic, 50 racks/hr.	9,175	810	9,985
0140	Automatic, 275 racks/hr.	30,200	3,450	33,650
0150	Cooler, beverage, reach-in, 6 ft. long	4,750	176	4,926
0160	Food warmer, counter, 1.65 kw	735		735
0170	Fryers, with submerger, single	1,600	151	1,751
0180	Double	2,625	211	2,836
0185	Ice maker, 1000 lb. per day, with bin	5,250	1,050	6,300
0190	Kettles, steam jacketed, 20 gallons	7,350	238	7,588
0200	Range, restaurant type, burners, 2 ovens and 24″ griddle	8,200	176	8,376
0210	Range hood, incl. carbon dioxide system, economy	4,500	350	4,850
0220	Deluxe	34,800	1,050	35,850

E1090 360	Food Service Equipment, S.F.	COST PER S.F.		
		MAT.	INST.	TOTAL
0110	Refrigerators, prefab, walk-in, 7′-6″ high, 6′ x 6′	146	18.10	164.10
0120	12′ x 20′	92	9.05	101.05

E1090 410	Residential Equipment, EACH	COST EACH		
		MAT.	INST.	TOTAL
0110	Arch. equip., appliances, range, cook top, 4 burner, economy	268	93.50	361.50
0120	Built in, single oven 30″ wide, economy	760	93.50	853.50
0130	Standing, single oven-21″ wide, economy	345	78.50	423.50
0140	Double oven-30″ wide, deluxe	1,800	78.50	1,878.50
0150	Compactor, residential, economy	530	99	629
0160	Deluxe	565	165	730
0170	Dishwasher, built-in, 2 cycles, economy	227	286	513
0180	4 or more cycles, deluxe	1,125	570	1,695
0190	Garbage disposer, sink type, economy	73.50	114	187.50
0200	Deluxe	211	114	325
0210	Refrigerator, no frost, 10 to 12 C.F., economy	490	78.50	568.50
0220	21 to 29 C.F., deluxe	3,250	261	3,511
0300	Washing machine, automatic	1,200	585	1,785

E1090 550	Darkroom Equipment, EACH	COST EACH		
		MAT.	INST.	TOTAL
0110	Darkroom equipment, developing tanks, 2′ x 4′, 5″ deep	5,525	525	6,050
0120	2′ x 9′, 10″ deep	9,375	700	10,075
0130	Combination, tray & tank sinks, washers & dry tables	10,600	2,350	12,950
0140	Dryers, dehumidified filtered air, 3′ x 2′ x 5′-8″ high	5,275	277	5,552

E10 Equipment

E1090 Other Equipment

E1090 550	Darkroom Equipment, EACH	COST EACH		
		MAT.	INST.	TOTAL
0150	Processors, manual, 16" x 20" print size	10,100	830	10,930
0160	Automatic, color print, deluxe	22,700	2,775	25,475
0170	Washers, round, 11" x 14" sheet size	3,375	525	3,900
0180	Square, 50" x 56" sheet size	5,325	1,325	6,650

E1090 610	School Equipment, EACH	COST EACH		
		MAT.	INST.	TOTAL
0110	School equipment, basketball backstops, wall mounted, wood, fixed	1,250	870	2,120
0120	Suspended type, electrically operated	5,350	1,675	7,025
0130	Bleachers-telescoping, manual operation, 15 tier, economy (per seat)	83.50	31	114.50
0140	Power operation, 30 tier, deluxe (per seat)	320	54.50	374.50
0150	Weight lifting gym, universal, economy	435	785	1,220
0160	Deluxe	16,200	1,575	17,775
0170	Scoreboards, basketball, 1 side, economy	2,450	735	3,185
0180	4 sides, deluxe	6,175	10,100	16,275
0800	Vocational shop equipment, benches, metal	405	198	603
0810	Wood	555	198	753
0820	Dust collector, not incl. ductwork, 6' diam.	3,300	530	3,830
0830	Planer, 13" x 6"	1,375	248	1,623

E1090 620	School Equipment, S.F.	COST PER S.F.		
		MAT.	INST.	TOTAL
0110	School equipment, gym mats, naugahyde cover, 2" thick	3.95		3.95
0120	Wrestling, 1" thick, heavy duty	6.05		6.05

E1090 810	Athletic, Recreational, and Therapeutic Equipment, EACH	COST EACH		
		MAT.	INST.	TOTAL
0050	Bolwling alley with gutters	583,000	125,000	708,000
0060	Combo. table and ball rack	7,500		7,500
0070	Bowling alley automatic scorer	131,000		131,000
0110	Sauna, prefabricated, incl. heater and controls, 7' high, 6' x 4'	5,100	755	5,855
0120	10' x 12'	12,300	1,675	13,975
0130	Heaters, wall mounted, to 200 C.F.	665		665
0140	Floor standing, to 1000 C.F., 12500 W	2,775	187	2,962
0610	Shooting range incl. bullet traps, controls, separators, ceilings, economy	13,200	6,550	19,750
0620	Deluxe	40,700	11,100	51,800
0650	Sport court, squash, regulation, in existing building, economy			38,500
0660	Deluxe			42,900
0670	Racketball, regulation, in existing building, economy	41,700	9,400	51,100
0680	Deluxe	45,100	18,800	63,900
0700	Swimming pool equipment, diving stand, stainless steel, 1 meter	6,025	365	6,390
0710	3 meter	10,300	2,475	12,775
0720	Diving boards, 16 ft. long, aluminum	3,375	365	3,740
0730	Fiberglass	2,675	365	3,040
0740	Filter system, sand, incl. pump, 6000 gal./hr.	1,600	650	2,250
0750	Lights, underwater, 12 volt with transformer, 300W	268	560	828
0760	Slides, fiberglass with aluminum handrails & ladder, 6' high, straight	4,000	620	4,620
0780	12' high, straight with platform	5,625	825	6,450

E1090 820	Athletic, Recreational, and Therapeutic Equipment, S.F.	COST PER S.F.		
		MAT.	INST.	TOTAL
0110	Swimming pools, residential, vinyl liner, metal sides	11.60	6.80	18.40
0120	Concrete sides	13.90	12.40	26.30
0130	Gunite shell, plaster finish, 350 S.F.	26	26	52
0140	800 S.F.	21	14.90	35.90
0150	Motel, gunite shell, plaster finish	32	32	64
0160	Municipal, gunite shell, tile finish, formed gutters	122	56	178

E10 Equipment

E1090 Other Equipment

E1090 910	Other Equipment, EACH	COST EACH		
		MAT.	INST.	TOTAL
0200	Steam bath heater, incl. timer and head, single, to 140 C.F.	1,550	490	2,040
0210	Commercial size, to 2500 C.F.	8,825	730	9,555
0900	Wine Cellar, redwood, refrigerated, 6'-8" high, 6'W x 6'D	3,675	660	4,335
0910	6'W x 12'D	7,625	990	8,615

E20 Furnishings

E2010 Fixed Furnishings

E2010 310	Window Treatment, EACH	COST EACH		
		MAT.	INST.	TOTAL
0110	Furnishings, blinds, exterior, aluminum, louvered, 1'-4" wide x 3'-0" long	61	49.50	110.50
0120	1'-4" wide x 6'-8" long	123	55	178
0130	Hemlock, solid raised, 1'-4" wide x 3'-0" long	44.50	49.50	94
0140	1'-4" wide x 6'-9" long	72.50	55	127.50
0150	Polystyrene, louvered, 1'-3" wide x 3'-3" long	30.50	49.50	80
0160	1'-3" wide x 6'-8" long	51	55	106
0200	Interior, wood folding panels, louvered, 7" x 20" (per pair)	50.50	29	79.50
0210	18" x 40" (per pair)	185	29	214

E2010 320	Window Treatment, S.F.	COST PER S.F.		
		MAT.	INST.	TOTAL
0110	Furnishings, blinds-interior, venetian-aluminum, stock, 2" slats, economy	2.02	.84	2.86
0120	Custom, 1" slats, deluxe	7.85	1.13	8.98
0130	Vertical, PVC or cloth, T&B track, economy	5.95	1.08	7.03
0140	Deluxe	12.90	1.24	14.14
0150	Draperies, unlined, economy	3.05		3.05
0160	Lightproof, deluxe	9.15		9.15
0510	Shades, mylar, wood roller, single layer, non-reflective	6.70	.72	7.42
0520	Metal roller, triple layer, heat reflective	11.95	.72	12.67
0530	Vinyl, light weight, 4 ga.	.61	.72	1.33
0540	Heavyweight, 6 ga.	1.84	.72	2.56
0550	Vinyl coated cotton, lightproof decorator shades	2.31	.72	3.03
0560	Woven aluminum, 3/8" thick, light and fireproof	5.90	1.42	7.32

E2010 420	Fixed Floor Grilles and Mats, S.F.	COST PER S.F.		
		MAT.	INST.	TOTAL
0110	Floor mats, recessed, inlaid black rubber, 3/8" thick, solid	21.50	2.53	24.03
0120	Colors, 1/2" thick, perforated	25.50	2.53	28.03
0130	Link-including nosings, steel-galvanized, 3/8" thick	18.60	2.53	21.13
0140	Vinyl, in colors	22.50	2.53	25.03

E2010 510	Fixed Multiple Seating, EACH	COST EACH		
		MAT.	INST.	TOTAL
0110	Seating, painted steel, upholstered, economy	145	28.50	173.50
0120	Deluxe	465	35.50	500.50
0400	Seating, lecture hall, pedestal type, economy	182	45	227
0410	Deluxe	610	68.50	678.50
0500	Auditorium chair, veneer construction	193	45	238
0510	Fully upholstered, spring seat	232	45	277

E20 Furnishings

E2020 Moveable Furnishings

E2020 210	Furnishings/EACH	COST EACH		
		MAT.	INST.	TOTAL
0200	Hospital furniture, beds, manual, economy	800		800
0210	Deluxe	2,750		2,750
0220	All electric, economy	1,450		1,450
0230	Deluxe	4,525		4,525
0240	Patient wall systems, no utilities, economy, per room	1,125		1,125
0250	Deluxe, per room	2,100		2,100
0300	Hotel furnishings, standard room set, economy, per room	2,550		2,550
0310	Deluxe, per room	9,125		9,125
0500	Office furniture, standard employee set, economy, per person	505		505
0510	Deluxe, per person	2,250		2,250
0550	Posts, portable, pedestrian traffic control, economy	114		114
0560	Deluxe	340		340
0700	Restaurant furniture, booth, molded plastic, stub wall and 2 seats, economy	355	248	603
0710	Deluxe	1,225	330	1,555
0720	Upholstered seats, foursome, single-economy	650	99	749
0730	Foursome, double-deluxe	1,500	165	1,665

E2020 220	Furniture and Accessories, L.F.	COST PER L.F.		
		MAT.	INST.	TOTAL
0210	Dormitory furniture, desk top (built-in),laminated plastc, 24"deep, economy	40.50	19.80	60.30
0220	30" deep, deluxe	228	25	253
0230	Dressing unit, built-in, economy	223	82.50	305.50
0240	Deluxe	670	124	794
0310	Furnishings, cabinets, hospital, base, laminated plastic	258	99	357
0320	Stainless steel	475	99	574
0330	Counter top, laminated plastic, no backsplash	44.50	25	69.50
0340	Stainless steel	144	25	169
0350	Nurses station, door type, laminated plastic	299	99	398
0360	Stainless steel	570	99	669
0710	Restaurant furniture, bars, built-in, back bar	190	99	289
0720	Front bar	262	99	361
0910	Wardrobes & coatracks, standing, steel, single pedestal, 30" x 18" x 63"	103		103
0920	Double face rack, 39" x 26" x 70"	105		105
0930	Wall mounted rack, steel frame & shelves, 12" x 15" x 26"	76.50	7.05	83.55
0940	12" x 15" x 50"	47	3.69	50.69

F10 Special Construction

F1010 Special Structures

F1010 120	Air-Supported Structures, S.F.	COST PER S.F.		
		MAT.	INST.	TOTAL
0110	Air supported struc., polyester vinyl fabric, 24oz., warehouse, 5000 S.F.	26.50	.31	26.81
0120	50,000 S.F.	12.35	.25	12.60
0130	Tennis, 7,200 S.F.	24.50	.26	24.76
0140	24,000 S.F.	17.30	.26	17.56
0150	Woven polyethylene, 6 oz., shelter, 3,000 S.F.	7.25	.52	7.77
0160	24,000 S.F.	6.15	.26	6.41
0170	Teflon coated fiberglass, stadium cover, economy	60.50	.14	60.64
0180	Deluxe	71.50	.19	71.69
0190	Air supported storage tank covers, reinf. vinyl fabric, 12 oz., 400 S.F.	24.50	.44	24.94
0200	18,000 S.F.	7.75	.40	8.15

F1010 210	Pre-Engineered Structures, EACH	COST EACH		
		MAT.	INST.	TOTAL
0600	Radio towers, guyed, 40 lb. section, 50' high, 70 MPH basic wind speed	3,600	1,275	4,875
0610	90 lb. section, 400' high, wind load 70 MPH basic wind speed	46,700	13,900	60,600
0620	Self supporting, 60' high, 70 MPH basic wind speed	7,500	2,425	9,925
0630	190' high, wind load 90 MPH basic wind speed	44,600	9,750	54,350
0700	Shelters, aluminum frame, acrylic glazing, 8' high, 3' x 9'	3,475	1,125	4,600
0710	9' x 12'	8,050	1,750	9,800

F1010 320	Other Special Structures, S.F.	COST PER S.F.		
		MAT.	INST.	TOTAL
0110	Swimming pool enclosure, transluscent, freestanding, economy	16.25	4.96	21.21
0120	Deluxe	247	14.15	261.15
0510	Tension structures, steel frame, polyester vinyl fabric, 12,000 S.F.	12.70	2.28	14.98
0520	20,800 S.F.	12.90	2.06	14.96

F1010 330	Special Structures, EACH	COST EACH		
		MAT.	INST.	TOTAL
0110	Kiosks, round, 5' diam., 1/4" fiberglass wall, 8' high	6,375		6,375
0120	Rectangular, 5' x 9', 1" insulated dbl. wall fiberglass, 7'-6" high	11,000		11,000
0200	Silos, concrete stave, industrial, 12' diam., 35' high	46,600	20,500	67,100
0210	25' diam., 75' high	175,000	45,100	220,100
0220	Steel prefab, 30,000 gal., painted, economy	20,900	5,250	26,150
0230	Epoxy-lined, deluxe	43,300	10,500	53,800

F1010 340	Special Structures, S.F.	COST PER S.F.		
		MAT.	INST.	TOTAL
0110	Comfort stations, prefab, mobile on steel frame, economy	51		51
0120	Permanent on concrete slab, deluxe	315	45.50	360.50
0210	Domes, bulk storage, wood framing, wood decking, 50' diam.	42.50	1.67	44.17
0220	116' diam.	25.50	1.93	27.43
0230	Steel framing, metal decking, 150' diam.	30.50	11.10	41.60
0240	400' diam.	25	8.45	33.45
0250	Geodesic, wood framing, wood panels, 30' diam.	30.50	1.79	32.29
0260	60' diam.	23	1.18	24.18
0270	Aluminum framing, acrylic panels, 40' diam.			116
0280	Aluminum panels, 400' diam.			26
0310	Garden house, prefab, wood, shell only, 48 S.F.	36	4.96	40.96
0320	200 S.F.	41	20.50	61.50
0410	Greenhouse, shell-stock, residential, lean-to, 8'-6" long x 3'-10" wide	42.50	29	71.50
0420	Freestanding, 8'-6" long x 13'-6" wide	38	9.20	47.20
0430	Commercial-truss frame, under 2000 S.F., deluxe			34.50
0440	Over 5,000 S.F., economy			23.50
0450	Institutional-rigid frame, under 500 S.F., deluxe			84
0460	Over 2,000 S.F., economy			33.50

F1010 Special Structures

F1010 340	Special Structures, S.F.	COST PER S.F.		
		MAT.	INST.	TOTAL
0510	Hangar, prefab, galv. roof and walls, bottom rolling doors, economy	14.65	4.39	19.04
0520	Electric bifolding doors, deluxe	16.75	5.70	22.45

F10 Special Construction

F1020 Integrated Construction

F1020 110	Integrated Construction, EACH	COST EACH		
		MAT.	INST.	TOTAL
0110	Integrated ceilings, radiant electric, 2' x 4' panel, manila finish	99	22.50	121.50
0120	ABS plastic finish	112	43	155

F1020 120	Integrated Construction, S.F.	COST PER S.F.		
		MAT.	INST.	TOTAL
0110	Integrated ceilings, Luminaire, suspended, 5' x 5' modules, 50% lighted	4.32	11.85	16.17
0120	100% lighted	6.45	21.50	27.95
0130	Dimensionaire, 2' x 4' module tile system, no air bar	2.61	1.98	4.59
0140	With air bar, deluxe	3.83	1.98	5.81
0220	Pedestal access floors, stl pnls, no stringers, vinyl covering, >6000 S.F.	11.45	1.55	13
0230	With stringers, high pressure laminate covering, under 6000 S.F.	10.65	1.65	12.30
0240	Carpet covering, under 6000 S.F.	14.05	1.65	15.70
0250	Aluminum panels, no stringers, no covering	34.50	1.98	36.48

F1020 250	Special Purpose Room, EACH	COST EACH		
		MAT.	INST.	TOTAL
0110	Portable booth, acoustical, 27 db 1000 hz., 15 S.F. floor	3,975		3,975
0120	55 S.F. flr.	8,175		8,175

F1020 260	Special Purpose Room, S.F.	COST PER S.F.		
		MAT.	INST.	TOTAL
0110	Anechoic chambers, 7' high, 100 cps cutoff, 25 S.F.			2,550
0120	200 cps cutoff, 100 S.F.			1,325
0130	Audiometric rooms, under 500 S.F.	56.50	20	76.50
0140	Over 500 S.F.	53.50	16.50	70
0300	Darkrooms, shell, not including door, 240 S.F., 8' high	29.50	8.25	37.75
0310	64 S.F., 12' high	75	15.50	90.50
0510	Music practice room, modular, perforated steel, under 500 S.F.	34.50	14.15	48.65
0520	Over 500 S.F.	29.50	12.40	41.90

F1020 330	Special Construction, L.F.	COST PER L.F.		
		MAT.	INST.	TOTAL
0110	Spec. const., air curtains, shipping & receiving, 8'high x 5'wide, economy	385	116	501
0120	20' high x 8' wide, heated, deluxe	1,375	291	1,666
0130	Customer entrance, 10' high x 5' wide, economy	385	116	501
0140	12' high x 4' wide, heated, deluxe	720	145	865

F10 Special Construction

F1030 Special Construction Systems

F1030 120	Sound, Vibration, and Seismic Construction, S.F.	COST PER S.F.		
		MAT.	INST.	TOTAL
0020	Special construction, acoustical, enclosure, 4" thick, 8 psf panels	35.50	20.50	56
0030	Reverb chamber, 4" thick, parallel walls	50	25	75
0110	Sound absorbing panels, 2'-6" x 8', painted metal	13.30	6.90	20.20
0120	Vinyl faced	10.35	6.20	16.55
0130	Flexible transparent curtain, clear	8	8.10	16.10
0140	With absorbing foam, 75% coverage	11.20	8.10	19.30
0150	Strip entrance, 2/3 overlap	8.25	12.90	21.15
0160	Full overlap	6.95	15.15	22.10
0200	Audio masking system, plenum mounted, over 10,000 S.F.	.79	.25	1.04
0210	Ceiling mounted, under 5,000 S.F.	1.38	.47	1.85

F1030 210	Radiation Protection, EACH	COST EACH		
		MAT.	INST.	TOTAL
0110	Shielding, lead x-ray protection, radiography room, 1/16" lead, economy	8,900	3,375	12,275
0120	Deluxe	10,700	5,600	16,300

F1030 220	Radiation Protection, S.F.	COST PER S.F.		
		MAT.	INST.	TOTAL
0110	Shielding, lead, gypsum board, 5/8" thick, 1/16" lead	8	5.85	13.85
0120	1/8" lead	15.50	6.65	22.15
0130	Lath, 1/16" thick	7.90	6.25	14.15
0140	1/8" thick	15.90	7	22.90
0150	Radio frequency, copper, prefab screen type, economy	34.50	5.50	40
0160	Deluxe	39	6.85	45.85

F1030 910	Other Special Construction Systems, EACH	COST EACH		
		MAT.	INST.	TOTAL
0110	Disappearing stairways, folding, pine, 8'-6" ceiling	215	124	339
0120	9'-6" ceiling	248	124	372
0200	Fire escape, galvanized steel, 8'-0" to 10'-6" ceiling	1,675	990	2,665
0210	10'-6" to 13'-6" ceiling	2,200	990	3,190
0220	Automatic electric, wood, 8' to 9' ceiling	9,075	990	10,065
0230	Aluminum, 14' to 15' ceiling	10,400	1,425	11,825
0300	Fireplace prefabricated, freestanding or wall hung, economy	1,275	380	1,655
0310	Deluxe	3,425	550	3,975
0320	Woodburning stoves, cast iron, economy	1,250	760	2,010
0330	Deluxe	3,025	1,250	4,275

F10 Special Construction

F1040 Special Facilities

F1040 210	Ice Rinks, EACH	COAT EACH		
		MAT.	INST.	TOTAL
0100	Ice skating rink, 85' x 200', 55° system, 5 mos., 100 ton			655,500
0110	90° system, 12 mos., 135 ton			715,000
0120	Dash boards, acrylic screens, polyethylene coated plywood	143,000	33,500	176,500
0130	Fiberglass and aluminum construction	154,000	33,500	187,500

F1040 510	Liquid & Gas Storage Tanks, EACH	COST EACH		
		MAT.	INST.	TOTAL
0100	Tanks, steel, ground level, 100,000 gal.			228,000
0110	10,000,000 gal.			4,124,500
0120	Elevated water, 50,000 gal.		177,500	388,000
0130	1,000,000 gal.		828,500	1,811,000
0150	Cypress wood, ground level, 3,000 gal.	9,675	9,900	19,575
0160	Redwood, ground level, 45,000 gal.	73,500	26,800	100,300

F1040 910	Special Construction, EACH	COST EACH		
		MAT.	INST.	TOTAL
0110	Special construction, bowling alley incl. pinsetter, scorer etc., economy	44,300	9,900	54,200
0120	Deluxe	56,000	11,000	67,000
0130	For automatic scorer, economy, add	9,050		9,050
0140	Deluxe, add	10,900		10,900
0200	Chimney, metal, high temp. steel jacket, factory lining, 24" diameter	11,600	4,675	16,275
0210	60" diameter	84,000	20,300	104,300
0220	Poured concrete, brick lining, 10' diam., 200' high			1,600,000
0230	20' diam., 500' high			7,000,000
0240	Radial brick, 3'-6" I.D., 75' high			268,000
0250	7' I.D., 175' high			800,500
0300	Control tower, modular, 12' x 10', incl. instrumentation, economy			737,000
0310	Deluxe			1,028,500
0400	Garage costs, residential, prefab, wood, single car economy	5,600	990	6,590
0410	Two car deluxe	11,800	1,975	13,775
0500	Hangars, prefab, galv. steel, bottom rolling doors, economy (per plane)	14,200	4,850	19,050
0510	Electrical bi-folding doors, deluxe (per plane)	22,700	6,675	29,375

414

G1030 Site Earthwork

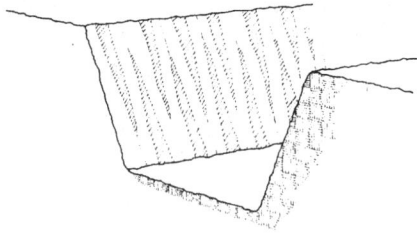

Trenching Systems are shown on a cost per linear foot basis. The systems include: excavation; backfill and removal of spoil; and compaction for various depths and trench bottom widths. The backfill has been reduced to accomodate a pipe of suitable diameter and bedding.

The slope for trench sides varies from none to 1:1.

The Expanded System Listing shows Trenching Systems that range from 2' to 12' in width. Depths range from 2' to 25'.

System Components	QUANTITY	UNIT	COST PER L.F.		
			EQUIP.	LABOR	TOTAL
SYSTEM G1030 805 1310					
TRENCHING COMMON EARTH, NO SLOPE, 2' WIDE, 2' DP, 3/8 C.Y. BUCKET					
Excavation, trench, hyd. backhoe, track mtd., 3/8 C.Y. bucket	.148	B.C.Y.	.32	.88	1.20
Backfill and load spoil, from stockpile	.153	L.C.Y.	.10	.26	.36
Compaction by vibrating plate, 6" lifts, 4 passes	.118	E.C.Y.	.03	.33	.36
Remove excess spoil, 8 C.Y. dump truck, 2 mile roundtrip	.040	L.C.Y.	.12	.13	.25
TOTAL			.57	1.60	2.17

G1030 805	Trenching Common Earth	COST PER L.F.		
		EQUIP.	LABOR	TOTAL
1310	Trenching, common earth, no slope, 2' wide, 2' deep, 3/8 C.Y. bucket	.57	1.60	2.17
1320	3' deep, 3/8 C.Y. bucket	.82	2.42	3.24
1330	4' deep, 3/8 C.Y. bucket	1.06	3.23	4.29
1340	6' deep, 3/8 C.Y. bucket	1.42	4.20	5.62
1350	8' deep, 1/2 C.Y. bucket	1.88	5.60	7.48
1360	10' deep, 1 C.Y. bucket	3.07	6.65	9.72
1400	4' wide, 2' deep, 3/8 C.Y. bucket	1.29	3.16	4.45
1410	3' deep, 3/8 C.Y. bucket	1.76	4.80	6.56
1420	4' deep, 1/2 C.Y. bucket	2.15	5.40	7.55
1430	6' deep, 1/2 C.Y. bucket	3.44	8.60	12.04
1440	8' deep, 1/2 C.Y. bucket	5.70	11.05	16.75
1450	10' deep, 1 C.Y. bucket	6.95	13.75	20.70
1460	12' deep, 1 C.Y. bucket	8.95	17.55	26.50
1470	15' deep, 1-1/2 C.Y. bucket	7.70	15.65	23.35
1480	18' deep, 2-1/2 C.Y. bucket	10.60	22	32.60
1520	6' wide, 6' deep, 5/8 C.Y. bucket w/trench box	7.65	12.70	20.35
1530	8' deep, 3/4 C.Y. bucket	10.15	16.75	26.90
1540	10' deep, 1 C.Y. bucket	9.75	17.35	27.10
1550	12' deep, 1-1/2 C.Y. bucket	10.45	18.65	29.10
1560	16' deep, 2-1/2 C.Y. bucket	14.75	23	37.75
1570	20' deep, 3-1/2 C.Y. bucket	18.85	28	46.85
1580	24' deep, 3-1/2 C.Y. bucket	22.50	33.50	56
1640	8' wide, 12' deep, 1-1/2 C.Y. bucket w/trench box	14.60	23.50	38.10
1650	15' deep, 1-1/2 C.Y. bucket	19.10	31	50.10
1660	18' deep, 2-1/2 C.Y. bucket	21.50	30.50	52
1680	24' deep, 3-1/2 C.Y. bucket	30.50	43	73.50
1730	10' wide, 20' deep, 3-1/2 C.Y. bucket w/trench box	24.50	41	65.50
1740	24' deep, 3-1/2 C.Y. bucket	37	49.50	86.50
1780	12' wide, 20' deep, 3-1/2 C.Y. bucket w/trench box	39	52	91
1790	25' deep, bucket	48	66.50	114.50
1800	1/2 to 1 slope, 2' wide, 2' deep, 3/8 C.Y. bucket	.82	2.42	3.24
1810	3' deep, 3/8 C.Y. bucket	1.36	4.26	5.62
1820	4' deep, 3/8 C.Y. bucket	2.01	6.50	8.51
1840	6' deep, 3/8 C.Y. bucket	3.43	10.55	13.98

416

G10 Site Preparation

G1030 Site Earthwork

G1030 805	Trenching Common Earth	COST PER L.F.		
		EQUIP.	LABOR	TOTAL
1860	8' deep, 1/2 C.Y. bucket	5.45	16.80	22.25
1880	10' deep, 1 C.Y. bucket	10.55	23.50	34.05
2300	4' wide, 2' deep, 3/8 C.Y. bucket	1.53	3.98	5.51
2310	3' deep, 3/8 C.Y. bucket	2.31	6.65	8.96
2320	4' deep, 1/2 C.Y. bucket	3.04	8.20	11.24
2340	6' deep, 1/2 C.Y. bucket	5.75	15.25	21
2360	8' deep, 1/2 C.Y. bucket	11.05	22.50	33.55
2380	10' deep, 1 C.Y. bucket	15.40	31.50	46.90
2400	12' deep, 1 C.Y. bucket	19.75	42.50	62.25
2430	15' deep, 1-1/2 C.Y. bucket	22	45.50	67.50
2460	18' deep, 2-1/2 C.Y. bucket	39.50	70	109.50
2840	6' wide, 6' deep, 5/8 C.Y. bucket w/trench box	11.30	18.90	30.20
2860	8' deep, 3/4 C.Y. bucket	16.40	28.50	44.90
2880	10' deep, 1 C.Y. bucket	15.70	29	44.70
2900	12' deep, 1-1/2 C.Y. bucket	19.90	38	57.90
2940	16' deep, 2-1/2 C.Y. bucket	33.50	54.50	88
2980	20' deep, 3-1/2 C.Y. bucket	47	75	122
3020	24' deep, 3-1/2 C.Y. bucket	67	103	170
3100	8' wide, 12' deep, 1-1/2 C.Y. bucket w/trench box	24.50	43	67.50
3120	15' deep, 1-1/2 C.Y. bucket	36	62.50	98.50
3140	18' deep, 2-1/2 C.Y. bucket	46.50	73	119.50
3180	24' deep, 3-1/2 C.Y. bucket	75	112	187
3270	10' wide, 20' deep, 3-1/2 C.Y. bucket w/trench box	47.50	86	133.50
3280	24' deep, 3-1/2 C.Y. bucket	82.50	122	204.50
3370	12' wide, 20' deep, 3-1/2 C.Y. bucket w/ trench box	69.50	100	169.50
3380	25' deep, 3-1/2 C.Y. bucket	96.50	143	239.50
3500	1 to 1 slope, 2' wide, 2' deep, 3/8 C.Y. bucket	1.06	3.24	4.30
3520	3' deep, 3/8 C.Y. bucket	2.98	7.25	10.23
3540	4' deep, 3/8 C.Y. bucket	2.98	9.75	12.73
3560	6' deep, 3/8 C.Y. bucket	3.43	10.55	13.98
3580	8' deep, 1/2 C.Y. bucket	6.80	21	27.80
3600	10' deep, 1 C.Y. bucket	18	40	58
3800	4' wide, 2' deep, 3/8 C.Y. bucket	1.76	4.80	6.56
3820	3' deep, 3/8 C.Y. bucket	2.86	8.45	11.31
3840	4' deep, 1/2 C.Y. bucket	3.93	11	14.93
3860	6' deep, 1/2 C.Y. bucket	8.10	22	30.10
3880	8' deep, 1/2 C.Y. bucket	16.50	33.50	50
3900	10' deep, 1 C.Y. bucket	24	49	73
3920	12' deep, 1 C.Y. bucket	35	71	106
3940	15' deep, 1-1/2 C.Y. bucket	36	76	112
3960	18' deep, 2-1/2 C.Y. bucket	54.50	96.50	151
4030	6' wide, 6' deep, 5/8 C.Y. bucket w/trench box	14.85	25.50	40.35
4040	8' deep, 3/4 C.Y. bucket	21.50	35	56.50
4050	10' deep, 1 C.Y. bucket	22.50	42.50	65
4060	12' deep, 1-1/2 C.Y. bucket	30.50	57.50	88
4070	16' deep, 2-1/2 C.Y. bucket	52.50	86	138.50
4080	20' deep, 3-1/2 C.Y. bucket	76	122	198
4090	24' deep, 3-1/2 C.Y. bucket	111	172	283
4500	8' wide, 12' deep, 1-1/2 C.Y. bucket w/trench box	34.50	62.50	97
4550	15' deep, 1-1/2 C.Y. bucket	52.50	94	146.50
4600	18' deep, 2-1/2 C.Y. bucket	70.50	113	183.50
4650	24' deep, 3-1/2 C.Y. bucket	119	181	300
4800	10' wide, 20' deep, 3-1/2 C.Y. bucket w/trench box	70	131	201
4850	24' deep, 3-1/2 C.Y. bucket	127	191	318
4950	12' wide, 20' deep, 3-1/2 C.Y. bucket w/ trench box	100	148	248
4980	25' deep, 3-1/2 C.Y. bucket	144	216	360

417

G1030 Site Earthwork

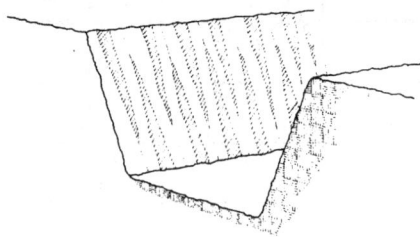

Trenching Systems are shown on a cost per linear foot basis. The systems include: excavation; backfill and removal of spoil; and compaction for various depths and trench bottom widths. The backfill has been reduced to accomodate a pipe of suitable diameter and bedding.

The slope for trench sides varies from none to 1:1.

The Expanded System Listing shows Trenching Systems that range from 2' to 12' in width. Depths range from 2' to 25'.

System Components	QUANTITY	UNIT	COST PER L.F.		
			EQUIP.	LABOR	TOTAL
SYSTEM G1030 806 1310					
TRENCHING LOAM & SANDY CLAY, NO SLOPE, 2' WIDE, 2' DP, 3/8 C.Y. BUCKET					
Excavation, trench, hyd. backhoe, track mtd., 3/8 C.Y. bucket	.148	B.C.Y.	.30	.81	1.11
Backfill and load spoil, from stockpile	.165	L.C.Y.	.11	.29	.40
Compaction by vibrating plate 18" wide, 6" lifts, 4 passes	.118	E.C.Y.	.03	.33	.36
Remove excess spoil, 8 C.Y. dump truck, 2 mile roundtrip	.042	L.C.Y.	.13	.14	.27
TOTAL			.57	1.60	2.17

G1030 806	Trenching Loam & Sandy Clay	COST PER L.F.		
		EQUIP.	LABOR	TOTAL
1310	Trenching, loam & sandy clay, no slope, 2' wide, 2' deep, 3/8 C.Y. bucket	.57	1.57	2.14
1320	3' deep, 3/8 C.Y. bucket	.89	2.59	3.48
1330	4' deep, 3/8 C.Y. bucket	1.04	3.16	4.20
1340	6' deep, 3/8 C.Y. bucket	1.52	3.75	5.27
1350	8' deep, 1/2 C.Y. bucket	2.02	4.97	6.99
1360	10' deep, 1 C.Y. bucket	2.21	5.35	7.56
1400	4' wide, 2' deep, 3/8 C.Y. bucket	1.29	3.10	4.39
1410	3' deep, 3/8 C.Y. bucket	1.75	4.69	6.44
1420	4' deep, 1/2 C.Y. bucket	2.13	5.30	7.43
1430	6' deep, 1/2 C.Y. bucket	3.66	7.70	11.36
1440	8' deep, 1/2 C.Y. bucket	5.55	10.90	16.45
1450	10' deep, 1 C.Y. bucket	5.25	11.15	16.40
1460	12' deep, 1 C.Y. bucket	6.60	13.90	20.50
1470	15' deep, 1-1/2 C.Y. bucket	8.05	16.15	24.20
1480	18' deep, 2-1/2 C.Y. bucket	9.85	17.45	27.30
1520	6' wide, 6' deep, 5/8 C.Y. bucket w/trench box	7.35	12.50	19.85
1530	8' deep, 3/4 C.Y. bucket	9.75	16.40	26.15
1540	10' deep, 1 C.Y. bucket	8.90	16.65	25.55
1550	12' deep, 1-1/2 C.Y. bucket	10.15	18.70	28.85
1560	16' deep, 2-1/2 C.Y. bucket	14.30	23	37.30
1570	20' deep, 3-1/2 C.Y. bucket	17.25	28	45.25
1580	24' deep, 3-1/2 C.Y. bucket	21.50	34	55.50
1640	8' wide, 12' deep, 1-1/4 C.Y. bucket w/trench box	14.35	23.50	37.85
1650	15' deep, 1-1/2 C.Y. bucket	17.55	30	47.55
1660	18' deep, 2-1/2 C.Y. bucket	21.50	33	54.50
1680	24' deep, 3-1/2 C.Y. bucket	29.50	44	73.50
1730	10' wide, 20' deep, 3-1/2 C.Y. bucket w/trench box	29.50	44	73.50
1740	24' deep, 3-1/2 C.Y. bucket	37	54.50	91.50
1780	12' wide, 20' deep, 3-1/2 C.Y. bucket w/trench box	36	52	88
1790	25' deep, bucket	46.50	67.50	114
1800	1/2:1 slope, 2' wide, 2' deep, 3/8 C.Y. BK	.80	2.37	3.17
1810	3' deep, 3/8 C.Y. bucket	1.33	4.16	5.49
1820	4' deep, 3/8 C.Y. bucket	1.98	6.35	8.33
1840	6' deep, 3/8 C.Y. bucket	3.69	9.40	13.09
1860	8' deep, 1/2 C.Y. bucket	5.85	15.05	20.90
1880	10' deep, 1 C.Y. bucket	7.55	18.90	26.45

G10 Site Preparation

G1030   Site Earthwork

G1030 806	Trenching Loam & Sandy Clay	COST PER L.F.		
		EQUIP.	LABOR	TOTAL
2300	4' wide, 2' deep, 3/8 C.Y. bucket	1.52	3.90	5.42
2310	3' deep, 3/8 C.Y. bucket	2.28	6.50	8.78
2320	4' deep, 1/2 C.Y. bucket	2.99	8.05	11.04
2340	6' deep, 1/2 C.Y. bucket	6.15	13.70	19.85
2360	8' deep, 1/2 C.Y. bucket	10.75	22	32.75
2380	10' deep, 1 C.Y. bucket	11.50	25.50	37
2400	12' deep, 1 C.Y. bucket	19.30	42.50	61.80
2430	15' deep, 1-1/2 C.Y. bucket	23	47.50	70.50
2460	18' deep, 2-1/2 C.Y. bucket	39	71	110
2840	6' wide, 6' deep, 5/8 C.Y. bucket w/trench box	10.75	18.95	29.70
2860	8' deep, 3/4 C.Y. bucket	15.75	28	43.75
2880	10' deep, 1 C.Y. bucket	15.90	31.50	47.40
2900	12' deep, 1-1/2 C.Y. bucket	19.80	38.50	58.30
2940	16' deep, 2-1/2 C.Y. bucket	32.50	55	87.50
2980	20' deep, 3-1/2 C.Y. bucket	45.50	76	121.50
3020	24' deep, 3-1/2 C.Y. bucket	64.50	104	168.50
3100	8' wide, 12' deep, 1-1/2 C.Y. bucket w/trench box	24	43	67
3120	15' deep, 1-1/2 C.Y. bucket	32.50	61	93.50
3140	18' deep, 2-1/2 C.Y. bucket	45	74	119
3180	24' deep, 3-1/2 C.Y. bucket	72	114	186
3270	10' wide, 20' deep, 3-1/2 C.Y. bucket w/trench box	58	92	150
3280	24' deep, 3-1/2 C.Y. bucket	80	124	204
3320	12' wide, 20' deep, 3-1/2 C.Y. bucket w/trench box	64	100	164
3380	25' deep, 3-1/2 C.Y. bucket w/trench box	87.50	134	221.50
3500	1:1 slope, 2' wide, 2' deep, 3/8 C.Y. bucket	1.04	3.16	4.20
3520	3' deep, 3/8 C.Y. bucket	1.86	5.95	7.81
3540	4' deep, 3/8 C.Y. bucket	2.91	9.50	12.41
3560	6' deep, 3/8 C.Y. bucket	3.69	9.40	13.09
3580	8' deep, 1/2 C.Y. bucket	9.70	25	34.70
3600	10' deep, 1 C.Y. bucket	12.85	32.50	45.35
3800	2' deep, 3/8 C.Y. bucket	1.75	4.69	6.44
3820	4' wide, 3' deep, 3/8 C.Y. bucket	2.82	8.25	11.07
3840	4' deep, 1/2 C.Y. bucket	3.88	10.80	14.68
3860	6' deep, 1/2 C.Y. bucket	8.65	19.65	28.30
3880	8' deep, 1/2 C.Y. bucket	15.95	33.50	49.45
3900	10' deep, 1 C.Y. bucket	17.75	40	57.75
3920	12' deep, 1 C.Y. bucket	25.50	57	82.50
3940	15' deep, 1-1/2 C.Y. bucket	37.50	78.50	116
3960	18' deep, 2-1/2 C.Y. bucket	53.50	98	151.50
4030	6' wide, 6' deep, 5/8 C.Y. bucket w/trench box	14.15	25.50	39.65
4040	8' deep, 3/4 C.Y. bucket	22	39	61
4050	10' deep, 1 C.Y. bucket	23	46	69
4060	12' deep, 1-1/2 C.Y. bucket	29.50	58	87.50
4070	16' deep, 2-1/2 C.Y. bucket	51	87	138
4080	20' deep, 3-1/2 C.Y. bucket	73.50	124	197.50
4090	24' deep, 3-1/2 C.Y. bucket	107	174	281
4500	8' wide, 12' deep, 1-1/4 C.Y. bucket w/trench box	33.50	63	96.50
4550	15' deep, 1-1/2 C.Y. bucket	48	91.50	139.50
4600	18' deep, 2-1/2 C.Y. bucket	68	114	182
4650	24' deep, 3-1/2 C.Y. bucket	115	184	299
4800	10' wide, 20' deep, 3-1/2 C.Y. bucket w/trench box	86	140	226
4850	24' deep, 3-1/2 C.Y. bucket	123	194	317
4950	12' wide, 20' deep, 3-1/2 C.Y. bucket w/trench box	92	148	240
4980	25' deep, bucket	139	219	358

G1030 Site Earthwork

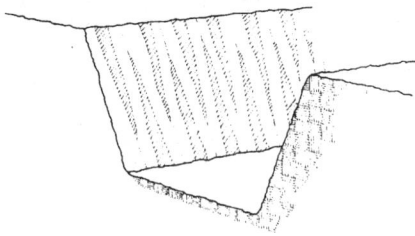

Trenching Systems are shown on a cost per linear foot basis. The systems include: excavation; backfill and removal of spoil; and compaction for various depths and trench bottom widths. The backfill has been reduced to accomodate a pipe of suitable diameter and bedding.

The slope for trench sides varies from none to 1:1.

The Expanded System Listing shows Trenching Systems that range from 2' to 12' in width. Depths range from 2' to 25'.

System Components			COST PER L.F.		
	QUANTITY	UNIT	EQUIP.	LABOR	TOTAL
SYSTEM G1030 807 1310					
TRENCHING SAND & GRAVEL, NO SLOPE, 2' WIDE, 2' DEEP, 3/8 C.Y. BUCKET					
Excavation, trench, hyd. backhoe, track mtd., 3/8 C.Y. bucket	.148	B.C.Y.	.29	.80	1.09
Backfill and load spoil, from stockpile	.140	L.C.Y.	.09	.24	.33
Compaction by vibrating plate 18" wide, 6" lifts, 4 passes	.118	E.C.Y.	.03	.33	.36
Remove excess spoil, 8 C.Y. dump truck, 2 mile roundtrip	.035	L.C.Y.	.11	.12	.23
TOTAL			.52	1.49	2.01

G1030 807	Trenching Sand & Gravel	COST PER L.F.		
		EQUIP.	LABOR	TOTAL
1310	Trenching, sand & gravel, no slope, 2' wide, 2' deep, 3/8 C.Y. bucket	.52	1.49	2.01
1320	3' deep, 3/8 C.Y. bucket	.84	2.49	3.33
1330	4' deep, 3/8 C.Y. bucket	.97	3	3.97
1340	6' deep, 3/8 C.Y. bucket	1.45	3.59	5.04
1350	8' deep, 1/2 C.Y. bucket	1.93	4.76	6.69
1360	10' deep, 1 C.Y. bucket	2.06	5	7.06
1400	4' wide, 2' deep, 3/8 C.Y. bucket	1.18	2.92	4.10
1410	3' deep, 3/8 C.Y. bucket	1.62	4.44	6.06
1420	4' deep, 1/2 C.Y. bucket	1.97	5	6.97
1430	6' deep, 1/2 C.Y. bucket	3.48	7.35	10.83
1440	8' deep, 1/2 C.Y. bucket	5.25	10.35	15.60
1450	10' deep, 1 C.Y. bucket	4.93	10.55	15.48
1460	12' deep, 1 C.Y. bucket	6.20	13.15	19.35
1470	15' deep, 1-1/2 C.Y. bucket	7.60	15.25	22.85
1480	18' deep, 2-1/2 C.Y. bucket	9.30	16.40	25.70
1520	6' wide, 6' deep, 5/8 C.Y. bucket w/trench box	7	11.80	18.80
1530	8' deep, 3/4 C.Y. bucket	9.25	15.60	24.85
1540	10' deep, 1 C.Y. bucket	8.40	15.75	24.15
1550	12' deep, 1-1/2 C.Y. bucket	9.60	17.60	27.20
1560	16' deep, 2 C.Y. bucket	13.70	22	35.70
1570	20' deep, 3-1/2 C.Y. bucket	16.40	26	42.40
1580	24' deep, 3-1/2 C.Y. bucket	20.50	32	52.50
1640	8' wide, 12' deep, 1-1/2 C.Y. bucket w/trench box	13.50	22.50	36
1650	15' deep, 1-1/2 C.Y. bucket	16.45	28.50	44.95
1660	18' deep, 2-1/2 C.Y. bucket	20.50	31	51.50
1680	24' deep, 3-1/2 C.Y. bucket	28	41	69
1730	10' wide, 20' deep, 3-1/2 C.Y. bucket w/trench box	28	41	69
1740	24' deep, 3-1/2 C.Y. bucket	35	51	86
1780	12' wide, 20' deep, 3-1/2 C.Y. bucket w/trench box	33.50	48.50	82
1790	25' deep, bucket	44	63	107
1800	1/2:1 slope, 2' wide, 2' deep, 3/8 CY bk	.75	2.25	3
1810	3' deep, 3/8 C.Y. bucket	1.24	3.96	5.20
1820	4' deep, 3/8 C.Y. bucket	1.85	6.05	7.90
1840	6' deep, 3/8 C.Y. bucket	3.52	9	12.52
1860	8' deep, 1/2 C.Y. bucket	5.60	14.40	20
1880	10' deep, 1 C.Y. bucket	7.05	17.80	24.85

G10 Site Preparation

G1030 Site Earthwork

G1030 807	Trenching Sand & Gravel	COST PER L.F.		
		EQUIP.	LABOR	TOTAL
2300	4' wide, 2' deep, 3/8 C.Y. bucket	1.40	3.68	5.08
2310	3' deep, 3/8 C.Y. bucket	2.12	6.15	8.27
2320	4' deep, 1/2 C.Y. bucket	2.78	7.65	10.43
2340	6' deep, 1/2 C.Y. bucket	5.90	13.10	19
2360	8' deep, 1/2 C.Y. bucket	10.25	21	31.25
2380	10' deep, 1 C.Y. bucket	10.90	24	34.90
2400	12' deep, 1 C.Y. bucket	18.30	40.50	58.80
2430	15' deep, 1-1/2 C.Y. bucket	21.50	44.50	66
2460	18' deep, 2-1/2 C.Y. bucket	37	67	104
2840	6' wide, 6' deep, 5/8 C.Y. bucket w/trench box	10.25	17.95	28.20
2860	8' deep, 3/4 C.Y. bucket	15	26.50	41.50
2880	10' deep, 1 C.Y. bucket	15.10	29.50	44.60
2900	12' deep, 1-1/2 C.Y. bucket	18.80	36	54.80
2940	16' deep, 2 C.Y. bucket	31.50	52	83.50
2980	20' deep, 3-1/2 C.Y. bucket	43	71	114
3020	24' deep, 3-1/2 C.Y. bucket	61.50	97.50	159
3100	8' wide, 12' deep, 1-1/4 C.Y. bucket w/trench box	23	41	64
3120	15' deep, 1-1/2 C.Y. bucket	31	57.50	88.50
3140	18' deep, 2-1/2 C.Y. bucket	42.50	69.50	112
3180	24' deep, 3-1/2 C.Y. bucket	68.50	107	175.50
3270	10' wide, 20' deep, 3-1/2 C.Y. bucket w/trench box	55	86.50	141.50
3280	24' deep, 3-1/2 C.Y. bucket	76	117	193
3370	12' wide, 20' deep, 3-1/2 C.Y. bucket w/trench box	61	96	157
3380	25' deep, bucket	88	134	222
3500	1:1 slope, 2' wide, 2' deep, 3/8 CY bk	1.64	3.68	5.32
3520	3' deep, 3/8 C.Y. bucket	1.74	5.65	7.39
3540	4' deep, 3/8 C.Y. bucket	2.73	9.05	11.78
3560	6' deep, 3/8 C.Y. bucket	3.52	9	12.52
3580	8' deep, 1/2 C.Y. bucket	9.25	24	33.25
3600	10' deep, 1 C.Y. bucket	12.05	30.50	42.55
3800	4' wide, 2' deep, 3/8 C.Y. bucket	1.62	4.44	6.06
3820	3' deep, 3/8 C.Y. bucket	2.62	7.85	10.47
3840	4' deep, 1/2 C.Y. bucket	3.60	10.25	13.85
3860	6' deep, 1/2 C.Y. bucket	8.30	18.85	27.15
3880	8' deep, 1/2 C.Y. bucket	15.20	31.50	46.70
3900	10' deep, 1 C.Y. bucket	16.80	38	54.80
3920	12' deep, 1 C.Y. bucket	24.50	54	78.50
3940	15' deep, 1-1/2 C.Y. bucket	35.50	74	109.50
3960	18' deep, 2-1/2 C.Y. bucket	50.50	92	142.50
4030	6' wide, 6' deep, 5/8 C.Y. bucket w/trench box	13.45	24	37.45
4040	8' deep, 3/4 C.Y. bucket	21	37.50	58.50
4050	10' deep, 1 C.Y. bucket	21.50	43.50	65
4060	12' deep, 1-1/2 C.Y. bucket	28	54.50	82.50
4070	16' deep, 2 C.Y. bucket	49	82.50	131.50
4080	20' deep, 3-1/2 C.Y. bucket	70	116	186
4090	24' deep, 3-1/2 C.Y. bucket	102	163	265
4500	8' wide, 12' deep, 1-1/2 C.Y. bucket w/trench box	32	59.50	91.50
4550	15' deep, 1-1/2 C.Y. bucket	45	86	131
4600	18' deep, 2-1/2 C.Y. bucket	64.50	108	172.50
4650	24' deep, 3-1/2 C.Y. bucket	109	172	281
4800	10' wide, 20' deep, 3-1/2 C.Y. bucket w/trench box	81.50	131	212.50
4850	24' deep, 3-1/2 C.Y. bucket	117	182	299
4950	12' wide, 20' deep, 3-1/2 C.Y. bucket w/trench box	87.50	139	226.50
4980	25' deep, bucket	132	206	338

G1030   Site Earthwork

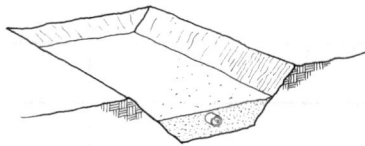

The Pipe Bedding System is shown for various pipe diameters. Compacted bank sand is used for pipe bedding and to fill 12″ over the pipe. No backfill is included. Various side slopes are shown to accommodate different soil conditions. Pipe sizes vary from 6″ to 84″ diameter.

System Components	QUANTITY	UNIT	COST PER L.F.		
			MAT.	INST.	TOTAL
SYSTEM G1030 815 1440					
PIPE BEDDING, SIDE SLOPE 0 TO 1, 1′ WIDE, PIPE SIZE 6″ DIAMETER					
Borrow, bank sand, 2 mile haul, machine spread	.067	C.Y.	.77	.43	1.20
Compaction, vibrating plate	.067	C.Y.		.14	.14
TOTAL			.77	.57	1.34

G1030 815	Pipe Bedding	COST PER L.F.		
		MAT.	INST.	TOTAL
1440	Pipe bedding, side slope 0 to 1, 1′ wide, pipe size 6″ diameter	.77	.57	1.34
1460	2′ wide, pipe size 8″ diameter	1.71	1.27	2.98
1480	Pipe size 10″ diameter	1.76	1.31	3.07
1500	Pipe size 12″ diameter	1.80	1.34	3.14
1520	3′ wide, pipe size 14″ diameter	2.97	2.18	5.15
1540	Pipe size 15″ diameter	3	2.21	5.21
1560	Pipe size 16″ diameter	3.04	2.25	5.29
1580	Pipe size 18″ diameter	3.11	2.30	5.41
1600	4′ wide, pipe size 20″ diameter	4.48	3.32	7.80
1620	Pipe size 21″ diameter	4.53	3.35	7.88
1640	Pipe size 24″ diameter	4.64	3.43	8.07
1660	Pipe size 30″ diameter	4.75	3.52	8.27
1680	6′ wide, pipe size 32″ diameter	8.30	6.15	14.45
1700	Pipe size 36″ diameter	8.50	6.30	14.80
1720	7′ wide, pipe size 48″ diameter	11.10	8.20	19.30
1740	8′ wide, pipe size 60″ diameter	13.85	10.20	24.05
1760	10′ wide, pipe size 72″ diameter	20	14.75	34.75
1780	12′ wide, pipe size 84″ diameter	27	20	47
2140	Side slope 1/2 to 1, 1′ wide, pipe size 6″ diameter	1.63	1.21	2.84
2160	2′ wide, pipe size 8″ diameter	2.70	2	4.70
2180	Pipe size 10″ diameter	2.93	2.17	5.10
2200	Pipe size 12″ diameter	3.13	2.32	5.45
2220	3′ wide, pipe size 14″ diameter	4.49	3.32	7.81
2240	Pipe size 15″ diameter	4.61	3.41	8.02
2260	Pipe size 16″ diameter	4.75	3.52	8.27
2280	Pipe size 18″ diameter	5	3.70	8.70
2300	4′ wide, pipe size 20″ diameter	6.65	4.91	11.56
2320	Pipe size 21″ diameter	6.80	5	11.80
2340	Pipe size 24″ diameter	7.30	5.40	12.70
2360	Pipe size 30″ diameter	8.15	6.05	14.20
2380	6′ wide, pipe size 32″ diameter	12	8.90	20.90
2400	Pipe size 36″ diameter	12.85	9.50	22.35
2420	7′ wide, pipe size 48″ diameter	17.60	13	30.60
2440	8′ wide, pipe size 60″ diameter	23	16.90	39.90
2460	10′ wide, pipe size 72″ diameter	32	23.50	55.50
2480	12′ wide, pipe size 84″ diameter	42.50	31.50	74
2620	Side slope 1 to 1, 1′ wide, pipe size 6″ diameter	2.48	1.84	4.32
2640	2′ wide, pipe size 8″ diameter	3.73	2.75	6.48

G10 Site Preparation

G1030 Site Earthwork

G1030 815	Pipe Bedding	COST PER L.F.		
		MAT.	INST.	TOTAL
2660	Pipe size 10" diameter	4.09	3.01	7.10
2680	Pipe size 12" diameter	4.49	3.32	7.81
2700	3' wide, pipe size 14" diameter	6	4.45	10.45
2720	Pipe size 15" diameter	6.25	4.60	10.85
2740	Pipe size 16" diameter	6.45	4.78	11.23
2760	Pipe size 18" diameter	6.95	5.15	12.10
2780	4' wide, pipe size 20" diameter	8.80	6.50	15.30
2800	Pipe size 21" diameter	9.05	6.70	15.75
2820	Pipe size 24" diameter	9.90	7.30	17.20
2840	Pipe size 30" diameter	11.55	8.55	20.10
2860	6' wide, pipe size 32" diameter	15.75	11.65	27.40
2880	Pipe size 36" diameter	17.20	12.70	29.90
2900	7' wide, pipe size 48" diameter	24	17.80	41.80
2920	8' wide, pipe size 60" diameter	32	23.50	55.50
2940	10' wide, pipe size 72" diameter	44	32.50	76.50
2960	12' wide, pipe size 84" diameter	58	43	101

G2010 Roadways

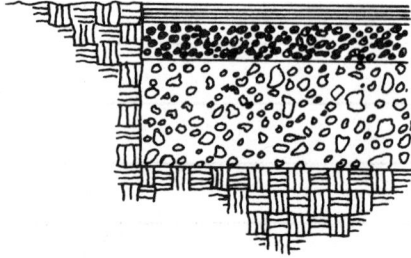

The Roadway System includes surveying; 8″ of compacted bank-run gravel; fine grading with a grader and roller; and bituminous wearing course. Roadway Systems are shown on a cost per linear foot basis.

The Expanded System Listing shows Roadway Systems with widths that vary from 20′ to 34′; for various pavement thicknesses.

System Components	QUANTITY	UNIT	COST PER L.F.		
			MAT.	INST.	TOTAL
SYSTEM G2010 210 1500					
ROADWAYS, BITUMINOUS CONC. PAVING, 2-1/2″ THICK, 20′ WIDE					
4 man surveying crew	.010	Ea.		21.27	21.27
Bank gravel, 2 mi. haul, place and spread by dozer	.494	C.Y.	14.82	3.17	17.99
Compaction, granular material to 98%	.494	C.Y.		2.82	2.82
Grading, fine grade, 3 passes with grader plus rolling	2.220	S.Y.		10.07	10.07
Bituminous, paving, wearing course 2″ - 2-1/2″	2.220	S.Y.	20.87	3.29	24.16
Curbs, granite, split face, straight, 5″ x 16″	2.000	L.F.	26.50	13.84	40.34
Painting lines, reflectorized, 4″ wide	1.000	L.F.	.28	.15	.43
TOTAL			62.47	54.61	117.08

G2010 210	Roadway Pavement	COST PER L.F.		
		MAT.	INST.	TOTAL
1500	Roadways, bituminous conc. paving, 2-1/2″ thick, 20′ wide	62.50	54.50	117
1520	24′ wide	69.50	56.50	126
1540	26′ wide	73	58	131
1560	28′ wide	76.50	60	136.50
1580	30′ wide	80.50	61.50	142
1600	32′ wide	84	63	147
1620	34′ wide	87.50	65	152.50
1800	3″ thick, 20′ wide	66.50	53.50	120
1820	24′ wide	74.50	56.50	131
1840	26′ wide	78.50	58.50	137
1860	28′ wide	82.50	60.50	143
1880	30′ wide	86.50	62.50	149
1900	32′ wide	90.50	64	154.50
1920	34′ wide	94.50	65.50	160
2100	4″ thick, 20′ wide	73	53.50	126.50
2120	24′ wide	82	56.50	138.50
2140	26′ wide	87	58.50	145.50
2160	28′ wide	91.50	60	151.50
2180	30′ wide	96	62.50	158.50
2200	32′ wide	101	64	165

G20 Site Improvements

G2010 Roadways

G2010 210	Roadway Pavement	COST PER L.F.		
		MAT.	INST.	TOTAL
2220	34' wide	105	65.50	170.50
2390				
2400	5" thick, 20' wide	87	54	141
2420	24' wide	99.50	57.50	157
2440	26' wide	105	60	165
2460	28' wide	111	61.50	172.50
2480	30' wide	117	63	180
2500	32' wide	123	65.50	188.50
2520	34' wide	129	67	196
3000	6" thick, 20' wide	95	55.50	150.50
3020	24' wide	109	59.50	168.50
3040	26' wide	115	62	177
3060	28' wide	122	63.50	185.50
3080	30' wide	129	66	195
3100	32' wide	136	68	204
3120	34' wide	143	69.50	212.50
3300	8" thick 20' wide	122	56	178
3320	24' wide	141	60.50	201.50
3340	26' wide	150	62	212
3360	28' wide	159	64.50	223.50
3380	30' wide	169	66.50	235.50
3400	32' wide	179	68.50	247.50
3420	34' wide	188	71	259
3600	12" thick 20' wide	153	59.50	212.50
3620	24' wide	178	64.50	242.50
3640	26' wide	190	67	257
3660	28' wide	203	69.50	272.50
3700	32' wide	228	74.50	302.50
3720	34' wide	241	76.50	317.50

G2010 305	Curbs & Berms	COST PER L.F.		
		MAT.	INST.	TOTAL
1000	Bituminous, curbs, 8" wide, 6" high	1.36	1.89	3.25
1100	8" high	1.57	2.10	3.67
1500	Berm, 12" wide, 3" to 6" high	1.75	2.71	4.46
1600	1-1/2" to 4" high	1.06	1.89	2.95
2000	Concrete, curb, 6" wide, 18" high, cast-in-place	5.30	5.70	11
2100	Precast	11.55	5.70	17.25
2500	Curb and gutter, monolithic, 6" high, 24" wide	19.05	7.60	26.65
2600	30" wide	21	8.40	29.40
3000	Granite, curb, 4-1/2" wide, 12" high precast	6.65	13.30	19.95
3100	5" wide, 16" high	13.25	6.90	20.15
3200	6" wide, 18" high	17.40	7.65	25.05
3290				

G2020 Parking Lots

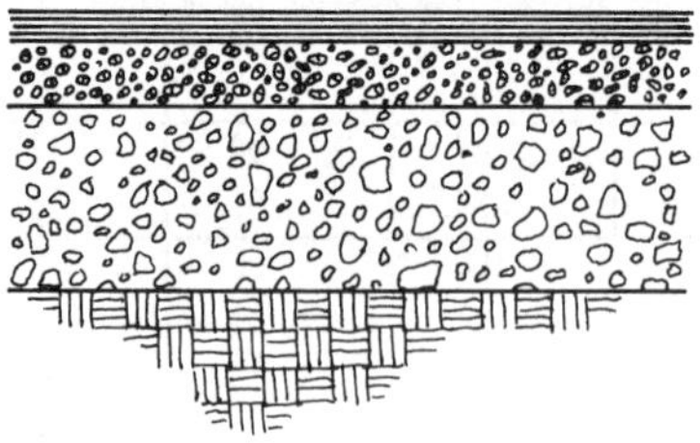

The Parking Lot System includes: compacted bank-run gravel; fine grading with a grader and roller; and bituminous concrete wearing course. All Parking Lot systems are on a cost per car basis. There are three basic types of systems: 90° angle, 60° angle, and 45° angle. The gravel base is compacted to 98%. Final stall design and lay-out of the parking lot with precast bumpers, sealcoating and white paint is also included.

The Expanded System Listing shows the three basic parking lot types with various depths of both gravel base and wearing course. The gravel base depths range from 6″ to 10″. The bituminous paving wearing course varies from a depth of 3″ to 6″.

System Components	QUANTITY	UNIT	COST PER CAR MAT.	INST.	TOTAL
SYSTEM G2020 210 1500					
PARKING LOT, 90° ANGLE PARKING, 3″ BITUMINOUS PAVING, 6″ GRAVEL BASE					
Surveying crew for layout, 4 man crew	.020	Day		42.54	42.54
Borrow, bank run gravel, haul 2 mi., spread w/dozer, no compaction	7.223	L.C.Y.	216.69	46.30	262.99
Grading, fine grade 3 passes with motor grader	43.333	S.Y.		82.06	82.06
Compact w/ vibrating plate, 8″ lifts, granular mat'l. to 98%	7.223	E.C.Y.		41.31	41.31
Bituminous paving, 3″ thick	43.333	S.Y.	487.50	71.93	559.43
Seal coating, petroleum resistant under 1,000 S.Y.	43.333	S.Y.	58.93	49.40	108.33
Painting lines on pavement, parking stall, white	1.000	Ea.	6.60	6.67	13.27
Precast concrete parking bar, 6″ x 10″ x 6′-0″	1.000	Ea.	64.50	16.55	81.05
TOTAL			834.22	356.76	1,190.98

G2020 210	Parking Lots	COST PER CAR MAT.	INST.	TOTAL
1500	Parking lot, 90° angle parking, 3″ bituminous paving, 6″ gravel base	835	355	1,190
1520	8″ gravel base	905	385	1,290
1540	10″ gravel base	980	415	1,395
1560	4″ bituminous paving, 6″ gravel base (RG2020 -500)	960	355	1,315
1580	8″ gravel base	1,025	385	1,410
1600	10″ gravel base	1,100	415	1,515
1620	6″ bituminous paving, 6″ gravel base	1,225	370	1,595
1640	8″ gravel base	1,300	400	1,700
1661	10″ gravel base	1,375	430	1,805
1800	60° angle parking, 3″ bituminous paving, 6″ gravel base	835	355	1,190
1820	8″ gravel base	905	385	1,290
1840	10″ gravel base	980	415	1,395
1860	4″ bituminous paving, 6″ gravel base	960	355	1,315
1880	8″ gravel base	1,025	385	1,410
1900	10″ gravel base	1,100	415	1,515
1920	6″ bituminous paving, 6″ gravel base	1,225	370	1,595
1940	8″ gravel base	1,300	400	1,700
1961	10″ gravel base	1,375	430	1,805
2200	45° angle parking, 3″ bituminous paving, 6″ gravel base	855	365	1,220
2220	8″ gravel base	930	395	1,325
2240	10″ gravel base	1,000	425	1,425
2260	4″ bituminous paving, 6″ gravel base	980	365	1,345
2280	8″ gravel base	1,050	395	1,445
2300	10″ gravel base	1,125	425	1,550
2320	6″ bituminous paving, 6″ gravel base	1,275	380	1,655
2340	8″ gravel base	1,350	410	1,760
2361	10″ gravel base	1,425	625	2,050

G20 Site Improvements

G2030 Pedestrian Paving

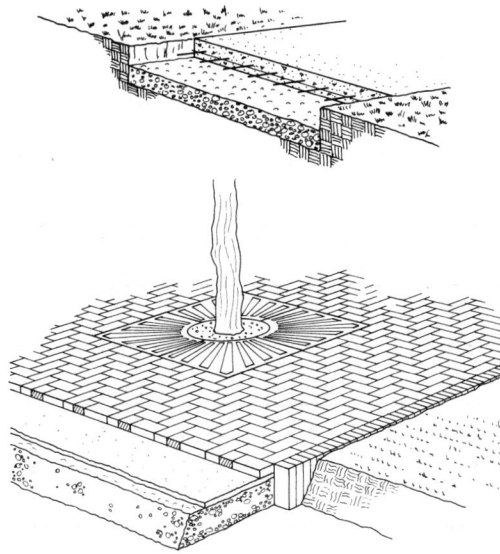

The Bituminous and Concrete Sidewalk Systems include excavation, hand grading and compacted gravel base. Pavements are shown for two conditions of each of the following variables; pavement thickness, gravel base thickness and pavement width. Costs are given on a linear foot basis.

The Plaza Systems listed include several brick and tile paving surfaces on two different bases: gravel and slab on grade. The type of bedding for the pavers depends on the base being used, and alternate bedding may be desirable. Also included in the paving costs are edging and precast grating costs and where concrete bases are involved, expansion joints. Costs are given on a square foot basis.

G2030 110	Bituminous Sidewalks	COST PER L.F.		
		MAT.	INST.	TOTAL
1580	Bituminous sidewalk, 1" thick paving, 4" gravel base, 3' width	2.22	3.76	5.98
1600	4' width	2.95	4.09	7.04
1640	6" gravel base, 3' width	2.79	3.95	6.74
1660	4' width	3.70	4.35	8.05
2120	2" thick paving, 4" gravel base, 3' width	3.32	4.55	7.87
2140	4' width	4.42	5.05	9.47
2180	6" gravel base, 3' width	3.89	4.74	8.63
2201	4' width	5.15	5.35	10.50

G2030 120	Concrete Sidewalks	COST PER L.F.		
		MAT.	INST.	TOTAL
1580	Concrete sidewalk, 4" thick, 4" gravel base, 3' wide	6.65	10.10	16.75
1600	4' wide	8.85	12.50	21.35
1640	6" gravel base, 3' wide	7.25	10.30	17.55
1660	4' wide	9.60	12.75	22.35
2120	6" thick concrete, 4" gravel base, 3' wide	9.75	11.55	21.30
2140	4' wide	13	14.30	27.30
2180	6" gravel base, 3' wide	10.30	11.75	22.05
2201	4' wide	13.75	15.15	28.90

G2030 150	Brick & Tile Plazas	COST PER S.F.		
		MAT.	INST.	TOTAL
2050	Brick pavers, 4" x 8" x 2-1/4", aggregate base, course washed sand bedding	8	5.10	13.10
2100	Slab on grade, asphalt bedding	10.45	7.40	17.85
3550	Concrete paving stone, 4" x 8" x 2-1/2", gravel base, sand bedding	4.62	3.56	8.18
3600	Slab on grade, asphalt bedding	6.55	5.40	11.95
4050	Concrete patio blocks, 8" x 16" x 2", gravel base, sand bedding	4.42	4.66	9.08
4100	Slab on grade, asphalt bedding	6.70	6.85	13.55
6050	Granite pavers, 3-1/2" x 3-1/2" x 3-1/2", gravel base, sand bedding	13.85	11.40	25.25
6101	Slab on grade, mortar bedding	16.15	15.20	31.35

G2040 Site Development

G2040 105	Fence & Guardrails	COST PER L.F.		
		MAT.	INST.	TOTAL
1000	Fence, chain link, 2" post, 1-5/8" rail, 9 ga. galv. wire, 5' high	14.30	4.62	18.92
1050	6' high w/barb. wire	16.80	5.80	22.60
1100	6 ga. galv. wire, 6' high	25.50	5.55	31.05
1150	6' high w/ barb. wire	26.50	5.80	32.30
1200	1-3/8" rail, 11 ga., galv. wire, 10' high	15.60	12.55	28.15
1250	12' high	18.40	14	32.40
1300	9 ga. vinyl covered wire, 10' high	18.45	12.55	31
1350	12' high	30.50	14	44.50
1500	1-5/8" post, 1-3/8" rail, 11 ga. galv. wire, 3' high	4.86	2.77	7.63
1550	4' high	5.60	3.47	9.07
1600	9 ga. vinyl covered wire, 3' high	5.25	2.77	8.02
1650	4' high	6.10	3.47	9.57
2000	Guardrail, corrugated steel, galvanized steel posts	24.50	2.80	27.30
2100	Timber, 6" x 8" posts, 4" x 8" rails	28.50	2.48	30.98
2200	Steel box beam, 6" x 6" rails	35.50	19.85	55.35
2250	6" x 8" rails	47	11.05	58.05
3000	Barrier, median, precast concrete, single face	57	10.55	67.55
3050	Double face	65	11.75	76.75

G2040 810	Flagpoles	COST EACH		
		MAT.	INST.	TOTAL
0110	Flagpoles, on grade, aluminum, tapered, 20' high	1,075	575	1,650
0120	70' high	8,675	1,425	10,100
0130	Fiberglass, tapered, 23' high	1,200	575	1,775
0140	59' high	6,950	1,275	8,225
0150	Concrete, internal halyard, 20' high	1,725	455	2,180
0160	100' high	17,100	1,150	18,250

G2040 950	Other Site Development, EACH	COST EACH		
		MAT.	INST.	TOTAL
0110	Grandstands, permanent, closed deck, steel, economy (per seat)			60.50
0120	Deluxe (per seat)			121
0130	Composite design, economy (per seat)			143
0140	Deluxe (per seat)			286

G3010 Water Supply

G3010 110	Water Distribution Piping	COST PER L.F.		
		MAT.	INST.	TOTAL
2000	Piping, excav. & backfill excl., ductile iron class 250, mech. joint			
2130	4" diameter	16.85	15.15	32
2150	6" diameter	19.55	19	38.55
2160	8" diameter	21.50	23	44.50
2170	10" diameter	29.50	27	56.50
2180	12" diameter	36	29	65
2210	16" diameter	50	41.50	91.50
2220	18" diameter	63	44	107
3000	Tyton joint			
3130	4" diameter	12.65	7.60	20.25
3150	6" diameter	16.15	9.10	25.25
3160	8" diameter	22.50	15.15	37.65
3170	10" diameter	31.50	16.70	48.20
3180	12" diameter	33.50	19	52.50
3210	16" diameter	59.50	27	86.50
3220	18" diameter	66	30	96
3230	20" diameter	72.50	34	106.50
3250	24" diameter	86	39.50	125.50
4000	Copper tubing, type K			
4050	3/4" diameter	9.65	2.63	12.28
4060	1" diameter	12.60	3.29	15.89
4080	1-1/2" diameter	20.50	3.98	24.48
4090	2" diameter	31.50	4.58	36.08
4110	3" diameter	64.50	7.85	72.35
4130	4" diameter	106	11.10	117.10
4150	6" diameter	269	20.50	289.50
5000	Polyvinyl chloride class 160, S.D.R. 26			
5130	1-1/2" diameter	.37	.98	1.35
5150	2" diameter	.56	1.08	1.64
5160	4" diameter	1.82	3.51	5.33
5170	6" diameter	3.63	4.16	7.79
5180	8" diameter	6.25	5.05	11.30
6000	Polyethylene 160 psi, S.D.R. 7			
6050	3/4" diameter	.52	1.41	1.93
6060	1" diameter	.81	1.52	2.33
6080	1-1/2" diameter	1.76	1.64	3.40
6090	2" diameter	3.04	2.02	5.06

G30 Site Mechanical Utilities

G3020 Sanitary Sewer

G3020 110	Drainage & Sewage Piping	COST PER L.F.		
		MAT.	INST.	TOTAL
2000	Piping, excavation & backfill excluded, PVC, plain			
2130	4" diameter	1.83	3.51	5.34
2150	6" diameter	3.74	3.76	7.50
2160	8" diameter	7.75	3.92	11.67
2900	Box culvert, precast, 8' long			
3000	6' x 3'	315	28.50	343.50
3020	6' x 7'	475	32	507
3040	8' x 3'	430	30	460
3060	8' x 8'	585	40	625
3080	10' x 3'	640	36	676
3100	10' x 8'	730	50	780
3120	12' x 3'	625	40	665
3140	12' x 8'	1,075	59.50	1,134.50
4000	Concrete, nonreinforced			
4150	6" diameter	6.70	10.45	17.15
4160	8" diameter	7.35	12.40	19.75
4170	10" diameter	8.15	12.85	21
4180	12" diameter	10	13.85	23.85
4200	15" diameter	11.70	15.45	27.15
4220	18" diameter	14.35	19.30	33.65
4250	24" diameter	21.50	27.50	49
4400	Reinforced, no gasket			
4580	12" diameter	11.20	18.50	29.70
4600	15" diameter	16.45	18.50	34.95
4620	18" diameter	19.20	21	40.20
4650	24" diameter	31.50	27.50	59
4670	30" diameter	46	43.50	89.50
4680	36" diameter	59	53	112
4690	42" diameter	85.50	58	143.50
4700	48" diameter	124	65.50	189.50
4720	60" diameter	222	87	309
4730	72" diameter	247	105	352
4740	84" diameter	280	131	411
4800	With gasket			
4980	12" diameter	15.55	10.20	25.75
5000	15" diameter	19.30	10.70	30
5020	18" diameter	26	11.30	37.30
5050	24" diameter	39.50	12.60	52.10
5070	30" diameter	55.50	43.50	99
5080	36" diameter	70.50	53	123.50
5090	42" diameter	114	52.50	166.50
5100	48" diameter	142	65.50	207.50
5120	60" diameter	198	84	282
5130	72" diameter	276	105	381
5140	84" diameter	335	174	509
5700	Corrugated metal, alum. or galv. bit. coated			
5760	8" diameter	13.95	8.45	22.40
5770	10" diameter	16.70	10.70	27.40
5780	12" diameter	20	13.25	33.25
5800	15" diameter	24.50	13.85	38.35
5820	18" diameter	31	14.60	45.60
5850	24" diameter	38	17.35	55.35
5870	30" diameter	50.50	32	82.50
5880	36" diameter	74	32	106
5900	48" diameter	113	38	151
5920	60" diameter	146	55.50	201.50
5930	72" diameter	217	93	310
6000	Plain			

G30 Site Mechanical Utilities

G3020 Sanitary Sewer

G3020 110	Drainage & Sewage Piping	COST PER L.F.		
		MAT.	INST.	TOTAL
6060	8" diameter	11.40	7.80	19.20
6070	10" diameter	12.50	9.90	22.40
6080	12" diameter	14.25	12.60	26.85
6100	15" diameter	18.15	12.60	30.75
6120	18" diameter	24	13.55	37.55
6140	24" diameter	35	15.85	50.85
6170	30" diameter	44.50	29	73.50
6180	36" diameter	72.50	29	101.50
6200	48" diameter	97	34.50	131.50
6220	60" diameter	153	53.50	206.50
6230	72" diameter	214	130	344
6300	Steel or alum. oval arch, coated & paved invert			
6400	15" equivalent diameter	43.50	13.85	57.35
6420	18" equivalent diameter	56	18.50	74.50
6450	24" equivalent diameter	80.50	22	102.50
6470	30" equivalent diameter	98	27.50	125.50
6480	36" equivalent diameter	146	38	184
6490	42" equivalent diameter	176	42	218
6500	48" equivalent diameter	195	50.50	245.50
6600	Plain			
6700	15" equivalent diameter	23.50	12.35	35.85
6720	18" equivalent diameter	28	15.85	43.85
6750	24" equivalent diameter	45	18.50	63.50
6770	30" equivalent diameter	56	35	91
6780	36" equivalent diameter	93	35	128
6790	42" equivalent diameter	110	41.50	151.50
6800	48" equivalent diameter	96.50	50.50	147
8000	Polyvinyl chloride SDR 35			
8130	4" diameter	1.83	3.51	5.34
8150	6" diameter	3.74	3.76	7.50
8160	8" diameter	7.75	3.92	11.67
8170	10" diameter	12.25	5.20	17.45
8180	12" diameter	13.95	5.35	19.30
8200	15" diameter	13.30	7.15	20.45

G3030 Storm Sewer

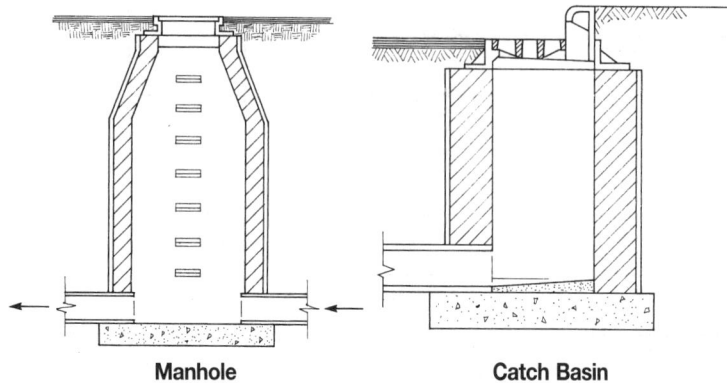

Manhole

Catch Basin

The Manhole and Catch Basin System includes: excavation with a backhoe; a formed concrete footing; frame and cover; cast iron steps and compacted backfill.

The Expanded System Listing shows manholes that have a 4′, 5′ and 6′ inside diameter riser. Depths range from 4′ to 14′. Construction material shown is either concrete, concrete block, precast concrete, or brick.

System Components	QUANTITY	UNIT	COST PER EACH		
			MAT.	INST.	TOTAL
SYSTEM G3030 210 1920					
MANHOLE/CATCH BASIN, BRICK, 4′ I.D. RISER, 4′ DEEP					
Excavation, hydraulic backhoe, 3/8 C.Y. bucket	14.815	C.Y.		102.97	102.97
Trim sides and bottom of excavation	64.000	S.F.		53.12	53.12
Forms in place, manhole base, 4 uses	20.000	SFCA	15.40	90.80	106.20
Reinforcing in place footings, #4 to #7	.019	Ton	30.88	20.90	51.78
Concrete, 3000 psi	.925	C.Y.	102.68		102.68
Place and vibrate concrete, footing, direct chute	.925	C.Y.		41.66	41.66
Catch basin or MH, brick, 4′ ID, 4′ deep	1.000	Ea.	440	880	1,320
Catch basin or MH steps; heavy galvanized cast iron	1.000	Ea.	14.25	12.30	26.55
Catch basin or MH frame and cover	1.000	Ea.	274	202.50	476.50
Fill, granular	12.954	C.Y.	179.41		179.41
Backfill, spread with wheeled front end loader	12.954	C.Y.		27.07	27.07
Air tamp, add	12.954	C.Y.		99.62	99.62
TOTAL			1,056.62	1,530.94	2,587.56

G3030 210		Manholes & Catch Basins	COST PER EACH		
			MAT.	INST.	TOTAL
1920	Manhole/catch basin, brick, 4′ I.D. riser, 4′ deep		1,050	1,550	2,600
1940		6′ deep	1,400	2,125	3,525
1960		8′ deep	1,825	2,925	4,750
1980		10′ deep	2,325	3,625	5,950
3000		12′ deep	3,000	3,925	6,925
3020		14′ deep	3,775	5,500	9,275
3200	Block, 4′ I.D. riser, 4′ deep		995	1,250	2,245
3220		6′ deep	1,300	1,750	3,050
3240		8′ deep	1,675	2,400	4,075
3260		10′ deep	1,950	2,975	4,925
3280		12′ deep	2,400	3,775	6,175
3300		14′ deep	2,925	4,600	7,525
4620	Concrete, cast-in-place, 4′ I.D. riser, 4′ deep		1,225	2,125	3,350
4640		6′ deep	1,700	2,850	4,550
4660		8′ deep	2,300	4,100	6,400
4680		10′ deep	2,750	5,100	7,850
4700		12′ deep	3,375	6,325	9,700
4720		14′ deep	4,050	7,550	11,600
5820	Concrete, precast, 4′ I.D. riser, 4′ deep		1,575	1,125	2,700
5840		6′ deep	2,025	1,500	3,525

G30 Site Mechanical Utilities

G3030 Storm Sewer

G3030 210	Manholes & Catch Basins	COST PER EACH		
		MAT.	INST.	TOTAL
5860	8' deep	2,475	2,100	4,575
5880	10' deep	3,025	2,575	5,600
5900	12' deep	3,750	3,200	6,950
5920	14' deep	4,550	4,075	8,625
6000	5' I.D. riser, 4' deep	1,675	1,250	2,925
6020	6' deep	2,225	1,750	3,975
6040	8' deep	2,825	2,325	5,150
6060	10' deep	3,575	2,975	6,550
6080	12' deep	4,400	3,775	8,175
6100	14' deep	5,300	4,625	9,925
6200	6' I.D. riser, 4' deep	2,425	1,650	4,075
6220	6' deep	3,150	2,175	5,325
6240	8' deep	3,925	3,075	7,000
6260	10' deep	4,975	3,875	8,850
6280	12' deep	6,075	4,900	10,975
6300	14' deep	7,275	5,900	13,175

G3060 Fuel Distribution

G3060 110	Gas Service Piping	COST PER L.F.		
		MAT.	INST.	TOTAL
2000	Piping, excavation & backfill excluded, polyethylene			
2070	1-1/4" diam, SDR 10	2.74	3.42	6.16
2090	2" diam, SDR 11	3.40	3.82	7.22
2110	3" diam, SDR 11.5	7.10	4.56	11.66
2130	4" diam, SDR 11	16.20	8.65	24.85
2150	6" diam, SDR 21	50.50	9.25	59.75
2160	8" diam, SDR 21	69	11.20	80.20
3000	Steel, schedule 40, plain end, tarred & wrapped			
3060	1" diameter	8.30	7.65	15.95
3090	2" diameter	13	8.15	21.15
3110	3" diameter	21.50	8.85	30.35
3130	4" diameter	28	14	42
3140	5" diameter	40.50	16.20	56.70
3150	6" diameter	50	19.85	69.85
3160	8" diameter	79	25.50	104.50
3170	10" diameter	100	35.50	135.50
3180	12" diameter	105	44.50	149.50
3190	14" diameter	112	47.50	159.50
3200	16" diameter	122	51.50	173.50
3210	18" diameter	157	55	212
3220	20" diameter	244	59.50	303.50
3230	24" diameter	279	71.50	350.50

G40 Site Electrical Utilities

G4020 Site Lighting

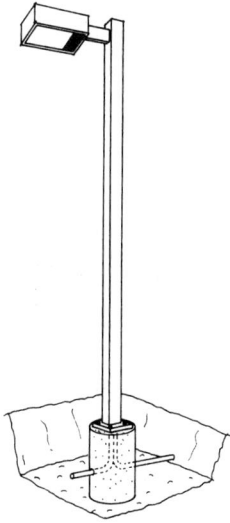

The Site Lighting System includes the complete unit from foundation to electrical fixtures. Each system includes: excavation; concrete base; backfill by hand; compaction with a plate compacter; pole of specified material; all fixtures; and lamps.

The Expanded System Listing shows Site Lighting Systems that use one of three types of lamps: high pressure sodium; mercury vapor; and metal halide. Systems are listed for 400-watt and 1000-watt lamps. Pole height varies from 20′ to 40′. There are four types of poles listed: aluminum, fiberglass, steel and wood.

G4020 110	Site Lighting	COST EACH		
		MAT.	INST.	TOTAL
2320	Site lighting, high pressure sodium, 400 watt, aluminum pole, 20′ high	1,625	1,200	2,825
2360	40′ high	3,025	1,875	4,900
2920	Wood pole, 20′ high	965	1,150	2,115
2960	40′ high	1,275	1,800	3,075
3120	1000 watt, aluminum pole, 20′ high	1,700	1,225	2,925
3160	40′ high	3,100	1,900	5,000
3520	Wood pole, 20′ high	1,050	1,175	2,225
3560	40′ high	1,350	1,825	3,175
5820	Metal halide, 400 watt, aluminum pole, 20′ high	1,600	1,200	2,800
5860	40′ high	3,000	1,875	4,875
7620	1000 watt, aluminum pole, 20′ high	1,675	1,225	2,900
7661	40′ high	3,075	1,900	4,975

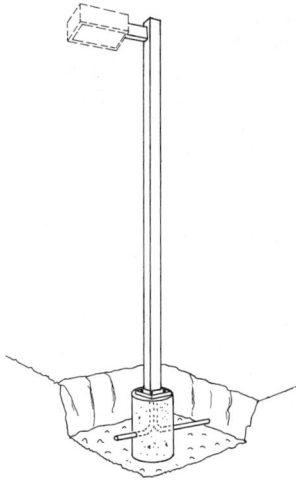

Table G4020 210 Procedure for Calculating Floodlights Required for Various Footcandles
Poles should not be spaced more than 4 times the fixture mounting height for good light distribution.

Estimating Chart
Select Lamp type.

Determine total square feet.

Chart will show quantity of fixtures to provide 1 footcandle initial, at intersection of lines. Multiply fixture quantity by desired footcandle level.

Chart based on use of wide beam luminaires in an area whose dimensions are large compared to mounting height and is approximate only.

To maintain 1 footcandle over a large area use these watts per square foot:

Incandescent	0.15
Metal Halide	0.032
Mercury Vapor	0.05
High Pressure Sodium	0.024

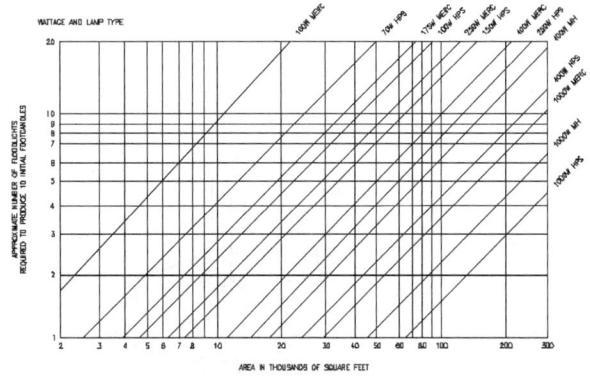

System Components			COST EACH		
	QUANTITY	UNIT	MAT.	INST.	TOTAL
SYSTEM G4020 210 0200					
LIGHT POLES, ALUMINUM, 20' HIGH, 1 ARM BRACKET					
Aluminum light pole, 20', no concrete base	1.000	Ea.	950	524	1,474
Bracket arm for Aluminum light pole	1.000	Ea.	125	70	195
Excavation by hand, pits to 6' deep, heavy soil or clay	2.368	C.Y.		232.06	232.06
Footing, concrete incl forms, reinforcing, spread, under 1 C.Y.	.465	C.Y.	80.91	80.96	161.87
Backfill by hand	1.903	C.Y.		67.56	67.56
Compaction vibrating plate	1.903	C.Y.		9.21	9.21
TOTAL			1,155.91	983.79	2,139.70

G4020 210	Light Pole (Installed)		COST EACH		
			MAT.	INST.	TOTAL
0200	Light pole, aluminum, 20' high, 1 arm bracket		1,150	985	2,135
0240	2 arm brackets		1,275	985	2,260
0280	3 arm brackets	RD5020 -250	1,400	1,025	2,425
0320	4 arm brackets		1,525	1,025	2,550
0360	30' high, 1 arm bracket		2,075	1,225	3,300
0400	2 arm brackets		2,200	1,225	3,425
0440	3 arm brackets		2,325	1,275	3,600
0480	4 arm brackets		2,450	1,275	3,725
0680	40' high, 1 arm bracket		2,550	1,650	4,200
0720	2 arm brackets		2,675	1,650	4,325
0760	3 arm brackets		2,800	1,700	4,500
0800	4 arm brackets		2,925	1,700	4,625
0840	Steel, 20' high, 1 arm bracket		1,400	1,050	2,450
0880	2 arm brackets		1,500	1,050	2,550
0920	3 arm brackets		1,525	1,075	2,600
0960	4 arm brackets		1,650	1,075	2,725
1000	30' high, 1 arm bracket		1,600	1,325	2,925
1040	2 arm brackets		1,725	1,325	3,050
1080	3 arm brackets		1,725	1,375	3,100
1120	4 arm brackets		1,850	1,375	3,225
1320	40' high, 1 arm bracket		2,100	1,775	3,875
1360	2 arm brackets		2,200	1,775	3,975
1400	3 arm brackets		2,225	1,825	4,050
1440	4 arm brackets		2,350	1,825	4,175

Reference Section

All the reference information is in one section, making it easy to find what you need to know . . . and easy to use the book on a daily basis. This section is visually identified by a vertical gray bar on the page edges.

In this section, you'll see the background that relates to the reference numbers that appeared in the Assembly Cost Sections. You'll find reference tables, explanations, and estimating information that support how we derive the systems data. Also included are alternate pricing methods, technical data, and estimating procedures, along with information on design and economy in construction.

Also in this Reference Section, we've included Historical Cost Indexes for cost comparison over time; City Cost Indexes and Location Factors for adjusting costs to the region you are in; Square Foot Costs; Reference Aids; Estimating Forms; and an explanation of all Abbreviations used in the book.

Table of Contents

General: A spread footing is used to convert a concentrated load (from one superstructure column, or substructure grade beams) into an allowable area load on supporting soil.

Because of punching action from the column load, a spread footing is usually thicker than strip footings which support wall loads. One or two story commercial or residential buildings should have no less than 1′ thick spread footings. Heavier loads require no less than 2′ thick. Spread footings may be square, rectangular or octagonal in plan.

Spread footings tend to minimize excavation and foundation materials, as well as labor and equipment. Another advantage is that footings and soil conditions can be readily examined. They are the most widely used type of footing, especially in mild climates and for buildings of four stories or under. This is because they are usually more economical than other types, if suitable soil and site conditions exist.

They are used when suitable supporting soil is located within several feet of the surface or line of subsurface excavation. Suitable soil types include sands and gravels, gravels with a small amount of clay or silt, hardpan, chalk, and rock. Pedestals may be used to bring the column base load down to the top of footing. Alternately, undesirable soil between underside of footing and top of bearing level can be removed and replaced with lean concrete mix or compacted granular material.

Depth of footing should be below topsoil, uncompacted fill, muck, etc. It must be lower than frost penetration (see local code or Table L1030-502 in Section L) but should be above the water table. It must not be at the ground surface because of potential surface erosion. If the ground slopes, approximately three horizontal feet of edge protection must remain. Differential footing elevations may overlap soil stresses or cause excavation problems if clear spacing between footings is less than the difference in depth.

Other footing types are usually used for the following reasons:
- A. Bearing capacity of soil is low.
- B. Very large footings are required, at a cost disadvantage.
- C. Soil under footing (shallow or deep) is very compressible, with probability of causing excessive or differential settlement.
- D. Good bearing soil is deep.
- E. Potential for scour action exists.
- F. Varying subsoil conditions within building perimeter.

Cost of spread footings for a building is determined by:
1. The soil bearing capacity.
2. Typical bay size.
3. Total load (live plus dead) per S.F. for roof and elevated floor levels.
4. The size and shape of the building.
5. Footing configuration. Does the building utilize outer spread footings or are there continuous perimeter footings only or a combination of spread footings plus continuous footings?

COST DETERMINATION

1. Determine Soil Bearing Capacity by a known value or using Table A1010-121 as a guide.

Table A1010-121   Soil Bearing Capacity in Kips per S.F.

Bearing Material	Typical Allowable Bearing Capacity
Hard sound rock	120 KSF
Medium hard rock	80
Hardpan overlaying rock	24
Compact gravel and boulder-gravel; very compact sandy gravel	20
Soft rock	16
Loose gravel; sandy gravel; compact sand; very compact sand-inorganic silt	12
Hard dry consolidated clay	10
Loose coarse to medium sand; medium compact fine sand	8
Compact sand-clay	6
Loose fine sand; medium compact sand-inorganic silts	4
Firm or stiff clay	3
Loose saturated sand-clay; medium soft clay	2

Table A1010-122 Working Load Determination

2. Determine Bay Size in S.F.

3. Determine Load to the Footing

Type Load	Reference	Working Loads			
		Roof	Floor	Column	Total
1. Total Load	Floor System	PSF	PSF		
2. Whole Bay Load/Level	(Line 1) x Bay Area ÷ 1000	Kips	Kips		
3. Load to Column	[(Roof + (Floor x No. Floors)]	Kips	Kips		Kips
4. Column Weight	Systems B1010 201 to B1010 208			Kips	Kips
5. Fireproofing	System B1010 720			Kips	Kips
6. Total Load to Footing	Roof + Floor + Column				Kips

Working loads represent the highest actual load a structural member is designed to support.

4. Determine Size and Shape of Building

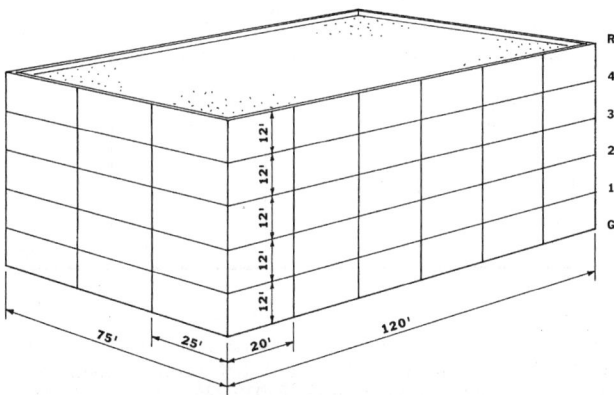

Figure A1010-123

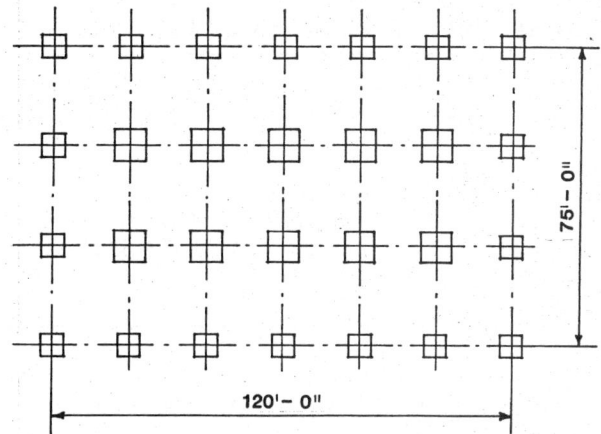

Figure A1010-124

5. Footing Configuration Possibilities

a. Building has spread footings throughout (including exterior and corners).

b. Building has continuous perimeter footings to support exterior columns, corners and exterior walls (as well as interior spread footings).

c. Building has both perimeter continuous footings for walls plus exterior spread footings for column loads. If there are interior spread footings also, use "b".

    1. Figure continuous footings from tables in Section A1010-110.

    2. Add total dollars for spread footings and continuous footings to determine total foundation dollars.

Table A1010-126 Working Load Determination

Tabulate interior footing costs and add continuous footing costs from Systems A1010 210 and A1010 110.

Example: Spread Footing Cost Determination

Office Building .... 5 Story
Story Height .... 12′
Dimensions .... 75′ x 120′
Bay Size .... 20′ x 25′
Soil Bearing Capacity .... 6 KSF
Foundation .... Spread Footings
Slab on Grade .... 5″ Reinforced
Code .... BOCA
Construction .... From Example RB1010-112
Using concrete flat plate

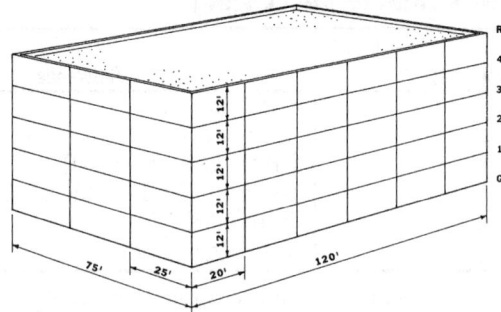

Figure A1010-125

Determine load to footings Table A1010-122 or Example RB1010-112.

Type Load	Reference	Working Loads			
		Roof	Floor	Column	Total
Determine floor system	System in Div. B1010 (Example: Flat Plate)				
1. Total Load	System # B1010 223	146PSF	188PSF		
2. Whole Bay Load/Level	Bay Area x Line 1/1000	73K	94K		
3. Load To Column	Roof + (Floor x No. Floors)	73K	276K		449K
4. Column Weight	Systems B1010 201 to B1010 208			25K	25K
5. Fireproofing	N/A			—	
6. Load to Footing	Roof + Floor + Column				474K

1. Total load working from System B1010 223: 40# superimposed load for roof and 75# superimposed load for floors.
2. Bay Area 20′ x 25′ = 500 S.F. x Line 1
   500 x 146/1000 = 73K roof
   500 x 188/1000 = 94K floor
3. Roof + (floor x no. floors)
   73K + (4 x 94K) = 449K

4. Enter System B1010 203 with a total working load of 474K and 12′ story height. A 16″ square tied column with 6 KSI concrete is most economical. Minimum column size from System B1010 223 is 20″. Use the larger. Column weight = 394 #/V.L.F. x 12′ = approx. 5K.
5. No fireproofing is required for concrete column.
6. Add roof, floors and column loads for a total load to each interior footing.

Footing Cost: Enter System A1010 210 with total working load and allowable soil pressure. Determine the cost per footing using the closest higher load on the table. Add the total costs for interior, exterior, and corner footings.

Table A1010-127 Typical Spread Footing, Square

EXTERIOR FOOTINGS

Exterior vs. Interior Footings

System A1010 210 contains the cost for individual spread interior footings for various loading and soil bearing conditions. To determine the loads for exterior footings:

1. Multiply whole bay working load by 60% for half bay footings and by 45% for quarter bay (or corner) footings. This will give the approximate column load for half and quarter bay footings.
2. Enter System A1010 210 and find cost of footing with loads determined in step 1.
3. Determine the number of each type of footing on the job and multiply the cost per footing by the appropriate quantity. This will give the total dollar cost for exterior and interior spread footings.

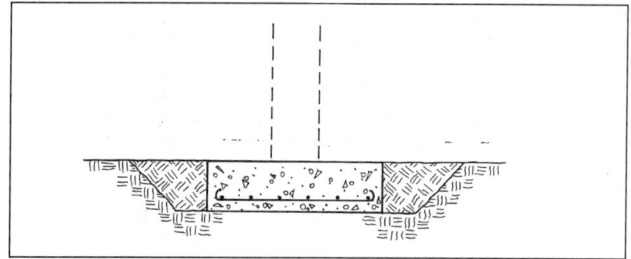

Example: Using soil capacity, 6KSF
Total working load from Table A1010-126 .... = 474K
Half Bay Footings .... 474K x 60% = 284K
Corner Footings .... 474K x 45% = 213K

Table A1010-128 Spread Footing Cost

Footing Type	Working Load (Kips)	Total Quantity	System A1010 210 Unit Cost	Total
Interior	474	10	$2,525	$25,250
Exterior	284	14	1,380	19,320
Corner	213	4	1,380	5,520
Total				$50,090

Table A1010-141 Section Thru Typical Strip Footing

General: Strip (continuous) footings are a type of spread footing, but are used to convert the lineal type wall into an allowable area load on supporting soil.

Many of the general comments on isolated spread footings (Section A1010 210) apply here as well.

Strip footings may be used under walls of concrete, brick, block or stone. They are constructed of continuously placed concrete, 2000 psi or greater. Normally they are not narrower than twice the wall thickness, nor are they thinner than the wall thickness. Plain concrete footings should be at least 8″ thick, and reinforced footings require at least 6″ of thickness above the bottom reinforcement. Where there is no basement, the bottom of footing is usually 3′ to 5′ below finished exterior grade, and 12″ below the average frost penetration (see Figure A1010-141), resting on undisturbed soil. In some cases strip footings serve as leveling pads (stone walls, etc.).

Method of construction is affected by reinforcement and soil type. If steel mats and/or dowels are present and require accurate placement (frequently for reinforced concrete or reinforced masonry walls), then side forms are required, as they are in wet or sandy soil. If there is no reinforcement or only longitudinal reinforcement, and the soil will "stand" forms might be eliminated and economically should be.

Preferably, strip footings are used on sand and gravel. The presence of a small amount of clay or dense silty sand in gravel is acceptable, as is bearing on rock or chalk. They are sometimes used in uniform, firm or stiff clays with little nearby ground vegetation, and placed at least 3-1/2′ into the clay. If the clay is sloped, there is potential for downhill creep. Footings up to 3-1/2′ wide are sometimes used in soft or silty clays, but settlement must be expected and provision made.

Sloping terrain requires stepping the footing to maintain depth, since a sloped footing will create horizontal thrust which may distress the structure. Steps should be at least 2′ long horizontally and each vertical step no greater than three-fourths the horizontal distance between steps. Vertical risers must be at least 6″ thick and of footing width. Potential for erosion of surrounding soil must be considered.

Strip footings should be used under walls constructed of unreinforced concrete, or of units bonded with mortar. They are also used when bearing capacity of soil is inadequate to support wall thickness alone.

Footings require horizontal reinforcement at right angles to the wall line when side projection of footings beyond face of wall is greater than 1/2 to 2/3 footing thickness. Horizontal reinforcement parallel to wall line should be used if soil bearing capacity is low, or soil compressibility is variable, or the footing spans pipe trenches, etc. Vertical reinforcing dowels depend upon wall design.

Alternate foundation types include bored piles, grade beams, and mat. Short, bored compacted concrete piles are sometimes competitive with strip footings, especially in shrinkable clay soils where they may provide an added factor of safety. When soil bearing capacity is inadequate, grade beams can be used to transfer wall loads across the inadequate ground to column support foundations. Mat footings replace both strip and isolated spread footings. They may eliminate differential settlement problems.

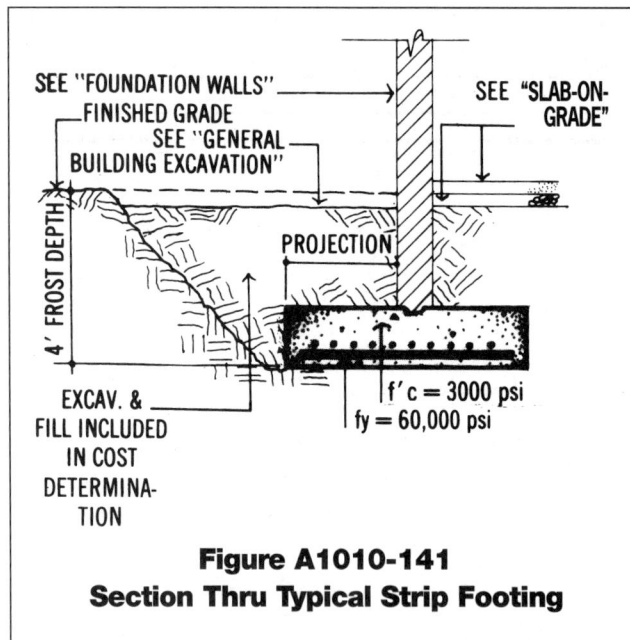

**Figure A1010-141
Section Thru Typical Strip Footing**

Example: Strip Footing Cost Determination for Bearing Wall

Office Building . . . . 5 Story
Story Height . . . . 12 VLF
Dimensions . . . . 50′ x 80′ = 4000 S.F.G.
Exterior Wall . . . . .12″ Brick and Block
Interior Wall . . . . .8″ Block
Floor . . . . .Prestressed Concrete Slabs + 2″ Topping
Roof . . . . .Prestressed Concrete Slabs, No Topping
Soil Bearing Capacity . . . . .6 KSF
Superimposed Load:

    floor . . . . 100 PSF
    roof . . . . .40 PSF

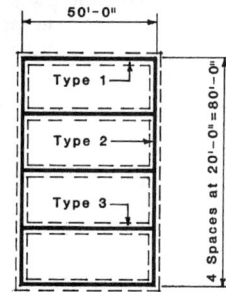

Figure A1010-142 Footing Plan

A. Total load to each footing

Table A1010-143 Strip Footing, Load Determination

		Working Loads			
Type Load	Reference	Roof	Floor	Wall	Total
1. Total Load	System B1010 229	90 PSF	175 PSF	95 PSF	
2. Load Span or Height	Building Design	10 LF	10 LF	12 VLF	
3. Whole Bay Load/Level	(Line 1 x Line 2) ÷ 1000	.9 KIPS	1.75 KIPS	1.1 KIPS	
4. Load to Footing/L.F.	Roof + (Floor + Wall)	.9 KLF	7.0 KLF	5.9 KLF	13.8 KLF

1. Total load from System B1010-229.
2. Roof and floor load spans for exterior strip footings are 1/2 total span of 20′ or 10′ (wall height for our example is 12′).
3. Multiply Line 1 by Line 2 in each column and divide by 1,000 Lb/Kip.
4. Add 1 roof + 4 elevated floors + 5.33 stories of wall (includes .33 added for a frost wall) for the total load to the footing.

B. Enter System A1010-110 with a load of 13.8 KLF and interpolate between line items 2700 and 3100.

$$\text{Interpolating: } \$39.50 + \left[\frac{13.8 - 11.1}{14.8 - 11.1} \times (\$45.50 - \$39.50) \right] = \$43.80$$

C. If the building design incorporates other strip footing loads,
   repeat steps A and B for each type of footing load.

Table A1010-144 Cost Determination, Strip Footings

Description	Strip Footing Type or Size	Total Length Each Type (Ft.)	Cost From System A1010 110 ($/L.F.)	Cost x Length ($/Each Type)	Total Cost All Types
End Bearing Wall	1	100	$43.80	$ 4,380	
Non Bearing Wall	2	160	39.50	$ 6,320	
Interior Bearing Wall	3	150	90.50	$13,575	$24,275

Note: Type 2 wall: use "wall" loads only in Step A above.
    Type 3 wall: same as Step A above with load span increased to 20′.
    Interpolated values from System A1010 110 are rounded to $.05.

General: The function of a reinforced concrete pile cap is to transfer superstructure load from isolated column or pier to each pile in its supporting cluster. To do this, the cap must be thick and rigid, with all piles securely embedded into and bonded to it.

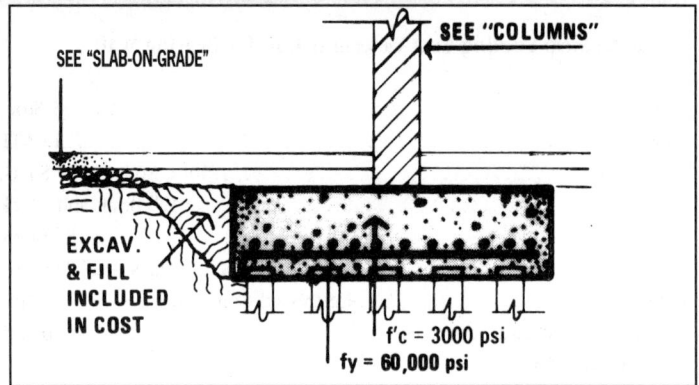

Figure A1010-331 Section Through Pile Cap

Table A1010-332   Concrete Quantities for Pile Caps

Load	Number of Piles @ 3'-0" O.C. Per Footing Cluster									
Working (K)	2 (CY)	4 (CY)	6 (CY)	8 (CY)	10 (CY)	12 (CY)	14 (CY)	16 (CY)	18 (CY)	20 (CY)
50	(.9)	(1.9)	(3.3)	(4.9)	(5.7)	(7.8)	(9.9)	(11.1)	(14.4)	(16.5)
100	(1.0)	(2.2)	(3.3)	(4.9)	(5.7)	(7.8)	(9.9)	(11.1)	(14.4)	(16.5)
200	(1.0)	(2.2)	(4.0)	(4.9)	(5.7)	(7.8)	(9.9)	(11.1)	(14.4)	(16.5)
400	(1.1)	(2.6)	(5.2)	(6.3)	(7.4)	(8.2)	(13.7)	(11.1)	(14.4)	(16.5)
800		(2.9)	(5.8)	(7.5)	(9.2)	(13.6)	(17.6)	(15.9)	(19.7)	(22.1)
1200			(5.8)	(8.3)	(9.7)	(14.2)	(18.3)	(20.4)	(21.2)	(22.7)
1600				(9.8)	(11.4)	(14.5)	(19.5)	(20.4)	(24.6)	(27.2)
2000				(9.8)	(11.4)	(16.6)	(24.1)	(21.7)	(26.0)	(28.8)
3000						(17.5)		(26.5)	(30.3)	(32.9)
4000								(30.2)	(30.7)	(36.5)

Table A1010-333   Concrete Quantities for Pile Caps

Load	Number of Piles @ 4'-6" O.C. Per Footing					
Working (K)	2 (CY)	3 (CY)	4 (CY)	5 (CY)	6 (CY)	7 (CY)
50	(2.3)	(3.6)	(5.6)	(11.0)	(13.7)	(12.9)
100	(2.3)	(3.6)	(5.6)	(11.0)	(13.7)	(12.9)
200	(2.3)	(3.6)	(5.6)	(11.0)	(13.7)	(12.9)
400	(3.0)	(3.6)	(5.6)	(11.0)	(13.7)	(12.9)
800			(6.2)	(11.5)	(14.0)	(12.9)
1200				(13.0)	(13.7)	(13.4)
1600						(14.0)

General: Piles are column-like shafts which receive superstructure loads, overturning forces, or uplift forces. They receive these loads from isolated column or pier foundations (pile caps), foundation walls, grade beams, or foundation mats. The piles then transfer these loads through shallower poor soil strata to deeper soil of adequate support strength and acceptable settlement with load.

Be sure that other foundation types aren't better suited to the job. Consider ground and settlement, as well as loading, when reviewing. Piles usually are associated with difficult foundation problems and substructure condition. Ground conditions determine type of pile (different pile types have been developed to suit ground conditions). Decide each case by technical study, experience, and sound engineering judgment—not rules of thumb. A full investigation of ground conditions, early, is essential to provide maximum information for professional foundation engineering and an acceptable structure.

Piles support loads by end bearing and friction. Both are generally present; however, piles are designated by their principal method of load transfer to soil.

Boring should be taken at expected pile locations. Ground strata (to bedrock or depth of 1-1/2 building width) must be located and identified with appropriate strengths and compressibilities. The sequence of strata determines if end-bearing or friction piles are best suited. See **Table A1020-105** for site investigation costs.

End-bearing piles have shafts which pass through soft strata or thin hard strata and tip bear on bedrock or penetrate some distance into a dense, adequate soil (sand or gravel).

Friction piles have shafts which may be entirely embedded in cohesive soil (moist clay), and develop required support mainly by adhesion or "skin-friction" between soil and shaft area.

Piles pass through soil by either one of two ways:
1. Displacement piles force soil out of the way. This may cause compaction, ground heaving, remolding of sensitive soils, damage to adjacent structures, or hard driving.
2. Non-displacement piles have either a hole bored and the pile cast or placed in the hole, or open-ended pipe (casing) driven and the soil core removed. They tend to eliminate heaving or lateral pressure damage to adjacent structures of piles. Steel "HP" piles are considered of small displacement.

Placement of piles (attitude) is most often vertical; however, they are sometimes battered (placed at a small angle from vertical) to advantageously resist lateral loads. Seldom are piles installed singly but rather in clusters (or groups). Codes require a minimum of three piles per major column load or two per foundation wall or grade beam. Single pile capacity is limited by pile structural strength or support strength of soil. Support capacity of a pile cluster is almost always less than the sum of its individual pile capacities due to overlapping of bearing the friction stresses. See **Table A1020-101** for minimum pile spacing requirements.

Large rigs for heavy, long piles create large soil surface loads and additional expense on weak ground. See **Table A1020-103** for percent of cost increases.

Fewer piles create higher costs per pile. See **Table A1020-102** for effect of mobilization.

Pile load tests are frequently required by code, ground situation, or pile type. See **Table A1020-104** for costs. Test load is twice the design load.

Table A1020-101   Min. Pile Spacing by Pile Type

Type Support		Min. Pile Spacing	
		X's Butt Diameter	Foot
End	Bedrock	2	2'-0"
Bearing	Hard Strata	2.5	2'-6"
Friction		3 to 5	3'-6"

Table A1020-102   Add cost for Mobilization (Setup and Removal)

Job Size	Cost	
	$/Job	$/L.F. Pile
Small (12,000 L.F.)	$13,200	$1.10
Large (25,000 L.F.)	22,000	.88

Table A1020-103 Add Costs for Special Soil Conditions

Special Conditions	Add % of Total
Soft, damp ground	40%
Swampy, wet ground	40%
Barge-mounted drilling rig	30%

Table A1020-104 Testing Costs, if Required, Any Pile Type

Test Weight (Tons)	$/Test
50-100T	$18,700
100-200T	24,200
150-300T	30,800
200-400T	33,000

Table A1020-105 Cost/L.F. of Boring Types (4") and Per Job Cost of Other Items

Type Soil	Type Boring	Sample	$/L.F. Boring	Total $
Earth	Auger	None	$28.00	
	Cased	Yes	62.00	
Rock, "BX"	Core	None	65.50	
	Cased	Yes	94.00	
Filed survey, mobilization, demobilization & engineering report				$2,819

445

General: Caissons, as covered in this section, are drilled cylindrical foundation shafts which function primarily as short column-like compression members. They transfer superstructure loads through inadequate soils to bedrock or hard stratum. They may be either reinforced or unreinforced and either straight or belled out at the bearing level.

Shaft diameters range in size from 20″ to 84″ with the most usual sizes beginning at 34″. If inspection of bottom is required, the minimum diameter practical is 30″. If handwork is required (in addition to mechanical belling, etc.) the minimum diameter is 32″. The most frequently used shaft diameter is probably 36″ with a 5′ or 6′ bell diameter. The maximum bell diameter practical is three times the shaft diameter.

Plain concrete is commonly used, poured directly against the excavated face of soil. Permanent casings add to cost and economically should be avoided. Wet or loose strata are undesirable. The associated installation sometimes involves a mudding operation with bentonite clay slurry to keep walls of excavation stable (costs not included here).

Reinforcement is sometimes used, especially for heavy loads. It is required if uplift, bending moment, or lateral loads exist. A small amount of reinforcement is desirable at the top portion of each caisson, even if the above conditions theoretically are not present. This will provide for construction eccentricities and other possibilities. Reinforcement, if present, should extend below the soft strata. Horizontal reinforcement is not required for belled bottoms.

There are three basic types of caisson bearing details:

1. Belled, which are generally recommended to provide reduced bearing pressure on soil. These are not for shallow depths or poor soils. Good soils for belling include most clays, hardpan, soft shale, and decomposed rock.

Soils requiring handwork include hard shale, limestone, and sandstone.

Soils not recommended include sand, gravel, silt, and igneous rock. Compact sand and gravel above water table may stand. Water in the bearing strata is undesirable.

2. Straight shafted, which have no bell but the entire length is enlarged to permit safe bearing pressures. They are most economical for light loads on high bearing capacity soil.

3. Socketed (or keyed), which are used for extremely heavy loads. They involve sinking the shaft into rock for combined friction and bearing support action. Reinforcement of shaft is usually necessary. Wide flange cores are frequently used here.

Advantages include:

A. Shafts can pass through soils that piles cannot

B. No soil heaving or displacement during installation

C. No vibration during installation

D. Less noise than pile driving

E. Bearing strata can be visually inspected & tested

Uses include:

A. Situations where unsuitable soil exists to moderate depth

B. Tall structures

C. Heavy structures

D. Underpinning (extensive use)

See **Table A1010-121** for Soil Bearing Capacities.

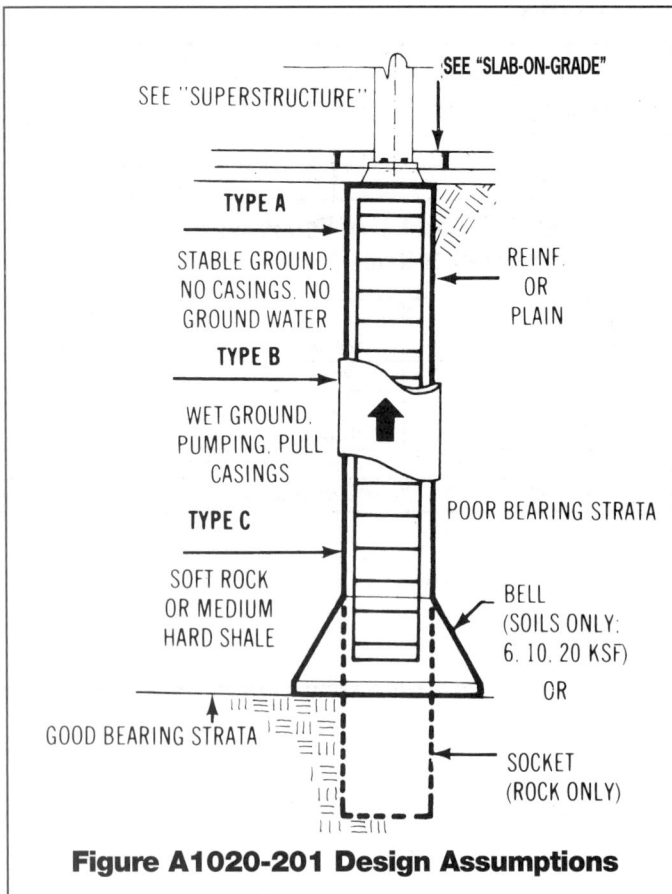

Figure A1020-201 Design Assumptions

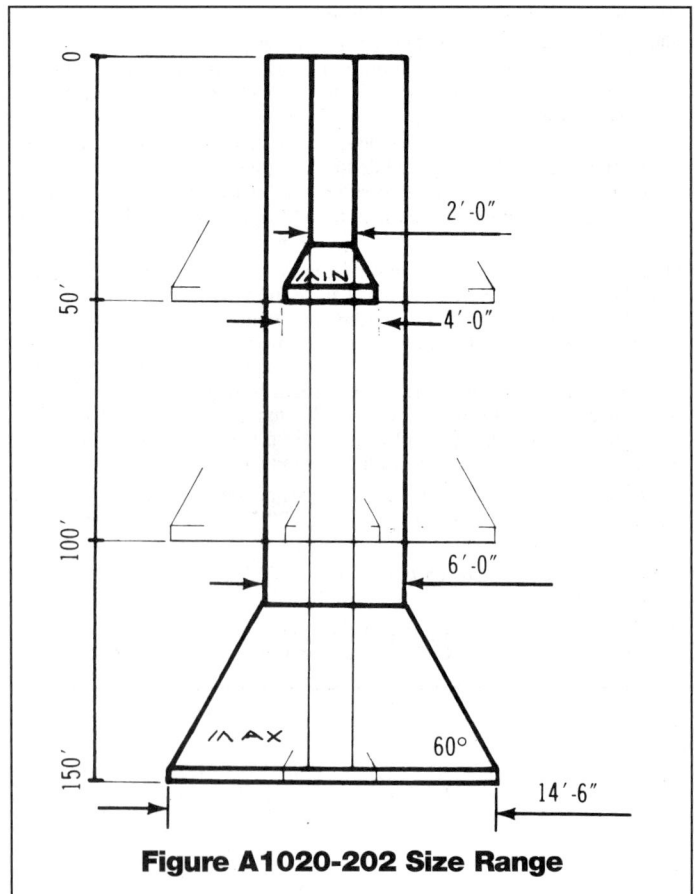

Figure A1020-202 Size Range

446

Table A1020-231   Grade Beam Detail

General: Grade beams are stiff self-supporting structural members, partly exposed above grade. They support wall loads and carry them across unacceptable soil to column footings, support piles or caissons.

They should be deep enough to be below frost depth. Therefore, they are more frequently used in mild climates.

They must be stiff enough to prevent cracking of the supported wall. Usually, they are designed as simply supported beams, so as to minimize effects of unequal footing settlement.

Heavy column loads and light wall loads tend to make grade beams an economical consideration. Light column loads but heavy wall loads tend to make strip footings more economical.

Conditions which favor other foundation types include:

A. Deep frost penetration requiring deep members

B. Good soil bearing near surface

C. Large live loads to grade beam

D. Ground floor 2′ or 3′ above finished grade requiring wall depth beams (not included here).

E. Varying ground elevations

F. Basement, requiring walls

Design Assumption: See Figure A1020-231
    Concrete placed by chute
    Forms, four uses
    Simply supported
    Max. deflection = L/480
    Design span = Bay width less 2′

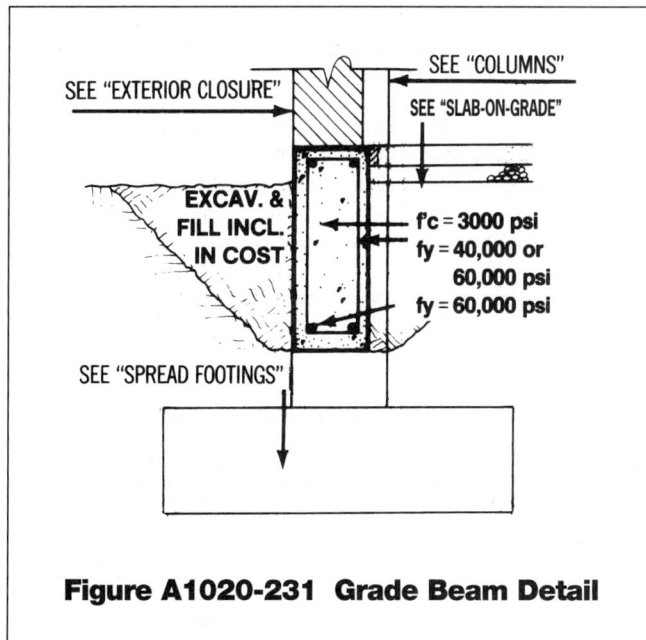

Figure A1020-231 Grade Beam Detail

Table A1030-202 Thickness and Loading Assumptions by Type of Use

General: Grade slabs are classified on the basis of use. Thickness is generally controlled by the heaviest concentrated load supported. If load area is greater than 80 sq. in., soil bearing may be important. The base granular fill must be a uniformly compacted material of limited capillarity, such as gravel or crushed rock. Concrete is placed on this surface or the vapor barrier on top of base.

Grade slabs are either single or two course floors. Single course are widely used. Two course floors have a subsequent wear resistant topping.

Reinforcement is provided to maintain tightly closed cracks.

Control joints limit crack locations and provide for differential horizontal movement only. Isolation joints allow both horizontal and vertical differential movement.

Use of Table: Determine appropriate type of slab (A, B, C, or D) by considering type of use or amount of abrasive wear of traffic type.

Determine thickness by maximum allowable wheel load or uniform load, opposite 1st column thickness. Increase the controlling thickness if details require, and select either plain or reinforced slab thickness and type.

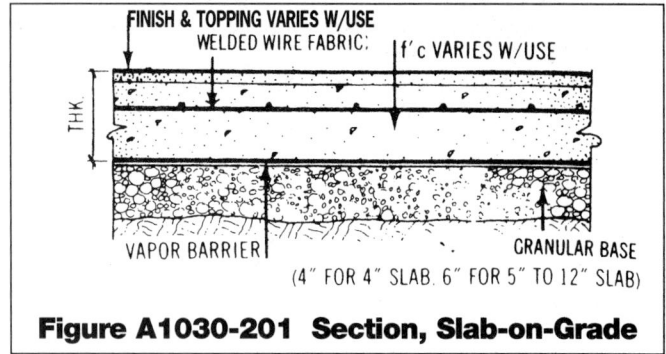

Figure A1030-201 Section, Slab-on-Grade

Table A1030-202 Thickness and Loading Assumptions by Type of Use

SLAB THICKNESS (IN.)	TYPE	A Non Little Foot Only Load* (K)	B Light Light Pneumatic Wheels Load* (K)	C Normal Moderate Solid Rubber Wheels Load* (K)	D Heavy Severe Steel Tires Load* (K)	◀ Slab I.D. ◀ Industrial ◀ Abrasion ◀ Type of Traffic / Max. Uniform Load to Slab ▼ (PSF)
4"	Reinf. Plain	4K				100
5"	Reinf. Plain	6K	4K			200
6"	Reinf. Plain		8K	6K	6K	500 to 800
7"	Reinf. Plain			9K	8K	1,500
8"	Reinf. Plain				11K	
10"	Reinf. Plain				14K	* Max. Wheel Load in Kips (incl. impact)
12"	Reinf. Plain					
D E S I G N A S S U M P T I O N S	Concrete, Chuted	f'c = 3.5 KSI	4 KSI	4.5 KSI	Slab @ 3.5 KSI	ASSUMPTIONS BY SLAB TYPE
	Topping			1" Integral	1" Bonded	
	Finish	Steel Trowel	Steel Trowel	Steel Trowel	Screed & Steel Trowel	
	Compacted Granular Base	4" deep for 4" slab thickness / 6" deep for 5" slab thickness & greater				
	Vapor Barrier	6 mil polyethylene				ASSUMPTIONS FOR ALL SLAB TYPES
	Forms & Joints	Allowances included				
	Reinforcement	WWF As required ≥ 60,000 psi				

Table A2020-212   Minimum Wall Reinforcement Weight (PSF)

General: Foundation wall heights depend upon depth of frost penetration, basement configuration or footing requirements. For certain light load conditions, good soil bearing and adequate wall thickness, strip footings may not be required.

Thickness requirements are based upon code requirements (usually 8″ min. for foundation, exposed basement, fire and party walls), unsupported height or length, configuration of wall above, and structural load considerations.

Shrinkage and temperature forces make longitudinal reinforcing desirable, while high vertical and lateral loads make vertical reinforcing necessary. Earthquake design also necessitates reinforcing.

Figure A2020-211 shows the range in wall sizes for which costs are provided in System A2020 110. Other pertinent information is also given.

Design Assumptions: Reinforced concrete wall designs are based upon ACI 318-71-14, Empirical Design of Walls. Plain concrete wall designs are based upon nonreinforced load bearing masonry criteria (fm = 1700 psi).

Earthquake, lateral or shear load requirements may add to reinforcement costs, and are not included.

Minimum Wall Reinforcement: Table below lists the weight of wall for minimum quantities of reinforcing steel for various wall thicknesses.

Design Assumptions:
Reinforcing steel is A 615, grade 60 (fy = 60 k).
Total percent of steel area is 0.20% for horizontal steel, 0.12% for vertical steel.
For steel at both faces, mats are identical.
For single layer, mat is in center of wall.
For grade 40 steel, add 25% to weights and costs.
If other than the minimum amount of steel area is required, factor weights directly with increased percentage.

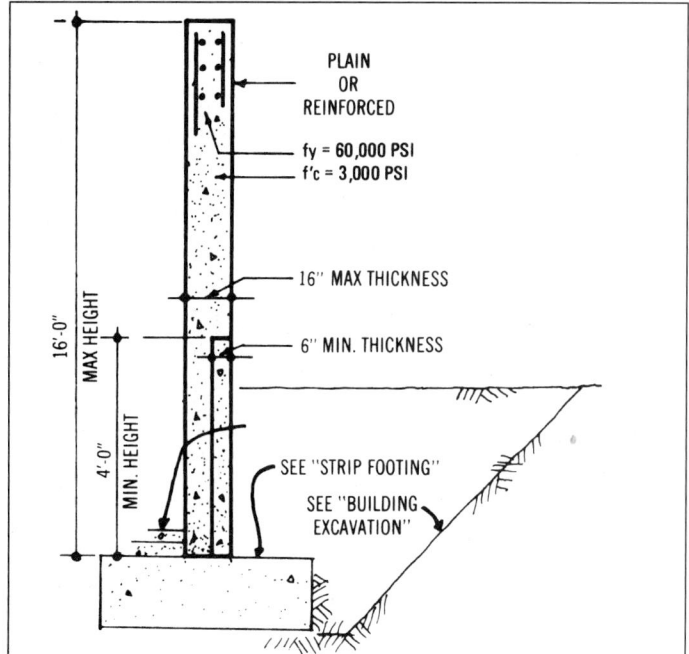

Figure A2020-211
Concrete Foundation Walls
Size Range for System A2020 110

Note: Excavation and fill costs are not included in System A2020 110, but are provided in STRIP FOOTING COST, System A1010 110

Table A2020-212 Minimum Wall Reinforcement Weight (PSF)

Location	Wall Thickness	Horizontal Steel				Vertical Steel				Horizontal & Vertical Steel
		Bar Size	Spacing C/C	Sq. In. per L.F.	Total Wt. per S.F.	Bar Size	Spacing C/C	Sq. In. per L.F.	Total Wt. per S.F.	Total Wt. per S.F.
Both Faces	10″	#4	18″	.13	.891#	#3	18″	.07	.501#	1.392#
	12″	#4	16″	.15	1.002	#3	15″	.09	.602	1.604
	14″	#4	14″	.17	1.145	#3	13″	.10	.694	1.839
	16″	#4	12″	.20	1.336	#3	11″	.12	.820	2.156
	18″	#5	17″	.22	1.472	#4	18″	.13	.891	2.363
One Face	6″	#3	9″	.15	.501	#3	18″	.07	.251	.752
	8″	#4	12″	.20	.668	#3	11″	.12	.410	1.078
	10″	#5	15″	.25	.834	#4	16″	.15	.501	1.335

Table A2020-213 Weight of Steel Reinforcing per S.F. in Walls (PSF)

Reinforced Weights: Table below lists the weight per S.F. for reinforcing steel in walls.

Design Assumptions: Reinforcing Steel is A615 grade 60 (for costs).

Weights will be correct for any grade steel reinforcing. For bars in two directions, add weights for each size and spacing.

C/C Spacing in Inches	Bar size								
	#3 Wt. (PSF)	#4 Wt. (PSF)	#5 Wt. (PSF)	#6 Wt. (PSF)	#7 Wt. (PSF)	#8 Wt. (PSF)	#9 Wt. (PSF)	#10 Wt. (PSF)	#11 Wt. (PSF)
2"	2.26	4.01	6.26	9.01	12.27				
3"	1.50	2.67	4.17	6.01	8.18	0.68	13.60	17.21	21.25
4"	1.13	2.01	3.13	4.51	6.13	8.10	10.20	12.91	15.94
5"	.90	1.60	2.50	3.60	4.91	6.41	8.16	10.33	12.75
6"	.752	1.34	2.09	3.00	4.09	5.34	6.80	8.61	10.63
8"	.564	1.00	1.57	2.25	3.07	4.01	5.10	6.46	7.97
10"	.451	.802	1.25	1.80	2.45	3.20	4.08	5.16	6.38
12"	.376	.668	1.04	1.50	2.04	2.67	3.40	4.30	5.31
18"	.251	.445	.695	1.00	1.32	1.78	2.27	2.86	3.54
24"	.188	.334	.522	.751	1.02	1.34	1.70	2.15	2.66
30"	.150	.267	.417	.600	.817	1.07	1.36	1.72	2.13
36"	.125	.223	.348	.501	.681	.890	1.13	1.43	1.77
42"	.107	.191	.298	.429	.584	.763	.97	1.17	1.52
48"	.094	.167	.261	.376	.511	.668	.85	1.08	1.33

Selecting a Floor System:

1. Determine size of building — total S.F. and occupancy
   (i.e., office building, dormitory, etc.).
2. Determine number of floors and S.F./floor.
3. Select a possible bay size and layout.
4. From Table L1010-101 determine the minimum design live load.
   Add partition load, ceiling load and miscellaneous
   loads such as mechanical, light fixtures and flooring.

Example:

Live Load Office Building	50 PSF	BOCA Code Table L1010-101
Partitions	20	Assume using Tables L1010-201
Ceiling	5	through L1010-226
	75 PSF	Total Superimposed Load

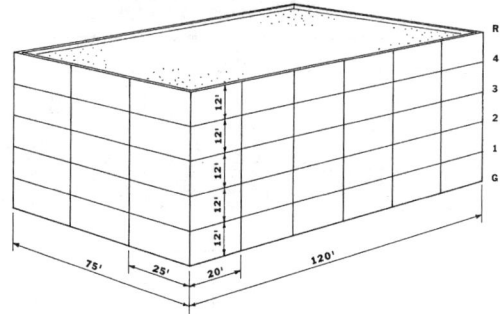

Suspended Floors					
Building Dimension (Ft.)	Suspended Floor (S.F.)	Number of Suspended Floors	Total Suspended Floors (S.F.)	Bay Size (Ft.)	Superimposed Load PSF
75 x 120	9,000	4	36,000	20 x 25	75

5. Enter Tables with 75 PSF superimposed load and 20 x 25 bay size.

Table B1010-223	C.I.P. Flat Plate	$15.20 /S.F.
Table B1010-256	Composite Beam, Deck & Slab	19.95 /S.F.
Table B1010-250	Steel Joists & Beams on Col.	17.76 /S.F.
	(col. not included)	

It appears, from the costs listed above, that the most economical system for framing the suspended floors with 20′ x 25′ bays, 75 PSF superimposed load, is the open web joists, beams, slab form and 2-1/2″ concrete slabs. However, the system will require a fireproof ceiling for fireproofing the structure, which will ultimately raise the cost.

The concrete flat plate may be more economical if no hung ceiling is required and the soil capacity is sufficient to justify the additional load to the foundations.

Some of the factors affecting the selection of a floor system are:
1. Location
2. Owner's preference
3. Fire code
4. Availability of materials
5. Subsoil conditions
6. Economy
7. Clear span
8. Acoustical characteristics
9. Contractor capability

Table B1010-101 Comparative Costs ($/S.F.) of Floor Systems/Type (Table Number), Bay Size, & Load

Bay Size	Cast-In-Place Concrete						Precast Concrete			Structural Steel					Wood	
	1 Way BM & Slab B1010-219	2 Way BM & Slab B1010-220	Flat Slab B1010-222	Flat Plate B1010-223	Joist Slab B1010-226	Waffle Slab B1010-227	Beams & Hollow Core Slabs B1010-236	Beams & Hollow Core Slabs Topped B1010-238	Beams & Double Tees Topped B1010-239	Bar Joists on Cols. & Brg. Walls B1010-248	Bar Joists & Beams on cols. B1010-250	Composite Beams & C.I.P. Slab B1010-252	Composite Deck & Slab, W Shapes B1010-254	Composite Beam & DK., Lt. Wt. Slab B1010-256	Wood Beams & Joists B1010-264	Laminated Wood Beams & Joists B1010-265
Superimposed Load = 40 PSF																
15 x 15	15.40	14.60	13.05	12.60	16.05	—	—	—	—	—	—	—	—	—	12.51	13.40
15 x 20	15.50	15.35	13.50	13.50	16.20	—	—	—	—	11.53	13.65	—	19.30	—	15.78	13.40
20 x 20	15.75	16.20	13.95	13.50	16.25	16.70	21.31	24.00	—	11.79	14.69	—	20.75	—	14.93	13.61
20 x 25	16.05	17.05	15.05	14.90	16.40	16.95	20.63	22.95	—	13.61	16.87	22.00	21.05	19.35	—	—
25 x 25	16.15	17.50	15.45	15.20	16.20	17.35	21.63	24.65	—	14.41	17.95	22.65	23.45	18.95	—	—
25 x 30	16.50	18.40	16.50	—	17.00	17.65	20.73	23.35	22.35	14.84	18.70	24.55	23.45	19.20	—	—
30 x 30	18.35	19.95	17.45	—	17.55	18.50	21.28	24.75	23.75	14.71	18.73	24.50	25.70	19.35	—	—
30 x 35	19.10	21.50	18.65	—	18.15	18.85	21.16	24.35	—	16.48	21.00	25.55	28.10	20.20	—	—
35 x 35	20.55	22.60	18.95	—	18.35	19.75	22.26	24.90	—	16.92	21.55	26.60	28.60	21.30	—	—
35 x 40	21.00	23.95	—	—	19.05	20.40	22.40	25.75	21.00	—	—	28.80	29.50	23.40	—	—
40 x 40	—	—	—	—	19.70	21.20	23.40	26.75	23.55	—	—	—	—	—	—	—
40 x 45	—	—	—	—	20.50	21.75	—	—	—	—	—	—	—	—	—	—
40 x 50	—	—	—	—	—	—	—	—	22.15	—	—	—	—	—	—	—
Superimposed Load = 75 PSF																
15 x 15	15.60	14.85	13.25	12.65	16.10	—	—	—	—	—	—	—	—	—	15.53	16.63
15 x 20	16.10	16.65	14.05	14.10	16.90	—	—	—	—	12.52	15.54	—	21.05	—	19.41	18.30
20 x 20	17.15	17.60	14.65	14.20	17.15	17.00	22.41	25.10	—	13.38	16.90	—	23.20	—	19.19	18.05
20 x 25	17.60	19.30	16.05	15.20	17.30	17.45	20.63	23.95	—	15.57	17.76	24.55	25.65	19.95	—	—
25 x 25	17.55	19.05	16.30	15.90	17.25	17.95	22.38	24.65	—	15.36	19.50	26.00	26.75	20.65	—	—
25 x 30	17.75	20.10	17.60	—	17.65	18.25	21.32	24.70	22.35	15.75	20.00	27.65	28.05	20.35	—	—
30 x 30	19.85	21.80	18.60	—	18.15	19.00	22.18	25.85	23.75	16.92	21.55	27.30	30.00	20.75	—	—
30 x 35	20.15	22.70	19.90	—	18.45	18.85	22.21	24.50	—	19.35	24.20	29.40	31.45	21.75	—	—
35 x 35	22.40	23.20	20.40	—	19.35	20.35	23.16	26.80	—	20.75	26.05	30.40	32.95	23.95	—	—
35 x 40	22.80	24.95	—	—	19.95	21.20	—	—	22.22	—	—	32.75	33.95	25.15	—	—
40 x 40	—	—	—	—	20.25	22.15	—	—	24.20	—	—	—	—	—	—	—
40 x 45	—	—	—	—	20.70	22.70	—	—	—	—	—	—	—	—	—	—
40 x 50	—	—	—	—	—	—	—	—	22.25	—	—	—	—	—	—	—
Superimposed Load = 125 PSF																
15 x 15	15.90	15.70	13.75	13.05	16.40	—	—	—	—	—	—	—	—	—	22.65	24.25
15 x 20	16.80	18.05	14.70	14.95	17.35	—	—	—	—	14.83	18.50	—	24.15	—	27.20	27.90
20 x 20	17.95	18.20	15.70	15.00	17.45	17.40	—	—	—	16.74	19.30	—	26.05	—	36.55	26.90
20 x 25	18.75	19.70	17.25	16.20	18.25	17.80	—	—	—	16.68	20.80	27.55	30.25	24.20	—	—
25 x 25	20.50	20.65	17.45	16.70	19.25	18.55	—	—	—	18.00	22.75	29.65	32.10	21.70	—	—
25 x 30	20.60	21.90	18.35	—	19.05	18.80	—	—	—	19.15	24.10	32.45	31.40	23.55	—	—
30 x 30	21.25	23.15	19.35	—	19.30	19.35	—	—	—	21.30	26.95	32.35	35.05	24.70	—	—
30 x 35	22.65	24.95	20.65	—	19.40	20.00	—	—	—	22.00	27.45	36.15	37.50	25.45	—	—
35 x 35	24.10	26.10	21.05	—	19.45	20.85	—	—	—	24.35	28.65	35.45	38.70	27.85	—	—
35 x 40	24.35	26.40	—	—	19.95	22.05	—	—	—	—	—	38.00	40.10	28.45	—	—
40 x 40	—	—	—	—	21.20	22.40	—	—	—	—	—	—	—	—	—	—
40 x 45	—	—	—	—	21.50	23.70	—	—	—	—	—	—	—	—	—	—
40 x 50	—	—	—	—	—	—	—	—	—	—	—	—	—	—	—	—
Superimposed Load = 200 PSF																
15 x 15	16.65	16.85	14.35	—	17.05	—	—	—	—	—	—	—	—	—	43.35	42.05
15 x 20	18.30	19.55	15.15	—	18.05	—	—	—	—	—	—	—	29.35	—	39.10	37.20
20 x 20	19.75	19.95	16.05	—	18.35	18.50	—	—	—	—	—	—	30.65	—	—	39.30
20 x 25	20.35	21.65	17.90	—	19.30	18.75	—	—	—	—	—	33.90	33.30	26.85	—	—
25 x 25	22.40	23.90	18.10	—	20.05	19.15	—	—	—	—	—	35.10	36.45	28.70	—	—
25 x 30	22.55	24.15	19.30	—	20.25	20.45	—	—	—	—	—	37.60	38.70	28.80	—	—
30 x 30	23.80	25.05	20.55	—	20.40	21.30	—	—	—	—	—	39.45	46.40	29.30	—	—
30 x 35	24.20	26.65	—	—	21.05	22.15	—	—	—	—	—	42.65	44.95	29.35	—	—
35 x 35	26.30	27.50	—	—	20.95	22.60	—	—	—	—	—	43.85	49.55	32.05	—	—
35 x 40	26.50	28.45	—	—	21.45	24.10	—	—	—	—	—	47.55	50.65	34.40	—	—
40 x 40	—	—	—	—	—	—	—	—	—	—	—	—	—	—	—	—
40 x 45	—	—	—	—	—	—	—	—	—	—	—	—	—	—	—	—
40 x 50	—	—	—	—	—	—	—	—	—	—	—	—	—	—	—	—

Design Assumptions:

Bay Size 20′ x 25′ = 500 S.F.

Use Concrete Flat Plate

Roof Load:	Superimposed Load	40 PSF
	Total Load	146 PSF
Floor Load:	Superimposed Load	75 PSF
	Total Load	188 PSF

Minimum Column size 20″

ROOF Total Roof Load .146 KSF x 500 S.F. =	73 K	
	73 K	
Column Load 1.67′ x 1.67′ x .15 KCF x 12′ =	5 K	
	78 K	
FOURTH Total Floor Load .188 K x 500 S.F. =	94 K	
	172 K	
Column	5 K	
	177 K	
THIRD Total Floor Load	94 K	
	271 K	
Column	5 K	
	276 K	
SECOND Total Floor Load	94 K	
	370 K	
Column	5 K	
	375 K	
FIRST Total Floor Load	94 K	
	469 K	
Column	5 K	
	474 K	
GROUND		
	474 K	
Total Load		

Description:

1. Multiply roof load x bay area.
2. Show total load at top of column.
3. Multiply est. column weight x floor to floor height.
4. Add to roof load.
5. Multiply floor load x bay area.
6. Add to roof and column load.
7. Multiply est. column weight x floor to floor height.
8. Add to total loads above.
9. Repeat steps above for remainder of floors & columns.
10. Enter the minimum reinforced portion of the table with total load on the column and the minimum allowable column size for the selected cast-in-place floor system.

    If the total load on the column does not exceed the allowable load shown, use the cost per L.F. multiplied by the length of columns required to obtain the column cost.
11. If the total load on the column exceeds the allowable working load shown in the minimum reinforced portion of the table enter the first portion of the table with the total load on the column and the minimum allowable column size from the selected cast-in-place floor system.

    Select a cost per L.F. for bottom level columns by total load or minimum allowable column size.

    Select a cost per L.F. for top level columns using the column size required for bottom level columns from the minimum reinforced portion of the table.

$$\frac{\text{Bottom \& Top Col. Costs/L.F.}}{2} = \text{Average Column Cost/L.F.}$$

$$\text{Column Cost} = \text{Average Col. Cost/L.F. x Length of Cols. Req'd.}$$

Description: Below is an example of steel column determination when roof and floor loads are known.

Calculation of Total Loads

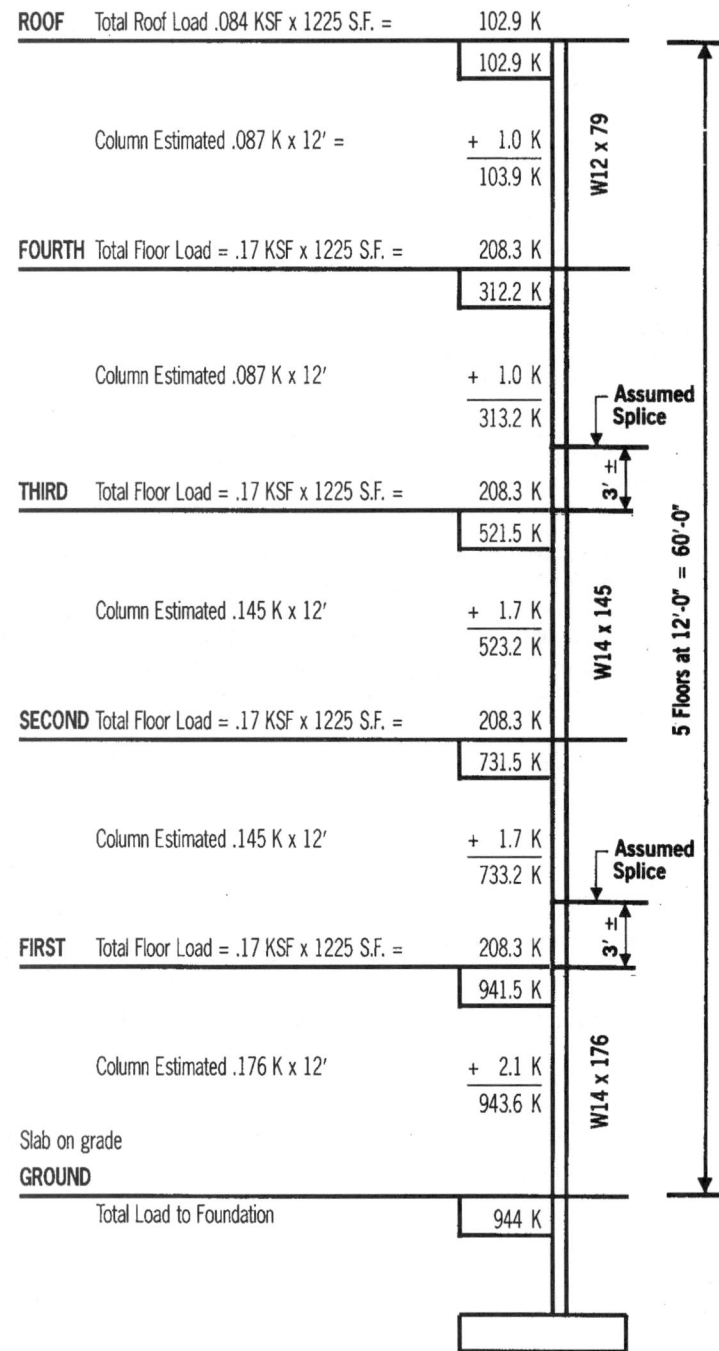

ROOF	Total Roof Load .084 KSF x 1225 S.F. =	102.9 K
		102.9 K
	Column Estimated .087 K x 12' =	+ 1.0 K
		103.9 K
FOURTH	Total Floor Load = .17 KSF x 1225 S.F. =	208.3 K
		312.2 K
	Column Estimated .087 K x 12'	+ 1.0 K
		313.2 K
THIRD	Total Floor Load = .17 KSF x 1225 S.F. =	208.3 K
		521.5 K
	Column Estimated .145 K x 12'	+ 1.7 K
		523.2 K
SECOND	Total Floor Load = .17 KSF x 1225 S.F. =	208.3 K
		731.5 K
	Column Estimated .145 K x 12'	+ 1.7 K
		733.2 K
FIRST	Total Floor Load = .17 KSF x 1225 S.F. =	208.3 K
		941.5 K
	Column Estimated .176 K x 12'	+ 2.1 K
		943.6 K
Slab on grade		
GROUND		
	Total Load to Foundation	944 K

W12 x 79

Assumed Splice — 3' ±H

5 Floors at 12'-0" = 60'-0"

W14 x 145

Assumed Splice — 3' ±H

W14 x 176

Design Assumptions:

Bay Size	35' x 35' = 1,225 S.F.		
Roof Load:	Superimposed Load	40	PSF
	Dead Load	44	PSF
	Total Load	84	PSF
Floor Load:	Superimposed Load	125	PSF
	Dead Load	45	PSF
	Total Load	170	PSF

Description:

1. Multiply roof load x bay area.
2. Show total load at top of column.
3. Multiply estimated column weight* x floor to floor height.
4. Add to roof load.
5. Multiply floor load x bay area.
6. Add to roof and column load.
7. Multiply estimated column weight x floor to floor height.
8. Add to total loads above.
9. Choose column from **System B1010 208** using unsupported height.
10. Interpolate or use higher loading to obtain cost/L.F.
11. Repeat steps above for remainder of floors and columns.
12. Multiply number of columns by the height of the column times the cost per foot to obtain the cost of each type of column.

* To Estimate Column Weight

Roof Load		.084 KSF
Floor Load x No Floors above Splice		
170 x 1		.170 KSF
	Total	.254 KSF

Total Load (KSF) x Bay Area (S.F.) = Load to Col.
.254 KSF x 1,225 KSF = 311 K

From **System B1010 208**, choose a column by:

Load,	Height,	Weight
400 K	10'	79 lb.

Table B1010-241   One, Two and Three Member Beams, Maximum Load for Various Spans

Description: Table below lists the maximum uniform load (W) or concentrated load (P) allowable for various size beams at various spans.

Design Assumptions: Fiber strength (f) is 1,000 psi.

Maximum deflection does not exceed 1/360 the span of the beam.

Modulus of elasticity (E) is 1,100,000 psi. Anything less will result in excessive deflection in the longer members so that spans must be reduced or member size increased.

The uniform loads (W) are in pounds per foot; concentrated loads (P), at the midpoint, are in pounds.

The span is in feet and is the unsupported clear span.

The member sizes are from 2″ x 6″ to 4″ x 12″ and include one, two and three pieces of each member size to arrive at spans and loading.

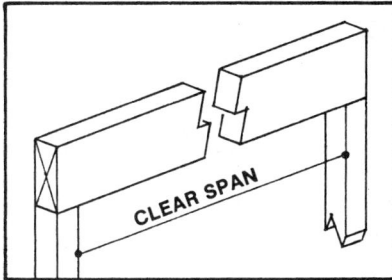

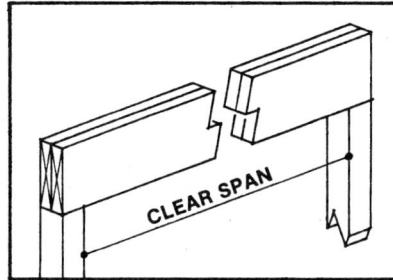

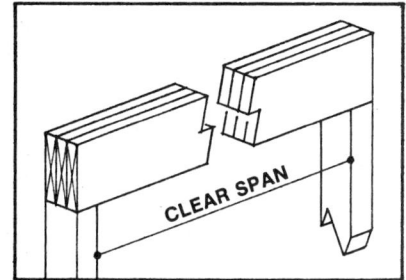

Individual Member		6′		8′		10′		12′		14′		16′		18′	
		W	P	W	P	W	P	W	P	W	P	W	P	W	P
Size	No.	#/L.F.	#	#/L.F.	#	#/L.F.	#	#/L.F.	#	#/L.F.	#	#/L.F.	#	#/L.F.	#
2″ x 6″	2	238	840	—	—										
	3	357	1260	—	—										
2″ x 8″	2	314	1460	235	1095										
	3	471	2190	353	1642	232	1164								
2″ x 10″	2	400	2376	300	1782	240	1426	—	—						
	3	600	3565	450	2673	360	2139	279	1679						
2″ x 12″	2	487	2925	365	2636	292	2109	243	1757	208	1479	—	—		
	3	731	4387	548	3955	548	3164	365	2636	313	2219	212	1699		
2″ x 14″	2	574	3445	430	3445	344	2926	287	2438	246	2090	215	1828	—	—
	3	861	5167	645	5167	516	4389	430	3657	369	3135	322	2743	243	2193
3″ x 6″	1	—	—	—	—										
	2	397	1400	200	882										
	3	595	2100	300	1323										
3″ x 8″	1	261	1216	—	—	—	—	—	—						
	2	523	2433	392	1825	258	1293	—	—						
	3	785	3650	589	2737	388	1940	224	1347						
3″ x 10″	1	334	1980	250	1485	200	1188	—	—	—	—				
	2	668	3961	500	2970	400	2376	310	1865	—	—	—	—		
	3	1002	5941	750	4456	600	3565	466	2798	293	2056				
3″ x 12″	1	406	2437	304	2197	243	1757	203	1464	—	—	—	—	—	—
	2	812	4875	608	4394	487	3515	406	2929	348	2466	236	1888	—	—
	3	1218	7312	912	6591	731	5273	609	4394	522	3699	354	2832	248	2237
4″ x 6″	1	278	980	—	—	—	—								
	2	556	1960	308	1235	—	—								
4″ x 8″	1	366	1703	274	1277	—	—	—	—						
	2	733	3406	549	2555	362	1811	209	1257						
4″ x 10″	1	467	2772	350	2079	280	1663	217	1306	—	—				
	2	935	5545	700	4159	560	3327	435	2612	274	1919				
4″ x 12″	1	568	3412	426	3076	341	2460	284	2050	243	1726	—	—		
	2	1137	6825	853	6152	682	4921	568	4100	487	3452	330	2643	232	2088

Table B1010-711 Maximum Floor Joist Spans

Description: Table below lists the maximum clear spans and the framing lumber quantity required per S.F. of floor.

Design Assumptions: Dead load = joist weight plus floor weight of 5 psf plus ceiling or partition load of 10 psf.

Maximum L.L. deflection is 1/360 of the clear span.

Modulus of elasticity is 1,100,000 psi.

Fiber strength (f_b) is 1,000 psi.

10% allowance has been added to framing quantities for overlaps, waste, double joists at openings, etc. 5% added to subfloor for waste.

Maximum span is in feet and is the unsupported clear span.

Floor Joist		Framing Lumber	Live Load in Pounds per Square Foot							
Size in Inches	Spacing C/C	B.F./S.F.	30	40	50	60	70	80	90	100
			Span	Span	Span	Span	Span	Span	Span	Span
2" x 6"	12"	1.10	9'-0"	8'-6"	8'-0"	7'-8"	7'-4"	7'-0"	6'-10"	6'-7"
	16"	.83	8'-4"	7'-9"	7'-4"	7'-0"	6'-8"	6'-4"	6'-0"	—
	24"	.55	7'-2"	6'-9"	6'-2"	5'-9"	—	—	—	—
2" x 8"	12"	1.46	12'-0"	11'-2"	10'-7"	10'-0"	9'-8"	9'-4"	9'-0"	8'-8"
	16"	1.10	10'-10"	10'-2"	9'-7"	9'-2"	8'-9"	8'-4"	7'-10"	7'-6"
	24"	.74	9'-6"	8'-10"	8'-2"	7'-6"	7'-0"	6'-9"	6'-5"	6'-0"
2" x 10"	12"	1.84	15'-4"	14'-4"	13'-6"	12'-10"	12'-4"	11'-10"	11'-6"	11'-2"
	16"	1.38	13'-10"	13'-0"	12'-4"	11'-8"	11'-2"	10'-7"	10'-0"	9'-7"
	24"	.91	12'-0"	11'-4"	10'-5"	9'-9"	9'-0"	8'-7"	8'-3"	7'-10"
2" x 12"	12"	2.20	18'-7"	17'-4"	16'-5"	15'-8"	15'-0"	14'-6"	14'-0"	13'-6"
	16"	1.65	16'-10"	15'-9"	14'-10"	14'-3"	13'-7"	12'-10"	12'-3"	11'-8"
	24"	1.10	14'-9"	13'-9"	12'-8"	11'-9"	11'-2"	10'-6"	10'-0"	9'-6"
2" x 14"	12"	2.56	21'-10"	20'-6"	19'-4"	18'-5"	17'-9"	17'-0"	16'-6"	15'-10"
	16"	1.93	19'-10"	18'-7"	17'-7"	16'-9"	16'-0"	15'-2"	14'-5"	13'-9"
	24"	1.29	17'-4"	16'-3"	15'-0"	13'-10"	13'-2"	12'-4"	11'-9"	11'-3"
3" x 6"	12"	1.65	10'-9"	10'-0"	9'-6"	9'-0"	8'-8"	8'-4"	8'-0"	7'-10"
	16"	1.24	9'-9"	9'-0"	8'-8"	8'-3"	7'-10"	7'-7"	7'-4"	7'-0"
	24"	.83	8'-6"	8'-0"	7'-6"	7'-2"	6'-10"	6'-8"	6'-4"	6'-0"
3" x 8"	12"	2.20	14'-2"	13'-3"	12'-6"	11'-10"	11'-6"	11'-0"	10'-8"	10'-4"
	16"	1.65	12'-10"	12'-0"	11'-4"	10'-10"	10'-5"	10'-0"	9'-9"	9'-5"
	24"	1.10	11'-3"	10'-6"	9'-10"	9'-6"	9'-0"	8'-9"	8'-3"	7'-10"
3" x 10"	12"	2.75	18'-0"	16'-10"	16'-0"	15'-4"	14'-8"	14'-0"	13'-8"	13'-3"
	16"	2.07	16'-5"	15'-4"	14'-6"	13'-10"	13'-3"	12'-10"	12'-4"	12'-0"
	24"	1.38	14'-4"	13'-5"	12'-8"	12'-0"	11'-7"	11'-0"	10'-7"	10'-0"
3" x 12"	12"	3.30	22'-0"	20'-7"	19'-6"	18'-7"	17'-10"	17'-0"	16'-7"	16'-0"
	16"	2.48	20'-0"	18'-9"	17'-8"	16'-10"	16'-2"	15'-7"	15'-0"	14'-7"
	24"	1.65	17'-6"	16'-4"	15'-6"	14'-9"	14'-0"	13'-7"	12'-10"	12'-4"
4" x 6"	12"	2.20	—	—	—	10'-0"	9'-9"	9'-4"	9'-0"	8'-9"
	16"	1.65	—	—	—	9'-2"	8'-10"	8'-6"	8'-3"	8'-0"
	24"	1.10	—	—	—	8'-0"	7'-9"	7'-4"	7'-2"	7'-0"
4" x 8"	12"	2.93	—	—	—	13'-4"	12'-10"	12'-4"	12'-0"	11'-7"
	16"	2.20	—	—	—	12'-0"	11'-8"	11'-3"	10'-10"	10'-6"
	24"	1.47	—	—	—	10'-7"	9'-2"	9'-9"	9'-6"	9'-2"
4" x 10"	12"	3.67	—	—	—	17'-0"	16'-4"	15'-9"	15'-3"	14'-9"
	16"	2.75	—	—	—	15'-6"	14'-7"	14'-4"	13'-10"	13'-5"
	24"	1.83	—	—	—	13'-6"	13'-0"	12'-6"	12'-0"	11'-9"

Table B1010-781   Decking Material Characteristics

Description: The table below lists the maximum spans for commonly used wood decking materials for various loading conditions.

Design Assumptions: Applied load is the total load, live plus dead.

Maximum deflection is 1/180 of the clear span which is not suitable if plaster ceilings will be supported by the roof or floor.

Modulus of elasticity (E) and fiber strength (F) are as shown in table to the right.

No allowance for waste has been included.

Deck Material	Modulus of Elasticity	Fiber Strength
Cedar	1,100,000 psi	1,000 psi
Douglas Fir	1,760,000	1,200
Hemlock	1,600,000	1,200
White Spruce	1,320,000	1,200

Assume Dead Load of wood plank is 33.3 lbs. per C.F.

Table B1010-782 Maximum Spans for Wood Decking

Nominal Thickness	Type Wood	B.F. per S.F.	Total Uniform Load per S.F.							
			40	50	70	80	100	150	200	250
2″	Cedar	2.40	6.5′	6′	5.7′	5.3′	5′	4.5′	4′	3.5′
	Douglas fir	2.40	8′	7′	6.7′	6.3′	6′	5′	4.5′	4′
	Hemlock	2.40	7.5′	7′	6.5′	6′	5.5′	5′	4.5′	4′
	White spruce	2.40	7′	6.5′	6′	5.5′	5′	4.5′	4′	3.5′
3″	Cedar	3.65	11′	10′	9′	8.5′	8′	7′	6.5′	6′
	Douglas fir	3.65	13′	12′	11′	10.5′	10′	8′	7.5′	7′
	Hemlock	3.65	12′	11′	10.5′	10′	9′	8′	7.5′	7′
	White spruce	3.65	11′	10′	9.5′	9′	8′	7.5′	7′	6′
4″	Cedar	4.65	15′	14′	13′	12′	11′	10′	9′	8′
	Douglas fir	4.65	18′	17′	15′	14′	13′	11′	10.5′	10′
	Hemlock	4.65	17′	16′	14′	13.5′	13′	11′	10′	9′
	White spruce	4.65	16′	15′	13′	12.5′	12′	10′	9′	8′
6″	Cedar	6.65	23′	21′	19′	18′	17′	15′	14′	13′
	Douglas fir	6.65	24′+	24′	23′	22′	21′	18′	16′	15′
	Hemlock	6.65	24′+	24′	22′	21′	20′	17′	16′	15′
	White spruce	6.65	24′	23′	21′	20′	18′	16′	14′	13′

Table B1020-511 Maximum Roof Joist or Rafter Spans

Description: Table below lists the maximum clear spans and the framing lumber quantity required per S.F. of roof. Spans and loads are based on flat roofs.

Design Assumptions: Dead load = Joist weight plus weight of roof sheathing at 2.5 psf plus either of two alternate roofing weights.

Maximum deflection is 1/360 of the clear span.

Modulus of elasticity is 1,100,000 psi.

Fiber strength (f) is 1,000 psi.

Allowance: 10% has been added to framing quantities for overlaps, waste, double joists at openings, etc.

Maximum span is measured horizontally in feet and is the clear span.

Note: To convert to inclined measurements or to add for overhangs, multiply quantities in Table B1020-511 by factors in Table B1020-512.

Joist			Live Load in Pounds per Square Foot							
			Group I Covering (2.5 psf)				Group II Covering (8.0 psf)			
Size in Inches	Center to Center Spacing	Framing B.F. per S.F.	15 Span	20 Span	30 Span	40 Span	20 Span	30 Span	40 Span	50 Span
2" x 4"	12"	0.73	7'-11"	7'-3"	6'-5"	5'-10"	6'-9"	6'-1"	5'-7"	5'-3"
	16"	0.55	7'-2"	6'-7"	5'-10"	5'-4"	6'-1"	5'-6"	5'-1"	4'-10"
	24"	0.37	6'-3"	5'-9"	5'-0"	4'-8"	5'-4"	4'-10"	4'-6"	4'-2"
2" x 6"	12"	1.10	12'-5"	11'-5"	10'-0"	9'-3"	10'-7"	9'-7"	8'-10"	8'-4"
	16"	0.83	11'-3"	10'-4"	9'-2"	8'-5"	9'-8"	8'-9"	8'-0"	7'-6"
	24"	0.55	9'-10"	9'-0"	8'-0"	7'-4"	8'-5"	7'-7"	7'-0"	6'-6"
2" x 8"	12"	1.46	16'-5"	15'-0"	13'-4"	12'-2"	14'-0"	12'-8"	11'-8"	11'-0"
	16"	1.10	14'-11"	13'-8"	12'-0"	11'-0"	12'-9"	11'-6"	10'-7"	11'-0"
	24"	0.74	13'-0"	12'-0"	10'-7"	9'-8"	11'-1"	10'-0"	9'-3"	10'-0"
2" x 10"	12"	1.84	20'-11"	19'-3"	17'-0"	19'-7"	17'-11"	16'-2"	14'-11"	14'-0"
	16"	1.38	19'-0"	17'-6"	15'-6"	14'-2"	16'-3"	14'-8"	13'-7"	12'-9"
	24"	0.91	16'-7"	15'-3"	13'-6"	12'-4"	14'-2"	12'-10"	11'-10"	11'-0"
2" x 12"	12"	2.20	25'-6"	23'-5"	20'-9"	18'-11"	21'-9"	19'-8"	18'-2"	17'-1"
	16"	1.65	23'-1"	21'-3"	18'-10"	17'-2"	19'-9"	17'-10"	16'-6"	15'-6"
	24"	1.10	20'-2"	18'-7"	16'-5"	15'-0"	17'-3"	15'-7"	14'-5"	13'-5"
2" x 14"	12"	2.56	30'-0"	27'-7"	24'-5"	22'-4"	25'-8"	23'-2"	21'-5"	20'-1"
	16"	1.93	27'-3"	25'-0"	22'-2"	20'-3"	23'-3"	21'-0"	19'-6"	18'-3"
	24"	1.29	23'-10"	21'-11"	19'-4"	17'-8"	20'-4"	18'-4"	17'-0"	15'-10"
3" x 6"	12"	1.65	14'-9"	13'-7"	12'-0"	11'-0"	12'-7"	11'-4"	10'-6"	9'-11"
	16"	1.24	13'-4"	12'-4"	10'-11"	9'-11"	11'-5"	10'-4"	9'-7"	9'-0"
	24"	.83	11'-9"	10'-9"	9'-6"	8'-8"	10'-0"	9'-0"	8'-4"	7'-10"
3" x 8"	12"	2.20	19'-5"	17'-11"	15'-10"	14'-6"	16'-7"	15'-0"	13'-11"	13'-0"
	16"	1.65	17'-8"	16'-9"	14'-4"	13'-2"	15'-1"	13'-7"	12'-7"	11'-10"
	24"	1.10	15'-5"	14'-2"	12'-6"	11'-6"	13'-2"	11'-11"	11'-0"	10'-4"
3" x 10"	12"	2.75	24'-10"	22'-10"	22'-2"	18'-6"	21'-9"	19'-2"	17'-9"	16'-8"
	16"	2.07	22'-6"	20'-9"	18'-4"	16'-9"	19'-9"	17'-4"	16'-1"	15'-1"
	24"	1.38	19'-8"	18'-1"	16'-0"	14'-8"	16'-10"	15'-2"	14'-0"	13'-2"
3" x 12"	12"	3.30	30'-2"	27'-9"	24'-6"	22'-5"	25'-10"	23'-4"	21'-6"	20'-3"
	16"	2.48	27'-5"	25'-3"	22'-3"	20'-4"	23'-5"	21'-2"	19'-7"	18'-4"
	24"	1.65	24'-0"	22'-0"	19'-6"	17'-10"	20'-6"	18'-6"	17'-1"	16'-0"

Table B1020-512 Factors for Converting Inclined to Horizontal

Rafters: The quantities shown in Table B1020-511 are for flat roofs. For inclined roofs using rafters, the quantities should be multiplied by the factors in Table B1020-512 to allow for the increased area of the inclined roofs.

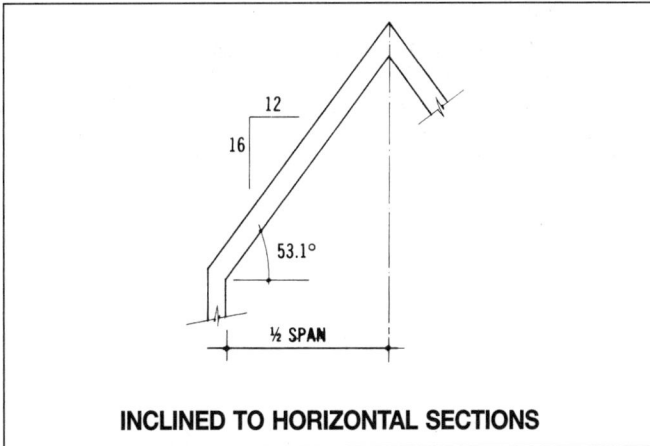

INCLINED TO HORIZONTAL SECTIONS

Roof Slope	Approx. Angle	Factor	Roof Slope	Approx. Angle	Factor
Flat	0°	1.000	12 in 12	45.0	1.414
1 in 12	4.8°	1.003	13 in 12	47.3	1.474
2 in 12	9.5°	1.014	14 in 12	49.4	1.537
3 in 12	14.0°	1.031	15 in 12	51.3	1.601
4 in 12	18.4°	1.054	16 in 12	53.1	1.667
5 in 12	22.6°	1.083	17 in 12	54.8	1.734
6 in 12	26.6°	1.118	18 in 12	56.3	1.803
7 in 12	30.3°	1.158	19 in 12	57.7	1.873
8 in 12	33.7°	1.202	20 in 12	59.0	1.943
9 in 12	36.9°	1.250	21 in 12	60.3	2.015
10 in 12	39.8°	1.302	22 in 12	61.4	2.088
11 in 12	42.5°	1.357	23 in 12	62.4	2.162

Table B1020-513 below can be used two ways:
1. Use 1/2 span with overhang on one side only.
2. Use whole span with total overhang from both sides.

Roof or Deck Overhangs: The quantities shown in table do not include cantilever overhangs. The S.F. quantities should be multiplied by the factors in Table B1020-513 to allow for the increased area of the overhang. All dimensions are horizontal. For inclined overhangs also multiply by factors in Table B1020-512 above.

Table B1020-513 Allowance Factors for Including Roof or Deck Overhangs

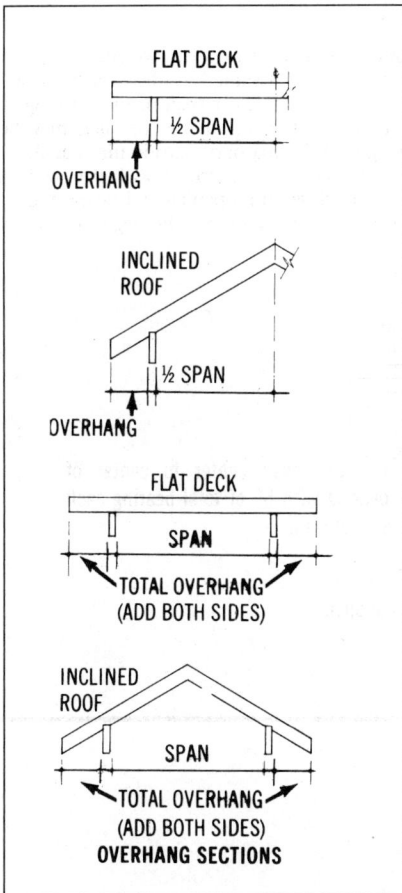

OVERHANG SECTIONS

Horizontal Span	Roof overhang measured horizontally							
	0'-6"	1'-0"	1'-6"	2'-0"	2'-6"	3'-0"	3'-6"	4'-0"
6'	1.083	1.167	1.250	1.333	1.417	1.500	1.583	1.667
7'	1.071	1.143	1.214	1.286	1.357	1.429	1.500	1.571
8'	1.063	1.125	1.188	1.250	1.313	1.375	1.438	1.500
9'	1.056	1.111	1.167	1.222	1.278	1.333	1.389	1.444
10'	1.050	1.100	1.150	1.200	1.250	1.300	1.350	1.400
11'	1.045	1.091	1.136	1.182	1.227	1.273	1.318	1.364
12'	1.042	1.083	1.125	1.167	1.208	1.250	1.292	1.333
13'	1.038	1.077	1.115	1.154	1.192	1.231	1.269	1.308
14'	1.036	1.071	1.107	1.143	1.179	1.214	1.250	1.286
15'	1.033	1.067	1.100	1.133	1.167	1.200	1.233	1.267
16'	1.031	1.063	1.094	1.125	1.156	1.188	1.219	1.250
17'	1.029	1.059	1.088	1.118	1.147	1.176	1.206	1.235
18'	1.028	1.056	1.083	1.111	1.139	1.167	1.194	1.222
19'	1.026	1.053	1.079	1.105	1.132	1.158	1.184	1.211
20'	1.025	1.050	1.075	1.100	1.125	1.150	1.175	1.200
21'	1.024	1.048	1.071	1.095	1.119	1.143	1.167	1.190
22'	1.023	1.045	1.068	1.091	1.114	1.136	1.159	1.182
23'	1.022	1.043	1.065	1.087	1.109	1.130	1.152	1.174
24'	1.021	1.042	1.063	1.083	1.104	1.125	1.146	1.167
25'	1.020	1.040	1.060	1.080	1.100	1.120	1.140	1.160
26'	1.019	1.038	1.058	1.077	1.096	1.115	1.135	1.154
27'	1.019	1.037	1.056	1.074	1.093	1.111	1.130	1.148
28'	1.018	1.036	1.054	1.071	1.089	1.107	1.125	1.143
29'	1.017	1.034	1.052	1.069	1.086	1.103	1.121	1.138
30'	1.017	1.033	1.050	1.067	1.083	1.100	1.117	1.133
32'	1.016	1.031	1.047	1.063	1.078	1.094	1.109	1.125

The "Exterior Closure" Division has many types of wall systems which are most commonly used in current United States and Canadian building construction.

Systems referenced have an illustration as well as description and cost breakdown. Total costs per square foot of each system are provided. Each system is itemized with quantities, extended and summed for material, labor, or total.

Exterior walls usually consist of several component systems. The weighted square foot cost of the enclosure is made up of individual component systems costs, prorated for their percent of wall area. For instance, a given wall system may take up 70% of the enclosure wall's composite area. The fenestration may make up the balance of 30%. The components should be factored by the proportion of area each contributes to the total.

Example: An enclosure wall is 100' long and 15' high.
The total wall area is therefore 1,500 S.F.
70% of the wall area is a brick & block system.
@ $32.00/S.F. (B2010-132-1240)

30% is a window system:

Framing	$27.10	(B2020-210-1700)
Glazing	29.15	(B2020-220-1400)
	$56.25	

Solution: Brick & block = 1,500 S.F. x .70 x $32.00 = $33,600
Windows = 1,500 S.F. x .30 x $56.25 = $25,313
Total enclosure wall cost = $58,913
or $39.28 per S.F. of wall

The square foot unit costs include the installing contractors markup for overhead and profit. However, fees for professional services must be added along with other items. See H1010, General Conditions, for these costs. Note that interior finishes (drywall, etc.) are included elsewhere.

There may be times when the wall systems costs will need to be expressed in units of square feet of floor area.

Example: A building 25' x 25' requires a wall 100' long and 15' high.

Cost of the wall is $39.28 per S.F. of wall.
The floor area is 625 S.F.

Solution: Wall cost per S.F of floor area

$$\frac{100 \text{ L.F. x } 15' \text{ x } \$39.28}{625 \text{ S.F.}} = \$94.26 \text{ per S.F. of floor}$$

Concrete Block Lintels

Design Assumptions:
Bearing length each end = 8 inches
f'c = 2,000 p.s.i.
Wall load = 300 lbs. per linear foot (p.l.f.)
Floor & roof load (includes wall load) = 100 p.l.f.
Weight of 8" deep lintel (included in wall load) = 50 p.l.f.
Weight of 16" deep lintel (included in wall load) = 100 p.l.f.

Load to be supported by the lintel is shown below. The weight of the masonry above the opening can be safely assumed as the weight of a triangular section whose height is one-half the clear span of the opening. Corbelling action of the masonry above the top of the opening may be counted upon to support the weight of the masonry outside the triangle. To the dead load of a wall must be added the uniform dead & live loads of floors & roof that frame into the wall above the opening and below the apex of the 45° triangle. Any load above the apex may be neglected.

6x8x8   8x8x8   6x16x8   8x16x8

Uniform Floor & Roof Load

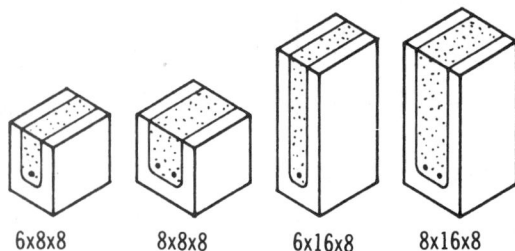

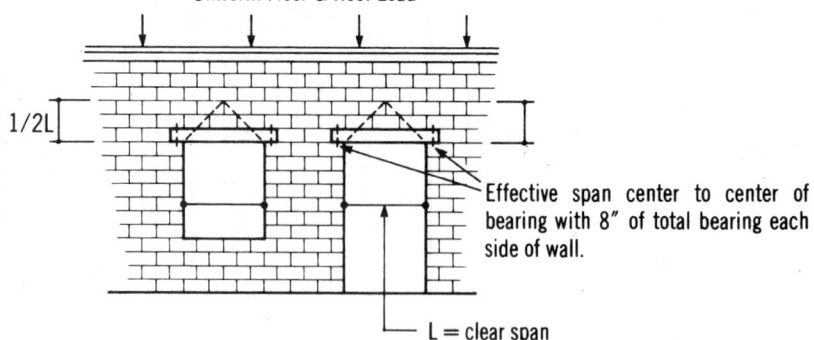

1/2L

Effective span center to center of bearing with 8" of total bearing each side of wall.

L = clear span

Table B2010-031 Example Using 14″ Masonry Cavity Wall with 1″ Plaster for the Exterior Closure

Total Heat Transfer is found using the equation
$Q = AU (T_2 — T_1)$ where:

Q = Heat flow, BTU per hour
A = Area, square feet
U = Overall heat transfer coefficient
$(T_2 — T_1)$ = Difference in temperature of the air on each side of the construction component in degrees Fahrenheit

Coefficients of Transmission ("U") are expressed in BTU per (hour) (square foot) (Fahrenheit degree difference in temperature between the air on two sides) and are based on 15 mph outside wind velocity.

The lower the U-value the higher the insulating value.
$U = 1/R$ where "R" is the summation of the resistances of air films, materials and air spaces that make up the assembly.

Figure B2010-031 Example Using 14″ Masonry Cavity Wall with 1″ Plaster for the Exterior Closure

Construction	Resistance R.
1. Outside surface (15 mph wind)	0.17
2. Face Brick (4″)	0.44
3. Air space (2″, 50° mean temp, 10° diff)	1.02
4. Concrete block (8″, lightweight)	2.12
5. Plaster (1″, sand aggregate)	0.18
6. Inside Surface (still air)	0.68
Total resistance	4.61

$U = 1/R = 1/4.61 =$ 0.22
Weight of system = 77 psf
I.S. = Initial System

Assume 20,000 S.F. of wall in Boston, MA (5600 degree days) Table L1030-501
Initial system effective $U = U_w \times M = 0.22 \times 0.905 = 0.20$
  (M = weight correction factor from Table L1030-203)
Modified system effective $U = U_w \times M = 0.067 \times 0.915 = 0.06$
Initial systems: $Q = 20{,}000 \times 0.20 \times (72° — 10°) = 248{,}000$ BTU/hr.
Modified system: $Q = 20{,}000 \times 0.06 \times (72° — 10°) = 74{,}400$ BTU/hr.

Reduction in BTU/hr. heat loss; $\dfrac{\text{I.S. - M.S.}}{\text{I.S.}} \times 100 = \dfrac{248{,}000 - 74{,}400}{248{,}000} \times 100 = 70\%$

Replace item 3 with 2″ smooth rigid polystyrene insulation and item 5 with 3/4″ furring and 1/2″ drywall.

Total resistance		4.61
Deduct 3. Airspace	1.02	
5. Plaster	0.18	
	1.20	
Difference 4.61 — 1.20 =		3.41
Add rigid polystyrene insulation		10.00
3/4″ air space		1.01
1/2″ gypsum board		0.45
Total resistance		14.87

$U = 1/R = 1/14.87 =$ 0.067
Weight of system = 70 psf
M.S. = Modified System

Example of Calculating Envelope Requirements per ASHRAE 90-75

Example: For Boston, Massachusetts, based on the envelope energy considerations of ASHRAE 90-75, determine the amount of glass that could be used in a five story 50′ x 80′ office building with a 12′ floor to floor height using both systems above.

Step #1. Look up Degree Days for Boston from Table L1030-501=5600

Step #2. Look up overall "U" value for walls (buildings over three story) from Table L1030-102 $U_o = 0.34$

Step #3. Look up weight modification factor for each system from Table L1030-203 $M_{i.s.} = 0.905$, $M_{m.s.} = 0.915$

Step #4. Calculate effective "U" values for each system
$U_{i.s.} = 0.22 \times 0.905 = 0.20$,
$U_{m.s.} = 0.067 \times 0.915 = 0.06$

Step #5. Look up R values for glass from Table L1030-401
Single glass R = 0.91, insulated glass with 1/2″ air space R = 2.04

Step #6. Calculate "U" values for glass
Single glass $U = 1/R = 1/.91 = 1.10$,
Insulated glass $U = 1/R = 1/2.04 = 0.49$

Step #7. From Table L1030-201 for single glass, enter table from left at $U_o = 0.34$ and proceed to the right to the intersection of the $U_w = 0.20$ and $U_w = 0.06$ lines. Project to the bottom and read 15% and 26% respectively.

Step #8. Similarly, from Table L1030-202 for insulated glass, read 28% and 47% respectively.

Step #9. Calculate area of exterior closure:
$[50 + 50 + 80 + 80] \times 5 \times 12 = 15{,}600$ SF

Step #10. Calculate maximum glass area for each situation:

Initial System
Single glass: $A = 15{,}600 \times .15 = 2{,}340$ SF
Double glass: $A = 15{,}600 \times .28 = 4{,}368$ SF

Modified System
$A = 15{,}600 \times .26 = 4{,}056$ SF
$A = 15{,}600 \times .47 = 7{,}332$ SF

Conclusion: In order to meet the envelope requirements of ASHRAE 90-75, the glass area is limited to the square footage shown in Step #10 for the two systems considered. If the maximum glass were used in any of the situations, then the heat loss through the walls would be based on $U_o = 0.34$ (Step 2).

For any system:

$$U = \frac{U \text{ Wall} \times \text{Area Wall} + U \text{ Glass} \times \text{Area Glass}}{\text{Total Area}}$$

461

Table B2010-032　Thermal Coefficients of Exterior Closures

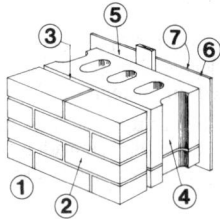

EXAMPLE:

Construction	Resistance R
1. Outside surface (15 MPH wind)	0.17
2. Common brick, 4″	0.80
3. Nonreflective air space, 0.75″	1.01
4. Concrete block, S&G aggregate, 8″	1.46
5. Nonreflective air space, 0.75″	1.01
6. Gypsum board, 0.5″	0.45
7. Inside surface (still air)	0.68
Total resistance	5.58

$U = 1/R = 1/5.58 = 0.18$

System weight = 97 psf

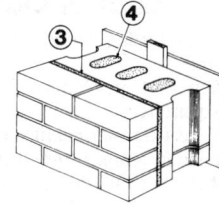

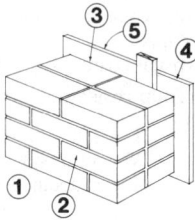

Substitution

Replace item 3 with perlite loose fill insulation and fill block cavities with the same.

Total resistance (Example)	**R =**	5.58
Deduct 3. Air space	1.01	
4. Concrete block	1.46	
	2.47	− 2.47
Difference		3.11
Add 0.75″ perlite cavity fill		2.03
Perlite filled 8″ block		2.94
Net total resistance		8.08

$U = 1/R = 1/8.08 = 0.12$

System weight = 97 psf

EXAMPLE:

Construction	Resistance R
1. Outside surface (15 MPH wind)	0.17
2. Common brick, 8″	1.60
3. Nonreflective air space, 0.75″	1.01
4. Gypsum board, 0.625″	0.56
5. Inside surface (still air)	0.68
Total resistance	4.02

$U = 1/R = 1/4.02 = 0.25$

System weight = 82 psf

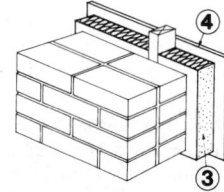

Substitution

Replace item 3 with 4″ blanket insulation and item 4 with 0.75″ Gypsum plaster (sand agg.).

Total resistance (Example)	**R =**	4.02
Deduct 3. Air space	1.01	
4. Gypsum board	0.56	
	1.57	− 1.57
Difference		2.45
ADD Blanket insulation		13.00
Gypsum plaster		0.14
Net total resistance		15.59

$U = 1/R = 1/15.59 = 0.06$

System weight = 87 psf

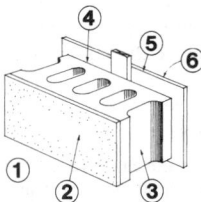

EXAMPLE:

Construction	Resistance R
1. Outside surface (15 MPH wind)	0.17
2. Cement stucco, 0.75″	0.15
3. Concrete block, 8″ light weight	2.12
4. Reflective air space, 0.75″	2.77
5. Gypsum board, 0.5″	0.45
6. Inside surface (still air)	0.68
Total resistance	6.34

$U = 1/R = 1/6.34 = 0.16$

System weight = 47 psf

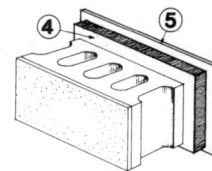

Substitution

Replace item 4 with 2″ insulation board and item 5 with 0.75″ Perlite plaster.

Total resistance (Example)	**R =**	6.34
Deduct 4. Air space	2.77	
5. Gypsum board	0.45	
	3.22	− 3.22
Difference		3.12
Add Insulation board 2″ (polyurethane)		12.50
Perlite plaster		1.34
Net total resistance		16.96

$U = 1/R = 1/16.96 = 0.06$

System weight = 48 psf

Table B2010-112   Partially and Fully Grouted Reinforced Concrete Masonry Wall Capacities Per L.F. (Kips & In-Kips)

Thk. T (Nom.) (in)	Length Or Height h' (Ft.)	Length Or Height h'/T (in/in)	Grouted Core & Rebar (spacing) (in O.C.)	Grouted Core & Rebar Rebar Size (@ d.)	Eccentric Loads 7.0 in-K/Ft. Inspection No (K/Ft.)	Eccentric Loads 7.0 in-K/Ft. Inspection Yes (K/Ft.)	Eccentric Loads 3.5 in-K/Ft. Inspection No (K/Ft.)	Eccentric Loads 3.5 in-K/Ft. Inspection Yes (K/Ft.)	Without Wind or Eccentric Loads Inspection No (K/Ft.)	Without Wind or Eccentric Loads Inspection Yes (K/Ft.)	With Wind Inspection No 15 psf (K/Ft.)	With Wind Inspection No 30 psf (K/Ft.)	With Wind Inspection Yes 15 & 30 (K/Ft.)	Allowable Wall Moments Not Wind or Earthquake Inspection No (in.-K/Ft.)	Allowable Wall Moments Not Wind or Earthquake Inspection Yes (in.-K/Ft.)
8" Conc. Block	8'	12	48"	#8	5.10	12.55	6.25	13.70	7.75	14.90	7.45	7.45	14.90	7.55	12.20
			32"	#5	5.45	13.35	6.65	14.60	7.90	15.80	7.90	7.90	15.80	6.45	9.60
			16"	↓	6.50	15.80	7.90	17.15	9.25	18.50	9.25	9.25	18.50	7.95	12.85
			8"	↓	10.10	23.40	11.70	25.00	13.30	26.55	13.30	13.30	26.55	10.20	17.15
	12'	18	48"	#8	4.70	11.60	5.80	12.65	6.85	13.75	6.85	6.85	13.75	7.55	12.20
			32"	#5	5.05	12.35	6.15	13.45	7.30	14.60	7.30	7.30	14.60	6.45	9.60
			16"	↓	6.00	14.55	7.30	15.85	8.55	17.10	8.55	8.55	17.10	7.95	12.85
			8"	↓	9.30	21.60	10.80	23.05	12.25	24.55	12.25	12.25	24.55	10.20	17.15
	16'	24	48"	#8	3.95	9.70	4.85	10.60	5.75	11.55	5.75	—	11.55	7.55	12.20
			32"	#5	4.20	12.25	5.15	11.30	6.10	12.25	6.10	—	12.25	6.45	9.60
			16"	↓	5.05	12.20	6.10	13.30	7.15	14.35	7.15	—	14.35	7.95	12.85
			8"	↓	7.80	18.10	9.05	19.35	10.30	20.60	10.30	9.65	20.60	10.20	17.15
10" Conc. Block	8'	9.6	48"	#8	7.25	16.45	8.20	17.40	9.15	—	9.15	9.15	18.35	13.10	21.05
			32"	↓	7.80	17.60	8.80	18.60	9.80	19.55	9.80	9.80	19.55	14.55	24.15
			16"	#5	9.45	21.15	10.55	22.25	11.65	23.35	11.65	11.65	23.35	13.90	22.35
			8"	↓	14.50	31.65	15.80	32.95	17.10	34.25	17.10	17.10	34.25	18.45	30.15
	12	14.4	48"	#8	7.05	16.00	8.00	16.90	8.90	17.85	8.90	8.90	17.85	13.10	21.05
			32"	↓	7.60	17.10	8.55	18.05	9.50	19.05	9.50	9.50	19.05	14.55	24.15
			16"	#5	9.20	20.55	10.25	21.60	11.35	22.70	11.35	11.35	22.70	13.90	22.35
			8"	↓	14.10	30.75	15.35	32.05	16.65	33.00	16.65	16.65	33.30	18.35	30.15
	16'	19.2	48"	#8	6.70	15.10	7.55	16.00	8.40	16.85	8.40	8.40	16.85	13.10	21.05
			32"	↓	7.15	16.15	8.05	17.05	9.00	17.95	9.00	9.00	17.95	14.55	24.15
			16"	#5	8.70	19.40	9.70	20.40	10.70	21.45	10.70	10.70	21.45	13.90	22.35
			8"	↓	13.35	29.05	14.50	30.25	15.70	31.45	15.70	15.70	31.45	18.35	30.15
	20'	24	48"	#8	6.05	13.65	6.80	14.45	7.60	15.20	7.60	—	15.20	13.10	21.05
			32"	↓	6.45	14.60	7.30	15.40	8.10	16.25	8.10	0.70	16.25	14.55	24.15
			16"	#5	7.85	17.50	8.75	18.45	9.65	19.35	9.65	0.25	19.35	13.90	22.35
			8"	↓	12.05	26.25	13.10	27.30	14.20	28.40	14.20	13.40	28.40	18.35	30.15
12" Conc. Block	8'	8	48"	#8	9.20	20.00	10.00	20.75	10.75	21.55	10.75	10.75	21.55	15.30	24.40
			32"	↓	9.90	21.50	10.75	22.30	11.55	23.15	11.55	11.55	23.15	17.10	28.10
			16"	#5	12.10	26.05	13.00	26.95	13.90	27.85	13.90	13.90	27.85	16.20	25.90
			8"	↓	18.60	39.35	19.65	40.40	20.75	41.50	20.75	20.75	41.50	21.45	35.10
	12'	12	48"	#8	9.00	19.55	9.75	20.35	10.55	21.10	10.55	10.55	21.10	15.30	24.40
			32"	↓	9.70	21.05	10.50	21.85	11.30	22.65	11.30	11.30	22.65	17.10	28.10
			16"	#5	11.85	25.50	12.75	26.40	13.65	27.30	13.65	13.65	27.30	16.20	25.90
			8"	↓	18.20	38.50	19.25	39.55	20.30	40.60	20.30	20.30	40.60	21.45	35.10
	16'	16	48"	#8	8.65	18.75	9.35	19.50	10.10	20.20	10.10	10.10	20.20	15.30	24.40
			32"	↓	9.30	20.15	10.05	20.90	10.85	21.70	10.85	10.85	21.70	17.10	28.10
			16"	#5	11.35	24.40	12.20	25.25	13.05	26.15	13.05	13.05	26.15	16.20	25.90
			8"	↓	17.45	36.90	18.45	37.90	19.45	38.90	19.45	19.45	38.90	21.45	35.10
	24'	24	48"	#8	7.05	15.35	7.65	15.95	8.30	16.55	8.30	—	16.55	15.30	24.40
			32"	↓	7.60	16.55	8.25	17.15	8.90	17.80	8.90	—	17.80	17.10	28.10
			16"	#5	9.30	20.00	10.00	20.70	10.70	21.45	10.40*	—	21.45	16.20	25.90
			8"	↓	14.30	30.25	15.10	31.10	15.95	31.90	15.55*	15.15	31.90	21.45	35.10

*Zone 3 only

Table B2010-114 Fully Grouted Reinforced Masonry Wall Capacities Per L.F. (Kips & In-Kips)

Thk. T (Nom.) (in)	Length or Height h' (Ft.)	h'/T (in/in)	Type Wall Brick Grout Conc. M.U.	Rebar Size and Spacing (in. O.C.)	Eccentric Loads 7.0 in-K/Ft. Insp. No (K/Ft.)	7.0 in-K/Ft. Insp. Yes (K/Ft.)	3.5 in-K/Ft. Insp. No (K/Ft.)	3.5 in-K/Ft. Insp. Yes (K/Ft.)	Without Wind or Eccentric Loads Insp. No (K/Ft.)	Insp. Yes (K/Ft.)	With Wind No 15 psf (K/Ft.)	No 30 psf (K/Ft.)	Yes 15 & 30 (K/Ft.)	Moments No (in.-K/Ft.)	Moments Yes (in.-K/Ft.)
8"	8'	12	4"0"4"	#5@32	10.10	23.35	11.70	24.50	13.30	26.55	13.30	13.30	26.55	6.90	9.65
			Solid CMU		12.75	28.70	14.35	30.30	15.95	31.90	15.95	15.95	31.90	7.80	9.75
	12'	18	4"0"4"		9.30	21.60	10.80	23.05	12.25	24.55	12.25	12.25	24.55	6.90	9.65
			Solid CMU		16.65	26.50	18.15	28.00	14.70	29.45	14.70	14.70	29.45	7.80	9.75
	16'	24	4"0"4"	↓	7.80	18.10	9.05	19.35	10.30	20.60	10.30	9.65	20.60	6.90	9.65
			Solid CMU		9.85	22.20	11.10	23.45	12.35	24.70	12.35	12.35	12.70	7.80	9.75
10"	8'	9.6	4"0"6"	#5@24	14.45	31.50	15.75	32.80	17.05	34.10	17.05	17.05	34.10	11.20	16.20
			Solid CMU		17.85	38.35	19.15	39.65	20.45	40.90	20.45	20.45	40.90	12.65	16.35
			4"2"4"		14.45	31.50	15.75	32.80	17.05	34.10	17.05	17.05	34.10	11.20	16.20
	12'	14.4	4"0"6"		13.90	30.35	15.15	31.60	16.40	32.80	16.40	16.40	32.80	11.20	16.20
			Solid CMU		17.20	36.90	18.45	38.15	19.70	39.40	19.70	19.70	39.40	12.65	16.35
			4"2"4"		13.90	30.35	15.15	31.60	16.40	32.80	16.40	16.40	32.80	11.20	16.20
	16'	19.2	4"0"6"		12.85	28.05	14.00	29.20	15.15	30.35	15.15	15.15	30.35	11.20	16.20
			Solid CMU		15.90	34.10	17.05	35.25	18.20	36.40	13.20	13.20	36.40	12.65	16.35
			4"2"4"		12.85	23.05	14.00	29.20	15.15	30.35	15.15	15.15	30.35	11.20	16.20
	20'	24	4"0"6"		11.15	24.25	12.10	25.25	13.10	26.20	13.10	12.40	26.20	11.20	16.20
			Solid CMU		13.75	29.50	14.75	30.50	15.75	31.50	15.75	15.75	31.50	12.65	16.35
			4"2"4"	↓	11.15	24.25	12.10	25.25	13.10	26.20	13.10	12.40	26.20	11.20	16.20
12"	8'	8	4"0"8"	#8@48	18.60	39.35	19.65	40.40	20.75	41.50	20.75	20.75	41.50	16.65	24.90
			Solid CMU		22.75	47.65	23.80	48.70	24.90	49.80	24.90	24.90	49.80	18.80	25.10
			4"2"6"	#5@20	18.60	39.35	19.65	40.40	20.75	41.50	20.75	20.75	41.50	11.90	18.75
	12'	12	4"0"8"	#8@48	18.20	38.50	19.25	39.55	20.30	40.60	20.30	20.30	40.60	16.65	24.90
			Solid CMU	↓	22.25	46.65	23.30	47.70	24.35	48.75	24.35	24.35	48.75	18.80	25.10
			4"2"6"	#5@20	18.20	38.50	19.25	39.55	20.30	40.60	20.30	20.30	40.60	11.90	18.75
	16'	16	4"0"8"	#8@48	17.45	36.90	18.45	37.90	19.45	38.90	19.45	19.45	38.90	16.65	24.90
			Solid CMU	↓	21.30	44.70	22.35	45.70	23.35	46.70	23.25	23.35	46.70	18.80	25.10
			4"2"6"	#5@20	17.45	36.90	18.45	37.90	19.45	38.90	19.45	19.45	38.90	11.90	18.75
	24'	24	4"0"8"	#8@48	14.30	30.25	15.10	31.10	15.95	31.90	15.95	15.15	31.90	16.65	24.90
			Solid CMU	↓	17.15	36.65	18.30	37.45	19.15	38.30	19.15	19.15	38.30	18.80	25.10
			4"2"6"	#5@20	14.30	30.25	15.10	31.10	15.95	31.90	15.95	15.15	31.90	11.90	18.75
16"	8'	6	4"0"12"	#8@32	26.40	54.40	27.20	55.20	28.00	56.05	28.00	28.00	56.05	31.20	49.10
			Solid CMU	↓	32.00	65.65	32.80	66.45	33.60	67.25	33.60	33.60	67.25	35.25	50.40
			4"2"10"	#5@15	26.40	54.40	27.20	55.20	28.00	56.05	28.00	28.00	56.05	14.75	24.10
	12'	9	4"0"12"	#8@32	26.15	53.95	26.95	54.75	27.75	55.55	27.75	27.75	55.55	31.20	49.10
			Solid CMU	↓	31.70	66.05	32.50	65.85	33.30	66.65	33.00	33.30	66.65	35.25	50.40
			4"2"10"	#5@15	26.15	53.95	26.95	54.75	27.75	55.55	27.75	27.75	55.55	14.75	24.10
	16'	12	4"0"12"	#8@32	25.70	53.05	26.50	53.80	28.30	54.60	27.30	27.30	54.60	31.20	49.10
			Solid CMU		31.20	63.95	31.95	64.75	32.75	65.50	32.75	32.75	65.50	35.25	50.40
			4"2"10"	↓	25.70	53.05	26.50	53.80	27.30	54.60	27.30	27.30	59.60	14.75	24.10
	32'	24	4"0"12"	#8@32	20.35	41.95	20.95	42.55	21.60	43.20	18.75*	18.75*	43.20	31.20	49.10
			Solid CMU	↓	24.65	50.60	25.30	51.20	25.90	51.85	24.55*	24.55*	51.85	35.25	50.40
			4"2"10"	#5@15	20.35	41.95	20.95	42.55	21.60	43.20	18.75*	18.75*	43.20	14.75	24.10

Column groupings: Earthquake Zones 1, 2 & 3 (Thk., Length or Height, Type Wall, Rebar). Allowable Vertical Wall Loads (Eccentric Loads; Without Wind or Eccentric Loads; With Wind). With Wind: No (15 psf, 30 psf), Yes (15 & 30). Allowable Wall Moments Without Vertical Wall Loads — Not Wind or Earthquake (Inspection No, Yes).

*Zone 3 Only

Reed Construction Data, Inc. is a leading provider of construction information and building information modeling (BIM) solutions. The company's portfolio of information products and services is designed specifically to help construction industry professionals advance their business with timely and accurate project, product, and cost data. Reed Construction Data is a division of Reed Business Information, a member of the Reed Elsevier PLC group of companies.

Cost Information

RSMeans, the undisputed market leader in construction costs, provides current cost and estimating information through its innovative MeansCostworks.com® web-based solution. In addition, RSMeans publishes annual cost books, estimating software, and a rich library of reference books. RSMeans also conducts a series of professional seminars and provides construction cost consulting for owners, manufacturers, designers, and contractors to sharpen personal skills and maximize the effective use of cost estimating and management tools.

Project Data

Reed Construction Data assembles one of the largest databases of public and private project data for use by contractors, distributors, and building product manufacturers in the U.S. and Canadian markets. In addition, Reed Construction Data is the North American construction community's premier resource for project leads and bid documents. Reed Bulletin and Reed CONNECT™ provide project leads and project data through all stages of construction for many of the country's largest public sector, commercial, industrial, and multi-family residential projects.

Research and Analytics

Reed Construction Data's forecasting tools cover most aspects of the construction business in the U.S. and Canada. With a vast network of resources, Reed Construction Data is uniquely qualified to give you the information you need to keep your business profitable.

SmartBIM Solutions

Reed Construction Data has emerged as the leader in the field of building information modeling (BIM) with products and services that have helped to advance the evolution of BIM. Through an in-depth SmartBIM Object Creation Program, BPMs can rely on Reed to create high-quality, real world objects embedded with superior cost data (RSMeans). In addition, Reed has made it easy for architects to manage these manufacturer-specific objects as well as generic objects in Revit with the SmartBIM Library.

SmartBuilding Index

The leading industry source for product research, product documentation, BIM objects, design ideas, and source locations. Search, select, and specify available building products with our online directories of manufacturer profiles, MANU-SPEC, SPEC-DATA, guide specs, manufacturer catalogs, building codes, historical project data, and BIM objects.

SmartBuilding Studio

A vibrant online resource library which brings together a comprehensive catalog of commercial interior finishes and products under a single standard for high-definition imagery, product data, and searchable attributes.

Associated Construction Publications (ACP)

Reed Construction Data's regional construction magazines cover the nation through a network of 14 regional magazines. Serving the construction market for more than 100 years, our magazines are a trusted source of news and information in the local and national construction communities.

Table B2010-116 Unreinforced Masonry Wall Capacities Per L.F. (Kips & In-Kips)

Thk. T (Nom.) (in)	Earthquake Zones 0 & 1 Only		Type of Wall	Allowable Vertical Wall Loads					Allowable Wall Moments (Without Vertical Wall Loads)			
	Length Or Height			Eccentric Loads		Without Wind or Eccentric Loads	With Wind		Not Wind or Earthquake Inspection		Wind or Earthquake Inspection	
	h' (Ft.)	h'/t (in/in)		7.0 (K/Ft.)	3.5 (K/Ft.)	(K/Ft.)	15 psf (K/Ft.)	30 psf (K/Ft.)	No (in-K/Ft.)	Yes (in-K/Ft.)	No (in-K/Ft.)	Yes (in-K/Ft.)
8"	8'	12	Solid Brick	4.85	7.50	10.15	10.15	10.15	1.15	2.30	1.55	3.10
			Solid CM Units	8.85	11.50	14.15	14.15	14.15	.70	1.40	.95	1.85
			Hollow CM Units	7.95	10.60	13.30	13.30	13.30	.70	1.40	.60	1.25
	10'	15	Solid Brick	4.70	7.30	9.85	9.85	9.85	1.15	2.30	1.55	3.10
			Solid CM Units	8.55	11.15	13.75	13.75	13.75	.70	1.40	.90	1.85
			Hollow CM Units	—	4.45	6.30	6.30	6.05	.45	.95	.60	1.25
	12'	18	Solid Brick	—	6.95	9.40	9.40	7.95	1.15	2.30	1.55	3.10
			Solid CM Units	8.15	10.60	13.10	13.10	12.90	.70	1.40	.90	1.85
			Hollow CM Units	—	4.25	6.00	6.00	4.75	.45	.95	.60	1.25
12"	8'	8	Solid Brick	12.30	14.10	15.90	15.90	15.90	2.70	5.40	3.60	7.20
			Solid CM Units	18.50	20.30	22.10	22.10	22.10	1.60	3.20	2.15	4.30
			Hollow CM Units	7.85	9.10	10.40	10.40	10.40	1.10	2.25	1.50	3.00
			Brick & Hollow CMU	5.50	7.80	10.05	10.05	10.05	1.35	2.70	1.80	3.60
	12'	12	Solid Brick	12.05	13.80	15.55	15.55	15.55	2.70	5.40	3.60	7.20
			Solid CM Units	13.45	14.75	16.05	16.05	16.05	1.60	3.20	2.15	4.30
			Hollow CM Units	7.65	8.90	10.15	10.15	10.15	1.10	2.25	1.50	3.00
			Brick & Hollow CMU	5.40	7.65	9.85	9.85	9.05	1.35	2.70	1.80	3.60
	16'	16	Solid Brick	11.55	13.20	14.90	14.90	14.35	2.70	5.40	3.60	7.20
			Solid CM Units	17.35	19.05	20.70	20.70	20.70	1.60	3.20	2.15	4.30
			Hollow CM Units	7.35	8.55	9.75	9.75	9.05	1.10	2.25	1.50	3.00
			Brick & Hollow CMU	5.20	7.30	9.45	9.10	—	1.35	2.70	1.80	3.60
16"	12'	9	Solid Brick	18.60	19.95	21.30	21.30	21.30	4.85	9.75	6.50	13.00
			Solid CM Units	26.95	28.30	29.60	29.60	29.60	2.90	5.85	3.90	7.80
			Hollow CM Units	10.85	12.10	13.30	13.30	13.30	1.50	3.05	2.00	4.05
			Brick & Hollow CMU	9.30	11.10	12.85	12.85	12.85	2.20	4.40	2.90	5.85
	16'	12	Solid Brick	18.30	19.60	20.90	20.90	20.90	4.85	9.75	6.50	13.00
			Solid CM Units	26.50	27.80	29.10	29.10	29.10	2.90	5.85	3.90	7.80
			Hollow CM Units	10.70	11.90	13.10	13.10	13.10	1.50	3.05	2.00	4.05
			Brick & Hollow CMU	9.15	10.90	12.65	12.65	11.10	2.20	4.40	2.90	5.85
	20'	15	Solid Brick	17.80	19.05	20.30	20.30	20.30	4.85	9.75	6.50	13.00
			Solid CM Units	25.75	27.00	28.25	28.25	28.25	2.90	5.85	3.90	7.80
			Hollow CM Units	10.40	11.55	12.70	12.70	10.95	1.50	3.05	2.00	4.05
			Brick & Hollow CMU	8.90	10.60	12.30	12.00	—	2.20	4.40	2.90	5.85
20"	12'	7.2	Solid Brick	24.80	25.85	26.90	26.90	26.90	7.70	15.40	10.25	20.50
			Solid CM Units	35.30	36.40	37.45	37.45	37.45	4.60	9.20	6.15	12.30
			Hollow CM Units	16.15	17.15	18.15	18.15	18.15	2.50	5.05	3.35	6.75
			Brick & Hollow CMU	16.10	17.25	18.45	18.45	18.45	4.05	8.10	5.40	10.80
	16'	9.6	Solid Brick	24.55	25.60	26.70	26.70	26.70	7.70	15.40	10.25	20.50
			Solid CM Units	35.00	36.10	37.15	37.15	37.15	4.60	9.20	6.15	12.30
			Hollow CM Units	16.00	17.00	18.00	18.00	18.00	2.50	5.05	3.35	6.75
			Brick & Hollow CMU	15.95	17.10	18.30	18.30	18.30	4.05	8.10	5.40	10.80
	24'	14.4	Solid Brick	23.70	24.70	25.75	25.75	25.75	7.70	15.40	10.25	20.50
			Solid CM Units	33.80	34.80	35.80	35.80	35.80	4.60	9.20	6.15	12.30
			Hollow CM Units	15.45	16.40	17.40	17.40	16.05	2.50	5.05	3.35	6.75
			Brick & Hollow CMU	15.40	16.50	17.65	17.65	15.25	4.05	8.10	5.40	10.80

Table B2010-202   Mix Proportions by Volume, Compressive Strength of Mortar

| Where Used | Mortar Type | Allowable Proportions by Volume | | | | Compressive Strength @ 28 days |
		Portland Cement	Masonry Cement	Hydrated Lime	Masonry Sand	
Plain Masonry	M	1	1	—	6	2500 psi
		1	—	1/4	3	
	S	1/2	1	—	4	1800 psi
		1	—	1/4 to 1/2	4	
	N	—	1	—	3	750 psi
		1	—	1/2 to 1-1/4	6	
	O	—	1	—	3	350 psi
		1	—	1-1/4 to 2-1/2	9	
	K	1	—	2-1/2 to 4	12	75 psi
Reinforced Masonry	PM	1	1	—	6	2500 psi
	PL	1	—	1/4 to 1/2	4	2500 psi

Note: The total aggregate should be between 2.25 to 3 times the sum of the cement and lime used.

Table B2010-203   Volume and Cost Per S.F. of Grout Fill for Concrete Block Walls

| Center to Center Spacing Grouted Cores | 6" C.M.U. Per S.F. | | | | 8" C.M.U. Per S.F. | | | | 12" C.M.U. Per S.F. | | | |
| | Volume in C.F. | | Cost | | Volume in C.F. | | Cost | | Volume in C.F. | | Cost | |
	40% Solid	75% Solid	40% Solid	75% Solid	40% Solid	75% Solid	40% Solid	75% Solid	40% Solid	75% Solid	40% Solid	75% Solid
All cores grouted solid	.27	.11	$2.71	$1.11	.36	.15	$3.62	$1.51	.55	.23	$5.53	$2.31
cores grouted 16" O.C.	.14	.06	1.41	.60	.18	.08	1.81	.80	.28	.12	2.81	1.21
cores grouted 24" O.C.	.09	.04	.90	.40	.12	.05	1.21	.50	.18	.08	1.81	.80
cores grouted 32" O.C.	.07	.03	.70	.30	.09	.04	.90	.40	.14	.06	1.41	.60
cores grouted 40" O.C.	.05	.02	.50	.20	.07	.03	.70	.30	.11	.05	1.11	.50
cores grouted 48" O.C.	.04	.02	.40	.20	.06	.03	.60	.30	.09	.04	.90	.40

Installed costs, including O&P, based on High-Lift Grouting method

Low-Lift Grouting is used when the wall is built to a maximum height of 5'. The grout is pumped or poured into the cores of the concrete block. The operation is repeated after each five additional feet of wall height has been completed.

High-Lift Grouting is used when the wall has been built to the full-story height. Some of the advantages are: the vertical reinforcing steel can be placed after the wall is completed; and the grout can be supplied by a ready-mix concrete supplier so that it may be pumped in a continuous operation.

B20 Exterior Enclosure   RB2010-430 Pre-Engin. Steel Building

Reference Tables

Pre-engineered Steel Buildings

These buildings are manufactured by many companies and normally erected by franchised dealers throughout the U.S. The four basic types are: Rigid Frames, Truss type, Post and Beam and the Sloped Beam type. Most popular roof slope is low pitch of 1″ in 12″. The minimum economical area of these buildings is about 3000 S.F. of floor area. Bay sizes are usually 20′ to 24′ but can go as high as 30′ with heavier girts and purlins. Eave heights are usually 12′ to 24′ with 18′ to 20′ most typical.

Material prices generally do not include floors, foundations, interior finishes or utilities. Costs assume at least three bays of 24′ each, a 1″ in 12″ roof slope, and they are based on 30 psf roof load and 20 psf wind load (wind load is a function of wind speed, building height, and terrain characteristics; this should be determined by a Registered Structural Engineer) and no unusual requirements. Costs include the structural frame, 26 ga. non-insulated colored corrugated or ribbed roofing or siding panels, fasteners, closures, trim and flashing but no allowance for insulation, doors, windows, skylights, gutters or downspouts. Very large projects would generally cost less for materials than the prices shown.

Conditions at the site, weather, shape and size of the building, and labor availability will affect the erection cost of the building.

Table B3010-011   Thermal Coefficients for Roof Systems

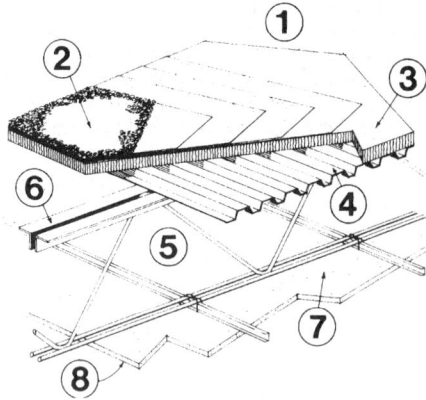

EXAMPLE:

Construction	Resistance R
1. Outside surface (15 MPH wind)	0.17
2. Built up roofing .375" thick	0.33
3. Rigid roof insulation, 2"	5.26
4. Metal decking	0.00
5. Air space (non reflective)	1.14
6. Structural members	0.00
7. Gypsum board ceiling 5/8" thick	0.56
8. Inside surface (still air)	0.61
Total resistance	8.07

$$U = 1/R = 1/8.07 = 0.12$$

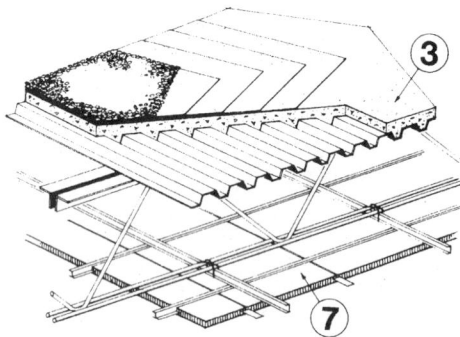

Substitution

Replace item 3 with Perlite lightweight concrete and item 7 with acoustical tile.

		R = 8.07
Total resistance (Example)		
Deduct 3. Rigid insulation	5.26	
7. Gypsum board ceiling	0.05	
	5.31	-5.31
Difference		2.76
Add Perlite concrete, 4"		4.32
Acoustical tile, 3/4"		1.56
Total resistance		8.64

$$U = 1/R = 1/8.64 = 0.12$$

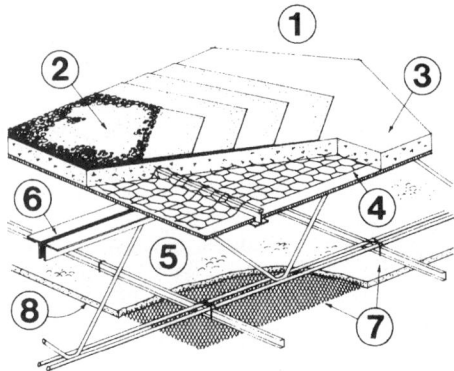

EXAMPLE

Construction	Resistance R
1. Outside surface (15 MPH wind)	0.17
2. Built up roofing .375" thick	0.33
3. Rigid roof insulation, 2" thick	1.20
4. Form bd., 1/2" thick	0.45
5. Air space (non reflective)	1.14
6. Structural members	0.00
7. Gypsum board ceiling 3/4" thick	0.14
8. Inside surface (still air)	0.61
Total resistance	4.04

$$U = 1/R = 1/4.04 = 0.25$$

(for modified system see next page)

Table B3010-011 Thermal Coefficients for Roof Systems

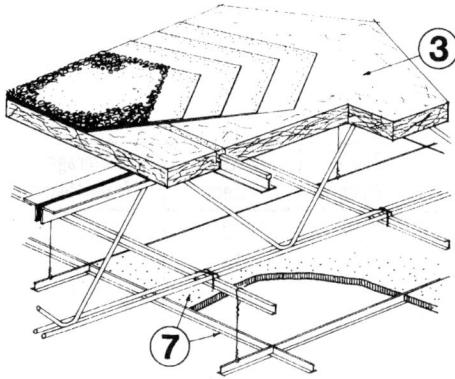

Substitution

Replace item 3 with Tectum board and item 7 with acoustical tile.

Total resistance (Example)		R =	4.04
Deduct 3. Gypsum concrete	1.20		
4. Form bd.	0.45		
7. Plaster	0.14		
	1.79	<-1.79>	
Difference		2.25	
Add Tectum bd. 3" thick		5.25	
Acoustical tile, 3/4"		1.56	
Total resistance		9.06	

$$U = 1/R = 1/9.06 = 0.11$$

(for limited system see previous page)

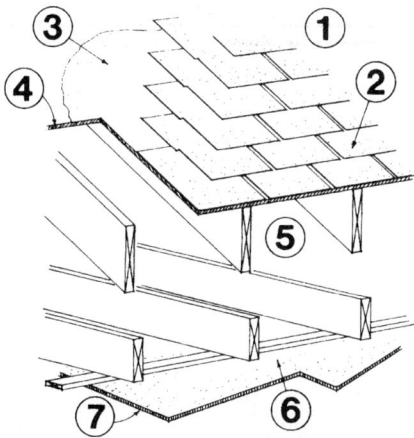

EXAMPLE:

Construction	Resistance R
1. Outside surface (15 MPH wind)	0.17
2. Asphalt shingle roof	0.44
3. Felt building paper	0.06
4. Plywood sheathing, 5/8" thick	0.78
5. Reflective air space, 3-1/2"	2.17
6. Gypsum board, 1/2" foil backed	0.45
7. Inside surface (still air)	0.61
Total resistance	4.68

$$U = 1/R = 1/4.68 = 0.21$$

Substitution

Replace item 2 with slate, item 5 with insulation and item 6 with gypsum plaster

Total resistance (Example)		R =	4.68
Deduct 2. Asphalt shingles	0.44		
5. Air space	2.17		
6. Gypsum board	0.45		
	3.06	<-3.06>	
Difference		1.62	
Add Slate, 1/4" thick		0.03	
Insulation, 6" blanket		19.00	
Gypsum sand plaster 3/4"		0.14	
Total resistance		20.79	

$$U = 1/R = 1/20.79 = 0.05$$

Table C2010-101   Typical Range of Risers for Various Story Heights

General Design: See Table L1010-101 for code requirements.
Maximum height between landings is 12′; usual stair angle is 20°
to 50° with 30° to 35° best. Usual relation of riser to treads is:

Riser + tread = 17.5
2x (Riser) + tread = 25
Riser x tread = 70 or 75

Maximum riser height is 7″ for commercial, 8-1/4″ for residential.
Usual riser height is 6-1/2″ to 7-1/4″.
Minimum tread width is 11″ for commercial, 9″ for residential.

Story Height	Minimum Risers	Maximum Riser Height	Tread Width	Maximum Risers	Minimum Riser Height	Tread Width	Average Risers	Average Riser Height	Tread Width
7′-6″	12	7.50″	10.00″	14	6.43″	11.07″	13	6.92″	10.58″
8′-0″	13	7.38	10.12	15	6.40	11.10	14	6.86	10.64
8′-6″	14	7.29	10.21	16	6.38	11.12	15	6.80	10.70
9′-0″	15	7.20	10.30	17	6.35	11.15	16	6.75	10.75
9′-6″	16	7.13	10.37	18	6.33	11.17	17	6.71	10.79
10′-0″	16	7.50	10.00	19	6.32	11.18	18	6.67	10.83
10′-6″	17	7.41	10.09	20	6.30	11.20	18	7.00	10.50
11′-0″	18	7.33	10.17	21	6.29	11.21	19	6.95	10.55
11′-6″	19	7.26	10.24	22	6.27	11.23	20	6.90	10.60
12′-0″	20	7.20	10.30	23	6.26	11.24	21	6.86	10.64
12′-6″	20	7.50	10.00	24	6.25	11.25	22	6.82	10.68
13′-0″	21	7.43	10.07	25	6.24	11.26	22	7.09	10.41
13′-6″	22	7.36	10.14	25	6.48	11.02	23	7.04	10.46
14′-0″	23	7.30	10.20	26	6.46	11.04	24	7.00	10.50

Table D1010-011 Elevator Hoistway Sizes

Elevator Type	Floors	Building Type	Capacity Lbs.	Capacity Passengers	Entry	Hoistway Width	Hoistway Depth	S.F. Area per Floor
Hydraulic	5	Apt./Small office	1500	10	S	6'-7"	4'-6"	29.6
			2000	13	S	7'-8"	4'-10"	37.4
	7	Average office/Hotel	2500	16	S	8'-4"	5'-5"	45.1
			3000	20	S	8'-4"	5'-11"	49.3
		Large office/Store	3500	23	S	8'-4"	6'-11"	57.6
		Freight light duty	2500		D	7'-2"	7'-10"	56.1
		Heavy duty	5000		D	10'-2"	10'-10"	110.1
			7500		D	10'-2"	12'-10"	131
			5000		D	10'-2"	10'-10"	110.1
			7500		D	10'-2"	12'-10"	131
			10000		D	10'-4"	14'-10"	153.3
		Hospital	3500		D	6'-10"	9'-2"	62.7
			4000		D	7'-4"	9'-6"	69.6
Electric Traction, high speed	High Rise	Apt./Small office	2000	13	S	7'-8"	5'-10"	44.8
			2500	16	S	8'-4"	6'-5"	54.5
			3000	20	S	8'-4"	6'-11"	57.6
			3500	23	S	8'-4"	7'-7"	63.1
		Store	3500	23	S	9'-5"	6'-10"	64.4
		Large office	4000	26	S	9'-4"	7'-6"	70
		Hospital	3500		D	7'-6"	9'-2"	69.4
			4000		D	7'-10"	9'-6"	58.8
Geared, low speed		Apartment	1200	8	S	6'-4"	5'-3"	33.2
			2000	13	S	7'-8"	5'-8"	43.5
			2500	16	S	8'-4"	6'-3"	52
		Office	3000	20	S	8'-4"	6'-9"	56
		Store	3500	23	S	9'-5"	6'-10"	64.4
						Add 4" width for multiple units		

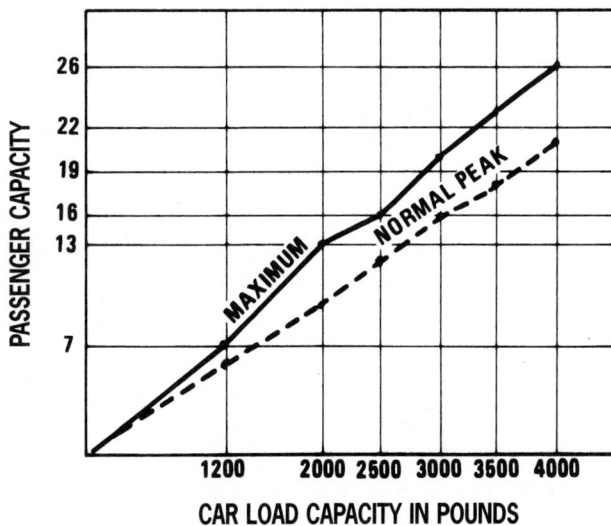

Figure D1010-012 Elevator Passenger Capacity During Peak Periods

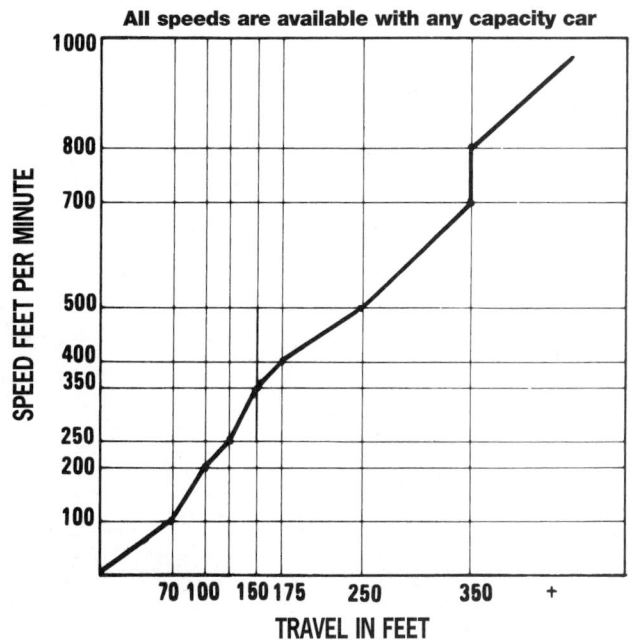

Figure D1010-013 Speed & Travel Chart

All speeds are available with any capacity car

Table D1010-014   Elevator Speed vs. Height Requirements

Building Type and Elevator Capacities	Travel Speeds in Feet per Minute								
	100 fpm	200 fpm	250 fpm	350 fpm	400 fpm	500 fpm	700 fpm	800 fpm	1000 fpm
Apartments 1200 lb. to 2500 lb.	to 70	to 100	to 125	to 150	to 175	to 250	to 350		
Department Stores 1200 lb. to 2500 lb.		100		125	175	250	350		
Hospitals 3500 lb. to 4000 lb.	70	100	125	150	175	250	350		
Office Buildings 2000 lb. to 4000 lb.		100	125	150		175	250	to 350	over 350

Note: Vertical transportation capacity may be determined by code occupancy requirements of the number of square feet per person divided into the total square feet of building type. If we are contemplating an office building, we find that the Occupancy Code Requirement is 100 S.F. per person. For a 20,000 S.F. building, we would have a legal capacity of two hundred people. Elevator handling capacity is subject to the five minute evacuation recommendation, but it may vary from 11 to 18%. Speed required is a function of the travel height, number of stops and capacity of the elevator.

Table D2010-031   Plumbing Approximations for Quick Estimating

Water Control

Water Meter; Backflow Preventer,

Shock Absorbers; Vacuum

Breakers; Mixer .... 10 to 15% of Fixtures

Pipe And Fittings .... 30 to 60% of Fixtures

> **Note:** Lower percentage for compact buildings or larger buildings with plumbing in one area.
> Larger percentage for large buildings with plumbing spread out.
> In extreme cases pipe may be more than 100% of fixtures.
> Percentages **do not** include special purpose or process piping.

Plumbing Labor

1 & 2 Story Residential ....Rough-in Labor = 80% of Materials

Apartment Buildings ....Rough-in Labor = 90 to 100% of Materials

Labor for handling and placing fixtures is approximately 25 to 30% of fixtures

Quality/Complexity Multiplier (for all installations)

Economy installation, add ....0 to 5%

Good quality, medium complexity, add ....5 to 15%

Above average quality and complexity, add ....15 to 25%

Table D2010-032   Pipe Material Considerations

1. Malleable fittings should be used for gas service.
2. Malleable fittings are used where there are stresses/strains due to expansion and vibration.
3. Cast fittings may be broken as an aid to disassembling of heating lines frozen by long use, temperature and minerals.
4. Cast iron pipe is extensively used for underground and submerged service.
5. Type M (light wall) copper tubing is available in hard temper only and is used for nonpressure and less severe applications than K and L.
6. Type L (medium wall) copper tubing, available hard or soft for interior service.
7. Type K (heavy wall) copper tubing, available in hard or soft temper for use where conditions are severe. For underground and interior service.
8. Hard drawn tubing requires fewer hangers or supports but should not be bent. Silver brazed fittings are recommended, however soft solder is normally used.
9. Type DMV (very light wall) copper tubing designed for drainage, waste and vent plus other non-critical pressure services.

Table D2010-033   Domestic/Imported Pipe and Fittings Cost

The prices shown in this publication for steel/cast iron pipe and steel, cast iron, malleable iron fittings are based on domestic production sold at the normal trade discounts. The above listed items of foreign manufacture may be available at prices of 1/3 to 1/2 those shown. Some imported items after minor machining or finishing operations are being sold as domestic to further complicate the system.

Caution: Most pipe prices in this book also include a coupling and pipe hangers which for the larger sizes can add significantly to the per foot cost and should be taken into account when comparing "book cost" with quoted supplier's cost.

Table D2010-401 Minimum Plumbing Fixture Requirements

Minimum Plumbing Fixture Requirements

Type of Building or Occupancy (2)	Water Closets (14) (Fixtures per Person)		Urinals (5,10) (Fixtures per Person)		Lavatories (Fixtures per Person)		Bathtubs or Showers (Fixtures per Person)	Drinking Fountains (Fixtures per Person) (3, 13)
	Male	Female	Male	Female	Male	Female		
Assembly Places- Theatres, Auditoriums, Convention Halls, etc.-for permanent employee use	1: 1 - 15 2: 16 - 35 3: 36 - 55 Over 55, add 1 fixture for each additional 40 persons	1: 1 - 15 2: 16 - 35 3: 36 - 55	0: 1 - 9 1: 10 - 50 Add one fixture for each additional 50 males		1 per 40	1 per 40		
Assembly Places- Theatres, Auditoriums, Convention Halls, etc. - for public use	1: 1 - 100 2: 101 - 200 3: 201 - 400 Over 400, add 1 fixture for each additional 500 males and 1 for each additional 125 females	3: 1 - 50 4: 51 - 100 8: 101 - 200 11: 201 - 400	1: 1 - 100 2: 101 - 200 3: 201 - 400 4: 401 - 600 Over 600, add 1 fixture for each additional 300 males		1: 1 - 200 2: 201 - 400 3: 401 - 750 Over 750, add 1 fixture for each additional 500 persons	1: 1 - 200 2: 201 - 400 3: 401 - 750		1: 1 - 150 2: 151 - 400 3: 401 - 750 Over 750, add one fixture for each additional 500 persons
Dormitories (9) School or Labor	1 per 10 Add 1 fixture for each additional 25 males (over 10) and 1 for each additional 20 females (over 8)	1 per 8	1 per 25 Over 150, add 1 fixture for each additional 50 males		1 per 12 Over 12 add 1 fixture for each additional 20 males and 1 for each 15 additional females	1 per 12	1 per 8 For females add 1 bathtub per 30. Over 150, add 1 per 20	1 per 150 (12)
Dormitories- for Staff Use	1: 1 - 15 2: 16 - 35 3: 36 - 55 Over 55, add 1 fixture for each additional 40 persons	1: 1 - 15 3: 16 - 35 4: 36 - 55	1 per 50		1 per 40	1 per 40	1 per 8	
Dwellings: Single Dwelling Multiple Dwelling or Apartment House	1 per dwelling 1 per dwelling or apartment unit				1 per dwelling 1 per dwelling or apartment unit		1 per dwelling 1 per dwelling or apartment unit	
Hospital Waiting rooms	1 per room				1 per room			1 per 150 (12)
Hospitals- for employee use	1: 1 - 15 2: 16 - 35 3: 36 - 55 Over 55, add 1 fixture for each additional 40 persons	1: 1 - 15 3: 16 - 35 4: 36 - 55	0: 1 - 9 1: 10 - 50 Add 1 fixture for each additional 50 males		1 per 40	1 per 40		
Hospitals: Individual Room Ward Room	1 per room 1 per 8 patients				1 per room 1 per 10 patients		1 per room 1 per 20 patients	1 per 150 (12)
Industrial (6) Warehouses Workshops, Foundries and similar establishments- for employee use	1: 1 -10 2: 11 - 25 3: 26 - 50 4: 51 - 75 5: 76 - 100 Over 100, add 1 fixture for each additional 30 persons	1: 1 -10 2: 11 - 25 3: 26 - 50 4: 51 - 75 5: 76 - 100			Up to 100, per 10 persons Over 100, 1 per 15 persons (7, 8)		1 shower for each 15 persons exposed to excessive heat or to skin contamination with poisonous, infectious or irritating material	1 per 150 (12)
Institutional - Other than Hospitals or Penal Institutions (on each occupied floor)	1 per 25	1 per 20	0: 1 - 9 1: 10 - 50 Add 1 fixture for each additional 50 males		1 per 10	1 per 10	1 per 8	1 per 150 (12)
Institutional - Other than Hospitals or Penal Institutions (on each occupied floor)- for employee use	1: 1 - 15 2: 16 - 35 3: 36 - 55 Over 55, add 1 fixture for each additional 40 persons	1: 1 - 15 3: 16 - 35 4: 36 - 55	0: 1 - 9 1: 10 - 50 Add 1 fixture for each additional 50 males		1 per 40	1 per 40	1 per 8	1 per 150 (12)
Office or Public Buildings	1: 1 - 100 2: 101 - 200 3: 201 - 400 Over 400, add 1 fixture for each additional 500 males and 1 for each additional 150 females	3: 1 - 50 4: 51 - 100 8: 101 - 200 11: 201 - 400	1: 1 - 100 2: 101 - 200 3: 201 - 400 4: 401 - 600 Over 600, add 1 fixture for each additional 300 males		1: 1 - 200 2: 201 - 400 3: 401 - 750 Over 750, add 1 fixture for each additional 500 persons	1: 1 - 200 2: 201 - 400 3: 401 - 750		1 per 150 (12)
Office or Public Buildings - for employee use	1: 1 - 15 2: 16 - 35 3: 36 - 55 Over 55, add 1 fixture for each additional 40 persons	1: 1 - 15 3: 16 - 35 4: 36 - 55	0: 1 - 9 1: 10 - 50 Add 1 fixture for each additional 50 males		1 per 40	1 per 40		

Table D2010-401   Minimum Plumbing Fixture Requirements (cont.)

Minimum Plumbing Fixture Requirements

Type of Building or Occupancy	Water Closets (14) (Fixtures per Person)		Urinals (5, 10) (Fixtures per Person)		Lavatories (Fixtures per Person)		Bathtubs or Showers (Fixtures per Person)	Drinking Fountains (Fixtures per Person) (3, 13)
	Male	Female	Male	Female	Male	Female		
Penal Institutions - for employee use	1: 1 - 15 2: 16 - 35 3: 36 - 55	1: 1 - 15 3: 16 - 35 4: 36 - 55	0: 1 - 9 1: 10 - 50		1 per 40	1 per 40		1 per 150 (12)
	Over 55, add 1 fixture for each additional 40 persons		Add 1 fixture for each additional 50 males					
Penal Institutions - for prison use								
Cell	1 per cell				1 per cell			1 per cellblock floor
Exercise room	1 per exercise room		1 per exercise room		1 per exercise room			1 per exercise room
Restaurants, Pubs and Lounges (11)	1: 1 - 50 2: 51 - 150 3: 151 - 300	1: 1 - 50 2: 51 - 150 4: 151 - 300	1: 1 - 150		1: 1 - 150 2: 151 - 200 3: 201 - 400	1: 1 - 150 2: 151 - 200 3: 201 - 400		
	Over 300, add 1 fixture for each additional 200 persons		Over 150, add 1 fixture for each additional 150 males		Over 400, add 1 fixture for each additional 400 persons			
Schools - for staff use All Schools	1: 1 - 15 2: 16 - 35 3: 36 - 55	1: 1 - 15 3: 16 - 35 4: 36 - 55	1 per 50		1 per 40	1 per 40		
	Over 55, add 1 fixture for each additional 40 persons							
Schools - for student use: Nursery	1: 1 - 20 2: 21 - 50	1: 1 - 20 2: 21 - 50			1: 1 - 25 2: 26 - 50	1: 1 - 25 2: 26 - 50		1 per 150 (12)
	Over 50, add 1 fixture for each additional 50 persons				Over 50, add 1 fixture for each additional 50 persons			
Elementary	1 per 30	1 per 25	1 per 75		1 per 35	1 per 35		1 per 150 (12)
Secondary	1 per 40	1 per 30	1 per 35		1 per 40	1 per 40		1 per 150 (12)
Others (Colleges, Universities, Adult Centers, etc.	1 per 40	1 per 30	1 per 35		1 per 40	1 per 40		1 per 150 (12)
Worship Places Educational and Activities Unit	1 per 150	1 per 75	1 per 150		1 per 2 water closets			1 per 150 (12)
Worship Places Principal Assembly Place	1 per 150	1 per 75	1 per 150		1 per 2 water closets			1 per 150 (12)

Notes:
1. The figures shown are based upon one (1) fixture being the minimum required for the number of persons indicated or any fraction thereof.
2. Building categories not shown on this table shall be considered separately by the Administrative Authority.
3. Drinking fountains shall not be installed in toilet rooms.
4. Laundry trays. One (1) laundry tray or one (1) automatic washer standpipe for each dwelling unit or one (1) laundry trays or one (1) automatic washer standpipes, or combination thereof, for each twelve (12) apartments. Kitchen sinks, one (1) for each dwelling or apartment unit.
5. For each urinal added in excess of the minimum required, one water closet may be deducted. The number of water closets shall not be reduced to less than two-thirds (2/3) of the minimum requirement.
6. As required by ANSI Z4.1-1968, Sanitation in Places of Employment.
7. Where there is exposure to skin contamination with poisonous, infectious, or irritating materials, provide one (1) lavatory for each five (5) persons.
8. Twenty-four (24) lineal inches of wash sink or eighteen (18) inches of a circular basin, when provided with water outlets for such space shall be considered equivalent to one (1) lavatory.
9. Laundry trays, one (1) for each fifty (50) persons. Service sinks, one (1) for each hundred (100) persons.
10. General. In applying this schedule of facilities, consideration shall be given to the accessibility of the fixtures. Conformity purely on a numerical basis may not result in an installation suited to the need of the individual establishment. For example, schools should be provided with toilet facilities on each floor having classrooms.
  a. Surrounding materials, wall and floor space to a point two (2) feet in front of urinal lip and four (4) feet above the floor, and at least two (2) feet to each side of the urinal shall be lined with non-absorbent materials.
  b. Trough urinals shall be prohibited.
11. A restaurant is defined as a business which sells food to be consumed on the premises.
  a. The number of occupants for a drive-in restaurant shall be considered as equal to the number of parking stalls.
  b. Employee toilet facilities shall not to be included in the above restaurant requirements. Hand washing facilities shall be available in the kitchen for employees.
12. Where food is consumed indoors, water stations may be substituted for drinking fountains. Offices, or public buildings for use by more than six (6) persons shall have one (1) drinking fountain for the first one hundred fifty (150) persons and one additional fountain for each three hundred (300) persons thereafter.
13. There shall be a minimum of one (1) drinking fountain per occupied floor in schools, theaters, auditoriums, dormitories, offices of public building.
14. The total number of water closets for females shall be at least equal to the total number of water closets and urinals required for males.

Table D2010-403   Water Cooler Application

Type of Service	Requirement
Office, School or Hospital	12 persons per gallon per hour
Office, Lobby or Department Store	4 or 5 gallons per hour per fountain
Light manufacturing	7 persons per gallon per hour
Heavy manufacturing	5 persons per gallon per hour
Hot heavy manufacturing	4 persons per gallon per hour
Hotel	.08 gallons per hour per room
Theater	1 gallon per hour per 100 seats

Table D2020-101 Hot Water Consumption Rates

Type of Building	Size Factor	Maximum Hourly Demand	Average Day Demand
Apartment Dwellings	No. of Apartments: Up to 20 21 to 50 51 to 75 76 to 100 101 to 200 201 up	 12.0 Gal. per apt. 10.0 Gal. per apt. 8.5 Gal. per apt. 7.0 Gal. per apt. 6.0 Gal. per apt. 5.0 Gal. per apt.	 42.0 Gal. per apt. 40.0 Gal. per apt. 38.0 Gal. per apt. 37.0 Gal. per apt. 36.0 Gal. per apt. 35.0 Gal. per apt.
Dormitories	Men Women	3.8 Gal. per man 5.0 Gal. per woman	13.1 Gal. per man 12.3 Gal. per woman
Hospitals	Per bed	23.0 Gal. per patient	90.0 Gal. per patient
Hotels	Single room with bath Double room with bath	17.0 Gal. per unit 27.0 Gal. per unit	50.0 Gal. per unit 80.0 Gal. per unit
Motels	No. of units: Up to 20 21 to 100 101 Up	 6.0 Gal. per unit 5.0 Gal. per unit 4.0 Gal. per unit	 20.0 Gal. per unit 14.0 Gal. per unit 10.0 Gal. per unit
Nursing Homes		4.5 Gal. per bed	18.4 Gal. per bed
Office buildings		0.4 Gal. per person	1.0 Gal. per person
Restaurants	Full meal type Drive-in snack type	1.5 Gal./max. meals/hr. 0.7 Gal./max. meals/hr.	2.4 Gal. per meal 0.7 Gal. per meal
Schools	Elementary Secondary & High	0.6 Gal. per student 1.0 Gal. per student	0.6 Gal. per student 1.8 Gal. per student

For evaluation purposes, recovery rate and storage capacity are inversely proportional. Water heaters should be sized so that the maximum hourly demand anticipated can be met in addition to allowance for the heat loss from the pipes and storage tank.

Table D2020-102 Fixture Demands in Gallons Per Fixture Per Hour

Table below is based on 140° F final temperature except for dishwashers in public places (*) where 180° F water is mandatory.

Fixture	Apartment House	Club	Gym	Hospital	Hotel	Indust. Plant	Office	Private Home	School
Bathtubs	20	20	30	20	20			20	
Dishwashers, automatic	15	50-150*		50-150*	50-200*	20-100*		15	20-100*
Kitchen sink	10	20		20	30	20	20	10	20
Laundry, stationary tubs	20	28		28	28			20	
Laundry, automatic wash	75	75		100	150			75	
Private lavatory	2	2	2	2	2	2	2	2	2
Public lavatory	4	6	8	6	8	12	6		15
Showers	30	150	225	75	75	225	30	30	225
Service sink	20	20		20	30	20	20	15	20
Demand factor	0.30	0.30	0.40	0.25	0.25	0.40	0.30	0.30	0.40
Storage capacity factor	1.25	0.90	1.00	0.60	0.80	1.00	2.00	0.70	1.00

To obtain the probable maximum demand multiply the total demands for the fixtures (gal./fixture/hour) by the demand factor. The heater should have a heating capacity in gallons per hour equal to this maximum. The storage tank should have a capacity in gallons equal to the probable maximum demand multiplied by the storage capacity factor.

Table D3010-601 Collector Tilt for Domestic Hot Water

Optimum collector tilt is usually equal to the site latitude. Variations of plus or minus 10 degrees are acceptable, and orientation of 20 degrees on either side of true south is acceptable. However, local climate and collector type may influence the choice between east or west deviations.

Flat plate collectors consist of a number of components as follows: Insulation to reduce heat loss through the bottom and sides of the collector. The enclosure which contains all the components in this assembly is usually weatherproof and prevents dust, wind and water from coming in contact with the absorber plate. The cover plate usually consists of one or more layers of a variety of glass or plastic and reduces the reradiation. It creates an air space which traps the heat by reducing radiation losses between the cover plate and the absorber plate.

The absorber plate must have a good thermal bond with the fluid passages.

The absorber plate is usually metallic and treated with a surface coating which improves absorptivity. Black or dark paints or selective coatings are used for this purpose, and the design of this passage and plate combination helps determine a solar system's effectiveness.

Heat transfer fluid passage tubes are attached above and below or integral with an absorber plate for the purpose of transferring thermal energy from the absorber plate to a heat transfer medium. The heat exchanger is a device for transferring thermal energy from one fluid to another. The rule of thumb of space heating sizing is one S.F. of collector per 2.5 S.F. of floor space.

For domestic hot water the rule of thumb is 3/4 S.F. of collector for one gallon of water used per day, on an average use of twenty-five gallons per day per person, plus ten gallons per dishwasher or washing machine.

Reference Tables

Table D3020-011

Heating Systems

The basic function of a heating system is to bring an enclosed volume up to a desired temperature and then maintain that temperature within a reasonable range. To accomplish this, the selected system must have sufficient capacity to offset transmission losses resulting from the temperature difference on the interior and exterior of the enclosing walls in addition to losses due to cold air infiltration through cracks, crevices and around doors and windows. The amount of heat to be furnished is dependent upon the building size, construction, temperature difference, air leakage, use, shape, orientation and exposure. Air circulation is also an important consideration. Circulation will prevent stratification which could result in heat losses through uneven temperatures at various levels. For example, the most

efficient use of unit heaters can usually be achieved by circulating the space volume through the total number of units once every 20 minutes or 3 times an hour. This general rule must, of course, be adapted for special cases such as large buildings with low ratios of heat transmitting surface to cubical volume. The type of occupancy of a building will have considerable bearing on the number of heat transmitting units and the location selected. It is axiomatic, however, that the basis of any successful heating system is to provide the maximum amount of heat at the points of maximum heat loss such as exposed walls, windows, and doors. Large roof areas, wind direction, and wide doorways create problems of excessive heat loss and require special consideration and treatment.

Heat Transmission

Heat transfer is an important parameter to be considered during selection of the exterior wall style, material and window area. A high rate of transfer will permit greater heat loss during the wintertime with the resultant increase in heating energy costs and a greater rate of heat gain in the summer with proportionally greater cooling cost. Several terms are used to describe various aspects of heat transfer. However, for general estimating purposes this book lists U valves for systems of construction materials. U is the "overall heat transfer coefficient." It is defined as the heat flow per hour through one square foot when the temperature difference in the air on either side of the structure wall, roof, ceiling or floor is one degree Fahrenheit. The structural segment may be a single homogeneous material or a composite.

Total heat transfer is found using the following equation:

$$Q = AU(T_2 - T_1) \text{ where}$$

Q = Heat flow, BTU per hour
A = Area, square feet
U = Overall heat transfer coefficient
$(T_2 - T_1)$ = Difference in temperature of air on each side of the construction component. (Also abbreviated TD)

Note that heat can flow through all surfaces of any building and this flow is in addition to heat gain or loss due to ventilation, infiltration and generation (appliances, machinery, people).

Table D3020-012 Heating Approximations for Quick Estimating

Oil Piping & Boiler Room Piping:

Small System	20 to 30% of Boiler
Complex System with Pumps, Headers, Etc.	80 to 110% of Boiler

Breeching With Insulation:

Small	10 to 15% of Boiler
Large	15 to 25% of Boiler

Coils: ..... 15 to 30% of Containing Unit
Balancing (Independent) ..... 1/2% of H.V.A.C. Estimating

Quality/Complexity Adjustment: For all heating installations add these adjustments to the estimate to more closely allow for the equipment and conditions of the particular job under consideration.

Economy installation, add	0 to 5% of System
Good quality, medium complexity, add	5 to 15% of System
Above average quality and complexity, add	15 to 25% of System

Table D3020-021 Factor for Determining Heat Loss for Various Types of Buildings

General: While the most accurate estimates of heating requirements would naturally be based on detailed information about the building being considered, it is possible to arrive at a reasonable approximation using the following procedure:

1. Calculate the cubic volume of the room or building.
2. Select the appropriate factor from Table D3020-021 below. Note that the factors apply only to inside temperatures listed in the first column and to 0° F outside temperature.

3. If the building has bad north and west exposures, multiply the heat loss factor by 1.1.
4. If the outside design temperature is other than 0°F, multiply the factor from Table D3020-021 by the factor from Table D3020-022.
5. Multiply the cubic volume by the factor selected from Table D3020-021. This will give the estimated BTUH heat loss which must be made up to maintain inside temperature.

Building Type	Conditions	Qualifications	Loss Factor*
Factories & Industrial Plants General Office Areas at 70° F	One Story	Skylight in Roof	6.2
		No Skylight in Roof	5.7
	Multiple Story	Two Story	4.6
		Three Story	4.3
		Four Story	4.1
		Five Story	3.9
		Six Story	3.6
	All Walls Exposed	Flat Roof	6.9
		Heated Space Above	5.2
	One Long Warm Common Wall	Flat Roof	6.3
		Heated Space Above	4.7
	Warm Common Walls on Both Long Sides	Flat Roof	5.8
		Heated Space Above	4.1
Warehouses at 60° F	All Walls Exposed	Skylights in Roof	5.5
		No Skylight in Roof	5.1
		Heated Space Above	4.0
	One Long Warm Common Wall	Skylight in Roof	5.0
		No Skylight in Roof	4.9
		Heated Space Above	3.4
	Warm Common Walls on Both Long Sides	Skylight in Roof	4.7
		No Skylight in Roof	4.4
		Heated Space Above	3.0

*Note: This table tends to be conservative particularly for new buildings designed for minimum energy consumption.

Table D3020-022 Outside Design Temperature Correction Factor (for Degrees Fahrenheit)

Outside Design Temperature	50	40	30	20	10	0	−10	−20	−30
Correction Factor	0.29	0.43	0.57	0.72	0.86	1.00	1.14	1.28	1.43

Tables D3020-023 and D3020-024 provide a way to calculate heat transmission of various construction materials from their U values and the TD (Temperature Difference).

1. From the exterior Enclosure Division or elsewhere, determine U values for the construction desired.
2. Determine the coldest design temperature. The difference between this temperature and the desired interior temperature is the TD (temperature difference).

3. Enter Figure D3020-023 or D3020-024 at correct U value. Cross horizontally to the intersection with appropriate TD. Read transmission per square foot from bottom of figure.
4. Multiply this value of BTU per hour transmission per square foot of area by the total surface area of that type of construction.

Table D3020-023 Transmission of Heat

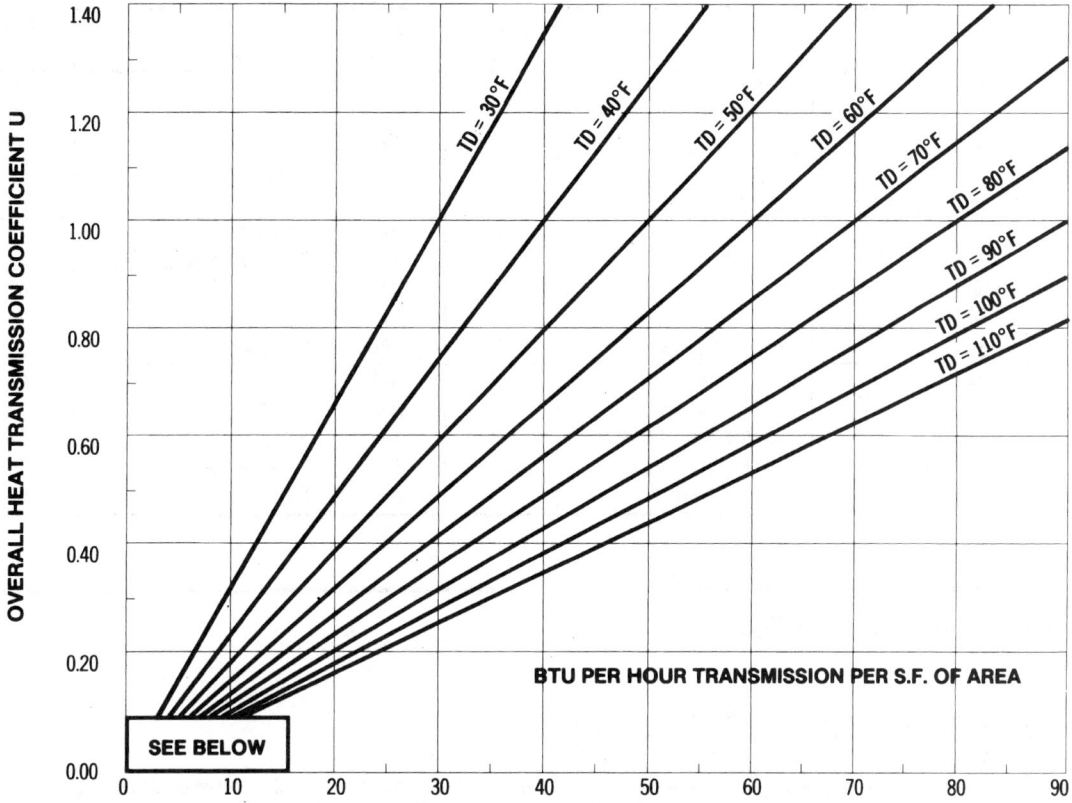

Table D3020-024 Transmission of Heat (Low Rate)

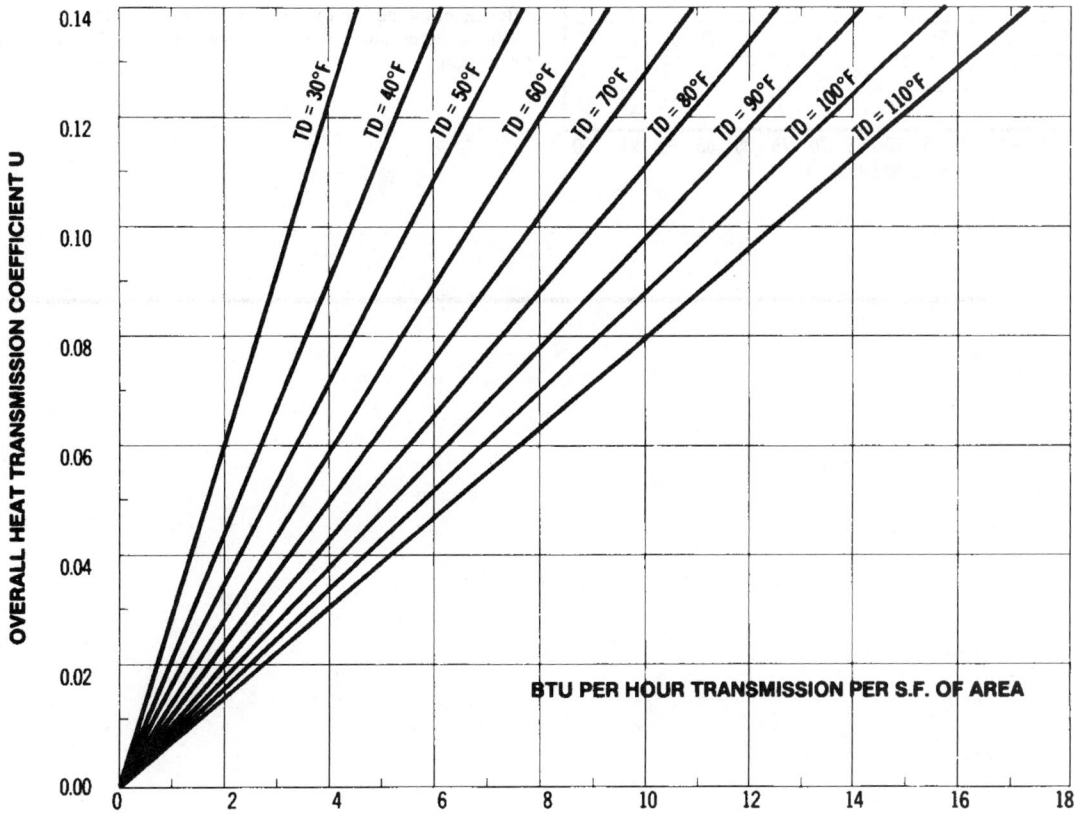

Table D3030-012 Air Conditioning Requirements

BTU's per hour per S.F. of floor area and S.F. per ton of air conditioning.

Type of Building	BTU/Hr per S.F.	S.F. per Ton	Type of Building	BTU/Hr per S.F.	S.F. per Ton	Type of Building	BTU/Hr per S.F.	S.F. per Ton
Apartments, Individual	26	450	Dormitory, Rooms	40	300	Libraries	50	240
Corridors	22	550	Corridors	30	400	Low Rise Office, Exterior	38	320
Auditoriums & Theaters	40	300/18*	Dress Shops	43	280	Interior	33	360
Banks	50	240	Drug Stores	80	150	Medical Centers	28	425
Barber Shops	48	250	Factories	40	300	Motels	28	425
Bars & Taverns	133	90	High Rise Office—Ext. Rms.	46	263	Office (small suite)	43	280
Beauty Parlors	66	180	Interior Rooms	37	325	Post Office, Individual Office	42	285
Bowling Alleys	68	175	Hospitals, Core	43	280	Central Area	46	260
Churches	36	330/20*	Perimeter	46	260	Residences	20	600
Cocktail Lounges	68	175	Hotel, Guest Rooms	44	275	Restaurants	60	200
Computer Rooms	141	85	Corridors	30	400	Schools & Colleges	46	260
Dental Offices	52	230	Public Spaces	55	220	Shoe Stores	55	220
Dept. Stores, Basement	34	350	Industrial Plants, Offices	38	320	Shop'g. Ctrs., Supermarkets	34	350
Main Floor	40	300	General Offices	34	350	Retail Stores	48	250
Upper Floor	30	400	Plant Areas	40	300	Specialty	60	200

*Persons per ton
12,000 BTU = 1 ton of air conditioning

Table D3030-013 Psychrometric Table

Dewpoint or Saturation Temperature (F)

	32	35	40	45	50	55	60	65	70	75	80	85	90	95	100
100	32	35	40	45	50	55	60	65	70	75	80	85	90	95	100
90	30	33	37	42	47	52	57	62	67	72	77	82	87	92	97
80	27	30	34	39	44	49	54	58	64	68	73	78	83	88	93
70	24	27	31	36	40	45	50	55	60	64	69	74	79	84	88
60	20	24	28	32	36	41	46	51	55	60	65	69	74	79	83
50	16	20	24	28	33	36	41	46	50	55	60	64	69	73	78
40	12	15	18	23	27	31	35	40	45	49	53	58	62	67	71
30	8	10	14	18	21	25	29	33	37	42	46	50	54	59	62
20	6	7	8	9	13	16	20	24	28	31	35	40	43	48	52
10	4	4	5	5	6	8	9	10	13	17	20	24	27	30	34

Dry bulb temperature (F)

This table shows the relationship between RELATIVE HUMIDITY, DRY BULB TEMPERATURE AND DEWPOINT. As an example, assume that the thermometer in a room reads 75° F, and we know that the relative humidity is 50%. The chart shows the dewpoint temperature to be 55°. That is, any surface colder than 55° F will "sweat" or collect condensing moisture. This surface could be the outside of an uninsulated chilled water pipe in the summertime, or the inside surface of a wall or deck in the wintertime. After determining the extreme ambient parameters, the table at the left is useful in determining which surfaces need insulation or vapor barrier protection.

Table D3030-014　Recommended Ventilation Air Changes

Table below lists range of time in minutes per change for various types of facilities.

Assembly Halls	2-10	Dance Halls	2-10	Laundries	1-3
Auditoriums	2-10	Dining Rooms	3-10	Markets	2-10
Bakeries	2-3	Dry Cleaners	1-5	Offices	2-10
Banks	3-10	Factories	2-5	Pool Rooms	2-5
Bars	2-5	Garages	2-10	Recreation Rooms	2-10
Beauty Parlors	2-5	Generator Rooms	2-5	Sales Rooms	2-10
Boiler Rooms	1-5	Gymnasiums	2-10	Theaters	2-8
Bowling Alleys	2-10	Kitchens-Hospitals	2-5	Toilets	2-5
Churches	5-10	Kitchens-Restaurant	1-3	Transformer Rooms	1-5

CFM air required for changes = Volume of room in cubic feet ÷ Minutes per change.

Table D3030-015　Ductwork

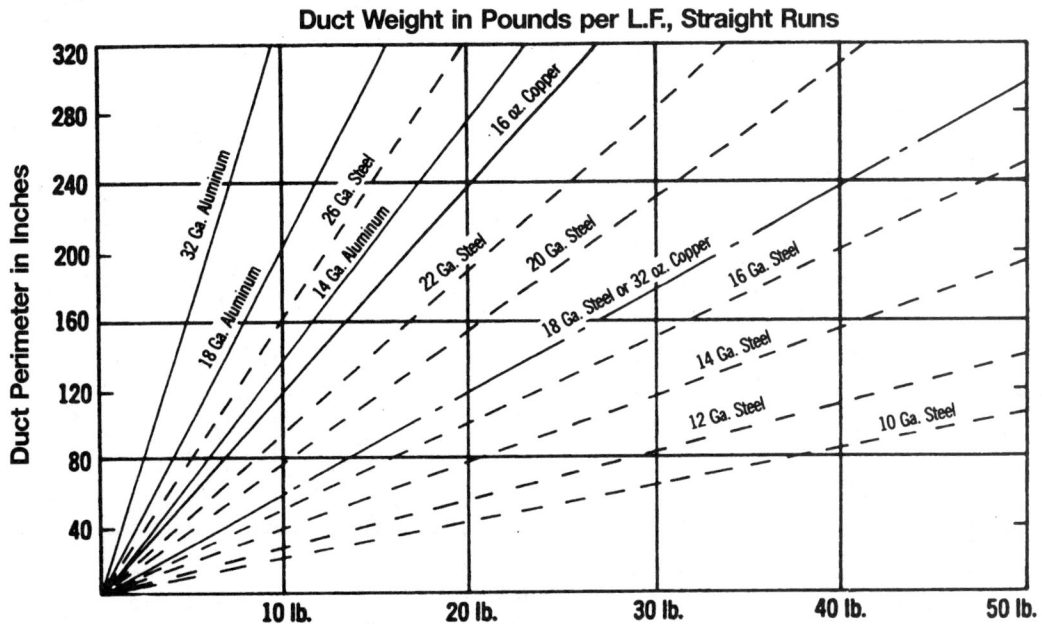

Duct Weight in Pounds per L.F., Straight Runs

Add to the above for fittings; 90° elbow is 3 L.F.; 45° elbow is 2.5 L.F.; offset is 4 L.F.; transition offset is 6 L.F.; square-to-round transition is 4 L.F.; 90° reducing elbow is 5 L.F. For bracing and waste, add 20% to aluminum and copper, 15% to steel.

Table D3030-017　Ductwork Packages (per Ton of Cooling)

System	Sheet Metal	Insulation	Diffusers	Return Register
Roof Top Unit Single Zone	120 Lbs.	52 S.F.	1	1
Roof Top Unit Multizone	240 Lbs.	104 S.F.	2	1
Self-contained Air or Water Cooled	108 Lbs.	—	2	—
Split System Air Cooled	102 Lbs.	—	2	—

Systems reflect most common usage. Refer to system graphics for duct layout.

Table D3030-018　Quality/Complexity Adjustment for Air Conditioning Systems

Economy installation, add	0 to 5%
Good quality, medium complexity, add	5 to 15%
Above average quality and complexity, add	15 to 25%

Add the above adjustments to the estimate to more closely allow for the equipment and conditions of the particular job under consideration.

Table D3030-019   Sheet Metal Calculator (Weight in Lb./Ft. of Length)

Gauge	26	24	22	20	18	16	Gauge	26	24	22	20	18	16
Wt.-Lb./S.F.	.906	1.156	1.406	1.656	2.156	2.656	Wt.-Lb./S.F.	.906	1.156	1.406	1.656	2.156	2.656
SMACNA Max. Dimension – Long Side		30"	54"	84"	85" Up		SMACNA Max. Dimension – Long Side		30"	54"	84"	85" Up	
Sum-2 sides							Sum-2 Sides						
2	.3	.40	.50	.60	.80	.90	56	9.3	12.0	14.0	16.2	21.3	25.2
3	.5	.65	.80	.90	1.1	1.4	57	9.5	12.3	14.3	16.5	21.7	25.7
4	.7	.85	1.0	1.2	1.5	1.8	58	9.7	12.5	14.5	16.8	22.0	26.1
5	.8	1.1	1.3	1.5	1.9	2.3	59	9.8	12.7	14.8	17.1	22.4	26.6
6	1.0	1.3	1.5	1.7	2.3	2.7	60	10.0	12.9	15.0	17.4	22.8	27.0
7	1.2	1.5	1.8	2.0	2.7	3.2	61	10.2	13.1	15.3	17.7	23.2	27.5
8	1.3	1.7	2.0	2.3	3.0	3.6	62	10.3	13.3	15.5	18.0	23.6	27.9
9	1.5	1.9	2.3	2.6	3.4	4.1	63	10.5	13.5	15.8	18.3	24.0	28.4
10	1.7	2.2	2.5	2.9	3.8	4.5	64	10.7	13.7	16.0	18.6	24.3	28.8
11	1.8	2.4	2.8	3.2	4.2	5.0	65	10.8	13.9	16.3	18.9	24.7	29.3
12	2.0	2.6	3.0	3.5	4.6	5.4	66	11.0	14.1	16.5	19.1	25.1	29.7
13	2.2	2.8	3.3	3.8	4.9	5.9	67	11.2	14.3	16.8	19.4	25.5	30.2
14	2.3	3.0	3.5	4.1	5.3	6.3	68	11.3	14.6	17.0	19.7	25.8	30.6
15	2.5	3.2	3.8	4.4	5.7	6.8	69	11.5	14.8	17.3	20.0	26.2	31.1
16	2.7	3.4	4.0	4.6	6.1	7.2	70	11.7	15.0	17.5	20.3	26.6	31.5
17	2.8	3.7	4.3	4.9	6.5	7.7	71	11.8	15.2	17.8	20.6	27.0	32.0
18	3.0	3.9	4.5	5.2	6.8	8.1	72	12.0	15.4	18.0	20.9	27.4	32.4
19	3.2	4.1	4.8	5.5	7.2	8.6	73	12.2	15.6	18.3	21.2	27.7	32.9
20	3.3	4.3	5.0	5.8	7.6	9.0	74	12.3	15.8	18.5	21.5	28.1	33.3
21	3.5	4.5	5.3	6.1	8.0	9.5	75	12.5	16.1	18.8	21.8	28.5	33.8
22	3.7	4.7	5.5	6.4	8.4	9.9	76	12.7	16.3	19.0	22.0	28.9	34.2
23	3.8	5.0	5.8	6.7	8.7	10.4	77	12.8	16.5	19.3	22.3	29.3	34.7
24	4.0	5.2	6.0	7.0	9.1	10.8	78	13.0	16.7	19.5	22.6	29.6	35.1
25	4.2	5.4	6.3	7.3	9.5	11.3	79	13.2	16.9	19.8	22.9	30.0	35.6
26	4.3	5.6	6.5	7.5	9.9	11.7	80	13.3	17.1	20.0	23.2	30.4	36.0
27	4.5	5.8	6.8	7.8	10.3	12.2	81	13.5	17.3	20.3	23.5	30.8	36.5
28	4.7	6.0	7.0	8.1	10.6	12.6	82	13.7	17.5	20.5	23.8	31.2	36.9
29	4.8	6.2	7.3	8.4	11.0	13.1	83	13.8	17.8	20.8	24.1	31.5	37.4
30	5.0	6.5	7.5	8.7	11.4	13.5	84	14.0	18.0	21.0	24.4	31.9	37.8
31	5.2	6.7	7.8	9.0	11.8	14.0	85	14.2	18.2	21.3	24.7	32.3	38.3
32	5.3	6.9	8.0	9.3	12.2	14.4	86	14.3	18.4	21.5	24.9	32.7	38.7
33	5.5	7.1	8.3	9.6	12.5	14.9	87	14.5	18.6	21.8	25.2	33.1	39.2
34	5.7	7.3	8.5	9.9	12.9	15.3	88	14.7	18.8	22.0	25.5	33.4	39.6
35	5.8	7.5	8.8	10.2	13.3	15.8	89	14.8	19.0	22.3	25.8	33.8	40.1
36	6.0	7.8	9.0	10.4	13.7	16.2	90	15.0	19.3	22.5	26.1	34.2	40.5
37	6.2	8.0	9.3	10.7	14.1	16.7	91	15.2	19.5	22.8	26.4	34.6	41.0
38	6.3	8.2	9.5	11.0	14.4	17.1	92	15.3	19.7	23.0	26.7	35.0	41.4
39	6.5	8.4	9.8	11.3	14.8	17.6	93	15.5	19.9	23.3	27.0	35.3	41.9
40	6.7	8.6	10.0	11.6	15.2	18.0	94	15.7	20.1	23.5	27.3	35.7	42.3
41	6.8	8.8	10.3	11.9	15.6	18.5	95	15.8	20.3	23.8	27.6	36.1	42.8
42	7.0	9.0	10.5	12.2	16.0	18.9	96	16.0	20.5	24.0	27.8	36.5	43.2
43	7.2	9.2	10.8	12.5	16.3	19.4	97	16.2	20.8	24.3	28.1	36.9	43.7
44	7.3	9.5	11.0	12.8	16.7	19.8	98	16.3	21.0	24.5	28.4	37.2	44.1
45	7.5	9.7	11.3	13.1	17.1	20.3	99	16.5	21.2	24.8	28.7	37.6	44.6
46	7.7	9.9	11.5	13.3	17.5	20.7	100	16.7	21.4	25.0	29.0	38.0	45.0
47	7.8	10.1	11.8	13.6	17.9	21.2	101	16.8	21.6	25.3	29.3	38.4	45.5
48	8.0	10.3	12.0	13.9	18.2	21.6	102	17.0	21.8	25.5	29.6	38.8	45.9
49	8.2	10.5	12.3	14.2	18.6	22.1	103	17.2	22.0	25.8	29.9	39.1	46.4
50	8.3	10.7	12.5	14.5	19.0	22.5	104	17.3	22.3	26.0	30.2	39.5	46.8
51	8.5	11.0	12.8	14.8	19.4	23.0	105	17.5	22.5	26.3	30.5	39.9	47.3
52	8.7	11.2	13.0	15.1	19.8	23.4	106	17.7	22.7	26.5	30.7	40.3	47.7
53	8.8	11.4	13.3	15.4	20.1	23.9	107	17.8	22.9	26.8	31.0	40.7	48.2
54	9.0	11.6	13.5	15.7	20.5	24.3	108	18.0	23.1	27.0	31.3	41.0	48.6
55	9.2	11.8	13.8	16.0	20.9	24.8	109	18.2	23.3	27.3	31.6	41.4	49.1
							110	18.3	23.5	27.5	31.9	41.8	49.5

Example: If duct is 34" x 20" x 15' long, 34" is greater than 30" maximum, for 24 ga. so must be 22 ga. 34" + 20" = 54" going across from 54" find 13.5 lb. per foot. 13.5 x 15' = 202.5 lbs.

For S.F. of surface area 202.5 ÷ 1.406 = 144 S.F.
Note: Figures include an allowance for scrap.

Table D4010-102   System Classification

System Classification
Rules for installation of sprinkler systems vary depending on the classification of occupancy falling into one of three categories as follows:

Light Hazard Occupancy
The protection area allotted per sprinkler should not exceed 225 S.F., with the maximum distance between lines and sprinklers on lines being 15'. The sprinklers do not need to be staggered. Branch lines should not exceed eight sprinklers on either side of a cross main. Each large area requiring more than 100 sprinklers and without a sub-dividing partition should be supplied by feed mains or risers sized for ordinary hazard occupancy.
Maximum system area = 52,000 S.F.

Included in this group are:

Churches	Nursing Homes
Clubs	Offices
Educational	Residential
Hospitals	Restaurants
Institutional	Theaters and Auditoriums
Libraries	(except stages and prosceniums)
(except large stack rooms)	Unused Attics
Museums	

Ordinary Hazard Occupancy
The protection area allotted per sprinkler shall not exceed 130 S.F. of noncombustible ceiling and 130 S.F. of combustible ceiling. The maximum allowable distance between sprinkler lines and sprinklers on line is 15'. Sprinklers shall be staggered if the distance between heads exceeds 12'. Branch lines should not exceed eight sprinklers on either side of a cross main.
Maximum system area = 52,000 S.F.

Included in this group are:

Group 1	Group 2
Automotive Parking and Showrooms	Cereal Mills
Bakeries	Chemical Plants—Ordinary
Beverage manufacturing	Confectionery Products
Canneries	Distilleries
Dairy Products Manufacturing/Processing	Dry Cleaners
Electronic Plans	Feed Mills
Glass and Glass Products Manufacturing	Horse Stables
Laundries	Leather Goods Manufacturing
Restaurant Service Areas	Libraries—Large Stack Room Areas
	Machine Shops
	Metal Working
	Mercantile
	Paper and Pulp Mills
	Paper Process Plants
	Piers and Wharves
	Post Offices
	Printing and Publishing
	Repair Garages
	Stages
	Textile Manufacturing
	Tire Manufacturing
	Tobacco Products Manufacturing
	Wood Machining
	Wood Product Assembly

Extra Hazard Occupancy
The protection area allotted per sprinkler shall not exceed 100 S.F. of noncombustible ceiling and 100 S.F. of combustible ceiling. The maximum allowable distance between lines and between sprinklers on lines is 12'. Sprinklers on alternate lines shall be staggered if the distance between sprinklers on lines exceeds 8'. Branch lines should not exceed six sprinklers on either side of a cross main.
Maximum system area:
Design by pipe schedule = 25,000 S.F.
Design by hydraulic calculation = 40,000 S.F.

Included in this group are:

Group 1	Group 2
Aircraft hangars	Asphalt Saturating
Combustible Hydraulic Fluid Use Area	Flammable Liquids Spraying
Die Casting	Flow Coating
Metal Extruding	Manufactured/Modular Home
Plywood/Particle Board Manufacturing	Building Assemblies (where
Printing (inks with flash points < 100 degrees F)	finished enclosure is present and has combustible interiors)
Rubber Reclaiming, Compounding, Drying, Milling, Vulcanizing	Open Oil Quenching
Saw Mills	Plastics Processing
Textile Picking, Opening, Blending, Garnetting, Carding, Combing of Cotton, Synthetics, Wood Shoddy, or Burlap	Solvent Cleaning
Upholstering with Plastic Foams	Varnish and Paint Dipping

Table D4020-301   Standpipe Systems

The basis for standpipe system design is National Fire Protection Association NFPA 14. However, the authority having jurisdiction should be consulted for special conditions, local requirements and approval.

Standpipe systems, properly designed and maintained, are an effective and valuable time saving aid for extinguishing fires, especially in the upper stories of tall buildings, the interior of large commercial or industrial malls, or other areas where construction features or access make the laying of temporary hose lines time consuming and/or hazardous. Standpipes are frequently installed with automatic sprinkler systems for maximum protection.

There are three general classes of service for standpipe systems:
Class I for use by fire departments and personnel with special training for heavy streams (2-1/2″ hose connections).
Class II for use by building occupants until the arrival of the fire department (1-1/2″ hose connector with hose).

Class III for use by either fire departments and trained personnel or by the building occupants (both 2-1/2″ and 1-1/2″ hose connections or one 2-1/2″ hose valve with an easily removable 2-1/2″ by 1-1/2″ adapter).

Standpipe systems are also classified by the way water is supplied to the system. The four basic types are:
Type 1: Wet standpipe system having supply valve open and water pressure maintained at all times.
Type 2: Standpipe system so arranged through the use of approved devices as to admit water to the system automatically by opening a hose valve.
Type 3: Standpipe system arranged to admit water to the system through manual operation of approved remote control devices located at each hose station.
Type 4: Dry standpipe having no permanent water supply.

Table D4020-302 NFPA 14 Basic Standpipe Design

Class	Design-Use	Pipe Size Minimums	Water Supply Minimums
Class I	2 1/2″ hose connection on each floor All areas within 150′ of an exit in every exit stairway Fire Department Trained Personnel	Height to 100′, 4″ dia. Heights above 100′, 6″ dia. (275′ max. except with pressure regulators 400′ max.)	For each standpipe riser 500 GPM flow For common supply pipe allow 500 GPM for first standpipe plus 250 GPM for each additional standpipe (2500 GPM max. total) 30 min. duration 65 PSI at 500 GPM
Class II	1 1/2″ hose connection with hose on each floor All areas within 130′ of hose connection measured along path of hose travel Occupant personnel	Height to 50′, 2″ dia. Height above 50′, 2 1/2″ dia.	For each standpipe riser 100 GPM flow For multiple riser common supply pipe 100 GPM 300 min. duration, 65 PSI at 100 GPM
Class III	Both of above. Class I valved connections will meet Class III with additional 2 1/2″ by 1 1/2″ adapter and 1 1/2″ hose.	Same as Class I	Same as Class I

*Note: Where 2 or more standpipes are installed in the same building or section of building they shall be interconnected at the bottom.

Combined Systems

Combined systems are systems where the risers supply both automatic sprinklers and 2-1/2″ hose connection outlets for fire department use. In such a system the sprinkler spacing pattern shall be in accordance with NFPA 13 while the risers and supply piping will be sized in accordance with NFPA 14. When the building is completely sprinklered the risers may be sized by hydraulic calculation. The minimum size riser for buildings not completely sprinklered is 6″.
The minimum water supply of a completely sprinklered, light hazard, high-rise occupancy building will be 500 GPM while the supply required for other types of completely sprinklered high-rise buildings is 1000 GPM.

General System Requirements

1. Approved valves will be provided at the riser for controlling branch lines to hose outlets.
2. A hose valve will be provided at each outlet for attachment of hose.
3. Where pressure at any standpipe outlet exceeds 100 PSI a pressure reducer must be installed to limit the pressure to 100 PSI. Note that

the pressure head due to gravity in 100′ of riser is 43.4 PSI. This must be overcome by city pressure, fire pumps, or gravity tanks to provide adequate pressure at the top of the riser.

4. Each hose valve on a wet system having linen hose shall have an automatic drip connection to prevent valve leakage from entering the hose.
5. Each riser will have a valve to isolate it from the rest of the system.
6. One or more fire department connections as an auxiliary supply shall be provided for each Class I or Class III standpipe system. In buildings having two or more zones, a connection will be provided for each zone.
7. There will be no shutoff valve in the fire department connection, but a check valve will be located in the line before it joins the system.
8. All hose connections street side will be identified on a cast plate or fitting as to purpose.

Table D4020-303 Quality/Complexity Adjustment for Sprinkler/Standpipe Systems

Economy installation, add . . . . 0 to 5%
Good quality, medium complexity, add . . . . 5 to 15%
Above average quality and complexity, add . . . . 15 to 25%

Figure D5010-111   Typical Overhead Service Entrance

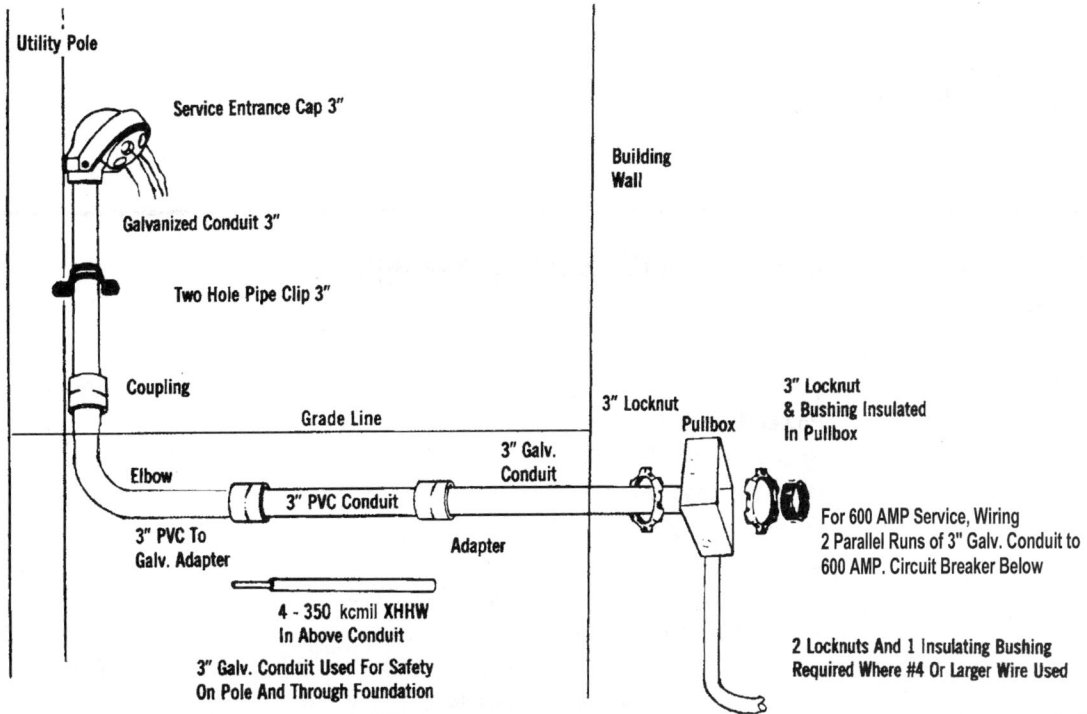

Utility Pole

Service Entrance Cap 3"

Galvanized Conduit 3"

Two Hole Pipe Clip 3"

Coupling

Building Wall

Grade Line

3" Locknut

3" Locknut & Bushing Insulated In Pullbox

Pullbox

Elbow

3" Galv. Conduit

3" PVC Conduit

Adapter

3" PVC To Galv. Adapter

4 - 350 kcmil XHHW In Above Conduit

3" Galv. Conduit Used For Safety On Pole And Through Foundation

For 600 AMP Service, Wiring 2 Parallel Runs of 3" Galv. Conduit to 600 AMP. Circuit Breaker Below

2 Locknuts And 1 Insulating Bushing Required Where #4 Or Larger Wire Used

Figure D5010-112   Typical Commercial Electric System

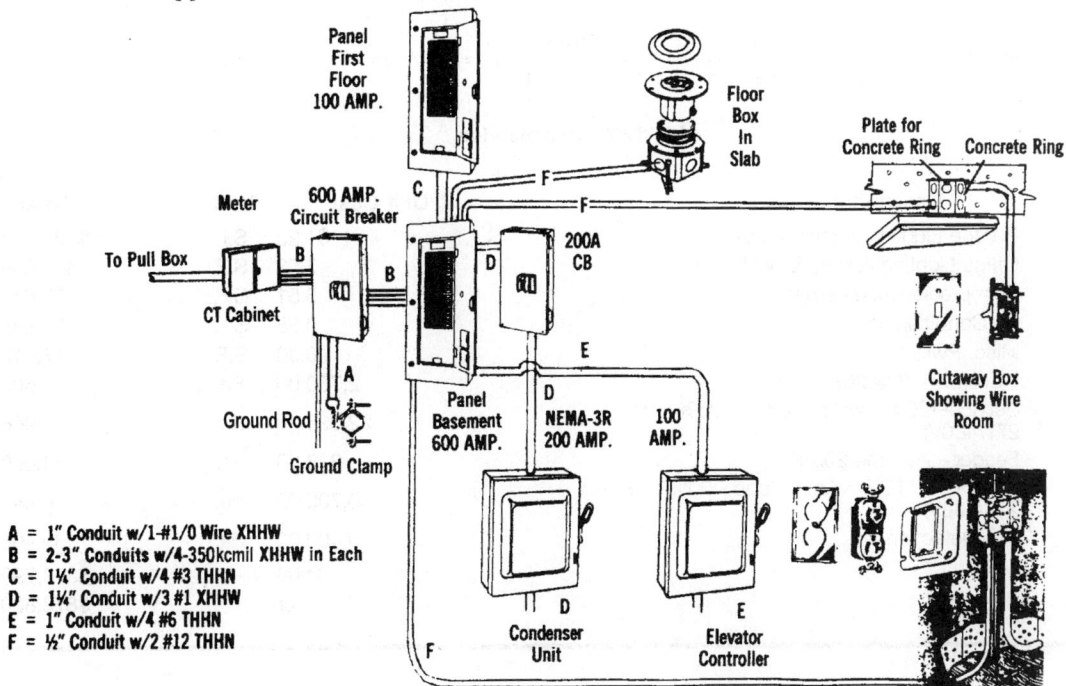

Panel First Floor 100 AMP.

Floor Box In Slab

Plate for Concrete Ring

Concrete Ring

Meter

600 AMP. Circuit Breaker

To Pull Box

CT Cabinet

200A CB

Cutaway Box Showing Wire Room

Ground Rod

Ground Clamp

Panel Basement 600 AMP.

NEMA-3R 200 AMP.

100 AMP.

Condenser Unit

Elevator Controller

A = 1" Conduit w/1-#1/0 Wire XHHW
B = 2-3" Conduits w/4-350kcmil XHHW in Each
C = 1¼" Conduit w/4 #3 THHN
D = 1¼" Conduit w/3 #1 XHHW
E = 1" Conduit w/4 #6 THHN
F = ½" Conduit w/2 #12 THHN

Figure D5010-113 Preliminary Procedure

1. Determine building size and use.
2. Develop total load in watts.
   a. Lighting
   b. Receptacle
   c. Air Conditioning
   d. Elevator
   e. Other power requirements
3. Determine best voltage available from utility company.
4. Determine cost from tables for loads (a) thru (e) above.
5. Determine size of service from formulas (D5010-116).
6. Determine costs for service, panels, and feeders from tables.

Figure D5010-114 Office Building 90' x 210', 3 story, w/garage

Garage Area = 18,900 S.F.
Office Area = 56,700 S.F.
Elevator = 2 @ 125 FPM

Tables	Power Required	Watts
D5010-1151	Garage Lighting .5 Watt/S.F.	9,450
	Office Lighting 3 Watts/S.F.	170,100
D5010-1151	Office Receptacles 2 Watts/S.F.	113,400
D5010-1151, R262213-27	Low Rise Office A.C. 4.3 Watts/S.F.	243,810
D5010-1152, 1153	Elevators - 2 @ 20 HP = 2 @ 17,404 Watts/Ea.	34,808
D5010-1151	Misc. Motors + Power 1.2 Watts/S.F.	68,040
	Total	639,608 Watts

Voltage Available

277/480V, 3 Phase, 4 Wire

Formula

D5010-116

$$\text{Amperes} = \frac{\text{Watts}}{\text{Volts x Power Factor x 1.73}} = \frac{639,608}{480V \text{ x } .8 \text{ x } 1.73} = 963 \text{ Amps}$$

Use 1200 Amp Service

System	Description	Unit Cost	Unit	Total
D5020 210 0200	Garage Lighting (Interpolated)	$1.23	S.F.	$ 23,247
D5020 210 0540	Office Lighting (Using 32 W lamps)	7.52	S.F.	426,384
D5020 115 0880	Receptacle-Undercarpet	3.51	S.F.	199,017
D5020 140 0280	Air Conditioning	0.58	S.F.	32,886
D5020 135 0320	Misc. Pwr.	0.30	S.F.	17,010
D5020 145 2120	Elevators - 2 @ 20HP	2,750.00	Ea.	5,500
D5010 120 0480	Service-1200 Amp (add 25% for 277/480V)	22,325.00	Ea.	27,906
D5010 230 0480	Feeder - Assume 200 Ft.	373.00	Ft.	74,600
D5010 240 0320	Panels - 1200 Amp (add 20% for 277/480V)	30,700.00	Ea.	36,840
D5030 910 0400	Fire Detection	30,400.00	Ea.	30,400
		Total		$873,790
		or		$873,800

Table D5010-1151 Nominal Watts Per S.F. for Electric Systems for Various Building Types

Type Construction	1. Lighting	2. Devices	3. HVAC	4. Misc.	5. Elevator	Total Watts
Apartment, luxury high rise	2	2.2	3	1		
Apartment, low rise	2	2	3	1		
Auditorium	2.5	1	3.3	.8		
Bank, branch office	3	2.1	5.7	1.4		
Bank, main office	2.5	1.5	5.7	1.4		
Church	1.8	.8	3.3	.8		
College, science building	3	3	5.3	1.3		
College, library	2.5	.8	5.7	1.4		
College, physical education center	2	1	4.5	1.1		
Department store	2.5	.9	4	1		
Dormitory, college	1.5	1.2	4	1		
Drive-in donut shop	3	4	6.8	1.7		
Garage, commercial	.5	.5	0	.5		
Hospital, general	2	4.5	5	1.3		
Hospital, pediatric	3	3.8	5	1.3		
Hotel, airport	2	1	5	1.3		
Housing for the elderly	2	1.2	4	1		
Manufacturing, food processing	3	1	4.5	1.1		
Manufacturing, apparel	2	1	4.5	1.1		
Manufacturing, tools	4	1	4.5	1.1		
Medical clinic	2.5	1.5	3.2	1		
Nursing home	2	1.6	4	1		
Office building, hi rise	3	2	4.7	1.2		
Office building, low rise	3	2	4.3	1.2		
Radio-TV studio	3.8	2.2	7.6	1.9		
Restaurant	2.5	2	6.8	1.7		
Retail store	2.5	.9	5.5	1.4		
School, elementary	3	1.9	5.3	1.3		
School, junior high	3	1.5	5.3	1.3		
School, senior high	2.3	1.7	5.3	1.3		
Supermarket	3	1	4	1		
Telephone exchange	1	.6	4.5	1.1		
Theater	2.5	1	3.3	.8		
Town Hall	2	1.9	5.3	1.3		
U.S. Post Office	3	2	5	1.3		
Warehouse, grocery	1	.6	0	.5		

Rule of Thumb: 1 KVA = 1 HP (Single Phase)

Three Phase:

Watts = 1.73 x Volts x Current x Power Factor x Efficiency

$$\text{Horsepower} = \frac{\text{Volts x Current x 1.73 x Power Factor}}{746 \text{ Watts}}$$

Table D5010-1152 Horsepower Requirements for Elevators with 3 Phase Motors

Type	Maximum Travel Height in Ft.	Travel Speeds in FPM	Capacity of Cars in Lbs.		
			1200	1500	1800
Hydraulic	70	70	10	15	15
		85	15	15	15
		100	15	15	20
		110	20	20	20
		125	20	20	20
		150	25	25	25
		175	25	30	30
		200	30	30	40
Geared Traction	300	200			
		350			
			2000	2500	3000
Hydraulic	70	70	15	20	20
		85	20	20	25
		100	20	25	30
		110	20	25	30
		125	25	30	40
		150	30	40	50
		175	40	50	50
		200	40	50	60
Geared Traction	300	200	10	10	15
		350	15	15	23
			3500	4000	4500
Hydraulic	70	70	20	25	30
		85	25	30	30
		100	30	40	40
		110	40	40	50
		125	40	50	50
		150	50	50	60
		175	60		
		200	60		
Geared Traction	300	200	15		23
		350	23		35

The power factor of electric motors varies from 80% to 90% in larger size motors. The efficiency likewise varies from 80% on a small motor to 90% on a large motor.

Table D5010-1153 Watts per Motor

90% Power Factor & Efficiency @ 200 or 460V			
HP	Watts	HP	Watts
10	9024	30	25784
15	13537	40	33519
20	17404	50	41899
25	21916	60	49634

Table D5010-116 Electrical Formulas

Ohm's Law

Ohm's Law is a method of explaining the relation existing between voltage, current, and resistance in an electrical circuit. It is practically the basis of all electrical calculations. The term "electromotive force" is often used to designate pressure in volts. This formula can be expressed in various forms.

To find the current in amperes:

$$\text{Current} = \frac{\text{Voltage}}{\text{Resistance}} \quad \text{or} \quad \text{Amperes} = \frac{\text{Volts}}{\text{Ohms}} \quad \text{or} \quad I = \frac{E}{R}$$

The flow of current in amperes through any circuit is equal to the voltage or electromotive force divided by the resistance of that circuit.

To find the pressure or voltage:

$$\text{Voltage} = \text{Current} \times \text{Resistance} \quad \text{or} \quad \text{Volts} = \text{Amperes} \times \text{Ohms}$$
$$\text{or} \quad E = I \times R$$

The voltage required to force a current through a circuit is equal to the resistance of the circuit multiplied by the current.

To find the resistance:

$$\text{Resistance} = \frac{\text{Voltage}}{\text{Current}} \quad \text{or} \quad \text{Ohms} = \frac{\text{Volts}}{\text{Amperes}} \quad \text{or} \quad R = \frac{E}{I}$$

The resistance of a circuit is equal to the voltage divided by the current flowing through that circuit.

Power Formulas

One horsepower = 746 watts   One kilowatt = 1000 watts

The power factor of electric motors varies from 80% to 90% in the larger size motors.

Single-Phase Alternating Current Circuits

Power in Watts = Volts x Amperes x Power Factor

To find current in amperes:

$$\text{Current} = \frac{\text{Watts}}{\text{Volts} \times \text{Power Factor}} \quad \text{or}$$

$$\text{Amperes} = \frac{\text{Watts}}{\text{Volts} \times \text{Power Factor}} \quad \text{or} \quad I = \frac{W}{E \times PF}$$

To find current of a motor, single phase:

$$\text{Current} = \frac{\text{Horsepower} \times 746}{\text{Volts} \times \text{Power Factor} \times \text{Efficiency}} \quad \text{or}$$

$$I = \frac{HP \times 746}{E \times PF \times \text{Eff.}}$$

To find horsepower of a motor, single phase:

$$\text{Horsepower} = \frac{\text{Volts} \times \text{Current} \times \text{Power Factor} \times \text{Efficiency}}{746 \text{ Watts}}$$

$$HP = \frac{E \times I \times PF \times \text{Eff.}}{746}$$

To find power in watts of a motor, single phase:

Watts = Volts x Current x Power Factor x Efficiency   or
Watts = E x I x PF x Eff.

To find single phase kVA:

$$1 \text{ Phase kVA} = \frac{\text{Volts} \times \text{Amps}}{1000}$$

Three-Phase Alternating Current Circuits

Power in Watts = Volts x Amperes x Power Factor x 1.73

To find current in amperes in each wire:

$$\text{Current} = \frac{\text{Watts}}{\text{Voltage} \times \text{Power Factor} \times 1.73} \quad \text{or}$$

$$\text{Amperes} = \frac{\text{Watts}}{\text{Volts} \times \text{Power Factor} \times 1.73} \quad \text{or} \quad I = \frac{W}{E \times PF \times 1.73}$$

To find current of a motor, 3 phase:

$$\text{Current} = \frac{\text{Horsepower} \times 746}{\text{Volts} \times \text{Power Factor} \times \text{Efficiency} \times 1.73} \quad \text{or}$$

$$I = \frac{HP \times 746}{E \times PF \times \text{Eff.} \times 1.73}$$

To find horsepower of a motor, 3 phase:

$$\text{Horsepower} = \frac{\text{Volts} \times \text{Current} \times 1.73 \times \text{Power Factor}}{746 \text{ Watts}}$$

$$HP = \frac{E \times I \times 1.73 \times PF}{746}$$

To find power in watts of a motor, 3 phase:

Watts = Volts x Current x 1.73 x Power Factor x Efficiency   or
Watts = E x I x 1.73 x PF x Eff.

To find 3 phase kVA:

$$3 \text{ phase kVA} = \frac{\text{Volts} \times \text{Amps} \times 1.73}{1000} \quad \text{or}$$

$$\text{kVA} = \frac{V \times A \times 1.73}{1000}$$

Power Factor (PF) is the percentage ratio of the measured watts (effective power) to the volt-amperes (apparent watts)

$$\text{Power Factor} = \frac{\text{Watts}}{\text{Volts} \times \text{Amperes}} \times 100\%$$

A Conceptual Estimate of the costs for a building when final drawings are not available can be quickly figured by using **Table D5010-117 Cost Per S.F. for Electrical Systems for Various Building Types.** The following definitions apply to this table.

1. **Service and Distribution:** This system includes the incoming primary feeder from the power company, main building transformer, metering arrangement, switchboards, distribution panel boards, stepdown transformers, and power and lighting panels. Items marked (*) include the cost of the primary feeder and transformer. In all other projects the cost of the primary feeder and transformer is paid for by the local power company.

2. **Lighting:** Includes all interior fixtures for decor, illumination, exit and emergency lighting. Fixtures for exterior building lighting are included, but parking area lighting is not included unless mentioned. See also Section D5020 for detailed analysis of lighting requirements and costs.

3. **Devices:** Includes all outlet boxes, receptacles, switches for lighting control, dimmers and cover plates.

4. **Equipment Connections:** Includes all materials and equipment for making connections for heating, ventilating and air conditioning, food service and other motorized items requiring connections.

5. **Basic Materials:** This category includes all disconnect power switches not part of service equipment, raceways for wires, pull boxes, junction boxes, supports, fittings, grounding materials, wireways, busways, and wire and cable systems.

6. **Special Systems:** Includes installed equipment only for the particular system such as fire detection and alarm, sound, emergency generator and others as listed in the table.

Table D5010-117 Cost per S.F. for Electric Systems for Various Building Types

Type Construction	1. Service & Distrib.	2. Lighting	3. Devices	4. Equipment Connections	5. Basic Materials	6. Special Systems Fire Alarm & Detection	Lightning Protection	Master TV Antenna
Apartment, luxury high rise	$2.04	$1.23	$ .91	$1.21	$ 3.61	$ .59		$.42
Apartment, low rise	1.18	1.03	.84	1.03	2.12	.50		
Auditorium	2.60	6.21	.76	1.73	4.35	.76		
Bank, branch office	3.12	6.90	1.23	1.78	4.18	2.17		
Bank, main office	2.34	3.75	.40	.75	4.46	1.12		
Church	1.63	3.81	.47	.45	2.14	1.12		
* College, science building	3.36	5.05	1.81	1.57	5.62	.98		
* College library	2.20	2.80	.36	.82	2.67	1.12		
* College, physical education center	3.46	3.91	.50	.67	2.08	.63		
Department store	1.15	2.70	.36	1.19	3.58	.50		
* Dormitory, college	1.55	3.47	.37	.75	3.51	.79		.51
Drive-in donut shop	4.34	1.37	1.77	1.79	5.75	—		
Garage, commercial	.59	1.40	.27	.55	1.23	—		
* Hospital, general	8.39	5.34	2.02	1.40	7.10	.69	$.24	
* Hospital, pediatric	7.35	7.95	1.75	5.15	13.20	.78		.61
* Hotel, airport	3.30	4.37	.37	.69	5.22	.64	.39	.57
Housing for the elderly	.93	1.04	.49	1.35	4.41	.78		.50
Manufacturing, food processing	2.09	5.45	.34	2.56	4.86	.49		
Manufacturing, apparel	1.39	2.86	.41	1.02	2.60	.43		
Manufacturing, tools	3.19	6.66	.38	1.20	4.35	.51		
Medical clinic	1.65	2.53	.61	2.01	3.77	.77		
Nursing home	2.18	4.31	.61	.54	4.35	1.06		.42
Office Building	2.92	5.84	.34	1.02	4.55	.56	.35	
Radio-TV studio	2.09	6.02	.91	1.87	5.37	.72		
Restaurant	7.81	5.71	1.11	2.84	6.46	.43		
Retail Store	1.65	3.04	.37	.69	2.03	—		
School, elementary	2.81	5.39	.75	.69	5.42	.68		.32
School, junior high	1.71	4.51	.37	1.26	4.31	.79		
* School, senior high	1.84	3.53	.64	1.64	4.80	.70		
Supermarket	1.88	3.05	.44	2.74	4.16	.35		
* Telephone Exchange	4.94	1.70	.43	1.54	3.27	1.30		
Theater	3.60	4.17	.77	2.38	4.22	.96		
Town Hall	2.21	3.30	.77	.85	5.48	.63		
* U.S.Post Office	6.46	4.25	.78	1.31	3.99	.63		
Warehouse, grocery	1.21	1.84	.28	.69	2.96	.41		

*Includes cost of primary feeder and transformer. Cont'd on next page.

COST ASSUMPTIONS:

Each of the projects analyzed in Table D5010-117 was bid within the last 10 years in the northeastern part of the United States. Bid prices have been adjusted to Jan. 1 levels. The list of projects is by no means all-inclusive, yet by carefully examining the various systems for a particular building type, certain cost relationships will emerge. The use of Square Foot/Cubic Costs, in the Reference Section, along with the Project Size Modifier, should produce a budget S.F. cost for the electrical portion of a job that is consistent with the amount of design information normally available at the conceptual estimate stage.

Table D5010-117 Cost per S.F. for Electric Systems for Various Building Types (cont.)

Type Construction	Intercom Systems	Sound Systems	Closed Circuit TV	Snow Melting	Emergency Generator	Security	Master Clock Sys.
Apartment, luxury high rise	$ .79						
Apartment, low rise	.57						
Auditorium		$2.07	$ .95		$1.55		
Bank, branch office	1.09		2.19			$1.85	
Bank, main office	.62		.48		1.25	1.01	$.44
Church	.77	.48				.30	
* College, science building	.80	1.02			1.75	.52	.50
* College, library		.95			.81	.51	
* College, physical education center		1.06				.53	
Department store					.38		
* Dormitory, college	1.07						
Drive-in donut shop							.23
Garage, commercial							.20
Hospital, general	.81		.35		2.14		
* Hospital, pediatric	5.51	.56	.60		1.34		
* Hotel, airport	.80				.80		
Housing for the elderly	.96						
Manufacturing, food processing		.37			2.78		
Manufacturing apparel		.49					
Manufacturing, tools		.62		$.41			
Medical clinic							
Nursing home	1.83				.71		
Office Building		.33			.71	.36	.20
Radio-TV studio	1.08				1.76		.76
Restaurant		.49					
Retail Store							
School, elementary		.35					.35
School, junior high		.88			.59		.60
* School, senior high	.73		.49		.81	.44	.45
Supermarket		.38			.75	.49	
* Telephone exchange					7.07	.28	
Theater		.71					
Town Hall							.35
* U.S. Post Office	.71			.19	.79		
Warehouse, grocery	.46						

*Includes cost of primary feeder and transformer. Cont'd on next page.

General: Variations in the following square foot costs are due to the type of structural systems of the buildings, geographical location, local electrical codes, designer's preference for specific materials and equipment, and the owner's particular requirements.

Table D5010-117 Cost per S.F. for Total Electric Systems for Various Building Types (cont.)

Type Construction	Basic Description	Total Floor Area in Square Feet	Total Cost per Square Foot for Total Electric Systems
Apartment building, luxury high rise	All electric, 18 floors, 86 1 B.R., 34 2 B.R.	115,000	$10.81
Apartment building, low rise	All electric, 2 floors, 44 units, 1 & 2 B.R.	40,200	7.26
Auditorium	All electric, 1200 person capacity	28,000	20.97
Bank, branch office	All electric, 1 floor	2,700	24.50
Bank, main office	All electric, 8 floors	54,900	16.62
Church	All electric, incl. Sunday school	17,700	11.16
*College, science building	All electric, 3-1/2 floors, 47 rooms	27,500	22.97
*College, library	All electric	33,500	12.24
*College, physical education center	All electric	22,000	12.83
Department store	Gas heat, 1 floor	85,800	9.85
*Dormitory, college	All electric, 125 rooms	63,000	12.03
Drive-in donut shop	Gas heat, incl. parking area lighting	1,500	24.25
Garage, commercial	All electric	52,300	4.24
*Hospital, general	Steam heat, 4 story garage, 300 beds	540,000	28.49
*Hospital, pediatric	Steam heat, 6 stories	278,000	44.81
Hotel, airport	All electric, 625 guest rooms	536,000	17.16
Housing for the elderly	All electric, 7 floors, 100 1 B.R. units	67,000	10.46
Manufacturing, food processing	Electric heat, 1 floor	9,600	18.95
Manufacturing, apparel	Electric heat, 1 floor	28,000	9.19
Manufacturing, tools	Electric heat, 2 floors	42,000	17.32
Medical clinic	Electric heat, 2 floors	22,700	11.34
Nursing home	Gas heat, 3 floors, 60 beds	21,000	16.01
Office building	All electric, 15 floors	311,200	17.18
Radio-TV studio	Electric heat, 3 floors	54,000	20.59
Restaurant	All electric	2,900	24.85
Retail store	All electric	3,000	7.78
School, elementary	All electric, 1 floor	39,500	16.76
School, junior high	All electric, 1 floor	49,500	15.02
*School, senior high	All electric, 1 floor	158,300	16.08
Supermarket	Gas heat	30,600	14.23
*Telephone exchange	Gas heat, 300 kW emergency generator	24,800	20.53
Theater	Electric heat, twin cinema	14,000	16.80
Town Hall	All electric	20,000	13.59
*U.S. Post Office	All electric	495,000	19.09
Warehouse, grocery	All electric	96,400	7.86

*Includes cost of primary feeder and transformer.

D5010-120   Electric Circuit Voltages

General: The following method provides the user with a simple non-technical means of obtaining comparative costs of wiring circuits. The circuits considered serve the electrical loads of motors, electric heating, lighting and transformers, for example, that require low voltage 60 Hertz alternating current.

The method used here is suitable only for obtaining estimated costs. It is **not** intended to be used as a substitute for electrical engineering design applications.

Conduit and wire circuits can represent from twenty to thirty percent of the total building electrical cost. By following the described steps and using the tables the user can translate the various types of electric circuits into estimated costs.

Wire Size: Wire size is a function of the electric load which is usually listed in one of the following units:

1. Amperes (A)
2. Watts (W)
3. Kilowatts (kW)
4. Volt amperes (VA)
5. Kilovolt amperes (kVA)
6. Horsepower (HP)

These units of electric load must be converted to amperes in order to obtain the size of wire necessary to carry the load. To convert electric load units to amperes one must have an understanding of the voltage classification of the power source and the voltage characteristics of the electrical equipment or load to be energized. The seven A.C. circuits commonly used are illustrated in Figures D5010-121 thru D5010-127 showing the transformer load voltage and the point of use voltage at the point on the circuit where the load is connected. The difference between the source and point of use

voltages is attributed to the circuit voltage drop and is considered to be approximately 4%.

Motor Voltages: Motor voltages are listed by their point of use voltage and not the power source voltage.

For example: 460 volts instead of 480 volts
200 instead of 208 volts
115 volts instead of 120 volts

Lighting and Heating Voltages: Lighting and heating equipment voltages are listed by the power source voltage and not the point of wire voltage.

For example: 480, 277, 120 volt lighting
480 volt heating or air conditioning unit
208 volt heating unit

Transformer Voltages: Transformer primary (input) and secondary (output) voltages are listed by the power source voltage.

For example: Single phase 10 kVA
Primary 240/480 volts
Secondary 120/240 volts

In this case, the primary voltage may be 240 volts with a 120 volts secondary or may be 480 volts with either a 120V or a 240V secondary.

For example: Three phase 10 kVA
Primary 480 volts
Secondary 208Y/120 volts

In this case the transformer is suitable for connection to a circuit with a 3 phase 3 wire or 3 phase 4 wire circuit with a 480 voltage. This application will provide a secondary circuit of 3 phase 4 wire with 208 volts between phase wires and 120 volts between any phase wire and the neutral (white) wire.

Figure D5010-121

3 Wire, 1 Phase, 120/240 Volt System

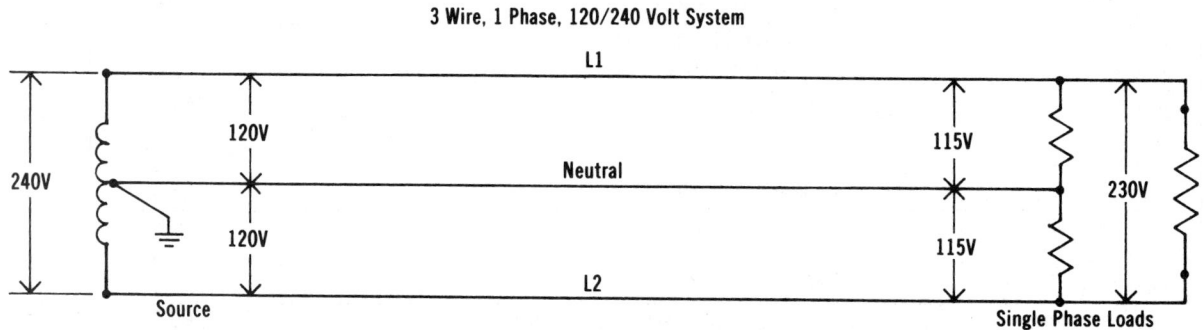

Figure D5010-122

4 wire, 3 Phase, 208Y/120 Volt System

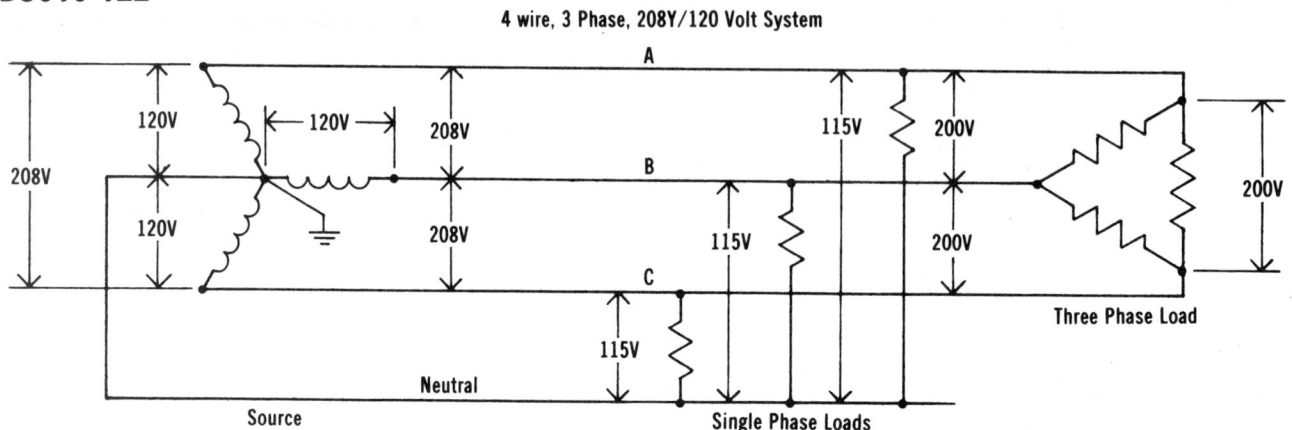

Figure D5010-123　Electric Circuit Voltages (cont.)

3 Wire, 3 Phase 240 Volt System

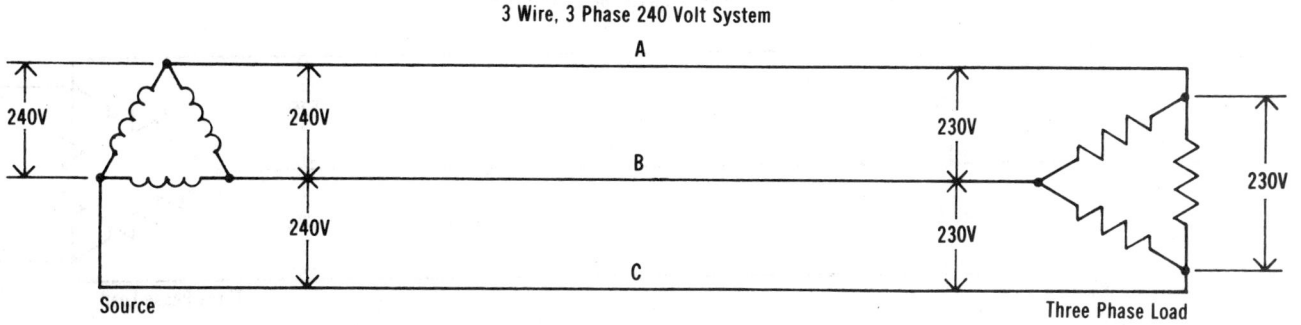

Figure D5010-124

4 Wire, 3 Phase, 240/120 Volt System

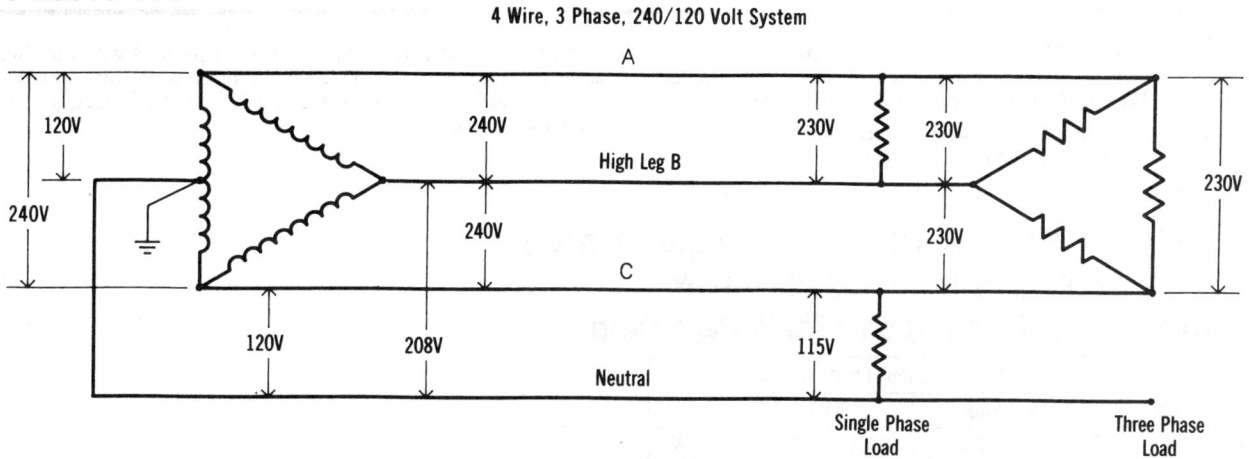

Figure D5010-125

3 Wire, 3 Phase 480 Volt System

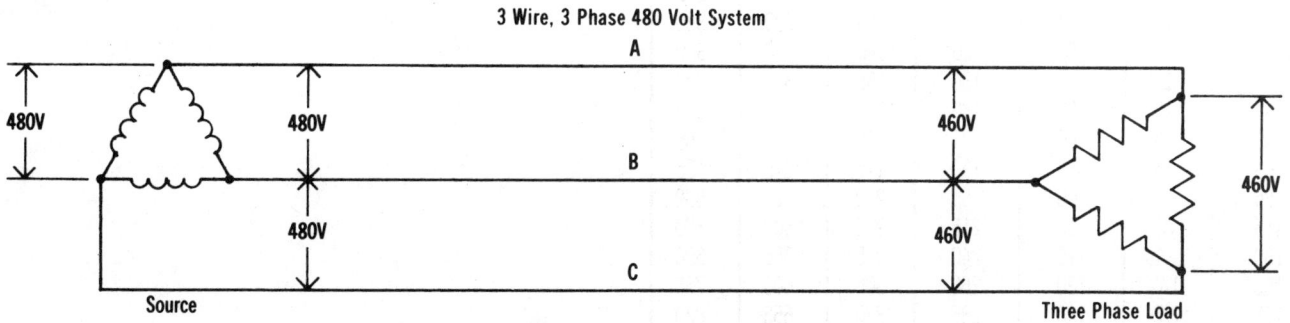

Figure D5010-126

4 Wire, 3 Phase, 480Y/277 Volt System

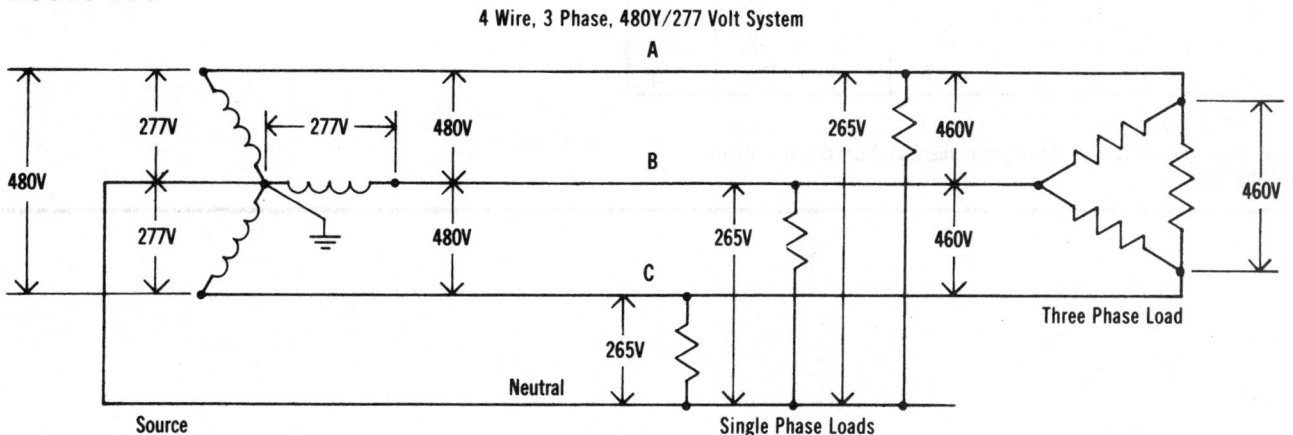

Figure D5010-127   Electric Circuit Voltages (cont.)

3 Wire, 3 Phase, 600 Volt System

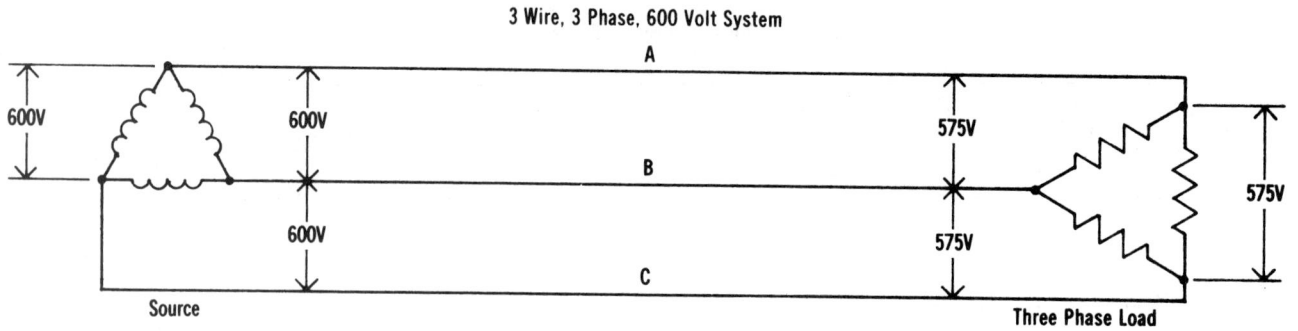

Maximum Circuit Length: Table D5010-130 indicates typical maximum installed length a circuit can have and still maintain an adequate voltage level at the point of use. The circuit length is similar to the conduit length.

If the circuit length for an ampere load and a copper wire size exceeds the length obtained from Table D5010-130, use the next largest wire size to compensate voltage drop.

Example: A 130 ampere load at 480 volts, 3 phase, 3 wire with No. 1 wire can be run a maximum of 555 L.F. and provide satisfactory operation. If the same load is to be wired at the end of a 625 L.F. circuit, then a larger wire must be used.

Table D5010-130   Maximum Circuit Length (approximate) for Various Power Requirements Assuming THW, Copper Wire @ 75°C, Based Upon a 4% Voltage Drop

Amperes	Wire Size	Maximum Circuit Length in Feet				
		2 Wire, 1 Phase		3 Wire, 3 Phase		
		120V	240V	240V	480V	600V
15	14*	50	105	120	240	300
	14	50	100	120	235	295
20	12*	60	125	145	290	360
	12	60	120	140	280	350
30	10*	65	130	155	305	380
	10	65	130	150	300	375
50	8	60	125	145	285	355
65	6	75	150	175	345	435
85	4	90	185	210	425	530
115	2	110	215	250	500	620
130	1	120	240	275	555	690
150	1/0	130	260	305	605	760
175	2/0	140	285	330	655	820
200	3/0	155	315	360	725	904
230	4/0	170	345	395	795	990
255	250	185	365	420	845	1055
285	300	195	395	455	910	1140
310	350	210	420	485	975	1220
380	500	245	490	565	1130	1415

*Solid Conductor

Note: The circuit length is the one-way distance between the origin and the load.

RD5010-141　Minimum Copper and Aluminum Wire Size Allowed for Various Types of Insulation

Minimum Wire Sizes									
	Copper		Aluminum			Copper		Aluminum	
Amperes	THW THWN or XHHW	THHN XHHW *	THW XHHW	THHN XHHW *	Amperes	THW THWN or XHHW	THHN XHHW *	THW XHHW	THHN XHHW *
15A	#14	#14	#12	#12	195	3/0	2/0	250kcmil	4/0
20	#12	#12	#10	#10	200	3/0	3/0	250kcmil	4/0
25	#10	#10	#10	#10	205	4/0	3/0	250kcmil	4/0
30	#10	#10	#8	#8	225	4/0	3/0	300kcmil	250kcmil
40	#8	#8	#8	#8	230	4/0	4/0	300kcmil	250kcmil
45	#8	#8	#6	#8	250	250kcmil	4/0	350kcmil	300kcmil
50	#8	#8	#6	#6	255	250kcmil	4/0	400kcmil	300kcmil
55	#6	#8	#4	#6	260	300kcmil	4/0	400kcmil	350kcmil
60	#6	#6	#4	#6	270	300kcmil	250kcmil	400kcmil	350kcmil
65	#6	#6	#4	#4	280	300kcmil	250kcmil	500kcmil	350kcmil
75	#4	#6	#3	#4	285	300kcmil	250kcmil	500kcmil	400kcmil
85	#4	#4	#2	#3	290	350kcmil	250kcmil	500kcmil	400kcmil
90	#3	#4	#2	#2	305	350kcmil	300kcmil	500kcmil	400kcmil
95	#3	#4	#1	#2	310	350kcmil	300kcmil	500kcmil	500kcmil
100	#3	#3	#1	#2	320	400kcmil	300kcmil	600kcmil	500kcmil
110	#2	#3	1/0	#1	335	400kcmil	350kcmil	600kcmil	500kcmil
115	#2	#2	1/0	#1	340	500kcmil	350kcmil	600kcmil	500kcmil
120	#1	#2	1/0	1/0	350	500kcmil	350kcmil	700kcmil	500kcmil
130	#1	#2	2/0	1/0	375	500kcmil	400kcmil	700kcmil	600kcmil
135	1/0	#1	2/0	1/0	380	500kcmil	400kcmil	750kcmil	600kcmil
150	1/0	#1	3/0	2/0	385	600kcmil	500kcmil	750kcmil	600kcmil
155	2/0	1/0	3/0	3/0	420	600kcmil	500kcmil		700kcmil
170	2/0	1/0	4/0	3/0	430		500kcmil		750kcmil
175	2/0	2/0	4/0	3/0	435		600kcmil		750kcmil
180	3/0	2/0	4/0	4/0	475		600kcmil		

*Dry Locations Only

Notes:

1. Size #14 to 4/0 is in AWG units (American Wire Gauge).
2. Size 250 to 750 is in kcmil units (Thousand Circular Mils).
3. Use next higher ampere value if exact value is not listed in table.
4. For loads that operate continuously increase ampere value by 25% to obtain proper wire size.
5. Refer to Table D5010-130 for the maximum circuit length for the various size wires.
6. Table D5010-141 has been written for estimating purpose only, based on ambient temperature of 30°C (86° F); for ambient temperature other than 30°C (86° F), ampacity correction factors will be applied.

Table D5010-142 Conductors in Conduit

Table below lists maximum number of conductors for various sized conduit using THW, TW or THWN insulations.

Copper Wire Size	Conduit Size																										
	½″			¾″			1″			1-¼″			1-½″			2″			2-½″			3″		3-½″		4″	
	TW	THW	THWN	TW	THW	THWN	TW	THW	THWN	TW	THW	THWN	TW	THW	THWN	TW	THW	THWN	TW	THW	THWN	THW	THWN	THW	THWN	THW	THWN
#14	9	6	13	15	10	24	25	16	39	44	29	69	60	40	94	99	65	154	142	93		143		192			
#12	7	4	10	12	8	18	19	13	29	35	24	51	47	32	70	78	53	114	111	76	164	117		157			
#10	5	4	6	9	6	11	15	11	18	26	19	32	36	26	44	60	43	73	85	61	104	95	160	127		163	
#8	2	1	3	4	3	5	7	5	9	12	10	16	17	13	22	28	22	36	40	32	51	49	79	66	106	85	136
#6		1	1		2	4		4	6		7	11		10	15		16	26		23	37	36	57	48	76	62	98
#4		1	1		1	2		3	4		5	7		7	9		12	16		17	22	27	35	36	47	47	60
#3		1	1		1	1		2	3		4	6		6	8		10	13		15	19	23	29	31	39	40	51
#2		1	1		1	1		2	3		4	5		5	7		9	11		13	16	20	25	27	33	34	43
#1					1	1		1	1		3	3		4	5		6	8		9	12	14	18	19	25	25	32
1/0					1	1		1	1		2	3		3	4		5	7		8	10	12	15	16	21	21	27
2/0					1	1		1	1		1	2		3	3		5	6		7	8	10	13	14	17	18	22
3/0					1	1		1	1		1	1		2	3		4	5		6	7	9	11	12	14	15	18
4/0						1		1	1		1	1		1	2		3	4		5	6	7	9	10	12	13	15
250 MCM								1	1		1	1		1	1		2	3		4	4	6	7	8	10	10	12
300								1	1		1	1		1	1		2	3		3	4	5	6	7	8	9	11
350									1		1	1		1	1		1	2		3	3	4	5	6	7	8	9
400											1	1		1	1		1	1		2	3	4	5	5	6	7	8
500											1	1		1	1		1	1		1	2	3	4	4	5	6	7
600												1		1	1		1	1		1	1	3	3	4	4	5	5
700														1	1		1	1		1	1	2	3	3	4	4	5
750 ▼														1	1		1	1		1	1	2	2	3	3	4	4

Table D5010-146A Metric Equivalent, Conduit

U.S. vs. European Conduit – Approximate Equivalents			
United States		European	
Trade Size	Inside Diameter Inch/MM	Trade Size	Inside Diameter MM
½	.622/15.8	11	16.4
¾	.824/20.9	16	19.9
1	1.049/26.6	21	25.5
1¼	1.380/35.0	29	34.2
1½	1.610/40.9	36	44.0
2	2.067/52.5	42	51.0
2½	2.469/62.7		
3	3.068/77.9		
3½	3.548/90.12		
4	4.026/102.3		
5	5.047/128.2		
6	6.065/154.1		

Table D5010-146B Metric Equivalent, Wire

U.S. vs. European Wire – Approximate Equivalents			
United States		European	
Size AWG or MCM	Area Cir. Mils.(CM) MM^2	Size MM^2	Area Cir. Mils.
18	1620/.82	.75	1480
16	2580/1.30	1.0	1974
14	4110/2.08	1.5	2961
12	6530/3.30	2.5	4935
10	10,380/5.25	4	7896
8	16,510/8.36	6	11,844
6	26,240/13.29	10	19,740
4	41,740/21.14	16	31,584
3	52,620/26.65	25	49,350
2	66,360/33.61	–	–
1	83,690/42.39	35	69,090
1/0	105,600/53.49	50	98,700
2/0	133,100/67.42	–	–
3/0	167,800/85.00	70	138,180
4/0	211,600/107.19	95	187,530
250	250,000/126.64	120	236,880
300	300,000/151.97	150	296,100
350	350,000/177.30	–	–
400	400,000/202.63	185	365,190
500	500,000/253.29	240	473,760
600	600,000/303.95	300	592,200
700	700,000/354.60	–	–
750	750,000/379.93	–	–

D5010-147 Concrete for Conduit Encasement

Table below lists C.Y. of concrete for 100 L.F. of trench. Conduits separation center to center should meet 7.5" (N.E.C.).

Number of Conduits	1	2	3	4	6	8	9	Number of Conduits
Trench Dimension	11.5" x 11.5"	11.5" x 19"	11.5" x 27"	19" x 19"	19" x 27"	19" x 38"	27" x 27"	Trench Dimension
Conduit Diameter 2.0"	3.29	5.39	7.64	8.83	12.51	17.66	17.72	Conduit Diameter 2.0"
2.5"	3.23	5.29	7.49	8.62	12.19	17.23	17.25	2.5"
3.0"	3.15	5.13	7.24	8.29	11.71	16.59	16.52	3.0"
3.5"	3.08	4.97	7.02	7.99	11.26	15.98	15.84	3.5"
4.0"	2.99	4.80	6.76	7.65	10.74	15.30	15.07	4.0"
5.0"	2.78	4.37	6.11	6.78	9.44	13.57	13.12	5.0"
6.0"	2.52	3.84	5.33	5.74	7.87	11.48	10.77	6.0"

Table D5010-148 Size Required and Weight (Lbs./1000 L.F.) of Aluminum and Copper THW Wire by Ampere Load

Amperes	Copper Size	Aluminum Size	Copper Weight	Aluminum Weight
15	14	12	24	11
20	12	10	33	17
30	10	8	48	39
45	8	6	77	52
65	6	4	112	72
85	4	2	167	101
100	3	1	205	136
115	2	1/0	252	162
130	1	2/0	324	194
150	1/0	3/0	397	233
175	2/0	4/0	491	282
200	3/0	250	608	347
230	4/0	300	753	403
255	250	400	899	512
285	300	500	1068	620
310	350	500	1233	620
335	400	600	1396	772
380	500	750	1732	951

Table D5010-150 Transformer Weight (Lbs.) by kVA

Oil Filled 3 Phase 5/15 kV To 480/277			
kVA	Lbs.	kVA	Lbs.
150	1800	1000	6200
300	2900	1500	8400
500	4700	2000	9700
750	5300	3000	15000

Dry 240/480 To 120/240 Volt			
1 Phase		3 Phase	
kVA	Lbs.	kVA	Lbs.
1	23	3	90
2	36	6	135
3	59	9	170
5	73	15	220
7.5	131	30	310
10	149	45	400
15	205	75	600
25	255	112.5	950
37.5	295	150	1140
50	340	225	1575
75	550	300	1870
100	670	500	2850
167	900	750	4300

Table D5010-149 Weight (Lbs./L.F.) of 4 Pole Aluminum and Copper Bus Duct by Ampere Load

Amperes	Aluminum Feeder	Copper Feeder	Aluminum Plug-In	Copper Plug-In
225			7	7
400			8	13
600	10	10	11	14
800	10	19	13	18
1000	11	19	16	22
1350	14	24	20	30
1600	17	26	25	39
2000	19	30	29	46
2500	27	43	36	56
3000	30	48	42	73
4000	39	67		
5000		78		

Table D5010-151 Generator Weight (Lbs.) by kW

3 Phase 4 Wire 277/480 Volt			
Gas		Diesel	
kW	Lbs.	kW	Lbs.
7.5	600	30	1800
10	630	50	2230
15	960	75	2250
30	1500	100	3840
65	2350	125	4030
85	2570	150	5500
115	4310	175	5650
170	6530	200	5930
		250	6320
		300	7840
		350	8220
		400	10750
		500	11900

499

Table D5010-152A Conduit Weight Comparisons (Lbs. per 100 ft.) Empty

Type	1/2"	3/4"	1"	1-1/4"	1-1/2"	2"	2-1/2"	3"	3-1/2"	4"	5"	6"
Rigid Aluminum	28	37	55	72	89	119	188	246	296	350	479	630
Rigid Steel	79	105	153	201	249	332	527	683	831	972	1314	1745
Intermediate Steel (IMC)	60	82	116	150	182	242	401	493	573	638		
Electrical Metallic Tubing (EMT)	29	45	65	96	111	141	215	260	365	390		
Polyvinyl Chloride, Schedule 40	16	22	32	43	52	69	109	142	170	202	271	350
Polyvinyl Chloride Encased Burial						38		67	88	105	149	202
Fibre Duct Encased Burial						127		164	180	206	400	511
Fibre Duct Direct Burial						150		251	300	354		
Transite Encased Burial						160		240	290	330	450	550
Transite Direct Burial						220		310		400	540	640

Table D5010-152B Conduit Weight Comparisons (Lbs. per 100 ft.) with Maximum Cable Fill*

Type	1/2"	3/4"	1"	1-1/4"	1-1/2"	2"	2-1/2"	3"	3-1/2"	4"	5"	6"
Rigid Galvanized Steel (RGS)	104	140	235	358	455	721	1022	1451	1749	2148	3083	4343
Intermediate Steel (IMC)	84	113	186	293	379	611	883	1263	1501	1830		
Electrical Metallic Tubing (EMT)	54	116	183	296	368	445	641	930	1215	1540		

*Conduit & Heaviest Conductor Combination

D5010-153 Weight Comparisons of Common Size Cast Boxes in Lbs.

Size NEMA 4 or 9	Cast Iron	Cast Aluminum	Size NEMA 7	Cast Iron	Cast Aluminum
6" x 6" x 6"	17	7	6" x 6" x 6"	40	15
8" x 6" x 6"	21	8	8" x 6" x 6"	50	19
10" x 6" x 6"	23	9	10" x 6" x 6"	55	21
12" x 12" x 6"	52	20	12" x 6" x 6"	100	37
16" x 16" x 6"	97	36	16" x 16" x 6"	140	52
20" x 20" x 6"	133	50	20" x 20" x 6"	180	67
24" x 18" x 8"	149	56	24" x 18" x 8"	250	93
24" x 24" x 10"	238	88	24" x 24" x 10"	358	133
30" x 24" x 12"	324	120	30" x 24" x 10"	475	176
36" x 36" x 12"	500	185	30" x 24" x 12"	510	189

General: Lighting and electric heating loads are expressed in watts and kilowatts.

Cost Determination:

The proper ampere values can be obtained as follows:
1. Convert watts to kilowatts
   (watts 1000 ÷ kilowatts)
2. Determine voltage rating of equipment.
3. Determine whether equipment is single phase or three phase.
4. Refer to Table D5010-158 to find ampere value from kW, Ton and Btu/hr. values.
5. Determine type of wire insulation – TW, THW, THWN.
6. Determine if wire is copper or aluminum.
7. Refer to Table D5010-141 to obtain copper or aluminum wire size from ampere values.

Notes:
1. Phase refers to single phase, 2 wire circuits.
2. Phase refers to three phase, 3 wire circuits.
3. For circuits which operate continuously for 3 hours or more, multiply the ampere values by 1.25 for a given kW requirement.
4. For kW ratings not listed, add ampere values.

8. Next refer to Table D5010-142 for the proper conduit size to accommodate the number and size of wires in each particular case.
9. Next refer to unit price data for the per linear foot cost of the conduit.
10. Next refer to unit price data for the per linear foot cost of the wire. Multiply cost of wire per L.F. x number of wires in the circuits to obtain total wire cost per L.F..
11. Add values obtained in Step 9 and 10 for total cost per linear foot for conduit and wire x length of circuit = Total Cost.

For example: Find the ampere value of 9 kW at 208 volt, single phase.

$$
\begin{array}{ll}
4\ kW & = 19.2A \\
5\ kW & = 24.0A \\
\hline
9\ kW & = 43.2A
\end{array}
$$

5. "Length of Circuit" refers to the one way distance of the run, not to the total sum of wire lengths.

Table D5010-158   Ampere Values as Determined by kW Requirements, BTU/HR or Ton, Voltage and Phase Values

			120V	208V		240V		277V	480V
kW	Ton	BTU/HR	1 Phase	1 Phase	3 Phase	1 Phase	3 Phase	1 Phase	3 Phase
0.5	.1422	1,707	4.2A	2.4A	1.4A	2.1A	1.2A	1.8A	0.6A
0.75	.2133	2,560	6.2	3.6	2.1	3.1	1.9	2.7	.9
1.0	.2844	3,413	8.3	4.9	2.8	4.2	2.4	3.6	1.2
1.25	.3555	4,266	10.4	6.0	3.5	5.2	3.0	4.5	1.5
1.5	.4266	5,120	12.5	7.2	4.2	6.3	3.1	5.4	1.8
2.0	.5688	6,826	16.6	9.7	5.6	8.3	4.8	7.2	2.4
2.5	.7110	8,533	20.8	12.0	7.0	10.4	6.1	9.1	3.1
3.0	.8532	10,239	25.0	14.4	8.4	12.5	7.2	10.8	3.6
4.0	1.1376	13,652	33.4	19.2	11.1	16.7	9.6	14.4	4.8
5.0	1.4220	17,065	41.6	24.0	13.9	20.8	12.1	18.1	6.1
7.5	2.1331	25,598	62.4	36.0	20.8	31.2	18.8	27.0	9.0
10.0	2.8441	34,130	83.2	48.0	27.7	41.6	24.0	36.5	12.0
12.5	3.5552	42,663	104.2	60.1	35.0	52.1	30.0	45.1	15.0
15.0	4.2662	51,195	124.8	72.0	41.6	62.4	37.6	54.0	18.0
20.0	5.6883	68,260	166.4	96.0	55.4	83.2	48.0	73.0	24.0
25.0	7.1104	85,325	208.4	120.2	70.0	104.2	60.0	90.2	30.0
30.0	8.5325	102,390		144.0	83.2	124.8	75.2	108.0	36.0
35.0	9.9545	119,455		168.0	97.1	145.6	87.3	126.0	42.1
40.0	11.3766	136,520		192.0	110.8	166.4	96.0	146.0	48.0
45.0	12.7987	153,585			124.8	187.5	112.8	162.0	54.0
50.0	14.2208	170,650			140.0	208.4	120.0	180.4	60.0
60.0	17.0650	204,780			166.4		150.4	216.0	72.0
70.0	19.9091	238,910			194.2		174.6		84.2
80.0	22.7533	273,040			221.6		192.0		96.0
90.0	25.5975	307,170					225.6		108.0
100.0	28.4416	341,300							120.0

General: Control transformers are listed in VA. Step-down and power transformers are listed in kVA.

Cost Determination:
1. Convert VA to kVA. Volt amperes (VA) ÷ 1000 = Kilovolt amperes (kVA).
2. Determine voltage rating of equipment.
3. Determine whether equipment is single phase or three phase.
4. Refer to Table D5010-161 to find ampere value from kVA value.
5. Determine type of wire insulation – TW, THW, THWN.
6. Determine if wire is copper or aluminum.
7. Refer to Table D5010-161 to obtain copper or aluminum wire size from ampere values.
8. Next refer to Table D5010-142 for the proper conduit size to accommodate the number and size of wires in each particular case.

9. Next refer to unit price data for the per linear foot cost of the conduit.
10. Next refer to unit price data for the per linear foot cost of the wire. Multiply cost of wire per L.F. x number of wires in the circuits to obtain total wire cost.
11. Add values obtained in Step 9 and 10 for total cost per linear foot for conduit and wire x length of circuit = Total Cost.

Example: A transformer rated 10 kVA 480 volts primary, 240 volts secondary, 3 phase has the capacity to furnish the following:
1. Primary amperes = 10 kVA x 1.20 = 12 amperes (from Table D5010-161)
2. Secondary amperes = 10 kVA x 2.40 = 24 amperes (from Table D5010-161)

Note: Transformers can deliver generally 125% of their rated kVA. For instance, a 10 kVA rated transformer can safely deliver 12.5 kVA.

Table D5010-161   Multiplier Values for kVA to Amperes Determined by Voltage and Phase Values

Volts	Multiplier for Circuits	
	2 Wire, 1 Phase	3 Wire, 3 Phase
115	8.70	
120	8.30	
230	4.30	2.51
240	4.16	2.40
200	5.00	2.89
208	4.80	2.77
265	3.77	2.18
277	3.60	2.08
460	2.17	1.26
480	2.08	1.20
575	1.74	1.00
600	1.66	0.96

General: Motors can be powered by any of the seven systems shown in Figure D5010-121 thru Figure D5010-127 provided the motor voltage characteristics are compatible with the power system characteristics.

Cost Determination:

Motor Amperes for the various size H.P. and voltage are listed in Table D5010-171. To find the amperes, locate the required H.P. rating and locate the amperes under the appropriate circuit characteristics.

For example:

  A. 100 H.P., 3 phase, 460 volt motor = 124 amperes (Table D5010-171)

  B. 10 H.P., 3 phase, 200 volt motor = 32.2 amperes (Table D5010-171)

Motor Wire Size: After the amperes are found in Table D5010-171 the amperes must be increased 25% to compensate for power losses. Next refer to Table D5010-141. Find the appropriate insulation column for copper or aluminum wire to determine the proper wire size.

For example:

  A. 100 H.P., 3 phase, 460 volt motor has an ampere value of 124 amperes from Table D5010-171

  B. 124A x 1.25 = 155 amperes

  C. Refer to Table D5010-141 for THW or THWN wire insulations to find the proper wire size. For a 155 ampere load using copper wire a size 2/0 wire is needed.

  D. For the 3 phase motor three wires of 2/0 size are required.

Conduit Size: To obtain the proper conduit size for the wires and type of insulation used, refer to Table D5010-142.

For example: For the 100 H.P., 460V, 3 phase motor, it was determined that three 2/0 wires are required. Assuming THWN insulated copper wire, use Table D5010-142 to determine that three 2/0 wires require 1-1/2" conduit.

Material Cost of the conduit and wire system depends on:
1. Wire size required
2. Copper or aluminum wire
3. Wire insulation type selected
4. Steel or plastic conduit
5. Type of conduit raceway selected.

Labor Cost of the conduit and wire system depends on:
1. Type and size of conduit
2. Type and size of wires installed
3. Location and height of installation in building or depth of trench
4. Support system for conduit.

Table D5010-171 Ampere Values Determined by Horsepower, Voltage and Phase Values

H.P.	Single Phase		Three Phase			
	115V	230V	200 V	230V	460V	575V
1/6	4.4A	2.2A				
1/4	5.8	2.9				
1/3	7.2	3.6				
1/2	9.8	4.9	2.3A	2.0A	1.0A	0.8A
3/4	13.8	6.9	3.2	2.8	1.4	1.1
1	16	8	4.1	3.6	1.8	1.4
1-1/2	20	10	6.0	5.2	2.6	2.1
2	24	12	7.8	6.8	3.4	2.7
3	34	17	11.0	9.6	4.8	3.9
5			17.5	15.2	7.6	6.1
7-1/2			25.3	22	11	9
10			32.2	28	14	11
15			48.3	42	21	17
20			62.1	54	27	22
25			78.2	68	34	27
30			92.0	80	40	32
40			119.6	104	52	41
50			149.5	130	65	52
60			177	154	77	62
75			221	192	96	77
100			285	248	124	99
125			359	312	156	125
150			414	360	180	144
200			552	480	240	192

Cost Determination (cont.)

Magnetic starters, switches, and motor connection:

To complete the cost picture from H.P. to Costs additional items must be added to the cost of the conduit and wire system to arrive at a total cost.

1. Assembly D5020 160 Magnetic Starters Installed Cost lists the various size starters for single phase and three phase motors.
2. Assembly D5020 165 Heavy Duty Safety Switches Installed Cost lists safety switches required at the beginning of a motor circuit and also one required in the vicinity of the motor location.
3. Assembly D5020 170 Motor Connection lists the various costs for single and three phase motors.

Worksheet to obtain total motor wiring costs:

It is assumed that the motors or motor driven equipment are furnished and installed under other sections for this estimate and the following work is done under this section:

1. Conduit
2. Wire (add 10% for additional wire beyond conduit ends for connections to switches, boxes, starters, etc.)
3. Starters
4. Safety switches
5. Motor connections

Table D5010-172

				Cost	
Item	Type	Size	Quantity	Unit	Total
Wire					
Conduit					
Switch					
Starter					
Switch					
Motor Connection					
Other					
Total Cost					

Worksheet for Motor Circuits

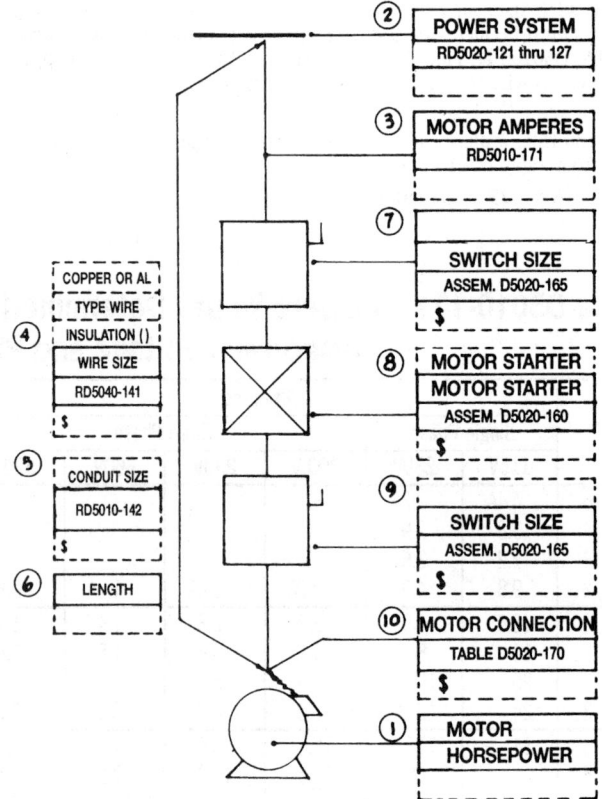

Table D5010-173 Maximum Horsepower for Starter Size by Voltage

Starter	Maximum HP (3φ)			
Size	208V	240V	480V	600V
00	1½	1½	2	2
0	3	3	5	5
1	7½	7½	10	10
2	10	15	25	25
3	25	30	50	50
4	40	50	100	100
5		100	200	200
6		200	300	300
7		300	600	600
8		450	900	900
8L		700	1500	1500

General: The cost of the lighting portion of the electrical costs is dependent upon:
1. The footcandle requirement of the proposed building.
2. The type of fixtures required.
3. The ceiling heights of the building.
4. Reflectance value of ceilings, walls and floors.
5. Fixture efficiencies and spacing vs. mounting height ratios.

Footcandle Requirements: See Table D5020-204 for Footcandle and Watts per S.F. determination.

Table D5020-201   IESNA* Recommended Illumination Levels in Footcandles

Commercial Buildings			Industrial Buildings		
Type	Description	Footcandles	Type	Description	Footcandles
Bank	Lobby	50	Assembly Areas	Rough bench & machine work	50
	Customer Areas	70		Medium bench & machine work	100
	Teller Stations	150		Fine bench & machine work	500
	Accounting Areas	150	Inspection Areas	Ordinary	50
Offices	Routine Work	100		Difficult	100
	Accounting	150		Highly Difficult	200
	Drafting	200	Material Handling	Loading	20
	Corridors, Halls, Washrooms	30		Stock Picking	30
Schools	Reading or Writing	70		Packing, Wrapping	50
	Drafting, Labs, Shops	100	Stairways	Service Areas	20
	Libraries	70	Washrooms	Service Areas	20
	Auditoriums, Assembly	15	Storage Areas	Inactive	5
	Auditoriums, Exhibition	30		Active, Rough, Bulky	10
Stores	Circulation Areas	30		Active, Medium	20
	Stock Rooms	30		Active, Fine	50
	Merchandise Areas, Service	100	Garages	Active Traffic Areas	20
	Self-Service Areas	200		Service & Repair	100

*IESNA - Illuminating Engineering Society of North America

505

Table D5020-202 General Lighting Loads by Occupancies

Type of Occupancy	Unit Load per S.F. (Watts)
Armories and Auditoriums	1
Banks	5
Barber Shops and Beauty Parlors	3
Churches	1
Clubs	2
Court Rooms	2
*Dwelling Units	3
Garages — Commercial (storage)	½
Hospitals	2
*Hotels and Motels, including apartment houses without provisions for cooking by tenants	2
Industrial Commercial (Loft) Buildings	2
Lodge Rooms	1½
Office Buildings	5
Restaurants	2
Schools	3
Stores	3
Warehouses (storage)	¼
*In any of the above occupancies except one-family dwellings and individual dwelling units of multi-family dwellings:	
Assembly Halls and Auditoriums	1
Halls, Corridors, Closets	½
Storage Spaces	¼

Table D5020-203 Lighting Limit (Connected Load) for Listed Occupancies: New Building Proposed Energy Conservation Guideline

Type of Use	Maximum Watts per S.F.
Interior	3.00
Category A: Classrooms, office areas, automotive mechanical areas, museums, conference rooms, drafting rooms, clerical areas, laboratories, merchandising areas, kitchens, examining rooms, book stacks, athletic facilities.	
Category B: Auditoriums, waiting areas, spectator areas, restrooms, dining areas, transportation terminals, working corridors in prisons and hospitals, book storage areas, active inventory storage, hospital bedrooms, hotel and motel bedrooms, enclosed shopping mall concourse areas, stairways.	1.00
Category C: Corridors, lobbies, elevators, inactive storage areas.	0.50
Category D: Indoor parking.	0.25
Exterior	
Category E: Building perimeter: wall-wash, facade, canopy.	5.00 (per linear foot)
Category F: Outdoor parking.	0.10

Table D5020-204   Procedure for Calculating Footcandles and Watts Per Square Foot

1. Initial footcandles = No. of fixtures × lamps per fixture × lumens per lamp × coefficient of utilization ÷ square feet
2. Maintained footcandles = initial footcandles × maintenance factor
3. Watts per square foot = No. of fixtures × lamps × (lamp watts + ballast watts) ÷ square feet

Example: To find footcandles and watts per S.F. for an office 20′ x 20′ with 11 fluorescent fixtures each having 4–40 watt C.W. lamps.

Based on good reflectance and clean conditions:

Lumens per lamp = 40 watt cool white at 3150 lumens per lamp RD5020-250, Table RD5020-251

Coefficient of utilization = .42 (varies from .62 for light colored areas to .27 for dark)

Maintenance factor = .75 (varies from .80 for clean areas with good maintenance to .50 for poor)

Ballast loss = 8 watts per lamp. (Varies with manufacturer. See manufacturers' catalog.)

1. Initial footcandles:

$$\frac{11 \times 4 \times 3150 \times .42}{400} = \frac{58,212}{400} = 145 \text{ footcandles}$$

2. Maintained footcandles:

$$145 \times .75 = 109 \text{ footcandles}$$

3. Watts per S.F.

$$\frac{11 \times 4 \,(40 + 8)}{400} = \frac{2,112}{400} = 5.3 \text{ watts per S.F.}$$

Table D5020-205   Approximate Watts Per Square Foot for Popular Fixture Types

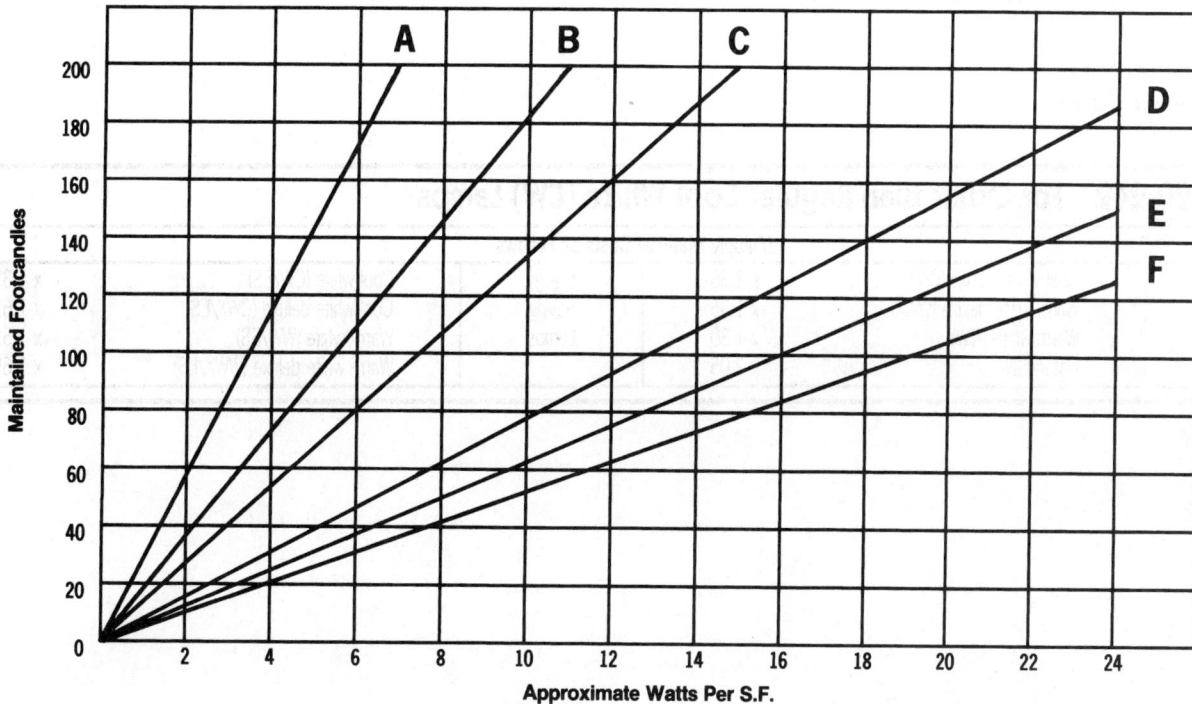

Due to the many variables involved, use for preliminary estimating only:
  a. Fluorescent – industrial System D5020 208
  b. Fluorescent – lens unit System D5020 208 Fixture types B & C
  c. Fluorescent – louvered unit
  d. Incandescent – open reflector System D5020 214, Type D
  e. Incandescent – lens unit System D5020 214, Type A
  f. Incandescent – down light System D5020 214, Type B

Table D5020-241   Comparison - Cost of Operation of High Intensity Discharge Lamps

Lamp Type	Wattage	Life (Hours)	[1] Circuit Wattage	Average Initial Lumens	[2] L.L.D.	[3] Mean Lumens
M.V.	100 DX	24,000	125	4,000	61%	2,440
L.P.S.	SOX-35	18,000	65	4,800	100%	4,800
H.P.S.	LU-70	12,000	84	5,800	90%	5,220
M.H.	No Equivalent					
M.V.	175 DX	24,000	210	8,500	66%	5,676
M.V.	250 DX	24,000	295	13,000	66%	7,986
L.P.S.	SOX-55	18,000	82	8,000	100%	8,000
H.P.S.	LU-100	12,000	120	9,500	90%	8,550
M.H.	No Equivalent					
M.V.	400 DX	24,000	465	24,000	64%	14,400
L.P.S.	SOX-90	18,000	141	13,500	100%	13,500
H.P.S.	LU-150	16,000	188	16,000	90%	14,400
M.H.	MH-175	7,500	210	14,000	73%	10,200
M.V.	No Equivalent					
L.P.S.	SOX-135	18,000	147	22,500	100%	22,500
H.P.S.	LU-250	20,000	310	25,500	92%	23,205
M.H.	MH-250	7,500	295	20,500	78%	16,000
M.V.	1000 DX	24,000	1,085	63,000	61%	37,820
L.P.S.	SOX-180	18,000	248	33,000	100%	33,000
H.P.S.	LU-400	20,000	480	50,000	90%	45,000
M.H.	MH-400	15,000	465	34,000	72%	24,600
H.P.S.	LU-1000	15,000	1,100	140,000	91%	127,400

1. Includes ballast losses and average lamp watts
2. Lamp lumen depreciation (% of initial light output at 70% rated life)
3. Lamp lumen output at 70% rated life (L.L.D. x initial)

M.V. = Mercury Vapor

L.P.S. = Low Pressure Sodium

H.P.S. = High Pressure Sodium

L.H. = Metal Halide

Table D5020-242   For Other than Regular Cool White (CW) Lamps

	Multiply Material Costs as Follows:					
Regular Lamps	Cool white deluxe (CWX)	x 1.35	Energy Saving Lamps	Cool white (CW/ES)	x 1.35
	Warm white deluxe (WWX)	x 1.35		Cool white deluxe (CWX/ES)	x 1.65
	Warm white (WW)	x 1.30		Warm white (WW/ES)	x 1.55
	Natural (N)	x 2.05		Warm white deluxe (WWX/ES)	x 1.65

Table D5020-251 Lamp Comparison Chart with Enclosed Floodlight, Ballast, & Lamp for Pole

Type	Watts	Initial Lumens	Lumens per Watt	Lumens @ 40% Life	Life (Hours)
Incandescent	150	2,880	19	85	750
	300	6,360	21	84	750
	500	10,850	22	80	1,000
	1,000	23,740	24	80	1,000
	1,500	34,400	23	80	1,000
Tungsten	500	10,950	22	97	2,000
Halogen	1,500	35,800	24	97	2,000
Fluorescent	40	3,150	79	88	20,000
Cool	110	9,200	84	87	12,000
White	215	16,000	74	81	12,000
Deluxe	250	12,100	48	86	24,000
Mercury	400	22,500	56	85	24,000
	1,000	63,000	63	75	24,000
Metal	175	14,000	80	77	7,500
Halide	400	34,000	85	75	15,000
	1,000	100,000	100	83	10,000
	1,500	155,000	103	92	1,500
High	70	5,800	83	90	20,000
Pressure	100	9,500	95	90	20,000
Sodium	150	16,000	107	90	24,000
	400	50,000	125	90	24,000
	1,000	140,000	140	90	24,000
Low	55	4,600	131	98	18,000
Pressure	90	12,750	142	98	18,000
Sodium	180	33,000	183	98	18,000

Color: High Pressure Sodium — Slightly Yellow
   Low Pressure Sodium — Yellow
   Mercury Vapor — Green-Blue
   Metal Halide — Blue White

Note: Pole not included.

Table D5020-602 Central Air Conditioning Watts per S.F., BTU's per Hour per S.F. of Floor Area and S.F. per Ton of Air Conditioning

Type Building	Watts per S.F.	BTUH per S.F.	S.F. per Ton	Type Building	Watts per S.F.	BTUH per S.F.	S.F. per Ton	Type Building	Watts per S.F.	BTUH per S.F.	S.F. per Ton
Apartments, Individual	3	26	450	Dormitory, Rooms	4.5	40	300	Libraries	5.7	50	240
Corridors	2.5	22	550	Corridors	3.4	30	400	Low Rise Office, Ext.	4.3	38	320
Auditoriums & Theaters	3.3	40	300/18*	Dress Shops	4.9	43	280	Interior	3.8	33	360
Banks	5.7	50	240	Drug Stores	9	80	150	Medical Centers	3.2	28	425
Barber Shops	5.5	48	250	Factories	4.5	40	300	Motels	3.2	28	425
Bars & Taverns	15	133	90	High Rise Off.-Ext. Rms.	5.2	46	263	Office (small suite)	4.9	43	280
Beauty Parlors	7.6	66	180	Interior Rooms	4.2	37	325	Post Office, Int. Office	4.9	42	285
Bowling Alleys	7.8	68	175	Hospitals, Core	4.9	43	280	Central Area	5.3	46	260
Churches	3.3	36	330/20*	Perimeter	5.3	46	260	Residences	2.3	20	600
Cocktail Lounges	7.8	68	175	Hotels, Guest Rooms	5	44	275	Restaurants	6.8	60	200
Computer Rooms	16	141	85	Public Spaces	6.2	55	220	Schools & Colleges	5.3	46	260
Dental Offices	6	52	230	Corridors	3.4	30	400	Shoe Stores	6.2	55	220
Dept. Stores, Basement	4	34	350	Industrial Plants, Offices	4.3	38	320	Shop'g. Ctrs., Sup. Mkts.	4	34	350
Main Floor	4.5	40	300	General Offices	4	34	350	Retail Stores	5.5	48	250
Upper Floor	3.4	30	400	Plant Areas	4.5	40	300	Specialty Shops	6.8	60	200

*Persons per ton

12,000 BTUH = 1 ton of air conditioning

G10 Site Preparation — RG1010-010 Information

The Building Site Work Division is divided into five sections:

G10 Site Preparation
G20 Site Improvements
G30 Site Mechanical Utilities
G40 Site Electrical Utilities
G90 Other Site Construction

Clear and Grub, Demolition and Landscaping must be added to these costs as they apply to individual sites.

The Site Preparation section includes typical bulk excavation, trench excavation, and backfill. To this must be added sheeting, dewatering, drilling and blasting, traffic control, and any unusual items relating to a particular site.

The Site Improvements section, including Parking Lot and Roadway construction, has quantities and prices for typical designs with gravel base and asphaltic concrete pavement. Different climate, soil conditions, material availability, and owner's requirements can easily be adapted to these parameters.

The Site Electrical Utilities sections tabulate the installed price for pipe. Excavation and backfill costs for the pipe are derived from the Site Preparation section.

By following the examples and illustrations of typical site work conditions in this division and adapting them to your particular site, an accurate price, including O&P, can be determined.

Table G1010-011 Site Preparation

Description	Unit	Total Costs
Clear and Grub Average	Acre	$8,975
Bulk Excavation with Front End Loader	C.Y.	1.96
Spread and Compact Dumped Material	C.Y.	2.58
Fine Grade and Seed	S.Y.	2.98

G10 Site Preparation — RG1030-400 Excavation, General

Bulk excavation on building jobs today is done entirely by machine. Hand work is used only for squaring out corners, fine grading, cutting out trenches and footings below the general excavating grade, and making minor miscellaneous cuts. Hand trimming of material other than sand and gravel is generally done with picks and shovels.

The following figures on machine excavating apply to basement pits, large footings and other excavation for buildings. On grading excavation this performance can be bettered by 33% to 50%. The figures presume loading directly into trucks. Allowance has been made for idle time due to truck delay and moving about on the job. Figures include no water problems or slow down due to sheeting and bracing operations. They also do not include truck spotters or labor for cleanup.

Table G1030-401 Bulk Excavation Cost, 15′ Deep

1½ C.Y. Hydraulic Excavator Backhoe, Four Tandem Trucks with a Two Mile round Trip Haul Distance

Item			Daily Cost	Labor & Equipment	Unit Price (C.Y.)
Equipment operator @	$ 63.95	per hr. x 8 hrs.	$ 511.60		
Laborer @	$ 55.30	per hr. x 8 hrs.	442.40	$2,526.80 Labor	$3.51 Labor
Four truck drivers @	$ 49.15	per hr. x 8 hrs.	1,572.80		
1½ C.Y. backhoe @	$865.80	per day	$ 865.80	$2,997.80 Equipment	$4.16 Equipment
Four 16 ton trucks @	$533.00	per day	2,132.00		
Production = 720 C.Y. per day			$5,524.60	$5,524.60 Labor & Equip.	$7.67 Labor & Equip.

Table G1030-402 Total Cost Per C.Y. Excavation for Various Size Jobs

No. of C.Y. to be Excavated	Cost Per Cubic Yard	Mobilizing & Demobilizing Cost*	Total Cost Per Cubic Yard
500 C.Y.	$7.67	$1.68	$9.35
1,000		.84	$8.51
1,500		.56	$8.23
2,000		.42	$8.09
3,000		.28	$7.95
5,000		.17	$7.84
10,000		.08	$7.75
15,000		.06	$7.73

*$420.00 for Mobilizing or Demobilizing 1½ C.Y. Hydraulic Excavator

Table G1030-403 Trench Bottom Widths for Various Outside Diameters of Buried Pipe

Outside Diameter	Trench Bottom Width
24 inches	4.1 feet
30	4.9
36	5.6
42	6.3
48	6.3
60	8.5
72	10.0
84	11.4

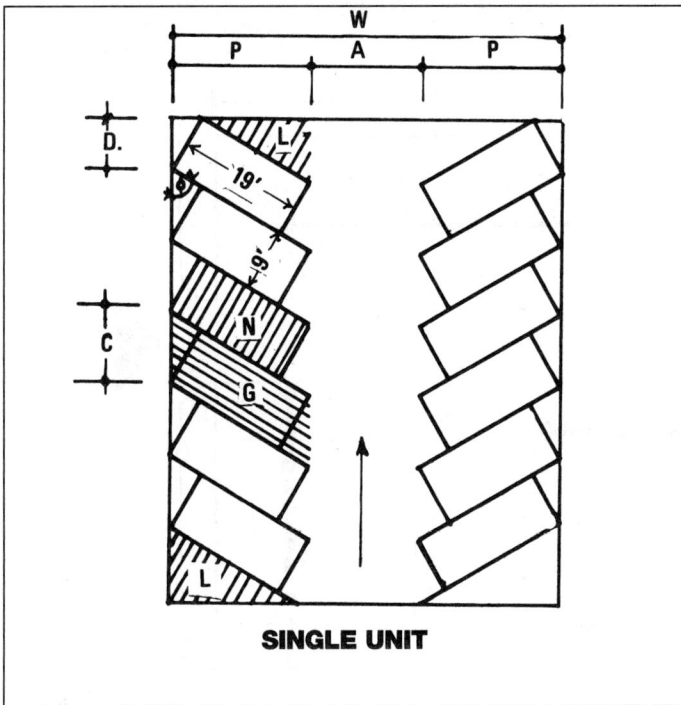

SINGLE UNIT

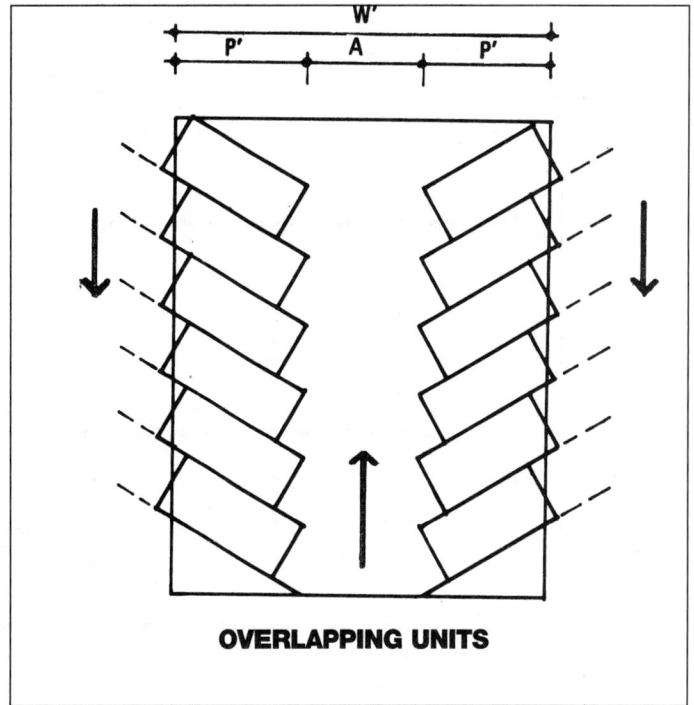

OVERLAPPING UNITS

Table G2020-502   Layout Data Based on 9' x 19' Parking Stall Size

φ	P	A	C	W	N	G	D	L	P'	W'
Angle of Stall	Parking Depth	Aisle Width	Curb Length	Width Overall	Net Car Area	Gross Car Area	Distance Last Car	Lost Area	Parking Depth	Width Overall
90°	19'.	24'	9'.	62'.	171 S.F.	171 S.F.	9'.	0	19'.	62'.
60°	21'.8	18'	10.4'	60'.	171 S.F.	217 S.F.	7.8'	205 S.F.	18.8'	55.5'
45°	19.8'	13'	12.8'	52.7'	171 S.F.	252 S.F.	6.4'	286 S.F.	16.6'	46.2'

Note: Square foot per car areas do not include the area of the travel lane.

90° Stall Angle: The main reason for use of this stall angle is to achieve the highest car capacity. This may be sound reasoning for employee lots with all day parking, but in most (in & out) lots there is difficulty in entering the stalls and no traffic lane direction. This may outweigh the advantage of high capacity.

60° Stall Angle: This layout is used most often due to the ease of entering and backing out, also the traffic aisle may be smaller.

45° Stall Angle: Requires a small change of direction from the traffic aisle to the stall, so the aisle may be reduced in width.

162'-6"

13 SPACES

CATCH BASIN

9 SPACES

5 SPACES

9 SPACES

MAN HOLE

CATCH BASIN

14 SPACES

120'-0'

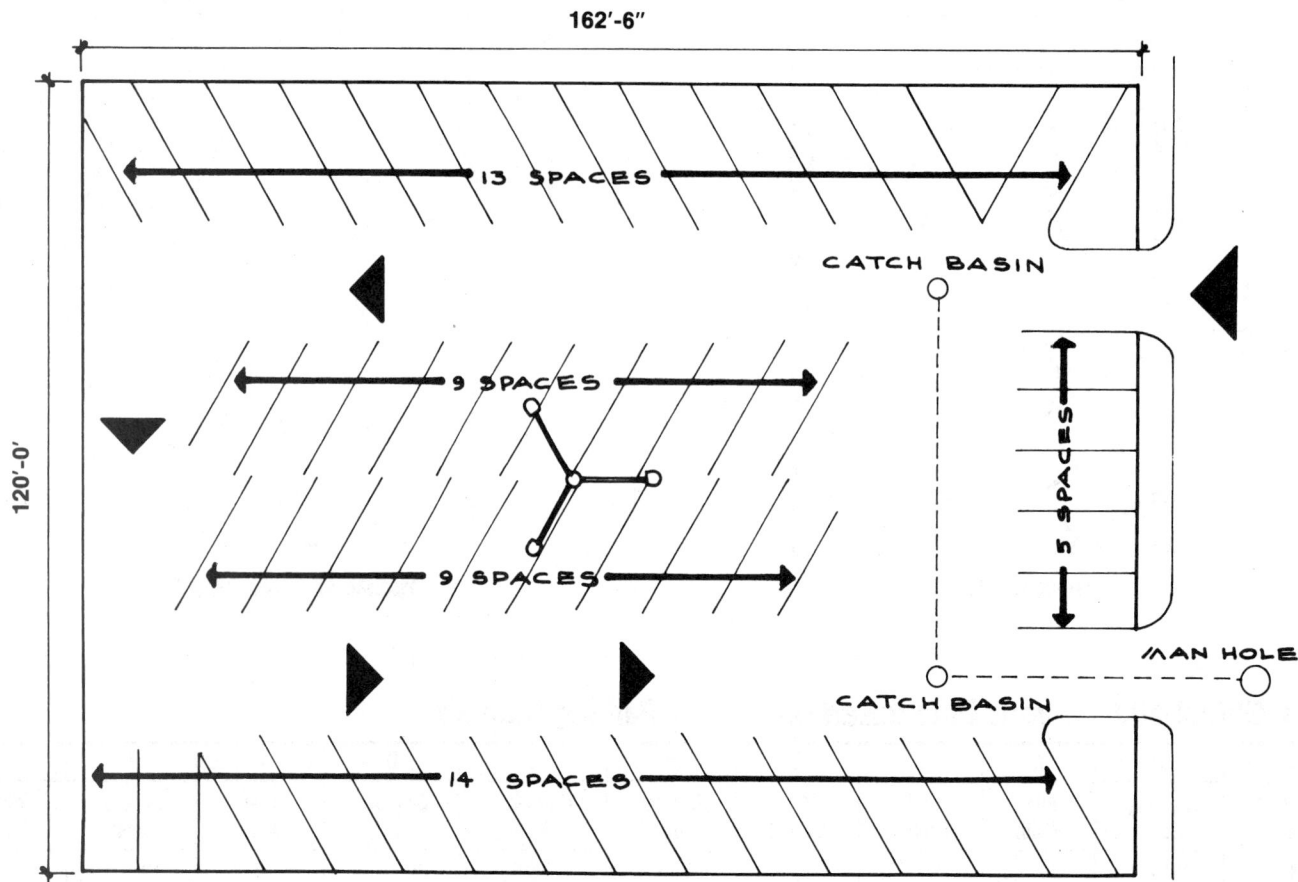

PARKING LOT PLAN (50 CAR)

Preliminary Design Data: The space required for parking and maneuvering is between 300 and 400 S.F. per car, depending upon engineering layout and design.

Ninety degree (90°) parking, with a central driveway and two rows of parked cars, will provide the best economy.

Diagonal parking is easier than 90° for the driver and reduces the necessary driveway width, but requires more total space.

Table H1010-101   General Contractor's Overhead

There are two distinct types of overhead on a construction project: Project Overhead and Main Office Overhead. Project Overhead includes those costs at a construction site not directly associated with the installation of construction materials. Examples of Project Overhead costs include the following:

1. Superintendent
2. Construction office and storage trailers
3. Temporary sanitary facilities
4. Temporary utilities
5. Security fencing
6. Photographs
7. Clean up
8. Performance and payment bonds

The above Project Overhead items are also referred to as General Requirements and therefore are estimated in Division 1. Division 1 is the first division listed in the CSI MasterFormat but it is usually the last division estimated. The sum of the costs in Divisions 1 through 49 is referred to as the sum of the direct costs.

All construction projects also include indirect costs. The primary components of indirect costs are the contractor's Main Office Overhead and profit. The amount of the Main Office Overhead expense varies depending on the the following:

1. Owner's compensation
2. Project managers and estimator's wages
3. Clerical support wages
4. Office rent and utilities
5. Corporate legal and accounting costs
6. Advertising
7. Automobile expenses
8. Association dues
9. Travel and entertainment expenses

These costs are usually calculated as a percentage of annual sales volume. This percentage can range from 35% for a small contractor doing less than $500,000 to 5% for a large contractor with sales in excess of $100 million.

Table H1010-102   Main Office Expense

A General Contractor's main office expense consists of many items not detailed in the front portion of the book. The percentage of main office expense declines with increased annual volume of the contractor. Typical main office expense ranges from 2% to 20% with the median about 7.2%

of total volume. This equals about 7.7% of direct costs. The following are approximate percentages of total overhead for different items usually included in a General Contractor's main office overhead. With different accounting procedures, these percentages may vary.

Item	Typical Range			Average
Managers', clerical and estimators' salaries	40 %	to	55 %	48%
Profit sharing, pension and bonus plans	2	to	20	12
Insurance	5	to	8	6
Estimating and project management (not including salaries)	5	to	9	7
Legal, accounting and data processing	0.5	to	5	3
Automobile and light truck expense	2	to	8	5
Depreciation of overhead capital expenditures	2	to	6	4
Maintenance of office equipment	0.1	to	1.5	1
Office rental	3	to	5	4
Utilities including phone and light	1	to	3	2
Miscellaneous	5	to	15	8
Total				100%

Reference Tables

Table H1010-201   Architectural Fees

Tabulated below are typical percentage fees by project size, for good professional architectural service. Fees may vary from those listed depending upon degree of design difficulty and economic conditions in any particular area.

Rates can be interpolated horizontally and vertically. Various portions of the same project requiring different rates should be adjusted

proportionately. For alterations, add 50% to the fee for the first $500,000 of project cost and add 25% to the fee for project cost over $500,000.

Architectural fees tabulated below include Structural, Mechanical and Electrical Engineering Fees. They do not include the fees for special consultants such as kitchen planning, security, acoustical, interior design, etc.

Building Types	Total Project Size in Thousands of Dollars						
	100	250	500	1,000	5,000	10,000	50,000
Factories, garages, warehouses, repetitive housing	9.0%	8.0%	7.0%	6.2%	5.3%	4.9%	4.5%
Apartments, banks, schools, libraries, offices, municipal buildings	12.2	12.3	9.2	8.0	7.0	6.6	6.2
Churches, hospitals, homes, laboratories, museums, research	15.0	13.6	12.7	11.9	9.5	8.8	8.0
Memorials, monumental work, decorative furnishings	—	16.0	14.5	13.1	10.0	9.0	8.3

Table H1010-202   Engineering Fees

Typical **Structural Engineering Fees** based on type of construction and total project size. These fees are included in Architectural Fees.

Type of Construction	Total Project Size (in thousands of dollars)			
	$500	$500-$1,000	$1,000-$5,000	Over $5000
Industrial buildings, factories & warehouses	Technical payroll times 2.0 to 2.5	1.60%	1.25%	1.00%
Hotels, apartments, offices, dormitories, hospitals, public buildings, food stores		2.00%	1.70%	1.20%
Museums, banks, churches and cathedrals		2.00%	1.75%	1.25%
Thin shells, prestressed concrete, earthquake resistive		2.00%	1.75%	1.50%
Parking ramps, auditoriums, stadiums, convention halls, hangars & boiler houses		2.50%	2.00%	1.75%
Special buildings, major alterations, underpinning & future expansion		Add to above 0.5%	Add to above 0.5%	Add to above 0.5%

For complex reinforced concrete or unusually complicated structures, add 20% to 50%.

Table H1010-203   Mechanical and Electrical Fees

Typical **Mechanical and Electrical Engineering Fees** based on the size of the subcontract. The fee structure for both are shown below. These fees are included in Architectural Fees.

Type of Construction	Subcontract Size							
	$25,000	$50,000	$100,000	$225,000	$350,000	$500,000	$750,000	$1,000,000
Simple structures	6.4%	5.7%	4.8%	4.5%	4.4%	4.3%	4.2%	4.1%
Intermediate structures	8.0	7.3	6.5	5.6	5.1	5.0	4.9	4.8
Complex structures	10.1	9.0	9.0	8.0	7.5	7.5	7.0	7.0

For renovations, add 15% to 25% to applicable fee.

Table H1010-204   Construction Time Requirements

Table below is average construction time in months for different types of building projects. Table at right is the construction time in months for different size projects. Design time runs 25% to 40% of construction time.

Type Building	Construction Time	Project Value	Construction Time
Industrial Buildings	12 Months	Under $1,400,000	10 Months
Commercial Buildings	15 Months	Up to $3,800,000	15 Months
Research & Development	18 Months	Up to $19,000,000	21 Months
Institutional Buildings	20 Months	over $19,000,000	28 Months

Table H1010-301   Builder's Risk Insurance

Builder's Risk Insurance is insurance on a building during construction. Premiums are paid by the owner or the contractor. Blasting, collapse and underground insurance would raise total insurance costs above those listed. Floater policy for materials delivered to the job runs $.75 to $1.25 per $100 value. Contractor equipment insurance runs $.50 to $1.50 per $100 value. Insurance for miscellaneous tools to $1,500 value runs from $3.00 to $7.50 per $100 value.

Tabulated below are New England Builder's Risk insurance rates in dollars per $100 value for $1,000 deductible. For $25,000 deductible, rates can be reduced 13% to 34%. On contracts over $1,000,000, rates may be lower than those tabulated. Policies are written annually for the total completed value in place. For "all risk" insurance (excluding flood, earthquake and certain other perils) add $.025 to total rates below.

Coverage	Frame Construction (Class 1)			Brick Construction (Class 4)			Fire Resistive (Class 6)		
	Range		Average	Range		Average	Range		Average
Fire Insurance	$.350 to	$ .850	$.600	$.158 to	$.189	$.174	$.052 to	$.080	$.070
Extended Coverage	.115 to	.200	.158	.080 to	.105	.101	.081 to	.105	.100
Vandalism	.012 to	.016	.014	.008 to	.011	.011	.008 to	.011	.010
Total Annual Rate	$.477 to	$1.066	$.772	$.246 to	$.305	$.286	$.141 to	$.196	$.180

Table H1010-302   Performance Bond

This table shows the cost of a Performance Bond for a construction job scheduled to be completed in 12 months. Add 1% of the premium cost per month for jobs requiring more than 12 months to complete. The rates are "standard" rates offered to contractors that the bonding company considers financially sound and capable of doing the work. Preferred rates

are offered by some bonding companies based upon financial strength of the contractor. Actual rates vary from contractor to contractor and from bonding company to bonding company. Contractors should prequalify through a bonding agency before submitting a bid on a contract that requires a bond.

Contract Amount	Building Construction Class B Projects			Highways & Bridges					
				Class A New Construction			Class A-1 Highway Resurfacing		
First $ 100,000 bid	$25.00 per M			$15.00 per M			$9.40 per M		
Next 400,000 bid	$ 2,500 plus	$15.00	per M	$ 1,500 plus	$10.00	per M	$ 940 plus	$7.20	per M
Next 2,000,000 bid	8,500 plus	10.00	per M	5,500 plus	7.00	per M	3,820 plus	5.00	per M
Next 2,500,000 bid	28,500 plus	7.50	per M	19,500 plus	5.50	per M	15,820 plus	4.50	per M
Next 2,500,000 bid	47,250 plus	7.00	per M	33,250 plus	5.00	per M	28,320 plus	4.50	per M
Over 7,500,000 bid	64,750 plus	6.00	per M	45,750 plus	4.50	per M	39,570 plus	4.00	per M

Table H1010-401   Workers' Compensation Insurance Rates by Trade

The table below tabulates the national averages for Workers' Compensation insurance rates by trade and type of building. The average "Insurance Rate" is multiplied by the "% of Building Cost" for each trade. This produces the "Workers' Compensation Cost" by % of total labor cost, to be added for each trade by building type to determine the weighted average Workers' Compensation rate for the building types analyzed.

Trade	Insurance Rate (% Labor Cost)			% of Building Cost			Workers' Compensation		
	Range		Average	Office Bldgs.	Schools & Apts.	Mfg.	Office Bldgs.	Schools & Apts.	Mfg.
Excavation, Grading, etc.	4.2 % to	17.5%	10.0%	4.8%	4.9%	4.5%	0.48%	0.49%	0.45%
Piles & Foundations	5.9 to	40.3	20.1	7.1	5.2	8.7	1.43	1.05	1.75
Concrete	5.1 to	26.8	14.6	5.0	14.8	3.7	0.73	2.16	0.54
Masonry	4.8 to	43.3	14.4	6.9	7.5	1.9	0.99	1.08	0.27
Structural Steel	5.9 to	104.1	37.9	10.7	3.9	17.6	4.06	1.48	6.67
Miscellaneous & Ornamental Metals	3.4 to	22.8	10.7	2.8	4.0	3.6	0.30	0.43	0.39
Carpentry & Millwork	5.9 to	53.2	17.8	3.7	4.0	0.5	0.66	0.71	0.09
Metal or Composition Siding	5.9 to	38	16.6	2.3	0.3	4.3	0.38	0.05	0.71
Roofing	5.9 to	77.1	31.2	2.3	2.6	3.1	0.72	0.81	0.97
Doors & Hardware	4.9 to	32	11.6	0.9	1.4	0.4	0.10	0.16	0.05
Sash & Glazing	5.9 to	32.4	13.9	3.5	4.0	1.0	0.49	0.56	0.14
Lath & Plaster	3.3 to	43.9	13.6	3.3	6.9	0.8	0.45	0.94	0.11
Tile, Marble & Floors	3.1 to	17.9	9.1	2.6	3.0	0.5	0.24	0.27	0.05
Acoustical Ceilings	2.6 to	45.6	10.6	2.4	0.2	0.3	0.25	0.02	0.03
Painting	4.7 to	29.6	12.5	1.5	1.6	1.6	0.19	0.20	0.20
Interior Partitions	5.9 to	53.2	17.8	3.9	4.3	4.4	0.69	0.77	0.78
Miscellaneous Items	2.1 to	168.2	16.0	5.2	3.7	9.7	0.83	0.59	1.55
Elevators	2.8 to	11.8	6.6	2.1	1.1	2.2	0.14	0.07	0.15
Sprinklers	2.5 to	14.3	7.8	0.5	—	2.0	0.04	—	0.16
Plumbing	2.9 to	12.4	7.8	4.9	7.2	5.2	0.38	0.56	0.41
Heat., Vent., Air Conditioning	4.3 to	23	11.6	13.5	11.0	12.9	1.57	1.28	1.50
Electrical	2.8 to	11.5	6.5	10.1	8.4	11.1	0.66	0.55	0.72
Total	2.1 % to	168.2%	—	100.0%	100.0%	100.0%	15.78%	14.23%	17.69%

Overall Weighted Average   15.90%

Workers' Compensation Insurance Rates by States

The table below lists the weighted average Workers' Compensation base rate for each state with a factor comparing this with the national average of 15.5%.

State	Weighted Average	Factor	State	Weighted Average	Factor	State	Weighted Average	Factor
Alabama	24.2%	156	Kentucky	18.4%	119	North Dakota	13.7%	88
Alaska	21.7	140	Louisiana	28.3	183	Ohio	14.8	95
Arizona	9.5	61	Maine	15.4	99	Oklahoma	14.6	94
Arkansas	13.2	85	Maryland	16.7	108	Oregon	13.5	87
California	19.6	126	Massachusetts	12.5	81	Pennsylvania	13.8	89
Colorado	13.2	85	Michigan	17.3	112	Rhode Island	21.2	137
Connecticut	21.0	135	Minnesota	25.7	166	South Carolina	18.3	118
Delaware	15.1	97	Mississippi	19.1	123	South Dakota	19.1	123
District of Columbia	13.7	88	Missouri	17.2	111	Tennessee	15.1	97
Florida	13.8	89	Montana	16.6	107	Texas	12.2	79
Georgia	26.2	169	Nebraska	23.0	148	Utah	12.3	79
Hawaii	17.0	110	Nevada	11.6	75	Vermont	18.0	116
Idaho	10.8	70	New Hampshire	20.0	129	Virginia	11.7	75
Illinois	21.6	139	New Jersey	13.4	86	Washington	9.8	63
Indiana	5.9	38	New Mexico	18.5	119	West Virginia	9.2	59
Iowa	11.5	74	New York	12.6	81	Wisconsin	15.0	97
Kansas	8.9	57	North Carolina	18.9	122	Wyoming	6.5	42

Weighted Average for U.S. is   15.9% of payroll = 100%

Rates in the following table are the base or manual costs per $100 of payroll for Workers' Compensation in each state. Rates are usually applied to straight time wages only and not to premium time wages and bonuses.

The weighted average skilled worker rate for 35 trades is 15.5%. For bidding purposes, apply the full value of Workers' Compensation directly to total labor costs, or if labor is 38%, materials 42% and overhead and profit 20% of total cost, carry 38/80 x 15.5% =7.4% of cost (before overhead and profit) into overhead. Rates vary not only from state to state but also with the experience rating of the contractor.

Rates are the most current available at the time of publication.

Table H1010-403 Workers' Compensation Insurance Rates by Trade and State

State	Carpentry — 3 stories or less 5651	Carpentry — interior cab. work 5437	Carpentry — general 5403	Concrete Work — NOC 5213	Concrete Work — flat (flr., sdwk.) 5221	Electrical Wiring — inside 5190	Excavation — earth NOC 6217	Excavation — rock 6217	Glaziers 5462	Insulation Work 5479	Lathing 5443	Masonry 5022	Painting & Decorating 5474	Pile Driving 6003	Plastering 5480	Plumbing 5183	Roofing 5551	Sheet Metal Work (HVAC) 5538	Steel Erection — door & sash 5102	Steel Erection — inter, ornam. 5102	Steel Erection — structure 5040	Steel Erection — NOC 5057	Tile Work — (interior ceramic) 5348	Waterproofing 9014	Wrecking 5701
AL	37.98	22.82	32.46	16.20	11.27	10.68	11.16	11.16	21.26	16.28	11.93	27.00	25.81	32.65	18.85	12.27	58.14	22.96	11.01	11.01	55.68	25.37	13.31	6.56	55.68
AK	14.73	12.27	17.44	11.07	9.71	8.72	12.14	12.14	32.36	27.78	9.74	43.27	21.95	32.40	43.85	9.97	35.12	8.61	9.39	9.39	43.01	33.00	6.76	6.61	43.01
AZ	11.60	6.45	15.40	8.56	4.45	4.59	4.90	4.90	6.93	12.48	6.61	6.51	6.33	11.45	5.78	5.24	13.38	7.38	10.66	10.66	21.10	15.34	3.17	2.59	21.10
AR	14.82	7.45	16.26	12.17	6.48	5.02	7.76	7.76	9.82	16.49	5.93	9.93	11.50	16.48	16.10	5.15	23.09	13.09	6.77	6.77	32.38	25.50	6.08	3.74	32.38
CA	31.95	31.95	31.95	11.89	11.89	7.98	13.79	13.79	15.74	12.50	11.24	16.21	17.08	15.27	20.53	12.03	44.43	15.60	12.46	12.46	17.17	20.88	8.54	17.08	20.88
CO	13.57	8.92	12.33	11.57	7.87	5.28	10.34	10.34	9.75	16.84	6.02	14.01	9.02	14.31	9.58	7.43	26.09	11.61	9.31	9.31	32.62	17.58	7.06	5.03	17.58
CT	15.67	14.66	28.76	24.79	10.09	8.01	12.91	12.91	22.98	15.94	21.94	22.24	17.61	22.88	14.48	11.16	41.95	15.61	18.57	18.57	42.72	23.45	11.58	5.41	42.72
DE	15.17	15.17	11.46	11.51	9.49	5.89	9.32	9.32	13.11	11.46	13.11	12.77	26.77	18.75	13.11	7.87	26.71	9.78	12.28	12.28	26.77	12.28	9.29	12.77	26.77
DC	11.01	12.78	10.19	9.66	9.14	6.01	12.54	12.54	17.02	8.18	9.40	11.97	7.51	12.11	12.48	10.38	19.21	8.93	10.61	10.61	37.21	16.49	17.86	4.00	37.21
FL	13.05	10.47	14.39	15.31	6.97	6.64	8.30	8.30	9.91	9.24	5.98	11.32	9.73	38.24	18.70	6.75	22.19	12.68	8.85	8.85	27.89	14.56	6.55	5.17	27.89
GA	35.68	20.54	25.67	16.85	13.26	10.82	17.38	17.38	18.77	23.64	15.09	21.06	27.77	27.36	22.83	11.4	54.71	20.76	19.93	19.93	64.78	42.21	12.42	8.50	64.78
HI	17.38	11.62	28.20	13.74	12.27	7.10	7.73	7.73	20.55	21.19	10.94	17.87	11.33	20.15	15.61	6.06	33.24	8.02	11.11	11.11	31.71	21.70	9.78	11.54	31.71
ID	9.16	6.04	10.35	11.25	6.13	3.70	5.72	5.72	8.70	7.67	8.89	7.59	8.45	12.13	13.37	5.36	30.27	8.89	5.72	5.72	29.16	10.28	9.79	4.75	29.16
IL	24.93	14.46	20.78	26.59	12.10	9.62	10.06	10.06	19.18	17.48	15.59	19.27	11.62	34.85	15.88	11.07	31.06	14.96	16.05	16.05	72.39	25.69	14.02	4.84	72.39
IN	9.65	4.92	7.29	5.10	3.22	2.84	4.87	4.87	6.28	7.44	2.62	4.84	4.72	6.44	3.72	2.91	10.88	4.45	3.43	3.43	12.12	7.04	3.09	2.47	12.12
IA	10.43	10.02	10.34	12.12	7.00	4.96	6.27	6.27	9.42	7.06	5.58	9.61	6.65	9.78	7.33	6.46	22.58	7.52	6.15	6.15	30.21	35.21	7.42	3.87	35.21
KS	10.05	7.85	9.89	7.19	6.42	3.37	4.41	4.41	9.22	7.45	4.68	6.91	7.57	11.31	6.25	4.85	15.06	7.14	4.82	4.82	26.19	13.21	6.35	3.46	13.21
KY	18.72	14.07	25.30	13.39	8.00	5.43	9.15	9.15	17.37	17.59	11.24	9.12	13.18	27.72	14.41	7.06	38.25	19.58	12.32	12.32	54.03	22.56	13.67	5.74	56.16
LA	22.02	24.93	53.17	26.43	15.61	9.88	17.49	17.49	20.14	21.43	24.51	28.33	29.58	31.25	22.37	8.64	77.12	21.20	18.98	18.98	51.76	24.21	13.77	13.78	66.41
ME	13.79	10.40	28.55	20.94	8.74	6.31	9.05	9.05	18.14	11.82	7.90	14.76	12.73	16.70	11.79	10.56	22.87	11.61	9.73	9.73	31.67	23.11	7.00	7.12	31.67
MD	15.05	11.06	14.11	15.67	6.67	9.24	9.20	9.20	20.03	15.02	8.83	11.43	7.48	20.09	15.02	6.11	34.74	13.54	9.51	9.51	59.03	28.60	7.60	5.36	28.60
MA	6.80	5.60	11.46	19.51	6.57	3.20	4.19	4.19	8.96	10.06	6.35	10.81	4.79	14.68	5.08	3.98	32.80	5.15	7.61	7.61	45.71	36.69	6.21	2.09	25.88
MI	18.40	11.39	20.64	19.45	8.95	4.72	10.20	10.20	12.41	14.75	13.59	15.54	13.64	40.27	15.18	6.84	30.67	10.00	10.16	10.16	40.27	22.21	11.11	4.87	40.27
MN	19.53	19.80	41.48	14.65	15.73	7.53	14.95	14.95	15.68	11.90	22.05	19.18	16.95	24.63	22.05	10.88	70.18	14.23	10.58	10.58	104.13	33.30	14.55	6.68	36.03
MS	21.07	11.11	22.29	15.49	9.26	6.28	11.91	11.91	13.07	14.85	7.91	14.09	13.14	35.34	31.72	9.80	44.11	16.91	14.23	14.23	36.11	25.03	9.84	4.18	36.11
MO	29.99	11.16	13.98	18.91	10.39	7.64	9.17	9.17	11.13	15.17	7.93	15.01	12.53	19.12	17.02	10.05	29.98	13.19	11.31	11.31	34.56	35.84	11.18	6.51	34.56
MT	17.12	11.15	22.57	12.37	12.79	6.39	15.54	15.54	10.67	32.03	11.06	13.75	9.61	25.18	10.02	8.75	37.61	11.53	8.70	8.70	27.45	15.73	7.53	7.46	15.73
NE	24.65	17.47	21.80	25.07	15.13	10.38	16.90	16.90	23.70	31.50	12.20	22.22	20.07	22.50	21.67	12.35	35.22	19.42	16.05	16.05	45.40	35.27	11.10	6.38	47.05
NV	15.25	7.14	11.58	10.76	8.23	7.17	9.77	9.77	10.19	8.12	5.27	7.50	7.97	13.33	7.30	5.98	14.08	18.10	8.67	8.67	25.10	21.54	5.39	5.55	21.54
NH	25.92	13.36	18.90	26.79	14.17	6.94	14.52	14.52	13.07	21.36	9.63	24.34	7.77	18.99	11.71	10.13	49.23	13.59	15.27	15.27	52.24	18.83	11.37	6.09	52.24
NJ	14.34	8.94	14.34	16.01	9.99	4.76	8.35	8.35	8.55	12.65	11.64	13.09	11.92	16.03	11.64	6.17	37.77	6.49	9.46	9.46	23.72	13.59	6.99	5.72	22.31
NM	18.91	7.15	20.11	17.36	10.05	7.51	11.20	11.20	21.26	13.72	8.39	15.34	11.64	21.86	13.44	8.13	36.37	13.47	13.69	13.69	55.78	38.81	6.59	6.29	55.78
NY	14.72	6.45	12.67	16.02	11.64	6.03	8.29	8.29	10.97	7.98	12.10	16.02	9.89	12.77	9.03	6.88	27.98	10.46	8.84	8.84	23.45	12.72	6.66	6.04	8.73
NC	18.76	14.56	15.36	18.15	6.88	11.50	11.70	11.70	15.21	16.00	15.28	12.51	12.90	17.50	15.68	9.16	29.70	15.90	10.44	10.44	79.66	26.21	9.28	5.55	79.66
ND	11.17	11.17	11.17	6.42	6.42	3.63	6.64	6.64	11.17	11.17	7.51	7.71	6.61	22.78	7.51	5.22	23.53	5.22	22.78	22.78	22.78	22.78	11.17	23.53	10.98
OH	11.04	8.52	11.68	12.33	9.99	6.14	8.70	8.70	7.86	16.94	45.64	12.71	12.37	19.06	3.28	6.99	31.63	9.71	7.59	7.59	29.23	16.59	9.60	6.42	16.59
OK	13.42	9.17	11.76	11.76	6.34	5.73	10.83	10.83	16.07	19.47	8.45	10.02	8.47	20.25	12.19	7.09	21.13	9.05	13.76	13.76	39.72	23.97	6.50	5.71	39.72
OR	15.43	8.10	15.69	14.22	8.97	5.14	9.33	9.33	15.22	11.80	7.60	14.07	10.63	16.17	12.73	5.50	26.30	9.41	7.93	7.93	26.94	18.78	11.04	4.54	26.94
PA	13.18	13.18	11.38	14.21	10.46	5.95	8.27	8.27	11.47	11.38	11.47	12.22	13.31	15.78	11.47	7.33	27.42	7.84	14.30	14.30	21.31	14.30	8.02	12.22	21.37
RI	19.53	11.65	18.07	18.23	16.24	4.43	10.38	10.38	12.85	22.78	11.97	25.11	24.13	37.66	17.25	8.52	33.92	10.29	14.07	14.07	59.49	37.50	14.36	7.70	59.49
SC	22.54	15.41	20.91	14.80	7.92	10.92	13.16	13.16	16.12	14.69	11.32	12.88	16.52	19.48	16.02	11.47	45.24	12.95	12.99	12.99	32.57	26.39	9.62	6.68	47.05
SD	27.77	8.13	29.58	25.56	9.55	5.52	11.01	11.01	12.16	12.05	7.70	11.31	9.53	21.32	12.44	11.75	20.73	12.48	8.98	8.98	80.86	40.98	8.74	4.98	80.86
TN	15.68	13.11	16.06	13.89	7.97	7.35	13.68	13.68	13.09	11.80	7.30	15.60	12.63	24.89	16.02	7.66	22.25	12.77	7.17	7.17	29.18	23.47	10.15	4.65	29.18
TX	11.25	8.89	11.25	9.00	7.26	6.15	8.54	8.54	9.65	12.87	6.36	10.91	8.98	22.43	8.98	6.75	19.59	12.84	9.09	9.09	29.68	13.37	7.07	6.69	9.96
UT	15.74	8.88	12.97	9.81	8.47	4.57	11.59	11.59	14.70	12.49	6.35	11.25	8.37	12.47	7.26	6.18	24.99	12.64	7.16	7.16	25.67	14.67	6.80	4.27	18.37
VT	15.70	10.75	17.37	17.27	7.21	6.98	12.59	12.59	19.75	17.78	8.33	15.82	11.62	16.74	10.78	10.44	36.43	11.59	12.51	12.51	44.95	42.80	11.24	7.28	44.95
VA	11.05	8.02	9.16	11.05	5.31	5.52	7.57	7.57	10.25	9.90	14.34	9.62	8.99	10.75	10.22	5.83	22.52	7.65	8.37	8.37	33.52	19.66	4.74	3.04	33.52
WA	8.48	8.48	8.48	7.32	7.32	2.81	6.69	6.69	12.27	9.77	8.43	9.09	10.93	16.12	9.95	4.17	16.36	4.27	7.60	7.60	7.60	7.60	8.45	16.36	7.60
WV	10.65	7.30	10.23	8.24	4.41	4.70	6.22	6.22	7.89	7.25	6.11	7.98	7.90	11.90	7.93	4.90	18.05	5.74	5.58	5.58	24.17	13.20	5.65	2.66	17.34
WI	9.99	11.82	14.45	11.15	9.69	6.02	8.19	8.19	10.44	13.58	6.25	15.31	13.03	16.47	9.96	6.24	37.36	7.21	9.36	9.36	28.23	44.93	13.80	4.86	28.23
WY	5.87	5.87	5.87	5.87	5.87	5.87	5.87	5.87	5.87	5.87	5.87	5.87	5.87	5.87	5.87	5.87	5.87	5.87	5.87	5.87	5.87	5.87	5.87	5.87	5.87
AVG.	16.63	11.62	17.80	14.58	9.14	6.46	10.01	10.01	13.89	14.44	10.63	14.37	12.49	20.09	13.60	7.84	31.18	11.57	10.74	10.74	37.94	23.15	9.13	6.69	34.13

Table H1010-404   Workers' Compensation (Canada in Canadian dollars)

Province		Alberta	British Columbia	Manitoba	Ontario	New Brunswick	Newfndld. & Labrador	Northwest Territories	Nova Scotia	Prince Edward Island	Quebec	Saskat-chewan	Yukon
Carpentry—3 stories or less	Rate	7.32	4.74	4.36	4.35	3.79	10.08	3.84	7.82	5.96	15.05	6.23	8.89
	Code	42143	721028	40102	723	4226	4226	4-41	4226	401	80110	B1317	202
Carpentry—interior cab. work	Rate	2.18	4.90	4.36	4.35	4.40	4.69	3.84	4.52	3.71	15.05	3.51	8.89
	Code	42133	721021	40102	723	4279	4270	4-41	4274	402	80110	B11-27	202
CARPENTRY—general	Rate	7.32	4.74	4.36	4.35	3.79	4.69	3.84	7.82	5.96	15.05	6.23	8.89
	Code	42143	721028	40102	723	4226	4299	4-41	4226	401	80110	B1317	202
CONCRETE WORK—NOC	Rate	4.41	4.81	7.46	16.02	3.79	10.08	3.84	4.52	5.96	17.99	6.24	4.67
	Code	42104	721010	40110	748	4224	4224	4-41	4224	401	80100	B13-14	203
CONCRETE WORK—flat (flr. sidewalk)	Rate	4.41	4.81	7.46	16.02	3.79	10.08	3.84	4.52	5.96	17.99	6.24	4.67
	Code	42104	721010	40110	748	4224	4224	4-41	4224	401	80100	B13-14	203
ELECTRICAL Wiring—inside	Rate	2.13	1.77	2.51	3.79	2.17	3.01	3.52	2.31	3.71	6.12	3.51	4.67
	Code	42124	721019	40203	704	4261	4261	4-46	4261	402	80170	B11-05	206
EXCAVATION—earth NOC	Rate	2.51	3.56	3.76	4.55	2.62	4.37	3.64	3.34	3.82	7.13	3.93	4.67
	Code	40604	721031	40706	711	4214	4214	4-43	4214	404	80030	R11-06	207
EXCAVATION—rock	Rate	2.51	3.56	3.76	4.55	2.62	4.37	3.64	3.34	3.82	7.13	3.93	4.67
	Code	40604	721031	40706	711	4214	4214	4-43	4214	404	80030	R11-06	207
GLAZIERS	Rate	3.16	3.62	4.36	8.90	4.55	6.81	3.84	5.21	3.71	17.08	6.23	4.67
	Code	42121	715020	40109	751	4233	4233	4-41	4233	402	80150	B13-04	212
INSULATION WORK	Rate	2.57	8.42	4.36	8.90	4.55	6.81	3.84	5.21	5.96	15.05	5.98	8.89
	Code	42184	721029	40102	751	4234	4234	4-41	4234	401	80110	B12-07	202
LATHING	Rate	5.59	6.78	4.36	4.35	4.40	4.69	3.84	4.52	3.71	15.05	6.23	8.89
	Code	42135	721033	40102	723	4273	4279	4-41	4271	402	80110	B13-16	202
MASONRY	Rate	4.41	6.78	4.36	11.15	4.55	6.81	3.84	5.21	5.96	17.99	6.23	8.89
	Code	42102	721037	40102	741	4231	4231	4-41	4231	401	80100	B13-18	202
PAINTING & DECORATING	Rate	4.35	4.94	3.18	6.75	4.40	4.69	3.84	4.52	3.71	15.05	5.98	8.89
	Code	42111	721041	40105	719	4275	4275	4-41	4275	402	80110	B12-01	202
PILE DRIVING	Rate	4.41	5.18	3.76	6.34	3.79	9.78	3.64	4.64	5.96	7.13	6.23	8.89
	Code	42159	722004	40706	732	4221	4221	4-43	4221	401	80030	B13-10	202
PLASTERING	Rate	5.59	6.78	5.08	6.75	4.40	4.69	3.84	4.52	3.71	15.05	5.98	8.89
	Code	42135	721042	40108	719	4271	4271	4-41	4271	402	80110	B12-21	202
PLUMBING	Rate	2.13	3.11	3.05	4.02	2.32	3.37	3.52	2.43	3.71	6.93	3.51	4.67
	Code	42122	721043	40204	707	4241	4241	4-46	4241	402	80160	B11-01	214
ROOFING	Rate	8.12	8.68	7.02	12.98	4.55	10.08	3.84	10.57	5.96	23.01	6.23	8.89
	Code	42118	721036	40403	728	4236	4236	4-41	4236	401	80130	B13-20	202
SHEET METAL WORK (HVAC)	Rate	2.13	3.11	7.02	4.02	2.32	3.37	3.52	2.43	3.71	6.93	3.51	4.67
	Code	42117	721043	40402	707	4244	4244	4-46	4244	402	80160	B11-07	208
STEEL ERECTION—door & sash	Rate	2.57	14.07	11.32	16.02	3.79	10.08	3.84	5.21	5.96	25.32	6.23	8.89
	Code	42106	722005	40502	748	4227	4227	4-41	4227	401	80080	B13-22	202
STEEL ERECTION—inter., ornam.	Rate	2.57	14.07	11.32	16.02	3.79	10.08	3.84	5.21	5.96	25.32	6.23	8.89
	Code	42106	722005	40502	748	4227	4227	4-41	4227	401	80080	B13-22	202
STEEL ERECTION—structure	Rate	2.57	14.07	11.32	16.02	3.79	10.08	3.84	5.21	5.96	25.32	6.23	8.89
	Code	42106	722005	40502	748	4227	4227	4-41	4227	401	80080	B13-22	202
STEEL ERECTION—NOC	Rate	2.57	14.07	11.32	16.02	3.79	10.08	3.84	5.21	5.96	25.32	6.23	8.89
	Code	42106	722005	40502	748	4227	4227	4-41	4227	401	80080	B13-22	202
TILE WORK—inter. (ceramic)	Rate	3.63	5.62	1.97	6.75	4.40	10.08	3.84	4.52	3.71	15.05	6.23	8.89
	Code	42113	721054	40103	719	4276	4276	4-41	4276	402	80110	B13-01	202
WATERPROOFING	Rate	4.35	4.93	4.36	4.35	4.55	4.69	3.84	5.21	3.71	23.01	5.98	8.89
	Code	42139	721016	40102	723	4239	4299	4-41	4239	402	80130	B12-17	202
WRECKING	Rate	2.51	4.59	6.59	16.02	2.62	4.39	3.64	3.34	5.96	15.05	6.23	8.89
	Code	40604	721005	40106	748	4211	4211	4-43	4211	401	80110	B13-09	202

General Conditions — RH1020-300 Overhead & Miscellaneous Data

Unemployment Taxes and Social Security Taxes

State Unemployment Tax rates vary not only from state to state, but also with the experience rating of the contractor. The Federal Unemployment Tax rate is 6.2% of the first $7,000 of wages. This is reduced by a credit of up to 5.4% for timely payment to the state. The minimum Federal Unemployment Tax is 0.8% after all credits.

Social Security (FICA) for 2009 is estimated at time of publication to be 7.65% of wages up to $102,000.

General Conditions — RH1020-400 Overtime

Overtime

One way to improve the completion date of a project or eliminate negative float from a schedule is to compress activity duration times. This can be achieved by increasing the crew size or working overtime with the proposed crew.

To determine the costs of working overtime to compress activity duration times, consider the following examples. Below is an overtime efficiency and cost chart based on a five, six, or seven day week with an eight through twelve hour day. Payroll percentage increases for time and one half and double time are shown for the various working days.

| Days per Week | Hours per Day | Production Efficiency | | | | | Payroll Cost Factors | |
		1 Week	2 Weeks	3 Weeks	4 Weeks	Average 4 Weeks	@ 1-1/2 Times	@ 2 Times
5	8	100%	100%	100%	100%	100 %	100 %	100 %
	9	100	100	95	90	96.25	105.6	111.1
	10	100	95	90	85	91.25	110.0	120.0
	11	95	90	75	65	81.25	113.6	127.3
	12	90	85	70	60	76.25	116.7	133.3
6	8	100	100	95	90	96.25	108.3	116.7
	9	100	95	90	85	92.50	113.0	125.9
	10	95	90	85	80	87.50	116.7	133.3
	11	95	85	70	65	78.75	119.7	139.4
	12	90	80	65	60	73.75	122.2	144.4
7	8	100	95	85	75	88.75	114.3	128.6
	9	95	90	80	70	83.75	118.3	136.5
	10	90	85	75	65	78.75	121.4	142.9
	11	85	80	65	60	72.50	124.0	148.1
	12	85	75	60	55	68.75	126.2	152.4

General Conditions — RH1030-100 General

Sales Tax by State

State sales tax on materials is tabulated below (5 states have no sales tax). Many states allow local jurisdictions, such as a county or city, to levy additional sales tax.

Some projects may be sales tax exempt, particularly those constructed with public funds.

State	Tax (%)	State	Tax (%)	State	Tax (%)	State	Tax (%)
Alabama	4	Illinois	6.25	Montana	0	Rhode Island	7
Alaska	0	Indiana	6	Nebraska	5.5	South Carolina	5
Arizona	5.6	Iowa	5	Nevada	6.5	South Dakota	4
Arkansas	6	Kansas	5.3	New Hampshire	0	Tennessee	7
California	7.25	Kentucky	6	New Jersey	7	Texas	6.25
Colorado	2.9	Louisiana	4	New Mexico	5	Utah	4.65
Connecticut	6	Maine	5	New York	4	Vermont	6
Delaware	0	Maryland	6	North Carolina	4.25	Virginia	5
District of Columbia	5.75	Massachusetts	5	North Dakota	5	Washington	6.5
Florida	6	Michigan	6	Ohio	5.5	West Virginia	6
Georgia	4	Minnesota	6.5	Oklahoma	4.5	Wisconsin	5
Hawaii	4	Mississippi	7	Oregon	0	Wyoming	4
Idaho	6	Missouri	4.225	Pennsylvania	6	Average	4.91 %

Table H1040-101   Steel Tubular Scaffolding

On new construction, tubular scaffolding is efficient up to 60' high or five stories. Above this it is usually better to use a hung scaffolding if construction permits. Swing scaffolding operations may interfere with tenants. In this case, the tubular is more practical at all heights.

In repairing or cleaning the front of an existing building, the cost of tubular scaffolding per S.F. of building front increases as the height increases above the first tier. The first tier cost is relatively high due to leveling and alignment.

The minimum efficient crew for erection is three workers. For heights over 50', a crew of four is more efficient. Use two or more on top and two at the bottom for handing up or hoisting. Four workers can erect and dismantle about nine frames per hour up to five stories. From five to eight stories, they will average six frames per hour. With 7' horizontal spacing, this will run about 400 S.F. and 265 S.F. of wall surface, respectively. Time for placing

planks must be added to the above. On heights above 50', five planks can be placed per labor-hour.

The cost per 1,000 S.F. of building front in the table below was developed by pricing the materials required for a typical tubular scaffolding system eleven frames long and two frames high. Planks were figured five wide for standing plus two wide for materials.

Frames are 5' wide and usually spaced 7' O.C. horizontally. Sidewalk frames are 6' wide. Rental rates will be lower for jobs over three months' duration.

For jobs under twenty-five frames, add 50% to rental cost. These figures do not include accessories which are listed separately below. Large quantities for long periods can reduce rental rates by 20%.

Item	Unit	Monthly Rent	Per 1,000 S. F. of Building Front	
			No. of Pieces	Rental per Month
5' Wide Standard Frame, 6'-4" High	Ea.	$ 5.05	24	$121.20
Leveling Jack & Plate		2.00	24	48.00
Cross Brace		1.00	44	44.00
Side Arm Bracket, 21"		2.00	12	24.00
Guardrail Post		1.00	12	12.00
Guardrail, 7' section		1.00	22	22.00
Stairway Section		40.00	2	80.00
Walk-Thru Frame Guardrail		2.50	2	5.00
			Total	$356.20
			Per C.S.F., 1 Use/Mo.	$ 35.62

Scaffolding is often used as falsework over 15' high during construction of cast-in-place concrete beams and slabs. Two-foot wide scaffolding is generally used for heavy beam construction. The span between frames depends upon the load to be carried, with a maximum span of 5'.

Heavy duty shoring frames with a capacity of 10,000#/leg can be spaced up to 10' O. C., depending upon form support design and loading.

Scaffolding used as horizontal shoring requires less than half the material required with conventional shoring.

On new construction, erection is done by carpenters.

Rolling towers supporting horizontal shores can reduce labor and speed the job. For maintenance work, catwalks with spans up to 70' can be supported by the rolling towers.

General: The following information on current city cost indexes is calculated for over 930 zip code locations in the United States and Canada. Index figures for both material and installation are based upon the 30 major city average of 100 and represent the cost relationship on July 1, 2008.

In addition to index adjustment, the user should consider:

1. productivity
2. management efficiency
3. competitive conditions
4. automation

5. restrictive union practices
6. unique local requirements
7. regional variations due to specific building codes

The weighted-average index is calculated from about 66 materials and equipment types and 21 building trades. The component contribution of these in a model building is tabulated in Table J1010-011 below.

If the systems component distribution of a building is unknown, the weighted-average index can be used to adjust the cost for any city.

Table J1010-011 Labor, Material, and Equipment Cost Distribution for Weighted Average Listed by System Division

Division No.	Building System	Percentage	Division No.	Building System	Percentage
A	Substructure	6.3%	D10	Services: Conveying	3.9%
B10	Shell: Superstructure	19.8	D20-40	Mechanical	22.1
B20	Exterior Closure	11.5	D50	Electrical	12.0
B30	Roofing	2.9	E	Equipment & Furnishings	2.1
C	Interior Construction	15.5	G	Site Work	3.9
				Total weighted average (Div. A-G)	100.0%

How to Use the Component Indexes: Table J1010-012 below shows how the average costs obtained from this book for each division should be adjusted for your particular building in your city. The example is a building adjusted for New York, NY. Indexes for other cities are tabulated in Division RJ1030. These indexes should also be used when compiling data on the form in Division K1010.

Table J1010-012 Adjustment of "Book" Costs to a Particular City

Systems Division Number	System Description	New York, NY							
		City Cost Index	Cost Factor	Book Cost	Adjusted Cost	City Cost Index	Cost Factor	Book Cost	Adjusted Cost
A	Substructure	142.3	1.423	$ 93,400	$ 132,900				
B10	Shell: Superstructure	126.9	1.269	258,100	327,500				
B20	Exterior Closure	137.7	1.377	174,900	240,800				
B30	Roofing	130.6	1.306	38,600	50,400				
C	Interior Construction	135.9	1.359	249,000	338,400				
D10	Services: Conveying	108.8	1.088	49,900	54,300				
D20-40	Mechanical	125.8	1.258	309,100	388,800				
D50	Electrical	140.4	1.404	172,900	242,800				
E	Equipment	104.7	1.047	26,200	27,400				
G	Site Work	124.0	1.240	73,500	91,100				
Total Cost (Div. A-G)				$1,445,600	$1,894,400				

Alternate Method: Use the weighted-average index for your city rather than the component indexes. **For example:**

(Total weighted-average index A-G)/100 x (total cost) = Adjusted city cost
New York, NY 1.307 x $1,445,600 = $1,889,400

City to City: To convert known or estimated costs from one city to another, cost indexes can be used as follows:

$$\text{Unknown City Cost} = \text{Known Cost} \times \frac{\text{Unknown City Index}}{\text{Known City Index}}$$

For example: If the building cost in Boston, MA, is $2,000,000, how much would a duplicated building cost in Los Angeles, CA?

$$\text{L.A. Cost} = \$2,000,000 \times \frac{(\text{Los Angeles})\ 108.3}{(\text{Boston})\ 115.4} = \$1,877,000$$

The table below lists both the Means City Cost Index based on January 1, 1993 = 100 as well as the computed value of an index based on January 1, 2009 costs. Since the January 1, 2009 figure is estimated, space is left to write in the actual index figures as they become available through either the quarterly *Means Construction Cost Indexes* or as printed in the *Engineering News-Record*. To compute the actual index based on January 1, 2009 = 100, divide the Quarterly City Cost Index for a particular year by the actual January 1, 2009 Quarterly City Cost Index. Space has been left to advance the index figures as the year progresses.

Table J1020-011 Historical Cost Indexes

Year	Historical Cost Index Jan. 1, 1993 = 100		Current Index Based on Jan. 1, 2009 = 100		Year	Historical Cost Index Jan. 1, 1993 = 100	Current Index Based on Jan. 1, 2009 = 100		Year	Historical Cost Index Jan. 1, 1993 = 100	Current Index Based on Jan. 1, 2009 = 100	
	Est.	Actual	Est.	Actual		Actual	Est.	Actual		Actual	Est.	Actual
Oct 2009					July 1994	104.4	60.3		July 1976	46.9	27.1	
July 2009					1993	101.7	54.7		1975	44.8	24.1	
April 2009					1992	99.4	53.5		1974	41.4	22.3	
Jan 2009	185.9		100.0	100.0	1991	96.8	52.1		1973	37.7	20.3	
July 2008		180.4	97.0		1990	94.3	50.7		1972	34.8	18.7	
2007		169.4	91.1		1989	92.1	49.6		1971	32.1	17.3	
2006		162.0	87.1		1988	89.9	48.3		1970	28.7	15.4	
2005		151.6	81.5		1987	87.7	47.2		1969	26.9	14.5	
2004		143.7	77.3		1986	84.2	45.3		1968	24.9	13.4	
2003		132.0	71.0		1985	82.6	44.4		1967	23.5	12.6	
2002		128.7	69.2		1984	82.0	44.1		1966	22.7	12.2	
2001		125.1	67.3		1983	80.2	43.1		1965	21.7	11.7	
2000		120.9	65.0		1982	76.1	41.0		1964	21.2	11.4	
1999		117.6	63.3		1981	70.0	37.6		1963	20.7	11.1	
1998		115.1	61.9		1980	62.9	33.8		1962	20.2	10.9	
1997		112.8	60.7		1979	57.8	31.1		1961	19.8	10.7	
1996		110.2	59.3		1978	53.5	28.8		1960	19.7	10.6	
1995		107.6	57.9		July 1977	49.5	26.6		1959	19.3	10.4	

To find the **current cost** from a project built previously in either the same city or a different city, the following formula is used:

$$\text{Present Cost (City X)} = \frac{\text{Current HCI} \times \text{CCI (City X)}}{\text{Previous HCI} \times \text{CCI (City Y)}} \times \text{Former Cost (City Y)}$$

For example: Find the construction cost of a building to be built in San Francisco, CA, as of January 1, 2009 when the identical building cost $500,000 in Boston, MA, on July 1, 1968.

$$\text{Jan. 1, 2009 (San Francisco)} = \frac{(\text{San Francisco})\ 185.9 \times 123.8}{(\text{Boston})\ 24.9 \times 115.4} \times \$500,000 = \$4,004,500$$

Note: The City Cost Indexes for Canada can be used to convert U.S. national averages to local costs in Canadian dollars.

To Project Future Construction Costs: Using the results of the last five years average percentage increase as a basis, an average increase of 5.1% could be used.

The historical index figures above are compiled from the Means Construction Index Service.

ALABAMA

DIV. NO.	BUILDING SYSTEMS	BIRMINGHAM MAT.	INST.	TOTAL	HUNTSVILLE MAT.	INST.	TOTAL	MOBILE MAT.	INST.	TOTAL	MONTGOMERY MAT.	INST.	TOTAL	TUSCALOOSA MAT.	INST.	TOTAL
A	Substructure	94.6	77.9	85.2	88.8	74.4	80.7	94.3	66.5	78.6	95.2	63.2	77.1	91.0	66.0	76.9
B10	Shell: Superstructure	99.0	83.4	93.2	99.0	81.0	92.4	98.8	71.1	88.7	98.8	73.8	89.6	98.8	75.3	90.2
B20	Exterior Closure	89.6	81.6	85.9	89.4	66.9	79.0	89.8	61.9	76.9	89.1	46.4	69.3	89.0	57.8	74.5
B30	Roofing	94.6	87.0	91.9	92.2	79.7	87.7	92.2	79.4	87.6	91.6	72.3	84.7	92.5	75.3	86.3
C	Interior Construction	96.6	70.6	85.7	96.5	65.4	83.4	99.2	59.1	82.3	98.4	48.6	77.5	96.7	52.1	77.9
D10	Services: Conveying	100.0	89.9	97.3	100.0	86.1	96.3	100.0	84.6	95.9	100.0	82.8	95.4	100.0	85.0	96.3
D20 - 40	Mechanical	99.9	67.3	86.9	99.8	62.1	84.8	99.8	63.7	85.4	99.9	40.6	76.3	99.9	40.4	76.2
D50	Electrical	108.6	60.1	85.0	100.5	67.3	84.3	100.5	58.4	80.0	103.4	71.0	87.6	100.0	60.1	80.5
E	Equipment & Furnishings	100.0	69.7	98.4	100.0	66.9	98.3	100.0	58.5	97.8	100.0	49.9	97.4	100.0	51.9	97.5
G	Site Work	87.7	91.6	90.3	83.5	91.1	88.5	95.5	87.3	90.1	93.8	87.3	89.6	84.1	90.5	88.3
A-G	WEIGHTED AVERAGE	98.3	74.9	88.5	97.0	71.1	86.1	97.9	65.9	84.5	98.0	59.2	81.7	97.0	60.5	81.7

ALASKA / ARIZONA

DIV. NO.	BUILDING SYSTEMS	ANCHORAGE MAT.	INST.	TOTAL	FAIRBANKS MAT.	INST.	TOTAL	JUNEAU MAT.	INST.	TOTAL	FLAGSTAFF MAT.	INST.	TOTAL	MESA/TEMPE MAT.	INST.	TOTAL
A	Substructure	154.8	114.4	132.0	142.2	115.7	127.2	147.1	114.4	128.6	96.3	78.8	86.4	101.4	74.1	86.0
B10	Shell: Superstructure	130.5	105.6	121.4	128.3	106.5	120.3	129.0	105.5	120.4	96.0	73.3	87.7	95.4	71.0	85.3
B20	Exterior Closure	154.6	114.5	136.0	154.1	115.9	136.3	150.2	114.5	133.7	124.9	63.3	96.3	111.3	51.3	83.5
B30	Roofing	181.5	110.7	156.2	179.2	113.9	155.8	179.8	110.7	155.1	99.0	68.0	87.9	107.6	58.0	89.8
C	Interior Construction	134.8	114.4	126.3	134.3	116.8	127.0	132.8	114.4	125.1	100.9	61.4	84.3	98.3	60.0	82.2
D10	Services: Conveying	100.0	109.7	102.6	100.0	110.4	102.8	100.0	109.7	102.6	100.0	83.8	95.7	100.0	78.5	94.2
D20 - 40	Mechanical	100.5	103.0	101.5	100.4	109.8	104.2	100.5	94.4	98.1	100.2	76.8	90.9	100.1	66.6	86.8
D50	Electrical	140.0	109.5	125.1	148.2	109.5	129.4	141.5	109.5	125.9	101.6	61.3	81.9	92.5	61.3	77.3
E	Equipment & Furnishings	100.0	113.5	100.7	100.0	115.6	100.8	100.0	113.5	100.7	100.0	64.7	98.1	100.0	63.9	98.1
G	Site Work	141.9	126.4	131.7	128.6	126.5	127.2	141.1	126.4	131.4	81.1	102.3	95.0	87.0	105.0	98.8
A-G	WEIGHTED AVERAGE	128.3	110.3	120.8	127.6	112.6	121.3	127.0	108.5	119.2	101.4	71.6	88.9	99.2	66.7	85.6

ARIZONA / ARKANSAS

DIV. NO.	BUILDING SYSTEMS	PHOENIX MAT.	INST.	TOTAL	PRESCOTT MAT.	INST.	TOTAL	TUCSON MAT.	INST.	TOTAL	FORT SMITH MAT.	INST.	TOTAL	JONESBORO MAT.	INST.	TOTAL
A	Substructure	101.2	81.1	89.9	94.8	66.9	79.1	98.3	80.8	88.4	90.9	63.4	75.3	88.8	69.2	77.7
B10	Shell: Superstructure	96.7	75.3	88.8	95.8	63.3	83.9	95.5	74.5	87.8	101.0	63.7	87.3	94.8	73.5	87.0
B20	Exterior Closure	104.9	66.2	87.0	109.1	51.7	82.5	103.7	63.5	85.0	92.3	56.8	75.8	90.9	52.6	73.1
B30	Roofing	107.5	68.5	93.6	97.6	55.6	82.6	108.9	64.9	93.2	99.2	51.3	82.1	102.6	57.9	86.6
C	Interior Construction	100.3	66.5	86.1	99.6	48.4	78.1	95.5	64.3	82.4	95.7	46.5	75.0	93.0	56.7	77.7
D10	Services: Conveying	100.0	84.2	95.8	100.0	82.0	95.2	100.0	84.3	95.8	100.0	74.3	93.1	100.0	56.8	88.4
D20 - 40	Mechanical	100.1	76.9	90.9	100.2	71.2	88.7	100.1	68.8	87.7	100.1	48.1	79.5	100.3	48.1	79.6
D50	Electrical	100.3	64.5	82.9	101.3	61.2	81.8	94.8	58.8	77.3	95.8	75.3	85.8	103.8	52.2	78.7
E	Equipment & Furnishings	100.0	69.1	98.4	100.0	48.1	97.3	100.0	68.9	98.4	100.0	37.8	96.7	100.0	55.2	97.6
G	Site Work	87.4	106.3	99.8	70.5	100.9	90.5	83.1	105.8	98.0	81.2	91.1	87.7	102.4	95.7	98.0
A-G	WEIGHTED AVERAGE	99.9	74.0	89.0	99.2	63.7	84.3	98.0	70.6	86.5	97.4	60.1	81.8	96.9	60.2	81.5

ARKANSAS / CALIFORNIA

DIV. NO.	BUILDING SYSTEMS	LITTLE ROCK MAT.	INST.	TOTAL	PINE BLUFF MAT.	INST.	TOTAL	TEXARKANA MAT.	INST.	TOTAL	ANAHEIM MAT.	INST.	TOTAL	BAKERSFIELD MAT.	INST.	TOTAL
A	Substructure	91.8	72.9	81.1	85.4	72.8	78.2	92.2	59.4	73.7	98.9	120.6	111.2	106.7	119.8	114.1
B10	Shell: Superstructure	97.4	70.5	87.5	97.9	70.2	87.8	93.2	61.6	81.6	105.1	112.9	107.9	102.2	112.2	105.9
B20	Exterior Closure	91.0	58.2	75.8	103.7	58.2	82.6	89.9	37.6	65.6	101.6	114.9	107.8	108.0	115.9	111.7
B30	Roofing	98.5	55.7	83.2	98.0	55.7	82.9	98.9	47.4	80.5	102.6	119.1	108.5	100.5	110.6	104.1
C	Interior Construction	95.4	69.0	84.3	91.1	69.0	81.8	95.2	47.8	75.3	106.0	122.2	112.8	106.0	118.1	111.1
D10	Services: Conveying	100.0	74.3	93.1	100.0	74.3	93.1	100.0	40.1	83.9	100.0	113.1	103.5	100.0	113.7	103.7
D20 - 40	Mechanical	100.1	66.4	86.7	100.1	50.8	80.5	100.1	35.6	74.5	100.1	115.5	106.3	100.1	99.9	100.0
D50	Electrical	100.5	76.6	88.9	95.1	76.6	86.1	97.0	42.0	70.2	91.5	105.2	98.2	104.2	97.6	101.0
E	Equipment & Furnishings	100.0	72.3	98.5	100.0	72.3	98.5	100.0	47.7	97.2	100.0	126.5	101.4	100.0	125.9	101.4
G	Site Work	88.7	91.1	90.3	83.2	91.1	88.4	96.7	88.0	91.0	103.3	109.6	107.4	108.6	108.0	108.2
A-G	WEIGHTED AVERAGE	97.1	69.9	85.7	96.9	66.6	84.2	95.9	48.8	76.1	101.5	114.7	107.0	103.3	109.4	105.8

CALIFORNIA

DIV. NO.	BUILDING SYSTEMS	FRESNO MAT.	INST.	TOTAL	LOS ANGELES MAT.	INST.	TOTAL	OAKLAND MAT.	INST.	TOTAL	OXNARD MAT.	INST.	TOTAL	REDDING MAT.	INST.	TOTAL
A	Substructure	101.7	118.2	111.0	106.0	120.4	114.1	122.6	124.9	123.9	106.5	119.7	113.9	114.7	117.7	116.4
B10	Shell: Superstructure	105.6	111.3	107.7	104.6	112.3	107.5	103.5	118.4	109.0	102.2	112.3	105.9	107.7	110.6	108.8
B20	Exterior Closure	112.3	116.2	114.1	101.4	120.3	110.2	134.8	126.3	130.8	110.1	110.3	110.2	115.4	109.5	112.7
B30	Roofing	96.4	113.1	102.4	104.3	119.8	109.9	107.4	126.1	114.1	105.8	115.7	109.3	106.3	111.0	108.0
C	Interior Construction	109.3	125.2	116.0	103.6	122.6	111.6	109.9	138.7	122.0	106.2	121.4	112.6	108.2	122.1	114.1
D10	Services: Conveying	100.0	126.2	107.1	100.0	112.5	103.4	100.0	127.8	107.5	100.0	113.4	103.6	100.0	126.2	107.1
D20 - 40	Mechanical	100.3	111.6	104.8	100.3	115.5	106.3	100.3	143.0	117.3	100.2	115.6	106.3	100.2	103.5	101.5
D50	Electrical	90.7	94.5	92.5	100.4	114.5	107.3	104.9	130.0	117.1	96.4	99.5	97.9	99.4	98.1	98.8
E	Equipment & Furnishings	100.0	126.0	101.4	100.0	126.8	101.4	100.0	145.6	102.4	100.0	126.0	101.4	100.0	126.0	101.4
G	Site Work	108.2	107.4	107.7	99.6	110.3	106.6	142.5	105.7	118.3	108.5	106.5	107.2	112.2	106.4	108.4
A-G	WEIGHTED AVERAGE	103.2	112.6	107.2	102.2	116.7	108.3	108.8	129.5	117.5	102.9	112.8	107.0	105.8	109.8	107.5

CALIFORNIA

DIV. NO.	BUILDING SYSTEMS	RIVERSIDE MAT.	INST.	TOTAL	SACRAMENTO MAT.	INST.	TOTAL	SAN DIEGO MAT.	INST.	TOTAL	SAN FRANCISCO MAT.	INST.	TOTAL	SAN JOSE MAT.	INST.	TOTAL
A	Substructure	103.6	120.2	113.0	104.9	119.5	113.2	104.9	108.9	107.2	128.6	127.4	127.9	119.9	124.7	122.6
B10	Shell: Superstructure	106.4	112.5	108.6	97.1	112.0	102.6	103.8	106.1	104.7	109.4	121.0	113.7	103.8	119.9	109.7
B20	Exterior Closure	98.3	111.1	104.2	121.4	116.9	119.3	104.4	110.8	107.4	135.8	134.3	135.1	125.3	128.4	126.7
B30	Roofing	102.8	114.6	107.1	112.9	115.3	113.7	108.5	106.4	107.7	110.0	138.8	120.3	102.1	137.7	114.8
C	Interior Construction	106.2	122.2	113.0	115.0	123.9	118.7	104.0	109.6	106.4	113.2	141.7	125.2	100.7	140.4	117.4
D10	Services: Conveying	100.0	113.1	103.5	100.0	126.9	107.2	100.0	112.2	103.3	100.0	128.4	107.6	100.0	127.1	107.3
D20 - 40	Mechanical	100.1	115.5	106.2	100.1	112.4	105.0	100.3	115.8	105.6	100.3	164.9	126.0	100.2	145.6	118.3
D50	Electrical	91.6	96.3	93.9	99.6	103.7	101.6	97.7	104.1	100.8	104.9	162.2	132.8	103.1	141.0	121.6
E	Equipment & Furnishings	100.0	126.1	101.4	100.0	126.4	101.4	100.0	108.1	102.5	100.0	147.4	102.5	100.0	146.3	102.5
G	Site Work	101.8	108.0	105.9	114.4	110.9	112.1	105.9	104.2	104.8	144.3	111.3	122.6	141.6	103.4	116.5
A-G	WEIGHTED AVERAGE	101.6	112.7	106.3	104.9	114.6	108.9	102.4	108.8	105.1	111.2	141.3	123.8	106.0	132.5	117.1

DIV. NO.	BUILDING SYSTEMS	CALIFORNIA									COLORADO					
		SANTA BARBARA			STOCKTON			VALLEJO			COLORADO SPRINGS			DENVER		
		MAT.	INST.	TOTAL	MAT.	INST.	TOTAL	MAT.	INST.	TOTAL	MAT.	INST.	TOTAL	MAT.	INST.	TOTAL
A	Substructure	106.3	119.9	114.0	106.3	118.2	113.0	112.0	124.9	119.3	101.0	84.0	91.4	98.1	86.7	91.7
B10	Shell: Superstructure	102.5	112.4	106.1	103.7	111.5	106.5	101.3	116.5	106.9	101.5	84.1	95.1	103.1	84.6	96.3
B20	Exterior Closure	108.6	114.4	111.3	110.0	110.1	110.0	104.8	121.8	112.7	108.1	77.9	94.1	105.2	80.4	93.7
B30	Roofing	101.9	111.1	105.2	105.8	112.3	108.1	110.1	123.4	114.8	100.0	79.5	92.6	98.9	75.3	90.4
C	Interior Construction	107.0	117.8	111.6	106.6	121.5	112.9	116.2	138.5	125.6	99.6	78.2	90.6	101.2	83.3	93.6
D10	Services: Conveying	100.0	113.4	103.6	100.0	126.2	107.1	100.0	126.6	107.1	100.0	88.4	96.9	100.0	89.2	97.1
D20 - 40	Mechanical	100.2	115.5	106.3	100.2	103.7	101.6	100.2	117.5	107.1	100.1	75.1	90.2	100.0	86.7	94.7
D50	Electrical	88.4	100.3	94.2	99.1	111.7	105.2	95.6	116.5	105.8	104.9	83.3	94.3	106.7	85.4	96.3
E	Equipment & Furnishings	100.0	126.0	101.4	100.0	126.8	101.4	100.0	145.0	102.4	100.0	83.5	99.1	100.0	83.8	99.1
G	Site Work	108.4	108.1	108.2	106.0	107.3	106.9	114.8	110.2	111.8	91.1	94.2	93.1	90.3	102.4	98.2
A-G	**WEIGHTED AVERAGE**	101.9	112.9	106.5	103.5	112.0	107.0	104.1	121.6	111.4	101.5	81.0	92.9	101.8	85.5	95.0

DIV. NO.	BUILDING SYSTEMS	COLORADO												CONNECTICUT		
		FORT COLLINS			GRAND JUNCTION			GREELEY			PUEBLO			BRIDGEPORT		
		MAT.	INST.	TOTAL	MAT.	INST.	TOTAL	MAT.	INST.	TOTAL	MAT.	INST.	TOTAL	MAT.	INST.	TOTAL
A	Substructure	108.1	80.9	92.8	114.4	81.8	96.0	96.2	64.5	78.3	105.5	84.1	93.4	100.4	118.1	110.4
B10	Shell: Superstructure	101.7	79.4	93.6	101.5	79.3	93.3	99.3	68.8	88.1	100.8	84.8	94.9	95.9	122.7	105.7
B20	Exterior Closure	114.1	58.1	88.1	129.6	59.7	97.2	104.9	42.6	76.0	107.2	76.5	93.0	112.6	130.5	120.9
B30	Roofing	99.7	66.5	87.8	103.7	62.6	89.0	99.1	56.7	83.9	103.3	79.2	94.6	94.5	126.9	106.1
C	Interior Construction	98.6	71.4	87.2	104.8	74.7	92.2	97.9	48.5	77.1	101.5	80.0	92.4	102.6	123.5	111.4
D10	Services: Conveying	100.0	87.0	96.5	100.0	88.4	96.9	100.0	83.3	95.5	100.0	89.4	97.1	100.0	110.6	102.9
D20 - 40	Mechanical	100.0	79.1	91.7	99.9	65.2	86.1	100.0	74.1	89.7	99.9	68.0	87.2	100.2	114.4	105.9
D50	Electrical	101.2	85.3	93.5	94.7	61.7	78.7	101.2	85.3	93.5	95.7	77.6	86.9	97.0	109.9	103.3
E	Equipment & Furnishings	100.0	78.1	98.8	100.0	76.9	98.8	100.0	47.6	97.2	100.0	83.2	99.1	100.0	119.0	101.0
G	Site Work	101.7	97.0	98.6	120.7	99.6	106.9	89.0	95.6	93.4	112.2	93.3	99.8	97.3	104.3	101.9
A-G	**WEIGHTED AVERAGE**	102.2	77.3	91.7	104.9	72.0	91.1	99.7	67.0	86.0	101.3	79.0	91,9	100.3	118.6	108.0

DIV. NO.	BUILDING SYSTEMS	CONNECTICUT														
		BRISTOL			HARTFORD			NEW BRITAIN			NEW HAVEN			NORWALK		
		MAT.	INST.	TOTAL	MAT.	INST.	TOTAL	MAT.	INST.	TOTAL	MAT.	INST.	TOTAL	MAT.	INST.	TOTAL
A	Substructure	96.6	117.9	108.7	97.5	117.9	109.1	97.5	117.9	109.1	98.5	115.5	108.1	99.4	120.6	111.4
B10	Shell: Superstructure	95.1	122.4	105.1	99.0	122.4	107.6	92.7	122.4	103.6	93.1	121.2	103.4	95.7	124.0	106.1
B20	Exterior Closure	109.1	129.9	118.8	109.4	129.9	118.9	110.1	129.9	119.3	126.1	130.5	128.1	109.0	132.0	119.6
B30	Roofing	94.7	122.7	104.7	95.2	122.7	105.0	94.7	123.2	104.9	94.7	121.5	104.3	94.7	128.6	106.8
C	Interior Construction	102.7	121.5	110.6	102.8	121.5	110.7	102.7	121.5	110.6	102.7	123.5	111.4	102.7	122.8	111.1
D10	Services: Conveying	100.0	110.6	102.9	100.0	110.6	102.9	100.0	110.6	102.9	100.0	110.6	102.9	100.0	110.6	102.9
D20 - 40	Mechanical	100.2	114.4	105.8	100.1	114.4	105.8	100.2	114.4	105.8	100.2	114.4	105.8	100.2	114.5	105.9
D50	Electrical	97.0	109.6	103.2	96.9	110.5	103.5	97.1	109.6	103.2	97.0	109.6	103.1	97.0	108.3	102.5
E	Equipment & Furnishings	100.0	119.0	101.0	100.0	119.0	101.0	100.0	119.0	101.0	100.0	119.0	101.0	100.0	118.9	101.0
G	Site Work	96.3	104.3	101.6	94.1	104.3	100.8	96.5	104.3	101.6	96.4	104.8	101.9	97.0	104.3	101.8
A-G	**WEIGHTED AVERAGE**	99.6	118.0	107.3	100.5	118.1	107.9	99.2	118.0	107.1	101.0	118.0	108.2	99.9	119.0	107.9

DIV. NO.	BUILDING SYSTEMS	CONNECTICUT						D.C.			DELAWARE			FLORIDA		
		STAMFORD			WATERBURY			WASHINGTON			WILMINGTON			DAYTONA BEACH		
		MAT.	INST.	TOTAL	MAT.	INST.	TOTAL	MAT.	INST.	TOTAL	MAT.	INST.	TOTAL	MAT.	INST.	TOTAL
A	Substructure	100.4	120.8	111.9	100.3	118.0	110.3	120.5	87.6	101.9	98.1	100.1	99.2	95.7	81.1	87.4
B10	Shell: Superstructure	95.9	124.2	106.3	95.9	122.5	105.7	102.6	98.7	101.2	99.6	107.5	102.5	99.8	87.6	95.3
B20	Exterior Closure	109.4	132.0	119.9	109.4	130.5	119.2	108.5	86.0	98.0	107.5	98.0	103.1	91.0	73.2	82.8
B30	Roofing	94.7	128.6	106.8	94.7	123.2	104.9	105.2	84.5	97.8	99.7	104.3	101.3	94.9	79.8	89.4
C	Interior Construction	102.7	122.8	111.1	102.6	123.5	111.4	109.1	85.8	99.3	97.3	101.8	99.2	101.7	78.8	92.1
D10	Services: Conveying	100.0	110.6	102.9	100.0	110.6	102.9	100.0	101.4	100.4	100.0	105.8	101.5	100.0	84.0	95.7
D20 - 40	Mechanical	100.2	114.5	105.9	100.2	114.4	105.9	100.1	93.8	97.6	100.2	112.4	105.1	99.9	69.7	87.8
D50	Electrical	97.0	154.5	125.0	96.6	109.9	103.1	103.4	101.5	102.5	105.3	108.1	106.6	99.2	62.9	81.6
E	Equipment & Furnishings	100.0	121.2	101.1	100.0	119.0	101.0	100.0	81.6	99.0	100.0	103.9	100.2	100.0	79.6	98.9
G	Site Work	97.6	104.4	102.0	96.9	104.3	101.8	110.8	91.3	98.0	91.4	108.9	102.9	113.2	89.1	97.3
A-G	**WEIGHTED AVERAGE**	100.0	125.4	110.7	99.9	118.5	107.7	104.6	92.8	99.7	100.6	105.9	102.8	99.1	76.5	89.6

DIV. NO.	BUILDING SYSTEMS	FLORIDA														
		FORT LAUDERDALE			JACKSONVILLE			MELBOURNE			MIAMI			ORLANDO		
		MAT.	INST.	TOTAL	MAT.	INST.	TOTAL	MAT.	INST.	TOTAL	MAT.	INST.	TOTAL	MAT.	INST.	TOTAL
A	Substructure	96.0	74.2	83.7	96.2	67.0	79.7	107.5	82.3	93.3	101.5	75.0	86.5	112.8	80.9	94.8
B10	Shell: Superstructure	100.2	84.3	94.4	98.7	71.9	88.9	109.1	88.4	101.5	105.7	84.7	98.0	108.3	86.9	100.4
B20	Exterior Closure	89.4	69.1	80.0	91.0	58.8	76.0	92.2	78.4	85.8	89.1	72.3	81.3	94.3	72.2	84.1
B30	Roofing	94.9	84.9	91.3	95.0	66.2	84.7	95.2	86.9	92.2	103.1	81.1	95.2	95.9	80.2	90.3
C	Interior Construction	100.0	65.8	85.6	101.7	56.8	82.8	100.4	82.2	92.7	100.5	66.7	86.3	102.3	78.8	92.4
D10	Services: Conveying	100.0	89.1	97.1	100.0	82.0	95.2	100.0	85.6	96.1	100.0	89.8	97.3	100.0	84.0	95.7
D20 - 40	Mechanical	99.9	67.2	86.9	99.9	48.1	79.3	99.9	71.9	88.7	99.9	72.1	88.8	99.8	64.3	85.7
D50	Electrical	99.2	75.5	87.7	98.9	68.2	84.0	100.5	72.9	87.1	107.1	76.3	92.1	107.0	40.4	74.6
E	Equipment & Furnishings	100.0	65.9	98.2	100.0	58.0	97.8	100.0	80.1	98.9	100.0	66.0	98.2	100.0	81.1	99.0
G	Site Work	100.5	78.3	85.9	113.3	88.3	96.9	121.7	88.3	99.8	104.0	78.5	87.3	116.5	88.0	97.8
A-G	**WEIGHTED AVERAGE**	98.5	73.6	88.0	98.8	63.1	83.8	102.0	80.0	92.7	101.2	75.4	90.3	103.1	72.0	90.0

DIV. NO.	BUILDING SYSTEMS	FLORIDA														
		PANAMA CITY			PENSACOLA			ST. PETERSBURG			TALLAHASSEE			TAMPA		
		MAT.	INST.	TOTAL	MAT.	INST.	TOTAL	MAT.	INST.	TOTAL	MAT.	INST.	TOTAL	MAT.	INST.	TOTAL
A	Substructure	101.6	55.0	75.3	115.3	64.3	86.5	103.4	62.8	80.4	101.5	58.7	77.3	101.5	80.4	89.6
B10	Shell: Superstructure	100.5	58.8	85.2	104.4	70.0	91.8	104.0	69.7	91.4	95.6	66.8	85.0	102.8	88.3	97.5
B20	Exterior Closure	98.0	45.0	73.4	106.0	55.2	82.4	111.3	51.1	83.3	92.4	49.3	72.4	91.6	73.6	83.2
B30	Roofing	95.2	47.5	78.2	95.2	59.2	82.3	94.7	56.4	81.0	100.2	78.3	92.4	95.0	92.5	94.1
C	Interior Construction	100.4	43.1	76.3	99.6	55.4	81.0	99.9	50.6	79.2	102.6	45.5	78.6	101.7	78.3	91.9
D10	Services: Conveying	100.0	47.5	85.9	100.0	46.8	85.7	100.0	57.9	88.7	100.0	75.6	93.5	100.0	84.8	95.9
D20 - 40	Mechanical	99.9	37.8	75.2	99.9	57.1	82.8	99.9	55.6	82.3	99.8	42.2	76.9	99.9	89.3	95.7
D50	Electrical	97.9	41.8	70.6	101.9	58.2	80.6	99.5	47.0	73.9	106.8	36.7	72.6	99.1	47.0	73.8
E	Equipment & Furnishings	100.0	42.8	97.0	100.0	56.8	97.1	100.0	48.6	97.2	100.0	43.3	97.0	100.0	83.9	99.1
G	Site Work	127.4	86.0	100.1	127.6	87.5	101.3	114.5	87.4	96.7	112.0	87.2	95.7	114.6	88.1	97.2
A-G	**WEIGHTED AVERAGE**	100.3	48.6	78.6	102.9	61.3	85.4	102.3	58.1	83.7	99.6	53.0	80.1	100.1	78.7	91.1

City Cost Indexes — RJ1030-010 Building Systems

DIV. NO.	BUILDING SYSTEMS	ALBANY MAT.	INST.	TOTAL	ATLANTA MAT.	INST.	TOTAL	AUGUSTA MAT.	INST.	TOTAL	COLUMBUS MAT.	INST.	TOTAL	MACON MAT.	INST.	TOTAL
								GEORGIA								
A	Substructure	97.0	62.9	77.7	106.9	81.7	92.7	102.9	64.6	81.3	96.9	70.6	82.0	96.7	69.9	81.5
B10	Shell: Superstructure	99.9	74.2	90.5	96.4	81.2	90.8	94.6	65.0	83.7	99.5	81.0	92.7	95.9	78.9	89.7
B20	Exterior Closure	90.6	48.2	70.9	89.8	74.5	82.7	87.4	47.7	69.0	90.2	65.2	78.6	94.5	47.6	72.7
B30	Roofing	92.2	64.7	82.4	94.2	78.1	88.4	93.8	58.2	81.0	92.3	66.9	83.2	90.9	70.4	83.6
C	Interior Construction	99.7	46.2	77.2	100.1	76.2	90.0	96.9	52.0	78.0	99.8	60.5	83.3	94.5	55.9	78.3
D10	Services: Conveying	100.0	86.5	96.4	100.0	86.9	96.5	100.0	60.1	89.3	100.0	86.5	96.4	100.0	84.1	95.7
D20-40	Mechanical	99.8	65.6	86.2	100.0	78.0	91.2	100.0	73.6	89.5	99.9	55.9	82.4	99.9	65.7	86.3
D50	Electrical	94.5	58.3	76.9	98.7	76.7	88.0	99.6	68.7	84.5	94.6	61.4	78.4	93.1	63.8	78.8
E	Equipment & Furnishings	100.0	44.9	97.1	100.0	76.9	98.8	100.0	55.4	97.6	100.0	60.5	97.9	100.0	59.5	97.9
G	Site Work	100.9	80.7	87.6	103.9	94.2	97.5	100.3	90.9	94.1	100.7	81.4	88.1	100.7	93.7	96.1
A-G	WEIGHTED AVERAGE	97.9	62.0	82.9	98.2	79.2	90.2	96.9	64.3	83.2	97.8	66.8	84.8	96.5	66.6	83.9

DIV. NO.	BUILDING SYSTEMS	SAVANNAH MAT.	INST.	TOTAL	VALDOSTA MAT.	INST.	TOTAL	HONOLULU MAT.	INST.	TOTAL	STATES & POSS., GUAM MAT.	INST.	TOTAL	BOISE MAT.	INST.	TOTAL
			GEORGIA						HAWAII					IDAHO		
A	Substructure	106.9	62.4	81.7	96.8	57.4	74.6	149.8	121.3	133.7	177.2	85.4	125.3	95.9	82.4	88.3
B10	Shell: Superstructure	98.5	71.0	88.4	98.5	64.0	85.8	140.3	115.3	131.1	152.9	81.9	126.8	100.3	76.6	91.6
B20	Exterior Closure	90.4	56.4	74.6	97.4	50.8	75.8	116.3	124.5	120.1	147.0	51.8	102.8	120.2	65.1	94.7
B30	Roofing	95.0	58.9	82.1	92.5	66.7	83.2	115.5	121.9	117.8	123.7	76.7	106.9	95.4	73.5	87.5
C	Interior Construction	101.2	53.6	81.2	95.8	44.3	74.2	120.4	133.9	126.1	155.2	60.9	115.6	96.5	72.6	86.5
D10	Services: Conveying	100.0	62.3	89.9	100.0	62.3	89.9	100.0	113.3	103.6	100.0	96.2	99.0	100.0	70.9	92.2
D20-40	Mechanical	99.9	59.6	83.9	99.9	40.8	76.4	100.2	107.3	103.0	102.8	44.9	79.8	99.9	69.2	87.7
D50	Electrical	97.4	60.5	79.4	92.7	36.0	65.1	111.7	121.3	116.4	159.7	48.1	105.4	97.4	77.1	87.5
E	Equipment & Furnishings	100.0	52.2	97.5	100.0	42.7	97.0	100.0	138.1	102.0	100.0	56.9	97.7	100.0	76.0	98.7
G	Site Work	103.6	80.9	88.7	110.6	80.6	90.9	136.6	107.2	117.3	170.3	103.8	126.6	79.4	101.3	93.8
A-G	WEIGHTED AVERAGE	98.8	62.0	83.3	97.8	51.0	78.2	118.7	118.7	118.7	138.1	64.3	107.1	100.6	74.8	89.8

DIV. NO.	BUILDING SYSTEMS	LEWISTON MAT.	INST.	TOTAL	POCATELLO MAT.	INST.	TOTAL	TWIN FALLS MAT.	INST.	TOTAL	CHICAGO MAT.	INST.	TOTAL	DECATUR MAT.	INST.	TOTAL
			IDAHO								ILLINOIS					
A	Substructure	106.2	83.6	93.4	98.2	82.3	89.2	100.8	55.6	75.3	105.2	138.5	124.0	97.8	104.3	101.5
B10	Shell: Superstructure	97.2	82.0	91.6	106.9	76.6	95.8	107.4	59.0	89.7	93.4	139.7	110.4	95.8	105.6	99.4
B20	Exterior Closure	130.3	82.1	107.9	116.8	68.4	94.4	125.8	36.2	84.2	97.1	150.8	122.0	84.0	108.5	95.4
B30	Roofing	146.8	77.5	122.0	95.5	66.5	85.1	96.2	41.8	76.8	100.8	142.9	115.9	99.3	102.2	100.4
C	Interior Construction	140.9	67.1	109.9	97.9	72.8	87.3	99.1	35.2	72.2	98.5	153.2	121.5	97.3	109.1	102.2
D10	Services: Conveying	100.0	77.3	93.9	100.0	70.9	92.2	100.0	42.7	84.6	100.0	123.2	106.2	100.0	93.7	98.3
D20-40	Mechanical	101.0	83.4	94.0	99.9	69.2	87.7	99.9	57.6	83.1	99.9	133.8	113.4	99.9	101.2	100.4
D50	Electrical	84.8	78.0	81.5	93.8	72.8	83.6	87.9	63.6	76.1	97.5	135.8	116.1	96.5	91.2	93.9
E	Equipment & Furnishings	100.0	57.0	97.7	100.0	75.8	98.7	100.0	31.8	96.4	100.0	152.7	102.8	100.0	107.6	100.4
G	Site Work	88.1	100.6	96.3	79.8	101.3	93.9	86.0	99.2	94.7	97.7	95.4	96.2	89.4	94.9	93.0
A-G	WEIGHTED AVERAGE	109.0	80.4	97.0	101.6	74.5	90.2	102.5	54.0	82.1	97.9	138.3	114.9	96.2	102.5	98.8

DIV. NO.	BUILDING SYSTEMS	EAST ST. LOUIS MAT.	INST.	TOTAL	JOLIET MAT.	INST.	TOTAL	PEORIA MAT.	INST.	TOTAL	ROCKFORD MAT.	INST.	TOTAL	SPRINGFIELD MAT.	INST.	TOTAL
								ILLINOIS								
A	Substructure	96.3	106.5	102.1	105.3	127.8	118.0	97.6	104.3	101.4	97.8	114.7	107.3	92.6	103.3	98.7
B10	Shell: Superstructure	93.0	113.6	100.5	91.9	129.7	105.7	95.5	108.4	100.2	95.6	119.6	104.4	96.3	105.9	99.8
B20	Exterior Closure	78.3	114.9	95.3	98.3	139.2	117.3	102.7	108.6	105.5	91.2	111.1	100.4	86.4	109.3	97.0
B30	Roofing	93.5	108.3	98.8	100.7	134.7	112.8	99.4	109.6	103.0	102.1	120.2	108.6	99.1	106.3	101.7
C	Interior Construction	91.7	110.6	99.6	97.0	135.6	113.2	97.4	109.8	102.6	97.4	119.6	106.7	97.4	107.5	101.7
D10	Services: Conveying	100.0	95.2	98.7	100.0	119.8	105.3	100.0	101.4	100.4	100.0	100.1	100.0	100.0	94.0	98.4
D20-40	Mechanical	99.9	96.5	98.5	100.0	125.8	110.3	99.9	105.9	102.3	100.0	111.9	104.7	99.8	102.1	100.7
D50	Electrical	93.5	100.0	96.7	96.3	118.0	106.9	95.6	95.1	95.4	96.0	122.8	109.0	101.8	96.9	99.4
E	Equipment & Furnishings	100.0	109.0	100.5	100.0	131.9	101.7	100.0	106.2	100.3	100.0	118.6	101.0	100.0	107.2	100.4
G	Site Work	103.9	97.6	99.8	97.5	93.7	95.0	97.0	94.0	95.0	96.3	95.4	95.7	95.1	94.9	95.0
A-G	WEIGHTED AVERAGE	93.9	105.7	98.8	97.4	126.9	109.8	98.2	104.9	101.0	97.1	115.0	104.7	97.0	103.4	99.7

DIV. NO.	BUILDING SYSTEMS	ANDERSON MAT.	INST.	TOTAL	BLOOMINGTON MAT.	INST.	TOTAL	EVANSVILLE MAT.	INST.	TOTAL	FORT WAYNE MAT.	INST.	TOTAL	GARY MAT.	INST.	TOTAL
								INDIANA								
A	Substructure	93.1	82.4	87.0	91.4	80.3	85.1	90.9	92.3	91.7	96.3	80.2	87.2	95.6	102.9	99.7
B10	Shell: Superstructure	99.1	83.8	93.5	94.2	76.1	87.6	88.1	85.0	87.0	99.8	81.9	93.2	99.5	101.4	100.2
B20	Exterior Closure	87.2	80.7	84.2	98.2	74.5	87.2	96.2	83.7	90.4	88.8	77.6	83.6	87.7	104.4	95.5
B30	Roofing	96.1	78.5	89.8	92.7	80.1	88.2	96.7	85.6	92.8	95.9	82.2	91.0	95.1	104.9	98.6
C	Interior Construction	92.8	79.1	87.0	98.3	78.1	89.8	93.8	81.6	88.6	92.7	74.5	85.1	92.4	107.1	98.6
D10	Services: Conveying	100.0	91.5	97.7	100.0	80.9	94.9	100.0	86.4	96.3	100.0	80.3	94.7	100.0	83.2	95.5
D20-40	Mechanical	99.9	74.0	89.6	99.7	78.3	91.2	99.9	80.6	92.2	99.9	73.7	89.5	99.9	100.5	100.2
D50	Electrical	83.2	93.0	88.0	98.6	78.8	89.0	93.8	87.3	90.6	84.0	77.4	80.8	96.0	97.4	96.7
E	Equipment & Furnishings	100.0	79.3	98.9	100.0	76.0	98.7	100.0	80.2	99.0	100.0	73.5	98.6	100.0	107.3	100.4
G	Site Work	85.4	94.5	91.4	76.2	95.0	88.6	80.8	121.5	107.5	87.2	94.3	91.8	86.0	97.4	93.5
A-G	WEIGHTED AVERAGE	94.8	82.5	89.6	96.9	78.8	89.3	94.4	86.5	91.1	95.3	78.4	88.2	96.3	101.5	98.5

DIV. NO.	BUILDING SYSTEMS	INDIANAPOLIS MAT.	INST.	TOTAL	MUNCIE MAT.	INST.	TOTAL	SOUTH BEND MAT.	INST.	TOTAL	TERRE HAUTE MAT.	INST.	TOTAL	CEDAR RAPIDS MAT.	INST.	TOTAL
								INDIANA							IOWA	
A	Substructure	91.1	87.0	88.8	95.9	82.7	88.4	94.3	82.6	87.7	89.2	92.1	90.9	103.8	82.6	91.8
B10	Shell: Superstructure	100.2	82.1	93.6	96.2	83.7	91.6	99.2	89.4	95.6	88.4	86.3	87.6	91.3	85.4	89.1
B20	Exterior Closure	94.6	83.4	89.4	94.5	80.7	88.1	83.2	80.1	81.8	103.5	81.4	93.2	98.5	77.2	88.6
B30	Roofing	94.4	84.1	90.7	95.1	78.6	89.2	91.1	84.5	88.7	96.8	80.8	91.1	101.8	78.4	93.4
C	Interior Construction	97.9	85.1	92.5	93.9	78.9	87.6	87.6	78.6	83.8	94.0	82.1	89.0	107.1	72.6	92.6
D10	Services: Conveying	100.0	92.8	98.1	100.0	90.8	97.5	100.0	80.7	94.8	100.0	94.7	98.6	100.0	85.9	96.2
D20-40	Mechanical	99.9	85.7	94.2	99.7	74.0	89.4	99.8	80.4	92.1	99.9	82.7	93.1	100.2	81.3	92.7
D50	Electrical	98.3	93.0	95.7	89.1	79.7	84.5	103.7	82.7	93.5	92.1	83.7	88.0	96.5	78.6	87.7
E	Equipment & Furnishings	100.0	84.0	99.2	100.0	78.7	98.9	100.0	77.5	98.8	100.0	79.6	98.9	100.0	76.2	98.7
G	Site Work	84.4	98.2	93.4	76.7	94.9	88.7	87.7	94.4	92.1	82.7	121.6	108.2	93.5	96.4	95.4
A-G	WEIGHTED AVERAGE	98.0	86.7	93.2	95.5	80.6	89.3	95.6	83.1	90.4	95.0	86.6	91.5	98.8	80.8	91.2

525

IOWA

DIV. NO.	BUILDING SYSTEMS	COUNCIL BLUFFS MAT.	INST.	TOTAL	DAVENPORT MAT.	INST.	TOTAL	DES MOINES MAT.	INST.	TOTAL	DUBUQUE MAT.	INST.	TOTAL	SIOUX CITY MAT.	INST.	TOTAL
A	Substructure	106.9	72.2	87.3	101.3	92.4	96.2	97.8	85.0	90.5	100.6	81.8	90.0	104.6	67.3	83.5
B10	Shell: Superstructure	95.2	77.5	88.7	90.8	95.3	92.5	90.4	85.6	88.7	89.1	84.5	87.4	91.2	71.5	84.0
B20	Exterior Closure	99.6	76.6	89.0	96.7	85.9	91.7	93.4	71.9	83.4	97.8	70.7	85.2	95.3	54.7	76.5
B30	Roofing	101.2	62.4	87.3	101.3	87.8	96.5	102.6	75.4	92.9	101.5	68.4	89.7	101.3	52.8	84.0
C	Interior Construction	101.1	52.6	80.7	104.6	89.1	98.1	102.8	67.4	87.9	103.8	64.2	87.1	105.5	60.8	86.7
D10	Services: Conveying	100.0	88.3	96.9	100.0	88.2	96.8	100.0	85.0	96.0	100.0	83.4	95.5	100.0	83.2	95.5
D20 - 40	Mechanical	100.2	77.8	91.3	100.2	88.0	95.4	100.1	70.5	88.4	100.2	73.0	89.4	100.2	72.4	89.1
D50	Electrical	101.6	82.4	92.3	93.9	87.3	90.7	109.1	72.8	91.4	100.2	75.0	87.9	96.5	71.5	84.3
E	Equipment & Furnishings	100.0	49.4	97.3	100.0	86.6	99.3	100.0	71.8	98.5	100.0	68.0	98.3	100.0	61.8	98.0
G	Site Work	98.8	92.3	94.5	91.7	99.7	96.9	85.2	99.6	94.6	91.8	93.5	92.9	101.5	95.8	97.8
A-G	**WEIGHTED AVERAGE**	99.6	74.6	89.1	97.7	90.2	94.5	98.2	76.7	89.2	97.9	75.8	88.6	98.4	68.8	86.0

IOWA (WATERLOO) / KANSAS

DIV. NO.	BUILDING SYSTEMS	WATERLOO MAT.	INST.	TOTAL	DODGE CITY MAT.	INST.	TOTAL	KANSAS CITY MAT.	INST.	TOTAL	SALINA MAT.	INST.	TOTAL	TOPEKA MAT.	INST.	TOTAL
A	Substructure	103.6	60.9	79.5	111.8	65.1	85.4	93.6	94.1	93.9	101.2	61.2	78.6	94.5	64.1	77.3
B10	Shell: Superstructure	91.3	71.7	84.1	96.0	70.4	86.6	96.7	96.9	96.7	93.8	70.9	85.4	96.6	77.8	89.7
B20	Exterior Closure	95.3	58.1	78.0	107.7	53.3	82.5	101.6	94.0	98.1	109.2	47.7	80.7	96.7	62.1	80.7
B30	Roofing	101.1	45.2	81.1	100.0	56.5	84.4	98.5	98.2	98.4	99.5	52.9	82.9	100.5	63.7	87.4
C	Interior Construction	103.0	45.6	78.9	96.7	58.1	80.5	95.0	95.9	95.4	96.0	47.2	75.4	96.4	47.0	75.6
D10	Services: Conveying	100.0	79.1	94.4	100.0	43.1	84.7	100.0	70.9	92.2	100.0	56.8	88.4	100.0	61.5	89.7
D20 - 40	Mechanical	100.2	39.9	76.2	99.9	59.4	83.8	99.9	90.2	96.0	99.9	59.3	83.8	99.9	63.9	85.6
D50	Electrical	96.5	64.9	81.1	96.7	71.8	84.5	101.8	98.7	100.3	96.4	71.8	84.4	108.4	72.2	90.7
E	Equipment & Furnishings	100.0	43.1	97.0	100.0	66.2	98.2	100.0	100.2	100.0	100.0	51.5	97.4	100.0	43.9	97.0
G	Site Work	92.3	94.8	93.9	111.2	90.5	97.6	93.2	85.0	87.8	101.3	90.9	94.5	95.2	88.8	91.0
A-G	**WEIGHTED AVERAGE**	97.8	58.4	81.2	99.9	64.0	84.8	98.3	93.7	96.4	98.7	61.6	83.1	98.9	66.1	85.1

KANSAS (WICHITA) / KENTUCKY

DIV. NO.	BUILDING SYSTEMS	WICHITA MAT.	INST.	TOTAL	BOWLING GREEN MAT.	INST.	TOTAL	LEXINGTON MAT.	INST.	TOTAL	LOUISVILLE MAT.	INST.	TOTAL	OWENSBORO MAT.	INST.	TOTAL
A	Substructure	91.8	65.0	76.6	82.6	93.4	88.7	89.9	85.7	87.5	91.5	85.7	88.2	85.1	95.5	91.0
B10	Shell: Superstructure	96.1	73.5	87.8	90.3	88.9	89.8	93.2	86.1	90.6	99.4	86.6	94.7	84.3	88.0	85.7
B20	Exterior Closure	93.9	60.3	78.3	96.6	76.3	87.2	92.2	70.8	82.3	92.7	81.6	87.6	102.3	75.7	90.0
B30	Roofing	99.0	58.5	84.6	85.1	89.5	86.7	97.1	92.6	95.5	91.0	80.7	87.3	96.8	80.2	90.9
C	Interior Construction	95.8	55.0	78.7	92.7	82.7	88.5	94.9	73.5	85.9	93.8	80.6	88.2	91.2	77.7	85.5
D10	Services: Conveying	100.0	59.7	89.2	100.0	67.9	91.4	100.0	96.1	99.0	100.0	94.4	98.5	100.0	91.9	97.8
D20 - 40	Mechanical	99.8	57.5	83.0	99.9	85.3	94.1	99.9	68.9	87.6	99.9	83.4	93.4	99.9	73.0	89.2
D50	Electrical	103.2	71.8	87.9	92.8	79.4	86.2	93.2	76.6	85.1	104.3	79.4	92.2	92.2	80.4	86.5
E	Equipment & Furnishings	100.0	49.8	97.3	100.0	82.4	99.1	100.0	74.3	98.6	100.0	82.2	99.1	100.0	71.3	98.5
G	Site Work	93.9	90.7	91.8	68.8	96.3	86.8	77.2	99.3	91.7	66.8	96.1	86.0	81.0	122.0	107.9
A-G	**WEIGHTED AVERAGE**	97.6	64.9	83.9	93.6	84.6	89.8	95.1	78.5	88.1	97.1	83.9	91.6	93.4	83.2	89.1

LOUISIANA

DIV. NO.	BUILDING SYSTEMS	ALEXANDRIA MAT.	INST.	TOTAL	BATON ROUGE MAT.	INST.	TOTAL	LAKE CHARLES MAT.	INST.	TOTAL	MONROE MAT.	INST.	TOTAL	NEW ORLEANS MAT.	INST.	TOTAL
A	Substructure	95.7	54.5	72.4	97.5	68.4	81.1	99.1	64.5	79.6	95.4	59.8	75.2	99.2	72.9	84.3
B10	Shell: Superstructure	92.6	61.6	81.2	103.1	69.5	90.8	99.0	65.9	86.9	92.4	63.5	81.8	109.7	73.0	96.3
B20	Exterior Closure	101.7	50.8	78.1	101.2	55.3	79.9	100.3	53.1	78.4	99.2	52.6	77.6	100.8	61.6	82.6
B30	Roofing	99.3	48.1	81.0	101.3	58.8	86.1	102.8	56.5	86.2	99.3	52.3	82.5	103.6	62.3	88.9
C	Interior Construction	95.3	43.8	73.7	102.1	64.2	86.1	102.4	55.1	82.5	95.2	44.8	74.0	103.1	67.3	88.1
D10	Services: Conveying	100.0	62.3	89.9	100.0	79.3	94.4	100.0	79.4	94.5	100.0	56.7	88.4	100.0	81.4	95.0
D20 - 40	Mechanical	100.1	55.4	82.3	99.9	54.4	81.8	100.0	55.9	82.5	100.1	49.1	79.8	100.0	66.1	86.5
D50	Electrical	93.4	55.6	75.0	107.6	62.9	85.8	95.3	61.1	78.7	95.2	56.7	76.5	96.3	68.6	82.8
E	Equipment & Furnishings	100.0	38.6	96.8	100.0	67.6	98.3	100.0	55.4	97.6	100.0	39.0	96.8	100.0	68.2	98.3
G	Site Work	103.4	88.6	93.7	115.9	86.5	96.6	117.5	86.3	97.0	103.4	88.1	93.3	120.2	89.9	100.3
A-G	**WEIGHTED AVERAGE**	97.0	56.0	79.8	102.2	63.8	86.1	100.1	61.1	83.7	96.9	56.0	79.7	102.9	69.6	88.9

LOUISIANA (SHREVEPORT) / MAINE

DIV. NO.	BUILDING SYSTEMS	SHREVEPORT MAT.	INST.	TOTAL	AUGUSTA MAT.	INST.	TOTAL	BANGOR MAT.	INST.	TOTAL	LEWISTON MAT.	INST.	TOTAL	PORTLAND MAT.	INST.	TOTAL
A	Substructure	96.3	57.2	74.2	83.8	84.2	84.0	79.8	82.2	81.1	85.8	82.2	83.8	98.3	82.2	89.2
B10	Shell: Superstructure	90.2	62.5	80.1	83.2	82.9	83.1	81.9	79.6	81.0	85.6	79.6	83.4	89.3	79.6	85.7
B20	Exterior Closure	94.0	52.8	74.9	112.9	48.1	82.9	119.0	53.6	88.7	107.9	53.6	82.7	108.0	53.6	82.8
B30	Roofing	98.6	50.4	81.4	94.4	56.8	81.0	94.8	56.0	80.9	94.7	56.0	80.8	97.2	56.0	82.5
C	Interior Construction	96.1	46.8	75.4	99.6	77.9	90.4	98.8	75.2	88.9	101.9	75.2	90.7	101.9	75.2	90.6
D10	Services: Conveying	100.0	72.7	92.7	100.0	53.0	87.4	100.0	67.9	91.4	100.0	67.9	91.4	100.0	67.9	91.4
D20 - 40	Mechanical	100.0	55.3	82.3	100.0	69.1	87.7	100.2	67.9	87.3	100.2	67.9	87.3	100.0	67.9	87.2
D50	Electrical	102.6	68.7	86.1	98.3	80.1	89.5	95.3	80.0	87.9	97.2	80.0	88.8	103.7	80.0	92.2
E	Equipment & Furnishings	100.0	42.7	97.0	100.0	95.0	99.7	100.0	88.0	99.4	100.0	88.0	99.4	100.0	88.0	99.4
G	Site Work	104.8	87.9	93.7	81.6	100.5	94.0	80.7	98.1	92.1	77.7	98.1	91.1	79.8	98.1	91.8
A-G	**WEIGHTED AVERAGE**	96.8	59.2	81.0	96.1	74.3	86.9	95.8	73.7	86.5	96.4	73.7	86.9	98.5	73.7	88.1

MARYLAND / MASSACHUSETTS

DIV. NO.	BUILDING SYSTEMS	BALTIMORE MAT.	INST.	TOTAL	HAGERSTOWN MAT.	INST.	TOTAL	BOSTON MAT.	INST.	TOTAL	BROCKTON MAT.	INST.	TOTAL	FALL RIVER MAT.	INST.	TOTAL
A	Substructure	96.4	84.2	89.5	83.4	76.9	79.7	98.3	133.9	118.4	95.6	127.3	113.5	93.8	124.6	111.2
B10	Shell: Superstructure	97.4	90.1	94.7	91.6	83.3	88.6	97.5	132.6	110.4	94.6	127.0	106.5	94.2	120.9	104.0
B20	Exterior Closure	100.9	76.2	89.4	97.0	75.5	87.0	119.9	152.6	135.0	117.7	143.3	129.6	117.6	142.2	129.0
B30	Roofing	97.6	80.5	91.5	96.9	68.6	86.8	94.8	146.1	113.2	94.9	139.8	111.0	94.9	133.7	108.8
C	Interior Construction	94.9	78.4	88.0	92.0	74.4	84.6	99.5	144.5	118.4	98.2	130.9	111.9	98.1	129.8	111.4
D10	Services: Conveying	100.0	86.9	96.5	100.0	89.9	97.3	100.0	119.8	105.3	100.0	118.0	104.8	100.0	118.8	105.1
D20 - 40	Mechanical	99.9	88.5	95.4	99.9	87.9	95.1	100.2	131.5	112.6	100.2	108.5	103.5	100.2	108.3	103.4
D50	Electrical	98.2	91.7	95.0	95.4	81.4	88.6	96.3	137.1	116.2	95.4	99.5	97.4	95.2	99.4	97.3
E	Equipment & Furnishings	100.0	80.4	99.0	100.0	80.3	99.0	100.0	136.7	101.9	100.0	118.8	101.0	100.0	118.6	101.0
G	Site Work	97.6	96.8	97.0	87.5	94.2	91.9	86.1	106.8	99.7	83.5	103.9	96.9	82.4	103.8	96.4
A-G	**WEIGHTED AVERAGE**	98.2	86.0	93.1	94.9	81.6	89.3	100.5	135.9	115.4	99.2	120.7	108.2	99.0	118.9	107.3

DIV. NO.	BUILDING SYSTEMS	MASSACHUSETTS														
		HYANNIS			LAWRENCE			LOWELL			NEW BEDFORD			PITTSFIELD		
		MAT.	INST.	TOTAL	MAT.	INST.	TOTAL	MAT.	INST.	TOTAL	MAT.	INST.	TOTAL	MAT.	INST.	TOTAL
A	Substructure	84.4	124.3	107.0	95.9	124.7	112.2	90.8	124.8	110.0	87.2	124.6	108.3	91.2	108.2	100.8
B10	Shell: Superstructure	89.4	120.5	100.8	92.8	122.7	103.8	91.6	121.6	102.6	92.8	120.9	103.1	91.5	105.1	96.5
B20	Exterior Closure	114.0	142.3	127.1	117.4	139.6	127.7	105.5	141.6	122.3	117.2	142.2	128.8	105.8	113.4	109.3
B30	Roofing	94.5	134.6	108.9	94.9	139.3	110.8	94.7	139.4	110.7	94.9	133.7	108.8	94.7	109.3	99.9
C	Interior Construction	94.2	129.3	108.9	98.0	130.1	111.5	101.7	129.8	113.5	98.1	129.8	111.4	101.7	108.9	104.7
D10	Services: Conveying	100.0	118.0	104.8	100.0	118.0	104.8	100.0	118.0	104.8	100.0	118.8	105.1	100.0	93.5	98.2
D20 - 40	Mechanical	100.2	108.2	103.4	100.2	112.9	105.3	100.2	122.9	109.2	100.2	108.3	103.4	100.2	87.5	95.1
D50	Electrical	93.5	99.4	96.4	96.7	125.4	110.7	97.1	125.4	110.9	96.3	99.4	97.8	97.1	90.8	94.1
E	Equipment & Furnishings	100.0	118.7	101.0	100.0	120.0	101.1	100.0	120.0	101.1	100.0	118.6	101.0	100.0	103.2	100.2
G	Site Work	81.2	103.8	96.1	84.3	103.9	97.1	82.8	103.9	96.6	80.9	103.8	95.9	84.0	101.9	95.7
A-G	WEIGHTED AVERAGE	96.2	118.7	105.7	98.9	123.6	109.3	97.7	125.7	109.5	98.4	118.9	107.0	97.8	101.0	99.1

DIV. NO.	BUILDING SYSTEMS	MASSACHUSETTS						MICHIGAN								
		SPRINGFIELD			WORCESTER			ANN ARBOR			DEARBORN			DETROIT		
		MAT.	INST.	TOTAL	MAT.	INST.	TOTAL	MAT.	INST.	TOTAL	MAT.	INST.	TOTAL	MAT.	INST.	TOTAL
A	Substructure	92.9	109.0	102.0	92.6	127.7	112.5	87.3	111.1	100.8	86.2	115.5	102.8	92.7	115.5	105.6
B10	Shell: Superstructure	93.9	105.2	98.0	93.9	125.1	105.3	86.7	118.2	99.3	86.5	121.6	99.3	90.6	111.6	98.3
B20	Exterior Closure	105.6	115.9	110.4	105.4	143.2	122.9	93.0	112.7	102.1	92.9	118.2	104.7	92.8	117.5	104.3
B30	Roofing	94.7	110.9	100.5	94.7	131.2	107.7	102.6	111.7	105.8	101.2	121.7	108.5	99.3	121.7	107.3
C	Interior Construction	101.7	109.4	105.0	101.7	133.2	114.9	92.5	111.4	100.5	92.5	120.0	104.1	94.0	120.0	104.9
D10	Services: Conveying	100.0	94.8	98.6	100.0	98.3	99.5	100.0	106.5	101.8	100.0	107.5	102.0	100.0	107.5	102.0
D20 - 40	Mechanical	100.2	101.9	100.9	100.2	105.4	102.3	99.9	99.5	99.7	99.9	110.7	104.2	99.9	112.8	105.0
D50	Electrical	97.2	90.8	94.1	97.2	95.7	96.5	95.2	76.9	86.3	95.2	116.0	105.3	96.5	115.9	106.0
E	Equipment & Furnishings	100.0	103.3	100.2	100.0	123.5	101.2	100.0	107.4	100.4	100.0	119.3	101.0	100.0	119.3	101.0
G	Site Work	83.3	102.1	95.6	83.2	103.8	96.8	78.9	96.0	90.2	78.6	96.2	90.2	91.2	98.5	96.0
A-G	WEIGHTED AVERAGE	98.4	104.5	101.0	98.3	118.9	107.0	93.7	104.5	98.2	93.5	115.5	102.7	95.3	114.2	103.2

DIV. NO.	BUILDING SYSTEMS	MICHIGAN														
		FLINT			GRAND RAPIDS			KALAMAZOO			LANSING			MUSKEGON		
		MAT.	INST.	TOTAL	MAT.	INST.	TOTAL	MAT.	INST.	TOTAL	MAT.	INST.	TOTAL	MAT.	INST.	TOTAL
A	Substructure	86.5	101.5	95.0	91.3	81.0	85.5	91.5	91.8	91.7	90.1	101.1	96.3	90.4	87.0	88.5
B10	Shell: Superstructure	87.0	111.2	95.8	93.6	80.3	88.7	92.4	88.9	91.1	86.0	110.6	95.1	90.5	85.1	88.5
B20	Exterior Closure	93.0	101.0	96.7	92.3	48.9	72.1	95.3	85.3	90.7	90.2	97.0	93.3	94.0	73.6	82.6
B30	Roofing	100.3	96.0	98.7	94.3	59.3	81.8	92.6	87.2	90.7	101.4	93.7	98.7	91.8	77.3	86.6
C	Interior Construction	92.2	93.1	92.6	93.7	64.0	81.2	92.5	85.3	89.4	94.6	94.7	94.7	90.8	76.1	84.6
D10	Services: Conveying	100.0	94.5	98.5	100.0	97.3	99.3	100.0	100.9	100.2	100.0	94.7	98.6	100.0	100.6	100.2
D20 - 40	Mechanical	99.9	101.4	100.5	99.9	56.3	82.6	99.9	82.3	92.9	99.8	99.8	99.8	99.8	84.2	93.6
D50	Electrical	95.2	99.3	97.2	103.2	53.6	79.1	92.6	81.8	87.3	101.2	77.2	89.5	93.2	79.1	86.3
E	Equipment & Furnishings	100.0	93.2	99.6	100.0	68.1	98.3	100.0	85.2	99.2	100.0	92.5	99.6	100.0	75.4	98.7
G	Site Work	70.4	95.2	86.7	82.4	81.6	81.9	81.3	82.3	82.0	86.3	94.7	91.8	78.9	82.2	81.1
A-G	WEIGHTED AVERAGE	93.4	100.8	96.5	96.1	65.2	83.1	94.8	85.6	91.0	94.4	96.9	95.5	93.5	81.4	88.4

DIV. NO.	BUILDING SYSTEMS	MICHIGAN			MINNESOTA											
		SAGINAW			DULUTH			MINNEAPOLIS			ROCHESTER			SAINT PAUL		
		MAT.	INST.	TOTAL	MAT.	INST.	TOTAL	MAT.	INST.	TOTAL	MAT.	INST.	TOTAL	MAT.	INST.	TOTAL
A	Substructure	85.8	96.4	91.8	104.8	107.5	106.3	107.1	123.7	116.5	105.8	105.6	105.7	107.3	118.8	113.8
B10	Shell: Superstructure	86.6	107.5	94.3	92.6	114.9	100.8	95.2	130.7	108.3	92.9	119.3	102.6	92.6	127.6	105.4
B20	Exterior Closure	93.2	91.1	92.2	97.0	115.5	105.6	98.4	131.7	113.9	96.2	117.8	106.2	104.0	131.3	116.7
B30	Roofing	101.0	87.6	96.2	102.5	115.4	107.1	102.5	135.8	114.4	102.7	100.1	101.8	103.7	133.4	114.3
C	Interior Construction	91.2	86.8	89.4	98.7	116.5	106.1	101.1	136.4	115.9	99.1	106.3	102.2	96.5	127.3	109.5
D10	Services: Conveying	100.0	92.6	98.0	100.0	92.4	98.0	100.0	101.7	100.5	100.0	96.1	98.9	100.0	101.3	100.4
D20 - 40	Mechanical	99.9	84.2	93.6	99.8	102.4	100.9	99.8	117.1	106.7	99.8	98.0	99.1	99.8	112.5	104.8
D50	Electrical	92.9	72.8	83.1	102.5	96.9	99.7	101.9	115.3	108.5	100.6	85.7	93.4	100.1	102.4	101.2
E	Equipment & Furnishings	100.0	84.6	99.2	100.0	114.4	100.8	100.0	136.1	101.9	100.0	102.8	100.1	100.0	122.8	101.2
G	Site Work	73.0	94.3	87.0	92.5	102.6	99.1	91.4	107.6	102.1	90.5	101.5	97.7	95.2	103.4	100.5
A-G	WEIGHTED AVERAGE	93.0	89.9	91.7	98.2	108.2	102.4	99.3	124.2	109.8	98.1	104.7	100.9	98.6	118.7	107.0

DIV. NO.	BUILDING SYSTEMS	MINNESOTA			MISSISSIPPI											
		ST. CLOUD			BILOXI			GREENVILLE			JACKSON			MERIDIAN		
		MAT.	INST.	TOTAL	MAT.	INST.	TOTAL	MAT.	INST.	TOTAL	MAT.	INST.	TOTAL	MAT.	INST.	TOTAL
A	Substructure	95.4	117.9	108.1	102.2	68.1	82.9	103.9	74.7	87.4	97.3	70.8	82.3	98.8	74.1	84.8
B10	Shell: Superstructure	91.5	125.7	104.0	98.4	76.6	90.4	96.6	82.5	91.4	97.3	78.0	90.2	95.7	80.1	89.9
B20	Exterior Closure	94.9	124.9	108.8	90.3	54.0	73.8	111.6	58.9	87.1	90.9	58.2	75.7	88.5	64.2	77.2
B30	Roofing	101.1	113.9	105.7	92.4	63.2	81.9	92.5	70.8	84.7	94.2	69.6	85.4	92.1	72.7	85.2
C	Interior Construction	92.2	126.7	106.7	100.2	61.4	83.9	97.1	74.0	87.4	99.5	67.3	86.0	96.6	73.4	86.8
D10	Services: Conveying	100.0	95.6	98.8	100.0	70.3	92.0	100.0	72.7	92.7	100.0	72.7	92.7	100.0	74.7	93.2
D20 - 40	Mechanical	99.5	115.8	106.0	99.9	50.0	80.0	99.9	57.4	83.0	99.9	63.4	85.4	99.9	65.9	86.4
D50	Electrical	100.8	102.4	101.6	99.9	55.5	78.3	99.2	73.2	86.5	104.6	73.2	89.3	98.9	58.9	79.4
E	Equipment & Furnishings	100.0	121.4	101.1	100.0	64.4	98.1	100.0	78.7	98.9	100.0	69.3	98.4	100.0	75.0	98.7
G	Site Work	84.6	105.2	98.2	103.2	89.2	94.0	107.0	89.3	95.4	98.4	89.3	92.4	98.8	89.6	92.8
A-G	WEIGHTED AVERAGE	95.8	117.5	104.9	98.5	62.6	83.4	100.0	70.9	87.8	98.5	69.9	86.5	96.8	70.9	85.9

DIV. NO.	BUILDING SYSTEMS	MISSOURI														
		CAPE GIRARDEAU			COLUMBIA			JOPLIN			KANSAS CITY			SPRINGFIELD		
		MAT.	INST.	TOTAL	MAT.	INST.	TOTAL	MAT.	INST.	TOTAL	MAT.	INST.	TOTAL	MAT.	INST.	TOTAL
A	Substructure	96.6	82.9	88.8	93.7	85.1	88.9	101.0	76.4	87.1	94.3	105.8	100.8	97.5	79.5	87.3
B10	Shell: Superstructure	94.2	93.1	93.8	90.2	100.7	94.1	96.3	79.6	90.2	100.9	111.4	104.8	94.6	92.1	93.7
B20	Exterior Closure	104.9	77.9	92.4	104.1	89.2	97.2	92.1	61.9	78.1	96.2	108.3	101.8	91.7	79.4	86.0
B30	Roofing	102.1	78.3	93.6	93.6	81.0	89.1	96.6	67.9	86.4	96.3	106.0	99.7	98.2	71.4	88.6
C	Interior Construction	94.4	72.9	85.4	93.2	69.4	83.2	97.8	63.2	83.2	98.1	108.0	102.3	98.1	68.6	85.7
D10	Services: Conveying	100.0	58.8	88.9	100.0	98.9	99.7	100.0	69.8	91.9	100.0	98.2	99.5	100.0	91.5	97.7
D20 - 40	Mechanical	99.9	89.9	95.8	99.9	91.3	96.5	100.1	52.6	81.2	100.0	104.1	101.6	99.9	68.3	87.4
D50	Electrical	100.9	108.3	104.5	96.0	85.3	90.8	90.9	68.8	80.1	101.2	105.2	103.1	99.7	73.1	86.7
E	Equipment & Furnishings	100.0	73.4	98.6	100.0	60.6	97.9	100.0	67.8	98.3	100.0	107.6	100.4	100.0	71.9	98.5
G	Site Work	93.0	93.4	93.3	102.5	95.2	97.7	104.2	98.7	100.6	97.9	93.2	94.8	95.9	91.2	92.8
A-G	WEIGHTED AVERAGE	98.2	87.4	93.7	96.3	88.1	92.9	97.1	68.0	84.9	99.2	106.1	102.1	97.3	77.6	89.1

Cost Indexes

MISSOURI / MONTANA

DIV. NO.	BUILDING SYSTEMS	ST. JOSEPH MAT.	INST.	TOTAL	ST. LOUIS MAT.	INST.	TOTAL	BILLINGS MAT.	INST.	TOTAL	BUTTE MAT.	INST.	TOTAL	GREAT FALLS MAT.	INST.	TOTAL
A	Substructure	94.0	91.9	92.8	95.3	106.4	101.6	104.8	74.5	87.7	113.9	76.3	92.7	117.0	74.7	93.1
B10	Shell: Superstructure	97.8	99.5	98.5	98.1	114.3	104.0	105.9	78.0	95.7	101.6	79.6	93.5	105.6	78.2	95.6
B20	Exterior Closure	96.0	92.6	94.4	95.1	111.8	102.9	106.8	70.1	89.8	101.8	73.2	88.6	106.7	71.9	90.6
B30	Roofing	96.6	92.6	95.2	102.0	108.0	104.1	101.4	68.2	89.5	101.3	72.8	91.1	101.8	67.2	89.5
C	Interior Construction	98.0	82.1	91.3	96.6	104.4	99.9	101.7	63.7	85.8	100.5	71.2	88.2	104.3	68.2	89.1
D10	Services: Conveying	100.0	95.4	98.8	100.0	104.1	101.1	100.0	74.9	93.2	100.0	75.4	93.4	100.0	76.3	93.6
D20 - 40	Mechanical	100.1	90.1	96.1	99.9	108.7	103.4	100.1	76.1	90.6	100.2	69.9	88.2	100.2	74.3	89.9
D50	Electrical	99.0	84.0	91.7	103.6	111.7	107.6	95.1	75.5	85.6	102.4	72.6	87.9	95.2	72.5	84.1
E	Equipment & Furnishings	100.0	74.9	98.7	100.0	103.2	100.2	100.0	65.1	98.2	100.0	69.6	98.4	100.0	65.1	98.2
G	Site Work	99.5	88.2	92.1	92.5	97.1	95.5	91.0	97.6	95.3	98.8	96.6	97.4	102.3	97.3	99.0
A-G	WEIGHTED AVERAGE	98.3	90.2	94.9	98.6	108.8	102.9	101.9	74.6	90.4	101.6	75.0	90.4	103.1	74.8	91.2

MONTANA / NEBRASKA

DIV. NO.	BUILDING SYSTEMS	HELENA MAT.	INST.	TOTAL	MISSOULA MAT.	INST.	TOTAL	GRAND ISLAND MAT.	INST.	TOTAL	LINCOLN MAT.	INST.	TOTAL	NORTH PLATTE MAT.	INST.	TOTAL
A	Substructure	106.0	73.9	87.9	94.1	73.1	82.2	111.0	81.2	94.2	94.3	75.7	83.8	111.3	79.8	93.5
B10	Shell: Superstructure	102.7	77.8	93.6	95.5	79.6	89.7	94.5	84.0	90.6	95.0	79.2	89.2	94.7	83.7	90.7
B20	Exterior Closure	103.2	70.2	87.9	105.7	69.0	88.7	100.7	84.4	93.1	95.3	73.5	85.2	93.5	94.6	94.0
B30	Roofing	101.5	73.6	91.5	100.6	69.4	89.4	99.1	78.8	91.8	98.5	75.5	90.3	99.0	83.6	93.5
C	Interior Construction	102.7	64.9	86.8	100.3	64.5	85.2	93.4	86.8	90.6	96.0	70.7	85.4	93.2	84.7	89.6
D10	Services: Conveying	100.0	89.4	97.1	100.0	86.9	96.5	100.0	88.8	97.0	100.0	88.8	97.0	100.0	81.9	95.1
D20 - 40	Mechanical	100.1	70.3	88.3	100.2	66.6	86.8	99.9	80.1	92.1	99.8	80.1	92.0	99.9	78.2	91.3
D50	Electrical	101.0	72.5	87.1	100.3	76.7	88.8	89.9	74.0	82.1	103.8	74.0	89.3	91.8	89.9	90.9
E	Equipment & Furnishings	100.0	64.9	98.1	100.0	60.6	97.9	100.0	83.4	99.1	100.0	61.7	98.0	100.0	83.9	99.2
G	Site Work	99.2	96.5	97.4	80.9	96.4	91.1	100.5	90.4	93.9	90.0	90.4	90.3	101.5	89.9	93.9
A-G	WEIGHTED AVERAGE	101.8	73.5	89.9	99.0	73.2	88.2	97.3	82.4	91.0	97.6	77.1	89.0	96.8	85.0	91.8

NEBRASKA / NEVADA / NEW HAMPSHIRE

DIV. NO.	BUILDING SYSTEMS	OMAHA MAT.	INST.	TOTAL	CARSON CITY MAT.	INST.	TOTAL	LAS VEGAS MAT.	INST.	TOTAL	RENO MAT.	INST.	TOTAL	MANCHESTER MAT.	INST.	TOTAL
A	Substructure	98.8	78.6	87.4	104.6	97.2	100.4	100.2	112.7	107.3	106.1	98.8	102.0	98.5	96.8	97.6
B10	Shell: Superstructure	95.6	76.8	88.7	97.4	98.7	97.9	101.6	111.7	105.3	98.2	101.9	99.5	95.0	92.1	94.0
B20	Exterior Closure	97.3	79.7	89.1	115.2	83.7	100.6	112.9	106.0	109.7	115.7	83.3	100.7	108.6	85.6	97.9
B30	Roofing	93.4	82.9	89.7	99.5	89.5	95.9	115.6	103.7	111.3	100.9	89.5	96.4	94.2	91.7	93.3
C	Interior Construction	108.1	68.5	91.4	96.9	87.9	93.1	98.4	109.7	103.1	97.8	88.5	93.9	101.7	85.4	94.9
D10	Services: Conveying	100.0	87.2	96.6	100.0	112.0	103.2	100.0	103.7	101.0	100.0	112.0	103.2	100.0	68.6	91.6
D20 - 40	Mechanical	99.7	80.0	91.8	99.9	87.3	94.9	100.0	110.2	104.1	99.9	87.5	95.0	100.0	85.0	94.1
D50	Electrical	103.6	83.2	93.7	98.8	99.4	99.1	99.0	128.0	113.1	94.5	99.4	96.9	96.7	83.4	90.2
E	Equipment & Furnishings	100.0	71.8	98.5	100.0	93.6	99.7	100.0	110.4	100.5	100.0	93.6	99.7	100.0	79.0	98.9
G	Site Work	79.6	90.7	86.9	70.1	103.3	91.9	65.5	106.7	92.6	64.1	103.3	89.8	82.8	98.4	93.1
A-G	WEIGHTED AVERAGE	99.5	78.8	90.8	99.9	93.1	97.1	101.1	112.0	105.7	99.8	93.9	97.3	99.1	87.7	94.3

NEW HAMPSHIRE / NEW JERSEY

DIV. NO.	BUILDING SYSTEMS	NASHUA MAT.	INST.	TOTAL	PORTSMOUTH MAT.	INST.	TOTAL	CAMDEN MAT.	INST.	TOTAL	ELIZABETH MAT.	INST.	TOTAL	JERSEY CITY MAT.	INST.	TOTAL
A	Substructure	91.7	96.8	94.6	82.3	99.0	91.7	85.5	116.7	103.1	85.8	118.3	104.1	85.5	118.2	104.0
B10	Shell: Superstructure	93.6	92.1	93.0	88.7	94.7	90.9	92.0	110.4	98.7	89.6	114.7	98.8	92.1	113.3	99.9
B20	Exterior Closure	108.5	85.6	97.9	103.8	88.3	96.6	105.5	118.9	111.7	115.2	119.5	117.2	104.3	119.3	111.2
B30	Roofing	94.9	91.7	93.7	94.5	116.2	102.2	94.4	120.2	103.6	95.1	126.2	106.2	94.4	126.2	105.8
C	Interior Construction	101.9	85.4	95.0	99.6	86.4	94.1	102.1	123.2	111.0	103.9	126.1	113.2	102.1	126.1	112.2
D10	Services: Conveying	100.0	68.6	91.6	100.0	70.6	92.1	100.0	108.7	102.3	100.0	117.6	104.7	100.0	117.6	104.7
D20 - 40	Mechanical	100.2	85.0	94.2	100.2	87.7	95.2	100.2	113.9	105.6	100.2	120.6	108.3	100.2	122.8	109.2
D50	Electrical	97.0	83.4	90.4	95.2	83.4	89.5	97.5	134.0	115.3	95.0	137.5	115.6	98.8	141.1	119.4
E	Equipment & Furnishings	100.0	79.0	98.9	100.0	79.3	98.9	100.0	120.9	101.1	100.0	121.3	101.1	100.0	121.4	101.1
G	Site Work	84.7	98.4	93.7	79.7	99.7	92.9	87.0	103.8	98.1	100.0	104.7	103.1	87.0	104.6	98.5
A-G	WEIGHTED AVERAGE	98.6	87.7	94.0	95.9	90.1	93.5	97.8	117.8	106.2	98.6	121.5	108.2	97.8	122.2	108.0

NEW JERSEY / NEW MEXICO

DIV. NO.	BUILDING SYSTEMS	NEWARK MAT.	INST.	TOTAL	PATERSON MAT.	INST.	TOTAL	TRENTON MAT.	INST.	TOTAL	ALBUQUERQUE MAT.	INST.	TOTAL	FARMINGTON MAT.	INST.	TOTAL
A	Substructure	99.3	118.3	110.0	98.6	118.2	109.7	91.4	117.1	105.9	103.6	80.5	90.5	106.7	80.5	91.9
B10	Shell: Superstructure	94.6	114.7	102.0	91.0	114.6	99.7	89.5	110.5	97.2	102.2	82.0	94.8	101.0	82.0	94.1
B20	Exterior Closure	109.4	119.5	114.1	107.3	119.5	112.9	103.7	118.8	110.7	111.1	67.2	90.7	116.8	67.2	93.8
B30	Roofing	94.6	126.2	105.9	94.5	120.4	104.1	90.5	120.3	101.2	100.3	76.2	91.7	100.7	76.2	91.9
C	Interior Construction	105.9	126.1	114.4	105.9	126.1	114.4	102.3	125.2	111.9	97.4	68.5	85.3	97.9	68.5	85.6
D10	Services: Conveying	100.0	117.6	104.7	100.0	117.6	104.7	100.0	108.5	102.3	100.0	76.0	93.5	100.0	76.0	93.5
D20 - 40	Mechanical	100.2	120.6	108.3	100.2	122.8	109.2	100.3	121.8	108.8	100.0	72.3	89.0	99.9	72.3	88.9
D50	Electrical	99.0	141.1	119.5	98.8	137.5	117.6	101.4	139.2	119.8	88.6	73.8	81.4	86.8	73.8	80.5
E	Equipment & Furnishings	100.0	121.4	101.1	100.0	121.3	101.1	100.0	121.3	101.1	100.0	69.2	98.4	100.0	69.2	98.4
G	Site Work	102.2	104.7	103.8	98.6	104.6	102.5	87.1	104.5	98.5	79.6	108.0	98.2	86.8	108.0	100.7
A-G	WEIGHTED AVERAGE	100.5	122.0	109.5	99.3	121.8	108.8	97.6	120.6	107.3	99.8	76.0	89.8	100.3	76.0	90.1

NEW MEXICO / NEW YORK

DIV. NO.	BUILDING SYSTEMS	LAS CRUCES MAT.	INST.	TOTAL	ROSWELL MAT.	INST.	TOTAL	SANTA FE MAT.	INST.	TOTAL	ALBANY MAT.	INST.	TOTAL	BINGHAMTON MAT.	INST.	TOTAL
A	Substructure	97.6	70.8	82.4	101.8	77.7	88.2	115.6	80.5	95.8	87.9	96.6	92.8	101.2	88.8	94.2
B10	Shell: Superstructure	98.7	70.1	88.3	99.9	75.6	91.0	102.8	82.0	95.2	93.5	99.9	95.8	93.6	101.6	96.5
B20	Exterior Closure	92.8	63.6	79.3	121.6	66.2	95.9	112.1	67.2	91.3	102.2	97.7	100.1	99.8	83.1	92.1
B30	Roofing	86.6	70.2	80.7	96.4	76.2	92.1	100.2	76.2	91.6	89.5	90.6	89.9	103.3	79.2	94.7
C	Interior Construction	100.2	65.7	85.7	96.4	67.3	84.1	96.9	68.5	85.0	95.9	85.3	91.5	94.5	79.8	88.3
D10	Services: Conveying	100.0	72.4	92.6	100.0	76.0	93.5	100.0	76.0	93.5	100.0	94.5	98.5	100.0	95.4	98.8
D20 - 40	Mechanical	100.2	71.7	88.9	99.9	72.0	88.8	99.9	72.3	88.9	100.0	93.0	97.3	100.4	79.0	91.9
D50	Electrical	88.0	57.1	72.9	87.5	73.8	80.9	101.6	73.8	88.1	106.9	93.9	100.6	102.2	86.8	94.7
E	Equipment & Furnishings	100.0	67.6	98.3	100.0	69.2	98.4	100.0	69.2	98.4	100.0	84.8	99.2	100.0	78.8	98.9
G	Site Work	90.0	91.4	90.9	90.9	108.0	102.1	85.7	108.0	100.3	72.8	105.0	94.0	95.1	95.0	95.1
A-G	WEIGHTED AVERAGE	97.0	68.5	85.0	100.3	74.3	89.3	102.0	76.0	91.1	97.4	94.7	96.3	98.1	86.9	93.4

NEW YORK

DIV. NO.	BUILDING SYSTEMS	BUFFALO			HICKSVILLE			NEW YORK			RIVERHEAD			ROCHESTER		
		MAT.	INST.	TOTAL	MAT.	INST.	TOTAL	MAT.	INST.	TOTAL	MAT.	INST.	TOTAL	MAT.	INST.	TOTAL
A	Substructure	104.9	109.7	107.6	97.9	147.5	125.9	110.7	166.6	142.3	99.6	147.5	126.6	96.8	97.0	96.9
B10	Shell: Superstructure	96.5	103.1	98.9	99.8	144.9	116.3	108.4	158.7	126.9	100.7	144.9	116.9	92.5	100.7	95.5
B20	Exterior Closure	106.8	114.3	110.3	111.5	157.8	133.0	110.7	169.0	137.7	113.4	157.8	134.0	114.6	95.9	105.9
B30	Roofing	99.2	108.0	102.4	106.3	148.4	121.3	109.5	168.5	130.6	106.9	148.4	121.7	95.0	94.9	95.0
C	Interior Construction	93.7	113.4	102.0	95.8	147.2	117.4	103.6	180.5	135.9	96.4	147.2	117.8	95.7	93.4	94.8
D10	Services: Conveying	100.0	106.8	101.8	100.0	127.0	107.3	100.0	132.7	108.8	100.0	127.0	107.3	100.0	87.9	96.7
D20 - 40	Mechanical	99.9	97.1	98.8	99.8	146.9	118.5	100.2	164.6	125.8	99.9	146.9	118.6	99.9	89.6	95.8
D50	Electrical	99.8	98.5	99.1	102.2	149.3	125.1	109.4	173.1	140.4	103.8	149.3	126.0	101.0	86.2	93.8
E	Equipment & Furnishings	100.0	112.6	100.7	100.0	142.3	102.2	100.0	188.6	104.7	100.0	142.3	102.2	100.0	92.1	99.6
G	Site Work	96.5	99.4	98.4	114.9	121.9	119.5	128.8	121.5	124.0	116.3	121.9	120.0	72.7	105.1	93.9
A-G	WEIGHTED AVERAGE	99.0	104.8	101.5	101.2	146.4	120.1	106.1	164.6	130.7	102.0	146.4	120.6	98.4	94.1	96.6

NEW YORK

DIV. NO.	BUILDING SYSTEMS	SCHENECTADY			SYRACUSE			UTICA			WATERTOWN			WHITE PLAINS		
		MAT.	INST.	TOTAL	MAT.	INST.	TOTAL	MAT.	INST.	TOTAL	MAT.	INST.	TOTAL	MAT.	INST.	TOTAL
A	Substructure	94.2	97.1	95.9	96.9	95.7	96.2	90.3	91.7	91.1	98.7	98.2	98.4	99.5	134.3	119.1
B10	Shell: Superstructure	95.2	100.1	97.0	95.3	99.2	96.7	92.9	96.8	94.3	93.9	100.6	96.4	96.7	135.2	110.8
B20	Exterior Closure	101.9	99.0	100.5	104.3	94.4	99.7	101.2	86.8	94.5	109.4	97.9	104.1	103.0	135.4	118.0
B30	Roofing	90.2	91.5	90.6	100.2	94.4	98.2	90.0	89.6	89.8	90.2	96.2	92.4	109.9	140.4	120.8
C	Interior Construction	96.8	85.6	92.1	95.6	88.1	92.4	96.8	86.3	92.4	95.5	91.5	93.8	96.2	137.7	113.7
D10	Services: Conveying	100.0	95.1	98.7	100.0	99.0	99.7	100.0	90.4	97.4	100.0	95.4	98.8	100.0	122.0	105.9
D20 - 40	Mechanical	100.2	93.9	97.7	100.2	86.9	94.9	100.2	81.7	92.9	100.2	76.0	90.6	100.4	127.3	111.1
D50	Electrical	102.2	93.9	98.2	102.2	91.0	96.8	99.7	90.4	95.2	102.2	90.9	96.7	99.0	137.5	117.8
E	Equipment & Furnishings	100.0	84.9	99.2	100.0	87.0	99.3	100.0	86.7	99.3	100.0	91.0	99.5	100.0	129.3	101.6
G	Site Work	73.2	105.1	94.2	93.9	104.6	100.9	71.2	102.5	91.8	79.5	105.1	96.3	120.6	114.9	116.9
A-G	WEIGHTED AVERAGE	97.7	93.6	96.7	98.7	93.1	96.4	96.7	89.5	93.6	96.4	92.2	95.8	99.8	132.8	113.6

		NEW YORK			NORTH CAROLINA											
DIV. NO.	BUILDING SYSTEMS	YONKERS			ASHEVILLE			CHARLOTTE			DURHAM			FAYETTEVILLE		
		MAT.	INST.	TOTAL	MAT.	INST.	TOTAL	MAT.	INST.	TOTAL	MAT.	INST.	TOTAL	MAT.	INST.	TOTAL
A	Substructure	109.0	134.2	123.3	102.3	52.9	74.4	110.7	53.8	78.5	103.9	55.7	76.7	106.0	56.9	78.3
B10	Shell: Superstructure	105.2	135.2	116.2	91.1	63.0	80.8	96.3	63.1	84.1	102.3	65.4	88.8	106.1	66.0	91.4
B20	Exterior Closure	110.6	135.6	122.2	97.7	43.9	72.7	100.9	47.5	76.1	100.4	41.3	73.0	98.2	45.6	73.8
B30	Roofing	110.2	140.4	121.0	102.2	45.4	81.9	99.7	46.5	80.7	103.0	48.3	83.4	102.1	45.8	81.9
C	Interior Construction	101.5	137.7	116.7	98.1	41.6	74.4	100.9	43.5	76.7	101.5	42.7	76.8	98.5	43.0	75.2
D10	Services: Conveying	100.0	122.0	105.9	100.0	80.0	94.6	100.0	80.3	94.7	100.0	74.3	93.1	100.0	75.3	93.4
D20 - 40	Mechanical	100.4	127.3	111.1	100.2	41.5	76.9	99.9	41.7	76.8	100.3	40.2	76.4	100.1	41.4	76.8
D50	Electrical	106.8	137.5	121.7	97.7	40.2	69.7	100.3	52.3	76.9	98.6	50.9	75.4	97.2	46.2	72.4
E	Equipment & Furnishings	100.0	129.3	101.5	100.0	39.6	96.8	100.0	42.3	96.9	100.0	41.4	96.9	100.0	41.3	96.9
G	Site Work	128.2	114.7	119.3	106.7	77.3	87.4	110.5	77.4	88.8	106.9	84.8	92.4	106.0	84.9	92.2
A-G	WEIGHTED AVERAGE	104.7	132.8	116.5	97.6	49.6	77.5	100.2	52.2	80.0	101.1	51.7	80.4	101.1	52.1	80.5

		NORTH CAROLINA												NORTH DAKOTA		
DIV. NO.	BUILDING SYSTEMS	GREENSBORO			RALEIGH			WILMINGTON			WINSTON-SALEM			BISMARCK		
		MAT.	INST.	TOTAL	MAT.	INST.	TOTAL	MAT.	INST.	TOTAL	MAT.	INST.	TOTAL	MAT.	INST.	TOTAL
A	Substructure	103.1	56.3	76.7	112.3	58.2	81.7	102.4	54.2	75.1	104.8	53.8	76.0	103.4	62.6	80.3
B10	Shell: Superstructure	97.2	65.5	85.6	97.1	66.7	86.0	90.9	64.9	81.4	95.6	62.0	83.3	94.8	67.5	84.8
B20	Exterior Closure	99.5	41.7	72.7	99.9	46.2	75.0	92.9	42.6	69.6	99.6	42.2	73.0	97.4	58.1	79.2
B30	Roofing	102.7	45.8	82.4	101.9	46.2	81.9	102.2	45.8	82.1	102.7	45.0	82.1	101.7	49.2	82.9
C	Interior Construction	101.7	42.4	76.7	99.6	44.0	76.2	98.5	42.3	74.8	101.7	41.9	76.5	107.1	46.9	81.8
D10	Services: Conveying	100.0	80.1	94.6	100.0	75.7	93.5	100.0	75.1	93.3	100.0	79.8	94.6	100.0	85.4	96.1
D20 - 40	Mechanical	100.2	41.5	76.8	100.0	39.6	76.0	100.2	41.2	76.7	100.2	40.1	76.3	100.3	60.9	84.6
D50	Electrical	97.5	40.4	69.7	104.9	40.1	73.3	98.0	38.8	69.2	97.5	44.0	71.5	103.0	66.2	85.1
E	Equipment & Furnishings	100.0	41.0	96.9	100.0	41.7	96.9	100.0	40.3	96.8	100.0	40.5	96.9	100.0	41.6	96.9
G	Site Work	106.7	84.9	92.4	110.9	85.0	93.9	108.0	77.3	87.9	107.1	84.9	92.5	94.7	97.2	96.3
A-G	WEIGHTED AVERAGE	99.8	50.7	79.2	100.7	51.3	80.0	97.2	49.6	77.2	99.5	50.0	78.7	100.1	62.9	84.5

		NORTH DAKOTA									OHIO					
DIV. NO.	BUILDING SYSTEMS	FARGO			GRAND FORKS			MINOT			AKRON			CANTON		
		MAT.	INST.	TOTAL	MAT.	INST.	TOTAL	MAT.	INST.	TOTAL	MAT.	INST.	TOTAL	MAT.	INST.	TOTAL
A	Substructure	99.0	63.1	78.7	105.9	59.1	79.5	106.3	67.7	84.5	94.6	98.4	96.7	95.1	91.6	93.1
B10	Shell: Superstructure	95.6	67.2	85.2	95.0	63.8	83.6	95.2	71.8	86.6	92.7	88.9	91.3	92.8	81.1	88.5
B20	Exterior Closure	108.3	45.8	79.3	104.4	62.8	85.1	102.9	64.7	85.2	94.5	94.6	94.5	94.1	83.6	89.3
B30	Roofing	100.3	50.3	82.4	102.6	48.7	83.3	102.4	51.9	84.3	104.4	97.9	102.1	105.3	94.6	101.5
C	Interior Construction	107.2	46.0	81.5	108.0	38.7	78.9	107.4	60.9	87.9	103.9	95.7	100.5	101.0	83.6	93.7
D10	Services: Conveying	100.0	85.4	96.1	100.0	41.4	84.2	100.0	85.4	96.1	100.0	87.8	96.7	100.0	65.8	90.8
D20 - 40	Mechanical	100.4	64.6	86.1	100.4	39.8	76.3	100.4	60.4	84.5	99.9	95.5	98.2	99.9	81.4	92.6
D50	Electrical	104.9	66.3	86.1	97.4	62.5	80.4	100.6	67.4	84.4	97.5	87.7	92.7	96.8	88.2	92.6
E	Equipment & Furnishings	100.0	42.2	96.9	100.0	37.8	96.7	100.0	60.5	97.9	100.0	95.8	99.8	100.0	85.0	99.2
G	Site Work	94.1	97.2	96.1	102.6	94.2	97.1	100.1	97.2	98.2	89.3	103.1	98.3	89.4	103.1	98.4
A-G	WEIGHTED AVERAGE	101.4	62.0	84.9	100.8	55.1	81.6	100.9	67.3	86.8	97.8	93.7	96.1	97.3	85.0	92.1

OHIO

DIV. NO.	BUILDING SYSTEMS	CINCINNATI			CLEVELAND			COLUMBUS			DAYTON			LORAIN		
		MAT.	INST.	TOTAL	MAT.	INST.	TOTAL	MAT.	INST.	TOTAL	MAT.	INST.	TOTAL	MAT.	INST.	TOTAL
A	Substructure	83.5	90.1	87.2	93.7	103.9	99.5	90.5	90.2	90.3	80.0	88.5	84.8	91.9	99.2	96.0
B10	Shell: Superstructure	95.0	87.0	92.0	93.7	93.6	93.6	91.8	84.7	89.2	93.7	80.9	89.0	92.6	90.0	91.7
B20	Exterior Closure	87.0	91.4	89.0	95.2	105.1	99.8	95.6	91.5	93.7	87.0	86.8	86.9	105.3	99.7	102.7
B30	Roofing	98.8	93.9	97.0	103.2	111.5	106.2	100.1	93.6	97.8	104.3	89.6	99.0	105.3	104.4	105.0
C	Interior Construction	98.4	85.8	93.1	98.7	102.7	100.4	96.9	84.6	91.7	99.0	80.6	91.3	100.9	97.8	99.6
D10	Services: Conveying	100.0	93.2	98.2	100.0	103.8	101.0	100.0	94.9	98.6	100.0	92.2	97.9	100.0	101.2	100.3
D20 - 40	Mechanical	100.0	87.0	94.9	99.9	105.5	102.1	99.7	89.8	95.8	101.0	85.2	94.7	99.9	84.1	93.6
D50	Electrical	94.7	80.0	87.5	97.3	104.2	100.7	98.1	85.1	91.8	93.1	83.3	88.3	97.0	91.6	94.4
E	Equipment & Furnishings	100.0	82.9	99.1	100.0	101.1	100.1	100.0	81.4	99.0	100.0	78.5	98.9	100.0	93.1	99.6
G	Site Work	75.1	105.5	95.0	89.2	104.7	99.4	89.2	100.3	96.5	74.3	105.4	94.7	88.6	102.6	97.8
A-G	WEIGHTED AVERAGE	95.3	88.1	92.3	97.2	102.6	99.5	96.3	88.6	93.0	95.2	85.5	91.1	96.9	93.6	95.5

DIV. NO.	BUILDING SYSTEMS	OHIO SPRINGFIELD MAT.	INST.	TOTAL	TOLEDO MAT.	INST.	TOTAL	YOUNGSTOWN MAT.	INST.	TOTAL	OKLAHOMA ENID MAT.	INST.	TOTAL	LAWTON MAT.	INST.	TOTAL
A	Substructure	81.2	89.3	85.8	90.5	101.6	96.8	94.0	96.0	95.1	96.0	57.4	74.2	94.1	62.2	76.1
B10	Shell: Superstructure	93.9	82.6	89.8	91.7	95.0	92.9	92.6	86.2	90.3	95.7	59.4	82.4	98.4	62.8	85.3
B20	Exterior Closure	86.6	86.4	86.5	99.7	99.5	99.6	93.9	93.1	93.5	97.1	55.6	77.8	92.3	56.1	75.5
B30	Roofing	104.2	89.5	98.9	101.8	108.4	104.2	105.5	94.9	101.7	99.5	55.8	83.9	99.3	57.9	84.5
C	Interior Construction	98.1	82.4	91.5	95.6	101.7	98.2	100.9	89.8	96.2	94.3	40.0	71.5	96.6	50.5	77.2
D10	Services: Conveying	100.0	91.9	97.8	100.0	89.6	97.2	100.0	87.6	96.7	100.0	69.4	91.8	100.0	69.4	91.8
D20-40	Mechanical	101.0	84.8	94.6	99.7	98.4	99.2	99.9	86.1	94.5	100.1	63.6	85.6	100.1	63.6	85.6
D50	Electrical	93.1	82.2	87.8	98.0	105.0	101.4	97.0	86.0	91.6	95.5	66.6	81.4	97.1	66.6	82.3
E	Equipment & Furnishings	100.0	80.7	99.0	100.0	103.5	100.2	100.0	88.1	99.4	100.0	34.1	96.5	100.0	50.2	97.4
G	Site Work	74.6	103.7	93.7	88.5	99.6	95.8	89.1	103.7	98.7	109.4	91.7	97.8	104.4	92.8	96.8
A-G	WEIGHTED AVERAGE	95.1	85.7	91.2	96.5	99.7	97.9	97.2	89.7	94.1	97.4	59.7	81.6	97.8	62.5	83.0

DIV. NO.	BUILDING SYSTEMS	OKLAHOMA MUSKOGEE MAT.	INST.	TOTAL	OKLAHOMA CITY MAT.	INST.	TOTAL	TULSA MAT.	INST.	TOTAL	OREGON EUGENE MAT.	INST.	TOTAL	MEDFORD MAT.	INST.	TOTAL
A	Substructure	89.3	46.9	65.3	95.6	60.7	75.9	94.4	59.2	74.6	104.2	101.2	102.5	106.3	101.2	103.4
B10	Shell: Superstructure	94.7	49.1	78.0	100.2	61.4	86.0	98.1	67.3	86.8	94.0	97.8	95.4	93.8	97.8	95.2
B20	Exterior Closure	98.1	48.0	74.8	92.2	57.3	76.0	92.0	58.2	76.3	109.6	98.5	104.4	114.8	98.5	107.2
B30	Roofing	99.1	40.4	78.1	96.7	57.7	82.8	99.1	54.5	83.2	105.4	90.9	100.2	106.0	93.3	101.5
C	Interior Construction	95.7	33.6	69.6	95.9	44.9	74.5	96.3	44.8	74.7	103.1	97.2	100.6	106.1	96.6	102.1
D10	Services: Conveying	100.0	68.8	91.6	100.0	70.2	92.0	100.0	70.6	92.1	100.0	101.6	100.4	100.0	101.6	100.4
D20-40	Mechanical	100.1	25.8	70.6	100.0	64.6	86.0	100.1	53.1	81.4	100.2	99.5	99.9	100.2	99.5	99.9
D50	Electrical	95.2	30.7	63.8	103.0	66.6	85.3	97.1	40.1	69.4	99.3	95.1	97.2	102.8	85.0	94.1
E	Equipment & Furnishings	100.0	31.8	96.4	100.0	41.1	96.9	100.0	41.0	96.9	100.0	99.7	100.0	100.0	99.2	100.0
G	Site Work	95.6	88.0	90.6	103.9	93.3	97.0	102.0	89.5	93.8	107.4	105.1	105.8	115.9	105.1	108.8
A-G	WEIGHTED AVERAGE	96.9	41.6	73.7	98.7	61.7	83.1	97.6	56.3	80.3	100.7	98.4	99.7	102.3	97.0	100.1

DIV. NO.	BUILDING SYSTEMS	OREGON PORTLAND MAT.	INST.	TOTAL	SALEM MAT.	INST.	TOTAL	PENNSYLVANIA ALLENTOWN MAT.	INST.	TOTAL	ALTOONA MAT.	INST.	TOTAL	ERIE MAT.	INST.	TOTAL
A	Substructure	106.3	101.4	103.5	98.0	101.3	99.9	91.5	106.2	99.8	96.6	88.8	92.2	96.9	92.5	94.4
B10	Shell: Superstructure	95.3	98.3	96.4	93.0	98.1	94.9	94.5	114.3	101.8	90.0	99.6	93.5	90.9	102.0	95.0
B20	Exterior Closure	109.5	102.7	106.4	111.9	102.7	107.7	101.7	100.9	101.3	93.2	78.6	86.4	87.0	91.4	89.0
B30	Roofing	105.2	99.4	103.1	102.8	94.4	99.8	100.2	116.8	106.2	99.4	88.7	95.6	99.6	94.9	98.0
C	Interior Construction	102.5	97.6	100.4	102.4	97.2	100.2	94.3	107.5	99.8	90.4	79.1	85.6	91.9	87.3	89.9
D10	Services: Conveying	100.0	101.6	100.4	100.0	101.6	100.4	100.0	101.7	100.5	100.0	95.6	98.8	100.0	100.3	100.1
D20-40	Mechanical	100.1	104.3	101.8	100.0	99.6	99.9	100.2	109.1	103.8	99.8	84.8	93.8	99.8	86.4	94.5
D50	Electrical	99.6	99.8	99.7	107.7	95.1	101.6	101.3	96.3	98.9	90.6	98.0	94.2	92.4	85.6	89.1
E	Equipment & Furnishings	100.0	100.0	100.0	100.0	99.7	100.0	100.0	114.3	100.8	100.0	82.5	99.1	100.0	87.4	99.3
G	Site Work	110.6	105.1	107.0	104.0	105.1	104.7	92.5	104.1	100.2	95.9	103.8	101.1	92.5	104.5	100.4
A-G	WEIGHTED AVERAGE	101.1	101.0	101.0	101.0	99.1	100.2	97.7	106.5	101.4	94.3	89.4	92.2	94.2	92.0	93.3

DIV. NO.	BUILDING SYSTEMS	PENNSYLVANIA HARRISBURG MAT.	INST.	TOTAL	PHILADELPHIA MAT.	INST.	TOTAL	PITTSBURGH MAT.	INST.	TOTAL	READING MAT.	INST.	TOTAL	SCRANTON MAT.	INST.	TOTAL
A	Substructure	95.3	93.5	94.3	106.4	125.4	117.2	98.0	97.0	97.4	88.6	95.1	92.3	93.8	92.4	93.0
B10	Shell: Superstructure	96.9	105.2	99.9	103.1	130.5	113.2	94.6	107.9	99.5	96.6	104.6	99.5	96.5	105.1	99.6
B20	Exterior Closure	99.8	89.8	95.2	102.5	134.5	117.3	92.9	96.4	94.5	103.0	92.8	98.3	101.9	95.7	99.0
B30	Roofing	105.5	107.3	106.1	99.4	133.4	111.6	102.6	96.2	100.3	99.0	109.7	102.8	100.1	98.3	99.5
C	Interior Construction	93.7	86.4	90.7	99.8	138.5	115.5	93.7	95.9	94.6	96.0	86.9	92.1	95.5	88.4	92.6
D10	Services: Conveying	100.0	89.8	97.3	100.0	117.9	104.8	100.0	99.3	99.8	100.0	98.8	99.7	100.0	100.1	100.0
D20-40	Mechanical	100.1	91.6	96.7	100.0	127.6	111.0	99.9	93.8	97.5	100.2	105.6	102.3	100.2	88.8	95.7
D50	Electrical	100.0	84.4	92.4	98.7	137.5	117.6	96.7	98.0	97.3	100.9	88.0	94.6	101.3	95.8	98.7
E	Equipment & Furnishings	100.0	82.4	99.1	100.0	137.4	102.0	100.0	92.9	99.6	100.0	81.8	99.0	100.0	79.3	98.9
G	Site Work	84.3	102.6	96.3	102.1	97.6	99.2	107.4	105.0	105.8	98.5	107.8	104.6	93.1	104.0	100.3
A-G	WEIGHTED AVERAGE	97.9	93.1	95.9	101.0	130.0	113.2	96.9	98.7	97.6	98.5	97.6	98.1	98.5	95.2	97.1

DIV. NO.	BUILDING SYSTEMS	PENNSYLVANIA YORK MAT.	INST.	TOTAL	PUERTO RICO SAN JUAN MAT.	INST.	TOTAL	RHODE ISLAND PROVIDENCE MAT.	INST.	TOTAL	SOUTH CAROLINA CHARLESTON MAT.	INST.	TOTAL	COLUMBIA MAT.	INST.	TOTAL
A	Substructure	87.8	93.7	91.1	120.5	35.5	72.4	95.0	111.8	104.5	98.9	59.8	76.8	103.1	59.6	78.5
B10	Shell: Superstructure	93.1	105.3	97.6	109.4	30.4	80.4	94.4	110.8	100.4	90.7	68.4	82.5	91.7	67.8	82.9
B20	Exterior Closure	103.5	90.3	97.4	105.6	16.4	64.2	116.0	121.2	118.4	103.1	45.9	76.6	100.7	44.0	74.4
B30	Roofing	99.3	107.6	102.3	130.8	22.8	92.2	93.0	113.1	100.2	102.4	46.9	82.5	101.4	45.8	81.6
C	Interior Construction	91.5	86.6	89.4	175.7	17.1	109.0	97.9	115.2	105.2	102.0	48.5	79.5	100.5	45.5	77.4
D10	Services: Conveying	100.0	90.0	97.3	100.0	19.9	78.5	100.0	96.7	99.1	100.0	72.2	92.5	100.0	72.2	92.5
D20-40	Mechanical	100.2	91.9	96.9	104.1	13.7	68.1	100.0	107.2	102.9	100.2	44.8	78.1	99.9	39.6	76.0
D50	Electrical	94.9	84.4	89.8	122.6	13.8	69.6	95.9	94.5	95.2	97.2	103.6	100.3	101.3	53.0	77.8
E	Equipment & Furnishings	100.0	82.4	99.1	100.0	16.5	95.6	100.0	108.9	100.5	100.0	46.2	97.2	100.0	43.2	97.0
G	Site Work	83.0	102.6	95.9	120.3	82.9	95.7	81.7	133.3	95.9	99.3	84.9	89.9	100.6	84.9	90.3
A-G	WEIGHTED AVERAGE	96.1	93.3	94.9	120.1	23.9	79.7	98.8	109.1	103.2	98.4	62.2	83.2	98.6	53.3	79.6

DIV. NO.	BUILDING SYSTEMS	SOUTH CAROLINA FLORENCE MAT.	INST.	TOTAL	GREENVILLE MAT.	INST.	TOTAL	SPARTANBURG MAT.	INST.	TOTAL	SOUTH DAKOTA ABERDEEN MAT.	INST.	TOTAL	PIERRE MAT.	INST.	TOTAL
A	Substructure	91.0	59.4	73.2	91.3	57.5	72.2	91.5	57.5	72.3	101.4	57.0	76.3	96.4	58.4	74.9
B10	Shell: Superstructure	87.4	67.9	80.3	88.1	64.5	79.4	88.3	64.5	79.6	99.6	59.4	84.9	98.5	63.0	85.5
B20	Exterior Closure	103.0	45.9	76.5	99.4	45.4	74.4	100.4	45.4	74.9	99.1	59.6	80.8	97.0	59.9	79.8
B30	Roofing	102.6	46.8	82.7	102.6	46.8	82.6	102.6	46.8	82.6	101.2	49.8	82.8	101.0	48.0	82.0
C	Interior Construction	97.9	45.9	76.0	99.0	47.4	77.3	99.5	47.4	77.6	102.2	44.6	78.0	102.4	44.0	77.9
D10	Services: Conveying	100.0	72.2	92.5	100.0	72.2	92.5	100.0	72.2	92.5	100.0	52.3	87.2	100.0	80.5	94.8
D20-40	Mechanical	100.2	39.7	76.1	100.2	38.0	75.5	100.2	38.0	75.5	100.1	47.9	79.4	100.0	47.5	79.2
D50	Electrical	95.2	50.4	73.4	97.3	54.5	76.4	97.3	54.5	76.4	97.9	59.7	79.3	100.9	51.8	77.0
E	Equipment & Furnishings	100.0	43.1	97.0	100.0	43.3	97.0	100.0	43.2	97.0	100.0	42.2	96.9	100.0	42.7	97.0
G	Site Work	109.7	84.9	93.4	104.4	84.5	91.3	104.2	84.5	91.2	90.5	95.1	93.5	89.4	95.1	93.2
A-G	WEIGHTED AVERAGE	96.6	53.3	78.4	96.7	52.9	78.3	96.9	52.9	78.4	99.9	56.3	81.6	99.5	56.5	81.4

DIV. NO.	BUILDING SYSTEMS	SOUTH DAKOTA RAPID CITY MAT.	INST.	TOTAL	SIOUX FALLS MAT.	INST.	TOTAL	TENNESSEE CHATTANOOGA MAT.	INST.	TOTAL	JACKSON MAT.	INST.	TOTAL	JOHNSON CITY MAT.	INST.	TOTAL
A	Substructure	97.8	57.7	75.2	95.0	59.9	75.2	94.8	58.0	74.0	97.7	50.2	70.9	83.8	57.2	68.8
B10	Shell: Superstructure	100.3	62.4	86.4	99.6	64.1	86.6	101.3	67.2	88.8	97.6	57.0	82.7	95.9	67.6	85.5
B20	Exterior Closure	97.8	56.2	78.5	94.9	58.4	78.0	94.3	42.3	70.2	100.8	33.2	69.5	114.9	42.4	81.3
B30	Roofing	101.6	51.1	83.5	102.7	52.4	84.7	95.3	56.6	81.5	91.8	37.3	72.3	90.1	56.5	78.1
C	Interior Construction	104.5	46.8	80.3	103.8	51.9	82.0	96.0	43.3	73.9	96.1	32.1	69.2	94.7	44.5	73.6
D10	Services: Conveying	100.0	80.9	94.9	100.0	80.5	94.8	100.0	47.2	85.8	100.0	52.8	87.3	100.0	54.5	87.8
D20-40	Mechanical	100.1	48.0	79.4	100.0	46.2	78.6	100.0	39.7	76.0	99.9	48.8	79.5	99.7	52.2	80.8
D50	Electrical	94.6	51.8	73.8	102.9	69.1	86.5	106.5	69.3	88.4	104.1	45.8	75.7	95.0	53.9	75.0
E	Equipment & Furnishings	100.0	39.0	96.8	100.0	46.3	97.2	100.0	42.7	97.0	100.0	33.4	96.5	100.0	42.8	97.0
G	Site Work	88.5	95.2	92.9	89.7	96.4	94.1	100.9	95.1	97.1	100.5	93.5	95.9	109.7	86.2	94.3
A-G	WEIGHTED AVERAGE	99.7	56.5	81.5	99.9	60.1	83.2	99.4	55.0	80.8	99.0	47.9	77.5	98.4	55.4	80.3

DIV. NO.	BUILDING SYSTEMS	TENNESSEE KNOXVILLE MAT.	INST.	TOTAL	MEMPHIS MAT.	INST.	TOTAL	NASHVILLE MAT.	INST.	TOTAL	TEXAS ABILENE MAT.	INST.	TOTAL	AMARILLO MAT.	INST.	TOTAL
A	Substructure	89.4	56.1	70.6	89.0	70.1	78.3	91.4	73.0	81.0	95.5	52.9	71.4	95.3	57.4	73.9
B10	Shell: Superstructure	101.0	67.3	88.6	101.6	78.8	93.3	102.1	78.8	93.5	101.2	58.0	85.4	101.1	59.9	86.0
B20	Exterior Closure	82.3	42.5	63.8	89.3	69.0	79.9	87.5	64.6	76.8	92.3	50.9	73.1	92.2	46.7	71.1
B30	Roofing	88.2	56.0	76.7	90.0	66.0	81.4	93.3	64.2	82.9	99.3	43.9	79.4	99.4	43.7	79.5
C	Interior Construction	90.7	44.5	71.3	93.9	57.5	78.6	98.4	65.9	84.8	94.8	44.0	73.4	94.3	49.3	75.4
D10	Services: Conveying	100.0	54.5	87.8	100.0	81.6	95.1	100.0	82.1	95.2	100.0	78.3	94.2	100.0	68.6	91.6
D20-40	Mechanical	99.7	52.2	80.8	99.9	66.6	86.6	99.9	78.9	91.5	100.1	39.9	76.2	100.0	48.3	79.5
D50	Electrical	100.8	58.3	80.1	103.2	68.7	86.4	103.0	65.4	84.7	97.3	43.8	71.2	101.9	68.6	85.7
E	Equipment & Furnishings	100.0	43.0	97.0	100.0	60.6	97.9	100.0	64.3	98.1	100.0	37.6	96.7	100.0	51.5	97.4
G	Site Work	88.4	86.2	87.0	89.7	91.1	90.6	96.7	98.4	97.8	102.5	88.6	93.4	103.3	89.5	94.2
A-G	WEIGHTED AVERAGE	95.8	55.8	79.0	97.5	70.1	86.0	98.5	73.6	88.0	98.2	50.8	78.3	98.6	56.8	81.0

DIV. NO.	BUILDING SYSTEMS	TEXAS AUSTIN MAT.	INST.	TOTAL	BEAUMONT MAT.	INST.	TOTAL	CORPUS CHRISTI MAT.	INST.	TOTAL	DALLAS MAT.	INST.	TOTAL	EL PASO MAT.	INST.	TOTAL
A	Substructure	92.2	59.1	73.5	95.3	57.3	73.8	101.5	50.4	72.7	96.4	64.0	78.1	90.7	52.5	69.1
B10	Shell: Superstructure	97.2	59.3	83.3	101.2	58.1	85.4	97.9	60.1	84.1	100.6	71.3	89.9	100.0	54.5	83.3
B20	Exterior Closure	85.7	51.0	69.6	95.7	55.5	77.1	85.2	49.1	68.4	94.3	59.0	77.9	91.2	46.5	70.5
B30	Roofing	86.9	50.1	73.7	104.6	53.9	86.5	95.2	41.7	76.1	93.1	63.6	82.6	95.7	51.4	79.8
C	Interior Construction	94.7	51.3	76.5	94.6	50.6	76.1	102.0	38.0	75.1	100.7	57.5	82.5	94.3	45.7	73.9
D10	Services: Conveying	100.0	68.3	91.5	100.0	84.9	96.0	100.0	83.8	95.6	100.0	82.0	95.2	100.0	67.7	91.3
D20-40	Mechanical	100.0	55.5	82.3	100.0	59.1	83.7	99.9	40.6	76.3	99.9	63.8	85.5	100.0	34.9	74.1
D50	Electrical	104.2	67.6	86.4	95.0	65.0	80.4	97.6	50.1	74.5	97.3	69.2	83.6	98.6	51.9	75.9
E	Equipment & Furnishings	100.0	57.7	97.8	100.0	48.4	97.3	100.0	35.0	96.6	100.0	59.8	97.9	100.0	44.9	97.1
G	Site Work	92.9	88.8	90.2	100.1	89.1	92.9	119.9	81.9	94.9	118.3	87.8	98.3	100.6	87.9	92.3
A-G	WEIGHTED AVERAGE	96.5	59.1	80.8	98.4	60.1	82.3	98.4	50.4	78.3	99.4	66.2	85.4	97.5	49.8	77.5

DIV. NO.	BUILDING SYSTEMS	TEXAS FORT WORTH MAT.	INST.	TOTAL	HOUSTON MAT.	INST.	TOTAL	LAREDO MAT.	INST.	TOTAL	LUBBOCK MAT.	INST.	TOTAL	ODESSA MAT.	INST.	TOTAL
A	Substructure	97.9	62.4	77.9	100.4	70.5	83.5	83.3	55.7	67.7	100.1	53.7	73.9	95.6	52.6	71.3
B10	Shell: Superstructure	98.9	64.0	86.1	104.7	78.0	94.9	95.9	55.6	81.2	104.3	65.2	90.0	100.7	57.1	84.7
B20	Exterior Closure	92.8	58.1	76.7	92.6	63.0	78.9	89.0	49.4	70.6	93.2	46.5	71.5	92.3	45.6	70.6
B30	Roofing	100.4	50.6	82.6	96.8	68.4	86.7	91.3	47.2	75.6	89.8	45.6	74.0	99.3	40.6	78.3
C	Interior Construction	94.8	55.3	78.2	102.1	64.0	86.1	95.0	37.6	70.9	100.8	37.3	74.1	94.8	36.1	70.1
D10	Services: Conveying	100.0	81.6	95.1	100.0	88.8	97.0	100.0	76.9	93.8	100.0	78.1	94.1	100.0	68.1	91.4
D20-40	Mechanical	100.0	53.6	81.6	99.9	71.9	88.8	99.9	36.6	74.7	99.6	45.9	78.2	100.1	33.4	73.6
D50	Electrical	98.1	60.3	79.7	97.2	68.5	83.2	99.3	64.9	82.6	95.8	41.2	69.2	97.4	38.3	68.7
E	Equipment & Furnishings	100.0	59.2	97.8	100.0	66.4	98.2	100.0	35.6	96.6	100.0	37.5	96.7	100.0	35.6	96.6
G	Site Work	104.8	89.2	94.5	120.5	85.2	97.3	91.7	88.6	89.7	126.9	83.8	98.6	102.5	89.0	93.6
A-G	WEIGHTED AVERAGE	98.0	60.7	82.4	100.7	71.2	88.3	95.8	51.7	77.3	100.1	51.1	79.5	98.1	46.3	76.3

DIV. NO.	BUILDING SYSTEMS	TEXAS SAN ANTONIO MAT.	INST.	TOTAL	WACO MAT.	INST.	TOTAL	WICHITA FALLS MAT.	INST.	TOTAL	UTAH LOGAN MAT.	INST.	TOTAL	OGDEN MAT.	INST.	TOTAL
A	Substructure	84.3	66.2	74.1	90.0	54.8	70.2	93.5	53.6	71.0	95.4	72.7	82.5	94.9	72.7	82.3
B10	Shell: Superstructure	96.7	64.4	84.9	100.0	56.9	84.2	100.7	58.8	85.3	99.7	69.9	88.8	100.2	69.9	89.1
B20	Exterior Closure	88.9	60.8	75.8	93.9	53.4	75.1	94.1	56.0	76.4	122.3	58.1	92.5	107.1	58.1	84.4
B30	Roofing	91.3	65.1	81.9	99.7	45.8	80.4	99.7	47.9	81.2	99.5	65.6	87.4	98.3	65.6	86.6
C	Interior Construction	96.2	55.2	79.0	94.6	35.8	69.9	94.9	45.8	74.3	94.3	53.5	77.1	93.3	53.5	76.6
D10	Services: Conveying	100.0	81.4	95.0	100.0	81.6	95.1	100.0	68.6	91.6	100.0	65.7	90.8	100.0	65.7	90.8
D20-40	Mechanical	99.9	67.0	86.8	100.1	54.1	81.9	100.1	45.6	78.5	99.9	69.5	87.8	99.9	69.5	87.8
D50	Electrical	99.8	64.9	82.8	97.4	71.9	85.0	99.7	60.7	80.7	94.5	66.8	81.0	94.9	66.8	81.2
E	Equipment & Furnishings	100.0	54.3	97.6	100.0	35.3	96.6	100.0	38.4	96.7	100.0	52.7	97.5	100.0	52.7	97.5
G	Site Work	91.3	92.4	92.0	101.1	89.1	93.2	101.9	88.6	93.2	89.0	101.9	97.5	78.6	101.9	93.9
A-G	WEIGHTED AVERAGE	96.2	65.4	83.3	97.8	56.8	80.6	98.5	55.3	80.3	100.3	67.3	86.4	98.4	67.3	85.3

DIV. NO.	BUILDING SYSTEMS	UTAH PROVO MAT.	INST.	TOTAL	SALT LAKE CITY MAT.	INST.	TOTAL	VERMONT BURLINGTON MAT.	INST.	TOTAL	RUTLAND MAT.	INST.	TOTAL	VIRGINIA ALEXANDRIA MAT.	INST.	TOTAL
A	Substructure	97.2	72.4	83.2	100.5	72.7	84.8	93.7	82.9	87.6	90.5	83.1	86.3	101.5	82.1	90.5
B10	Shell: Superstructure	98.8	70.0	88.2	105.2	70.0	92.3	95.3	77.8	88.9	93.6	78.2	87.9	97.9	88.9	94.6
B20	Exterior Closure	123.8	58.1	93.4	130.8	58.1	97.1	112.8	54.2	85.6	103.0	54.3	80.4	104.5	75.1	90.9
B30	Roofing	101.3	65.6	88.5	101.8	65.6	88.8	94.1	53.8	79.7	94.1	63.4	83.2	101.1	81.7	94.2
C	Interior Construction	96.6	55.1	79.2	94.2	55.1	77.8	100.5	55.4	81.6	100.5	55.4	81.5	102.1	78.3	92.1
D10	Services: Conveying	100.0	65.7	90.8	100.0	65.7	90.8	100.0	85.6	96.2	100.0	85.6	96.2	100.0	91.1	97.6
D20-40	Mechanical	99.9	69.5	87.8	100.0	69.5	87.9	100.0	64.9	86.1	100.2	65.0	86.2	100.2	88.5	95.5
D50	Electrical	95.1	69.7	82.8	98.2	69.7	84.4	99.2	62.8	81.5	96.9	62.8	80.3	97.6	98.2	97.9
E	Equipment & Furnishings	100.0	52.8	97.5	100.0	52.8	97.5	100.0	54.3	97.6	100.0	54.3	97.6	100.0	76.0	98.7
G	Site Work	86.0	100.6	95.6	78.6	101.9	93.9	77.7	96.8	90.2	77.5	96.8	90.2	118.4	88.7	98.9
A-G	WEIGHTED AVERAGE	100.8	67.8	87.0	102.9	67.9	88.2	99.3	67.7	86.1	97.5	68.1	85.2	100.7	86.0	94.5

Cost Indexes

DIV. NO.	BUILDING SYSTEMS	VIRGINIA														
		ARLINGTON			NEWPORT NEWS			NORFOLK			PORTSMOUTH			RICHMOND		
		MAT.	INST.	TOTAL	MAT.	INST.	TOTAL	MAT.	INST.	TOTAL	MAT.	INST.	TOTAL	MAT.	INST.	TOTAL
A	Substructure	104.2	79.6	90.3	104.0	74.1	87.1	110.4	74.3	90.0	102.6	69.1	83.7	105.0	70.2	85.4
B10	Shell: Superstructure	97.0	82.3	91.6	98.4	82.6	92.6	99.0	82.6	93.0	96.9	78.0	90.0	100.2	80.5	93.0
B20	Exterior Closure	115.1	71.4	94.8	103.5	60.9	83.8	106.0	60.9	85.1	105.8	60.1	84.6	102.7	63.3	84.4
B30	Roofing	102.9	72.3	92.0	102.4	61.0	87.6	99.7	61.0	85.9	102.4	58.1	86.5	101.3	60.4	86.7
C	Interior Construction	100.7	72.9	89.0	101.7	68.9	87.9	101.3	70.4	88.3	99.8	61.9	83.9	102.3	68.5	88.1
D10	Services: Conveying	100.0	81.5	95.0	100.0	81.6	95.0	100.0	81.6	95.0	100.0	73.8	92.9	100.0	80.7	94.8
D20 - 40	Mechanical	100.2	82.4	93.1	100.2	66.9	86.9	100.0	65.9	86.4	100.2	65.9	86.5	100.0	61.7	84.8
D50	Electrical	95.1	93.4	94.3	97.2	63.8	80.9	105.5	60.4	83.5	95.4	60.4	78.4	104.2	69.8	87.5
E	Equipment & Furnishings	100.0	75.6	98.7	100.0	74.7	98.7	100.0	74.5	98.7	100.0	58.3	97.8	100.0	66.6	98.2
G	Site Work	128.9	86.2	100.8	110.4	87.3	95.2	111.4	88.0	96.0	108.7	86.5	94.1	111.1	87.7	95.7
A-G	WEIGHTED AVERAGE	101.5	80.8	92.8	100.5	70.9	88.1	101.9	70.5	88.7	99.8	67.4	86.2	101.7	70.1	88.4

DIV. NO.	BUILDING SYSTEMS	VIRGINIA			WASHINGTON											
		ROANOKE			EVERETT			RICHLAND			SEATTLE			SPOKANE		
		MAT.	INST.	TOTAL	MAT.	INST.	TOTAL	MAT.	INST.	TOTAL	MAT.	INST.	TOTAL	MAT.	INST.	TOTAL
A	Substructure	111.1	71.7	88.8	104.5	101.8	103.0	106.4	86.5	95.1	107.1	104.6	105.7	109.5	86.5	96.5
B10	Shell: Superstructure	99.8	79.2	92.3	108.8	93.5	103.2	94.6	83.5	90.6	110.3	95.9	105.0	97.6	83.6	92.4
B20	Exterior Closure	103.8	62.7	84.7	111.9	96.3	104.6	107.1	79.6	94.4	110.7	99.9	105.7	107.6	81.7	95.6
B30	Roofing	102.3	59.4	86.9	105.0	93.2	100.8	151.7	80.2	126.2	105.0	97.4	102.3	148.4	80.7	124.2
C	Interior Construction	101.6	64.2	85.9	106.9	98.9	103.6	116.6	72.4	98.0	107.2	100.6	104.4	116.2	77.6	99.9
D10	Services: Conveying	100.0	72.0	92.5	100.0	97.7	99.4	100.0	75.6	93.4	100.0	101.4	100.4	100.0	75.6	93.4
D20 - 40	Mechanical	100.2	53.2	81.5	100.2	96.4	98.7	100.6	96.4	98.9	100.2	107.0	102.9	100.5	82.1	93.2
D50	Electrical	97.2	44.4	71.5	103.3	90.3	97.0	97.0	90.8	93.9	103.1	99.4	101.3	95.3	76.3	86.1
E	Equipment & Furnishings	100.0	70.6	98.4	100.0	101.4	100.1	100.0	79.9	98.9	100.0	102.1	100.1	100.0	79.2	98.9
G	Site Work	108.7	85.8	93.6	98.2	111.1	106.6	107.8	96.0	100.1	102.1	110.3	107.5	107.0	96.0	99.8
A-G	WEIGHTED AVERAGE	101.2	63.7	85.4	105.0	96.7	101.5	104.1	85.7	96.4	105.4	101.7	103.9	104.5	81.8	95.0

DIV. NO.	BUILDING SYSTEMS	WASHINGTON									WEST VIRGINIA					
		TACOMA			VANCOUVER			YAKIMA			CHARLESTON			HUNTINGTON		
		MAT.	INST.	TOTAL	MAT.	INST.	TOTAL	MAT.	INST.	TOTAL	MAT.	INST.	TOTAL	MAT.	INST.	TOTAL
A	Substructure	105.7	104.7	105.1	114.6	95.7	103.9	110.2	93.2	100.6	100.6	89.8	94.5	106.9	93.6	99.4
B10	Shell: Superstructure	110.0	95.0	104.5	109.4	91.4	102.8	109.5	87.1	101.3	98.0	93.1	96.2	99.2	96.0	98.0
B20	Exterior Closure	110.5	99.8	105.6	115.4	93.1	105.0	108.1	64.4	87.8	102.3	88.3	95.8	103.5	86.1	95.4
B30	Roofing	104.7	96.3	101.7	105.3	85.9	98.4	104.8	78.9	95.5	99.6	87.4	95.2	102.6	89.6	97.9
C	Interior Construction	106.4	100.6	104.0	103.5	85.6	95.9	105.8	88.0	98.3	101.5	86.8	95.3	100.9	88.8	95.8
D10	Services: Conveying	100.0	101.4	100.4	100.0	69.5	91.8	100.0	95.0	98.7	100.0	84.1	95.7	100.0	85.4	96.1
D20 - 40	Mechanical	100.2	98.8	99.7	100.3	100.9	100.6	100.2	93.0	97.4	100.0	83.0	93.3	100.2	84.9	94.1
D50	Electrical	103.1	97.1	100.2	109.7	97.6	103.8	106.1	90.8	98.6	102.5	88.0	95.4	97.2	95.8	96.5
E	Equipment & Furnishings	100.0	102.0	100.1	100.0	92.2	99.6	100.0	100.7	100.0	100.0	82.8	99.1	100.0	88.1	99.4
G	Site Work	101.4	111.4	107.9	112.5	100.9	104.9	104.5	109.6	107.9	104.0	89.0	94.2	107.7	89.9	96.0
A-G	WEIGHTED AVERAGE	105.1	99.6	102.8	106.5	93.8	101.1	105.3	88.0	98.0	100.4	87.8	95.1	100.6	90.3	96.3

DIV. NO.	BUILDING SYSTEMS	WEST VIRGINIA						WISCONSIN								
		PARKERSBURG			WHEELING			EAU CLAIRE			GREEN BAY			KENOSHA		
		MAT.	INST.	TOTAL	MAT.	INST.	TOTAL	MAT.	INST.	TOTAL	MAT.	INST.	TOTAL	MAT.	INST.	TOTAL
A	Substructure	102.3	89.1	94.9	102.3	88.8	94.6	98.1	97.4	97.7	100.8	94.7	97.4	105.1	102.6	103.7
B10	Shell: Superstructure	96.5	93.7	95.5	96.7	93.1	95.4	93.4	98.8	95.4	96.0	94.7	95.5	101.0	102.7	101.6
B20	Exterior Closure	101.1	83.9	93.1	112.2	83.1	98.7	92.8	94.5	93.6	106.2	93.9	100.5	98.7	107.5	102.8
B30	Roofing	102.4	85.3	96.3	102.8	84.3	96.2	100.2	83.5	94.2	102.2	81.6	94.8	99.9	96.0	98.5
C	Interior Construction	100.0	84.6	93.5	100.7	85.1	94.1	101.4	96.9	99.5	101.4	94.4	98.4	99.4	106.0	102.2
D10	Services: Conveying	100.0	84.1	95.7	100.0	95.0	98.6	100.0	86.3	96.3	100.0	85.5	96.1	100.0	102.1	100.6
D20 - 40	Mechanical	100.1	87.3	95.0	100.2	90.6	96.4	100.1	80.1	92.2	100.4	83.8	93.8	100.1	88.5	95.5
D50	Electrical	97.7	94.3	96.0	94.8	92.8	93.8	101.9	87.7	94.9	96.6	83.4	90.2	98.6	92.9	95.8
E	Equipment & Furnishings	100.0	81.6	99.0	100.0	82.8	99.1	100.0	94.5	99.7	100.0	94.3	99.7	100.0	103.8	100.2
G	Site Work	112.1	89.0	97.0	112.9	88.5	96.9	90.6	103.8	99.3	93.0	99.4	97.2	96.7	103.1	100.9
A-G	WEIGHTED AVERAGE	99.6	88.7	95.0	100.6	89.2	95.8	97.9	92.0	95.4	99.7	90.4	95.8	100.0	99.3	99.7

DIV. NO.	BUILDING SYSTEMS	WISCONSIN												WYOMING		
		LA CROSSE			MADISON			MILWAUKEE			RACINE			CASPER		
		MAT.	INST.	TOTAL	MAT.	INST.	TOTAL	MAT.	INST.	TOTAL	MAT.	INST.	TOTAL	MAT.	INST.	TOTAL
A	Substructure	90.6	96.1	93.7	97.4	97.9	97.7	97.6	107.0	102.9	97.7	103.6	101.0	103.9	66.2	82.6
B10	Shell: Superstructure	91.3	95.3	92.8	100.3	95.5	98.6	101.2	102.5	101.7	99.8	102.8	100.9	99.5	60.2	85.1
B20	Exterior Closure	89.4	93.8	91.5	99.0	102.8	100.8	100.3	115.2	107.2	100.0	107.5	103.4	108.5	41.0	77.2
B30	Roofing	99.7	83.5	93.9	95.6	90.9	93.9	97.8	111.4	102.7	99.5	97.5	98.8	97.6	54.9	82.3
C	Interior Construction	99.4	95.3	97.7	99.5	96.9	98.4	102.9	114.8	107.9	100.6	106.0	102.8	98.3	42.7	74.9
D10	Services: Conveying	100.0	87.5	96.6	100.0	87.0	96.5	100.0	103.8	101.0	100.0	102.1	100.6	100.0	84.6	95.9
D20 - 40	Mechanical	100.1	81.3	92.7	99.7	89.3	95.6	99.9	101.4	100.5	99.9	93.7	97.5	100.0	60.7	84.4
D50	Electrical	102.1	87.7	95.1	101.3	88.9	95.3	99.6	104.3	101.9	98.0	96.2	97.1	101.2	59.3	80.8
E	Equipment & Furnishings	100.0	94.6	99.7	100.0	94.7	99.7	100.0	113.6	100.7	100.0	103.8	100.2	100.0	42.6	97.0
G	Site Work	84.4	104.0	97.2	91.2	106.5	101.2	92.2	97.6	95.7	91.2	107.3	101.8	92.2	100.3	97.5
A-G	WEIGHTED AVERAGE	96.3	91.3	94.2	99.5	95.0	97.6	100.3	106.4	102.9	99.5	101.3	100.2	100.6	58.4	82.9

DIV. NO.	BUILDING SYSTEMS	WYOMING						CANADA								
		CHEYENNE			ROCK SPRINGS			CALGARY, ALBERTA			EDMONTON, ALBERTA			HALIFAX, NOVA SCOTIA		
		MAT.	INST.	TOTAL	MAT.	INST.	TOTAL	MAT.	INST.	TOTAL	MAT.	INST.	TOTAL	MAT.	INST.	TOTAL
A	Substructure	98.8	72.2	83.7	100.0	58.9	76.8	166.4	92.9	124.9	183.6	92.9	132.3	153.6	79.3	111.6
B10	Shell: Superstructure	99.6	65.4	87.1	96.9	56.2	82.0	154.6	88.7	130.4	157.7	88.6	132.4	127.5	78.1	109.4
B20	Exterior Closure	107.8	54.2	83.0	135.0	48.9	95.0	157.0	88.9	125.4	151.7	88.9	122.6	138.0	80.9	111.5
B30	Roofing	99.4	58.2	84.7	100.8	48.9	82.2	123.0	87.8	110.4	123.7	87.8	110.8	105.9	75.4	95.0
C	Interior Construction	99.4	59.5	82.6	100.2	40.3	75.0	113.5	90.2	103.7	114.8	89.5	104.2	99.6	73.5	88.7
D10	Services: Conveying	100.0	84.6	95.9	100.0	69.1	91.7	140.0	104.2	130.4	140.0	104.2	130.4	140.0	73.7	122.2
D20 - 40	Mechanical	99.9	59.9	84.0	99.9	57.3	83.0	100.1	87.6	95.1	100.1	87.6	95.2	99.2	76.1	90.0
D50	Electrical	97.6	72.7	85.5	93.3	60.4	77.3	125.2	84.7	105.5	118.5	84.7	102.1	128.4	74.6	102.2
E	Equipment & Furnishings	100.0	63.2	98.1	100.0	39.2	96.8	140.0	89.3	137.3	140.0	89.3	137.3	140.0	75.1	136.6
G	Site Work	86.1	100.3	95.4	84.4	99.0	94.0	126.6	104.8	112.3	153.9	104.8	121.7	104.0	99.2	100.8
A-G	WEIGHTED AVERAGE	100.0	65.9	85.7	102.0	56.5	82.9	130.7	89.9	113.6	131.8	89.8	114.1	119.0	78.0	101.8

CANADA

DIV. NO.	BUILDING SYSTEMS	HAMILTON, ONTARIO			KITCHENER, ONTARIO			LAVAL, QUEBEC			LONDON, ONTARIO			MONTREAL, QUEBEC		
		MAT.	INST.	TOTAL	MAT.	INST.	TOTAL	MAT.	INST.	TOTAL	MAT.	INST.	TOTAL	MAT.	INST.	TOTAL
A	Substructure	148.3	99.9	120.9	126.7	90.4	106.2	134.5	91.3	110.1	138.9	95.5	114.4	156.5	97.7	123.3
B10	Shell: Superstructure	132.8	95.2	119.0	119.0	90.5	108.5	113.2	88.7	104.2	128.7	92.0	115.3	131.9	95.5	118.5
B20	Exterior Closure	142.1	100.0	122.6	123.7	96.2	111.0	132.8	82.7	109.5	143.7	97.3	122.2	135.7	90.3	114.7
B30	Roofing	120.0	95.5	111.3	109.8	92.0	103.5	104.9	88.9	99.2	120.8	92.0	110.5	117.1	97.5	110.1
C	Interior Construction	112.5	95.0	105.1	102.3	88.2	96.4	104.7	87.2	97.3	115.0	89.6	104.3	106.4	95.1	101.6
D10	Services: Conveying	140.0	108.2	131.5	140.0	107.1	131.2	140.0	92.5	127.2	140.0	107.7	131.3	140.0	94.3	127.7
D20 - 40	Mechanical	100.0	91.9	96.8	99.2	88.9	95.1	99.2	92.7	96.6	100.0	89.0	95.6	100.1	94.4	97.8
D50	Electrical	128.8	96.1	112.9	124.1	93.4	109.1	123.5	74.4	99.6	123.3	93.4	108.8	129.8	86.1	108.5
E	Equipment & Furnishings	140.0	94.7	137.6	140.0	86.5	137.2	140.0	85.0	137.1	140.0	87.4	137.2	140.0	93.1	137.5
G	Site Work	117.9	111.1	113.4	104.5	104.7	104.6	98.7	101.0	100.2	117.5	104.9	109.2	105.9	101.0	102.6
A-G	WEIGHTED AVERAGE	123.4	96.9	112.3	114.5	92.2	105.1	114.5	87.6	103.2	122.0	93.3	110.0	121.7	93.8	110.0

CANADA

DIV. NO.	BUILDING SYSTEMS	OSHAWA, ONTARIO			OTTAWA, ONTARIO			QUEBEC, QUEBEC			REGINA, SASKATCHEWAN			SASKATOON, SASKATCHEWAN		
		MAT.	INST.	TOTAL	MAT.	INST.	TOTAL	MAT.	INST.	TOTAL	MAT.	INST.	TOTAL	MAT.	INST.	TOTAL
A	Substructure	157.4	90.8	119.8	148.2	96.1	118.8	139.6	98.0	116.1	139.7	71.4	101.1	129.5	71.4	96.7
B10	Shell: Superstructure	118.5	90.1	108.1	131.5	93.5	117.6	129.6	95.7	117.2	112.8	70.5	97.3	110.6	70.4	95.9
B20	Exterior Closure	133.8	93.2	114.9	141.4	96.0	120.4	139.0	92.8	117.6	127.3	63.3	97.6	126.8	63.3	97.3
B30	Roofing	110.8	85.2	101.6	123.4	91.8	112.1	118.5	97.7	111.1	105.8	63.5	90.7	104.7	62.2	89.5
C	Interior Construction	106.3	89.8	99.4	116.4	91.3	105.9	108.5	95.1	102.8	105.6	59.2	86.1	103.5	58.4	84.5
D10	Services: Conveying	140.0	107.3	131.2	140.0	104.8	130.5	140.0	94.4	127.7	140.0	71.4	121.6	140.0	71.4	121.6
D20 - 40	Mechanical	99.2	103.2	100.8	100.2	89.4	95.9	99.9	94.4	97.7	99.4	77.6	90.8	99.3	77.6	90.7
D50	Electrical	125.3	92.7	109.4	119.7	94.4	107.4	121.7	86.1	104.4	126.7	64.4	96.4	127.0	64.4	96.5
E	Equipment & Furnishings	140.0	86.1	137.2	140.0	91.1	137.4	140.0	93.2	137.5	140.0	57.6	135.6	140.0	57.6	135.6
G	Site Work	117.7	104.0	108.7	115.2	104.6	108.3	103.7	101.0	102.0	116.5	96.7	103.5	110.1	96.9	101.4
A-G	WEIGHTED AVERAGE	117.9	94.7	108.2	122.7	93.9	110.6	120.1	94.2	109.2	115.1	69.9	96.1	113.6	69.8	95.2

CANADA

DIV. NO.	BUILDING SYSTEMS	ST CATHARINES, ONTARIO			ST JOHNS, NEWFOUNDLAND			THUNDER BAY, ONTARIO			TORONTO, ONTARIO			VANCOUVER, BRITISH COLUMBIA		
		MAT.	INST.	TOTAL	MAT.	INST.	TOTAL	MAT.	INST.	TOTAL	MAT.	INST.	TOTAL	MAT.	INST.	TOTAL
A	Substructure	122.8	96.8	108.1	158.4	75.1	111.3	129.0	96.4	110.6	148.1	102.9	122.6	154.5	87.9	116.9
B10	Shell: Superstructure	110.4	94.2	104.5	119.0	74.0	102.5	111.9	93.4	105.1	132.0	97.9	119.5	163.4	85.6	134.9
B20	Exterior Closure	123.4	99.5	112.3	149.9	65.4	110.7	127.1	99.1	114.1	146.5	105.1	127.3	147.1	81.8	116.8
B30	Roofing	109.8	91.9	103.4	109.2	63.0	92.6	110.0	94.0	104.3	120.9	100.5	113.6	132.4	83.0	114.8
C	Interior Construction	100.0	93.1	97.1	108.2	62.3	88.9	102.3	87.1	95.9	113.8	100.0	108.0	113.7	79.5	99.3
D10	Services: Conveying	140.0	79.8	123.8	140.0	72.3	121.8	140.0	80.2	123.9	140.0	109.7	131.9	140.0	103.4	130.2
D20 - 40	Mechanical	99.2	90.6	95.8	99.6	64.2	85.5	99.2	90.7	95.8	99.9	95.9	98.3	99.9	73.3	89.3
D50	Electrical	126.3	94.5	110.8	122.9	65.8	95.1	124.1	92.7	108.8	122.0	96.8	109.8	124.7	75.0	100.5
E	Equipment & Furnishings	140.0	91.8	137.5	140.0	63.6	136.0	140.0	91.9	137.5	140.0	99.6	137.9	140.0	74.8	136.6
G	Site Work	105.7	101.9	103.2	120.6	99.4	106.7	111.0	101.8	105.0	138.7	105.5	116.9	123.5	107.3	112.8
A-G	WEIGHTED AVERAGE	112.3	94.2	104.7	119.9	69.2	98.6	113.5	92.9	104.9	123.6	99.8	113.6	131.2	82.0	110.5

CANADA

DIV. NO.	BUILDING SYSTEMS	WINDSOR, ONTARIO			WINNIPEG, MANITOBA											
		MAT.	INST.	TOTAL	MAT.	INST.	TOTAL	MAT.	INST.	TOTAL	MAT.	INST.	TOTAL	MAT.	INST.	TOTAL
A	Substructure	124.8	95.5	108.2	163.6	74.1	113.1									
B10	Shell: Superstructure	111.1	93.2	104.5	161.2	73.2	128.9									
B20	Exterior Closure	123.2	98.0	111.5	141.3	62.4	104.7									
B30	Roofing	109.9	92.5	103.6	114.0	68.2	97.6									
C	Interior Construction	101.3	90.4	96.7	109.7	64.1	90.5									
D10	Services: Conveying	140.0	79.5	123.7	140.0	72.6	121.9									
D20 - 40	Mechanical	99.2	90.2	95.6	99.9	65.6	86.3									
D50	Electrical	130.2	94.4	112.8	127.9	67.0	98.3									
E	Equipment & Furnishings	140.0	88.6	137.3	140.0	65.2	136.0									
G	Site Work	100.1	101.5	101.0	123.2	100.8	108.5									
A-G	WEIGHTED AVERAGE	113.0	93.2	104.7	129.6	69.6	104.4									

Costs shown in RSMeans cost data publications are based on national averages for materials and installation. To adjust these costs to a specific location, simply multiply the base cost by the factor and divide by 100 for that city. The data is arranged alphabetically by state and postal zip code numbers. For a city not listed, use the factor for a nearby city with similar economic characteristics.

STATE/ZIP	CITY	MAT.	INST.	TOTAL
ALABAMA				
350-352	Birmingham	98.3	74.9	88.5
354	Tuscaloosa	97.0	60.5	81.7
355	Jasper	97.2	54.0	79.0
356	Decatur	97.0	59.9	81.4
357-358	Huntsville	97.0	71.1	86.1
359	Gadsden	96.9	59.4	81.2
360-361	Montgomery	98.0	59.2	81.7
362	Anniston	96.3	52.9	78.1
363	Dothan	96.9	52.5	78.3
364	Evergreen	96.4	57.3	80.0
365-366	Mobile	97.9	65.9	84.5
367	Selma	96.6	54.4	78.9
368	Phenix City	97.3	59.4	81.4
369	Butler	96.8	55.7	79.5
ALASKA				
995-996	Anchorage	128.3	110.3	120.8
997	Fairbanks	127.6	112.6	121.3
998	Juneau	127.0	108.5	119.2
999	Ketchikan	139.6	108.1	126.4
ARIZONA				
850,853	Phoenix	99.9	74.0	89.0
852	Mesa/Tempe	99.2	66.7	85.6
855	Globe	99.5	63.1	84.2
856-857	Tucson	98.0	70.6	86.5
859	Show Low	99.5	64.2	84.7
860	Flagstaff	101.4	71.6	88.9
863	Prescott	99.2	63.7	84.3
864	Kingman	97.5	69.6	85.8
865	Chambers	97.5	64.4	83.6
ARKANSAS				
716	Pine Bluff	96.9	66.6	84.2
717	Camden	94.6	44.6	73.6
718	Texarkana	95.9	48.8	76.1
719	Hot Springs	93.8	47.9	74.5
720-722	Little Rock	97.1	69.9	85.7
723	West Memphis	96.0	59.6	80.8
724	Jonesboro	96.9	60.2	81.5
725	Batesville	94.7	54.3	77.7
726	Harrison	95.9	55.0	78.8
727	Fayetteville	93.4	54.7	77.1
728	Russellville	94.5	56.7	78.6
729	Fort Smith	97.4	60.1	81.8
CALIFORNIA				
900-902	Los Angeles	102.2	116.7	108.3
903-905	Inglewood	97.6	107.9	101.9
906-908	Long Beach	99.3	108.0	102.9
910-912	Pasadena	98.0	107.9	102.2
913-916	Van Nuys	101.2	107.9	104.0
917-918	Alhambra	100.2	107.9	103.4
919-921	San Diego	102.4	108.8	105.1
922	Palm Springs	99.2	106.8	102.4
923-924	San Bernardino	96.9	107.2	101.2
925	Riverside	101.6	112.7	106.3
926-927	Santa Ana	98.9	107.5	102.5
928	Anaheim	101.5	114.7	107.0
930	Oxnard	102.9	112.8	107.0
931	Santa Barbara	101.9	112.9	106.5
932-933	Bakersfield	103.3	109.4	105.8
934	San Luis Obispo	102.5	105.3	103.7
935	Mojave	99.6	104.3	101.6
936-938	Fresno	103.2	112.6	107.2
939	Salinas	103.3	116.4	108.8
940-941	San Francisco	111.2	141.3	123.8
942,956-958	Sacramento	104.9	114.6	108.9
943	Palo Alto	103.5	123.7	112.0
944	San Mateo	106.4	129.3	116.0
945	Vallejo	104.1	121.6	111.4
946	Oakland	108.8	129.5	117.5
947	Berkeley	108.3	123.8	114.8
948	Richmond	107.4	123.9	114.3
949	San Rafael	108.7	124.5	115.3
950	Santa Cruz	108.4	116.6	111.8

STATE/ZIP	CITY	MAT.	INST.	TOTAL
CALIFORNIA (CONT'D)				
951	San Jose	106.0	132.5	117.1
952	Stockton	103.5	112.0	107.0
953	Modesto	103.4	111.8	106.9
954	Santa Rosa	103.3	126.1	112.9
955	Eureka	104.7	109.6	106.7
959	Marysville	104.0	111.2	107.0
960	Redding	105.8	109.8	107.5
961	Susanville	104.6	110.0	106.8
COLORADO				
800-802	Denver	101.8	85.5	95.0
803	Boulder	98.6	82.0	91.6
804	Golden	100.7	80.8	92.4
805	Fort Collins	102.2	77.3	91.7
806	Greeley	99.7	67.0	86.0
807	Fort Morgan	99.2	80.6	91.4
808-809	Colorado Springs	101.5	81.0	92.9
810	Pueblo	101.3	79.0	91.9
811	Alamosa	102.4	75.3	91.0
812	Salida	102.2	76.2	91.3
813	Durango	102.8	76.0	91.6
814	Montrose	101.4	74.4	90.1
815	Grand Junction	104.9	72.0	91.1
816	Glenwood Springs	102.4	77.7	92.1
CONNECTICUT				
060	New Britain	99.2	118.0	107.1
061	Hartford	100.5	118.1	107.9
062	Willimantic	99.8	117.5	107.2
063	New London	95.8	117.1	104.8
064	Meriden	97.9	118.5	106.6
065	New Haven	101.0	118.0	108.2
066	Bridgeport	100.3	118.6	108.0
067	Waterbury	99.9	118.5	107.7
068	Norwalk	99.9	119.0	107.9
069	Stamford	100.0	125.4	110.7
D.C.				
200-205	Washington	104.6	92.8	99.7
DELAWARE				
197	Newark	100.5	105.8	102.7
198	Wilmington	100.6	105.9	102.8
199	Dover	101.2	105.8	103.1
FLORIDA				
320,322	Jacksonville	98.8	63.1	83.8
321	Daytona Beach	99.1	76.5	89.6
323	Tallahassee	99.6	53.0	80.1
324	Panama City	100.3	48.6	78.6
325	Pensacola	102.9	61.3	85.4
326,344	Gainesville	100.5	66.0	86.0
327-328,347	Orlando	103.1	72.0	90.0
329	Melbourne	102.0	80.0	92.7
330-332,340	Miami	101.2	75.4	90.3
333	Fort Lauderdale	98.5	73.6	88.0
334,349	West Palm Beach	97.2	69.8	85.7
335-336,346	Tampa	100.1	78.7	91.1
337	St. Petersburg	102.3	58.1	83.7
338	Lakeland	99.2	77.8	90.2
339,341	Fort Myers	98.6	70.1	86.6
342	Sarasota	100.4	71.1	88.1
GEORGIA				
300-303,399	Atlanta	98.2	79.2	90.2
304	Statesboro	97.9	48.1	77.0
305	Gainesville	96.5	63.2	82.5
306	Athens	95.9	66.9	83.7
307	Dalton	98.0	52.3	78.8
308-309	Augusta	96.9	64.3	83.2
310-312	Macon	96.5	66.6	83.9
313-314	Savannah	98.8	62.0	83.3
315	Waycross	97.7	57.2	80.7
316	Valdosta	97.8	51.0	78.2
317,398	Albany	97.9	62.0	82.9
318-319	Columbus	97.8	66.8	84.8

STATE/ZIP	CITY	MAT.	INST.	TOTAL
HAWAII				
967	Hilo	114.2	118.7	116.1
968	Honolulu	118.7	118.7	118.7
STATES & POSS.				
969	Guam	138.1	64.3	107.1
IDAHO				
832	Pocatello	101.6	74.5	90.2
833	Twin Falls	102.5	54.0	82.1
834	Idaho Falls	99.9	59.8	83.0
835	Lewiston	109.0	80.4	97.0
836-837	Boise	100.6	74.8	89.8
838	Coeur d'Alene	108.3	77.4	95.3
ILLINOIS				
600-603	North Suburban	97.4	123.7	108.5
604	Joliet	97.4	126.9	109.8
605	South Suburban	97.4	124.5	108.8
606-608	Chicago	97.9	138.3	114.9
609	Kankakee	93.6	107.0	99.2
610-611	Rockford	97.1	115.0	104.7
612	Rock Island	94.7	99.5	96.7
613	La Salle	95.9	106.3	100.3
614	Galesburg	95.7	99.0	97.1
615-616	Peoria	98.2	104.9	101.0
617	Bloomington	95.1	106.4	99.8
618-619	Champaign	98.4	105.0	101.2
620-622	East St. Louis	93.9	105.7	98.8
623	Quincy	94.9	94.8	94.9
624	Effingham	94.3	95.9	95.0
625	Decatur	96.2	102.5	98.8
626-627	Springfield	97.0	103.4	99.7
628	Centralia	92.4	100.5	95.8
629	Carbondale	92.1	94.1	92.9
INDIANA				
460	Anderson	94.8	82.5	89.6
461-462	Indianapolis	98.0	86.7	93.2
463-464	Gary	96.3	101.5	98.5
465-466	South Bend	95.6	83.1	90.4
467-468	Fort Wayne	95.3	78.4	88.2
469	Kokomo	92.5	80.8	87.6
470	Lawrenceburg	91.1	76.9	85.1
471	New Albany	92.4	74.4	84.9
472	Columbus	94.7	79.5	88.3
473	Muncie	95.5	80.6	89.3
474	Bloomington	96.9	78.8	89.3
475	Washington	92.7	82.4	88.4
476-477	Evansville	94.4	86.5	91.1
478	Terre Haute	95.0	86.6	91.5
479	Lafayette	94.3	82.1	89.2
IOWA				
500-503,509	Des Moines	98.2	76.7	89.2
504	Mason City	95.7	61.4	81.3
505	Fort Dodge	96.0	57.1	79.7
506-507	Waterloo	97.8	58.4	81.2
508	Creston	96.2	61.2	81.5
510-511	Sioux City	98.4	68.8	86.0
512	Sibley	96.9	47.6	76.2
513	Spencer	98.5	46.3	76.6
514	Carroll	95.6	51.4	77.0
515	Council Bluffs	99.6	74.6	89.1
516	Shenandoah	96.3	49.3	76.6
520	Dubuque	97.9	75.8	88.6
521	Decorah	96.5	48.8	76.5
522-524	Cedar Rapids	98.8	80.8	91.2
525	Ottumwa	96.6	69.3	85.1
526	Burlington	95.8	70.7	85.2
527-528	Davenport	97.7	90.2	94.5
KANSAS				
660-662	Kansas City	98.3	93.7	96.4
664-666	Topeka	98.9	66.1	85.1
667	Fort Scott	96.6	71.4	86.0
668	Emporia	96.7	59.7	81.2
669	Belleville	98.4	61.0	82.7
670-672	Wichita	97.6	64.9	83.9
673	Independence	98.1	63.3	83.5
674	Salina	98.7	61.6	83.1
675	Hutchinson	93.8	61.2	80.1
676	Hays	97.6	62.5	82.9
677	Colby	98.4	62.3	83.3

STATE/ZIP	CITY	MAT.	INST.	TOTAL
KANSAS (CONT'D)				
678	Dodge City	99.9	64.0	84.8
679	Liberal	97.5	62.2	82.7
KENTUCKY				
400-402	Louisville	97.1	83.9	91.6
403-405	Lexington	95.1	78.5	88.1
406	Frankfort	96.3	78.9	89.0
407-409	Corbin	92.3	65.8	81.2
410	Covington	92.8	99.3	95.5
411-412	Ashland	91.7	96.7	93.8
413-414	Campton	93.1	66.2	81.8
415-416	Pikeville	94.0	81.7	88.8
417-418	Hazard	92.4	58.7	78.3
420	Paducah	90.9	87.2	89.4
421-422	Bowling Green	93.6	84.6	89.8
423	Owensboro	93.4	83.2	89.1
424	Henderson	90.7	86.8	89.0
425-426	Somerset	90.6	70.3	82.1
427	Elizabethtown	90.1	83.6	87.3
LOUISIANA				
700-701	New Orleans	102.9	69.6	88.9
703	Thibodaux	100.4	63.7	85.0
704	Hammond	97.6	58.4	81.2
705	Lafayette	100.0	58.5	82.6
706	Lake Charles	100.1	61.1	83.7
707-708	Baton Rouge	102.2	63.8	86.1
710-711	Shreveport	96.8	59.2	81.0
712	Monroe	96.9	56.0	79.7
713-714	Alexandria	97.0	56.0	79.8
MAINE				
039	Kittery	93.0	74.4	85.2
040-041	Portland	98.5	73.7	88.1
042	Lewiston	96.4	73.7	86.9
043	Augusta	96.1	74.3	86.9
044	Bangor	95.8	73.7	86.5
045	Bath	94.2	74.4	85.9
046	Machias	93.7	73.9	85.4
047	Houlton	93.9	74.4	85.7
048	Rockland	92.9	74.5	85.2
049	Waterville	94.2	74.4	85.9
MARYLAND				
206	Waldorf	100.5	69.5	87.5
207-208	College Park	100.4	79.4	91.6
209	Silver Spring	99.7	75.5	89.5
210-212	Baltimore	98.2	86.0	93.1
214	Annapolis	98.7	80.2	90.9
215	Cumberland	94.0	80.4	88.3
216	Easton	95.5	42.8	73.4
217	Hagerstown	94.9	81.6	89.3
218	Salisbury	95.8	50.3	76.7
219	Elkton	93.1	62.9	80.4
MASSACHUSETTS				
010-011	Springfield	98.4	104.5	101.0
012	Pittsfield	97.8	101.0	99.1
013	Greenfield	95.7	102.3	98.4
014	Fitchburg	94.4	117.6	104.1
015-016	Worcester	98.3	118.9	107.0
017	Framingham	93.9	126.1	107.4
018	Lowell	97.7	125.7	109.5
019	Lawrence	98.9	123.6	109.3
020-022, 024	Boston	100.5	135.9	115.4
023	Brockton	99.2	120.7	108.2
025	Buzzards Bay	93.3	118.2	103.8
026	Hyannis	96.2	118.7	105.7
027	New Bedford	98.4	118.9	107.0
MICHIGAN				
480,483	Royal Oak	91.4	103.9	96.6
481	Ann Arbor	93.7	104.5	98.2
482	Detroit	95.3	114.2	103.2
484-485	Flint	93.4	100.8	96.5
486	Saginaw	93.0	89.9	91.7
487	Bay City	93.1	90.0	91.8
488-489	Lansing	94.4	96.9	95.5
490	Battle Creek	94.5	88.2	91.9
491	Kalamazoo	94.8	85.6	91.0
492	Jackson	92.7	91.6	92.3
493,495	Grand Rapids	96.1	65.2	83.1
494	Muskegon	93.5	81.4	88.4

STATE/ZIP	CITY	MAT.	INST.	TOTAL
MICHIGAN (CONT'D)				
496	Traverse City	92.3	69.1	82.5
497	Gaylord	93.3	72.1	84.4
498-499	Iron Mountain	95.3	83.2	90.2
MINNESOTA				
550-551	Saint Paul	98.6	118.7	107.0
553-555	Minneapolis	99.3	124.2	109.8
556-558	Duluth	98.2	108.2	102.4
559	Rochester	98.1	104.7	100.9
560	Mankato	96.0	103.2	99.0
561	Windom	94.8	77.6	87.6
562	Willmar	94.3	82.8	89.5
563	St. Cloud	95.8	117.5	104.9
564	Brainerd	95.9	99.5	97.4
565	Detroit Lakes	97.7	94.6	96.4
566	Bemidji	97.1	95.9	96.6
567	Thief River Falls	96.7	91.7	94.6
MISSISSIPPI				
386	Clarksdale	96.6	60.4	81.4
387	Greenville	100.0	70.9	87.8
388	Tupelo	97.9	63.1	83.2
389	Greenwood	97.8	59.9	81.9
390-392	Jackson	98.5	69.9	86.5
393	Meridian	96.8	70.9	85.9
394	Laurel	97.8	63.7	83.5
395	Biloxi	98.5	62.6	83.4
396	Mccomb	96.3	59.3	80.8
397	Columbus	97.7	61.0	82.3
MISSOURI				
630-631	St. Louis	98.6	108.8	102.9
633	Bowling Green	97.1	90.3	94.2
634	Hannibal	95.9	80.1	89.3
635	Kirksville	98.7	73.3	88.0
636	Flat River	98.1	91.1	95.1
637	Cape Girardeau	98.2	87.4	93.7
638	Sikeston	96.1	77.2	88.2
639	Poplar Bluff	95.6	77.7	88.1
640-641	Kansas City	99.2	106.1	102.1
644-645	St. Joseph	98.3	90.2	94.9
646	Chillicothe	95.1	68.5	83.9
647	Harrisonville	94.7	98.4	96.2
648	Joplin	97.1	68.0	84.9
650-651	Jefferson City	95.4	86.0	91.5
652	Columbia	96.3	88.1	92.9
653	Sedalia	95.4	81.7	89.6
654-655	Rolla	94.1	73.0	85.2
656-658	Springfield	97.3	77.6	89.1
MONTANA				
590-591	Billings	101.9	74.6	90.4
592	Wolf Point	101.1	71.5	88.7
593	Miles City	98.9	72.8	88.0
594	Great Falls	103.1	74.8	91.2
595	Havre	100.0	72.8	88.6
596	Helena	101.8	73.5	89.9
597	Butte	101.6	75.0	90.4
598	Missoula	99.0	73.2	88.2
599	Kalispell	97.9	72.0	87.0
NEBRASKA				
680-681	Omaha	99.5	78.8	90.8
683-685	Lincoln	97.6	77.1	89.0
686	Columbus	95.5	76.4	87.5
687	Norfolk	97.2	80.4	90.1
688	Grand Island	97.3	82.4	91.0
689	Hastings	96.5	84.8	91.6
690	Mccook	96.4	75.1	87.5
691	North Platte	96.8	85.0	91.8
692	Valentine	98.7	73.6	88.2
693	Alliance	98.6	71.6	87.3
NEVADA				
889-891	Las Vegas	101.1	112.0	105.7
893	Ely	99.5	72.0	88.0
894-895	Reno	99.8	93.9	97.3
897	Carson City	99.9	93.1	97.1
898	Elko	98.2	77.6	89.6
NEW HAMPSHIRE				
030	Nashua	98.6	87.7	94.0
031	Manchester	99.1	87.7	94.3

STATE/ZIP	CITY	MAT.	INST.	TOTAL
NEW HAMPSHIRE (CONT'D)				
032-033	Concord	97.0	83.8	91.5
034	Keene	95.1	53.4	77.6
035	Littleton	95.0	60.6	80.5
036	Charleston	94.6	50.8	76.2
037	Claremont	93.7	50.8	75.7
038	Portsmouth	95.9	90.1	93.5
NEW JERSEY				
070-071	Newark	100.5	122.0	109.5
072	Elizabeth	98.6	121.5	108.2
073	Jersey City	97.8	122.2	108.0
074-075	Paterson	99.3	121.8	108.8
076	Hackensack	97.3	122.3	107.8
077	Long Branch	97.1	121.2	107.2
078	Dover	97.6	122.0	107.8
079	Summit	97.6	121.5	107.6
080,083	Vineland	95.7	117.6	104.9
081	Camden	97.8	117.8	106.2
082,084	Atlantic City	96.3	117.6	105.2
085-086	Trenton	97.6	120.6	107.3
087	Point Pleasant	97.7	120.8	107.4
088-089	New Brunswick	98.1	121.4	107.9
NEW MEXICO				
870-872	Albuquerque	99.8	76.0	89.8
873	Gallup	99.9	76.0	89.9
874	Farmington	100.3	76.0	90.1
875	Santa Fe	102.0	76.0	91.1
877	Las Vegas	98.4	75.8	88.9
878	Socorro	97.9	76.0	88.7
879	Truth/Consequences	98.1	70.9	86.7
880	Las Cruces	97.0	68.5	85.0
881	Clovis	98.5	74.2	88.3
882	Roswell	100.3	74.3	89.3
883	Carrizozo	100.7	76.0	90.3
884	Tucumcari	99.3	74.2	88.8
NEW YORK				
100-102	New York	106.1	164.6	130.7
103	Staten Island	101.9	162.0	127.1
104	Bronx	100.0	162.0	126.0
105	Mount Vernon	99.8	132.8	113.6
106	White Plains	99.8	132.8	113.6
107	Yonkers	104.7	132.8	116.5
108	New Rochelle	100.1	132.8	113.8
109	Suffern	99.9	121.8	109.1
110	Queens	101.3	161.9	126.7
111	Long Island City	102.9	161.9	127.6
112	Brooklyn	103.1	161.9	127.8
113	Flushing	103.0	161.9	127.7
114	Jamaica	101.2	161.9	126.7
115,117,118	Hicksville	101.2	146.4	120.1
116	Far Rockaway	103.1	161.9	127.8
119	Riverhead	102.0	146.4	120.6
120-122	Albany	97.4	94.7	96.3
123	Schenectady	97.7	95.3	96.7
124	Kingston	100.8	112.5	105.7
125-126	Poughkeepsie	100.0	127.3	111.5
127	Monticello	99.3	115.7	106.2
128	Glens Falls	92.2	91.0	91.7
129	Plattsburgh	96.7	85.9	92.2
130-132	Syracuse	98.7	93.1	96.4
133-135	Utica	96.7	89.5	93.6
136	Watertown	98.4	92.2	95.8
137-139	Binghamton	98.1	86.9	93.4
140-142	Buffalo	99.0	104.8	101.5
143	Niagara Falls	96.7	101.5	98.7
144-146	Rochester	98.4	94.1	96.6
147	Jamestown	95.6	82.5	90.1
148-149	Elmira	95.5	84.4	90.8
NORTH CAROLINA				
270,272-274	Greensboro	99.8	50.7	79.2
271	Winston-Salem	99.5	50.0	78.7
275-276	Raleigh	100.7	51.3	80.0
277	Durham	101.1	51.7	80.4
278	Rocky Mount	96.9	41.6	73.7
279	Elizabeth City	97.7	43.0	74.7
280	Gastonia	98.5	49.6	78.0
281-282	Charlotte	100.2	52.2	80.0
283	Fayetteville	101.1	52.1	80.5
284	Wilmington	97.2	49.6	77.2
285	Kinston	95.2	43.3	73.4

STATE/ZIP	CITY	MAT.	INST.	TOTAL
NORTH CAROLINA (CONT'D)				
286	Hickory	95.5	45.9	74.7
287-288	Asheville	97.6	49.6	77.5
289	Murphy	96.3	35.7	70.9
NORTH DAKOTA				
580-581	Fargo	101.4	62.0	84.9
582	Grand Forks	100.8	55.1	81.6
583	Devils Lake	100.0	57.3	82.0
584	Jamestown	100.0	49.1	78.7
585	Bismarck	100.1	62.9	84.5
586	Dickinson	100.9	60.1	83.7
587	Minot	100.9	67.3	86.8
588	Williston	99.4	60.1	82.9
OHIO				
430-432	Columbus	96.3	88.6	93.0
433	Marion	92.6	83.5	88.8
434-436	Toledo	96.5	99.7	97.9
437-438	Zanesville	93.1	82.5	88.7
439	Steubenville	94.1	91.6	93.0
440	Lorain	96.9	93.6	95.5
441	Cleveland	97.2	102.6	99.5
442-443	Akron	97.8	93.7	96.1
444-445	Youngstown	97.2	89.7	94.1
446-447	Canton	97.3	85.0	92.1
448-449	Mansfield	94.4	88.3	91.9
450	Hamilton	95.0	85.8	91.1
451-452	Cincinnati	95.3	88.1	92.3
453-454	Dayton	95.2	85.5	91.1
455	Springfield	95.1	85.7	91.2
456	Chillicothe	93.8	91.4	92.8
457	Athens	96.7	75.9	88.0
458	Lima	97.0	85.3	92.1
OKLAHOMA				
730-731	Oklahoma City	98.7	61.7	83.1
734	Ardmore	95.2	61.1	80.9
735	Lawton	97.8	62.5	83.0
736	Clinton	96.6	59.7	81.1
737	Enid	97.4	59.7	81.6
738	Woodward	95.4	59.8	80.4
739	Guymon	96.4	32.0	69.4
740-741	Tulsa	97.6	56.3	80.3
743	Miami	94.1	64.5	81.7
744	Muskogee	96.9	41.6	73.7
745	Mcalester	93.8	52.4	76.5
746	Ponca City	94.4	60.4	80.1
747	Durant	94.4	60.4	80.2
748	Shawnee	95.9	57.8	79.9
749	Poteau	93.5	62.8	80.6
OREGON				
970-972	Portland	101.1	101.0	101.0
973	Salem	101.0	99.1	100.2
974	Eugene	100.7	98.4	99.7
975	Medford	102.3	97.0	100.1
976	Klamath Falls	102.2	96.7	99.9
977	Bend	101.0	98.5	100.0
978	Pendleton	96.1	99.0	97.3
979	Vale	93.8	90.3	92.3
PENNSYLVANIA				
150-152	Pittsburgh	96.9	98.7	97.6
153	Washington	93.8	98.4	95.7
154	Uniontown	94.1	95.5	94.7
155	Bedford	95.1	89.3	92.6
156	Greensburg	95.0	96.8	95.8
157	Indiana	93.9	95.6	94.6
158	Dubois	95.5	93.2	94.5
159	Johnstown	95.1	93.2	94.3
160	Butler	91.9	96.3	93.8
161	New Castle	91.9	94.7	93.1
162	Kittanning	92.4	98.6	95.0
163	Oil City	91.9	92.9	92.3
164-165	Erie	94.2	92.0	93.3
166	Altoona	94.3	89.4	92.2
167	Bradford	95.4	90.7	93.4
168	State College	94.9	90.1	92.9
169	Wellsboro	96.0	90.7	93.8
170-171	Harrisburg	97.9	93.1	95.9
172	Chambersburg	95.5	88.9	92.7
173-174	York	96.1	93.3	94.9
175-176	Lancaster	94.2	87.9	91.6

STATE/ZIP	CITY	MAT.	INST.	TOTAL
PENNSYLVANIA (CONT'D)				
177	Williamsport	92.9	81.1	87.9
178	Sunbury	95.0	92.2	93.8
179	Pottsville	94.1	92.3	93.4
180	Lehigh Valley	95.5	111.4	102.2
181	Allentown	97.7	106.5	101.4
182	Hazleton	94.9	92.4	93.9
183	Stroudsburg	94.8	98.8	96.5
184-185	Scranton	98.5	95.2	97.1
186-187	Wilkes-Barre	94.7	92.7	93.8
188	Montrose	94.4	93.7	94.1
189	Doylestown	94.4	118.6	104.5
190-191	Philadelphia	101.0	130.0	113.2
193	Westchester	97.2	119.9	106.8
194	Norristown	96.2	127.4	109.3
195-196	Reading	98.5	97.6	98.1
PUERTO RICO				
009	San Juan	120.1	23.9	79.7
RHODE ISLAND				
028	Newport	97.7	109.1	102.5
029	Providence	98.8	109.1	103.2
SOUTH CAROLINA				
290-292	Columbia	98.6	53.3	79.6
293	Spartanburg	96.9	52.9	78.4
294	Charleston	98.4	62.2	83.2
295	Florence	96.6	53.3	78.4
296	Greenville	96.7	52.9	78.3
297	Rock Hill	96.1	50.9	77.1
298	Aiken	97.1	70.0	85.7
299	Beaufort	97.9	46.2	76.2
SOUTH DAKOTA				
570-571	Sioux Falls	99.9	60.1	83.2
572	Watertown	98.1	55.3	80.1
573	Mitchell	97.1	55.2	79.5
574	Aberdeen	99.9	56.3	81.6
575	Pierre	99.5	56.5	81.4
576	Mobridge	97.7	55.0	79.8
577	Rapid City	99.7	56.5	81.5
TENNESSEE				
370-372	Nashville	98.5	73.6	88.0
373-374	Chattanooga	99.4	55.0	80.8
375,380-381	Memphis	97.5	70.1	86.0
376	Johnson City	98.4	55.4	80.3
377-379	Knoxville	95.8	55.8	79.0
382	McKenzie	97.1	55.2	79.5
383	Jackson	99.0	47.9	77.5
384	Columbia	95.6	57.0	79.4
385	Cookeville	97.0	58.1	80.6
TEXAS				
750	McKinney	98.7	51.5	78.8
751	Waxahachie	98.6	54.2	80.0
752-753	Dallas	99.4	66.2	85.4
754	Greenville	98.7	38.4	73.4
755	Texarkana	97.6	50.8	77.9
756	Longview	98.0	39.9	73.6
757	Tyler	98.8	54.2	80.1
758	Palestine	95.0	40.3	72.0
759	Lufkin	95.7	43.9	74.0
760-761	Fort Worth	98.0	60.7	82.4
762	Denton	97.6	48.1	76.8
763	Wichita Falls	98.5	55.3	80.3
764	Eastland	96.9	39.4	72.8
765	Temple	95.6	48.7	75.9
766-767	Waco	97.8	56.8	80.6
768	Brownwood	98.0	37.2	72.5
769	San Angelo	97.6	45.3	75.7
770-772	Houston	100.7	71.2	88.3
773	Huntsville	98.8	37.9	73.2
774	Wharton	100.2	42.2	75.8
775	Galveston	98.2	69.1	86.0
776-777	Beaumont	98.4	60.1	82.3
778	Bryan	95.4	62.6	81.6
779	Victoria	100.1	44.3	76.7
780	Laredo	95.8	51.7	77.3
781-782	San Antonio	96.2	65.4	83.3
783-784	Corpus Christi	98.4	50.4	78.3
785	McAllen	98.0	45.1	75.8
786-787	Austin	96.5	59.1	80.8

STATE/ZIP	CITY	MAT.	INST.	TOTAL
TEXAS (CONT'D)				
788	Del Rio	97.6	31.9	70.1
789	Giddings	94.9	39.5	71.7
790-791	Amarillo	98.6	56.8	81.0
792	Childress	97.6	49.6	77.4
793-794	Lubbock	100.1	51.1	79.5
795-796	Abilene	98.2	50.8	78.3
797	Midland	100.0	47.4	77.9
798-799,885	El Paso	97.5	49.8	77.5
UTAH				
840-841	Salt Lake City	102.9	67.9	88.2
842,844	Ogden	98.4	67.3	85.3
843	Logan	100.3	67.3	86.4
845	Price	100.7	46.5	78.0
846-847	Provo	100.8	67.8	87.0
VERMONT				
050	White River Jct.	96.3	56.6	79.7
051	Bellows Falls	94.8	64.0	81.9
052	Bennington	95.2	66.8	83.2
053	Brattleboro	95.6	69.0	84.4
054	Burlington	99.3	67.7	86.1
056	Montpelier	95.9	68.1	84.2
057	Rutland	97.5	68.1	85.2
058	St. Johnsbury	96.3	57.1	79.9
059	Guildhall	95.1	56.9	79.0
VIRGINIA				
220-221	Fairfax	100.2	82.2	92.6
222	Arlington	101.5	80.8	92.8
223	Alexandria	100.7	86.0	94.5
224-225	Fredericksburg	98.8	73.8	88.3
226	Winchester	99.4	66.5	85.6
227	Culpeper	99.3	72.0	87.9
228	Harrisonburg	99.6	68.2	86.4
229	Charlottesville	100.0	66.6	85.9
230-232	Richmond	101.7	70.1	88.4
233-235	Norfolk	101.9	70.5	88.7
236	Newport News	100.5	70.9	88.1
237	Portsmouth	99.8	67.4	86.2
238	Petersburg	99.9	70.1	87.4
239	Farmville	99.0	56.9	81.3
240-241	Roanoke	101.2	63.7	85.4
242	Bristol	98.7	57.4	81.4
243	Pulaski	98.4	54.7	80.1
244	Staunton	99.3	62.8	84.0
245	Lynchburg	99.4	68.1	86.2
246	Grundy	98.8	54.1	80.0
WASHINGTON				
980-981,987	Seattle	105.4	101.7	103.9
982	Everett	105.0	96.7	101.5
983-984	Tacoma	105.1	99.6	102.8
985	Olympia	103.1	99.5	101.6
986	Vancouver	106.5	93.8	101.1
988	Wenatchee	104.8	81.4	95.0
989	Yakima	105.3	88.0	98.0
990-992	Spokane	104.5	81.8	95.0
993	Richland	104.1	85.7	96.4
994	Clarkston	102.9	82.2	94.2
WEST VIRGINIA				
247-248	Bluefield	97.5	77.3	89.0
249	Lewisburg	99.3	82.0	92.0
250-253	Charleston	100.4	87.8	95.1
254	Martinsburg	98.9	76.5	89.5
255-257	Huntington	100.6	90.3	96.3
258-259	Beckley	97.3	86.0	92.6
260	Wheeling	100.6	89.2	95.8
261	Parkersburg	99.6	88.7	95.0
262	Buckhannon	98.9	88.9	94.7
263-264	Clarksburg	99.3	87.8	94.5
265	Morgantown	99.4	88.5	94.8
266	Gassaway	98.7	88.8	94.5
267	Romney	98.8	82.3	91.9
268	Petersburg	98.6	84.6	92.7
WISCONSIN				
530,532	Milwaukee	100.3	106.4	102.9
531	Kenosha	100.0	99.3	99.7
534	Racine	99.5	101.3	100.2
535	Beloit	99.3	94.6	97.3
537	Madison	99.5	95.0	97.6

STATE/ZIP	CITY	MAT.	INST.	TOTAL
WISCONSIN (CONT'D)				
538	Lancaster	96.9	90.3	94.1
539	Portage	95.6	94.3	95.0
540	New Richmond	95.5	93.6	94.7
541-543	Green Bay	99.7	90.4	95.8
544	Wausau	94.9	88.5	92.2
545	Rhinelander	98.0	88.7	94.1
546	La Crosse	96.3	91.3	94.2
547	Eau Claire	97.9	92.0	95.4
548	Superior	95.4	95.7	95.5
549	Oshkosh	95.5	89.9	93.2
WYOMING				
820	Cheyenne	100.0	65.9	85.7
821	Yellowstone Nat'l Park	97.5	58.6	81.2
822	Wheatland	98.6	58.6	81.8
823	Rawlins	100.2	58.4	82.6
824	Worland	98.1	56.5	80.6
825	Riverton	99.2	56.5	81.3
826	Casper	100.6	58.4	82.9
827	Newcastle	97.9	58.4	81.3
828	Sheridan	100.7	60.5	83.8
829-831	Rock Springs	102.0	56.5	82.9
CANADIAN FACTORS (reflect Canadian currency)				
ALBERTA				
	Calgary	130.7	89.9	113.6
	Edmonton	131.8	89.8	114.1
	Fort McMurray	127.7	92.6	112.9
	Lethbridge	121.9	92.0	109.3
	Lloydminster	116.4	88.8	104.8
	Medicine Hat	116.4	88.1	104.5
	Red Deer	116.9	88.1	104.8
BRITISH COLUMBIA				
	Kamloops	117.2	90.8	106.1
	Prince George	118.3	90.8	106.7
	Vancouver	131.2	82.0	110.5
	Victoria	118.3	78.8	101.7
MANITOBA				
	Brandon	116.4	76.1	99.5
	Portage la Prairie	116.5	74.7	98.9
	Winnipeg	129.6	69.6	104.4
NEW BRUNSWICK				
	Bathurst	114.7	67.8	95.0
	Dalhousie	114.6	67.8	95.0
	Fredericton	117.1	72.4	98.3
	Moncton	115.0	68.7	95.5
	Newcastle	114.7	67.8	95.0
	St. John	117.3	72.4	98.4
NEWFOUNDLAND				
	Corner Brook	119.7	68.2	98.0
	St. Johns	119.9	69.2	98.6
NORTHWEST TERRITORIES				
	Yellowknife	121.4	85.7	106.4
NOVA SCOTIA				
	Bridgewater	116.1	75.0	98.9
	Dartmouth	117.5	75.0	99.7
	Halifax	119.0	78.0	101.8
	New Glasgow	115.5	75.0	98.5
	Sydney	113.1	75.0	97.1
	Truro	115.5	75.0	98.5
	Yarmouth	115.4	75.0	98.5
ONTARIO				
	Barrie	119.2	93.0	108.2
	Brantford	118.4	96.9	109.4
	Cornwall	118.5	93.3	107.9
	Hamilton	123.4	96.9	112.3
	Kingston	119.4	93.4	108.5
	Kitchener	114.5	92.2	105.1
	London	122.0	93.3	110.0
	North Bay	118.4	91.2	107.0
	Oshawa	117.9	94.7	108.2
	Ottawa	122.7	93.9	110.6
	Owen Sound	119.4	91.2	107.6
	Peterborough	118.4	93.2	107.8
	Sarnia	117.9	97.5	109.3

STATE/ZIP	CITY	MAT.	INST.	TOTAL
ONTARIO (CONT'D)				
	Sault Ste Marie	112.9	92.6	104.4
	St. Catharines	112.3	94.2	104.7
	Sudbury	112.7	92.9	104.4
	Thunder Bay	113.5	92.9	104.9
	Timmins	118.6	91.2	107.1
	Toronto	123.6	99.8	113.6
	Windsor	113.0	93.2	104.7
PRINCE EDWARD ISLAND				
	Charlottetown	117.8	63.3	94.9
	Summerside	117.2	63.3	94.6
QUEBEC				
	Cap-de-la-Madeleine	115.3	87.8	103.8
	Charlesbourg	115.3	87.8	103.8
	Chicoutimi	114.0	93.5	105.4
	Gatineau	114.9	87.6	103.4
	Granby	115.1	87.6	103.5
	Hull	115.0	87.6	103.5
	Joliette	115.5	87.8	103.9
	Laval	114.5	87.6	103.2
	Montreal	121.7	93.8	110.0
	Quebec	120.1	94.2	109.2
	Rimouski	114.4	93.5	105.6
	Rouyn-Noranda	114.8	87.6	103.4
	Saint Hyacinthe	114.2	87.6	103.0
	Sherbrooke	115.2	87.6	103.6
	Sorel	115.4	87.8	103.8
	St. Jerome	114.8	87.6	103.4
	Trois Rivieres	115.4	87.8	103.8
SASKATCHEWAN				
	Moose Jaw	113.5	69.9	95.2
	Prince Albert	112.5	68.1	93.9
	Regina	115.1	69.9	96.1
	Saskatoon	113.6	69.8	95.2
YUKON				
	Whitehorse	113.2	68.9	94.6

Square Foot and Cubic Foot Building Costs

The cost figures in Division K1010 were derived from more than 11,200 projects contained in the Means Data Bank of Construction Costs, and include the contractor's overhead and profit, but do not include architectural fees or land costs. The figures have been adjusted to January of the current year. New projects are added to our files each year, and outdated projects are discarded. For this reason, certain costs may not show a uniform annual progression. In no case are all subdivisions of a project listed.

These projects were located throughout the U.S. and reflect a tremendous variation in S.F. and C.F. costs. This is due to differences, not only in labor and material costs, but also in individual owner's requirements. For instance, a bank in a large city would have different features than one in a rural area. This is true of all the different types of buildings analyzed. Therefore, caution should be exercised when using Division K1010 costs. For example, for court houses, costs in the data bank are local court house costs and will not apply to the larger, more elaborate federal court houses. As a general rule, the projects in the 1/4 column do not include any site work or equipment, while the projects in the 3/4 column may include both equipment and site work. The median figures do not generally include site work.

None of the figures "go with" any others. All individual cost items were computed and tabulated separately. Thus the sum of the median figures for Plumbing, HVAC, and Electrical will not normally total up to the total Mechanical and Electrical costs arrived at by separate analysis and tabulation of the projects.

Each building was analyzed as to total and component costs and percentages. The figures were arranged in ascending order with the results tabulated as shown. The 1/4 column shows that 25% of the projects had lower costs, 75% higher. The 3/4 column shows that 75% of the projects had lower costs,

25% had higher. The median column shows that 50% of the projects had lower costs, 50% had higher.

There are two times when square foot costs are useful. The first is in the conceptual stage when no details are available. Then square foot costs make a useful starting point. The second is after the bids are in and the costs can be worked back into their appropriate units for information purposes. As soon as details become available in the project design, the square foot approach should be discontinued and the project priced as to its particular components. When more precision is required or for estimating the replacement cost of specific buildings, the current edition of *RSMeans Square Foot Costs* should be used.

In using the figures in Division K1010, it is recommended that the median column be used for preliminary figures if no additional information is available. The median figures, when multiplied by the total city construction cost index figures (see City Cost Indexes) and then multiplied by the project size modifier in RK1010-050, should present a fairly accurate base figure, which would then have to be adjusted in view of the estimator's experience, local economic conditions, code requirements, and the owner's particular requirements. There is no need to factor the percentage figures as these should remain constant from city to city. All tabulations mentioning air conditioning had at least partial air conditioning.

The editors of this book would greatly appreciate receiving cost figures on one or more of your recent projects, which would then be included in the averages for next year. All cost figures received will be kept confidential except that they will be averaged with other similar projects to arrive at S.F. and C.F. cost figures for next year's book. See the last page of the book for details and the discount available for submitting one or more of your projects.

Table K1010-031 Unit Gross Area Requirements

The figures in the table below indicate typical ranges in square feet as a function of the "occupant" unit. This table is best used in the preliminary design stages to help determine the probable size

requirement for the total project. See RK1010-050 for the typical total size ranges for various types of buildings.

Building Type	Unit	Gross Area in S.F.		
		1/4	Median	3/4
Apartments	Unit	660	860	1,100
Auditorium & Play Theaters	Seat	18	25	38
Bowling Alleys	Lane		940	
Churches & Synagogues	Seat	20	28	39
Dormitories	Bed	200	230	275
Fraternity & Sorority Houses	Bed	220	315	370
Garages, Parking	Car	325	355	385
Hospitals	Bed	685	850	1,075
Hotels	Rental Unit	475	600	710
Housing for the Elderly	Unit	515	635	755
Housing, Public	Unit	700	875	1,030
Ice Skating Rinks	Total	27,000	30,000	36,000
Motels	Rental Unit	360	465	620
Nursing Homes	Bed	290	350	450
Restaurants	Seat	23	29	39
Schools, Elementary	Pupil	65	77	90
Junior High & Middle		85	110	129
Senior High		102	130	145
Vocational		110	135	195
Shooting Ranges	Point		450	
Theaters & Movies	Seat		15	

RK1010-050 Square Foot Project Size Modifier

One factor that affects the S.F. cost of a particular building is the size. In general, for buildings built to the same specifications in the same locality, the larger building will have the lower S.F. cost. This is due mainly to the decreasing contribution of the exterior walls plus the economy of scale usually achievable in larger buildings. The Area Conversion Scale shown below will give a factor to convert costs for the typical size building to an adjusted cost for the particular project.

The Square Foot Base Size lists the median costs, most typical project size in our accumulated data, and the range in size of the projects.

The Size Factor for your project is determined by dividing your project area in S.F. by the typical project size for the particular Building Type. With this factor, enter the Area Conversion Scale at the appropriate Size Factor and determine the appropriate cost multiplier for your building size.

Example: Determine the cost per S.F. for a 100,000 S.F. mid-rise apartment building.

$$\frac{\text{Proposed building area} = 100,000 \text{ S.F.}}{\text{Typical size from below} = 50,000 \text{ S.F.}} = 2.00$$

Enter Area Conversion scale at 2.0, intersect curve, read horizontally the appropriate cost multiplier of .94. Size adjusted cost becomes .94 x $107.00 = $101.00 based on national average costs.

Note: For Size Factors less than .50, the Cost Multiplier is 1.1
For Size Factors greater than 3.5, the Cost Multiplier is .90

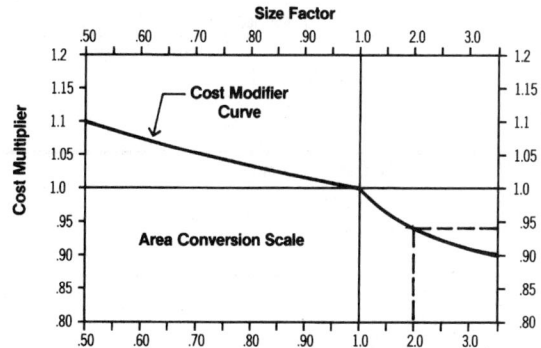

Square Foot Base Size							
Building Type	Median Cost per S.F.	Typical Size Gross S.F.	Typical Range Gross S.F.	Building Type	Median Cost per S.F.	Typical Size Gross S.F.	Typical Range Gross S.F.
Apartments, Low-Rise	$ 84.50	21,000	9,700 - 37,200	Jails	$ 257.00	40,000	5,500 - 145,000
Apartments, Mid-Rise	107.00	50,000	32,000 - 100,000	Libraries	161.00	12,000	7,000 - 31,000
Apartments, High-Rise	116.00	145,000	95,000 - 600,000	Living, Assisted	137.00	32,300	23,500 - 50,300
Auditoriums	141.00	25,000	7,600 - 39,000	Medical Clinics	146.00	7,200	4,200 - 15,700
Auto Sales	105.00	20,000	10,800 - 28,600	Medical Offices	138.00	6,000	4,000 - 15,000
Banks	189.00	4,200	2,500 - 7,500	Motels	102.00	40,000	15,800 - 120,000
Churches	130.00	17,000	2,000 - 42,000	Nursing Homes	142.00	23,000	15,000 - 37,000
Clubs, Country	132.00	6,500	4,500 - 15,000	Offices, Low-Rise	120.00	20,000	5,000 - 80,000
Clubs, Social	126.00	10,000	6,000 - 13,500	Offices, Mid-Rise	119.00	120,000	20,000 - 300,000
Clubs, YMCA	143.00	28,300	12,800 - 39,400	Offices, High-Rise	152.00	260,000	120,000 - 800,000
Colleges (Class)	152.00	50,000	15,000 - 150,000	Police Stations	190.00	10,500	4,000 - 19,000
Colleges (Science Lab)	242.00	45,600	16,600 - 80,000	Post Offices	141.00	12,400	6,800 - 30,000
College (Student Union)	179.00	33,400	16,000 - 85,000	Power Plants	1000.00	7,500	1,000 - 20,000
Community Center	134.00	9,400	5,300 - 16,700	Religious Education	120.00	9,000	6,000 - 12,000
Courthouses	180.00	32,400	17,800 - 106,000	Research	198.00	19,000	6,300 - 45,000
Dept. Stores	78.50	90,000	44,000 - 122,000	Restaurants	172.00	4,400	2,800 - 6,000
Dormitories, Low-Rise	143.00	25,000	10,000 - 95,000	Retail Stores	84.00	7,200	4,000 - 17,600
Dormitories, Mid-Rise	176.00	85,000	20,000 - 200,000	Schools, Elementary	125.00	41,000	24,500 - 55,000
Factories	76.50	26,400	12,900 - 50,000	Schools, Jr. High	129.00	92,000	52,000 - 119,000
Fire Stations	139.00	5,800	4,000 - 8,700	Schools, Sr. High	133.00	101,000	50,500 - 175,000
Fraternity Houses	131.00	12,500	8,200 - 14,800	Schools, Vocational	128.00	37,000	20,500 - 82,000
Funeral Homes	146.00	10,000	4,000 - 20,000	Sports Arenas	103.00	15,000	5,000 - 40,000
Garages, Commercial	93.50	9,300	5,000 - 13,600	Supermarkets	83.00	44,000	12,000 - 60,000
Garages, Municipal	119.00	8,300	4,500 - 12,600	Swimming Pools	194.00	20,000	10,000 - 32,000
Garages, Parking	50.50	163,000	76,400 - 225,300	Telephone Exchange	225.00	4,500	1,200 - 10,600
Gymnasiums	129.00	19,200	11,600 - 41,000	Theaters	120.00	10,500	8,800 - 17,500
Hospitals	228.00	55,000	27,200 - 125,000	Town Halls	137.00	10,800	4,800 - 23,400
House (Elderly)	115.00	37,000	21,000 - 66,000	Warehouses	60.00	25,000	8,000 - 72,000
Housing (Public)	106.00	36,000	14,400 - 74,400	Warehouse & Office	66.00	25,000	8,000 - 72,000
Ice Rinks	153.00	29,000	27,200 - 33,600				

K1010 | S.F. Costs

			UNIT	UNIT COSTS			% OF TOTAL		
				1/4	MEDIAN	3/4	1/4	MEDIAN	3/4
01	0010	**APARTMENTS Low Rise (1 to 3 story)**	S.F.	67	84.50	112			
	0020	Total project cost	C.F.	6	8	9.85			
	0100	Site work	S.F.	5.75	7.85	13.75	6.05%	10.55%	14.05%
	0500	Masonry		1.32	3.07	5.30	1.54%	3.67%	6.35%
	1500	Finishes		7.10	9.75	12.10	9.05%	10.75%	12.85%
	1800	Equipment		2.19	3.32	4.94	2.73%	4.03%	5.95%
	2720	Plumbing		5.20	6.70	8.55	6.65%	8.95%	10.05%
	2770	Heating, ventilating, air conditioning		3.33	4.10	6.05	4.20%	5.60%	7.60%
	2900	Electrical		3.88	5.15	6.90	5.20%	6.65%	8.40%
	3100	Total: Mechanical & Electrical	↓	13.45	17.10	21.50	15.90%	18.05%	23%
	9000	Per apartment unit, total cost	Apt.	62,500	95,500	140,500			
	9500	Total: Mechanical & Electrical	"	11,800	18,600	24,300			
02	0010	**APARTMENTS Mid Rise (4 to 7 story)**	S.F.	89	107	133			
	0020	Total project costs	C.F.	6.95	9.60	13.10			
	0100	Site work	S.F.	3.56	7.05	12.65	5.25%	6.70%	9.15%
	0500	Masonry		5.90	8.15	11.15	5.10%	7.25%	10.50%
	1500	Finishes		11.20	14.60	18.40	10.55%	13.45%	17.70%
	1800	Equipment		2.58	3.90	5.05	2.54%	3.48%	4.31%
	2500	Conveying equipment		1.91	2.44	2.96	1.94%	2.27%	2.69%
	2720	Plumbing		5.20	8.35	8.85	5.70%	7.20%	8.95%
	2900	Electrical		5.85	7.95	9.65	6.65%	7.20%	8.95%
	3100	Total: Mechanical & Electrical	↓	18.75	23.50	28.50	18.50%	21%	23%
	9000	Per apartment unit, total cost	Apt.	100,500	118,500	196,500			
	9500	Total: Mechanical & Electrical	"	19,000	22,000	27,800			
03	0010	**APARTMENTS High Rise (8 to 24 story)**	S.F.	101	116	139			
	0020	Total project costs	C.F.	9.80	11.40	14.55			
	0100	Site work	S.F.	3.66	5.90	8.30	2.58%	4.84%	6.15%
	0500	Masonry		5.85	10.60	13.20	4.74%	9.65%	11.05%
	1500	Finishes		11.20	14	16.50	9.75%	11.80%	13.70%
	1800	Equipment		3.24	3.99	5.30	2.78%	3.49%	4.35%
	2500	Conveying equipment		2.29	3.48	4.73	2.23%	2.78%	3.37%
	2720	Plumbing		7.40	8.75	12.25	6.80%	7.20%	10.45%
	2900	Electrical		6.95	8.75	11.85	6.45%	7.65%	8.80%
	3100	Total: Mechanical & Electrical	↓	21	26.50	32	17.95%	22.50%	24.50%
	9000	Per apartment unit, total cost	Apt.	105,000	115,500	160,000			
	9500	Total: Mechanical & Electrical	"	22,700	25,900	27,400			
04	0010	**AUDITORIUMS**	S.F.	104	141	203			
	0020	Total project costs	C.F.	6.55	9.15	13.10			
	2720	Plumbing	S.F.	6.35	9.20	11.15	5.85%	7.20%	8.70%
	2900	Electrical		8.10	11.70	15.75	6.80%	8.95%	11.30%
	3100	Total: Mechanical & Electrical	↓	23	46	56	24.50%	30.50%	31.50%
05	0010	**AUTOMOTIVE SALES**	S.F.	77.50	105	130			
	0020	Total project costs	C.F.	5.10	6.15	7.95			
	2720	Plumbing	S.F.	3.54	6.15	6.70	2.89%	6.05%	6.50%
	2770	Heating, ventilating, air conditioning		5.45	8.35	9	4.61%	10%	10.35%
	2900	Electrical		6.25	9.80	13.30	7.25%	9.80%	12.15%
	3100	Total: Mechanical & Electrical	↓	19.60	28	33.50	19.15%	20.50%	22%
06	0010	**BANKS**	S.F.	151	189	240			
	0020	Total project costs	C.F.	10.85	14.75	19.50			
	0100	Site work	S.F.	17.35	26.50	38	7.75%	12.45%	17%
	0500	Masonry		7.90	15.15	26.50	3.36%	6.95%	10.35%
	1500	Finishes		13.45	19.50	24	5.80%	8.45%	11.25%
	1800	Equipment		5.70	12.60	26	1.34%	5.95%	10.65%
	2720	Plumbing		4.76	6.80	9.95	2.82%	3.90%	4.93%
	2770	Heating, ventilating, air conditioning		9.05	12.10	16.10	4.86%	7.15%	8.50%
	2900	Electrical		14.35	19.20	25	8.20%	10.20%	12.15%
	3100	Total: Mechanical & Electrical	↓	33.50	45.50	53	16.25%	19.40%	23%
	3500	See also Divisions 11020 & 11030 (MF2004 11 16 00 & 11 17 00)								

K1010	S.F. Costs		UNIT	UNIT COSTS			% OF TOTAL		
				1/4	MEDIAN	3/4	1/4	MEDIAN	3/4
13	0010	**CHURCHES**	S.F.	102	130	168				13
	0020	Total project costs	C.F.	6.35	8	10.55			
	1800	Equipment	S.F.	1.22	2.93	6.25	.95%	2.11%	4.50%
	2720	Plumbing		3.99	5.55	8.20	3.51%	4.96%	6.25%
	2770	Heating, ventilating, air conditioning		9.30	12.15	17.20	7.50%	10%	12%
	2900	Electrical		8.60	11.80	15.90	7.35%	8.80%	11%
	3100	Total: Mechanical & Electrical	↓	26.50	34.50	46	18.30%	22%	24.50%
	3500	See also Division 11040 (MF2004 11 91 00)							
15	0010	**CLUBS, COUNTRY**	S.F.	110	132	167				15
	0020	Total project costs	C.F.	8.85	10.80	14.90			
	2720	Plumbing	S.F.	6.65	9.85	22.50	5.60%	7.90%	10%
	2900	Electrical		8.65	11.85	15.45	7%	8.95%	11%
	3100	Total: Mechanical & Electrical	↓	46	57.50	60.50	19%	26.50%	29.50%
17	0010	**CLUBS, SOCIAL Fraternal**	S.F.	87.50	126	169				17
	0020	Total project costs	C.F.	5.50	8.30	9.90			
	2720	Plumbing	S.F.	5.50	6.85	10.40	5.60%	6.90%	8.55%
	2770	Heating, ventilating, air conditioning		7.95	9.60	12.35	8.20%	9.25%	14.40%
	2900	Electrical		6.60	10.85	12.40	6.50%	9.50%	10.55%
	3100	Total: Mechanical & Electrical	↓	19.45	37	47	21%	23%	23.50%
18	0010	**CLUBS, Y.M.C.A.**	S.F.	110	143	179				18
	0020	Total project costs	C.F.	5.05	8.50	12.65			
	2720	Plumbing	S.F.	6.95	13.85	15.55	5.65%	7.60%	10.85%
	2900	Electrical		8.80	11	15.15	6.05%	7.60%	9.25%
	3100	Total: Mechanical & Electrical	↓	33.50	37.50	51.50	18.40%	21.50%	28.50%
19	0010	**COLLEGES Classrooms & Administration**	S.F.	111	152	201				19
	0020	Total project costs	C.F.	8.10	11.80	18.20			
	0500	Masonry	S.F.	8.15	15.35	18.85	5.65%	8.25%	10.50%
	2720	Plumbing		5.65	11.60	20.50	5.10%	6.60%	8.95%
	2900	Electrical		9.25	14.05	19.20	7.70%	9.85%	12%
	3100	Total: Mechanical & Electrical	↓	37	51.50	61.50	24%	28%	31.50%
21	0010	**COLLEGES Science, Engineering, Laboratories**	S.F.	207	242	280				21
	0020	Total project costs	C.F.	11.85	17.30	19.65			
	1800	Equipment	S.F.	11.50	26	28.50	2%	6.45%	12.65%
	2900	Electrical		17.05	23.50	37	7.10%	9.40%	12.10%
	3100	Total: Mechanical & Electrical	↓	63.50	75	116	28.50%	31.50%	41%
	3500	See also Division 11600 (MF2004 11 53 00)							
23	0010	**COLLEGES Student Unions**	S.F.	132	179	216				23
	0020	Total project costs	C.F.	7.35	9.65	11.90			
	3100	Total: Mechanical & Electrical	S.F.	49.50	53.50	63.50	23.50%	26%	29%
25	0010	**COMMUNITY CENTERS**	S.F.	109	134	181				25
	0020	Total project costs	C.F.	7.10	10.15	13.15			
	1800	Equipment	S.F.	2.64	4.46	7.10	1.48%	3.01%	5.45%
	2720	Plumbing		5.15	9	12.30	4.85%	7%	8.95%
	2770	Heating, ventilating, air conditioning		8.25	12.10	17.20	6.80%	10.35%	12.90%
	2900	Electrical		8.90	11.90	17.05	7.30%	9%	10.45%
	3100	Total: Mechanical & Electrical	↓	31	38	54	20%	25%	31%
28	0010	**COURTHOUSES**	S.F.	157	180	233				28
	0020	Total project costs	C.F.	12.05	14.40	18.15			
	2720	Plumbing	S.F.	7.45	10.45	15	5.95%	7.45%	8.20%
	2900	Electrical		16.70	18.50	27	8.90%	10.45%	11.55%
	3100	Total: Mechanical & Electrical	↓	42.50	58	64	22.50%	27.50%	30.50%
30	0010	**DEPARTMENT STORES**	S.F.	58	78.50	99				30
	0020	Total project costs	C.F.	3.11	4.03	5.50			
	2720	Plumbing	S.F.	1.81	2.28	3.47	1.82%	4.21%	5.90%
	2770	Heating, ventilating, air conditioning	↓	5.30	8.15	12.25	8.20%	9.10%	14.80%

		K1010 | S.F. Costs	UNIT	UNIT COSTS			% OF TOTAL			
				1/4	MEDIAN	3/4	1/4	MEDIAN	3/4	
30	2900	Electrical	S.F.	6.65	9.15	10.80	9.05%	12.15%	14.95%	30
	3100	Total: Mechanical & Electrical	↓	11.75	15	26.50	13.20%	21.50%	50%	
31	0010	DORMITORIES Low Rise (1 to 3 story)	S.F.	109	143	185				31
	0020	Total project costs	C.F.	6.20	10.05	15.05				
	2720	Plumbing	S.F.	6.60	8.85	11.20	8.05%	9%	9.65%	
	2770	Heating, ventilating, air conditioning		7	8.40	11.15	4.61%	8.05%	10%	
	2900	Electrical		7.15	10.95	14.85	6.25%	8.65%	9.50%	
	3100	Total: Mechanical & Electrical	↓	36.50	40.50	63.50	21.50%	24%	27%	
	9000	Per bed, total cost	Bed	46,700	52,000	111,000				
32	0010	DORMITORIES Mid Rise (4 to 8 story)	S.F.	135	176	218				32
	0020	Total project costs	C.F.	14.90	16.35	19.60				
	2900	Electrical	S.F.	14.35	16.30	22	8.20%	10.20%	11.95%	
	3100	Total: Mechanical & Electrical	"	40	41.50	81.50	25.50%	34.50%	37.50%	
	9000	Per bed, total cost	Bed	19,200	43,900	111,000				
34	0010	FACTORIES	S.F.	51	76.50	117				34
	0020	Total project costs	C.F.	3.28	4.89	8.10				
	0100	Site work	S.F.	5.85	10.65	16.85	6.95%	11.45%	17.95%	
	2720	Plumbing		2.76	5.15	8.50	3.73%	6.05%	8.10%	
	2770	Heating, ventilating, air conditioning		5.35	7.70	10.40	5.25%	8.45%	11.35%	
	2900	Electrical		6.35	10.05	15.35	8.10%	10.50%	14.20%	
	3100	Total: Mechanical & Electrical	↓	18.15	24.50	37	21%	28.50%	35.50%	
36	0010	FIRE STATIONS	S.F.	102	139	182				36
	0020	Total project costs	C.F.	5.95	8.15	10.85				
	0500	Masonry	S.F.	15	27	34.50	8.60%	11.65%	16.45%	
	1140	Roofing		3.31	8.95	10.20	1.90%	4.94%	5.05%	
	1580	Painting		2.57	3.84	3.93	1.37%	1.57%	2.07%	
	1800	Equipment		1.31	2.57	4.47	.74%	1.98%	3.54%	
	2720	Plumbing		5.70	9.15	13.25	5.85%	7.35%	9.45%	
	2770	Heating, ventilating, air conditioning		5.60	9.10	14.10	5.15%	7.40%	9.40%	
	2900	Electrical		7.20	12.55	16.95	6.85%	8.65%	10.60%	
	3100	Total: Mechanical & Electrical	↓	35.50	45.50	52.50	18.75%	23%	27%	
37	0010	FRATERNITY HOUSES & Sorority Houses	S.F.	102	131	179				37
	0020	Total project costs	C.F.	10.10	10.55	12.70				
	2720	Plumbing	S.F.	7.65	8.80	16.10	6.80%	8%	10.85%	
	2900	Electrical		6.70	14.45	17.70	6.60%	9.90%	10.65%	
	3100	Total: Mechanical & Electrical	↓	17.90	26	31		15.10%	15.90%	
38	0010	FUNERAL HOMES	S.F.	107	146	265				38
	0020	Total project costs	C.F.	10.95	12.15	23.50				
	2900	Electrical	S.F.	4.73	8.65	9.50	3.58%	4.44%	5.95%	
	3100	Total: Mechanical & Electrical	↓	16.75	24.50	34	12.90%	12.90%	12.90%	
39	0010	GARAGES, COMMERCIAL (Service)	S.F.	60.50	93.50	129				39
	0020	Total project costs	C.F.	3.99	5.90	8.55				
	1800	Equipment	S.F.	3.42	7.70	12	2.21%	4.62%	6.80%	
	2720	Plumbing		4.20	6.45	11.80	5.45%	7.85%	10.65%	
	2730	Heating & ventilating		5.50	7.30	9.90	5.25%	6.85%	8.20%	
	2900	Electrical		5.75	8.80	12.65	7.15%	9.25%	10.85%	
	3100	Total: Mechanical & Electrical	↓	12.70	24.50	36	12.35%	17.40%	26%	
40	0010	GARAGES, MUNICIPAL (Repair)	S.F.	89	119	167				40
	0020	Total project costs	C.F.	5.55	7.05	12.15				
	0500	Masonry	S.F.	8.35	16.35	25.50	5.60%	9.15%	12.50%	
	2720	Plumbing		4	7.65	14.45	3.59%	6.70%	7.95%	
	2730	Heating & ventilating		6.85	9.90	19.10	6.15%	7.45%	13.50%	
	2900	Electrical		6.60	10.35	14.90	6.65%	8.15%	11.15%	
	3100	Total: Mechanical & Electrical	↓	30.50	42	61	21.50%	25.50%	28.50%	

544

K1010 | Square Foot Costs

		K1010 | S.F. Costs	UNIT	UNIT COSTS			% OF TOTAL		
				1/4	MEDIAN	3/4	1/4	MEDIAN	3/4
41	0010	**GARAGES, PARKING**	S.F.	34.50	50.50	87				41
	0020	Total project costs	C.F.	3.26	4.42	6.45			
	2720	Plumbing	S.F.	.98	1.52	2.35	1.72%	2.70%	3.85%
	2900	Electrical		1.90	2.33	3.67	4.33%	5.20%	6.30%
	3100	Total: Mechanical & Electrical	↓	3.89	5.40	6.75	7%	8.90%	11.05%
	3200								
	9000	Per car, total cost	Car	14,600	18,400	23,400			
43	0010	**GYMNASIUMS**	S.F.	96.50	129	174				43
	0020	Total project costs	C.F.	4.81	6.50	8			
	1800	Equipment	S.F.	2.29	4.30	8.25	1.81%	3.30%	6.70%
	2720	Plumbing		6.10	7.25	9.35	4.65%	6.40%	7.75%
	2770	Heating, ventilating, air conditioning		6.55	10	20	5.15%	9.05%	11.10%
	2900	Electrical		7.40	10.10	13.30	6.60%	8.30%	10.30%
	3100	Total: Mechanical & Electrical	↓	27	36.50	43.50	19.75%	23.50%	29%
	3500	See also Division 11480 (MF2004 11 67 00)							
46	0010	**HOSPITALS**	S.F.	185	228	315				46
	0020	Total project costs	C.F.	14.05	17.45	25			
	1800	Equipment	S.F.	4.70	9.05	15.60	.96%	2.63%	5%
	2720	Plumbing		15.95	22.50	28.50	7.60%	9.10%	10.85%
	2770	Heating, ventilating, air conditioning		23.50	30	40.50	7.80%	12.95%	16.65%
	2900	Electrical		20	26.50	39	9.85%	11.55%	13.90%
	3100	Total: Mechanical & Electrical	↓	57	81.50	123	27%	33.50%	36.50%
	9000	Per bed or person, total cost	Bed	214,500	295,500	340,500			
	9900	See also Division 11700 (MF2004 11 71 00)							
48	0010	**HOUSING For the Elderly**	S.F.	91	115	142				48
	0020	Total project costs	C.F.	6.50	9	11.50			
	0100	Site work	S.F.	6.35	9.85	14.45	5.05%	7.90%	12.10%
	0500	Masonry		2.77	10.35	15.15	1.30%	6.05%	11%
	1800	Equipment		2.20	3.03	4.82	1.88%	3.23%	4.43%
	2510	Conveying systems		2.21	2.98	4.03	1.78%	2.20%	2.81%
	2720	Plumbing		6.75	8.60	10.85	8.15%	9.55%	10.50%
	2730	Heating, ventilating, air conditioning		3.47	4.91	7.35	3.30%	5.60%	7.25%
	2900	Electrical		6.80	9.20	11.80	7.30%	8.50%	10.25%
	3100	Total: Mechanical & Electrical	↓	23.50	28	37	18.10%	22.50%	29%
	9000	Per rental unit, total cost	Unit	84,500	99,000	110,500			
	9500	Total: Mechanical & Electrical	"	18,800	21,700	25,300			
50	0010	**HOUSING Public (Low Rise)**	S.F.	76.50	106	138				50
	0020	Total project costs	C.F.	6.80	8.50	10.55			
	0100	Site work	S.F.	9.75	14.05	22.50	8.35%	11.75%	16.50%
	1800	Equipment		2.08	3.39	5.15	2.26%	3.03%	4.24%
	2720	Plumbing		5.50	7.30	9.25	7.15%	9.05%	11.60%
	2730	Heating, ventilating, air conditioning		2.77	5.40	5.90	4.26%	6.05%	6.45%
	2900	Electrical		4.63	6.90	9.55	5.10%	6.55%	8.25%
	3100	Total: Mechanical & Electrical	↓	22	28.50	31.50	14.50%	17.55%	26.50%
	9000	Per apartment, total cost	Apt.	84,000	95,500	120,000			
	9500	Total: Mechanical & Electrical	"	17,900	22,100	24,500			
51	0010	**ICE SKATING RINKS**	S.F.	65.50	153	168				51
	0020	Total project costs	C.F.	4.81	4.92	5.65			
	2720	Plumbing	S.F.	2.44	4.58	4.69	3.12%	3.23%	5.65%
	2900	Electrical		7	10.75	11.35	6.30%	10.15%	15.05%
	3100	Total: Mechanical & Electrical	↓	11.65	16.50	20.50	18.95%	18.95%	18.95%
52	0010	**JAILS**	S.F.	199	257	330				52
	0020	Total project costs	C.F.	17.95	25	29.50			
	1800	Equipment	S.F.	7.75	23	39	2.80%	5.55%	10.35%
	2720	Plumbing		19.15	25.50	34	7%	8.90%	13.35%
	2770	Heating, ventilating, air conditioning		17.95	24	46.50	7.50%	9.45%	17.75%
	2900	Electrical	↓	20.50	27.50	35	8.20%	11.55%	14.95%

				UNIT COSTS			% OF TOTAL			
	K1010 | S.F. Costs		UNIT	1/4	MEDIAN	3/4	1/4	MEDIAN	3/4	
52	3100	Total: Mechanical & Electrical	S.F.	54	99	117	28%	30%	34%	52
53	0010	**LIBRARIES**	S.F.	126	161	207				53
	0020	Total project costs	C.F.	8.60	10.80	13.75				
	0500	Masonry	S.F.	9.85	17.40	29	5.80%	7.60%	11.80%	
	1800	Equipment		1.72	4.63	7	.37%	1.50%	4.16%	
	2720	Plumbing		4.65	6.75	9.15	3.38%	4.60%	5.70%	
	2770	Heating, ventilating, air conditioning		10.30	17.40	22.50	7.80%	10.95%	12.80%	
	2900	Electrical		12.85	16.65	21	8.30%	10.30%	11.95%	
	3100	Total: Mechanical & Electrical	↓	38	48	58.50	19.65%	23%	26.50%	
54	0010	**LIVING, ASSISTED**	S.F.	116	137	162				54
	0020	Total project costs	C.F.	9.80	11.45	13				
	0500	Masonry	S.F.	3.42	4.08	4.79	2.37%	3.16%	3.86%	
	1800	Equipment		2.65	3.08	3.94	2.12%	2.45%	2.66%	
	2720	Plumbing		9.75	13.05	13.55	6.05%	8.15%	10.60%	
	2770	Heating, ventilating, air conditioning		11.60	12.10	13.25	7.95%	9.35%	9.70%	
	2900	Electrical		11.40	12.60	14.55	9%	10%	10.70%	
	3100	Total: Mechanical & Electrical	↓	32	37.50	43	26%	29%	31.50%	
55	0010	**MEDICAL CLINICS**	S.F.	118	146	185				55
	0020	Total project costs	C.F.	8.70	11.25	14.95				
	1800	Equipment	S.F.	3.20	6.70	10.45	1.05%	2.94%	6.35%	
	2720	Plumbing		7.85	11.10	14.80	6.15%	8.40%	10.10%	
	2770	Heating, ventilating, air conditioning		9.40	12.30	18.10	6.65%	8.85%	11.35%	
	2900	Electrical		10.20	14.45	18.90	8.10%	10%	12.25%	
	3100	Total: Mechanical & Electrical	↓	32.50	44	60.50	22.50%	27%	33.50%	
	3500	See also Division 11700 (MF2004 11 71 00)								
57	0010	**MEDICAL OFFICES**	S.F.	112	138	169				57
	0020	Total project costs	C.F.	8.30	11.25	15.25				
	1800	Equipment	S.F.	3.68	7.25	10.35	.70%	5.10%	7.05%	
	2720	Plumbing		6.15	9.50	12.80	5.60%	6.80%	8.50%	
	2770	Heating, ventilating, air conditioning		7.45	10.75	14.20	6.10%	8%	9.70%	
	2900	Electrical		9.10	13.05	18.10	7.65%	9.80%	11.70%	
	3100	Total: Mechanical & Electrical	↓	26	35.50	51.50	19.35%	23%	29%	
59	0010	**MOTELS**	S.F.	70.50	102	133				59
	0020	Total project costs	C.F.	6.25	8.40	13.70				
	2720	Plumbing	S.F.	7.15	9.05	10.80	9.45%	10.60%	12.55%	
	2770	Heating, ventilating, air conditioning		4.34	6.50	11.60	5.60%	5.60%	10%	
	2900	Electrical		6.65	8.40	10.45	7.45%	9.05%	10.45%	
	3100	Total: Mechanical & Electrical	↓	22.50	28.50	48.50	18.50%	24%	25.50%	
	5000									
	9000	Per rental unit, total cost	Unit	35,800	68,000	73,500				
	9500	Total: Mechanical & Electrical	"	6,975	10,600	12,300				
60	0010	**NURSING HOMES**	S.F.	110	142	174				60
	0020	Total project costs	C.F.	8.65	10.80	14.80				
	1800	Equipment	S.F.	3.47	4.60	7.70	2.02%	3.62%	4.99%	
	2720	Plumbing		9.45	14.30	17.25	8.75%	10.10%	12.70%	
	2770	Heating, ventilating, air conditioning		9.95	15.10	20	9.70%	11.45%	11.80%	
	2900	Electrical		10.90	13.60	18.55	9.40%	10.55%	12.45%	
	3100	Total: Mechanical & Electrical	↓	26	36.50	61	26%	29.50%	30.50%	
	9000	Per bed or person, total cost	Bed	48,900	61,000	79,000				
61	0010	**OFFICES Low Rise (1 to 4 story)**	S.F.	93	120	155				61
	0020	Total project costs	C.F.	6.65	9.15	12.05				
	0100	Site work	S.F.	7.20	12.60	18.85	5.90%	9.70%	13.60%	
	0500	Masonry		3.46	7.25	13.15	2.62%	5.50%	8.45%	
	1800	Equipment		.99	1.93	5.25	.69%	1.50%	3.63%	
	2720	Plumbing		3.32	5.15	7.50	3.66%	4.50%	6.10%	
	2770	Heating, ventilating, air conditioning		7.35	10.25	15	7.20%	10.30%	11.70%	
	2900	Electrical	↓	7.55	10.75	15.20	7.45%	9.60%	11.35%	

			UNIT	UNIT COSTS			% OF TOTAL			
		K1010 | S.F. Costs		1/4	MEDIAN	3/4	1/4	MEDIAN	3/4	
61	3100	Total: Mechanical & Electrical	S.F.	20.50	27.50	41	18%	22%	27%	61
62	0010	**OFFICES Mid Rise (5 to 10 story)**	S.F.	98.50	119	157				62
	0020	Total project costs	C.F.	6.95	8.90	12.60				
	2720	Plumbing	S.F.	2.97	4.61	6.60	2.83%	3.74%	4.50%	
	2770	Heating, ventilating, air conditioning		7.45	10.65	17.05	7.65%	9.40%	11%	
	2900	Electrical		7.30	9.35	12.95	6.35%	7.80%	10%	
	3100	Total: Mechanical & Electrical		18.90	24.50	46.50	19.15%	21.50%	27.50%	
63	0010	**OFFICES High Rise (11 to 20 story)**	S.F.	121	152	187				63
	0020	Total project costs	C.F.	8.45	10.55	15.15				
	2900	Electrical	S.F.	7.35	8.95	13.30	5.80%	7.85%	10.50%	
	3100	Total: Mechanical & Electrical		23.50	31.50	53.50	16.90%	23.50%	34%	
64	0010	**POLICE STATIONS**	S.F.	145	190	240				64
	0020	Total project costs	C.F.	11.55	14.15	19.35				
	0500	Masonry	S.F.	13.60	24	30	6.70%	9.10%	11.35%	
	1800	Equipment		2.30	10.10	16.05	.98%	3.35%	6.70%	
	2720	Plumbing		8.10	16.15	20	5.65%	6.90%	10.75%	
	2770	Heating, ventilating, air conditioning		12.65	16.80	23	5.85%	10.55%	11.70%	
	2900	Electrical		15.85	22.50	30	9.80%	11.85%	14.80%	
	3100	Total: Mechanical & Electrical		52	62.50	84.50	25%	31.50%	32.50%	
65	0010	**POST OFFICES**	S.F.	114	141	180				65
	0020	Total project costs	C.F.	6.90	8.70	9.90				
	2720	Plumbing	S.F.	5.15	6.40	8.05	4.24%	5.30%	5.60%	
	2770	Heating, ventilating, air conditioning		8.05	9.95	11.05	6.65%	7.15%	9.35%	
	2900	Electrical		9.45	13.30	15.75	7.25%	9%	11%	
	3100	Total: Mechanical & Electrical		27.50	35.50	40.50	16.25%	18.80%	22%	
66	0010	**POWER PLANTS**	S.F.	795	1,000	1,925				66
	0020	Total project costs	C.F.	22	47.50	102				
	2900	Electrical	S.F.	56	119	177	9.30%	12.75%	21.50%	
	8100	Total: Mechanical & Electrical		140	455	1,025	32.50%	32.50%	52.50%	
67	0010	**RELIGIOUS EDUCATION**	S.F.	92	120	148				67
	0020	Total project costs	C.F.	5.10	7.35	9.15				
	2720	Plumbing	S.F.	3.85	5.45	7.70	4.40%	5.30%	7.10%	
	2770	Heating, ventilating, air conditioning		9.75	11	15.55	10.05%	11.45%	12.35%	
	2900	Electrical		7.30	10.30	13.65	7.60%	9.05%	10.35%	
	3100	Total: Mechanical & Electrical		29.50	39	47.50	22%	23%	27%	
69	0010	**RESEARCH Laboratories & Facilities**	S.F.	138	198	288				69
	0020	Total project costs	C.F.	10.40	20	24				
	1800	Equipment	S.F.	6.10	12.05	29.50	.94%	4.58%	8.80%	
	2720	Plumbing		13.85	17.70	28.50	6.15%	8.30%	10.80%	
	2770	Heating, ventilating, air conditioning		12.40	41.50	49.50	7.25%	16.50%	17.50%	
	2900	Electrical		16.15	26.50	43.50	9.55%	11.50%	15.40%	
	3100	Total: Mechanical & Electrical		49	92	132	29.50%	37%	42%	
70	0010	**RESTAURANTS**	S.F.	134	172	224				70
	0020	Total project costs	C.F.	11.25	14.75	19.35				
	1800	Equipment	S.F.	8.65	21	32	6.10%	13%	15.65%	
	2720	Plumbing		10.60	12.85	16.85	6.10%	8.15%	9%	
	2770	Heating, ventilating, air conditioning		13.45	18.60	24	9.20%	12%	12.40%	
	2900	Electrical		14.15	17.45	22.50	8.35%	10.55%	11.55%	
	3100	Total: Mechanical & Electrical		44	47	60.50	21%	24.50%	29.50%	
	9000	Per seat unit, total cost	Seat	4,900	6,550	7,750				
	9500	Total: Mechanical & Electrical	"	1,225	1,625	1,925				
72	0010	**RETAIL STORES**	S.F.	62.50	84	111				72
	0020	Total project costs	C.F.	4.23	6.05	8.40				
	2720	Plumbing	S.F.	2.26	3.78	6.45	3.26%	4.60%	6.80%	
	2770	Heating, ventilating, air conditioning		4.89	6.70	10.05	6.75%	8.75%	10.15%	
	2900	Electrical		5.65	7.70	11.10	7.25%	9.90%	11.65%	
	3100	Total: Mechanical & Electrical		14.95	19.20	26.50	17.05%	21%	23.50%	

		**K1010	S.F. Costs**	UNIT	UNIT COSTS			% OF TOTAL			
				1/4	MEDIAN	3/4	1/4	MEDIAN	3/4	
74	0010	**SCHOOLS Elementary**	S.F.	101	125	155				74
	0020	Total project costs	C.F.	6.65	8.55	11.05				
	0500	Masonry	S.F.	9.10	15.70	23.50	5.80%	11%	14.95%	
	1800	Equipment		2.78	4.70	8.75	1.89%	3.32%	4.71%	
	2720	Plumbing		5.85	8.30	11.05	5.70%	7.15%	9.35%	
	2730	Heating, ventilating, air conditioning		8.80	14	19.55	8.15%	10.80%	14.90%	
	2900	Electrical		9.60	12.70	15.95	8.40%	10.05%	11.70%	
	3100	Total: Mechanical & Electrical	▼	34	43	52.50	25%	27.50%	30%	
	9000	Per pupil, total cost	Ea.	11,700	17,400	50,000				
	9500	Total: Mechanical & Electrical	"	3,300	4,175	14,500				
76	0010	**SCHOOLS Junior High & Middle**	S.F.	103	129	157				76
	0020	Total project costs	C.F.	6.65	8.65	9.70				
	0500	Masonry	S.F.	13.10	16.90	19.75	8%	11.40%	14.30%	
	1800	Equipment		3.36	5.40	8.15	1.81%	3.26%	4.86%	
	2720	Plumbing		6.10	7.55	9.35	5.30%	6.80%	7.25%	
	2770	Heating, ventilating, air conditioning		12.20	14.85	26	8.90%	11.55%	14.20%	
	2900	Electrical		10.30	12.40	15.95	7.90%	9.35%	10.60%	
	3100	Total: Mechanical & Electrical	▼	33.50	43	53	23.50%	27%	29.50%	
	9000	Per pupil, total cost	Ea.	13,300	17,500	23,500				
78	0010	**SCHOOLS Senior High**	S.F.	108	133	167				78
	0020	Total project costs	C.F.	6.60	9.70	15.65				
	1800	Equipment	S.F.	2.87	6.75	9.50	1.88%	2.98%	4.80%	
	2720	Plumbing		6.15	9.20	16.80	5.60%	6.90%	8.30%	
	2770	Heating, ventilating, air conditioning		12.50	14.35	27.50	8.95%	11.60%	15%	
	2900	Electrical		10.95	14.25	21	8.65%	10.15%	11.95%	
	3100	Total: Mechanical & Electrical	▼	36.50	42.50	71.50	23.50%	26.50%	28.50%	
	9000	Per pupil, total cost	Ea.	10,300	21,000	26,200				
80	0010	**SCHOOLS Vocational**	S.F.	88.50	128	159				80
	0020	Total project costs	C.F.	5.50	7.90	10.90				
	0500	Masonry	S.F.	5.20	12.80	19.60	3.53%	6.70%	10.95%	
	1800	Equipment		2.77	6.90	9.60	1.24%	3.10%	4.26%	
	2720	Plumbing		5.65	8.45	12.40	5.40%	6.90%	8.55%	
	2770	Heating, ventilating, air conditioning		7.90	14.75	24.50	8.60%	11.90%	14.65%	
	2900	Electrical		9.20	12.05	16.60	8.45%	10.95%	13.20%	
	3100	Total: Mechanical & Electrical	▼	32	35.50	61	23.50%	27.50%	31%	
	9000	Per pupil, total cost	Ea.	12,300	33,000	49,200				
83	0010	**SPORTS ARENAS**	S.F.	77.50	103	159				83
	0020	Total project costs	C.F.	4.20	7.50	9.70				
	2720	Plumbing	S.F.	4.49	6.80	14.35	4.35%	6.35%	9.40%	
	2770	Heating, ventilating, air conditioning		9.65	11.40	15.85	8.80%	10.20%	13.55%	
	2900	Electrical		8.05	10.95	14.15	8.60%	9.90%	12.25%	
	3100	Total: Mechanical & Electrical	▼	20	35	47	21.50%	25%	27.50%	
85	0010	**SUPERMARKETS**	S.F.	71.50	83	97				85
	0020	Total project costs	C.F.	3.98	4.81	7.30				
	2720	Plumbing	S.F.	3.99	5.05	5.85	5.40%	6%	7.45%	
	2770	Heating, ventilating, air conditioning		5.85	7.80	9.50	8.60%	8.65%	9.60%	
	2900	Electrical		8.95	10.30	12.15	10.40%	12.45%	13.60%	
	3100	Total: Mechanical & Electrical	▼	23	25	32.50	20.50%	26.50%	31%	
86	0010	**SWIMMING POOLS**	S.F.	116	194	415				86
	0020	Total project costs	C.F.	9.25	11.55	12.60				
	2720	Plumbing	S.F.	10.70	12.20	16.60	4.80%	9.70%	20.50%	
	2900	Electrical		8.70	14.10	20.50	5.75%	6.95%	7.60%	
	3100	Total: Mechanical & Electrical	▼	21	55	73.50	11.15%	14.10%	23.50%	
87	0010	**TELEPHONE EXCHANGES**	S.F.	154	225	285				87
	0020	Total project costs	C.F.	9.55	15.35	21				
	2720	Plumbing	S.F.	6.50	10	14.65	4.52%	5.80%	6.90%	
	2770	Heating, ventilating, air conditioning	▼	15.05	30.50	37.50	11.80%	16.05%	18.40%	

		K1010 | S.F. Costs	UNIT	UNIT COSTS			% OF TOTAL			
				1/4	MEDIAN	3/4	1/4	MEDIAN	3/4	
87	2900	Electrical	S.F.	15.65	25	44	10.90%	14%	17.85%	87
	3100	Total: Mechanical & Electrical	↓	46	87.50	124	29.50%	33.50%	44.50%	
91	0010	**THEATERS**	S.F.	96.50	120	183				91
	0020	Total project costs	C.F.	4.46	6.60	9.70				
	2720	Plumbing	S.F.	3.22	3.49	14.25	2.92%	4.70%	6.80%	
	2770	Heating, ventilating, air conditioning		9.40	11.35	14.05	8%	12.25%	13.40%	
	2900	Electrical		8.45	11.40	23	8.05%	9.95%	12.25%	
	3100	Total: Mechanical & Electrical	↓	21.50	32.50	66.50	23%	26.50%	27.50%	
94	0010	**TOWN HALLS City Halls & Municipal Buildings**	S.F.	108	137	179				94
	0020	Total project costs	C.F.	9.85	11.80	16.60				
	2720	Plumbing	S.F.	4.50	8.40	15.50	4.31%	5.95%	7.95%	
	2770	Heating, ventilating, air conditioning		8.15	16.15	23.50	7.05%	9.05%	13.45%	
	2900	Electrical		10.25	14.60	19.95	8.05%	9.45%	11.65%	
	3100	Total: Mechanical & Electrical	↓	35.50	45	69	22%	26.50%	31%	
97	0010	**WAREHOUSES & Storage Buildings**	S.F.	40.50	60	86				97
	0020	Total project costs	C.F.	2.11	3.30	5.45				
	0100	Site work	S.F.	4.15	8.25	12.45	6.05%	12.95%	19.85%	
	0500	Masonry		2.42	5.70	12.35	3.73%	7.40%	12.30%	
	1800	Equipment		.65	1.39	7.80	.91%	1.82%	5.55%	
	2720	Plumbing		1.34	2.41	4.50	2.90%	4.80%	6.55%	
	2730	Heating, ventilating, air conditioning		1.53	4.32	5.80	2.41%	5%	8.90%	
	2900	Electrical		2.38	4.48	7.40	5.15%	7.20%	10.10%	
	3100	Total: Mechanical & Electrical	↓	6.65	10.20	20	12.75%	18.90%	26%	
99	0010	**WAREHOUSE & OFFICES Combination**	S.F.	49.50	66	90.50				99
	0020	Total project costs	C.F.	2.53	3.67	5.45				
	1800	Equipment	S.F.	.86	1.66	2.47	.52%	1.20%	2.40%	
	2720	Plumbing		1.91	3.39	4.96	3.74%	4.76%	6.30%	
	2770	Heating, ventilating, air conditioning		3.02	4.72	6.60	5%	5.65%	10.05%	
	2900	Electrical		3.31	4.93	7.75	5.75%	8%	10%	
	3100	Total: Mechanical & Electrical	↓	9.25	14.25	22.50	14.40%	19.95%	24.50%	

L1010-101   Minimum Design Live Loads in Pounds per S.F. for Various Building Codes

Occupancy	Description	Minimum Live Loads, Pounds per S.F.
		IBC
Access Floors	Office use	50
	Computer use	100
Armories		150
Assembly	Fixed seats	60
	Movable seats	100
	Platforms or stage floors	125
Commercial & Industrial	Light manufacturing	125
	Heavy manufacturing	250
	Light storage	125
	Heavy storage	250
	Stores, retail, first floor	100
	Stores, retail, upper floors	75
	Stores, wholesale	125
Dance halls	Ballrooms	100
Dining rooms	Restaurants	100
Fire escapes	Other than below	100
	Single family residential	40
Garages	Passenger cars only	40
Gymnasiums	Main floors and balconies	100
Hospitals	Operating rooms, laboratories	60
	Private room	40
	Corridors, above first floor	80
Libraries	Reading rooms	60
	Stack rooms	150
	Corridors, above first floor	80
Marquees		75
Office Buildings	Offices	50
	Lobbies	100
	Corridors, above first floor	80
Residential	Multi family private apartments	40
	Multi family, public rooms	100
	Multi family, corridors	100
	Dwellings, first floor	40
	Dwellings, second floor & habitable attics	30
	Dwellings, uninhabitable attics	20
	Hotels, guest rooms	40
	Hotels, public rooms	100
	Hotels, corridors serving public rooms	100
	Hotels, corridors	100
Schools	Classrooms	40
	Corridors	100
Sidewalks	Driveways, etc. subject to trucking	250
Stairs	Exits	100
Theaters	Aisles, corridors and lobbies	See Assembly
	Orchestra floors	↓
	Balconies	
	Stage floors	125
Yards	Terrace, pedestrian	100

UBC = Uniform Building Code, International Conference of Building Officials, 2006 Edition

Table L1010-201　Design Weight Per S.F. for Walls and Partitions

Type	Wall Thickness	Description	Weight Per S.F.	Type	Wall Thickness	Description	Weight Per S.F.
Brick	4″	Clay brick, high absorption	34 lb.	Clay tile	2″	Split terra cotta furring	10 lb.
		Clay brick, medium absorption	39			Non load bearing clay tile	11
		Clay brick, low absorption	46		3″	Split terra cotta furring	12
		Sand-lime brick	38			Non load bearing clay tile	18
		Concrete brick, heavy aggregate	46		4″	Non load bearing clay tile	20
		Concrete brick, light aggregate	33			Load bearing clay tile	24
	8″	Clay brick, high absorption	69		6″	Non load bearing clay tile	30
		Clay brick, medium absorption	79			Load bearing clay tile	36
		Clay brick, low absorption	89		8″	Non load bearing clay tile	36
		Sand-lime brick	74			Load bearing clay tile	42
		Concrete brick, heavy aggregate	89		12″	Non load bearing clay tile	46
		Concrete brick, light aggregate	68			Load bearing clay tile	58
	12″	Common brick	120	Gypsum block	2″	Hollow gypsum block	9.5
		Pressed brick	130			Solid gypsum block	12
		Sand-lime brick	105		3″	Hollow gypsum block	10
		Concrete brick, heavy aggregate	130			Solid gypsum block	18
		Concrete brick, light aggregate	98		4″	Hollow gypsum block	15
	16″	Clay brick, high absorption	134			Solid gypsum block	24
		Clay brick, medium absorption	155		5″	Hollow gypsum block	18
		Clay brick, low absorption	173		6″	Hollow gypsum block	24
		Sand-lime brick	138	Structural facing tile	2″	Facing tile	15
		Concrete brick, heavy aggregate	174		4″	Facing tile	25
		Concrete brick, light aggregate	130		6″	Facing tile	38
Concrete block	4″	Solid conc. block, stone aggregate	45	Glass	4″	Glass block	18
		Solid conc. block, lightweight	34		1″	Structural glass	15
		Hollow conc. block, stone aggregate	30	Plaster	1″	Gypsum plaster (1 side)	5
		Hollow conc. block, lightweight	20			Cement plaster (1 side)	10
	6″	Solid conc. block, stone aggregate	50			Gypsum plaster on lath	8
		Solid conc. block, lightweight	37			Cement plaster on lath	13
		Hollow conc. block, stone aggregate	42	Plaster partition (2 finished faces)	2″	Solid gypsum on metal lath	18
		Hollow conc. block, lightweight	30			Solid cement on metal lath	25
	8″	Solid conc. block, stone aggregate	67			Solid gypsum on gypsum lath	18
		Solid conc. block, lightweight	48			Gypsum on lath & metal studs	18
		Hollow conc. block, stone aggregate	55		3″	Gypsum on lath & metal studs	19
		Hollow conc. block, lightweight	38		4″	Gypsum on lath & metal studs	20
	10″	Solid conc. block, stone aggregate	84		6″	Gypsum on lath & wood studs	18
		Solid conc. block, lightweight	62	Concrete	6″	Reinf concrete, stone aggregate	75
		Hollow conc. block, stone aggregate	55			Reinf. concrete, lightweight	36-60
		Hollow conc. block, lightweight	38		8″	Reinf. concrete, stone aggregate	100
	12″	Solid conc. block, stone aggregate	108			Reinf. concrete, lightweight	48-80
		Solid conc. block, lightweight	72		10″	Reinf. concrete, stone aggregate	125
		Hollow conc. block, stone aggregate	85			Reinf. concrete, lightweight	60-100
		Hollow conc. block, lightweight	55		12″	Reinf concrete stone aggregate	150
Drywall	6″	Drywall on wood studs	10			Reinf concrete, lightweight	72-120

Table L1010-202　Design Weight per S.F. for Roof Coverings

Type		Description	Weight lb. Per S.F.	Type	Wall Thickness	Description	Weight lb. Per S.F.
Sheathing	Gypsum	1″ thick	4	Metal	Aluminum	Corr. & ribbed, .024″ to .040″	.4-.8
	Wood	¾″ thick	3		Copper	or tin	1.5-2.5
Insulation	per 1″	Loose	.5		Steel	Corrugated, 29 ga. to 12 ga.	.6-5.0
		Poured in place	2	Shingles	Asphalt	Strip shingles	1.7-2.8
		Rigid	1.5		Clay	Tile	8-16
Built-up	Tar & gravel	3 ply felt	5.5		Slate	¼″ thick	9.5
		5 ply felt	6.5		Wood		2-3

Table L1010-203   Design Weight per Square Foot for Floor Fills and Finishes

Type		Description	Weight lb. per S.F.	Type		Description	Weight lb. per S.F.
Floor fill	per 1"	Cinder fill Cinder concrete	5 9	Wood	Single 7/8"	On sleepers, light concrete fill On sleepers, stone concrete fill	16 25
		Lightweight concrete	3-9		Double 7/8"	On sleepers, light concrete fill	19
		Stone concrete	12			On sleepers, stone concrete fill	28
		Sand	8		3"	Wood block on mastic, no fill	15
		Gypsum	6			Wood block on ½" mortar	16
Terrazzo	1"	Terrazzo, 2" stone concrete	25		per 1"	Hardwood flooring (25/32")	4
Marble	and mortar	on stone concrete fill	33			Underlayment (Plywood per 1")	3
Resilient	1/16"-1/4"	Linoleum, asphalt, vinyl tile	2	Asphalt	1-1/2"	Mastic flooring	18
Tile	3/4"	Ceramic or quarry	10		2"	Block on ½" mortar	30

Table L1010-204   Design Weight Per Cubic Foot for Miscellaneous Materials

Type		Description	Weight lb. per C.F.	Type		Description	Weight lb. per C.F.
Bituminous	Coal, piled	Anthracite Bituminous Peat, turf, dry Coke	47-58 40-54 47 75	Masonry	Ashlar	Granite Limestone, crystalline Limestone, oolitic Marble	165 165 135 173
	Petroleum	Unrefined	54			Sandstone	144
		Refined Gasoline	50 42		Rubble, in mortar	Granite Limestone, crystalline	155 147
	Pitch		69			Limestone, oolitic	138
	Tar	Bituminous	75			Marble	156
Concrete	Plain	Stone aggregate	144			Sandstone & Bluestone	130
		Slag aggregate Expanded slag aggregate Haydite (burned clay agg.)	132 100 90		Brick	Pressed Common Soft	140 120 100
		Vermiculite & perlite, load bearing Vermiculite & perlite, non load bear	70-105 35-50		Cement	Portland, loose Portland set	90 183
	Reinforced	Stone aggregate	150		Lime	Gypsum, loose	53-64
		Slag aggregate	138		Mortar	Set	103
		Lightweight aggregates	30-120	Metals	Aluminum	Cast, hammered	165
Earth	Clay	Dry	63		Brass	Cast, rolled	534
		Damp, plastic	110		Bronze	7.9 to 14% Sn	509
		and gravel, dry	100		Copper	Cast, rolled	556
	Dry	Loose Packed	76 95		Iron	Cast, pig Wrought	450 480
	Moist	Loose	78		Lead		710
		Packed	96		Monel		556
	Mud	Flowing	108		Steel	Rolled	490
		Packed	115		Tin	Cast, hammered	459
	Riprap	Limestone	80-85		Zinc	Cast rolled	440
		Sandstone	90	Timber	Cedar	White or red	24.2
		Shale	105		Fir	Douglas	23.7
	Sand & gravel	Dry, loose	90-105			Eastern	25
		Dry, packed Wet	100-120 118-120		Maple	Hard White	44.5 33
Gases	Air	0C., 760 mm.	.0807		Oak	Red or Black	47.3
	Gas	Natural	.0385			White	47.3
Liquids	Alcohol	100%	49		Pine	White	26
	Water	4°C., maximum density Ice	62.5 56			Yellow, long leaf Yellow, short leaf	44 38
		Snow, fresh fallen	8		Redwood	California	26
		Sea water	64		Spruce	White or black	27

552

Table L1010-225   Design Weight Per S.F. for Structural Floor and Roof Systems

Type Slab		Description	Weight in Pounds per S.F. — Slab Depth in Inches											
			1"		2"		3"		4"		5"		6"	
Concrete Slab	Reinforced	Stone aggregate	1"	12.5	2"	25	3"	37.5	4"	50	5"	62.5	6"	75
		Lightweight sand aggregate		9.5		19		28.5		38		47.5		57
		All lightweight aggregate		9.0		18		27.0		36		45.0		54
	Plain, nonreinforced	Stone aggregate		12.0		24		36.0		48		60.0		72
		Lightweight sand aggregate		9.0		18		27.0		36		45.0		54
		All lightweight aggregate		8.5		17		25.5		34		42.5		51
Concrete Waffle	19" x 19"	5" wide ribs @ 24" O.C.	6+3	77	8+3	92	10+3	100	12+3	118				
			6+4 ½	96	8+4½	110	10+4½	119	12+4½	136				
	30" x 30"	6" wide ribs @ 36" O.C.	8+3	83	10+3	95	12+3	109	14+3	118	16+3	130	20+3	155
			8+4½	101	10+4½	113	12+4½	126	14+4½	137	16+4½	149	20+4½	173
Concrete Joist	20" wide form	5" wide rib	8+3	60	10+3	67	12+3	74	14+3	81				
		6" wide rib		63		70		78		86	16+3	94	20+3	111
		7" wide rib										99		118
		5" wide rib	8+4½	79	10+4½	85	12+4½	92	14+4½	99				
		6" wide rib		82		89		97		104	16+4½	113	20+4½	130
		7" wide rib										118		136
	30" wide form	5" wide rib	8+3	54	10+3	58	12+3	63	14+3	68				
		6" wide rib		56		61		67		72	16+3	78	20+3	91
		7" wide rib										83		96
		5" wide rib	8+4½	72	10+4½	77	12+4½	82	14+4½	87				
		6" wide rib		75		80		85		91	16+4½	97	20+4½	109
		7" wide rib										101		115
Wood Joists	Incl. Subfloor	12" O.C.	2x6	6	2x8	6	2x10	7	2x12	8	3x8	8	3x12	11
		16" O.C.		5		6		6		7		7		9

Table L1010-226   Superimposed Dead Load Ranges

Component	Load Range (PSF)
Ceiling	5-10
Partitions	20-30
Mechanical	4-8

Table L1010-301 Design Loads for Structures for Wind Load

Wind Loads: Structures are designed to resist the wind force from any direction. Usually 2/3 is assumed to act on the windward side, 1/3 on the leeward side.

For more than 1/3 openings, add 10 psf for internal wind pressure or 5 psf for suction, whichever is critical.

For buildings and structures, use psf values from Table L1010-301.

For glass over 4 S.F. use values in Table L1010-302 after determining 30' wind velocity from Table L1010-303.

Type Structure	Height Above Grade	Horizontal Load in Lb. per S.F.
Buildings	Up to 50 ft.	15
	50 to 100 ft.	20
	Over 100 ft.	20 + .025 per ft.
Ground signs & towers	Up to 50 ft.	15
	Over 50 ft.	20
Roof structures		30
Glass	See Table below	

Table L1010-302 Design Wind Load in PSF for Glass at Various Elevations (BOCA Code)

Height From Grade	Velocity in Miles per Hour and Design Load in Pounds per S.F.																				
	Vel.	PSF	Vel.	PSF	Vel.	PSF	Vel.	PSF	Vel.	PSF	Vel.	PSF	Vel.	PSF	Vel.	PSF	Vel.	PSF	Vel.	PSF	Vel.	PSF
To 10 ft.	42	6	46	7	49	8	52	9	55	10	59	11	62	12	66	14	69	15	76	19	83	22
10-20	52	9	58	11	61	11	65	14	70	16	74	18	79	20	83	22	87	24	96	30	105	35
20-30*	60	12	67	14	70	16	75	18	80	20	85	23	90	26	95	29	100	32	110	39	120	46
30-60	66	14	74	18	77	19	83	22	88	25	94	28	99	31	104	35	110	39	121	47	132	56
60-120	73	17	82	21	85	12	92	27	98	31	104	35	110	39	116	43	122	48	134	57	146	68
120-140	81	21	91	26	95	29	101	33	108	37	115	42	122	48	128	52	135	48	149	71	162	84
240-480	90	26	100	32	104	35	112	40	119	45	127	51	134	57	142	65	149	71	164	86	179	102
480-960	98	31	110	39	115	42	123	49	131	55	139	62	148	70	156	78	164	86	180	104	197	124
Over 960	98	31	110	39	115	42	123	49	131	55	139	62	148	70	156	78	164	86	180	104	197	124

*Determine appropriate wind at 30' elevation Fig. L1010-303 below.

Table L1010-303 Design Wind Velocity at 30 Ft. Above Ground

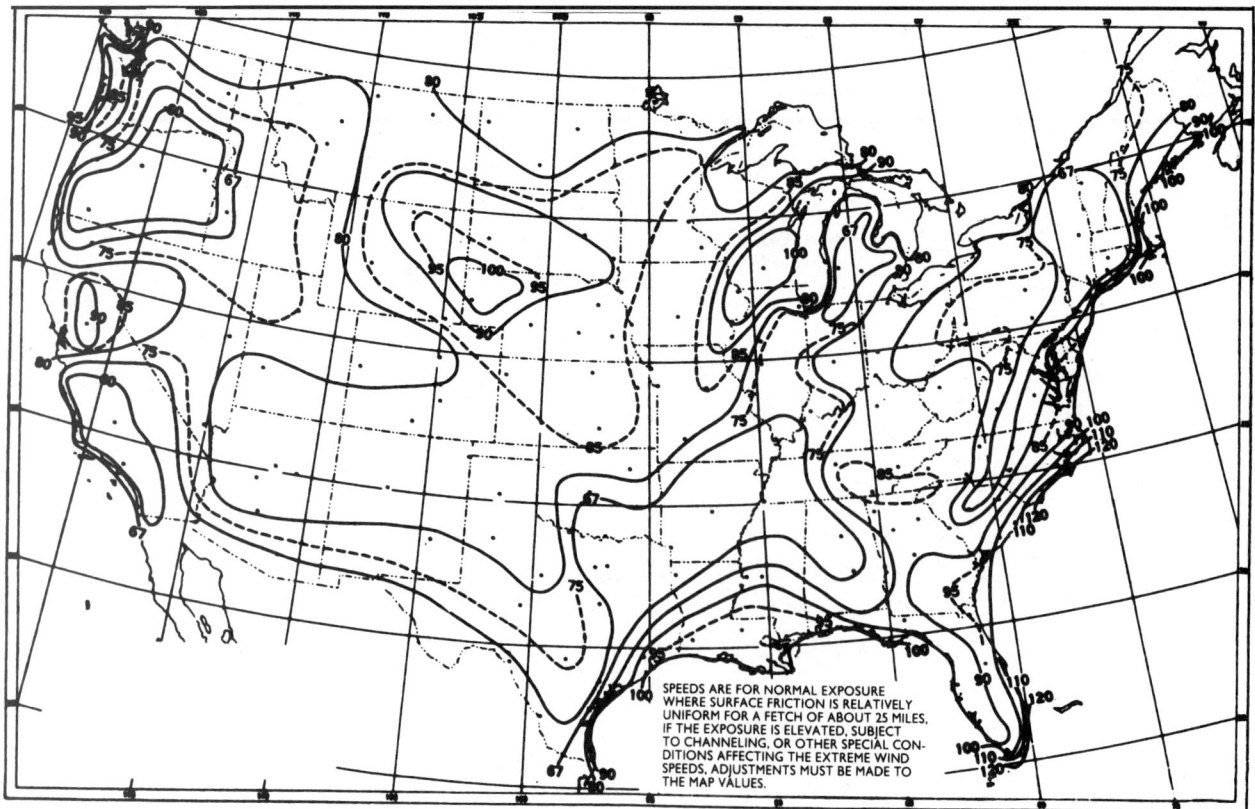

SPEEDS ARE FOR NORMAL EXPOSURE WHERE SURFACE FRICTION IS RELATIVELY UNIFORM FOR A FETCH OF ABOUT 25 MILES, IF THE EXPOSURE IS ELEVATED, SUBJECT TO CHANNELING, OR OTHER SPECIAL CONDITIONS AFFECTING THE EXTREME WIND SPEEDS, ADJUSTMENTS MUST BE MADE TO THE MAP VALUES.

Table L1010-401   Snow Load in Pounds Per Square Foot on the Ground

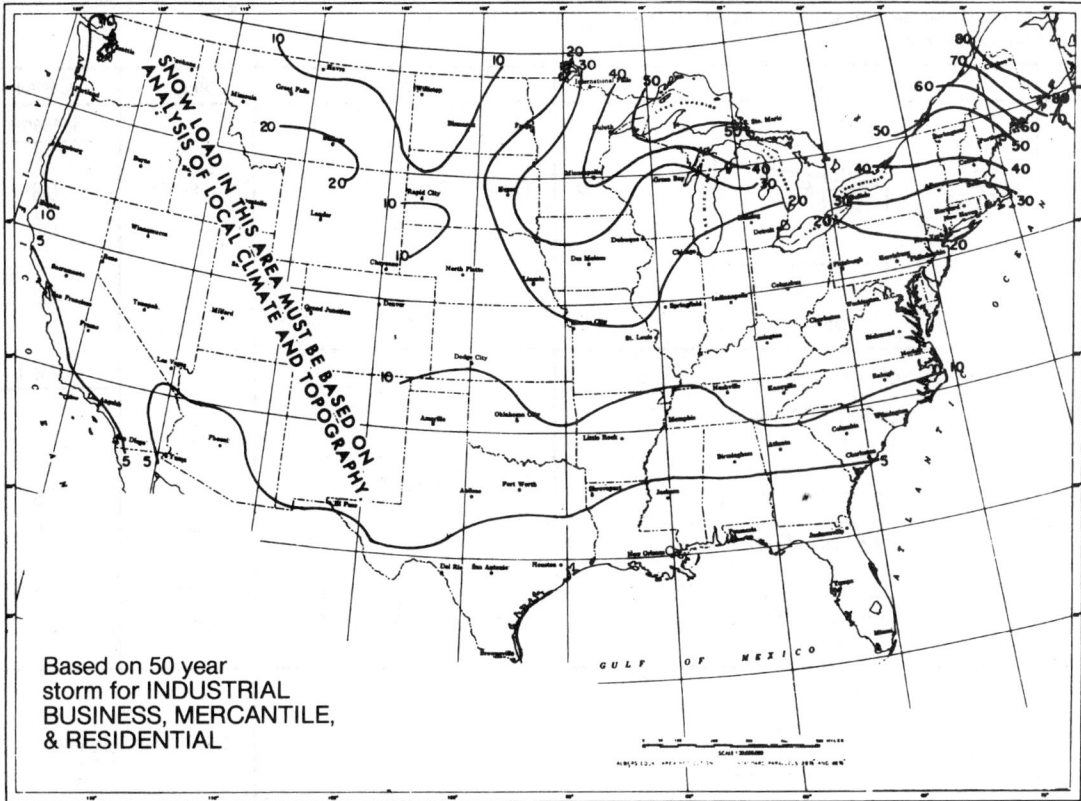

Based on 50 year
storm for INDUSTRIAL
BUSINESS, MERCANTILE,
& RESIDENTIAL

Table L1010-402   Snow Load in Pounds Per Square Foot on the Ground

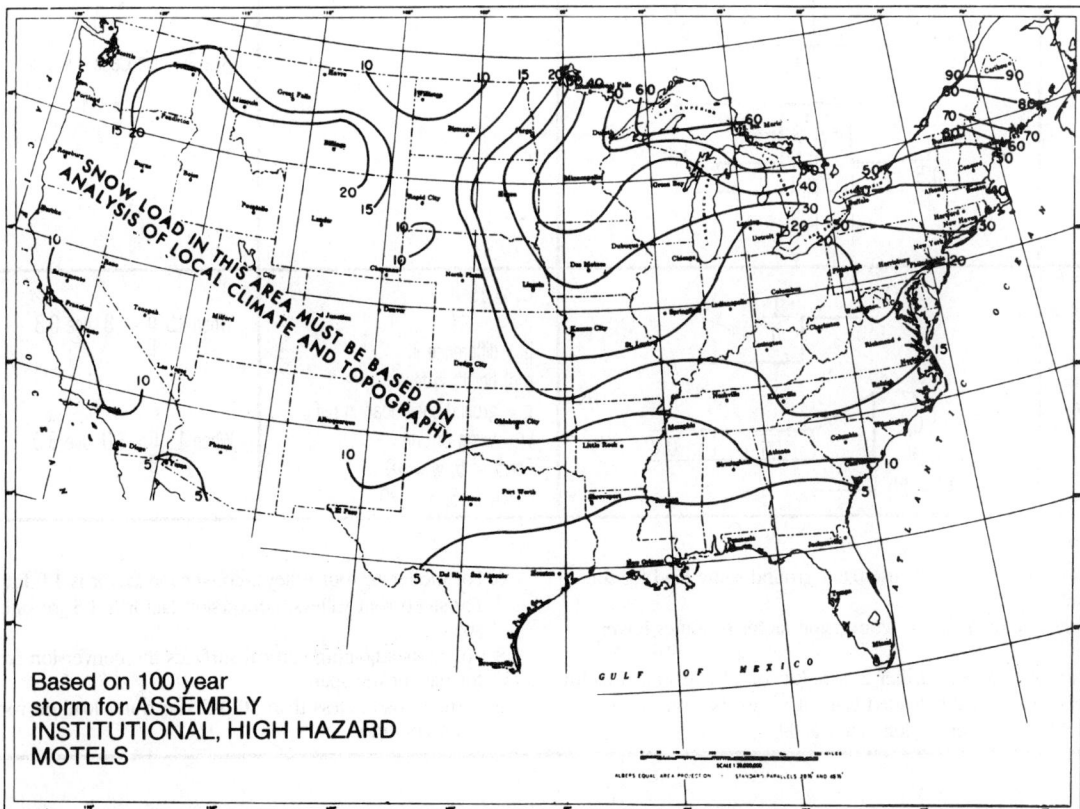

Based on 100 year
storm for ASSEMBLY
INSTITUTIONAL, HIGH HAZARD
MOTELS

555

Table L1010-403   Snow Loads

To convert the ground snow loads on the previous page to roof snow loads, the ground snow loads should be multiplied by the following factors depending upon the roof characteristics.

Note, in all cases $\dfrac{\alpha - 30}{50}$ is valid only for > 30 degrees.

Description	Sketch	Formula	Angle	Conversion Factor = C_f	
				Sheltered	Exposed
Simple flat and shed roofs	Load Diagram	For $\alpha > 30°$ $C_f = 0.8 - \dfrac{\alpha - 30}{50}$	0° to 30° 40° 50° 60° 70° to 90°	0.8 0.6 0.4 0.2 0	0.6 0.45 0.3 0.15 0
				Case I	Case II
Simple gable and hip roofs	Case I Load Diagram Case II Load Diagram	$C_f = 0.8 - \dfrac{\alpha - 30}{50}$ $C_f = 1.25\left(0.8 - \left(\dfrac{\alpha - 30}{50}\right)\right)$	10° 20° 30° 40° 50° 60°	0.8 0.8 0.8 0.6 0.4 0.2	— — 1.0 0.75 0.5 0.25
Valley areas of Two span roofs	Case I Case II	$\beta = \dfrac{\alpha_1 + \alpha_2}{2}$ $C_f = 0.8 - \dfrac{\alpha - 30}{50}$	$\beta \le 10°$ use Case I only $\beta > 10°$ $\beta < 20°$ use Case I & II $\beta \ge 20°$ use Case I, II & III		
Lower level of multi level roofs (or on an adjacent building not more then 15 ft. away)		$C_f = 15\dfrac{h}{g}$ h = difference in roof height in feet g = ground snow load in psf w = width of drift For h < 5, w = 10 h >15, w = 30	When $15\dfrac{h}{g} < .8$, use 0.8 When $15\dfrac{h}{g} > 3.0$, use 3.0		

Summary of above:

1. For flat roofs or roofs up to 30°, use 0.8 x ground snow load for the roof snow load.
2. For roof pitches in excess of 30°, conversion factor becomes lower than 0.8.
3. For exposed roofs there is a further 25% reduction of conversion factor.
4. For steep roofs a more highly loaded half span must be considered.
5. For shallow roof valleys conversion factor is 0.8.
6. For moderate roof valleys, conversion factor is 1.0 for half the span.
7. For steep roof valleys, conversion factor is 1.5 for one quarter of the span.
8. For roofs adjoining vertical surfaces the conversion factor is up to 3.0 for part of the span.
9. If snow load is less than 30 psf, use water load on roof for clogged drain condition.

Table L1020-101 Floor Area Ratios

Table below lists commonly used gross to net area and net to gross area ratios expressed in % for various building types.

Building Type	Gross to Net Ratio	Net to Gross Ratio	Building Type	Gross to Net Ratio	Net to Gross Ratio
Apartment	156	64	School Buildings (campus type)		
Bank	140	72	Administrative	150	67
Church	142	70	Auditorium	142	70
Courthouse	162	61	Biology	161	62
Department Store	123	81	Chemistry	170	59
Garage	118	85	Classroom	152	66
Hospital	183	55	Dining Hall	138	72
Hotel	158	63	Dormitory	154	65
Laboratory	171	58	Engineering	164	61
Library	132	76	Fraternity	160	63
Office	135	75	Gymnasium	142	70
Restaurant	141	70	Science	167	60
Warehouse	108	93	Service	120	83
			Student Union	172	59

The gross area of a building is the total floor area based on outside dimensions.

The net area of a building is the usable floor area for the function intended and excludes such items as stairways, corridors, and mechanical rooms.

In the case of a commercial building, it might be considered as the "leasable area."

Table L1020-201 Partition/Door Density

Building Type		Stories	Partition/Density	Doors	Description of Partition
Apartments		1 story	9 SF/LF	90 SF/door	Plaster, wood doors & trim
		2 story	8 SF/LF	80 SF/door	Drywall, wood studs, wood doors & trim
		3 story	9 SF/LF	90 SF/door	Plaster, wood studs, wood doors & trim
		5 story	9 SF/LF	90 SF/door	Plaster, metal studs, wood doors & trim
		6-15 story	8 SF/LF	80 SF/door	Drywall, metal studs, wood doors & trim
Bakery		1 story	50 SF/LF	500 SF/door	Conc. block, paint, door & drywall, wood studs
		2 story	50 SF/LF	500 SF/door	Conc. block, paint, door & drywall, wood studs
Bank		1 story	20 SF/LF	200 SF/door	Plaster, wood studs, wood doors & trim
		2-4 story	15 SF/LF	150 SF/door	Plaster, metal studs, wood doors & trim
Bottling Plant		1 story	50 SF/LF	500 SF/door	Conc. block, drywall, metal studs, wood trim
Bowling Alley		1 story	50 SF/LF	500 SF/door	Conc. block, wood & metal doors, wood trim
Bus Terminal		1 story	15 SF/LF	150 SF/door	Conc. block, ceramic tile, wood trim
Cannery		1 story	100 SF/LF	1000 SF/door	Drywall on metal studs
Car Wash		1 story	18 SF/LF	180 SF/door	Concrete block, painted & hollow metal door
Dairy Plant		1 story	30 SF/LF	300 SF/door	Concrete block, glazed tile, insulated cooler doors
Department Store		1 story	60 SF/LF	600 SF/door	Drywall, metal studs, wood doors & trim
		2-5 story	60 SF/LF	600 SF/door	30% concrete block, 70% drywall, wood studs
Dormitory		2 story	9 SF/LF	90 SF/door	Plaster, concrete block, wood doors & trim
		3-5 story	9 SF/LF	90 SF/door	Plaster, concrete block, wood doors & trim
		6-15 story	9 SF/LF	90 SF/door	Plaster, concrete block, wood doors & trim
Funeral Home		1 story	15 SF/LF	150 SF/door	Plaster on concrete block & wood studs, paneling
		2 story	14 SF/LF	140 SF/door	Plaster, wood studs, paneling & wood doors
Garage Sales & Service		1 story	30 SF/LF	300 SF/door	50% conc. block, 50% drywall, wood studs
Hotel		3-8 story	9 SF/LF	90 SF/door	Plaster, conc. block, wood doors & trim
		9-15 story	9 SF/LF	90 SF/door	Plaster, conc. block, wood doors & trim
Laundromat		1 story	25 SF/LF	250 SF/door	Drywall, wood studs, wood doors & trim
Medical Clinic		1 story	6 SF/LF	60 SF/door	Drywall, wood studs, wood doors & trim
		2-4 story	6 SF/LF	60 SF/door	Drywall, metal studs, wood doors & trim
Motel		1 story	7 SF/LF	70 SF/door	Drywall, wood studs, wood doors & trim
		2-3 story	7 SF/LF	70 SF/door	Concrete block, drywall on metal studs, wood paneling
Movie Theater	200-600 seats	1 story	18 SF/LF	180 SF/door	Concrete block, wood, metal, vinyl trim
	601-1400 seats		20 SF/LF	200 SF/door	Concrete block, wood, metal, vinyl trim
	1401-22000 seats		25 SF/LF	250 SF/door	Concrete block, wood, metal, vinyl trim
Nursing Home		1 story	8 SF/LF	80 SF/door	Drywall, metal studs, wood doors & trim
		2-4 story	8 SF/LF	80 SF/door	Drywall, metal studs, wood doors & trim
Office		1 story	20 SF/LF	200-500 SF/door	30% concrete block, 70% drywall on wood studs
		2 story	20 SF/LF	200-500 SF/door	30% concrete block, 70% drywall on metal studs
		3-5 story	20 SF/LF	200-500 SF/door	30% concrete block, 70% movable partitions
		6-10 story	20 SF/LF	200-500 SF/door	30% concrete block, 70% movable partitions
		11-20 story	20 SF/LF	200-500 SF/door	30% concrete block, 70% movable partitions
Parking Ramp (Open)		2-8 story	60 SF/LF	600 SF/door	Stair and elevator enclosures only
Parking garage		2-8 story	60 SF/LF	600 SF/door	Stair and elevator enclosures only
Pre-Engineered	Steel	1 story	0		
	Store	1 story	60 SF/LF	600 SF/door	Drywall on metal studs, wood doors & trim
	Office	1 story	15 SF/LF	150 SF/door	Concrete block, movable wood partitions
	Shop	1 story	15 SF/LF	150 SF/door	Movable wood partitions
	Warehouse	1 story	0		
Radio & TV Broadcasting		1 story	25 SF/LF	250 SF/door	Concrete block, metal and wood doors
& TV Transmitter		1 story	40 SF/LF	400 SF/door	Concrete block, metal and wood doors
Self Service Restaurant		1 story	15 SF/LF	150 SF/door	Concrete block, wood and aluminum trim
Cafe & Drive-in Restaurant		1 story	18 SF/LF	180 SF/door	Drywall, metal studs, ceramic & plastic trim
Restaurant with seating		1 story	25 SF/LF	250 SF/door	Concrete block, paneling, wood studs & trim
Supper Club		1 story	25 SF/LF	250 SF/door	Concrete block, paneling, wood studs & trim
Bar or Lounge		1 story	24 SF/LF	240 SF/door	Plaster or gypsum lath, wooded studs
Retail Store or Shop		1 story	60 SF/LF	600 SF/door	Drywall metal studs, wood doors & trim
Service Station	Masonry	1 story	15 SF/LF	150 SF/door	Concrete block, paint, door & drywall, wood studs
	Metal panel	1 story	15 SF/LF	150 SF/door	Concrete block, paint, door & drywall, wood studs
	Frame	1 story	15 SF/LF	150 SF/door	Drywall, wood studs, wood doors & trim
Shopping Center	(strip)	1 story	30 SF/LF	300 SF/door	Drywall, metal studs, wood doors & trim
	(group)	1 story	40 SF/LF	400 SF/door	50% concrete block, 50% drywall, wood studs
		2 story	40 SF/LF	400 SF/door	50% concrete block, 50% drywall, wood studs
Small Food Store		1 story	30 SF/LF	300 SF/door	Concrete block drywall, wood studs, wood trim
Store/Apt. above	Masonry	2 story	10 SF/LF	100 SF/door	Plaster, metal studs, wood doors & trim
	Frame	2 story	10 SF/LF	100 SF/door	Plaster, metal studs, wood doors & trim
	Frame	3 story	10 SF/LF	100 SF/door	Plaster, metal studs, wood doors & trim
Supermarkets		1 story	40 SF/LF	400 SF/door	Concrete block, paint, drywall & porcelain panel
Truck Terminal		1 story	0		
Warehouse		1 story	0		

Table L1020-301 Occupancy Determinations

Description		S.F. Required per Person IBC
Assembly Areas	Without fixed seats: Concentrated Unconcentrated (tables and chairs) Standing Space	7 15 5
Educational	Classrooms Shop Areas	20 50
Institutional	Outpatient Areas In-Patient Areas Sleeping Areas	100 240 120
Mercantile	Basement Ground Floor Upper Floors Storage	30 30 60 300
Office		100

IBC = International Building Code, 2006 Edition

** The occupancy load for assembly area with fixed seats shall be determined by the number of fixed seats installed.

Table L1020-302 Length of Exitway Access Travel (ft.)

Use Group	Without Fire Suppression System	With Fire Suppression System
Assembly	200	250
Business	200	300
Factory and industrial	300	400
High hazard	—	75
Institutional	200	250
Mercantile	200	250
Residential	200	250
Storage, low hazard	300	400
Storage, moderate hazard	200	250

Note: The maximum length of exitway access travel in unlimited area buildings shall be 400'.

Table L1020-303 Capacity per Unit Egress Width*

Use Group	Without Fire Suppression System (Inches per Person)*		With Fire Suppression System (Inches per Person)*	
	Stairways	Doors, Ramps and Corridors	Stairways	Doors, Ramps and Corridors
Assembly, Business, Educational, Factory Industrial, Mercantile, Residential, Storage	0.3	0.2	0.2	0.15
Institutional—1	0.3	0.2	0.2	0.15
Institutional—2	—	0.7	0.3	0.2
Institutional—3	0.3	0.2	0.2	0.15
High Hazard	0.7	0.4	0.3	0.2

* 1" = 25.4 mm

Table L1030-101 "U" Values for Type "A" Buildings

Type A buildings shall include:

    A1 Detached one and two family dwellings

    A2 All other residential buildings, three stories or less, including but not limited to:
        multi-family dwellings, hotels and motels.

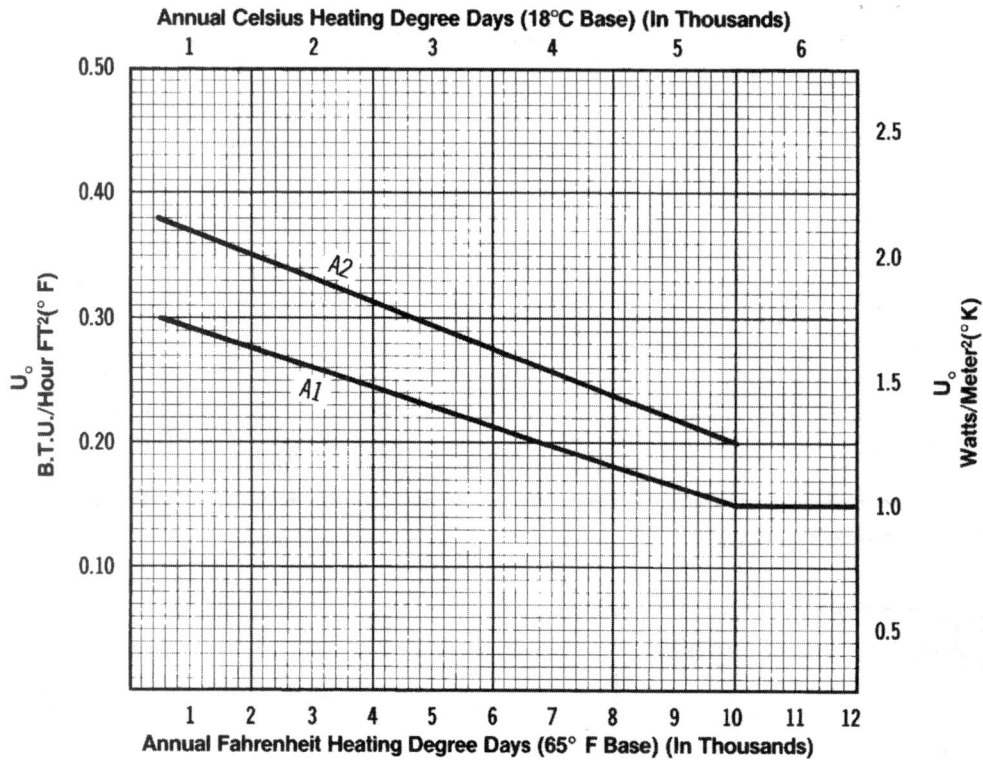

Table L1030-102 "U" Values for Type "B" Buildings

For all buildings Not Classified Type "A"

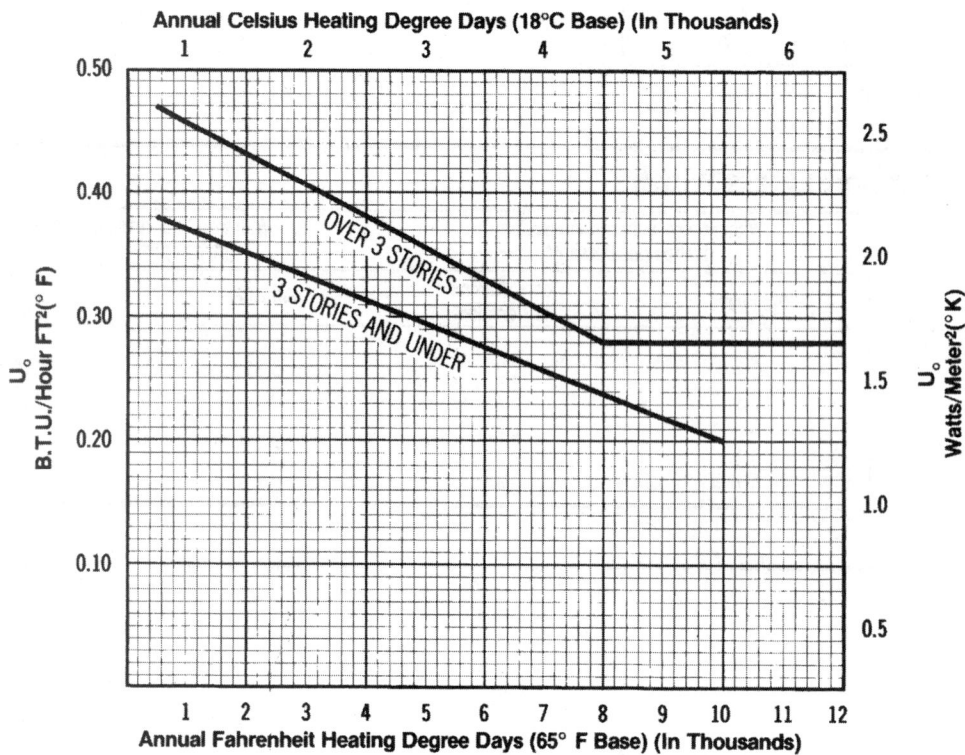

**Table L1030-201   Combinations of Wall and Single-Glazed Openings
(For Use With ASHRAE 90-75)**

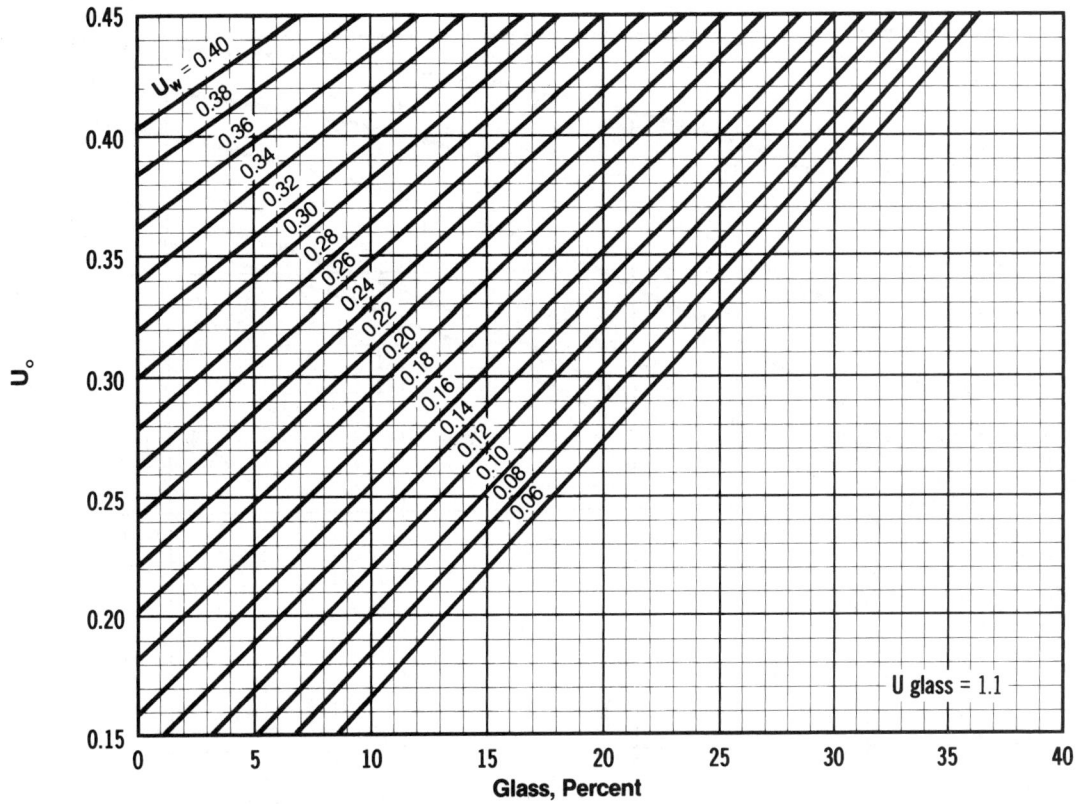

**Table L1030-202   Combinations of Wall and Double-Glazed Openings
(For Use With ASHRAE 90-75)**

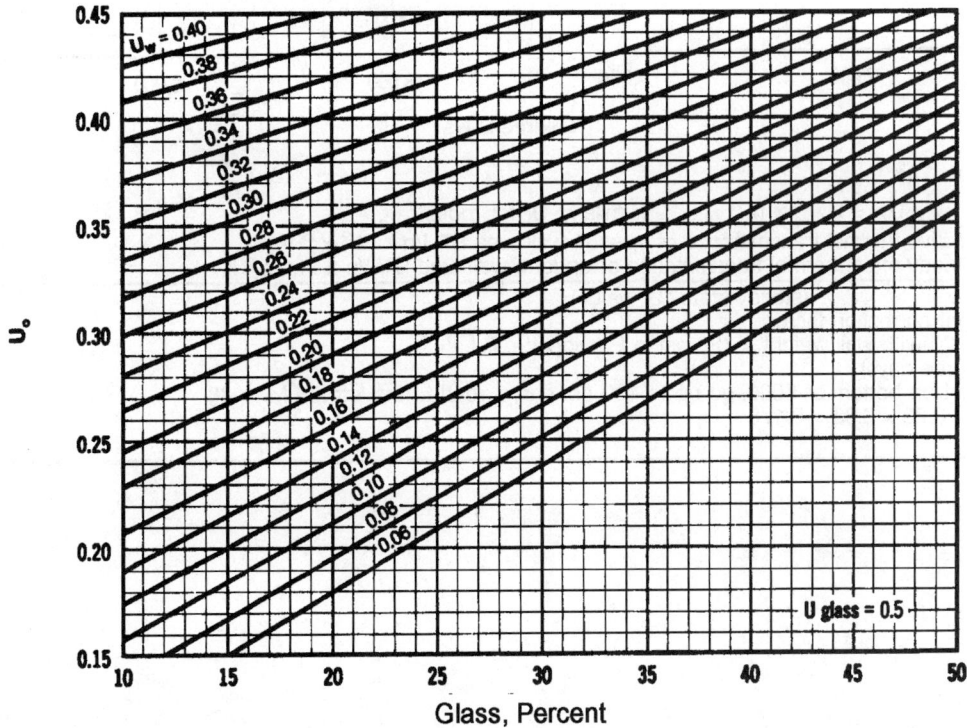

Table L1030-203 Influence of Wall and Roof Weight on "U" Value Correction for Heating Design

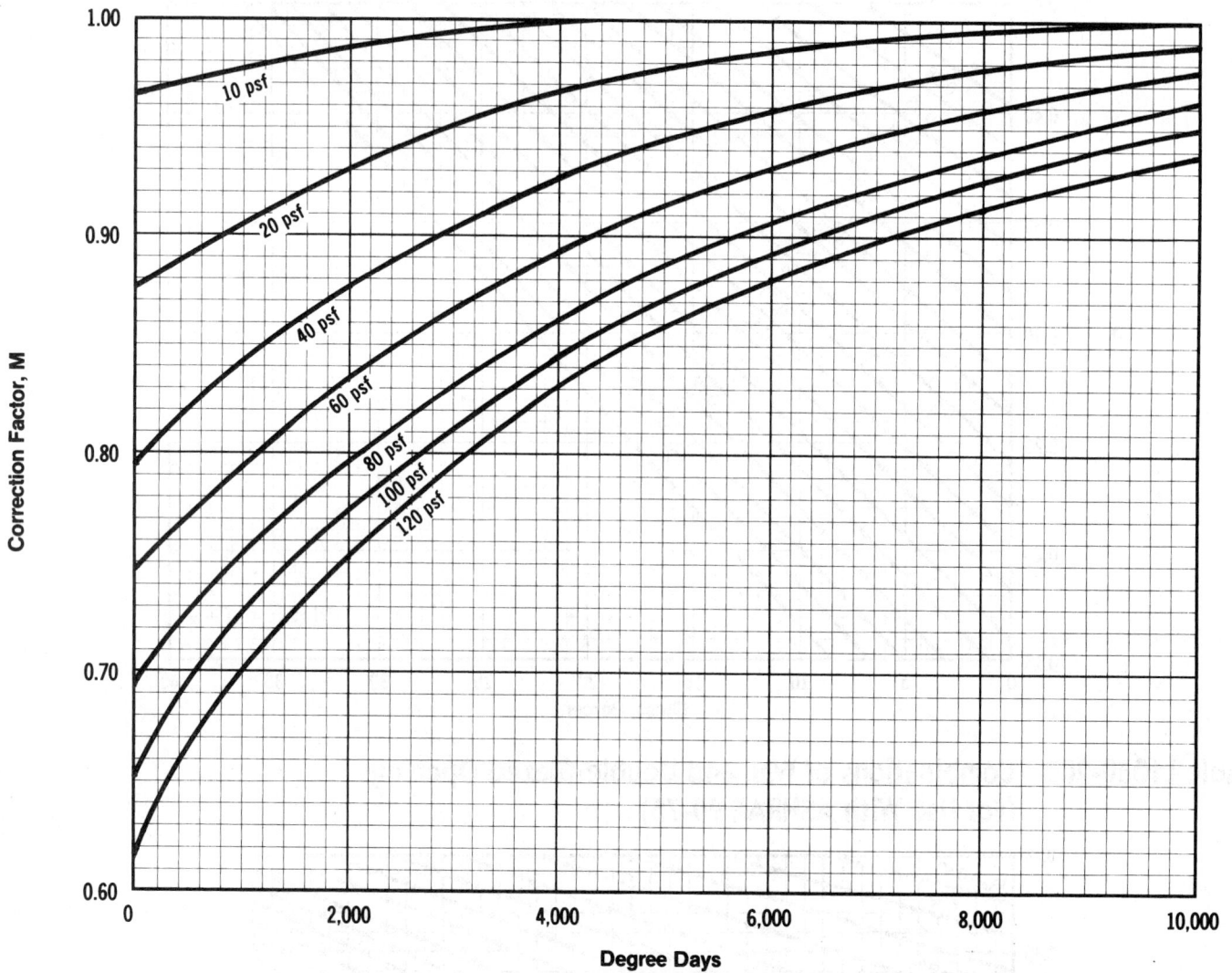

Effective "U" for walls: $Uw = Uw_{ss} \times M$ (similarly, Ur for roofs)

where Uw = effective thermal transmittance of opaque wall area BTU/h x Ft.^{2} x ° F

Uw_{ss} = steady state thermal transmittance of opaque wall area BTU/h x Ft.^{2} x ° F

(steady state "U" value)

M = weight correction factor

Example: Uw_{ss} = 0.20 with wall weight = 120 psf in Providence, R.I.

(6000 degree days)

Enter chart on bottom at 6000, go up to 120 psf curve, read to the left .88 M = .88

and Uw = 0.20 x 0.88 = 0.176

Table L1030-204   "U" Values for Type "A" Buildings

One and two family dwellings and residential buildings of three stories or less shall have an overall U value as follows:

A. For roof assemblies in which the finished interior surface is essentially the underside of the roof deck, such as exposed concrete slabs, joist slabs, cored slabs, or cathedral ceilings or wood-beam construction.

$$U_{or} = 0.08 \text{ Btu/hr./ft.}^2/^\circ \text{F. (all degree days)}$$

B. For roof-ceiling assemblies, such as roofs with finished ceilings attached to or suspended below the roof deck.

$$U_{or} = 0.05 \text{ for below 8,000 degree days}$$
$$= 0.04 \text{ for 8,000 degree days and greater.}$$

Table L1030-205   "U" Values for Type "B" Buildings*

For all buildings not classified type "A".

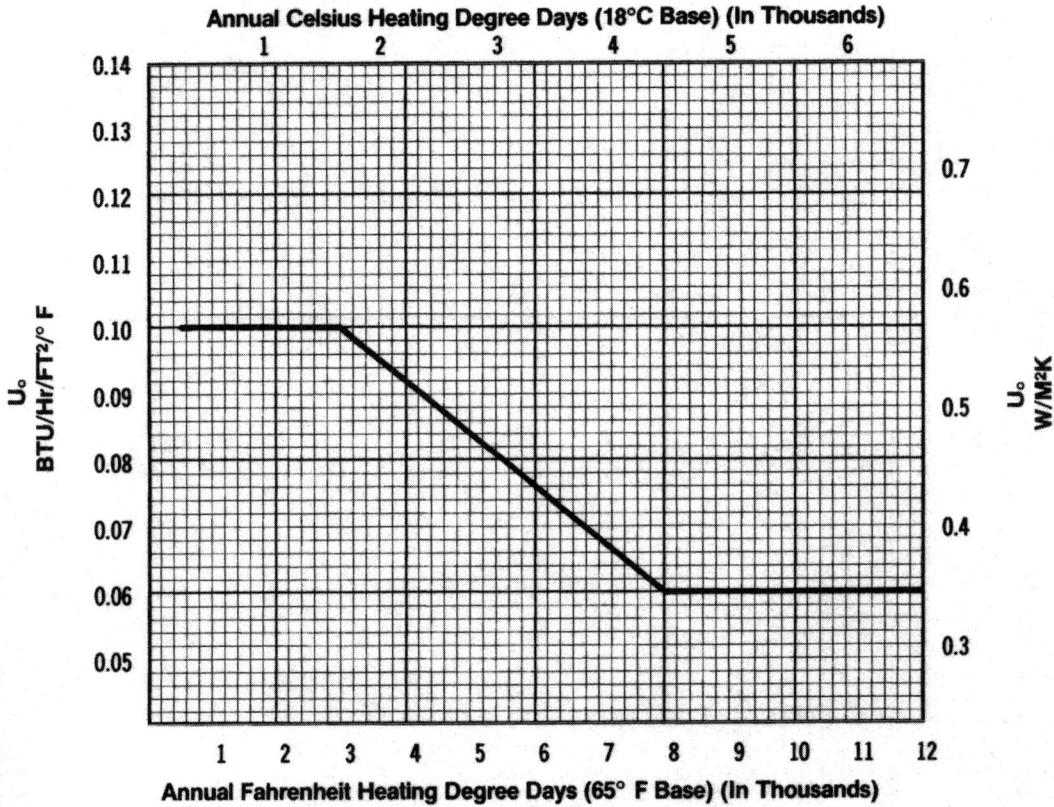

*Minimum requirements for thermal design, ASHRAE Standard 90-75

Table L1030-301   "U" Values for Floors Over Unheated Spaces

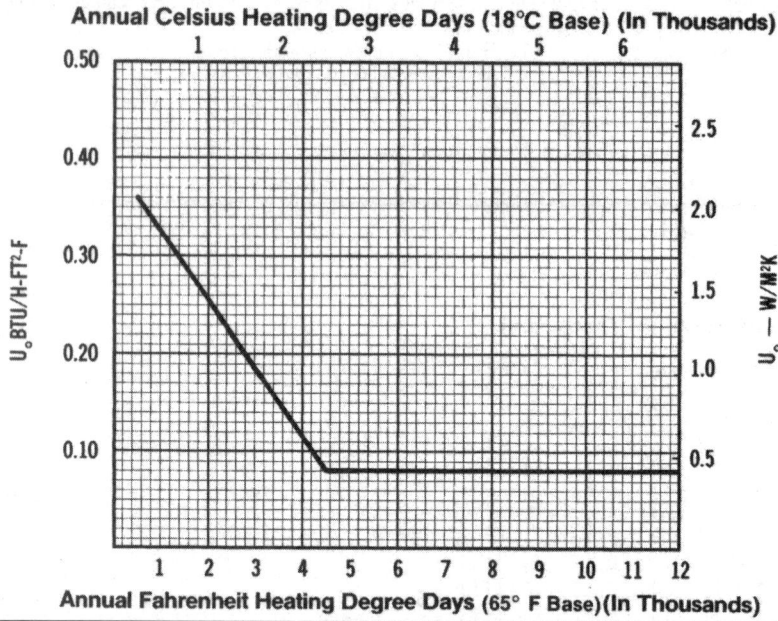

Annual Celsius Heating Degree Days (18°C Base) (In Thousands)

U_o BTU/H-FT²-F U_o — W/M²K

Annual Fahrenheit Heating Degree Days (65° F Base)(In Thousands)

Table L1030-302   "R" Values for Slabs on Grade

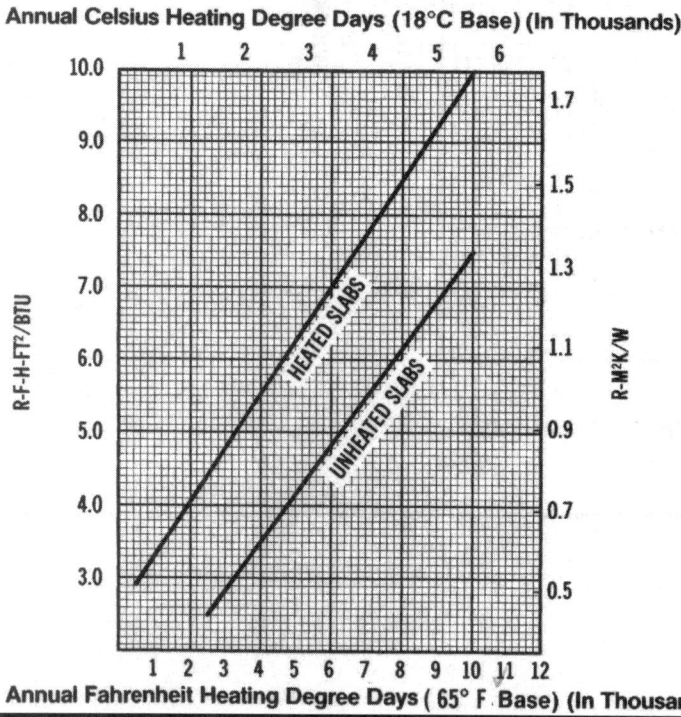

Annual Celsius Heating Degree Days (18°C Base) (In Thousands)

HEATED SLABS

UNHEATED SLABS

R-F-H-FT²/BTU   R-M²K/W

Annual Fahrenheit Heating Degree Days (65° F Base) (In Thousands)

Table L1030-303   Insulation Requirements for Slabs on Grade for all Buildings

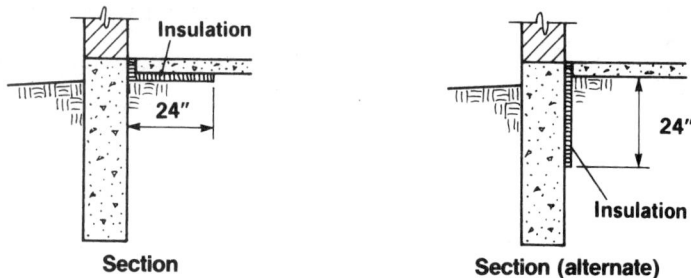

Insulation

24"

Section

24"

Insulation

Section (alternate)

Table L1030-401   Resistances ("R") of Building and Insulating Materials

Material	Wt./Lbs. per C.F.	R per Inch	R Listed Size
Air Spaces and Surfaces			
Enclosed non-reflective spaces, E=0.82			
50° F mean temp., 30°/10° F diff.			
.5"			.90/.91
.75"			.94/1.01
1.50"			.90/1.02
3.50"			.91/1.01
Inside vert. surface (still air)			0.68
Outside vert. surface (15 mph wind)			0.17
Building Boards			
Asbestos cement, 0.25" thick	120		0.06
Gypsum or plaster, 0.5" thick	50		0.45
Hardboard regular	50	1.37	
Tempered	63	1.00	
Laminated paper	30	2.00	
Particle board	37	1.85	
	50	1.06	
	63	0.85	
Plywood (Douglas Fir), 0.5" thick	34		0.62
Shingle backer, .375" thick	18		0.94
Sound deadening board, 0.5" thick	15		1.35
Tile and lay-in panels, plain or			
acoustical, 0.5" thick	18		1.25
Vegetable fiber, 0.5" thick	18		1.32
	25		1.14
Wood, hardwoods	48	0.91	
Softwoods	32	1.25	
Flooring Carpet with fibrous pad			2.08
With rubber pad			1.23
Cork tile, 1/8" thick			0.28
Terrazzo			0.08
Tile, resilient			0.05
Wood, hardwood, 0.75" thick			0.68
Subfloor, 0.75" thick			0.94
Glass			
Insulation, 0.50" air space			2.04
Single glass			0.91
Insulation Blanket or Batt, mineral, glass			
or rock fiber, approximate thickness			
3.0" to 3.5" thick			11
3.5" to 4.0" thick			13
6.0" to 6.5" thick			19
6.5" to 7.0" thick			22
8.5" to 9.0" thick			30
Boards			
Cellular glass	8.5	2.63	
Fiberboard, wet felted			
Acoustical tile	21	2.70	
Roof insulation	17	2.94	
Fiberboard, wet molded			
Acoustical tile	23	2.38	
Mineral fiber with resin binder	15	3.45	
Polystyrene, extruded,			
cut cell surface	1.8	4.00	
smooth skin surface	2.2	5.00	
	3.5	5.26	
Bead boards	1.0	3.57	
Polyurethane	1.5	6.25	
Wood or cane fiberboard, 0.5" thick			1.25

Material	Wt./Lbs. per C.F.	R per Inch	R Listed Size
Insulation Loose Fill			
Cellulose	2.3	3.13	
	3.2	3.70	
Mineral fiber, 3.75" to 5" thick	2-5		11
6.5" to 8.75" thick			19
7.5" to 10" thick			22
10.25" to 13.75" thick			30
Perlite	5-8	2.70	
Vermiculite	4-6	2.27	
Wood fiber	2-3.5	3.33	
Masonry Brick, Common	120	0.20	
Face	130	0.11	
Cement mortar	116	0.20	
Clay tile, hollow			
1 cell wide, 3" width			0.80
4" width			1.11
2 cells wide, 6" width			1.52
8" width			1.85
10" width			2.22
3 cells wide, 12" width			2.50
Concrete, gypsum fiber	51	0.60	
Lightweight	120	0.19	
	80	0.40	
	40	0.86	
Perlite	40	1.08	
Sand and gravel or stone	140	0.08	
Concrete block, lightweight			
3 cell units, 4"-15 lbs. ea.			1.68
6"-23 lbs. ea.			1.83
8"-28 lbs. ea.			2.12
12"-40 lbs. ea.			2.62
Sand and gravel aggregates,			
4"-20 lbs. ea.			1.17
6"-33 lbs. ea.			1.29
8"-38 lbs. ea.			1.46
12"-56 lbs. ea.			1.81
Plastering Cement Plaster,			
Sand aggregate	116	0.20	
Gypsum plaster, Perlite aggregate	45	0.67	
Sand aggregate	105	0.18	
Vermiculite aggregate	45	0.59	
Roofing			
Asphalt, felt, 15 lb.			0.06
Rolled roofing	70		0.15
Shingles	70		0.44
Built-up roofing .375" thick	70		0.33
Cement shingles	120		0.21
Vapor-permeable felt			0.06
Vapor seal, 2 layers of			
mopped 15 lb. felt			0.12
Wood, shingles 16"-7.5" exposure			0.87
Siding			
Aluminum or steel (hollow backed)			
oversheathing			0.61
With .375" insulating backer board			1.82
Foil backed			2.96
Wood siding, beveled, ½" x 8"			0.81

Table L1030-501 Weather Data and Design Conditions

City	Latitude (1) °	Latitude (1) 1'	Winter Temperatures (1) Med. of Annual Extremes	Winter Temperatures (1) 99%	Winter Temperatures (1) 97½%	Winter Degree Days (2)	Summer (Design Dry Bulb) Temperatures and Relative Humidity 1%	Summer (Design Dry Bulb) Temperatures and Relative Humidity 2½%	Summer (Design Dry Bulb) Temperatures and Relative Humidity 5%
UNITED STATES									
Albuquerque, NM	35	0	5.1	12	16	4,400	96/61	94/61	92/61
Atlanta, GA	33	4	11.9	17	22	3,000	94/74	92/74	90/73
Baltimore, MD	39	2	7	14	17	4,600	94/75	91/75	89/74
Birmingham, AL	33	3	13	17	21	2,600	96/74	94/75	92/74
Bismarck, ND	46	5	-32	-23	-19	8,800	95/68	91/68	88/67
Boise, ID	43	3	1	3	10	5,800	96/65	94/64	91/64
Boston, MA	42	2	-1	6	9	5,600	91/73	88/71	85/70
Burlington, VT	44	3	-17	-12	-7	8,200	88/72	85/70	82/69
Charleston, WV	38	2	3	7	11	4,400	92/74	90/73	87/72
Charlotte, NC	35	1	13	18	22	3,200	95/74	93/74	91/74
Casper, WY	42	5	-21	-11	-5	7,400	92/58	90/57	87/57
Chicago, IL	41	5	-8	-3	2	6,600	94/75	91/74	88/73
Cincinnati, OH	39	1	0	1	6	4,400	92/73	90/72	88/72
Cleveland, OH	41	2	-3	1	5	6,400	91/73	88/72	86/71
Columbia, SC	34	0	16	20	24	2,400	97/76	95/75	93/75
Dallas, TX	32	5	14	18	22	2,400	102/75	100/75	97/75
Denver, CO	39	5	-10	-5	1	6,200	93/59	91/59	89/59
Des Moines, IA	41	3	-14	-10	-5	6,600	94/75	91/74	88/73
Detroit, MI	42	2	-3	3	6	6,200	91/73	88/72	86/71
Great Falls, MT	47	3	-25	-21	-15	7,800	91/60	88/60	85/59
Hartford, CT	41	5	-4	3	7	6,200	91/74	88/73	85/72
Houston, TX	29	5	24	28	33	1,400	97/77	95/77	93/77
Indianapolis, IN	39	4	-7	-2	2	5,600	92/74	90/74	87/73
Jackson, MS	32	2	16	21	25	2,200	97/76	95/76	93/76
Kansas City, MO	39	1	-4	2	6	4,800	99/75	96/74	93/74
Las Vegas, NV	36	1	18	25	28	2,800	108/66	106/65	104/65
Lexington, KY	38	0	-1	3	8	4,600	93/73	91/73	88/72
Little Rock, AR	34	4	11	15	20	3,200	99/76	96/77	94/77
Los Angeles, CA	34	0	36	41	43	2,000	93/70	89/70	86/69
Memphis, TN	35	0	10	13	18	3,200	98/77	95/76	93/76
Miami, FL	25	5	39	44	47	200	91/77	90/77	89/77
Milwaukee, WI	43	0	-11	-8	-4	7,600	90/74	87/73	84/71
Minneapolis, MN	44	5	-22	-16	-12	8,400	92/75	89/73	86/71
New Orleans, LA	30	0	28	29	33	1,400	93/78	92/77	90/77
New York, NY	40	5	6	11	15	5,000	92/74	89/73	87/72
Norfolk, VA	36	5	15	20	22	3,400	93/77	91/76	89/76
Oklahoma City, OK	35	2	4	9	13	3,200	100/74	97/74	95/73
Omaha, NE	41	2	-13	-8	-3	6,600	94/76	91/75	88/74
Philadelphia, PA	39	5	6	10	14	4,400	93/75	90/74	87/72
Phoenix, AZ	33	3	27	31	34	1,800	109/71	107/71	105/71
Pittsburgh, PA	40	3	-1	3	7	6,000	91/72	88/71	86/70
Portland, ME	43	4	-10	-6	-1	7,600	87/72	84/71	81/69
Portland, OR	45	4	18	17	23	4,600	89/68	85/67	81/65
Portsmouth, NH	43	1	-8	-2	2	7,200	89/73	85/71	83/70
Providence, RI	41	4	-1	5	9	6,000	89/73	86/72	83/70
Rochester, NY	43	1	-5	1	5	6,800	91/73	88/71	85/70
Salt Lake City, UT	40	5	0	3	8	6,000	97/62	95/62	92/61
San Francisco, CA	37	5	36	38	40	3,000	74/63	71/62	69/61
Seattle, WA	47	4	22	22	27	5,200	85/68	82/66	78/65
Sioux Falls, SD	43	4	-21	-15	-11	7,800	94/73	91/72	88/71
St. Louis, MO	38	4	-3	3	8	5,000	98/75	94/75	91/75
Tampa, FL	28	0	32	36	40	680	92/77	91/77	90/76
Trenton, NJ	40	1	4	11	14	5,000	91/75	88/74	85/73
Washington, DC	38	5	7	14	17	4,200	93/75	91/74	89/74
Wichita, KS	37	4	-3	3	7	4,600	101/72	98/73	96/73
Wilmington, DE	39	4	5	10	14	5,000	92/74	89/74	87/73
ALASKA									
Anchorage	61	1	-29	-23	-18	10,800	71/59	68/58	66/56
Fairbanks	64	5	-59	-51	-47	14,280	82/62	78/60	75/59
CANADA									
Edmonton, Alta.	53	3	-30	-29	-25	11,000	85/66	82/65	79/63
Halifax, N.S.	44	4	-4	1	5	8,000	79/66	76/65	74/64
Montreal, Que.	45	3	-20	-16	-10	9,000	88/73	85/72	83/71
Saskatoon, Sask.	52	1	-35	-35	-31	11,000	89/68	86/66	83/65
St. John, Nwf.	47	4	1	3	7	8,600	77/66	75/65	73/64
Saint John, N.B.	45	2	-15	-12	-8	8,200	80/67	77/65	75/64
Toronto, Ont.	43	4	-10	-5	-1	7,000	90/73	87/72	85/71
Vancouver, B.C.	49	1	13	15	19	6,000	79/67	77/66	74/65
Winnipeg, Man.	49	5	-31	-30	-27	10,800	89/73	86/71	84/70

(1) Handbook of Fundamentals, ASHRAE, Inc., NY 1989
(2) Local Climatological Annual Survey, USDC Env. Science Services Administration, Asheville, NC

Table L1030-502   Maximum Depth of Frost Penetration in Inches

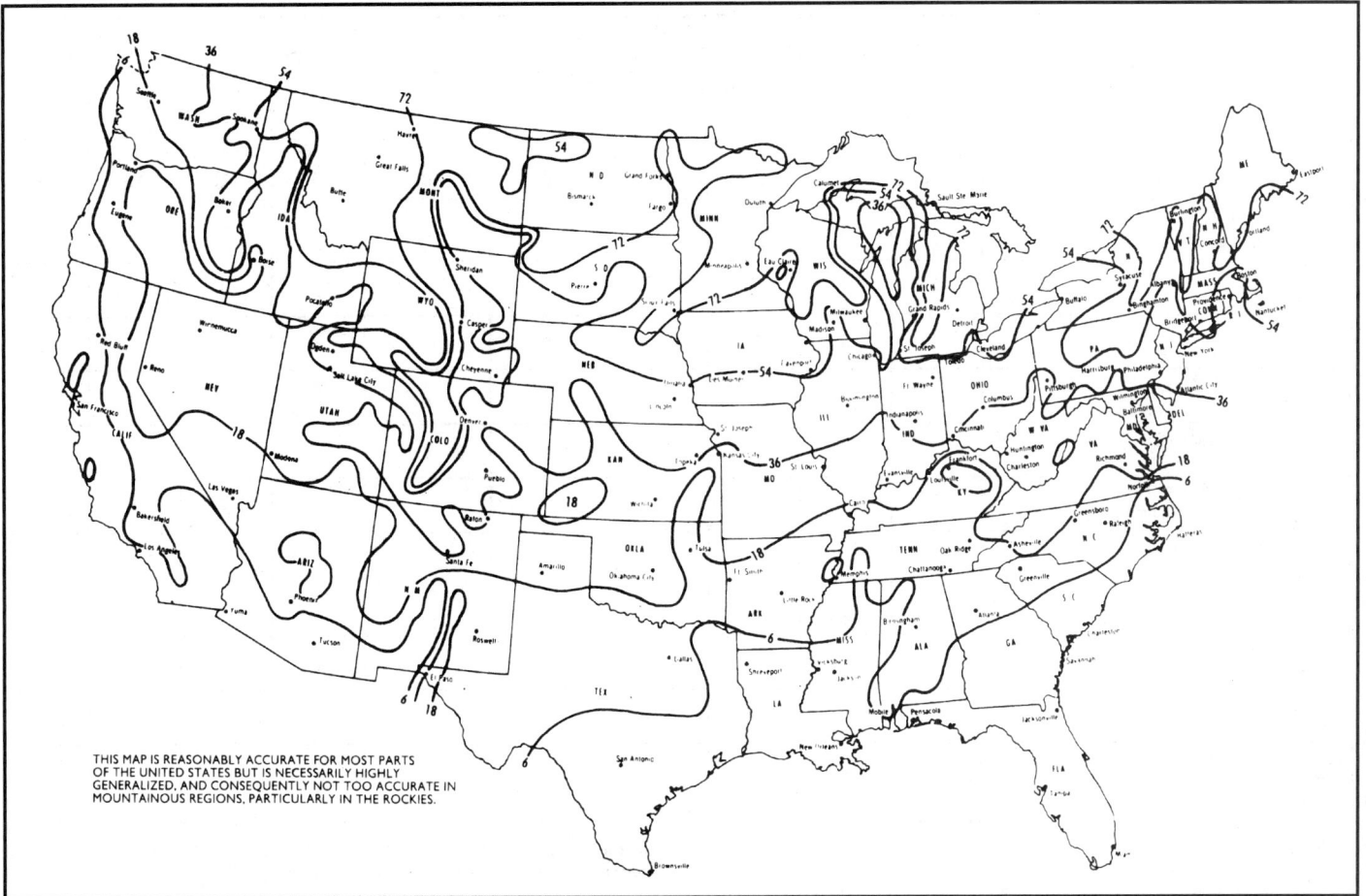

THIS MAP IS REASONABLY ACCURATE FOR MOST PARTS
OF THE UNITED STATES BUT IS NECESSARILY HIGHLY
GENERALIZED, AND CONSEQUENTLY NOT TOO ACCURATE IN
MOUNTAINOUS REGIONS, PARTICULARLY IN THE ROCKIES.

Table L1040-101   Fire-Resisting Ratings of Structural Elements (in hours)

Description of the Structural Element	No. 1 Fireproof		No. 2 Non Combustible			No. 3 Exterior Masonry Wall			No. 4 Frame	
			Protected		Unprotected	Heavy Timber	Ordinary		Protected	Unprotected
							Protected	Unprotected		
	1A	1B	2A	2B	2C	3A	3B	3C	4A	4B
Exterior, Bearing Walls	4	3	2	1½	1	2	2	2	1	1
Nonbearing Walls	2	2	1½	1	1	2	2	2	1	1
Interior Bearing Walls and Partitions	4	3	2	1	0	2	1	0	1	0
Fire Walls and Party Walls	4	3	2	2	2	2	2	2	2	2
Fire Enclosure of Exitways, Exit Hallways and Stairways	2	2	2	2	2	2	2	2	1	1
Shafts other than Exitways, Hallways and Stairways	2	2	2	2	2	2	2	2	1	1
Exitway access corridors and Vertical separation of tenant space	1	1	1	1	0	1	1	0	1	0
Columns, girders, trusses (other than roof trusses) and framing: Supporting more than one floor	4	3	2	1	0	—	1	0	1	0
Supporting one floor only	3	2	1½	1	0	—	1	0	1	0
Structural members supporting wall	3	2	1½	1	0	1	1	0	1	0
Floor construction including beams	3	2	1½	1	0	—	1	0	1	0
Roof construction including beams, trusses and framing arches and roof deck 15' or less in height to lowest member	2	1½	1	1	0	—	1	0	1	0

Note: a. Codes include special requirements and exceptions that are not included in the table above.

b. Each type of construction has been divided into sub-types which vary according to the degree of fire resistance required. Sub-types (A) requirements are more severe than those for sub-types (B).

c. Protected construction means all structural members are chemically treated, covered or protected so that the unit has the required fire resistance.

Type No. 1, Fireproof Construction — Buildings and structures of fireproof construction are those in which the walls, partitions, structural elements, floors, ceilings, and roofs, and the exitways are protected with approved noncombustible materials to afford the fire-resistance rating specified in Table L1040-101; except as otherwise specifically regulated. Fire-resistant treated wood may be used as specified.

Type No. 2, Noncombustible Construction — Buildings and structures of noncombustible construction are those in which the walls, partitions, structural elements, floors, ceilings, roofs and the exitways are approved noncombustible materials meeting the fire-resistance rating requirements specified in Table L1040-101; except as modified by the fire limit restrictions. Fire-retardant treated wood may be used as specified.

Type No. 3, Exterior Masonry Wall Construction — Buildings and structures of exterior masonry wall construction are those in which the exterior, fire and party walls are masonry or other approved noncombustible materials of the required fire-resistance rating and structural properties. The floors, roofs, and interior framing are wholly or partly wood or metal or other approved construction. The fire and party walls are ground-supported; except that girders and their supports, carrying walls of masonry shall be protected to afford the same degree of fire-resistance rating of the supported walls. All structural elements have the required fire-resistance rating specified in Table L1040-101.

Type No. 4, Frame Construction — Buildings and structures of frame construction are those in which the exterior walls, bearing walls, partitions, floor and roof construction are wholly or partly of wood stud and joist assemblies with a minimum nominal dimension of two inches or of other approved combustible materials. Fire stops are required at all vertical and horizontal draft openings in which the structural elements have required fire-resistance ratings specified in Table L1040-101.

Table L1040-201   Fire Grading of Building in Hours for Fire Walls and Fire Separation Walls

Fire Hazard for Fire Walls

The degree of fire hazard of buildings relating to their intended use is defined by "Fire Grading" the buildings. Such a grading system is listed in the Table L1040-201 below. This type of grading determines the requirements for fire walls and fire separation walls (exterior fire exposure). For mixed use occupancy, use the higher Fire Grading requirement of the components.

Building Type	Hours	Building Type	Hours
Businesses	2	Recreation Centers	2
Churches	1½	Residential Hotels	2
Factories	3	Residential, Multi-family Dwellings	1½
High Hazard*	4	Residential, One & Two Family Dwellings	¾
Industrial	3	Restaurants	2
Institutional, Incapacitated Occupants	2	Schools	1½
Institutional, Restrained Occupants	3	Storage, Low Hazard*	2
Lecture Halls	2	Storage, Moderate Hazard*	3
Mercantile	3	Terminals	2
Night Club	3	Theatres	3

Note: *The difference in "Fire Hazards" is determined by their occupancy and use.

High Hazard: Industrial and storage buildings in which the combustible contents might cause fires to be unusually intense or where explosives, combustible gases or flammable liquids are manufactured or stored.

Moderate Hazard: Mercantile buildings, industrial and storage buildings in which combustible contents might cause fires of moderate intensity.

Low Hazard: Business buildings that ordinarily do not burn rapidly.

Table L1040-202   Interior Finish Classification

Flame Spread for Interior Finishes

The flame spreadability of a material is the burning characteristic of the material relative to the fuel contributed by its combustion and the density of smoke developed. The flame spread classification of a material is based on a ten minute test on a scale of 0 to 100. Cement asbestos board is assigned a rating of 0 and select red oak flooring a rating of 100. The four classes are listed in Table L1040-202.

The flame spread ratings for interior finish walls and ceilings shall not be greater than the Class listed in the Table L1040-203 below.

Class of Material	Surface Burning Characteristics
I	0 to 25
II	26 to 75
III	76 to 200
IV	201 to 5000

Table L1040-203   Interior Finish Requirements by Class

Building Type	Vertical Exitways & Passageways	Corridors Providing Exitways	Rooms or Enclosed Spaces	Building Type	Vertical Exitways & Passageways	Corridors Providing Exitways	Rooms or Enclosed Space
Assembly Halls	I	I	II	Mercantile Walls	I	II	III
Businesses	I	II	III	Night Clubs	I	I	II
Churches	I	I	II	Residential Hotels	I	II	III
Factory	I	II	III	Residential, Multi-family	I	II	III
High Hazard	I	II	II	Residential, 1 & 2 Family	IV	IV	IV
Industrial	I	II	III	Restaurants	I	I	II
Institutional, Incapacitated	I	II	I	Storage, Low Hazard	I	II	III
Institutional, Restrained	I	I	I	Storage, Moderate Hazard	I	II	III
Mercantile Ceilings	I	II	II	Terminals	I	I	II

Description: This table is primarily for converting customary U.S. units in the left hand column to SI metric units in the right hand column.

In addition, conversion factors for some commonly encountered Canadian and non-SI metric units are included.

Table L1090-101   Metric Conversion Factors

	If You Know		Multiply By		To Find
Length	Inches	x	25.4[a]	=	Millimeters
	Feet	x	0.3048[a]	=	Meters
	Yards	x	0.9144[a]	=	Meters
	Miles (statute)	x	1.609	=	Kilometers
Area	Square inches	x	645.2	=	Square millimeters
	Square feet	x	0.0929	=	Square meters
	Square yards	x	0.8361	=	Square meters
Volume (Capacity)	Cubic inches	x	16,387	=	Cubic millimeters
	Cubic feet	x	0.02832	=	Cubic meters
	Cubic yards	x	0.7646	=	Cubic meters
	Gallons (U.S. liquids)[b]	x	0.003785	=	Cubic meters[c]
	Gallons (Canadian liquid)[b]	x	0.004546	=	Cubic meters[c]
	Ounces (U.S. liquid)[b]	x	29.57	=	Milliliters[c, d]
	Quarts (U.S. liquid)[b]	x	0.9464	=	Liters[c, d]
	Gallons (U.S. liquid)[b]	x	3.785	=	Liters[c, d]
Force	Kilograms force[d]	x	9.807	=	Newtons
	Pounds force	x	4.448	=	Newtons
	Pounds force	x	0.4536	=	Kilograms force[d]
	Kips	x	4448	=	Newtons
	Kips	x	453.6	=	Kilograms force[d]
Pressure, Stress, Strength (Force per unit area)	Kilograms force per square centimeter[d]	x	0.09807	=	Megapascals
	Pounds force per square inch (psi)	x	0.006895	=	Megapascals
	Kips per square inch	x	6.895	=	Megapascals
	Pounds force per square inch (psi)	x	0.07031	=	Kilograms force per square centimeter[d]
	Pounds force per square foot	x	47.88	=	Pascals
	Pounds force per square foot	x	4.882	=	Kilograms force per square meter[d]
Bending Moment Or Torque	Inch-pounds force	x	0.01152	=	Meter-kilograms force[d]
	Inch-pounds force	x	0.1130	=	Newton-meters
	Foot-pounds force	x	0.1383	=	Meter-kilograms force[d]
	Foot-pounds force	x	1.356	=	Newton-meters
	Meter-kilograms force[d]	x	9.807	=	Newton-meters
Mass	Ounces (avoirdupois)	x	28.35	=	Grams
	Pounds (avoirdupois)	x	0.4536	=	Kilograms
	Tons (metric)	x	1000	=	Kilograms
	Tons, short (2000 pounds)	x	907.2	=	Kilograms
	Tons, short (2000 pounds)	x	0.9072	=	Megagrams[e]
Mass per Unit Volume	Pounds mass per cubic foot	x	16.02	=	Kilograms per cubic meter
	Pounds mass per cubic yard	x	0.5933	=	Kilograms per cubic meter
	Pounds mass per gallon (U.S. liquid)[b]	x	119.8	=	Kilograms per cubic meter
	Pounds mass per gallon (Canadian liquid)[b]	x	99.78	=	Kilograms per cubic meter
Temperature	Degrees Fahrenheit	(F-32)/1.8		=	Degrees Celsius
	Degrees Fahrenheit	(F+459.67)/1.8		=	Degrees Kelvin
	Degrees Celsius	C+273.15		=	Degrees Kelvin

[a] The factor given is exact
[b] One U.S. gallon = 0.8327 Canadian gallon
[c] 1 liter = 1000 milliliters = 1000 cubic centimeters
  1 cubic decimeter = 0.001 cubic meter

[d] Metric but not SI unit
[e] Called "tonne" in England and "metric ton" in other metric countries

Table L1090-201   Metric Equivalents of Cement Content for Concrete Mixes

94 Pound Bags per Cubic Yard	Kilograms per Cubic Meter	94 Pound Bags per Cubic Yard	Kilograms per Cubic Meter
1.0	55.77	7.0	390.4
1.5	83.65	7.5	418.3
2.0	111.5	8.0	446.2
2.5	139.4	8.5	474.0
3.0	167.3	9.0	501.9
3.5	195.2	9.5	529.8
4.0	223.1	10.0	557.7
4.5	251.0	10.5	585.6
5.0	278.8	11.0	613.5
5.5	306.7	11.5	641.3
6.0	334.6	12.0	669.2
6.5	362.5	12.5	697.1

a. If you know the cement content in pounds per cubic yard, multiply by .5933 to obtain kilograms per cubic meter.

b. If you know the cement content in 94 pound bags per cubic yard, multiply by 55.77 to obtain kilograms per cubic meter.

Table L1090-202   Metric Equivalents of Common Concrete Strengths (to convert other psi values to megapascals, multiply by .006895)

U.S. Values psi	SI Value Megapascals	Non-SI Metric Value kgf/cm²*
2000	14	140
2500	17	175
3000	21	210
3500	24	245
4000	28	280
4500	31	315
5000	34	350
6000	41	420
7000	48	490
8000	55	560
9000	62	630
10,000	69	705

* kilograms force per square centimeter

Table L1090-203   Comparison of U.S. Customary Units and SI Units for Reinforcing Bars

U.S. Customary Units

Bar Designation No.[b]	Nominal Weight, lb/ft	Nominal Dimensions[a]			Deformation Requirements, in.		
		Diameter in.	Cross Sectional Area, in.²	Perimeter in.	Maximum Average Spacing	Minimum Average Height	Maximum Gap (Chord of 12-1/2% of Nominal Perimeter)
3	0.376	0.375	0.11	1.178	0.262	0.015	0.143
4	0.668	0.500	0.20	1.571	0.350	0.020	0.191
5	1.043	0.625	0.31	1.963	0.437	0.028	0.239
6	1.502	0.750	0.44	2.356	0.525	0.038	0.286
7	2.044	0.875	0.60	2.749	0.612	0.044	0.334
8	2.670	1.000	0.79	3.142	0.700	0.050	0.383
9	3.400	1.128	1.00	3.544	0.790	0.056	0.431
10	4.303	1.270	1.27	3.990	0.889	0.064	0.487
11	5.313	1.410	1.56	4.430	0.987	0.071	0.540
14	7.65	1.693	2.25	5.32	1.185	0.085	0.648
18	13.60	2.257	4.00	7.09	1.58	0.102	0.864

SI UNITS

Bar Designation No.[b]	Nominal Weight kg/m	Nominal Dimensions[a]			Deformation Requirements, mm		
		Diameter, mm	Cross Sectional Area, cm²	Perimeter, mm	Maximum Average Spacing	Minimum Average Height	Maximum Gap (Chord of 12-1/2% of Nominal Perimeter)
3	0.560	9.52	0.71	29.9	6.7	0.38	3.5
4	0.994	12.70	1.29	39.9	8.9	0.51	4.9
5	1.552	15.88	2.00	49.9	11.1	0.71	6.1
6	2.235	19.05	2.84	59.8	13.3	0.96	7.3
7	3.042	22.22	3.87	69.8	15.5	1.11	8.5
8	3.973	25.40	5.10	79.8	17.8	1.27	9.7
9	5.059	28.65	6.45	90.0	20.1	1.42	10.9
10	6.403	32.26	8.19	101.4	22.6	1.62	11.4
11	7.906	35.81	10.06	112.5	25.1	1.80	13.6
14	11.384	43.00	14.52	135.1	30.1	2.16	16.5
18	20.238	57.33	25.81	180.1	40.1	2.59	21.9

[a]Nominal dimensions of a deformed bar are equivalent to those of a plain round bar having the same weight per foot as the deformed bar.

[b]Bar numbers are based on the number of eighths of an inch included in the nominal diameter of the bars.

Table L1090-204   Metric Rebar Specification - ASTM A615-81

Grade 300 (300 MPa* = **43,560 psi; +8.7% vs. Grade 40**)				
Grade 400 (400 MPa* = **58,000 psi; –3.4% vs. Grade 60**)				
Bar No.	Diameter mm	Area mm²	Equivalent in.²	Comparison with U.S. Customary Bars
10	11.3	100	.16	Between #3 & #4
15	16.0	200	.31	#5 (.31 in.²)
20	19.5	300	.47	#6 (.44 in.²)
25	25.2	500	.78	#8 (.79 in.²)
30	29.9	700	1.09	#9 (1.00 in.²)
35	35.7	1000	1.55	#11 (1.56 in.²)
45	43.7	1500	2.33	#14 (2.25 in.²)
55	56.4	2500	3.88	#18 (4.00 in.²)

* MPa = megapascals   Grade 300 bars are furnished only in sizes 10 through 35

Estimating Form | Project Summary

PRELIMINARY ESTIMATE

PROJECT _____ TOTAL SITE AREA _____

BUILDING TYPE _____ OWNER _____

LOCATION _____ ARCHITECT _____

DATE OF CONSTRUCTION _____ ESTIMATED CONSTRUCTION PERIOD _____

BRIEF DESCRIPTION _____

_____

_____

TYPE OF PLAN _____ TYPE OF CONSTRUCTION _____

QUALITY _____ BUILDING CAPACITY _____

Floor			**Wall Areas**				
Below Grade Levels			Foundation Walls	L.F.	Ht.		S.F.
Area		S.F.	Frost Walls	L.F.	Ht.		S.F.
Area		S.F.	Exterior Closure		Total		S.F.
Total Area		S.F.	Comment				
Ground Floor			Fenestration		%		S.F.
Area		S.F.			%		S.F.
Area		S.F.	Exterior Wall		%		S.F.
Total Area		S.F.			%		S.F.
Supported Levels			**Site Work**				
Area		S.F.	Parking		S.F. (For		Cars)
Area		S.F.	Access Roads		L.F. (X		Ft. Wide)
Area		S.F.	Sidewalk		L.F. (X		Ft. Wide)
Area		S.F.	Landscaping		S.F. (		% Unbuilt Site)
Area		S.F.	**Building Codes**				
Total Area		S.F.	City		County		
Miscellaneous			National		Other		
Area		S.F.	**Loading**				
Area		S.F.	Roof	psf	Ground Floor		psf
Area		S.F.	Supported Floors	psf	Corridor		psf
Area		S.F.	Balcony	psf	Partition, allow		psf
Total Area		S.F.	Miscellaneous				psf
Net Finished Area		S.F.	Live Load Reduction				
Net Floor Area		S.F.	Wind				
Gross Floor Area		S.F.	Earthquake		Zone		
Roof			Comment				
Total Area		S.F.	Soil Type				
Comments			Bearing Capacity				K.S.F.
			Frost Depth				Ft.
Volume			**Frame**				
Depth of Floor System			Type		Bay Spacing		
Minimum		In.	Foundation, Standard				
Maximum		In.	Special				
Foundation Wall Height		Ft.	Substructure				
Floor to Floor Height		Ft.	Comment				
Floor to Ceiling Height		Ft.	Superstructure, Vertical		Horizontal		
Subgrade Volume		C.F.	Fireproofing	☐ Columns			Hrs.
Above Grade Volume		C.F.	☐ Girders	Hrs.	☐ Beams		Hrs.
Total Building Volume		C.F.	☐ Floor	Hrs.	☐ None		

Estimating Form

Systems Costs

ASSEMBLY NUMBER	DESCRIPTION	QTY	UNIT	TOTAL COST UNIT	TOTAL COST TOTAL	COST PER S.F.
A10	**Foundations**					
A20	**Basement Construction**					
B10	**Superstructure**					
B20	**Exterior Closure**					
B30	**Roofing**					

ASSEMBLY NUMBER	DESCRIPTION	QTY	UNIT	TOTAL COST		COST PER S.F.
				UNIT	TOTAL	
C	**Interior Construction**					
D10	**Conveying**					
D20	**Plumbing**					

ASSEMBLY NUMBER	DESCRIPTION	QTY	UNIT	TOTAL COST		COST PER S.F.
				UNIT	TOTAL	
D30	**HVAC**					
D40	**Fire Protection**					
D50	**Electrical**					

Estimating Form

Systems Costs

ASSEMBLY NUMBER	DESCRIPTION	QTY	UNIT	TOTAL COST		COST PER S.F.
				UNIT	TOTAL	
E	**Equipment & Furnishings**					
F	**Special Construction**					
G	**Sitework**					

Estimating Form

Systems Costs

Systems Costs Summary

Preliminary Estimate Cost Summary

PROJECT	TOTAL AREA	SHEET NO.
LOCATION	TOTAL VOLUME	ESTIMATE NO.
ARCHITECT	COST PER S.F.	DATE
OWNER	COST PER C.F.	NO OF STORIES
QUANTITIES BY	EXTENSIONS BY	CHECKED BY

DIV	DESCRIPTION	SUBTOTAL COST	COST/S.F.	PERCENTAGE
A	SUBSTRUCTURE			
B10	SHELL: SUPERSTRUCTURE			
B20	SHELL: EXTERIOR CLOSURE			
B30	SHELL: ROOFING			
C	INTERIOR CONSTRUCTION			
D10	SERVICES: CONVEYING			
D20	SERVICES: PLUMBING			
D30	SERVICES: HVAC			
D40	SERVICES: FIRE PROTECTION			
D50	SERVICES: ELECTRICAL			
E	EQUIPMENT & FURNISHINGS			
F	SPECIAL CONSTRUCTION			
G	SITEWORK			

BUILDING SUBTOTAL $ _____ -

Sales Tax ___ % x Subtotal /2 $ _____ -

General Conditions ___% x Subtotal $ _____ -

Subtotal "A" $ _____ -

Overhead ___% x Subtotal "A" $ _____ -

Subtotal "B" $ _____ -

Profit ___ % x Subtotal "B" $ _____ -

Subtotal "C" $ _____ -

Location Factor ____ % x Subtotal "C"   Localized Cost $ _____ -
   (Boston, MA)

Architects Fee____ x Localized Cost = $ _____ -
Contingency ____ x Localized Cost = $ _____ -

Project Total Cost $ _____ -

Square Foot Cost $ ____ / ____ S.F. =   **S.F. Cost** $ _____ -
Cubic Foot Cost $ ____ / ____ C.F. =   **C.F. Cost** $ _____ -

Abbreviations

A	Area Square Feet; Ampere	BTUH	BTU per Hour	Cwt.	100 Pounds	
ABS	Acrylonitrile Butadiene Stryrene;	B.U.R.	Built-up Roofing	C.W.X.	Cool White Deluxe	
	Asbestos Bonded Steel	BX	Interlocked Armored Cable	C.Y.	Cubic Yard (27 cubic feet)	
A.C.	Alternating Current;	°C	degree centegrade	C.Y./Hr.	Cubic Yard per Hour	
	Air-Conditioning;	c	Conductivity, Copper Sweat	Cyl.	Cylinder	
	Asbestos Cement;	C	Hundred; Centigrade	d	Penny (nail size)	
	Plywood Grade A & C	C/C	Center to Center, Cedar on Cedar	D	Deep; Depth; Discharge	
A.C.I.	American Concrete Institute	C-C	Center to Center	Dis., Disch.	Discharge	
AD	Plywood, Grade A & D	Cab.	Cabinet	Db.	Decibel	
Addit.	Additional	Cair.	Air Tool Laborer	Dbl.	Double	
Adj.	Adjustable	Calc	Calculated	DC	Direct Current	
af	Audio-frequency	Cap.	Capacity	DDC	Direct Digital Control	
A.G.A.	American Gas Association	Carp.	Carpenter	Demob.	Demobilization	
Agg.	Aggregate	C.B.	Circuit Breaker	d.f.u.	Drainage Fixture Units	
A.H.	Ampere Hours	C.C.A.	Chromate Copper Arsenate	D.H.	Double Hung	
A hr.	Ampere-hour	C.C.F.	Hundred Cubic Feet	DHW	Domestic Hot Water	
A.H.U.	Air Handling Unit	cd	Candela	DI	Ductile Iron	
A.I.A.	American Institute of Architects	cd/sf	Candela per Square Foot	Diag.	Diagonal	
AIC	Ampere Interrupting Capacity	CD	Grade of Plywood Face & Back	Diam., Dia	Diameter	
Allow.	Allowance	CDX	Plywood, Grade C & D, exterior	Distrib.	Distribution	
alt.	Altitude		glue	Div.	Division	
Alum.	Aluminum	Cefi.	Cement Finisher	Dk.	Deck	
a.m.	Ante Meridiem	Cem.	Cement	D.L.	Dead Load; Diesel	
Amp.	Ampere	CF	Hundred Feet	DLH	Deep Long Span Bar Joist	
Anod.	Anodized	C.F.	Cubic Feet	Do.	Ditto	
Approx.	Approximate	CFM	Cubic Feet per Minute	Dp.	Depth	
Apt.	Apartment	c.g.	Center of Gravity	D.P.S.T.	Double Pole, Single Throw	
Asb.	Asbestos	CHW	Chilled Water;	Dr.	Drive	
A.S.B.C.	American Standard Building Code		Commercial Hot Water	Drink.	Drinking	
Asbe.	Asbestos Worker	C.I.	Cast Iron	D.S.	Double Strength	
ASCE.	American Society of Civil Engineers	C.I.P.	Cast in Place	D.S.A.	Double Strength A Grade	
A.S.H.R.A.E.	American Society of Heating,	Circ.	Circuit	D.S.B.	Double Strength B Grade	
	Refrig. & AC Engineers	C.L.	Carload Lot	Dty.	Duty	
A.S.M.E.	American Society of Mechanical	Clab.	Common Laborer	DWV	Drain Waste Vent	
	Engineers	Clam	Common maintenance laborer	DX	Deluxe White, Direct Expansion	
A.S.T.M.	American Society for Testing and	C.L.F.	Hundred Linear Feet	dyn	Dyne	
	Materials	CLF	Current Limiting Fuse	e	Eccentricity	
Attchmt.	Attachment	CLP	Cross Linked Polyethylene	E	Equipment Only; East	
Avg., Ave.	Average	cm	Centimeter	Ea.	Each	
A.W.G.	American Wire Gauge	CMP	Corr. Metal Pipe	E.B.	Encased Burial	
AWWA	American Water Works Assoc.	C.M.U.	Concrete Masonry Unit	Econ.	Economy	
Bbl.	Barrel	CN	Change Notice	E.C.Y	Embankment Cubic Yards	
B&B	Grade B and Better;	Col.	Column	EDP	Electronic Data Processing	
	Balled & Burlapped	CO_2	Carbon Dioxide	EIFS	Exterior Insulation Finish System	
B.&S.	Bell and Spigot	Comb.	Combination	E.D.R.	Equiv. Direct Radiation	
B.&W.	Black and White	Compr.	Compressor	Eq.	Equation	
b.c.c.	Body-centered Cubic	Conc.	Concrete	EL	elevation	
B.C.Y.	Bank Cubic Yards	Cont.	Continuous; Continued, Container	Elec.	Electrician; Electrical	
BE	Bevel End	Corr.	Corrugated	Elev.	Elevator; Elevating	
B.F.	Board Feet	Cos	Cosine	EMT	Electrical Metallic Conduit;	
Bg. cem.	Bag of Cement	Cot	Cotangent		Thin Wall Conduit	
BHP	Boiler Horsepower;	Cov.	Cover	Eng.	Engine, Engineered	
	Brake Horsepower	C/P	Cedar on Paneling	EPDM	Ethylene Propylene Diene	
B.I.	Black Iron	CPA	Control Point Adjustment		Monomer	
Bit., Bitum.	Bituminous	Cplg.	Coupling	EPS	Expanded Polystyrene	
Bit., Conc.	Bituminous Concrete	C.P.M.	Critical Path Method	Eqhv.	Equip. Oper., Heavy	
Bk.	Backed	CPVC	Chlorinated Polyvinyl Chloride	Eqlt.	Equip. Oper., Light	
Bkrs.	Breakers	C.Pr.	Hundred Pair	Eqmd.	Equip. Oper., Medium	
Bldg.	Building	CRC	Cold Rolled Channel	Eqmm.	Equip. Oper., Master Mechanic	
Blk.	Block	Creos.	Creosote	Eqol.	Equip. Oper., Oilers	
Bm.	Beam	Crpt.	Carpet & Linoleum Layer	Equip.	Equipment	
Boil.	Boilermaker	CRT	Cathode-ray Tube	ERW	Electric Resistance Welded	
B.P.M.	Blows per Minute	CS	Carbon Steel, Constant Shear Bar	E.S.	Energy Saver	
BR	Bedroom		Joist	Est.	Estimated	
Brg.	Bearing	Csc	Cosecant	esu	Electrostatic Units	
Brhe.	Bricklayer Helper	C.S.F.	Hundred Square Feet	E.W.	Each Way	
Bric.	Bricklayer	CSI	Construction Specifications	EWT	Entering Water Temperature	
Brk.	Brick		Institute	Excav.	Excavation	
Brng.	Bearing	C.T.	Current Transformer	Exp.	Expansion, Exposure	
Brs.	Brass	CTS	Copper Tube Size	Ext.	Exterior	
Brz.	Bronze	Cu	Copper, Cubic	Extru.	Extrusion	
Bsn.	Basin	Cu. Ft.	Cubic Foot	f.	Fiber stress	
Btr.	Better	cw	Continuous Wave	F	Fahrenheit; Female; Fill	
Btu	British Thermal Unit	C.W.	Cool White; Cold Water	Fab.	Fabricated	

579

FBGS	Fiberglass
F.C.	Footcandles
f.c.c.	Face-centered Cubic
f'c.	Compressive Stress in Concrete; Extreme Compressive Stress
F.E.	Front End
FEP	Fluorinated Ethylene Propylene (Teflon)
F.G.	Flat Grain
F.H.A.	Federal Housing Administration
Fig.	Figure
Fin.	Finished
Fixt.	Fixture
Fl. Oz.	Fluid Ounces
Flr.	Floor
F.M.	Frequency Modulation; Factory Mutual
Fmg.	Framing
Fdn.	Foundation
Fori.	Foreman, Inside
Foro.	Foreman, Outside
Fount.	Fountain
fpm	Feet per Minute
FPT	Female Pipe Thread
Fr.	Frame
F.R.	Fire Rating
FRK	Foil Reinforced Kraft
FRP	Fiberglass Reinforced Plastic
FS	Forged Steel
FSC	Cast Body; Cast Switch Box
Ft.	Foot; Feet
Ftng.	Fitting
Ftg.	Footing
Ft lb.	Foot Pound
Furn.	Furniture
FVNR	Full Voltage Non-Reversing
FXM	Female by Male
Fy.	Minimum Yield Stress of Steel
g	Gram
G	Gauss
Ga.	Gauge
Gal., gal.	Gallon
gpm, GPM	Gallon per Minute
Galv.	Galvanized
Gen.	General
G.F.I.	Ground Fault Interrupter
Glaz.	Glazier
GPD	Gallons per Day
GPH	Gallons per Hour
GPM	Gallons per Minute
GR	Grade
Gran.	Granular
Grnd.	Ground
H	High Henry
H.C.	High Capacity
H.D.	Heavy Duty; High Density
H.D.O.	High Density Overlaid
H.D.P.E.	high density polyethelene
Hdr.	Header
Hdwe.	Hardware
Help.	Helper Average
HEPA	High Efficiency Particulate Air Filter
Hg	Mercury
HIC	High Interrupting Capacity
HM	Hollow Metal
HMWPE	high molecular weight polyethylene
H.O.	High Output
Horiz.	Horizontal
H.P.	Horsepower; High Pressure
H.P.F.	High Power Factor
Hr.	Hour
Hrs./Day	Hours per Day
HSC	High Short Circuit
Ht.	Height
Htg.	Heating
Htrs.	Heaters
HVAC	Heating, Ventilation & Air-Conditioning
Hvy.	Heavy
HW	Hot Water
Hyd.;Hydr.	Hydraulic
Hz.	Hertz (cycles)
I.	Moment of Inertia
IBC	International Building Code
I.C.	Interrupting Capacity
ID	Inside Diameter
I.D.	Inside Dimension; Identification
I.F.	Inside Frosted
I.M.C.	Intermediate Metal Conduit
In.	Inch
Incan.	Incandescent
Incl.	Included; Including
Int.	Interior
Inst.	Installation
Insul.	Insulation/Insulated
I.P.	Iron Pipe
I.P.S.	Iron Pipe Size
I.P.T.	Iron Pipe Threaded
I.W.	Indirect Waste
J	Joule
J.I.C.	Joint Industrial Council
K	Thousand; Thousand Pounds; Heavy Wall Copper Tubing, Kelvin
K.A.H.	Thousand Amp. Hours
KCMIL	Thousand Circular Mils
KD	Knock Down
K.D.A.T.	Kiln Dried After Treatment
kg	Kilogram
kG	Kilogauss
kgf	Kilogram Force
kHz	Kilohertz
Kip.	1000 Pounds
KJ	Kiljoule
K.L.	Effective Length Factor
K.L.F.	Kips per Linear Foot
Km	Kilometer
K.S.F.	Kips per Square Foot
K.S.I.	Kips per Square Inch
kV	Kilovolt
kVA	Kilovolt Ampere
K.V.A.R.	Kilovar (Reactance)
KW	Kilowatt
KWh	Kilowatt-hour
L	Labor Only; Length; Long; Medium Wall Copper Tubing
Lab.	Labor
lat	Latitude
Lath.	Lather
Lav.	Lavatory
lb.; #	Pound
L.B.	Load Bearing; L Conduit Body
L. & E.	Labor & Equipment
lb./hr.	Pounds per Hour
lb./L.F.	Pounds per Linear Foot
lbf/sq.in.	Pound-force per Square Inch
L.C.L.	Less than Carload Lot
L.C.Y.	Loose Cubic Yard
Ld.	Load
LE	Lead Equivalent
LED	Light Emitting Diode
L.F.	Linear Foot
L.F. Nose	Linear Foot of Stair Nosing
L.F. Rsr	Linear Foot of Stair Riser
Lg.	Long; Length; Large
L & H	Light and Heat
LH	Long Span Bar Joist
L.H.	Labor Hours
L.L.	Live Load
L.L.D.	Lamp Lumen Depreciation
lm	Lumen
lm/sf	Lumen per Square Foot
lm/W	Lumen per Watt
L.O.A.	Length Over All
log	Logarithm
L-O-L	Lateralolet
long.	longitude
L.P.	Liquefied Petroleum; Low Pressure
L.P.F.	Low Power Factor
LR	Long Radius
L.S.	Lump Sum
Lt.	Light
Lt. Ga.	Light Gauge
L.T.L.	Less than Truckload Lot
Lt. Wt.	Lightweight
L.V.	Low Voltage
M	Thousand; Material; Male; Light Wall Copper Tubing
M²CA	Meters Squared Contact Area
m/hr.; M.H.	Man-hour
mA	Milliampere
Mach.	Machine
Mag. Str.	Magnetic Starter
Maint.	Maintenance
Marb.	Marble Setter
Mat; Mat'l.	Material
Max.	Maximum
MBF	Thousand Board Feet
MBH	Thousand BTU's per hr.
MC	Metal Clad Cable
M.C.F.	Thousand Cubic Feet
M.C.F.M.	Thousand Cubic Feet per Minute
M.C.M.	Thousand Circular Mils
M.C.P.	Motor Circuit Protector
MD	Medium Duty
M.D.O.	Medium Density Overlaid
Med.	Medium
MF	Thousand Feet
M.F.B.M.	Thousand Feet Board Measure
Mfg.	Manufacturing
Mfrs.	Manufacturers
mg	Milligram
MGD	Million Gallons per Day
MGPH	Thousand Gallons per Hour
MH, M.H.	Manhole; Metal Halide; Man-Hour
MHz	Megahertz
Mi.	Mile
MI	Malleable Iron; Mineral Insulated
mm	Millimeter
Mill.	Millwright
Min., min.	Minimum, minute
Misc.	Miscellaneous
ml	Milliliter, Mainline
M.L.F.	Thousand Linear Feet
Mo.	Month
Mobil.	Mobilization
Mog.	Mogul Base
MPH	Miles per Hour
MPT	Male Pipe Thread

MRT	Mile Round Trip
ms	Millisecond
M.S.F.	Thousand Square Feet
Mstz.	Mosaic & Terrazzo Worker
M.S.Y.	Thousand Square Yards
Mtd., mtd.	Mounted
Mthe.	Mosaic & Terrazzo Helper
Mtng.	Mounting
Mult.	Multi; Multiply
M.V.A.	Million Volt Amperes
M.V.A.R.	Million Volt Amperes Reactance
MV	Megavolt
MW	Megawatt
MXM	Male by Male
MYD	Thousand Yards
N	Natural; North
nA	Nanoampere
NA	Not Available; Not Applicable
N.B.C.	National Building Code
NC	Normally Closed
N.E.M.A.	National Electrical Manufacturers Assoc.
NEHB	Bolted Circuit Breaker to 600V.
N.L.B.	Non-Load-Bearing
NM	Non-Metallic Cable
nm	Nanometer
No.	Number
NO	Normally Open
N.O.C.	Not Otherwise Classified
Nose.	Nosing
N.P.T.	National Pipe Thread
NQOD	Combination Plug-on/Bolt on Circuit Breaker to 240V.
N.R.C.	Noise Reduction Coefficient/ Nuclear Regulator Commission
N.R.S.	Non Rising Stem
ns	Nanosecond
nW	Nanowatt
OB	Opposing Blade
OC	On Center
OD	Outside Diameter
O.D.	Outside Dimension
ODS	Overhead Distribution System
O.G.	Ogee
O.H.	Overhead
O&P	Overhead and Profit
Oper.	Operator
Opng.	Opening
Orna.	Ornamental
OSB	Oriented Strand Board
O.S.&Y.	Outside Screw and Yoke
Ovhd.	Overhead
OWG	Oil, Water or Gas
Oz.	Ounce
P.	Pole; Applied Load; Projection
p.	Page
Pape.	Paperhanger
P.A.P.R.	Powered Air Purifying Respirator
PAR	Parabolic Reflector
Pc., Pcs.	Piece, Pieces
P.C.	Portland Cement; Power Connector
P.C.F.	Pounds per Cubic Foot
P.C.M.	Phase Contrast Microscopy
P.E.	Professional Engineer; Porcelain Enamel; Polyethylene; Plain End
Perf.	Perforated
PEX	Cross linked polyethylene
Ph.	Phase
P.I.	Pressure Injected
Pile.	Pile Driver
Pkg.	Package
Pl.	Plate
Plah.	Plasterer Helper
Plas.	Plasterer
Pluh.	Plumbers Helper
Plum.	Plumber
Ply.	Plywood
p.m.	Post Meridiem
Pntd.	Painted
Pord.	Painter, Ordinary
pp	Pages
PP, PPL	Polypropylene
P.P.M.	Parts per Million
Pr.	Pair
P.E.S.B.	Pre-engineered Steel Building
Prefab.	Prefabricated
Prefin.	Prefinished
Prop.	Propelled
PSF, psf	Pounds per Square Foot
PSI, psi	Pounds per Square Inch
PSIG	Pounds per Square Inch Gauge
PSP	Plastic Sewer Pipe
Pspr.	Painter, Spray
Psst.	Painter, Structural Steel
P.T.	Potential Transformer
P. & T.	Pressure & Temperature
Ptd.	Painted
Ptns.	Partitions
Pu	Ultimate Load
PVC	Polyvinyl Chloride
Pvmt.	Pavement
Pwr.	Power
Q	Quantity Heat Flow
Qt.	Quart
Quan., Qty.	Quantity
Q.C.	Quick Coupling
r	Radius of Gyration
R	Resistance
R.C.P.	Reinforced Concrete Pipe
Rect.	Rectangle
Reg.	Regular
Reinf.	Reinforced
Req'd.	Required
Res.	Resistant
Resi.	Residential
Rgh.	Rough
RGS	Rigid Galvanized Steel
R.H.W.	Rubber, Heat & Water Resistant; Residential Hot Water
rms	Root Mean Square
Rnd.	Round
Rodm.	Rodman
Rofc.	Roofer, Composition
Rofp.	Roofer, Precast
Rohe.	Roofer Helpers (Composition)
Rots.	Roofer, Tile & Slate
R.O.W.	Right of Way
RPM	Revolutions per Minute
R.S.	Rapid Start
Rsr	Riser
RT	Round Trip
S.	Suction; Single Entrance; South
SC	Screw Cover
SCFM	Standard Cubic Feet per Minute
Scaf.	Scaffold
Sch., Sched.	Schedule
S.C.R.	Modular Brick
S.D.	Sound Deadening
S.D.R.	Standard Dimension Ratio
S.E.	Surfaced Edge
Sel.	Select
S.E.R., S.E.U.	Service Entrance Cable
S.F.	Square Foot
S.F.C.A.	Square Foot Contact Area
S.F. Flr.	Square Foot of Floor
S.F.G.	Square Foot of Ground
S.F. Hor.	Square Foot Horizontal
S.F.R.	Square Feet of Radiation
S.F. Shlf.	Square Foot of Shelf
S4S	Surface 4 Sides
Shee.	Sheet Metal Worker
Sin.	Sine
Skwk.	Skilled Worker
SL	Saran Lined
S.L.	Slimline
Sldr.	Solder
SLH	Super Long Span Bar Joist
S.N.	Solid Neutral
S-O-L	Socketolet
sp	Standpipe
S.P.	Static Pressure; Single Pole; Self-Propelled
Spri.	Sprinkler Installer
spwg	Static Pressure Water Gauge
S.P.D.T.	Single Pole, Double Throw
SPF	Spruce Pine Fir
S.P.S.T.	Single Pole, Single Throw
SPT	Standard Pipe Thread
Sq.	Square; 100 Square Feet
Sq. Hd.	Square Head
Sq. In.	Square Inch
S.S.	Single Strength; Stainless Steel
S.S.B.	Single Strength B Grade
sst, ss	Stainless Steel
Sswk.	Structural Steel Worker
Sswl.	Structural Steel Welder
St.; Stl.	Steel
S.T.C.	Sound Transmission Coefficient
Std.	Standard
Stg.	Staging
STK	Select Tight Knot
STP	Standard Temperature & Pressure
Stpi.	Steamfitter, Pipefitter
Str.	Strength; Starter; Straight
Strd.	Stranded
Struct.	Structural
Sty.	Story
Subj.	Subject
Subs.	Subcontractors
Surf.	Surface
Sw.	Switch
Swbd.	Switchboard
S.Y.	Square Yard
Syn.	Synthetic
S.Y.P.	Southern Yellow Pine
Sys.	System
t.	Thickness
T	Temperature; Ton
Tan	Tangent
T.C.	Terra Cotta
T & C	Threaded and Coupled
T.D.	Temperature Difference
Tdd	Telecommunications Device for the Deaf
T.E.M.	Transmission Electron Microscopy
TFE	Tetrafluoroethylene (Teflon)
T. & G.	Tongue & Groove; Tar & Gravel
Th., Thk.	Thick
Thn.	Thin
Thrded	Threaded
Tilf.	Tile Layer, Floor
Tilh.	Tile Layer, Helper
THHN	Nylon Jacketed Wire
THW.	Insulated Strand Wire

Abbreviations

THWN	Nylon Jacketed Wire	USP	United States Primed	Wrck.	Wrecker
T.L.	Truckload	UTP	Unshielded Twisted Pair	W.S.P.	Water, Steam, Petroleum
T.M.	Track Mounted	V	Volt	WT., Wt.	Weight
Tot.	Total	V.A.	Volt Amperes	WWF	Welded Wire Fabric
T-O-L	Threadolet	V.C.T.	Vinyl Composition Tile	XFER	Transfer
T.S.	Trigger Start	VAV	Variable Air Volume	XFMR	Transformer
Tr.	Trade	VC	Veneer Core	XHD	Extra Heavy Duty
Transf.	Transformer	Vent.	Ventilation	XHHW, XLPE	Cross-Linked Polyethylene Wire
Trhv.	Truck Driver, Heavy	Vert.	Vertical		Insulation
Trlr	Trailer	V.F.	Vinyl Faced	XLP	Cross-linked Polyethylene
Trlt.	Truck Driver, Light	V.G.	Vertical Grain	Y	Wye
TTY	Teletypewriter	V.H.F.	Very High Frequency	yd	Yard
TV	Television	VHO	Very High Output	yr	Year
T.W.	Thermoplastic Water Resistant	Vib.	Vibrating	Δ	Delta
	Wire	V.L.F.	Vertical Linear Foot	%	Percent
UCI	Uniform Construction Index	Vol.	Volume	~	Approximately
UF	Underground Feeder	VRP	Vinyl Reinforced Polyester	Ø	Phase; diameter
UGND	Underground Feeder	W	Wire; Watt; Wide; West	@	At
U.H.F.	Ultra High Frequency	w/	With	#	Pound; Number
U.I.	United Inch	W.C.	Water Column; Water Closet	<	Less Than
U.L.	Underwriters Laboratory	W.F.	Wide Flange	>	Greater Than
Uld.	unloading	W.G.	Water Gauge	Z	zone
Unfin.	Unfinished	Wldg.	Welding		
URD	Underground Residential	W. Mile	Wire Mile		
	Distribution	W-O-L	Weldolet		
US	United States	W.R.	Water Resistant		

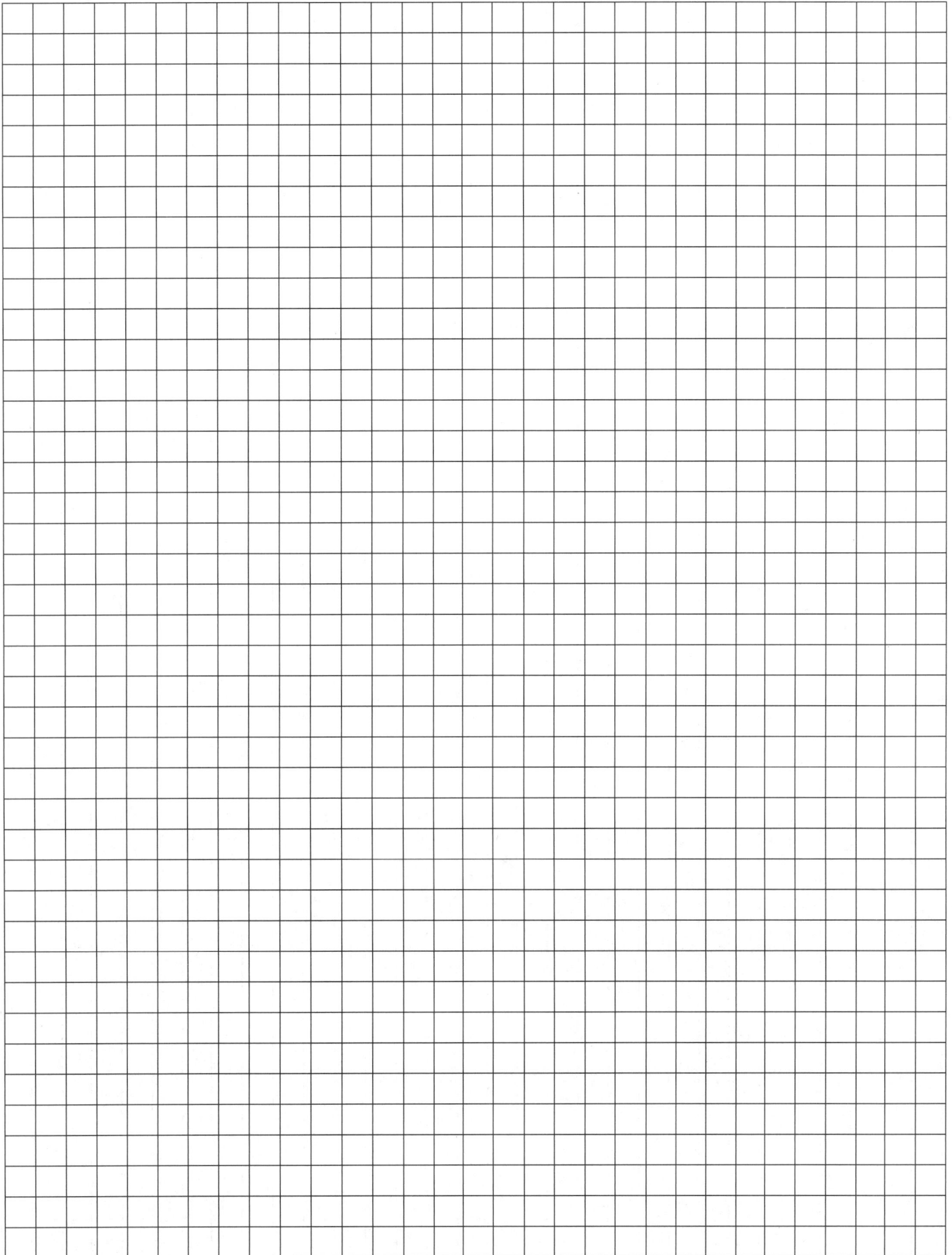

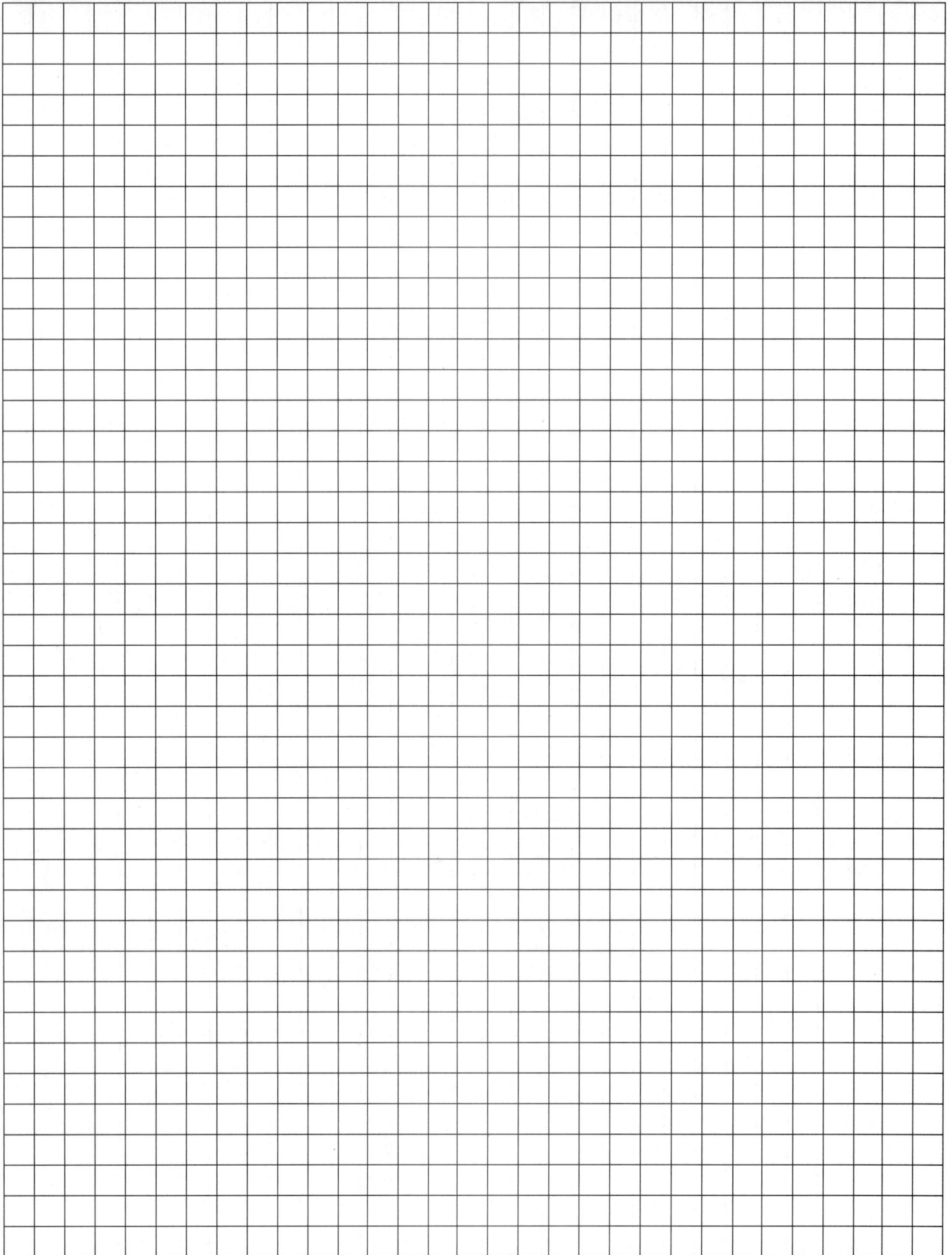

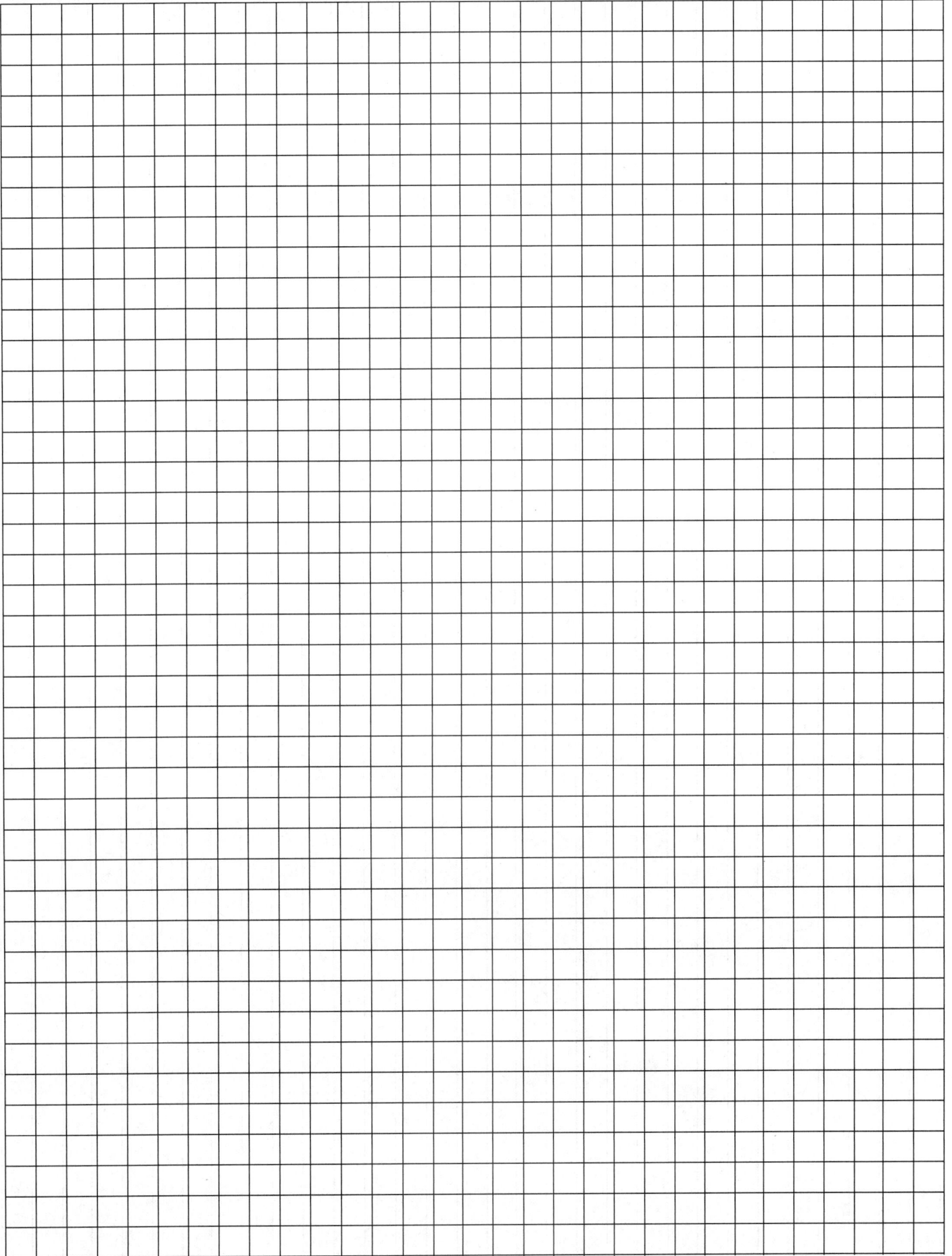

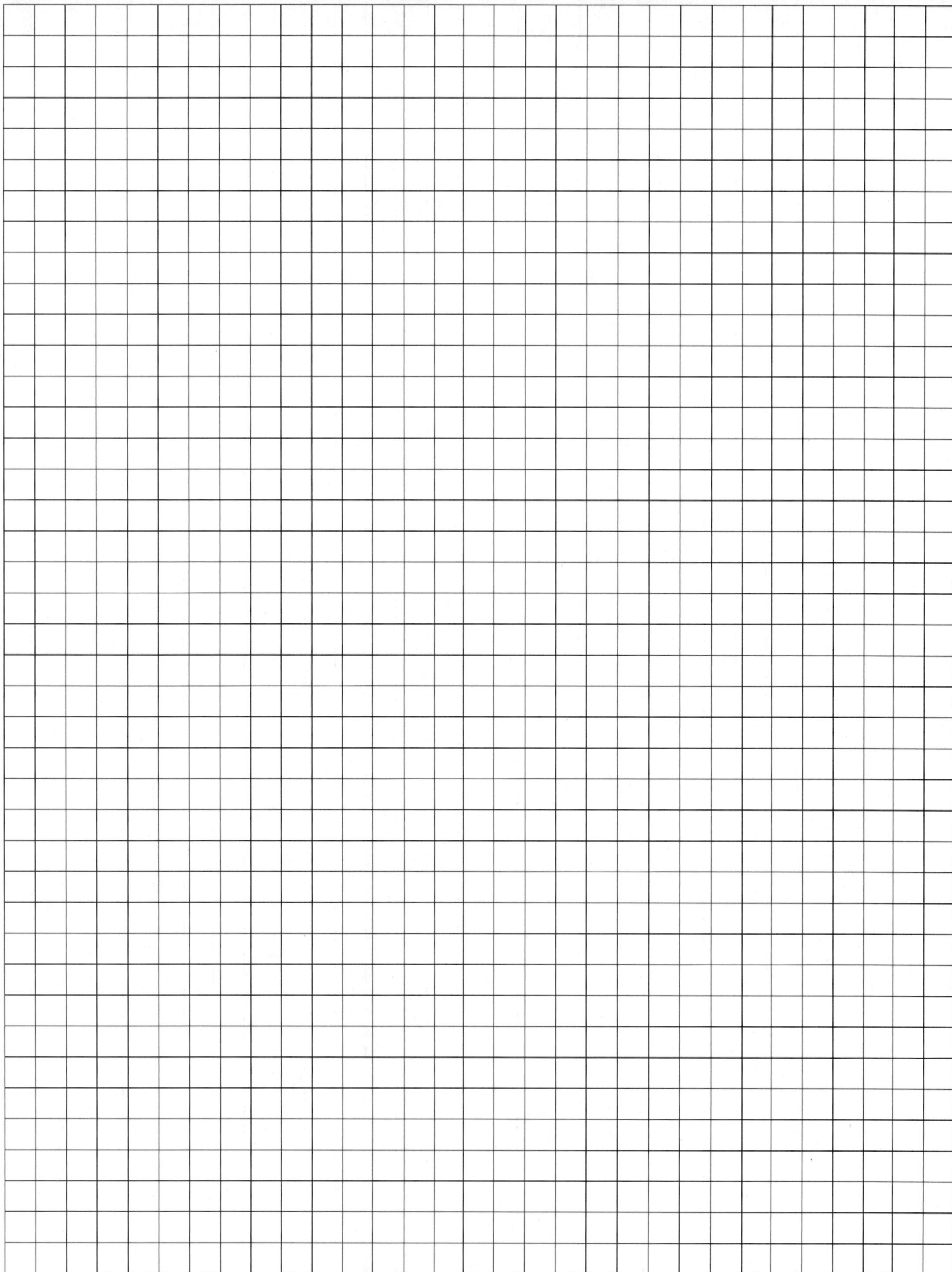

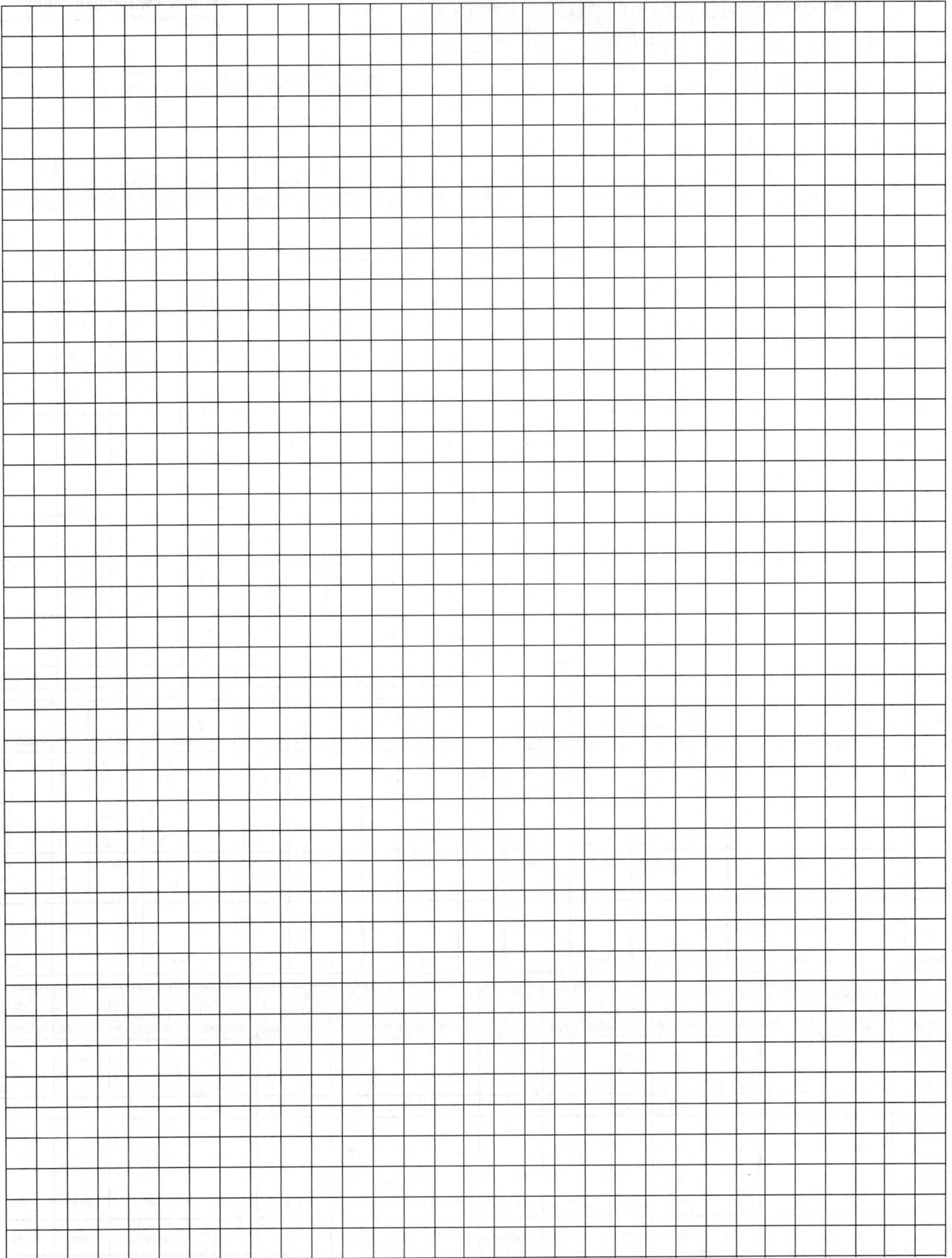

Reed Construction Data/RSMeans— a tradition of excellence in construction cost information and services since 1942

For more information visit the RSMeans website at www.rsmeans.com

Book Selection Guide

The following table provides definitive information on the content of each cost data publication. The number of lines of data provided in each unit price or assemblies division, as well as the number of reference tables and crews, is listed for each book. The presence of other elements such as equipment rental costs, historical cost indexes, city cost indexes, square foot models, or cross-referenced indexes is also indicated. You can use the table to help select the RSMeans book that has the quantity and type of information you most need in your work.

Unit Cost Divisions	Building Construction Costs	Mechanical	Electrical	Repair & Remodel	Square Foot	Site Work Landsc.	Assemblies	Interior	Concrete Masonry	Open Shop	Heavy Construction	Light Commercial	Facilities Construction	Plumbing	Residential
1	553	372	385	474		489		301	437	552	483	220	967	386	157
2	555	239	60	544		782		333	189	553	680	409	1022	262	217
3	1471	193	179	830		1321		266	1803	1467	1459	304	1386	162	309
4	878	24	0	677		683		557	1082	854	576	469	1125	0	385
5	1847	197	152	951		796		1023	751	1816	1045	809	1837	275	724
6	1870	78	80	1458		461		1191	310	1849	395	1562	1484	22	2076
7	1395	164	71	1418		506		513	439	1393	0	1091	1451	175	852
8	1941	61	10	1969		298		1677	673	1906	0	1389	2134	0	1319
9	1800	72	26	1606		292		1899	392	1761	0	1513	2035	54	1357
10	978	17	10	582		226		800	156	978	29	465	1072	233	212
11	998	209	160	476		126		848	28	983	0	223	1012	170	110
12	547	0	2	303		248		1691	12	536	19	349	1736	23	317
13	730	119	111	247		341		255	69	702	253	80	740	74	79
14	281	36	0	230		28		260	0	280	0	12	300	21	6
21	78	0	16	31		0		252	0	78	0	60	370	375	0
22	1090	6955	154	1049		1409		596	20	1069	1634	751	6824	8656	572
23	1189	7103	609	776		159		682	38	1103	110	649	5143	1860	425
26	1246	493	9825	991		743		1053	55	1234	527	1060	9761	438	557
27	75	0	200	34		9		67	0	75	35	48	198	0	4
28	87	59	94	64		0		69	0	72	0	42	109	45	27
31	1904	1162	1031	1225		2731		7	1635	1848	2804	1027	1964	1087	1033
32	737	46	0	603		3849		360	271	710	1366	353	1370	158	412
33	514	1020	429	199		1957		0	225	512	1883	114	1547	1189	112
34	99	0	20	4		134		0	31	55	113	0	119	0	0
35	18	0	0	0		166		0	0	18	281	0	83	0	0
41	52	0	0	24		7		22	0	52	30	0	59	14	0
44	98	90	0	0		32		0	0	23	32	0	98	105	0
Totals	23031	18709	13264	16765		17793		14722	8616	22479	13754	12999	45946	15784	11262

Assembly Divisions	Building Construction Costs	Mechanical	Electrical	Repair & Remodel	Square Foot	Site Work Landscape	Assemblies	Interior	Concrete Masonry	Open Shop	Heavy Construction	Light Commercial	Facilities Construction	Plumbing	Asm Div	Residential
A		15	0	182	150	530	598	0	536		570	154	24	0	1	374
B		0	0	808	2471	0	5577	329	1914		368	2014	143	0	2	217
C		0	0	630	864	0	1229	1567	146		0	765	242	0	3	588
D		1066	810	694	1779	0	2408	765	0		0	1270	1011	1068	4	861
E		0	0	85	260	0	297	5	0		0	257	5	0	5	392
F		0	0	0	123	0	124	0	0		0	123	3	0	6	358
G		528	294	311	202	2206	668	0	535		1194	200	293	685	7	301
															8	760
															9	80
															10	0
															11	0
															12	0
Totals		1609	1104	2710	5849	2736	10901	2666	3131		2132	4783	1721	1753		3931

Reference Section	Building Construction Costs	Mechanical	Electrical	Repair & Remodel	Square Foot	Site Work Landscape	Assemblies	Interior	Concrete Masonry	Open Shop	Heavy Construction	Light Commercial	Facilities Construction	Plumbing	Residential
Reference Tables	yes	yes	yes	yes	no	yes	yes	yes	yes	yes	yes	yes	yes	yes	yes
Models					105							46			32
Crews	482	482	482	462		482		482	482	460	482	460	462	482	460
Equipment Rental Costs	yes	yes	yes	yes		yes		yes	yes	yes	yes	yes	yes	yes	yes
Historical Cost Indexes	yes	yes	yes	yes		yes		yes	yes	yes	yes	yes	yes	yes	no
City Cost Indexes	yes	yes	yes	yes		yes		yes	yes	yes	yes	yes	yes	yes	yes

Annual Cost Guides

For more information
visit the RSMeans website
at www.rsmeans.com

RSMeans Building Construction Cost Data 2009

Offers you unchallenged unit price reliability in an easy-to-use arrangement. Whether used for complete, finished estimates or for periodic checks, it supplies more cost facts better and faster than any comparable source. Over 20,000 unit prices for 2009. The City Cost Indexes and Location Factors cover over 930 areas, for indexing to any project location in North America. Order and get *RSMeans Quarterly Update Service* FREE. You'll have year-long access to the RSMeans Estimating **HOTLINE** FREE with your subscription. Expert assistance when using RSMeans data is just a phone call away.

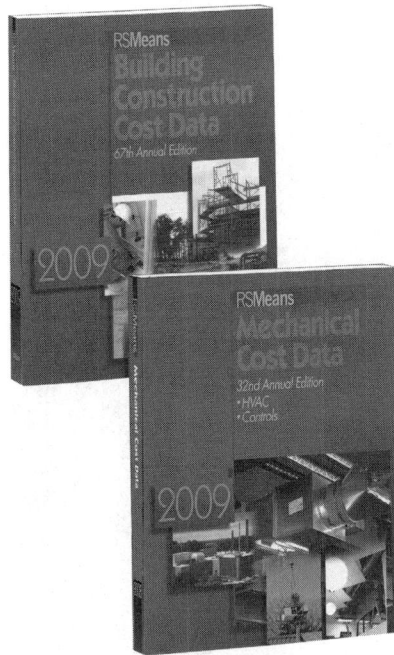

$154.95 per copy
Available Oct. 2008
Catalog no. 60019

RSMeans Mechanical Cost Data 2009
- **HVAC**
- **Controls**

Total unit and systems price guidance for mechanical construction. . . materials, parts, fittings, and complete labor cost information. Includes prices for piping, heating, air conditioning, ventilation, and all related construction.

Plus new 2009 unit costs for:

- Over 2,500 installed HVAC/controls assemblies components
- "On-site" Location Factors for over 930 cities and towns in the U.S. and Canada
- Crews, labor, and equipment

Unit prices organized according to MasterFormat 2004!

$154.95 per copy
Available Oct. 2008
Catalog no. 60029

RSMeans Plumbing Cost Data 2009

Comprehensive unit prices and assemblies for plumbing, irrigation systems, commercial and residential fire protection, point-of-use water heaters, and the latest approved materials. This publication and its companion, *RSMeans Mechanical Cost Data*, provide full-range cost estimating coverage for all the mechanical trades.

Now contains more lines of no-hub CI soil pipe fittings, more flange-type escutcheons, fiberglass pipe insulation in a full range of sizes for 2-1/2" and 3" wall thicknesses, 220 lines of grease duct, and much more.

$154.95 per copy
Available Oct. 2008
Catalog no. 60219

Unit prices organized according to MasterFormat 2004!

RSMeans Facilities Construction Cost Data 2009

For the maintenance and construction of commercial, industrial, municipal, and institutional properties. Costs are shown for new and remodeling construction and are broken down into materials, labor, equipment, and overhead and profit. Special emphasis is given to sections on mechanical, electrical, furnishings, site work, building maintenance, finish work, and demolition.

More than 43,000 unit costs, plus assemblies costs and a comprehensive Reference Section are included.

$368.95 per copy
Available Nov. 2008
Catalog no. 60209

For more information
visit the RSMeans website
at www.rsmeans.com

Annual Cost Guides

RSMeans Square Foot Costs 2009

It's Accurate and Easy To Use!

- **Updated price information** based on nationwide figures from suppliers, estimators, labor experts, and contractors

- "How-to-Use" sections, with **clear examples** of commercial, residential, industrial, and institutional structures

- Realistic graphics, offering true-to-life illustrations of building projects

- Extensive information on using square foot cost data, including sample estimates and alternate pricing methods

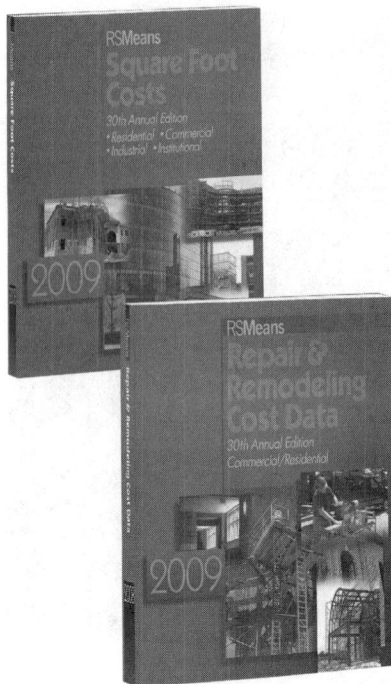

$168.95 per copy
Available Oct. 2008
Catalog no. 60059

RSMeans Repair & Remodeling Cost Data 2009

Commercial/Residential

Use this valuable tool to estimate commercial and residential renovation and remodeling.

Includes: New costs for hundreds of unique methods, materials, and conditions that only come up in repair and remodeling, PLUS:

- Unit costs for over 15,000 construction components
- Installed costs for over 90 assemblies
- Over 930 "on-site" localization factors for the U.S. and Canada.

Unit prices organized according to MasterFormat 2004!

$132.95 per copy
Available Nov. 2008
Catalog no. 60049

RSMeans Electrical Cost Data 2009

Pricing information for every part of electrical cost planning. More than 13,000 unit and systems costs with design tables; clear specifications and drawings; engineering guides; illustrated estimating procedures; complete labor-hour and materials costs for better scheduling and procurement; and the latest electrical products and construction methods.

- A variety of special electrical systems including cathodic protection

- Costs for maintenance, demolition, HVAC/mechanical, specialties, equipment, and more

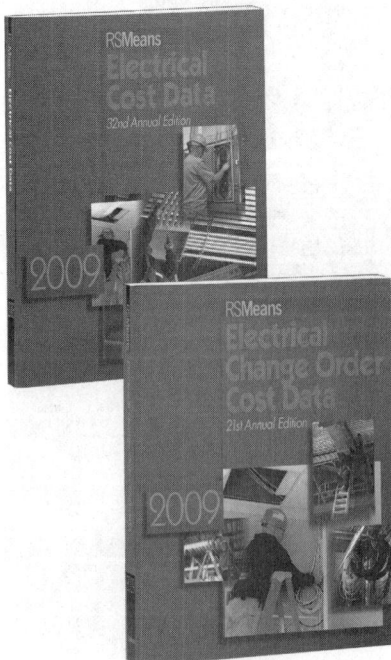

$154.95 per copy
Available Oct. 2008
Catalog no. 60039

Unit prices organized according to MasterFormat 2004!

RSMeans Electrical Change Order Cost Data 2009

RSMeans Electrical Change Order Cost Data provides you with electrical unit prices exclusively for pricing change orders—based on the recent, direct experience of contractors and suppliers. Analyze and check your own change order estimates against the experience others have had doing the same work. It also covers productivity analysis and change order cost justifications. With useful information for calculating the effects of change orders and dealing with their administration.

$154.95 per copy
Available Dec. 2008
Catalog no. 60239

RSMeans Assemblies Cost Data 2009

RSMeans Assemblies Cost Data takes the guesswork out of preliminary or conceptual estimates. Now you don't have to try to calculate the assembled cost by working up individual component costs. We've done all the work for you.

Presents detailed illustrations, descriptions, specifications, and costs for every conceivable building assembly—240 types in all—arranged in the easy-to-use UNIFORMAT II system. Each illustrated "assembled" cost includes a complete grouping of materials and associated installation costs, including the installing contractor's overhead and profit.

$245.95 per copy
Available Oct. 2008
Catalog no. 60069

RSMeans Open Shop Building Construction Cost Data 2009

The latest costs for accurate budgeting and estimating of new commercial and residential construction. . . renovation work. . . change orders. . . cost engineering.

RSMeans Open Shop "BCCD" will assist you to:
- Develop benchmark prices for change orders
- Plug gaps in preliminary estimates and budgets
- Estimate complex projects
- Substantiate invoices on contracts
- Price ADA-related renovations

Unit prices organized according to MasterFormat 2004!

$154.95 per copy
Available Dec. 2008
Catalog no. 60159

RSMeans Residential Cost Data 2009

Contains square foot costs for 30 basic home models with the look of today, plus hundreds of custom additions and modifications you can quote right off the page. With costs for the 100 residential systems you're most likely to use in the year ahead. Complete with blank estimating forms, sample estimates, and step-by-step instructions.

Now contains line items for cultured stone and brick, PVC trim lumber, and TPO roofing.

$132.95 per copy
Available Oct. 2008
Catalog no. 60179

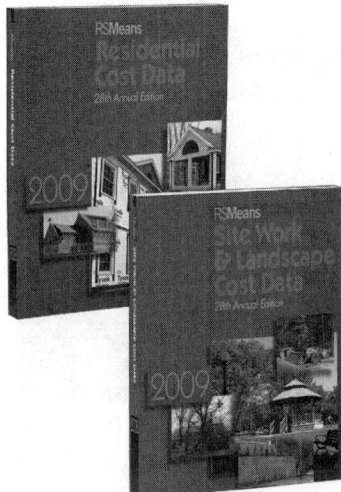

RSMeans Site Work & Landscape Cost Data 2009

Includes unit and assemblies costs for earthwork, sewerage, piped utilities, site improvements, drainage, paving, trees & shrubs, street openings/repairs, underground tanks, and more. Contains 57 tables of assemblies costs for accurate conceptual estimates.

Includes:
- Estimating for infrastructure improvements
- Environmentally-oriented construction
- ADA-mandated handicapped access
- Hazardous waste line items

Unit prices organized according to MasterFormat 2004!

$154.95 per copy
Available Nov. 2008
Catalog no. 60289

RSMeans Facilities Maintenance & Repair Cost Data 2009

RSMeans Facilities Maintenance & Repair Cost Data gives you a complete system to manage and plan your facility repair and maintenance costs and budget efficiently. Guidelines for auditing a facility and developing an annual maintenance plan. Budgeting is included, along with reference tables on cost and management, and information on frequency and productivity of maintenance operations.

The only nationally recognized source of maintenance and repair costs. Developed in cooperation with the Civil Engineering Research Laboratory (CERL) of the Army Corps of Engineers.

$336.95 per copy
Available Dec. 2008
Catalog no. 60309

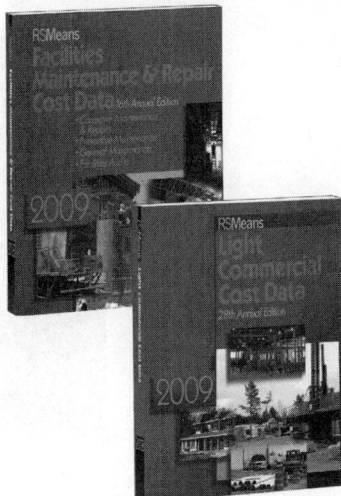

RSMeans Light Commercial Cost Data 2009

Specifically addresses the light commercial market, which is a specialized niche in the construction industry. Aids you, the owner/designer/contractor, in preparing all types of estimates—from budgets to detailed bids. Includes new advances in methods and materials.

Assemblies Section allows you to evaluate alternatives in the early stages of design/planning.

Over 11,000 unit costs ensure that you have the prices you need. . . when you need them.

Unit prices organized according to MasterFormat 2004!

$132.95 per copy
Available Nov. 2008
Catalog no. 60189

For more information
visit the RSMeans website
at www.rsmeans.com

Annual Cost Guides

RSMeans Concrete & Masonry Cost Data 2009

Provides you with cost facts for virtually all concrete/masonry estimating needs, from complicated formwork to various sizes and face finishes of brick and block—all in great detail. The comprehensive Unit Price Section contains more than 7,500 selected entries. Also contains an Assemblies [Cost] Section, and a detailed Reference Section that supplements the cost data.

Unit prices organized according to MasterFormat 2004!

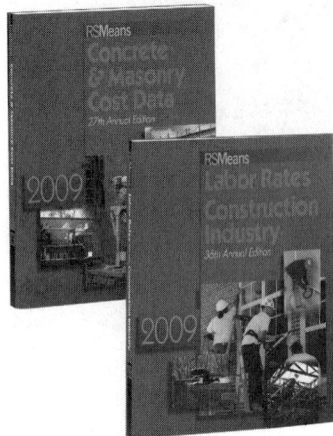

$139.95 per copy
Available Dec. 2008
Catalog no. 60119

RSMeans Labor Rates for the Construction Industry 2009

Complete information for estimating labor costs, making comparisons, and negotiating wage rates by trade for over 300 U.S. and Canadian cities. With 46 construction trades listed by local union number in each city, and historical wage rates included for comparison. Each city chart lists the county and is alphabetically arranged with handy visual flip tabs for quick reference.

$336.95 per copy
Available Dec. 2008
Catalog no. 60129

RSMeans Construction Cost Indexes 2009

What materials and labor costs will change unexpectedly this year? By how much?

• Breakdowns for 316 major cities
• National averages for 30 key cities
• Expanded five major city indexes
• Historical construction cost indexes

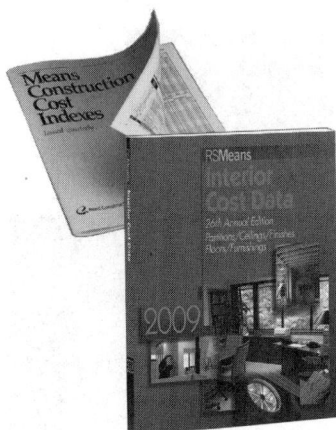

$335.00 per year (subscription)
Catalog no. 50149

$83.25 individual quarters
Catalog no. 60149 A,B,C,D

RSMeans Interior Cost Data 2009

Provides you with prices and guidance needed to make accurate interior work estimates. Contains costs on materials, equipment, hardware, custom installations, furnishings, and labor costs. . . for new and remodel commercial and industrial interior construction, including updated information on office furnishings, and reference information.

Unit prices organized according to MasterFormat 2004!

$154.95 per copy
Available Nov. 2008
Catalog no. 60099

RSMeans Heavy Construction Cost Data 2009

A comprehensive guide to heavy construction costs. Includes costs for highly specialized projects such as tunnels, dams, highways, airports, and waterways. Information on labor rates, equipment, and materials costs is included. Features unit price costs, systems costs, and numerous reference tables for costs and design.

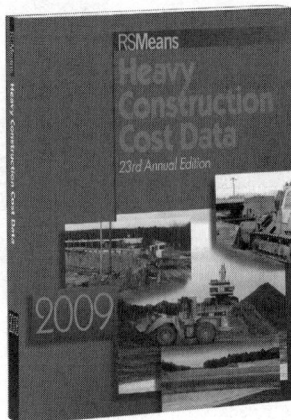

$154.95 per copy
Available Dec. 2008
Catalog no. 60169

Unit prices organized according to MasterFormat 2004!

Reference Books

For more information
visit the RSMeans website
at www.rsmeans.com

Value Engineering: Practical Applications

For Design, Construction, Maintenance & Operations

by Alphonse Dell'Isola, PE

A tool for immediate application—for engineers, architects, facility managers, owners, and contractors. Includes making the case for VE—the management briefing; integrating VE into planning, budgeting, and design; conducting life cycle costing; using VE methodology in design review and consultant selection; case studies; VE workbook; and a life cycle costing program on disk.

$79.95 per copy
Over 450 pages, illustrated, softcover
Catalog no. 67319A

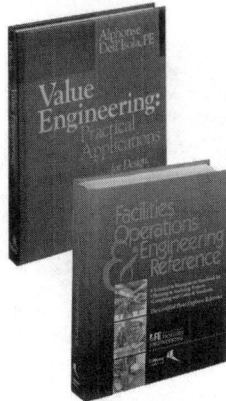

Facilities Operations & Engineering Reference

by the Association for Facilities Engineering and RSMeans

An all-in-one technical reference for planning and managing facility projects and solving day-to-day operations problems. Selected as the official certified plant engineer reference, this handbook covers financial analysis, maintenance, HVAC and energy efficiency, and more.

$54.98 per copy
Over 700 pages, illustrated, hardcover
Catalog no. 67318

The Building Professional's Guide to Contract Documents

3rd Edition

by Waller S. Poage, AIA, CSI, CVS

A comprehensive reference for owners, design professionals, contractors, and students

- Structure your documents for maximum efficiency.
- Effectively communicate construction requirements.
- Understand the roles and responsibilities of construction professionals.
- Improve methods of project delivery.

$64.95 per copy, 400 pages
Diagrams and construction forms, hardcover
Catalog no. 67261A

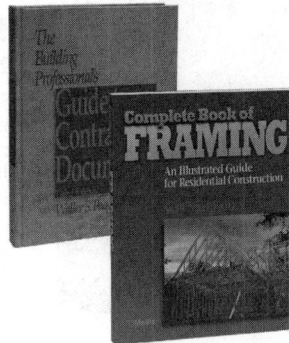

Complete Book of Framing

by Scot Simpson

This straightforward, easy-to-learn method will help framers, carpenters, and handy homeowners build their skills in rough carpentry and framing. Shows how to frame all the parts of a house: floors, walls, roofs, door & window openings, and stairs—with hundreds of color photographs and drawings that show every detail.

The book gives beginners all the basics they need to go from zero framing knowledge to a journeyman level.

$29.95 per copy
352 pages, softcover
Catalog no. 67353

Cost Planning & Estimating for Facilities Maintenance

In this unique book, a team of facilities management authorities shares their expertise on:

- Evaluating and budgeting maintenance operations
- Maintaining and repairing key building components
- Applying *RSMeans Facilities Maintenance & Repair Cost Data* to your estimating

Covers special maintenance requirements of the ten major building types.

$89.95 per copy
Over 475 pages, hardcover
Catalog no. 67314

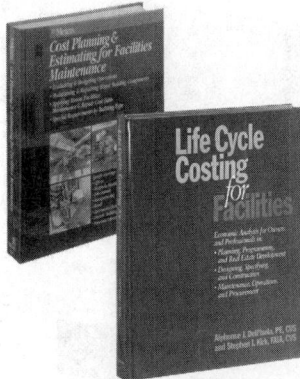

Life Cycle Costing for Facilities

by Alphonse Dell'Isola and Dr. Steven Kirk

Guidance for achieving higher quality design and construction projects at lower costs! Cost-cutting efforts often sacrifice quality to yield the cheapest product. Life cycle costing enables building designers and owners to achieve both. The authors of this book show how LCC can work for a variety of projects — from roads to HVAC upgrades to different types of buildings.

$99.95 per copy
450 pages, hardcover
Catalog no. 67341

Planning & Managing Interior Projects 2nd Edition

by Carol E. Farren, CFM

Expert guidance on managing renovation & relocation projects

This book guides you through every step in relocating to a new space or renovating an old one. From initial meeting through design and construction, to post-project administration, it helps you get the most for your company or client. Includes sample forms, spec lists, agreements, drawings, and much more!

$34.98 per copy
200 pages, softcover
Catalog no. 67245A

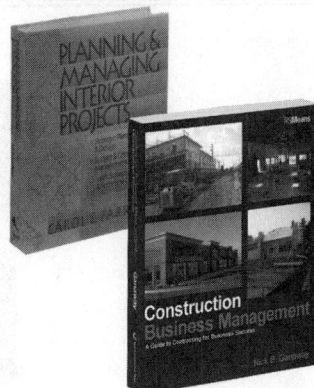

Construction Business Management

by Nick Ganaway

Only 43% of construction firms stay in business after four years. Make sure your company thrives with valuable guidance from a pro with 25 years of success as a commercial contractor. Find out what it takes to build all aspects of a business that is profitable, enjoyable, and enduring. With a bonus chapter on retail construction.

$49.95 per copy
200 pages, softcover
Catalog no. 67352

For more information
visit the RSMeans website
at www.rsmeans.com

Reference Books

Interior Home Improvement Costs 9th Edition

Updated estimates for the most popular remodeling and repair projects—from small, do-it-yourself jobs to major renovations and new construction. Includes: kitchens & baths; new living space from your attic, basement, or garage; new floors, paint, and wallpaper; tearing out or building new walls; closets, stairs, and fireplaces; new energy-saving improvements, home theaters, and more!

$24.95 per copy
250 pages, illustrated, softcover
Catalog no. 67308E

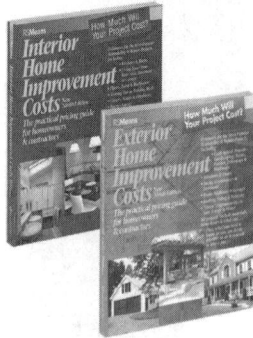

Exterior Home Improvement Costs 9th Edition

Updated estimates for the most popular remodeling and repair projects—from small, do-it-yourself jobs, to major renovations and new construction. Includes: curb appeal projects—landscaping, patios, porches, driveways, and walkways; new windows and doors; decks, greenhouses, and sunrooms; room additions and garages; roofing, siding, and painting; "green" improvements to save energy & water.

$24.95 per copy
Over 275 pages, illustrated, softcover
Catalog no. 67309E

Builder's Essentials: Plan Reading & Material Takeoff

For Residential and Light Commercial Construction

by Wayne J. DelPico

A valuable tool for understanding plans and specs, and accurately calculating material quantities. Step-by-step instructions and takeoff procedures based on a full set of working drawings.

$35.95 per copy
Over 420 pages, softcover
Catalog no. 67307

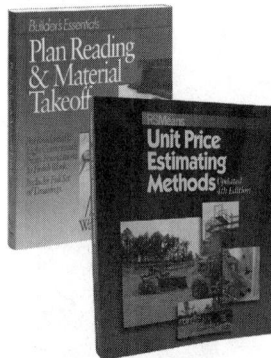

Means Unit Price Estimating Methods

New 4th Edition

This new edition includes up-to-date cost data and estimating examples, updated to reflect changes to the CSI numbering system and new features of RSMeans cost data. It describes the most productive, universally accepted ways to estimate, and uses checklists and forms to illustrate shortcuts and timesavers. A model estimate demonstrates procedures. A new chapter explores computer estimating alternatives.

$59.95 per copy
Over 350 pages, illustrated, softcover
Catalog no. 67303B

Total Productive Facilities Management

by Richard W. Sievert, Jr.

Today, facilities are viewed as strategic resources. . . elevating the facility manager to the role of asset manager supporting the organization's overall business goals. Now, Richard Sievert Jr., in this well-articulated guidebook, sets forth a new operational standard for the facility manager's emerging role. . . a comprehensive program for managing facilities as a true profit center.

$29.98 per copy
275 pages, softcover
Catalog no. 67321

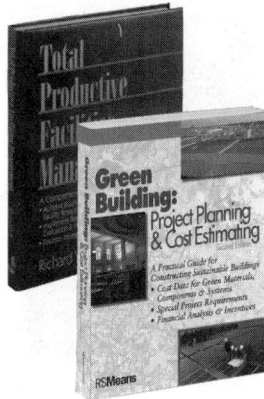

Green Building: Project Planning & Cost Estimating 2nd Edition

This new edition has been completely updated with the latest in green building technologies, design concepts, standards, and costs. Now includes a 2009 Green Building *CostWorks* CD with more than 300 green building assemblies and over 5,000 unit price line items for sustainable building. The new edition is also full-color with all new case studies—plus a new chapter on deconstruction, a key aspect of green building.

$129.95 per copy
350 pages, softcover
Catalog no. 67338A

Concrete Repair and Maintenance Illustrated

by Peter Emmons

Hundreds of illustrations show users how to analyze, repair, clean, and maintain concrete structures for optimal performance and cost effectiveness. From parking garages to roads and bridges to structural concrete, this comprehensive book describes the causes, effects, and remedies for concrete wear and failure. Invaluable for planning jobs, selecting materials, and training employees, this book is a must-have for concrete specialists, general contractors, facility managers, civil and structural engineers, and architects.

$69.95 per copy
300 pages, illustrated, softcover
Catalog no. 67146

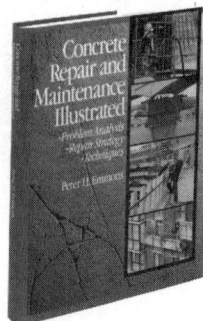

Reference Books

For more information
visit the RSMeans website
at www.rsmeans.com

Means Illustrated Construction Dictionary Condensed, 2nd Edition

The best portable dictionary for office or field use—an essential tool for contractors, architects, insurance and real estate personnel, facility managers, homeowners, and anyone who needs quick, clear definitions for construction terms. The second edition has been further enhanced with updates and hundreds of new terms and illustrations . . . in keeping with the most recent developments in the construction industry.

Now with a quick-reference Spanish section. Includes tools and equipment, materials, tasks, and more!

$59.95 per copy
Over 500 pages, softcover
Catalog no. 67282A

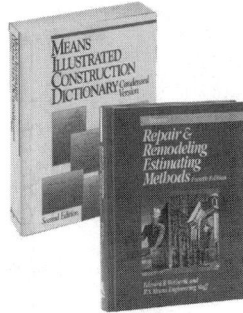

Means Repair & Remodeling Estimating 4th Edition
By Edward B. Wetherill and RSMeans

This important reference focuses on the unique problems of estimating renovations of existing structures, and helps you determine the true costs of remodeling through careful evaluation of architectural details and a site visit.

New section on disaster restoration costs.

$69.95 per copy
Over 450 pages, illustrated, hardcover
Catalog no. 67265B

Facilities Planning & Relocation
by David D. Owen

A complete system for planning space needs and managing relocations. Includes step-by-step manual, over 50 forms, and extensive reference section on materials and furnishings.

New lower price and user-friendly format.

$49.98 per copy
Over 450 pages, softcover
Catalog no. 67301

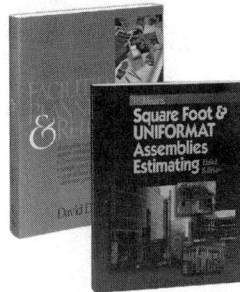

Means Square Foot & UNIFORMAT Assemblies Estimating Methods 3rd Edition

Develop realistic square foot and assemblies costs for budgeting and construction funding. The new edition features updated guidance on square foot and assemblies estimating using UNIFORMAT II. An essential reference for anyone who performs conceptual estimates.

$69.95 per copy
Over 300 pages, illustrated, softcover
Catalog no. 67145B

Means Electrical Estimating Methods 3rd Edition

Expanded edition includes sample estimates and cost information in keeping with the latest version of the CSI MasterFormat and UNIFORMAT II. Complete coverage of fiber optic and uninterruptible power supply electrical systems, broken down by components, and explained in detail. Includes a new chapter on computerized estimating methods. A practical companion to *RSMeans Electrical Cost Data.*

$64.95 per copy
Over 325 pages, hardcover
Catalog no. 67230B

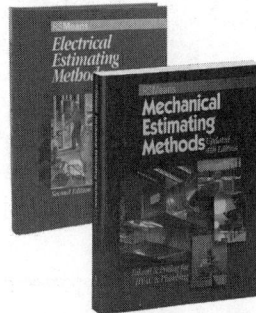

Means Mechanical Estimating Methods 4th Edition

Completely updated, this guide assists you in making a review of plans, specs, and bid packages, with suggestions for takeoff procedures, listings, substitutions, and pre-bid scheduling for all components of HVAC. Includes suggestions for budgeting labor and equipment usage. Compares materials and construction methods to allow you to select the best option.

$64.95 per copy
Over 350 pages, illustrated, softcover
Catalog no. 67294B

Means ADA Compliance Pricing Guide 2nd Edition
by Adaptive Environments and RSMeans

Completely updated and revised to the new 2004 *Americans with Disabilities Act Accessibility Guidelines*, this book features more than 70 of the most commonly needed modifications for ADA compliance. Projects range from installing ramps and walkways, widening doorways and entryways, and installing and refitting elevators, to relocating light switches and signage.

$79.95 per copy
Over 350 pages, illustrated, softcover
Catalog no. 67310A

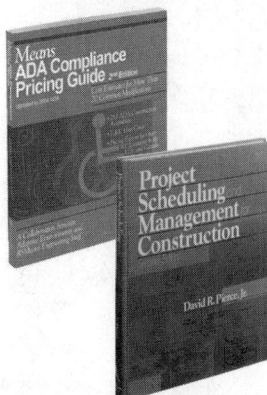

Project Scheduling & Management for Construction 3rd Edition
by David R. Pierce, Jr.

A comprehensive yet easy-to-follow guide to construction project scheduling and control—from vital project management principles through the latest scheduling, tracking, and controlling techniques. The author is a leading authority on scheduling, with years of field and teaching experience at leading academic institutions. Spend a few hours with this book and come away with a solid understanding of this essential management topic.

$64.95 per copy
Over 300 pages, illustrated, hardcover
Catalog no. 67247B

599

For more information
visit the RSMeans website
at www.rsmeans.com

Reference Books

The Practice of Cost Segregation Analysis

by Bruce A. Desrosiers and Wayne J. DelPico

This expert guide walks you through the practice of cost segregation analysis, which enables property owners to defer taxes and benefit from "accelerated cost recovery" through depreciation deductions on assets that are properly identified and classified.

With a glossary of terms, sample cost segregation estimates for various building types, key information resources, and updates via a dedicated website, this book is a critical resource for anyone involved in cost segregation analysis.

$99.95 per copy
Over 225 pages
Catalog no. 67345

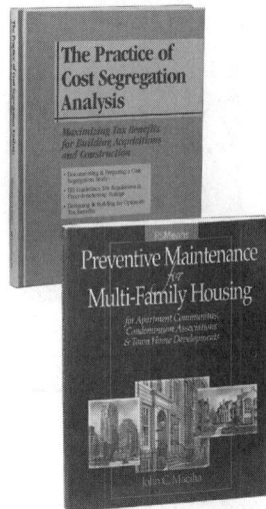

Preventive Maintenance for Multi-Family Housing

by John C. Maciha

Prepared by one of the nation's leading experts on multi-family housing.

This complete PM system for apartment and condominium communities features expert guidance, checklists for buildings and grounds maintenance tasks and their frequencies, a reusable wall chart to track maintenance, and a dedicated website featuring customizable electronic forms. A must-have for anyone involved with multi-family housing maintenance and upkeep.

$89.95 per copy
225 pages
Catalog no. 67346

How to Estimate with Means Data & CostWorks

New 3rd Edition

by RSMeans and Saleh A. Mubarak, Ph.D.

New 3rd Edition—fully updated with new chapters, plus new CD with updated *CostWorks* cost data and MasterFormat organization. Includes all major construction items—with more than 300 exercises and two sets of plans that show how to estimate for a broad range of construction items and systems—including general conditions and equipment costs.

$59.95 per copy
272 pages, softcover
Includes CostWorks CD
Catalog no. 67324B

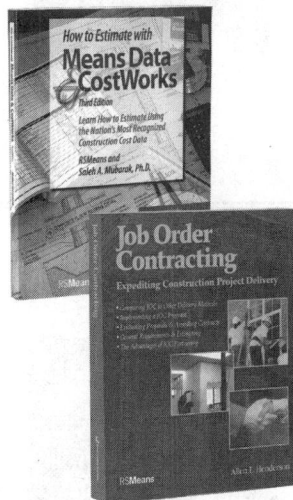

Job Order Contracting

Expediting Construction Project Delivery

by Allen Henderson

Expert guidance to help you implement JOC—fast becoming the preferred project delivery method for repair and renovation, minor new construction, and maintenance projects in the public sector and in many states and municipalities. The author, a leading JOC expert and practitioner, shows how to:

- Establish a JOC program
- Evaluate proposals and award contracts
- Handle general requirements and estimating
- Partner for maximum benefits

$89.95 per copy
192 pages, illustrated, hardcover
Catalog no. 67348

Builder's Essentials: Estimating Building Costs

For the Residential & Light Commercial Contractor

by Wayne J. DelPico

Step-by-step estimating methods for residential and light commercial contractors. Includes a detailed look at every construction specialty—explaining all the components, takeoff units, and labor needed for well-organized, complete estimates. Covers correctly interpreting plans and specifications, and developing accurate and complete labor and material costs.

$29.95 per copy
Over 400 pages, illustrated, softcover
Catalog no. 67343

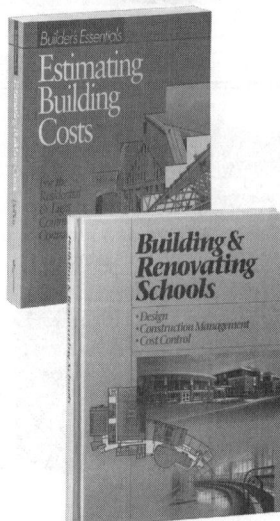

Building & Renovating Schools

This all-inclusive guide covers every step of the school construction process—from initial planning, needs assessment, and design, right through moving into the new facility. A must-have resource for anyone concerned with new school construction or renovation. With square foot cost models for elementary, middle, and high school facilities, and real-life case studies of recently completed school projects.

The contributors to this book—architects, construction project managers, contractors, and estimators who specialize in school construction—provide start-to-finish, expert guidance on the process.

$69.98 per copy
Over 425 pages, hardcover
Catalog no. 67342

Reference Books

For more information
visit the RSMeans website
at www.rsmeans.com

The Homeowner's Guide to Mold By Michael Pugliese

Expert guidance to protect your health and your home.

Mold, whether caused by leaks, humidity or flooding, is a real health and financial issue—for homeowners and contractors. This full-color book explains:
- Construction and maintenance practices to prevent mold
- How to inspect for and remove mold
- Mold remediation procedures and costs
- What to do after a flood
- How to deal with insurance companies

$21.95 per copy
144 pages, softcover
Catalog no. 67344

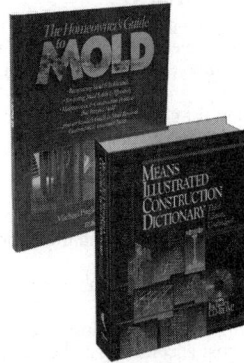

Means Landscape Estimating Methods

New 5th Edition

Answers questions about preparing competitive landscape construction estimates, with up-to-date cost estimates and the new MasterFormat classification system. Expanded and revised to address the latest materials and methods, including new coverage on approaches to green building. Includes:
- Step-by-step explanation of the estimating process
- Sample forms and worksheets that save time and prevent errors

$64.95 per copy
Over 350 pages, softcover
Catalog no. 67295C

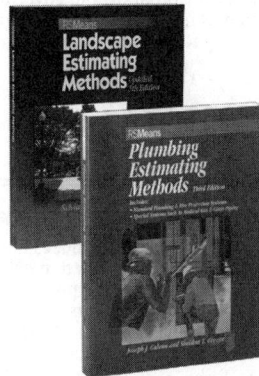

Understanding & Negotiating Construction Contracts

by Kit Werremeyer

Take advantage of the author's 30 years' experience in small-to-large (including international) construction projects. Learn how to identify, understand, and evaluate high risk terms and conditions typically found in all construction contracts—then negotiate to lower or eliminate the risk, improve terms of payment, and reduce exposure to claims and disputes. The author avoids "legalese" and gives real-life examples from actual projects.

$69.95 per copy
300 pages, softcover
Catalog no. 67350

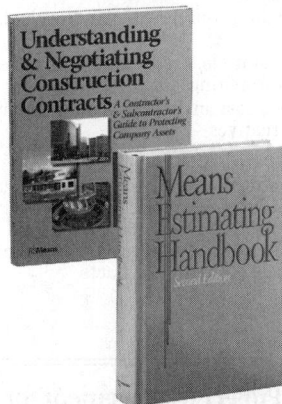

Means Spanish/English Construction Dictionary 2nd Edition

by RSMeans and the International Code Council

This expanded edition features thousands of the most common words and useful phrases in the construction industry with easy-to-follow pronunciations in both Spanish and English. Over 800 new terms, phrases, and illustrations have been added. It also features a new stand-alone "Safety & Emergencies" section, with colored pages for quick access.

Unique to this dictionary are the systems illustrations showing the relationship of components in the most common building systems for all major trades.

$23.95 per copy
Over 400 pages
Catalog no. 67327A

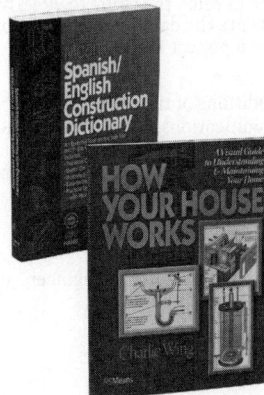

Means Illustrated Construction Dictionary

Unabridged 3rd Edition, with CD-ROM

Long regarded as the industry's finest, *Means Illustrated Construction Dictionary* is now even better. With the addition of over 1,000 new terms and hundreds of new illustrations, it is the clear choice for the most comprehensive and current information. The companion CD-ROM that comes with this new edition adds many extra features: larger graphics, expanded definitions, and links to both CSI MasterFormat numbers and product information.

$99.95 per copy
Over 790 pages, illustrated, hardcover
Catalog no. 67292A

Means Plumbing Estimating Methods 3rd Edition

by Joseph Galeno and Sheldon Greene

Updated and revised! This practical guide walks you through a plumbing estimate, from basic materials and installation methods through change order analysis. *Plumbing Estimating Methods* covers residential, commercial, industrial, and medical systems, and features sample takeoff and estimate forms and detailed illustrations of systems and components.

$29.98 per copy
330+ pages, softcover
Catalog no. 67283B

Means Estimating Handbook

2nd Edition

Updated Second Edition answers virtually any estimating technical question—all organized by CSI MasterFormat. This comprehensive reference covers the full spectrum of technical data required to estimate construction costs. The book includes information on sizing, productivity, equipment requirements, code-mandated specifications, design standards, and engineering factors.

$99.95 per copy
Over 900 pages, hardcover
Catalog no. 67276A

How Your House Works

by Charlie Wing

A must-have reference for every homeowner, handyman, and contractor—for repair, remodeling, and new construction. This book uncovers the mysteries behind just about every major appliance and building element in your house—from electrical, heating and AC, to plumbing, framing, foundations and appliances. Clear, "exploded" drawings show exactly how things should be put together and how they function—what to check if they don't work, and what you can do that might save you having to call in a professional.

$21.95 per copy
160 pages, softcover
Catalog no. 67351

For more information
visit the RSMeans website
at www.rsmeans.com

Professional Development

Means CostWorks® Training

This one-day course helps users become more familiar with the functionality of *Means CostWorks* program. Each menu, icon, screen, and function found in the program is explained in depth. Time is devoted to hands-on estimating exercises.

Some of what you'll learn:
- Search the database utilizing all navigation methods
- Export RSMeans Data to your preferred spreadsheet format
- View crews, assembly components, and much more!
- Automatically regionalize the database

You are required to bring your own laptop computer for this course.

When you register for this course you will receive an outline for your laptop requirements.

Unit Price Estimating

This interactive *two-day* seminar teaches attendees how to interpret project information and process it into final, detailed estimates with the greatest accuracy level.

The most important credential an estimator can take to the job is the ability to visualize construction, and estimate accurately.

Some of what you'll learn:
- Interpreting the design in terms of cost
- The most detailed, time-tested methodology for accurate pricing
- Key cost drivers—material, labor, equipment, staging, and subcontracts
- Understanding direct and indirect costs for accurate job cost accounting and change order management

Who should attend: Corporate and government estimators and purchasers, architects, engineers… and others needing to produce accurate project estimates.

Square Foot & Assemblies Estimating

This *two-day* course teaches attendees how to quickly deliver accurate square foot estimates using limited budget and design information.

Some of what you'll learn:
- How square foot costing gets the estimate done faster
- Taking advantage of a "systems" or "assemblies" format
- The RSMeans "building assemblies/square foot cost approach"
- How to create a reliable preliminary systems estimate using bare-bones design information

Who should attend: Facilities managers, facilities engineers, estimators, planners, developers, construction finance professionals… and others needing to make quick, accurate construction cost estimates at commercial, government, educational, and medical facilities.

Repair & Remodeling Estimating

This *two-day* seminar emphasizes all the underlying considerations unique to repair/remodeling estimating and presents the correct methods for generating accurate, reliable R&R project costs using the unit price and assemblies methods.

Some of what you'll learn:
- Estimating considerations—like labor-hours, building code compliance, working within existing structures, purchasing materials in smaller quantities, unforeseen deficiencies
- Identifying problems and providing solutions to estimating building alterations
- Rules for factoring in minimum labor costs, accurate productivity estimates, and allowances for project contingencies
- R&R estimating examples calculated using unit price and assemblies data

Who should attend: Facilities managers, plant engineers, architects, contractors, estimators, builders… and others who are concerned with the proper preparation and/or evaluation of repair and remodeling estimates.

Mechanical & Electrical Estimating

This *two-day* course teaches attendees how to prepare more accurate and complete mechanical/electrical estimates, avoiding the pitfalls of omission and double-counting, while understanding the composition and rationale within the RSMeans Mechanical/Electrical database.

Some of what you'll learn:
- The unique way mechanical and electrical systems are interrelated
- M&E estimates–conceptual, planning, budgeting, and bidding stages
- Order of magnitude, square foot, assemblies, and unit price estimating
- Comparative cost analysis of equipment and design alternatives

Who should attend: Architects, engineers, facilities managers, mechanical and electrical contractors… and others needing a highly reliable method for developing, understanding, and evaluating mechanical and electrical contracts.

Green Building Planning & Construction

In this *two-day* course, learn about tailoring a building and its placement on the site to the local climate, site conditions, culture, and community. Includes information to reduce resource consumption and augment resource supply.

Some of what you'll learn:
- Green technologies, materials, systems, and standards
- Energy efficiencies with energy modeling tools
- Cost vs. value of green products over their life cycle
- Low-cost strategies and economic incentives and funding
- Health, comfort, and productivity goals and techniques

Who should attend: Contractors, project managers, building owners, building officials, healthcare and insurance professionals.

Facilities Maintenance & Repair Estimating

This *two-day* course teaches attendees how to plan, budget, and estimate the cost of ongoing and preventive maintenance and repair for existing buildings and grounds.

Some of what you'll learn:
- The most financially favorable maintenance, repair, and replacement scheduling and estimating
- Auditing and value engineering facilities
- Preventive planning and facilities upgrading
- Determining both in-house and contract-out service costs
- Annual, asset-protecting M&R plan

Who should attend: Facility managers, maintenance supervisors, buildings and grounds superintendents, plant managers, planners, estimators… and others involved in facilities planning and budgeting.

Practical Project Management for Construction Professionals

In this *two-day* course, acquire the essential knowledge and develop the skills to effectively and efficiently execute the day-to-day responsibilities of the construction project manager.

Covers:
- General conditions of the construction contract
- Contract modifications: change orders and construction change directives
- Negotiations with subcontractors and vendors
- Effective writing: notification and communications
- Dispute resolution: claims and liens

Who should attend: Architects, engineers, owner's representatives, project managers.

Facilities Repair & Remodeling Estimating

In this *two-day* course, professionals working in facilities management can get help with their daily challenges to establish budgets for all phase of a project

Some of what you'll learn:
- Determine the full scope of a project
- Identify the scope of risks & opportunities
- Creative solutions to estimating issues
- Organizing estimates for presentation & discussion
- Special techniques for repair/remodel and maintenance projects
- Negotiating project change orders

Who should attend: Facility managers, engineers, contractors, facility trades-people, planners & project managers

Scheduling and Project Management

This *two-day* course teaches attendees the most current and proven scheduling and management techniques needed to bring projects in on time and on budget.

Some of what you'll learn:
- Crucial phases of planning and scheduling
- How to establish project priorities and develop realistic schedules and management techniques
- Critical Path and Precedence Methods
- Special emphasis on cost control

Who should attend: Construction project managers, supervisors, engineers, estimators, contractors... and others who want to improve their project planning, scheduling, and management skills.

Understanding & Negotiating Construction Contracts

In this *two-day* course, learn how to protect the assets of your company by justifying or eliminating commercial risk through negotiation of a contract's terms and conditions.

Some of what you'll learn:
- Myths and paradigms of contracts
- Scope of work, terms of payment, and scheduling
- Dispute resolution
- Why clients love CLAIMS!
- Negotiating issues and much more

Who should attend: Contractors & subcontractors, material suppliers, project managers, risk & insurance managers, procurement managers, owners and facility managers, corporate executives.

Site Work & Heavy Construction Estimating

This *two-day* course teaches attendees how to estimate earthwork, site utilities, foundations, and site improvements, using the assemblies and the unit price methods.

Some of what you'll learn:
- Basic site work and heavy construction estimating skills
- Estimating foundations, utilities, earthwork, and site improvements
- Correct equipment usage, quality control, and site investigation for estimating purposes

Who should attend: Project managers, design engineers, estimators, and contractors doing site work and heavy construction.

Assessing Scope of Work for Facility Construction Estimating

This *two-day* course is a practical training program that addresses the vital importance of understanding the SCOPE of projects in order to produce accurate cost estimates in a facility repair and remodeling environment.

Some of what you'll learn:
- Discussions of site visits, plans/specs, record drawings of facilities, and site-specific lists
- Review of CSI divisions, including means, methods, materials, and the challenges of scoping each topic
- Exercises in SCOPE identification and SCOPE writing for accurate estimating of projects
- Hands-on exercises that require SCOPE, take-off, and pricing

Who should attend: Corporate and government estimators, planners, facility managers, and others who need to produce accurate project estimates.

2009 RSMeans Seminar Schedule

Note: call for exact dates and details.

Location	Dates
Las Vegas, NV	March
Washington, DC	April
Denver, CO	May
San Francisco, CA	June
Washington, DC	September
Dallas, TX	September
Las Vegas, NV	October
Atlantic City, NJ	October
Orlando, FL	November
San Diego, CA	December

1-800-334-3509, ext. 5115

For more information
visit the RSMeans website
at www.rsmeans.com

Professional Development

Registration Information

Register early... Save up to $100! Register 30 days before the start date of a seminar and save $100 off your total fee. *Note: This discount can be applied only once per order. It cannot be applied to team discount registrations or any other special offer.*

How to register Register by phone today! The RSMeans toll-free number for making reservations is **1-800-334-3509, ext. 5115.**

Individual seminar registration fee - $935. *Means CostWorks®* **training registration fee - $375.** To register by mail, complete the registration form and return, with your full fee, to: RSMeans Seminars, 63 Smiths Lane, Kingston, MA 02364.

Government pricing All federal government employees save off the regular seminar price. Other promotional discounts cannot be combined with the government discount.

Team discount program for two to four seminar registrations. Call for pricing: 1-800-334-3509, ext. 5115

Multiple course discounts When signing up for two or more courses, call for pricing.

Refund policy Cancellations will be accepted up to ten business days prior to the seminar start. There are no refunds for cancellations received later than ten working days prior to the first day of the seminar. A $150 processing fee will be applied for all cancellations. Written notice of cancellation is required. Substitutions can be made at any time before the session starts. **No-shows are subject to the full seminar fee.**

AACE approved courses Many seminars described and offered here have been approved for 14 hours (1.4 recertification credits) of credit by the AACE International Certification Board toward meeting the continuing education requirements for recertification as a Certified Cost Engineer/Certified Cost Consultant.

AIA Continuing Education We are registered with the AIA Continuing Education System (AIA/CES) and are committed to developing quality learning activities in accordance with the CES criteria. Many seminars meet the AIA/CES criteria for Quality Level 2. AIA members may receive (14) learning units (LUs) for each two-day RSMeans course.

NASBA CPE sponsor credits We are part of the National Registry of CPE sponsors. Attendees may be eligible for (16) CPE credits.

Daily course schedule The first day of each seminar session begins at 8:30 a.m. and ends at 4:30 p.m. The second day begins at 8:00 a.m. and ends at 4:00 p.m. Participants are urged to bring a hand-held calculator, since many actual problems will be worked out in each session.

Continental breakfast Your registration includes the cost of a continental breakfast, and a morning and afternoon refreshment break. These informal segments allow you to discuss topics of mutual interest with other seminar attendees. (You are free to make your own lunch and dinner arrangements.)

Hotel/transportation arrangements RSMeans arranges to hold a block of rooms at most host hotels. To take advantage of special group rates when making your reservation, be sure to mention that you are attending the RSMeans seminar. You are, of course, free to stay at the lodging place of your choice. **(Hotel reservations and transportation arrangements should be made directly by seminar attendees.)**

Important Class sizes are limited, so please register as soon as possible.

Note: Pricing subject to change.

Registration Form

ADDS-1000

Call 1-800-334-3509, ext. 5115 to register or FAX this form 1-800-632-6732. Visit our Web site: www.rsmeans.com

Please register the following people for the RSMeans construction seminars as shown here. We understand that we must make our own hotel reservations if overnight stays are necessary.

☐ Full payment of $_____enclosed.

☐ Bill me

Please print name of registrant(s)

(To appear on certificate of completion)

_____

_____

_____

_____

P.O. #:_____
GOVERNMENT AGENCIES MUST SUPPLY PURCHASE ORDER NUMBER OR TRAINING FORM.

Firm name_____

Address_____

City/State/Zip_____

Telephone no._____ Fax no._____

E-mail address_____

Charge registration(s) to: ☐ MasterCard ☐ VISA ☐ American Express

Account no._____Exp. date_____

Cardholder's signature_____

Seminar name_____

Seminar City _____

Please mail check to: RSMeans Seminars, 63 Smiths Lane, P.O. Box 800, Kingston, MA 02364 USA